RSME Springer Series

Volume 5

Editor-in-Chief

José Bonet, Instituto Universitario de Matemática Pura y Aplicada (IUMPA), Universitat Politècnica de València, Valencia, Spain

Series Editors

Nicolas Andruskiewitsch, FaMAF - CIEM (CONICET), Universidad Nacional de Córdoba, Córdoba, Argentina

María Emilia Caballero, Instituto de Matemáticas, Universidad Nacional Autónoma de México, México, México

Pablo Mira, Departamento de Matematica Aplicada y Estadistica, Universidad Politécnica de Cartagena, Cartagena, Spain

Timothy G. Myers, Centre de Recerca Matemàtica, Barcelona, Spain

Marta Sanz-Solé, Department of Mathematics and Informatics, Barcelona Graduate School of Mathematics (BGSMath), Universitat de Barcelona, Barcelona, Spain

Karl Schwede, Department of Mathematics, University of Utah, Salt Lake City, USA

As of 2015, RSME - Real Sociedad Matemática Española - and Springer cooperate in order to publish works by authors and volume editors under the auspices of a co-branded series of publications including advanced textbooks, Lecture Notes, collections of surveys resulting from international workshops and Summer Schools, SpringerBriefs, monographs as well as contributed volumes and conference proceedings. The works in the series are written in English only, aiming to offer high level research results in the fields of pure and applied mathematics to a global readership of students, researchers, professionals, and policymakers.

More information about this series at http://www.springer.com/series/13759

Ben Elias • Shotaro Makisumi • Ulrich Thiel •
Geordie Williamson

Introduction to Soergel Bimodules

Real Sociedad Matemática
Española

Ben Elias
Department of Mathematics
University of Oregon, Fenton Hall
Eugene, OR, USA

Shotaro Makisumi
Department of Mathematics
Columbia University
New York, NY, USA

Ulrich Thiel
Department of Mathematics
University of Kaiserslautern
Kaiserslautern, Germany

Geordie Williamson
School of Mathematics and Statistics
University of Sydney
Sydney, Australia

ISSN 2509-8888　　　　　　　ISSN 2509-8896　(electronic)
RSME Springer Series
ISBN 978-3-030-48825-3　　　ISBN 978-3-030-48826-0　(eBook)
https://doi.org/10.1007/978-3-030-48826-0

© The Editor(s) (if applicable) and The Author(s), under exclusive licence to Springer Nature Switzerland AG 2020

This work is subject to copyright. All rights are solely and exclusively licensed by the Publisher, whether the whole or part of the material is concerned, specifically the rights of translation, reprinting, reuse of illustrations, recitation, broadcasting, reproduction on microfilms or in any other physical way, and transmission or information storage and retrieval, electronic adaptation, computer software, or by similar or dissimilar methodology now known or hereafter developed.

The use of general descriptive names, registered names, trademarks, service marks, etc. in this publication does not imply, even in the absence of a specific statement, that such names are exempt from the relevant protective laws and regulations and therefore free for general use.

The publisher, the authors, and the editors are safe to assume that the advice and information in this book are believed to be true and accurate at the date of publication. Neither the publisher nor the authors or the editors give a warranty, expressed or implied, with respect to the material contained herein or for any errors or omissions that may have been made. The publisher remains neutral with regard to jurisdictional claims in published maps and institutional affiliations.

This Springer imprint is published by the registered company Springer Nature Switzerland AG.
The registered company address is: Gewerbestrasse 11, 6330 Cham, Switzerland

Preface

The theory of Soergel bimodules began with the work of Wolfgang Soergel in the 1990s and 2000s [161, 162, 164, 165]. Soergel bimodules are certain bimodules over a polynomial ring, and the collection of all such bimodules forms a monoidal category under the tensor product. Soergel bimodules are fairly elementary objects, yet they have deep links to Lie theory and geometry and have a remarkably rich internal structure. The purpose of this book is to provide a comprehensive introduction to the theory of Soergel bimodules.

Let us begin with some historical context. The starting point of the whole story is the Kazhdan–Lusztig conjecture [108], the assertion that quantities of importance in Lie theory (Jordan–Hölder multiplicities of simple modules in Verma modules) can be calculated as the value at 1 of Kazhdan–Lusztig polynomials (certain inductively defined polynomials which depend only on the Weyl group). Soon thereafter, this conjecture was solved by Kazhdan–Lusztig [109], Beilinson–Bernstein [17], and Brylinski–Kashiwara [38] using an extraordinary array of geometric techniques (D-modules, the Riemann–Hilbert correspondence, ℓ-adic perverse sheaves, the Weil conjectures, etc.), most of which had only been developed in the decade preceding the conjecture! In the two decades following the Kazhdan–Lusztig conjecture, a new paradigm emerged, in which canonical advances like the Kazhdan–Lusztig basis provided answers to a growing number of problems in representation theory.

It was remarkable at the time that solving such a basic (and purely algebraic) question in Lie theory required such sophisticated geometric tools. As such, it was only natural to seek a simpler proof. In his landmark paper [161] from 1990, Soergel proposed a reformulation of the Kazhdan–Lusztig conjecture in terms of what are now known as Soergel modules. (Soergel modules can be obtained from Soergel bimodules by a simple process, but historically preceded them.) The original solution to the Kazhdan–Lusztig conjecture used D-modules as a bridge between Lie algebra representations on one side, and perverse sheaves of the flag variety on the other, via the so-called localization theorem. Soergel discovered a new bridge: both Lie algebra representations and perverse sheaves can be understood in terms of certain modules over the coinvariant algebra, which is the cohomology ring of the flag variety. Moreover, Soergel showed that the Kazhdan–Lusztig conjecture is equivalent to a very natural statement concerning Soergel modules (roughly speaking that the indecomposable Soergel modules are "not too big"). In his original paper, Soergel deduced this property from the decomposition theorem [19], a difficult and powerful geometric theorem concerning the decomposition behavior of constructible sheaves. Thus, although Soergel's proof was more elementary than earlier arguments, he still needed a deep geometric theorem to complete his proof.

Soergel's bridge is not merely an algebraic reformulation of the localization theorem. The localization theorem is an equivalence of Abelian categories, matching simple objects to simple objects and projectives to projectives. In contrast, Soergel proved that his modules are an algebraic model for projective Lie algebra representations, and for semisimple perverse sheaves. By swapping simples with projectives, Soergel actually proved that these categories are Koszul dual. Another important difference is the appearance of the Langlands dual flag variety in Soergel's work. By combining his results with the localization theorem, it can be shown that Langlands dual Lie algebras produce Koszul dual Abelian categories.

In subsequent years, Soergel developed the theory of (what are known today as) Soergel bimodules [162]. Soergel noticed that certain bimodules over a polynomial ring satisfy the relations of the Hecke algebra, which allowed him to produce a monoidal category that categorifies the Hecke algebra. In modern language, Soergel modules categorify the regular representation of the Hecke algebra, and Soergel bimodules categorify the Hecke algebra itself. Moreover, the Kazhdan–Lusztig conjecture is equivalent to the statement that the indecomposable Soergel bimodules categorify the Kazhdan–Lusztig basis of the Hecke algebra. In a remarkable paper from 2006 [165], Soergel pointed out that his theory works for any Coxeter system, whether or not it arises as a Weyl group. He provided an analogue of the Kazhdan–Lusztig conjecture in this setting, which we call the Soergel conjecture, but lacking a decomposition theorem, he was unable to prove it. Soergel also proved that the Soergel conjecture implies the Kazhdan–Lusztig positivity conjecture: for any Coxeter system, the coefficients of Kazhdan–Lusztig polynomials are positive.[1]

More recently, a "diagrammatic" theory of Soergel bimodules has emerged in the work of Elias–Williamson [59] (building on earlier work by Elias–Khovanov [53] in type A and by Elias [47] for dihedral groups). This concerns using generators and relations to find a presentation for the category of Soergel bimodules, using planar graphs to represent morphisms. In fact, although the presentation was found using Soergel bimodules, once defined, the resulting monoidal category can be considered independently of Soergel bimodules. This "diagrammatic Hecke category"[2] offers several advantages over the classical theory of Soergel bimodules:

---

[1] This was already known for crystallographic Coxeter systems (like Weyl and affine Weyl groups), thanks to the work of Kazhdan–Lusztig.

[2] In the literature, one increasingly encounters the term "Elias–Williamson diagrammatic category" for the diagrammatic Hecke category. For type A, one should use the term "Elias–Khovanov diagrammatic category."

- It reduces complicated polynomial computations in Soergel bimodules (which may encode still more complicated representation-theoretic or geometric computations) to (still complicated but essentially algorithmic) manipulations of planar diagrams.
- It categifies the Hecke algebra under very weak assumptions, even when the category of Soergel bimodules does not.
- It can be defined integrally, yielding an integral model for representation-theoretic and geometric categories that are defined over fields of various characteristics. This diagrammatic approach to Soergel bimodules has been a catalyst for a number of recent advances:
- The discovery of a Hodge theory of Soergel bimodules [57], and an algebraic proof of the decomposition theorem as it applies to the Hecke category. This led to a purely algebraic proof of Soergel's conjecture on the characteristics of indecomposable Soergel bimodules. In turn, this completed Soergel's agenda to provide an algebraic proof of the Kazhdan–Lusztig conjecture and prove the Kazhdan–Lusztig positivity conjecture.
- The discovery of counterexamples to the expected bounds in the long-standing Lusztig conjecture for modular representations of reductive groups [141].
- The further development of the Hodge theory of Soergel bimodules in its local and relative forms. This led to an algebraic proof of the Jantzen conjectures [179] following on earlier work by Soergel [166] and Kübel [115, 116] and to a proof of the unimodality of the structure constants of the Hecke algebra [58].
- A reproof of the geometric Satake equivalence in type A, and a surprising quantum version of this equivalence [49].
- A theory of categorical diagonalization, which aims to categorify the representation theory of the Hecke algebra [50, 51]. This has led to some remarkable conjectures relating the Hecke category in type A to the geometry of Hilbert schemes [78].
- An algebraic/diagrammatic approach to Koszul duality [5, 133], which was important in the proof of the Riche–Williamson conjecture on the characteristics of tilting modules in the modular representation theory of reductive groups [6].

The aim of this book is not to explain all the topics and advances mentioned above. Rather, our primary goal is to provide a solid introduction to the basic theory of Soergel bimodules, both classical (Part I) and diagrammatic (Part II). After working through these two parts, we hope that the reader will be fluent enough with the diagrammatics to carry out their own computations. We feel that such fluency is a crucial aspect of learning the theory and its applications. While some of the discoveries above do not logically rely on diagrammatics, it was the newfound ability to compute examples that made these questions manageable. Indeed, it is no exaggeration to say that the computational power afforded by diagrammatics was the key breakthrough that allowed many of the advances above to be made.

A secondary goal is to introduce the reader to the rapidly evolving field of categorical ("higher") representation theory, using Soergel bimodules as the main example. Interwoven in our discussion will be various techniques from the "toolbox" of higher representation theory. These include diagrammatics for monoidal categories, cellular categories, categorical cell modules, categorification theorems and their structure, algebraic Hodge theory, homotopy categories and their uses, categorical diagonalization, Koszul duality, and more. These tools are powerful and their potential has by no means been exhausted. We look forward to seeing applications of these tools to problems beyond the scope of this book!

Though our motivations, outlined above, have largely been drawn from representation theory and geometry, one particularly striking feature of Soergel bimodules is their additional generality. To any Coxeter group W, one can construct its Hecke category using Soergel bimodules or diagrammatics; only when W is crystallographic are there known connections to geometry or representation theory. Nonetheless, many of the results above (e.g. algebraic Hodge theory, Koszul duality) continue to apply even when W has no corresponding geometry. This mysterious phenomenon, the existence of nongeometric categories that behave as though there were some underlying geometry, has also appeared in several other contexts recently (e.g. nonrational polytopes and their "toric varieties" [30, 34, 98, 140], the Hodge theory of matroids [8], and the "spetses" idea put forward by Broué et al. [36]). One task of the twenty-first century mathematics will be to explain this phenomenon.

How to Read This Book

First, a word of advice for the reader: Do the exercises! There are hundreds of exercises scattered throughout the book, and the uninitiated reader will profit tremendously from attempting the exercises as they progress through the book.

This book consists of 27 chapters, organized into 5 parts. The Leitfaden on page xiii summarizes the logical structure of the chapters.

In Part I, we cover the basic theory of Coxeter groups and Hecke algebras, and the classical theory of Soergel bimodules.

In Part II, we cover the diagrammatic approach to Soergel bimodules, and the diagrammatic proof of the fact that Soergel bimodules categorify the Hecke algebra. Together, Parts I and II form the core of this book.

Part III places Soergel bimodules in context, describing the connections between them and the other incarnations of the Hecke category: category $\mathcal{O}$ in representation theory and perverse sheaves in geometry. This relies on the material covered in Part I. Chapter 16 is a stand-alone "user's guide to perverse sheaves," while the remainder develops the theory of category $\mathcal{O}$ and its "Soergel functor" to bimodules.

Part IV concerns the Hodge theory of Soergel bimodules, as developed in [57] (see also [150]). Much of this part can be read for its general ideas. Chapter 17 includes an elementary overview of the linear algebra involved. Chapter 18 describes the main results and gives the reader the big picture, while Chap. 20 features an extensive outline of the proof of Soergel's conjecture. In this regard, certain complexes of Soergel bimodules called Rouquier complexes play a pivotal role. Chapter 19 describes Rouquier complexes and the necessary homological algebra; after this point, the focus of the book gradually shifts toward homotopy categories rather than additive categories.

In turn, Part V offers a sampling of more advanced topics: cell theory and Hecke algebra representations (Chap. 22) and their connections to knot theory (Chap. 21) and categorical diagonalization (Chap. 23); diagrammatics for individual Soergel bimodules (Chap. 24); Koszul duality for category $\mathcal{O}$, reinterpreted using Soergel bimodules (Chaps. 25 and 26); and the situation in positive characteristic (Chap. 27). After completing the basics, many of the advanced topics can be read independently.

A book website with supplementary material (like Coxeter complexes and sample computations of Kazhdan–Lusztig basis elements in types A_3, B_3, and $\widetilde{A}_2$) can be found at

https://github.com/ulthiel/soergelbook.

Any future updates to this book will also appear here.

Much of the text should be accessible for beginning graduate students (or even advanced undergraduates, especially the core) who have the following prerequisites:

- Basic algebra: modules, bimodules, tensor products, linear algebra (bilinear forms, etc.).
- Basic category theory: equivalence of categories, isomorphisms of functors, adjunctions.
- Basic homological algebra (less essential for the first two parts): projective objects, Ext, derived category.

For some later chapters, it is helpful, though by no means necessary, to have a basic grasp of the following topics:

- Complex semisimple Lie algebras: root systems, Weyl groups, Coxeter groups.
- Cohomology: long exact sequences of cohomology, cohomology of simple spaces like projective spaces, the circle, etc.
- Algebraic geometry: proper maps, the fibers of a morphism of varieties.

Especially in the first two parts, we focus on the algebraic and combinatorial aspects of the theory of Soergel bimodules in order to quickly get the uninitiated reader up to speed. In particular, most chapters only mention the connections to geometry in passing. However, the entire theory is heavily motivated by geometry, and we have sought to include a glimpse of these connections in the later chapters. We hope that this book will whet the reader's appetite to explore them further.

Finally, did we mention that you should do the exercises?

Eugene, OR, USA	Ben Elias
New York, NY, USA	Shotaro Makisumi
Kaiserslautern, Germany	Ulrich Thiel
Sydney, Australia	Geordie Williamson
March 2020	

About This Book and Acknowledgements

This book grew out of lectures given at a graduate summer school at the Mathematical Sciences Research Institute in Berkeley in July 2017.[3] There were four previous incarnations of this summer school: a Master Class by BE and GW at the Centre for Quantum Geometry of Moduli Spaces in Aarhus in March 2013; a course given by GW at the University of Sydney at the end of 2013; a lecture series by BE at the Institute of Mathematical Sciences in Chennai in February 2014; and a WARTHOG[4] by BE at the University of Oregon in August 2014. Each of these previous lecture series covered the basic theory of Soergel bimodules and their Hodge theory. The 2-week format of the MSRI graduate summer school allowed us to cover this foundational material at a more leisurely pace, then discuss further topics and applications, many of which have only emerged since the time of the previous lecture series.

Lectures at the MSRI summer school were given by the authors, with help from Qi You and Leonardo Patimo. We thank them heartily for their contribution. Each chapter of this book was originally written by participant volunteers based on one lecture. A complete list of contributors can be found on page xxiii. Rather than simple note-taking, however, the participants were encouraged to struggle with the material to refine it into a book chapter, by furnishing details omitted in the lectures, by including more examples or exercises, etc. We thank all volunteers for their efforts, as well as all workshop participants for their questions and enthusiastic participation in the daily exercise sessions. Finally, we thank the MSRI staff for providing a fantastic workshop environment.

After each chapter had been contributed, significant work was necessary to combine them into a coherent whole. With a pedagogical mindset, we took considerable liberties in this process, shuffling around content between different

---

[3] Videos of most lectures are available on the MSRI website: http://www.msri.org/summer_schools/790.

[4] Workshop on Algebra and Representation Theory Held on Oregonian Grounds, organized by Nick Proudfoot.

chapters, splitting chapters in two, creating and rewriting and (rarely) removing sections. Soon enough the book was half-again as long as the original, and we needed to make a decision. Ultimately, we decided that this book should be a thorough reference on Soergel bimodules, the first of its kind, rather than an enhanced transcript of the activities of the workshop. This change in focus led to more reorganization of material, and the book grew longer still. We apologize to any contributor whose work was distributed or shattered by the process, but we hope it was for the greater readability of the book.

This is a fairly unusual textbook. We made little conscious effort to streamline the writing styles of the different contributors, so the chapters feel quite unique. Each chapter tended to recall the background material it needed (as one would in a lecture), and we kept most of that in place. Thus, there is more repetition of material than the average textbook, but we feel this is a feature rather than a bug: it enables the reader to read continuously for longer before getting lost, and it makes each chapter more independent of the others, allowing readers to read the book out of order (which is intended).

The process of combining the contributed chapters was begun while SM and GW were in residence at MSRI during Spring 2018, supported by the National Science Foundation under Grant No. 1440140. We would like to thank the QGM in Aarhus, the IMSc in Chennai, the University of Oregon, and the University of Sydney for hosting previous lecture series, as well as the participants for their active involvement in shaping the ideas of this book. We thank Cailan Li and Gunter Malle for their many useful comments on a preliminary version of this book. Thanks also to Columbia University, the University of Oxford, and the Max Planck Institute for Mathematics in Bonn, where much of the original work discussed here was done. BE was supported by NSF CAREER grant DMS-1553032 and by a Sloan Foundation Fellowship. UT greatly acknowledges the support by the Australian Research Council under the Grants DP160103897 and DE190101099.

Finally, it is a great pleasure to dedicate this book to Wolfgang Soergel. May his bimodules and barefootedness live long and prosper!

Leitfaden

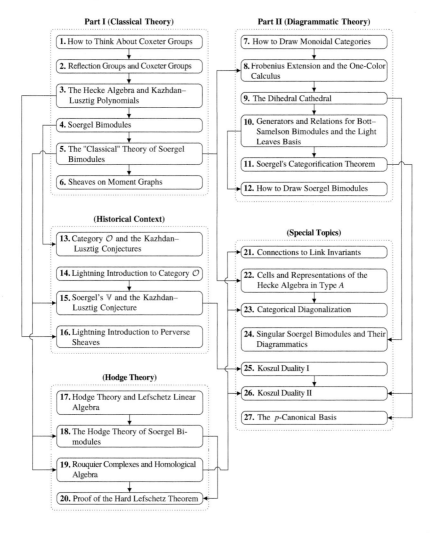

Contents

Part I The Classical Theory of Soergel Bimodules

1 How to Think About Coxeter Groups 3
   1.1   Coxeter Systems and Examples 3
          1.1.1   Definition of a Coxeter System 3
          1.1.2   Example: Type A 4
          1.1.3   Example: Type B 7
          1.1.4   Example: Type D 10
          1.1.5   Example: Dihedral Groups 10
          1.1.6   Coxeter Groups and Reflections 11
          1.1.7   The Geometric Representation and the
                  Classification of Finite Coxeter Groups 12
          1.1.8   Crystallographic Coxeter Systems 14
   1.2   Coxeter Group Fundamentals 14
          1.2.1   The Length Function 14
          1.2.2   The Descent Set.. 16
          1.2.3   The Exchange Condition 17
          1.2.4   The Longest Element 19
          1.2.5   Matsumoto's Theorem.................................... 19
          1.2.6   Bruhat Order 20
          1.2.7   Additional Exercises....................................... 21

2 Reflection Groups and Coxeter Groups 25
   2.1   Reflections and Affine Reflections 25
   2.2   Affine Reflection Groups 26
   2.3   Affine Reflection Groups are Coxeter Groups 27
   2.4   Expressions and Strolls 31
   2.5   Classification of Affine Reflection Groups......................... 34
   2.6   The Coxeter Complex 35

3 The Hecke Algebra and Kazhdan–Lusztig Polynomials 39
   3.1   The Hecke Algebra... 39

		3.1.1	The Standard Basis	40
		3.1.2	Inversion	42
	3.2	The Kazhdan–Lusztig Basis		44
		3.2.1	The Standard Form on H	45
	3.3	Existence of the Kazhdan–Lusztig Basis		48
		3.3.1	A Motivating Example	48
		3.3.2	Construction of the Kazhdan–Lusztig Basis	51
		3.3.3	The Kazhdan–Lusztig Presentation	54
		3.3.4	Deodhar's Formula	55
4	**Soergel Bimodules**			59
	4.1	Gradings		59
	4.2	Polynomials		61
		4.2.1	Invariant Polynomials	62
	4.3	Demazure Operators		64
	4.4	Bimodules and Tensor Products		66
	4.5	Bott–Samelson Bimodules		67
	4.6	Soergel Bimodules		69
	4.7	Examples of Soergel Bimodules		70
	4.8	A First Glimpse of Categorification		75
5	**The "Classical" Theory of Soergel Bimodules**			77
	5.1	Twisted Actions		77
	5.2	Standard Bimodules		78
	5.3	Soergel Bimodules and Standard Filtrations		80
	5.4	Localization		84
	5.5	Soergel's Categorification Theorem		86
	5.6	A Technical Wrinkle		89
	5.7	Realizations of a Coxeter System		90
	5.8	More Technicalities		96
6	**Sheaves on Moment Graphs**			99
	6.1	Roots in the Geometric Representation		99
	6.2	Bruhat Graphs		100
	6.3	Moment Graphs		104
	6.4	Sheaves on Moment Graphs		105
	6.5	The Braden–MacPherson Algorithm		107
	6.6	Stalks and Standard Bimodules		112
	6.7	A Functor to Sheaves on Moment Graphs		114
	6.8	Soergel's Conjecture and the Braden–MacPherson Algorithm		115

Part II Diagrammatic Hecke Category

7	**How to Draw Monoidal Categories**		119
	7.1	Linear Diagrams for Categories	119
	7.2	Planar Diagrams for 2-Categories	121
	7.3	Drawing Monoidal Categories	124

	7.4	The Temperley–Lieb Category	125
	7.5	More About Isotopy	129
8	**Frobenius Extensions and the One-Color Calculus**		**133**
	8.1	Frobenius Structures	133
		8.1.1 Frobenius Algebra Objects	133
		8.1.2 Diagrammatics for Frobenius Algebra Objects	135
		8.1.3 Playing with Isotopy	139
		8.1.4 Frobenius Extensions	140
	8.2	A Tale of One Color	143
		8.2.1 Frobenius Structure	144
		8.2.2 Additional Generators and Relations	145
		8.2.3 The Moral of the Tale	148
		8.2.4 A Direct Sum Decomposition, Diagrammatically	149
9	**The Dihedral Cathedral**		**151**
	9.1	A Tale of Two Colors	151
	9.2	The Temperley–Lieb 2-Category	154
	9.3	Jones–Wenzl Projectors	158
	9.4	Two-color Relations	162
10	**Generators and Relations for Bott–Samelson Bimodules and the Double Leaves Basis**		**167**
	10.1	Why Present $\mathbb{BS}$Bim?	167
	10.2	Generators and Relations	171
		10.2.1 A Diagrammatic Reminder	172
		10.2.2 An Isotopy Presentation of $\mathcal{H}_{BS}$	174
		10.2.3 Examples and Exercises	180
		10.2.4 A Presentation of $\mathcal{H}_{BS}$	182
		10.2.5 The Functor to Bimodules	185
		10.2.6 General Realizations	187
	10.3	Rex Moves and the 3-color Relations	188
	10.4	Light Leaves and Double Leaves	190
		10.4.1 Overview	191
		10.4.2 The Algorithm	193
		10.4.3 Diagrammatics and Bimodules	197
		10.4.4 Light Leaves and Localization	198
11	**The Soergel Categorification Theorem**		**201**
	11.1	Introduction	201
	11.2	Prelude: From $\mathcal{H}_{BS}$ to $\mathcal{H}$	204
		11.2.1 Graded Categories	204
		11.2.2 Additive Closure	208
		11.2.3 Karoubian Closure	209
		11.2.4 Karoubi Envelopes Are Krull–Schmidt	210
		11.2.5 Diagrammatics and Karoubi Envelopes	212
	11.3	Grothendieck Groups of Object-Adapted Cellular Categories	213

	11.3.1	Object-Adapted Cellular Categories	213
	11.3.2	First Properties	216
	11.3.3	The Main Example	217
	11.3.4	Classifying Indecomposables in Object-Adapted Cellular Categories	218
		Appendix 1: Krull–Schmidt Categories	224
		Appendix 2: Composition Forms, Cellular Forms, and Local Intersection Forms	229

12 How to Draw Soergel Bimodules ... 239
- 12.1 The 01-Basis ... 239
- 12.2 Commutative Ring Structure on a Bott–Samelson Bimodule ... 241
- 12.3 Trace and the Global Intersection Form ... 243
- 12.4 Bott–Samelson Bimodules and the Light Leaves Basis ... 246
- 12.5 Light Leaves Basis and the Standard Filtration on Bott–Samelson Bimodules ... 249
  - Appendix 1: A Crucial Positivity Result ... 251

Part III Historical Context: Category $\mathcal{O}$ and the Kazhdan–Lusztig Conjectures

13 Category $\mathcal{O}$ and the Kazhdan–Lusztig Conjectures ... 259
- 13.1 Introduction ... 259
- 13.2 The Verma Problem ... 260
  - 13.2.1 Verma Modules ... 260
  - 13.2.2 Category $\mathcal{O}$ and Its Mysteries ... 262
- 13.3 The Kazhdan–Lusztig Conjectures ... 263
  - 13.3.1 The Multiplicity Conjecture ... 263
  - 13.3.2 Positivity and Schubert Varieties ... 265
- 13.4 Two Proofs of the Kazhdan–Lusztig Conjecture ... 266

14 Lightning Introduction to Category $\mathcal{O}$ ... 271
- 14.1 Lie Algebra Basics ... 271
- 14.2 Category $\mathcal{O}$ ... 277
- 14.3 Duality in $\mathcal{O}$ ... 278
  - 14.3.1 Standard Filtrations and BGG Reciprocity ... 281
- 14.4 Blocks of Category $\mathcal{O}$ ... 282
- 14.5 Example: $\mathfrak{g} = \mathfrak{sl}_3(\mathbb{C})$ ... 288

15 Soergel's $\mathbb{V}$ Functor and the Kazhdan–Lusztig Conjecture ... 293
- 15.1 Brief Reminder on Category $\mathcal{O}$ ... 293
- 15.2 Translation Functors ... 295
  - 15.2.1 Tensor Products ... 295
  - 15.2.2 Definition of Translation Functors and First Properties ... 297
  - 15.2.3 Effect on Verma Modules ... 298
  - 15.2.4 Wall-Crossing Functors ... 301

		15.2.5 Effect on Projective Modules	303
	15.3	Soergel Modules	305
	15.4	Soergel's $\mathbb{V}$ Functor	307
	15.5	Soergel's Approach to the Kazhdan–Lusztig Conjecture	311

16 Lightning Introduction to Perverse Sheaves ... 315
    16.1 Motivation ... 315
    16.2 Stratified Spaces and Examples ... 316
        16.2.1 Stratified Resolutions and Schubert Varieties ... 319
        16.2.2 Constructible Sheaves and Pushforwards ... 322
        16.2.3 Perverse Sheaves ... 325
        16.2.4 The Decomposition Theorem ... 329
        16.2.5 Connection to the Hecke Algebra ... 329

Part IV   The Hodge Theory of Soergel Bimodules

17 Hodge Theory and Lefschetz Linear Algebra ... 335
    17.1 Introduction ... 335
    17.2 Hard Lefschetz ... 337
    17.3 Hodge–Riemann Bilinear Relations ... 340
    17.4 Lefschetz Lemmas ... 343

18 The Hodge Theory of Soergel Bimodules ... 347
    18.1 Introduction ... 347
    18.2 Overview and Preliminaries ... 348
        18.2.1 The Conjectures of Soergel and Kazhdan–Lusztig ... 348
        18.2.2 Duality and Invariant Forms ... 349
        18.2.3 The Main Theorem ... 356
    18.3 Outline of the Proof ... 356
        18.3.1 Step 1 ... 357
        18.3.2 Step 2 ... 359
    18.4 The Weak Lefschetz Problem ... 361
    18.5 From $\zeta = 0$ to $\zeta \gg 0$ ... 363
    18.6 From Local to Global Intersection Forms ... 365
    18.7 Hodge Theory of Matroids ... 366

19 Rouquier Complexes and Homological Algebra ... 369
    19.1 Motivation ... 369
    19.2 Some Homological Algebra ... 372
        19.2.1 Complexes and Homotopies ... 372
        19.2.2 Gaussian Elimination and Minimal Complexes ... 375
        19.2.3 Grothendieck Groups ... 378
    19.3 Rouquier Complexes and Categorification of the Braid Group ... 380
    19.4 Cohomology of Rouquier Complexes ... 387
    19.5 Perversity ... 389
    19.6 The Diagonal Miracle ... 391
        Appendix: More Homological Algebra ... 393

20	Proof of the Hard Lefschetz Theorem	401
	20.1 Introduction	401
	20.2 Preliminaries	402
	20.3 The Hodge–Riemann Relations for the Rouquier Complex	403
	20.4 Positivity of Breaking	407
	20.5 Sketch of the Proof of Hard Lefschetz	410
	Appendix: Some Historical Context and Geometric Intuition for the Proof of Soergel's Conjecture	413

Part V Special Topics

21	Connections to Link Invariants	421
	21.1 Temperley–Lieb Algebra	421
	21.2 Schur–Weyl Duality	425
	21.3 Trace and Link Invariants	426
	21.4 Quantum Groups and Link Invariants	430
	21.5 Ocneanu Trace and HOMFLYPT Polynomial	432
	21.6 Categorification of Braids and of the HOMFLYPT Invariant	435
	21.6.1 Rouquier Complexes	435
	21.6.2 Hochschild Homology	436
	21.6.3 Categorifying the Standard Trace	439

22	Cells and Representations of the Hecke Algebra in Type A	441
	22.1 Cells	441
	22.1.1 Cells for a Monoidal Category	442
	22.1.2 Cell Module Categories	446
	22.1.3 Cells for a Based Algebra	447
	22.2 Cells in Type A	448
	22.2.1 Young Diagrams and Tableaux	448
	22.2.2 The Robinson–Schensted Correspondence	449
	22.2.3 Cells in Type A	452
	22.2.4 The k-row Quotient of the Hecke Algebra	454
	22.3 Representations of the Hecke Algebra in Type A	455

23	Categorical Diagonalization	461
	23.1 Classical Linear Algebra	461
	23.2 Categorified Linear Algebra	464
	23.2.1 Eigenobjects	464
	23.2.2 Prediagonalizability	465
	23.2.3 Twisted Complexes	467
	23.2.4 Diagonalizability	468
	23.2.5 Smallness	469
	23.2.6 Lagrange Interpolation	469
	23.3 A Toy Example	471
	23.4 Diagonalizing the Full Twist	475
	23.4.1 Type A_1	475

		23.4.2 Type A	478
		23.4.3 The General Case	480
24	**Singular Soergel Bimodules and Their Diagrammatics**		**481**
	24.1	The Classical Theory of Singular Soergel Bimodules	481
		24.1.1 Bimodules and Functors	482
		24.1.2 Singular Soergel Bimodules	482
		24.1.3 Categorification Theorems	485
	24.2	One-Color Singular Diagrammatics	486
		24.2.1 Diagrammatics for a Frobenius Extension	486
		24.2.2 Relationship to the One-Color Soergel Calculus	490
		24.2.3 Relationship with the Temperley–Lieb 2-Category	492
	24.3	Singular Soergel Diagrammatics in General	495
		24.3.1 The Upgraded Chevalley Theorem, Part I	495
		24.3.2 The Upgraded Chevalley Theorem, Part II	500
		24.3.3 Diagrammatics for a Cube of Frobenius Extensions	503
		24.3.4 The Jones–Wenzl Relation	506
		24.3.5 Additional Relations, Applications, and Future Work	507
		24.3.6 Other Realizations	508
25	**Koszul Duality I**		**513**
	25.1	Introduction	513
		25.1.1 Morita Theory	514
	25.2	dg-Algebras	515
	25.3	dg-Morita Theory	517
	25.4	Koszul Duality for Polynomial Rings	520
	25.5	Review of the Kazhdan–Lusztig Conjecture	523
	25.6	Evidence of Koszul Duality in Category $\mathcal{O}$	525
26	**Koszul Duality II**		**529**
	26.1	Introduction	529
	26.2	Graded Category $\mathcal{O}_0$	530
		26.2.1 Desiderata	530
		26.2.2 Motivation from Soergel's $\mathbb{V}$ Functor	532
		26.2.3 Definition of Soergel Category $\mathcal{O}_0$	533
		26.2.4 Example: Soergel $\mathcal{O}_0$ in Type A_1	535
	26.3	Homological Properties of Soergel $\mathcal{O}_0$	538
		26.3.1 Highest Weight Structure	538
		26.3.2 Tilting Objects and the Realization Functor	540
		26.3.3 Ringel Duality	541
	26.4	Koszul Duality	541
		26.4.1 Statement	541
		26.4.2 Monodromy Action	542
		26.4.3 Wall-Crossing Functors	545
		26.4.4 Outline of the Proof of Theorem 26.26	546
	26.5	Some Odds and Ends	547

27	**The *p*-Canonical Basis**		549
	27.1	Introduction	549
	27.2	Definition of the *p*-Canonical Basis	550
	27.3	Computing the *p*-Canonical Basis	554
	27.4	Geometric Incarnation of the Hecke Category	559
		27.4.1 Parity Complexes on Flag Varieties	559
		27.4.2 Parity Complexes and the Hecke Category	561
	27.5	Modular Representation Theory of Reductive Groups	564
		27.5.1 Soergel's Modular Category $\mathcal{O}$	566
		27.5.2 The Riche–Williamson Conjecture	567
		27.5.3 This Is Not the End	569
References			571
Index			581

Contributors

Noah Arbesfeld Department of Mathematics, Columbia University, New York, NY, USA

Cihan Bahran School of Mathematics, University of Minnesota, Minneapolis, MN, USA

Elijah Bodish Department of Mathematics, University of Oregon, Fenton Hall, Eugene, OR, USA

Gordon C. Brown Department of Mathematics, University of Oklahoma, Norman, OK, USA

Charles Cain University at Buffalo SUNY, Buffalo, NY, USA

Anna Cepek Department of Mathematical Sciences, Montana State University, Bozeman, MT, USA

Alex Chandler Department of Mathematics, North Carolina State University, Raleigh, NC, USA

Christopher Chung Department of Mathematics, University of Virginia, Charlottesville, VA, USA

Gurbir Dhillon Department of Mathematics, Stanford University, Stanford, CA, USA

Ben Elias Department of Mathematics, University of Oregon, Fenton Hall, Eugene, OR, USA

Johannes Flake Department of Mathematics, Rutgers University, Piscataway, NJ, USA

Joel Gibson School of Mathematics and Statistics, University of Sydney, Sydney, NSW, Australia

Nicolle E. S. González Department of Mathematics, University of Southern California, Los Angeles, CA, USA

Jay Hathaway Department of Mathematics, University of Oregon, Fenton Hall, Eugene, OR, USA

Hang Huang University of Wisconsin Madison, Madison, WI, USA

Ilseung Jang Department of Mathematical Sciences, Seoul National University GwanAkRo 1, Gwanak-Gu, Seoul, Korea

Nachiket Karnick Department of Mathematics, Indiana University-Bloomington, Bloomington, IN, USA

Alexander Kerschl School of Mathematics and Statistics, University of Sydney, Sydney, NSW, Australia

Oscar Kivinen Department of Mathematics, UC Davis, Davis, CA, USA

Ethan Kowalenko University of California, Riverside, CA, USA

Christopher Leonard Department of Mathematics, University of Virginia, Charlottesville, VA, USA

Visu Makam Department of Mathematics, University of Michigan, Ann Arbor, MI, USA

Shotaro Makisumi Department of Mathematics, Columbia University, New York, NY, USA

Dmytro Matvieievskyi Department of Mathematics, Northeastern University, Boston, MA, USA

Leonardo Patimo Max Planck Institute for Mathematics, Bonn, Germany

You Qi Division of Physics, Mathematics and Astronomy, Caltech, Pasadena, CA, Germany

Sean Rogers Department of Mathematics, University of North Carolina, Chapel Hill, NC, USA

Anna Romanov Department of Mathematics, University of Utah, Salt Lake City, UT, USA

Seth Shelley-Abrahamson Department of Mathematics, Massachusetts Institute of Technology, Cambridge, MA, USA

Alexander Sistko Department of Mathematics, University of Iowa, Iowa, IA, USA

Andrew Stephens Department of Mathematics, University of Oregon, Eugene, OR, USA

Libby Taylor Georgia School of Mathematics, Institute of Technology, Atlanta, GA, USA

Sean Taylor Department of Mathematics, Louisiana State University, Baton Rouge, LA, USA

Ulrich Thiel Department of Mathematics, University of Kaiserslautern, Kaiserslautern, Germany

Kostiantyn Timchenko Department of Mathematics, University of Notre Dame, Notre Dame, IN, USA

Minh-Tam Trinh Department of Mathematics, University of Chicago, Chicago, IL, USA

Boris Tsvelikhovsky Department of Mathematics, Northeastern University, Boston, MA, USA

Dmitry Vagner Department of Mathematics, Duke University, Durham, NC, USA

Siddharth Venkatesh Department of Mathematics, Massachusetts Institute of Technology, Cambridge, MA, USA

Joshua Jeishing Wen Department of Mathematics, University of Illinois at Urbana-Champaign, Urbana, IL, USA

Geordie Williamson School of Mathematics and Statistics, University of Sydney, Sydney, Australia

Ziqing Xiang Department of Mathematics, University of Georgia, Athens, GA, USA

Yue Zhao Department of Mathematics, UC Davis, Davis, CA, USA

Part I
The Classical Theory of Soergel Bimodules

Chapter 1
How to Think About Coxeter Groups

This chapter is based on expanded notes of a lecture given by the authors and taken by

Yue Zhao and Ilseung Jang

Abstract This chapter gives a first introduction to the theory of Coxeter groups. In Sect. 1.1 we define Coxeter groups, discuss the Coxeter groups of type $ABCD$, and give the classification of finite Coxeter groups. We also discuss diagrammatic ways to present elements in type A and B. In Sect. 1.2 we discuss the fundamentals of Coxeter groups: the length function, descent sets, the exchange condition, the longest element, Matsumoto's theorem, and Bruhat order.

1.1 Coxeter Systems and Examples

1.1.1 Definition of a Coxeter System

Definition 1.1 A *Coxeter system* (W, S) is a group W and a finite set $S \subset W$ of generators of W, for which W admits a presentation of a very particular form. Namely, there must be a matrix $(m_{st})_{s,t \in S}$ satisfying $m_{ss} = 1$ for each $s \in S$, and $m_{st} = m_{ts} \in \{2, 3, \ldots\} \cup \{\infty\}$ for $s \neq t \in S$, such that

$$W = \langle s \in S \mid (st)^{m_{st}} = \text{id for any } s, t \in S \text{ with } m_{st} < \infty \rangle. \tag{1.1}$$

Here and throughout this book, the identity element of a Coxeter group is denoted by id. The *rank* of the Coxeter system (W, S) is defined as $|S|$.

Y. Zhao
Department of Mathematics, UC Davis, Davis, CA, USA

I. Jang
Department of Mathematical Sciences, Seoul National University GwanAkRo 1, Gwanak-Gu, Seoul, Korea

© The Editor(s) (if applicable) and The Author(s), under exclusive licence to Springer Nature Switzerland AG 2020
B. Elias et al., *Introduction to Soergel Bimodules*, RSME Springer Series 5, https://doi.org/10.1007/978-3-030-48826-0_1

One often sees these relations described in a different way. The relations for $s = t$ have the form

$$s^2 = \text{id}, \tag{1.2}$$

and are called the *quadratic relations*. The relation $(st)^{m_{st}} = \text{id}$ for $s \neq t \in S$ is equivalent (under the quadratic relations) to the *braid relation*

$$\underbrace{sts\cdots}_{m_{st}} = \underbrace{tst\cdots}_{m_{st}}. \tag{1.3}$$

Although it is not obvious from the presentation, m_{st} is precisely the order of the element st. When $m_{st} = \infty$, there is no corresponding relation between s and t.

The elements of S are often called *simple reflections*. Elements of W which are conjugate to elements of S are called *reflections*.

Remark 1.2 A group W for which there exists a Coxeter system (W, S) is called a *Coxeter group*. A Coxeter group can often be equipped with the structure of a Coxeter system in multiple different ways. For example, any conjugate of S could also be used as a set of simple reflections. However, it is even possible (see Exercise 1.20) for the same group to be described as a Coxeter group using two Coxeter systems with different ranks! In this book (and often implicitly in the literature), the object of study is a Coxeter system, not a Coxeter group.

For each $w \in W$, one can write $w = s_1 \cdots s_k$ for some $s_1, \ldots, s_k \in S$. We call the sequence $(s_1, \ldots, s_k)$ an *expression* for w, of *length k*. We often use a notational shorthand, writing $\underline{w}$ for the sequence $(s_1, \ldots, s_k)$, when the product $s_1 \cdots s_k$ is equal to w. That is, the notation $\underline{w}$ indicates both an element $w \in W$ and a particular choice of expression for w.

The *Coxeter graph* of a Coxeter system (W, S) is the following labeled graph, which efficiently encodes the data of the Coxeter system. Its vertex set is S, and vertices s and t are joined by an edge if $m_{st} > 2$. The edge is labeled by m_{st} if $m_{st} > 3$. Thus $m_{st} = 2$ if s and t are not connected, and $m_{st} = 3$ if they are joined by an unlabeled edge.

Remark 1.3 Let (W_1, S_1) and (W_2, S_2) be two Coxeter systems with Coxeter graphs X and Y, respectively. Then the Coxeter system whose Coxeter graph is the disjoint union $X \sqcup Y$ can be identified with the product system $(W_1 \times W_2, S_1 \times \{\text{id}\} \cup \{\text{id}\} \times S_2)$.

1.1.2 Example: Type A

Definition 1.4 The *Coxeter system of type* A_{n-1}, $n \geq 2$, is given by the following Coxeter graph:

$$\underset{s_1}{\bullet}\!\!-\!\!\underset{s_2}{\bullet}\!\!-\!\!\underset{s_3}{\bullet}\cdots\cdots\underset{s_{n-2}}{\bullet}\!\!-\!\!\underset{s_{n-1}}{\bullet}$$

1.1 Coxeter Systems and Examples

Thus, it has generating set $S = \{s_1, s_2, \ldots, s_{n-1}\}$ and relations

$$s_i^2 = \mathrm{id}, \qquad s_i s_{i+1} s_i = s_{i+1} s_i s_{i+1}, \qquad s_i s_j = s_j s_i \quad \text{if } |i - j| > 1. \tag{1.4}$$

This Coxeter group is isomorphic to the symmetric group S_n, the group of permutations of $\{1, 2, \ldots, n\}$. The generator s_i corresponds to the adjacent transposition $(i, i+1)$.

Exercise 1.5 Confirm that the set of all reflections agrees with the set of transpositions (i, j). For each transposition (i, j) with $i < j$, find an expression for (i, j) with length $2(j - i) - 1$.

Here are three common ways to describe a permutation $w \in S_n$.

- The *one-line notation* for w is the sequence $w(1)w(2) \cdots w(n)$. For example, $w = 4\,1\,3\,2$ means that $w(1) = 4$, $w(2) = 1$, $w(3) = 3$, and $w(4) = 2$. This notation will become useful in Sect. 22.2.2 when we discuss the Robinson–Schensted correspondence.
- *Cycle notation* describes w as the product of its disjoint cycles. Considering the same permutation as above, one writes $w = (1, 4, 2)(3)$ in cycle notation. Cycle notation is useful for understanding conjugacy classes, but will not appear frequently in this book.
- A third approach, *strand diagram notation*, will prove more useful for us. A permutation $w \in S_n$ is depicted as a diagram with n strands, in which each strand connects from i on the bottom line to $w(i)$ on the top line. For example, the following diagram describes $w = 4\,1\,3\,2$.

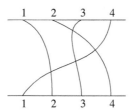

Stacking diagrams vertically multiplies permutations: note that xy is x on top of y.

Of the three notations, it is the strand diagram notation which relates most closely to the Coxeter presentation of S_n. In fact, a strand diagram represents not a permutation $w \in S_n$ but an expression $\underline{w}$ for a permutation. Any strand diagram which is suitably generic (see Remark 1.10) can be built by vertically stacking *crossings*, which are diagrams with a single crossing that correspond to the simple reflections $(i, i+1)$. Thus the length of the expression is the number of crossings. For example, the strand diagram

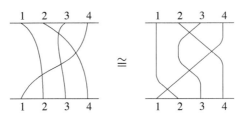

corresponds to the expression $\underline{w} = (s_2, s_3, s_2, s_1)$. Here is a different expression for the same permutation:

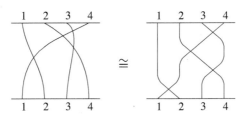

which corresponds to $\underline{w} = (s_3, s_2, s_3, s_1)$. The equivalence signs here represent generic isotopy, see Remark 1.10.

Exercise 1.6 In the previous exercise you found an expression for the transposition (i, j) of length $2(j-i)-1$ (when $i < j$). Draw this expression as a strand diagram, and think about what other expressions for (i, j) of this length you can find. Can you find any shorter expressions for (i, j)?

The Coxeter relations (1.4) become the following manipulations of strand diagrams, representing equalities in the symmetric group.

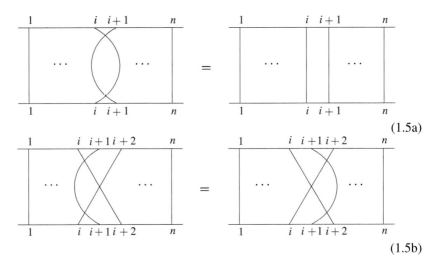

(1.5a)

(1.5b)

1.1 Coxeter Systems and Examples

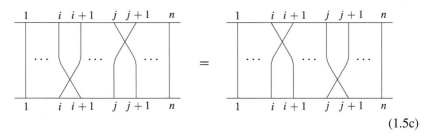

$$(1.5c)$$

Exercise 1.7 Verify that S_n is indeed given by the above presentation for $n = 2, 3$ and 4. You should attempt this using diagrammatic arguments only: try to show that the diagrammatic manipulations (1.5) above are sufficient to equate any two strand diagrams for the same permutation.

Exercise 1.8 If you are feeling courageous, attempt the previous exercise for all n. (This is tricky, and it is a good idea to finish reading this chapter first.)

Remark 1.9 The simple reflections of the symmetric group S_n are traditionally denoted by $s_1, s_2, \ldots, s_{n-1}$. This conflicts with the notation for an expression in a general Coxeter system, written as $(s_1, s_2, \ldots, s_m)$ where each s_i is an arbitrary simple reflection. Our preference in this book is for the second convention (as we are usually interested in general Coxeter groups). However sometimes we will use the first convention when dealing with the symmetric group. It should be clear from context when we are doing so.

Remark 1.10 A strand diagram is *generic* if it has only transverse intersections with no triple intersections, and if no two crossings occur at the same height. We obtain an expression from a generic strand diagram by reading off the crossings in order of their height. Thus an isotopy of diagrams which does not change the order of the crossings will not change the resulting expression, and we typically think of two diagrams which are isotopic in this way as being the same. An isotopy which does not create triple intersections but does change the order of the crossings will produce a different expression, related to the original by applications of the braid relation (1.5c). If the precise expression is not relevant then a non-generic strand diagram can often be much more compact to draw, see e.g. (1.14). In this case, one can perturb the diagram slightly to obtain a generic strand diagram if desired.

1.1.3 Example: Type B

Definition 1.11 The *Coxeter system of type* B_n, $n \geq 2$, is given by the following Coxeter graph:

$$\underset{t}{\bullet} \overset{4}{\rule{1cm}{0.4pt}} \underset{s_1}{\bullet} \rule{1cm}{0.4pt} \underset{s_2}{\bullet} \cdots\cdots\cdots \underset{s_{n-2}}{\bullet} \rule{1cm}{0.4pt} \underset{s_{n-1}}{\bullet}$$

This Coxeter group is isomorphic to the *signed symmetric group*, denoted by SS_n, the group of permutations w of $\{\pm 1, \pm 2, \ldots, \pm n\}$ such that $w(-i) = -w(i)$. This isomorphism sends t to the transposition $(-1, 1)$, and s_i to the product of transpositions $(i, i+1)(-i, -(i+1))$.

Exercise 1.12 Verify that $s_1 t s_1 t = t s_1 t s_1$ in SS_n. Verify that the Coxeter group of type B_n is isomorphic to SS_n for $n = 2$ and 3.

As SS_n is evidently a subgroup of S_{2n}, one can use one-line notation, cycle notation, or strand diagram notation to describe elements of SS_n. There are two more reasonable diagrammatic approaches.

The first approach uses *dotted strand diagrams*, where a dot corresponds to a sign flip. For example, the diagram

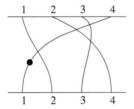

corresponds to the signed permutation w sending $1 \mapsto -4, 2 \mapsto 1, 3 \mapsto 3$, and $4 \mapsto 2$. Therefore it also sends $-1 \mapsto 4, -2 \mapsto -1$, etc. In terms of simple reflections, one has $w = s_3 s_2 s_3 s_1 t$.

To give another example, the braid relation $t s_1 t s_1 = s_1 t s_1 t$ could be drawn as follows.

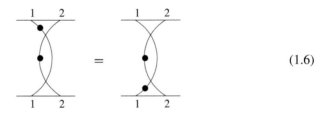 (1.6)

Exercise 1.13 Write down diagrammatic relations for dotted strand diagrams, akin to the relations in (1.5). What happens when two dots meet? How do dots interact with crossings? Does the braid relation in (1.6) follow from simpler relations?

Dotted strand diagrams have the advantage of being easy to draw and easy to multiply. However, they obfuscate the Coxeter structure. The simple reflections do correspond to nice elementary dotted strand diagrams: s_i to a crossing, and t to a dot on the first strand. However, when a dot appears on a strand which is not the leftmost strand, it does not correspond to a simple reflection. Thus a dotted strand

1.1 Coxeter Systems and Examples

diagram corresponds to an expression only when all dots appear on the first strand. This is true for the diagrams above, but not for the following diagram.

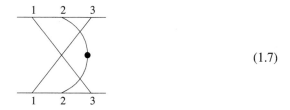
(1.7)

Exercise 1.14 Let x be the signed permutation in (1.7). Find an expression for x (i.e. find a dotted strand diagram for x where dots only appear on the first strand). Compare x with $s_2 x s_2$: which one has a dotted strand diagram with fewer crossings? Which one has a shorter expression?

The second method uses *mirror strand diagrams* or *mirror pictures*, which are a special kind of $2n$-strand diagram. Each strand connects i and $w(i)$, but the picture is required to be symmetric across a central axis (the mirror). The following is the mirror picture of the same signed permutation $w = s_3 s_2 s_1 s_3 t$.

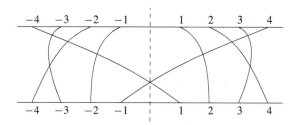

It is the mirror strand diagrams which best reflect the Coxeter structure. Each mirror pair of crossings corresponds to a simple reflection s_i, while a crossing which occurs on top of the mirror (the dashed line in the picture) corresponds to the simple reflection t. The length of the expression can be computed by counting the number of crossings on or to the right of the mirror. The disadvantage of mirror pictures is that they are hard to draw by hand.

Using mirror pictures, the braid relation $t s_1 t s_1 = s_1 t s_1 t$ appears as follows:

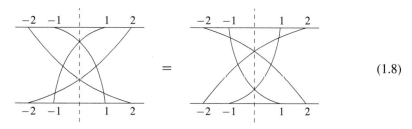
(1.8)

Remark 1.15 Mirror strand diagrams are typically not generic in the sense of Remark 1.10, but they have their own notion of genericity which we leave to the reader.

1.1.4 Example: Type D

Definition 1.16 The *Coxeter system of type* D_n, $n \geq 4$, is given by the following Coxeter graph:

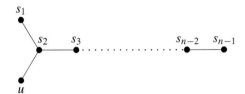

The Coxeter group of type D_n is isomorphic to the *even-signed symmetric group*, denoted by ESS_n, which is a subgroup of the signed symmetric group SS_n. A signed permutation w is in the subgroup ESS_n whenever the number of sign flips, i.e. the cardinality of $\{i > 0 \mid w(i) < 0\}$, is even. The generators of ESS_n are $\{u = ts_1t, s_1, \ldots, s_{n-1}\}$.

Neither diagrammatic method for drawing signed permutations is truly effective at describing the Coxeter structure on ESS_n. However, mirror pictures give a reasonable approximation.

Exercise 1.17 Draw the mirror picture for u. What kinds of mirror pictures represent expressions in type D? How does one compute the length of an expression from the picture?

1.1.5 Example: Dihedral Groups

Definition 1.18 The *Coxeter system of type* $I_2(m)$, $m > 2$, is given by the following Coxeter graph:

$$\overset{s}{\bullet} \overset{m}{} \overset{t}{\bullet}$$

Thus $I_2(m)$ has two generators s and t satisfying $(st)^m = \text{id}$. Note that $I_2(3) = A_2$ and $I_2(4) = B_2$. We also use $I_2(m)$ to denote the group itself, and call it a *dihedral group*.

1.1 Coxeter Systems and Examples

Exercise 1.19 Consider the group Dih_m of symmetries of the regular m-gon. Show that Dih_m admits a presentation $\langle r, \sigma \mid r^m = \sigma^2 = (r\sigma)^2 = \text{id}\rangle$, where r denotes a rotation and σ denotes the reflection. Show that Dih_m and $I_2(m)$ are isomorphic groups under the map sending s to σ and t to $r\sigma$. Conclude that the order of $I_2(m)$ is $2m$. What symmetries of the m-gon do the Coxeter generators correspond to?

Exercise 1.20 Find two Coxeter presentations of different ranks for the dihedral group with 12 elements.

Definition 1.21 The *Coxeter system of type* $I_2(\infty)$ is given by the following Coxeter graph:

$$\underset{s}{\bullet} \overset{\infty}{\text{———}} \underset{t}{\bullet}$$

We also use $I_2(\infty)$ to denote the group itself, and call it an *infinite dihedral group*.

Thus, $I_2(\infty)$ is freely generated by two involutions s and t, satisfying no other relations. It is an example of an infinite Coxeter group.

1.1.6 Coxeter Groups and Reflections

Let $O(\mathbb{R}^n)$ be the group of orthogonal transformations of the Euclidean space $\mathbb{R}^n$. A *reflection* is an orthogonal transformation $s \in O(\mathbb{R}^n)$ whose fixed subspace is a hyperplane. It is a beautiful theorem of Coxeter that any finite subgroup W of $O(\mathbb{R}^n)$ which is generated by reflections admits a Coxeter presentation (see e.g. [85, §1.9]).

Definition 1.22 Any nonzero vector $v \in \mathbb{R}^n$ gives rise to an element $s_v \in O(\mathbb{R}^n)$ which fixes H_v, the hyperplane perpendicular to v, and sends v to $-v$. We call s_v the *reflection along* v, or the *reflection through* H_v.

Exercise 1.23 Verify that the reflection along v can be expressed using the formula

$$s_v(x) = x - 2\frac{(x, v)}{(v, v)} v. \tag{1.9}$$

Here $(-, -)$ represents the inner product on $\mathbb{R}^n$. Verify that $(s_v x, s_v y) = (x, y)$ for any vectors $x, y \in \mathbb{R}^n$.

Exercise 1.24 In the previous section we saw that the symmetric group S_n is isomorphic to a Coxeter group of type A_{n-1}. Consider $\mathbb{R}^n = \bigoplus_{1 \le i \le n} \mathbb{R} e_i$ with its standard Euclidean structure (i.e. such that $e_1, e_2, \ldots, e_n$ is an orthonormal basis). Consider the action of S_n on $\mathbb{R}^n$ via permutation of coordinates. Show that S_n acts via orthogonal transformations of $\mathbb{R}^n$. Show that each transposition (i, j) acts as $s_{e_i - e_j}$.

Exercise 1.25 In Sect. 1.1.1 we defined the reflections in a Coxeter group, and in Exercise 1.5 you verified that the reflections in S_n are precisely the transpositions (i, j). In the previous exercise you verified that the transpositions (i, j) act as reflections on $\mathbb{R}^n$. Are there any other elements of S_n which also act as reflections?

Exercise 1.26 The Coxeter groups of types B_n and D_n both act on the subset $\{\pm e_1, \ldots, \pm e_n\} \subset \mathbb{R}^n$. Use this to embed these Coxeter groups into $O(\mathbb{R}^n)$. Which vectors (up to scalars) correspond to the simple reflections?

1.1.7 The Geometric Representation and the Classification of Finite Coxeter Groups

Conversely, any finite Coxeter group can be embedded in some orthogonal group, by means of its geometric representation.

Definition 1.27 The *geometric representation* of a Coxeter system (W, S) is the representation V of W defined as follows. Let V be the real vector space with basis $\{\alpha_s \mid s \in S\}$ indexed by S. Equip V with the symmetric bilinear form $(-, -)$ determined by

$$(\alpha_s, \alpha_t) = -\cos\frac{\pi}{m_{st}}. \tag{1.10}$$

When $m_{st} = \infty$ we use the convention that $\frac{\pi}{m_{st}} = 0$. Note that $(\alpha_s, \alpha_s) = 1$. One may define an action of W on V, where each $s \in S$ acts by reflection along α_s. That is,

$$s(\lambda) = \lambda - 2(\lambda, \alpha_s)\alpha_s. \tag{1.11}$$

Note that the geometric representation is defined for any Coxeter group, finite or infinite. However, the form $(-, -)$ is not positive definite in general (but wait for Theorem 1.34). We have the following important fact about the geometric representation.

Proposition 1.28 *For any Coxeter system, the geometric representation is faithful.*

Since the proof requires some more development, we will omit it and refer the reader to [85, Chapter 5].

Exercise 1.29 Recall the action of S_n on $\mathbb{R}^n$ considered in Exercise 1.24. Consider the S_n-stable subspace

$$U = \{(v_1, \ldots, v_n) \in \mathbb{R}^n \mid \sum v_i = 0\} \subset \mathbb{R}^n.$$

1.1 Coxeter Systems and Examples

Show that the map

$$\alpha_{s_i} \mapsto \frac{1}{\sqrt{2}}(e_i - e_{i+1})$$

yields an isomorphism of the geometric representation of S_n with U.

Exercise 1.30 In Exercise 1.19 we saw a realization of the dihedral group $I_2(m)$ as the symmetries of a regular m-gon. If this m-gon is centered at the origin of $\mathbb{R}^2$ this yields a representation of $I_2(m)$ as a group of orthogonal transformations of a 2-dimensional vector space. Show that this representation agrees with the geometric representation of $I_2(m)$.

Exercise 1.31 When the form $(-,-)$ is positive definite so that we can discuss lengths and angles in V, confirm that α_s is a unit vector for each s, and that the hyperplanes H_s and H_t (perpendicular to α_s and α_t, respectively) meet at an angle $\frac{\pi}{m_{st}}$. Thus st is rotation by $\frac{2\pi}{m_{st}}$, and has order m_{st}.

Remark 1.32 Suppose one is given two lines H_s and H_t in the Euclidean plane $\mathbb{R}^2$, which meet at an irrational angle θ (i.e. an angle which is not a rational multiple of π). We can choose unit vectors α_s and α_t, orthogonal to the lines H_s and H_t respectively, such that

$$(\alpha_s, \alpha_t) = -\cos\theta. \tag{1.12}$$

Then, letting s and t be the corresponding reflections, the operator st has infinite order, giving us an action of the infinite dihedral group $I_2(\infty)$ on the plane. However, this is not the geometric representation of $I_2(\infty)$.

Exercise 1.33 In the geometric representation of $I_2(\infty)$, confirm that the bilinear form $(-,-)$ is not positive definite. What is the kernel? Is the form positive semi-definite?

Theorem 1.34 *Let (W, S) be a Coxeter system. Then W is finite if and only if the form (1.10) is positive definite. This is the case if and only if the Coxeter graph is a finite disjoint union Coxeter graphs from the following list:*

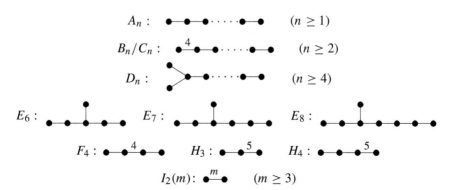

Remark 1.35 It is common to refer to this classification by type, saying "type A" to refer to the family of all symmetric groups. The Coxeter system of type B_n can also be called the Coxeter system of type C_n, to match conventions in Lie theory. For future use, note that *type $X \times Y$* refers to the product system of type X and type Y, whose Coxeter graph is their disjoint union, see Remark 1.3. For example, type $A_1 \times A_2$ refers to the Coxeter group $S_2 \times S_3$.

1.1.8 Crystallographic Coxeter Systems

Let us introduce some terminology, mostly for the literacy of the reader.

Definition 1.36 A Coxeter system (W, S) is called *crystallographic* if $m_{st} \in \{2, 3, 4, 6, \infty\}$ for all $s \neq t \in S$.

Crystallographic Coxeter groups are special for two reasons. The first is that a variant of the geometric representation can be defined over $\mathbb{Z}$ rather than $\mathbb{R}$. The second is that crystallographic Coxeter groups can be related to the geometry of Kac–Moody groups, while arbitrary Coxeter groups have no related geometry. Both of these special features will eventually have a role to play in this book, but we say nothing more for now.

By the classification theorem of the previous section, the finite crystallographic Coxeter systems are those of types A, B/C, D, E, F, and $I_2(6)$. Note that $I_2(6)$ is usually called G_2.

1.2 Coxeter Group Fundamentals

In this section we give some of the basic definitions and the fundamental theorems in the theory of Coxeter groups. Our goal is not to give proofs or a detailed exposition of these results, as there are many nice references in the literature, such as [85]. Instead, we aim to give some intuition for these concepts.

1.2.1 The Length Function

As noted in Sect. 1.1.1, each element $w \in W$ admits an expression $\underline{w} = (s_1, \ldots, s_k)$ for some $s_1, \ldots, s_k \in S$, since S is a generating set for W. The length $\ell(\underline{w})$ of this expression is k.

Definition 1.37 The *length* of w, denoted by $\ell(w)$, is the minimal k for which w admits an expression of length k. Any expression for w with this minimal length $\ell(w)$ is called a *reduced expression*. In particular, $\ell(w) = 0$ if and only if $w = \text{id}$.

1.2 Coxeter Group Fundamentals

Exercise 1.38 Show that W admits a *sign representation*: an action of W on $\mathbb{R}$ where each simple reflection acts by -1. Deduce that, for any two expressions for the same element $w \in W$, their lengths have the same parity (i.e. they are both even or both odd). Deduce also that $\ell(ws) \neq \ell(w)$, for any $w \in W$ and $s \in S$.

Example 1.39 Let $\underline{w}$ be an expression for $w \in S_n$ and consider the corresponding strand diagram. If two strands cross twice, then one can remove these two crossings and obtain a shorter expression for the same permutation. Here is an example:

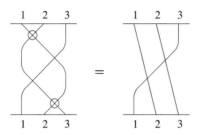

An *inversion* of w is a pair $\{i, j\}$ with $1 \leq i < j \leq n$, such that $w(i) > w(j)$. If $\{i, j\}$ is an inversion of w, then it is clear that the strand with i on bottom and the strand with j on bottom must cross at least once in any expression for w. Thus $\ell(w)$ is at least the number of inversions of w. It is not difficult to argue that, if no two strands cross twice, then the length of this expression is precisely the number of inversions, and therefore this expression is reduced. We thus conclude that $\underline{w}$ is reduced if and only if any two strands of its strand diagram cross each other at most once. In this case, $\ell(w)$ is the number of crossings in such a diagram.

Here is an example. Let $W = S_4$ and $\underline{w} = (s_1, s_2, s_1, s_3, s_2, s_1)$:

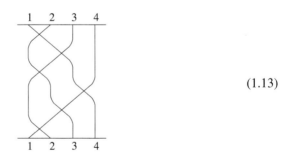

(1.13)

Then $\underline{w}$ is a reduced expression with $\ell(w) = 6$. This can be seen by checking that no strands cross each other twice in the above diagram. In fact, the element $w = s_1 s_2 s_1 s_3 s_2 s_1$ is the longest element of S_4 (see Sect. 1.2.4 below), whose set of inversions consists of all pairs $\{i, j\}$ with $1 \leq i < j \leq n$.

Example 1.40 Consider the expression (s_1, t, s_1) in the type B_3 Coxeter system. The following is its mirror picture:

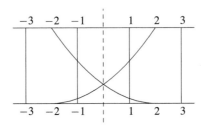

In general, in the mirror picture an expression is reduced if no two strands cross more than once, in which case one obtains the length by counting the number of crossings on and to the right of the mirror. Thus $\ell(s_1 t s_1) = 3$.

As noted in Sect. 1.1.3, it is difficult to read off the length from a dotted strand diagram. For instance, the following is a dotted strand diagram for the length 3 element $s_1 t s_1$:

Proposition 1.41 *The length function has the following properties (for $s \in S$ and $w, w' \in W$):*

1. $\ell(w) = 1$ if only if $w \in S$,
2. $\ell(w) = \ell(w^{-1})$,
3. $\ell(ww') \leq \ell(w) + \ell(w')$,
4. $\ell(ww') \geq \ell(w) - \ell(w')$,
5. $\ell(ws) = \ell(w) \pm 1$.

Proof Properties 1, 2, and 3 are straightforward. Applying property 3 to ww' and $(w')^{-1}$ yields $\ell(w) \leq \ell(ww') + \ell((w')^{-1})$. Using property 2 and rewriting the inequality gives property 4. When $w' = s$, properties 3 and 4 imply property 5. □

Similar arguments show that $\ell(ww') \geq \ell(w') - \ell(w)$, and that $\ell(sw) = \ell(w) \pm 1$.

1.2.2 The Descent Set

In Proposition 1.41 we showed that either $\ell(ws) = \ell(w) + 1$ or $\ell(ws) = \ell(w) - 1$, for $w \in W$ and $s \in S$. We say colloquially that "right multiplication by s brings w either up or down".

1.2 Coxeter Group Fundamentals

Definition 1.42 Given $w \in W$, its *right descent set* $\mathcal{R}(w)$ is the set $\{s \in S \mid \ell(ws) < \ell(w)\}$. Its *left descent set* $\mathcal{L}(w)$ is the set $\{s \in S \mid \ell(sw) < \ell(w)\}$.

Example 1.43 In type A, the right and left descent sets of $w \in S_n$ are easy to determine from the strand diagram of a reduced expression $\underline{w}$. The simple reflection s_i is in $\mathcal{R}(w)$ if the strands with bottom label i and $i+1$ eventually cross in $\underline{w}$. After all, if they do not cross, then adding this crossing s_i will still produce a reduced expression, so $\ell(ws_i) = \ell(w) + 1$. If they do cross already, then adding s_i will produce a double crossing, hence a non-reduced expression. Then $\ell(ws_i) < \ell(w) + 1$, meaning that $\ell(ws_i) = \ell(w) - 1$. Similarly, $s_i \in \mathcal{L}(w)$ if the strands with top label i and $i+1$ eventually cross in $\underline{w}$.

As an example, let $W = S_6$, and consider the reduced expression

$$\underline{w} = (s_1, s_3, s_5, s_2, s_4, s_1, s_3, s_2, s_4, s_3, s_5),$$

corresponding to the following strand diagram:

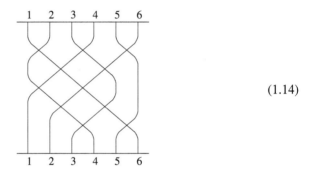

(1.14)

Then $\mathcal{L}(w) = \{s_1, s_2, s_3, s_5\}$ and $\mathcal{R}(w) = \{s_2, s_3, s_5\}$.

Exercise 1.44 Find a similar description of the left and right descent sets in type B using mirror pictures.

1.2.3 The Exchange Condition

Theorem 1.45 (Exchange Condition) *Let* $\underline{w} = (s_1, s_2, \ldots, s_k)$ *be a reduced expression for w, and $t \in S$. If $\ell(wt) < \ell(w)$ then there exists i such that $1 \leq i \leq k$ and $wt = s_1 s_2 \cdots \widehat{s_i} \cdots s_k$.*

Example 1.46 In type A the idea behind the exchange condition has already been explained. When $\ell(wt) < \ell(w)$, it is because adding the crossing t will re-cross two strands which have already crossed in $\underline{w}$. That previous crossing is s_i. By removing the new crossing t and the previous crossing s_i, one obtains an expression for wt.

Example 1.47 Let $W = S_4$, $w = s_2s_1s_3s_2s_1$ and $t = s_2$. Then $wt = s_2\widehat{s_1}s_3s_2s_1$. We can verify this diagrammatically as follows:

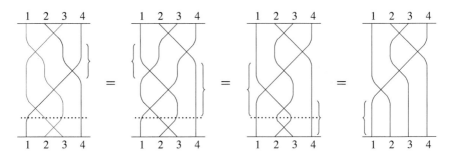

Here, the offending (red) crossing in the reduced expression for w was "pushed down" by braid relations, until it reached the bottom of the diagram, where it could be used to cancel t by the quadratic relation. In the penultimate diagram one witnesses a reduced expression for w which ends in t.

Exercise 1.48 For the example in (1.14), for each $t \in \mathcal{R}(w)$, find the crossing s_i which should be removed to produce wt. Then verify (by applying the relations) that $wt = s_1s_2 \cdots \widehat{s_i} \cdots s_k$.

The following corollary of the exchange condition is a truly crucial idea.

Corollary 1.49 *For $w \in W$, the descent set $\mathcal{R}(w)$ is equal to the set*

$$\{t \in S \mid w \text{ admits a reduced expression ending in } t\}.$$

Proof Let $\ell(w) = k$. If w admits a reduced expression ending in t, so that $w = s_1s_2 \cdots s_{k-1}t$, then $wt = s_1s_2 \cdots s_{k-1}$ by the quadratic relation. Thus $\ell(wt) \leq \ell(w)-1$, and $t \in \mathcal{R}(w)$. Conversely, if $t \in \mathcal{R}(w)$, then $wt = s_1 \cdots \widehat{s_i} \cdots s_k$ for some i, by the exchange condition. Then $w = wtt = s_1 \cdots \widehat{s_i} \cdots s_k t$ is an expression of length k, and is thus a reduced expression ending in t. □

In type A the corollary states that if adjacent strands (i.e. with labels j and $j+1$ on bottom) are going to cross at some point in a permutation, we can choose a reduced expression such that they are the first strands to cross.

We state one more famous corollary of the exchange condition.

Corollary 1.50 (Deletion Condition) *Let $\underline{w} = (s_1, s_2, \ldots, s_k)$ be an expression for $w \in W$ with $\ell(w) < k$. Then there exist $i < j$ such that $w = s_1s_2 \cdots \widehat{s_i} \cdots \widehat{s_j} \cdots s_k$.*

Again, in type A, the deletion condition is the now-familiar idea that any non-reduced expression has two crossings (of the same two strands) which can be canceled to leave a shorter reduced expression.

1.2.4 The Longest Element

Proposition 1.51 *Let (W, S) be a Coxeter system. Then there exists an element $w_0 \in W$ with $\mathcal{R}(w_0) = S$ if and only if W is finite. Such an element w_0 is unique, satisfies $\ell(w) < \ell(w_0)$ for all $w \neq w_0 \in W$, and is called the* longest element *of W. It also satisfies, and is determined by, the condition $\mathcal{L}(w_0) = S$.*

Example 1.52 We have already seen the longest element of S_4 in (1.13). In general, the longest element of S_n is the permutation sending $i \mapsto n+1-i$, and its inversion set is all pairs $\{i, j\}$ with $1 \leq i < j \leq n$.

Exercise 1.53 What is the longest element in type B? In type D?

Corollary 1.49 says that, whenever $t \in S$ is in the right descent set of w, then w has a reduced expression ending in t. This result can be upgraded as follows.

Proposition 1.54 *Let (W, S) be a Coxeter system (possibly infinite). Choose $w \in W$, and let $I \subset \mathcal{R}(w)$. Then I generates a finite Coxeter group inside W, with longest element w_I. Moreover, w has a reduced expression ending in a reduced expression for w_I.*

For instance, when $I = \{t\}$, this recovers Corollary 1.49. When $I = \{s, t\}$ and $m_{st} = 4$, this says that any element of W with a reduced expression ending in s and a reduced expression ending in t also must have a reduced expression ending in $stst$.

Exercise 1.55 For the example in (1.14), letting $I = \mathcal{R}(w)$, find a reduced expression for w ending in a reduced expression for w_I.

1.2.5 Matsumoto's Theorem

Let $(s_{i_1}, \ldots, s_{i_k})$ and $(s_{j_1}, \ldots, s_{j_k})$ be two arbitrary expressions of the same length. If we can apply a sequence of braid relations to obtain $(s_{j_1}, \ldots, s_{j_k})$ from $(s_{i_1}, \ldots, s_{i_k})$ we say that they are *related by braid relations*. The following beautiful theorem will be used throughout this book:

Theorem 1.56 (Matsumoto's Theorem [136], See Also [76, §1.2]) *Any two reduced expressions for $w \in W$ are related by braid relations.*

Remark 1.57 In the next chapter we will see a sketch of a geometric argument as to why this theorem is true.

The following exercise introduces a useful aspect of Coxeter theory which is underrepresented in the literature.

Exercise 1.58 The *reduced expression graph*, usually abbreviated *rex graph*, for a fixed element $w \in W$ is the graph defined as follows. Its vertices are the reduced expressions for w. There is an edge between two reduced expressions if they differ by a single application of a braid relation (it helps to label the edge with the number m_{st} associated to this braid relation).

1. Let W have type $A_1 \times A_1 \times A_1$. Draw the rex graph for every element of W. Verify that only the longest element has a cycle in its rex graph.
2. Repeat this exercise when W has type $A_1 \times A_2$.
3. Draw the rex graph for the longest element of S_4. What cycles appear?
4. (A challenge!) Do any cycles appear for other elements of S_4? Draw the rex graph for the longest elements in type B_3 and H_3. It might save space to identify two vertices if they are connected by an edge with $m_{st} = 2$.

1.2.6 Bruhat Order

In the exchange condition, we have seen that when $w = s_1 s_2 \cdots s_k$, $t \in S$, and $\ell(wt) < \ell(w)$, then $wt = s_1 \cdots \widehat{s_i} \cdots s_k$ for some i. However, not every i with $1 \leq i \leq k$ can appear in this way. Let T denote the set of all reflections, i.e. the set of all elements of W conjugate to an element of S. It is quite easy to see that any element of the form $s_1 \cdots \widehat{s_i} \cdots s_k$, for any $1 \leq i \leq k$, can be expressed as wt where $t \in T$. After all, the element

$$t = s_k s_{k-1} \cdots s_{i+1} s_i s_{i+1} \cdots s_{k-1} s_k$$

is a conjugate of s_i, and will suffice.

Definition 1.59 For $x, y \in W$ we write $x \to y$ if $\ell(x) < \ell(y)$ and $xt = y$ for some $t \in T$. The *Bruhat graph* is the directed graph with vertices the elements of W and arrows given by the relation $\to$. The *Bruhat order* on W is the partial order $\leq$ obtained as the transitive closure of the relation $\to$. Thus, $x \leq y$ if and only if there is a directed path in the Bruhat graph from x to y; equivalently, if there exists a chain $x = x_0, x_1, \ldots, x_m = y$ such that $\ell(x_i) < \ell(x_{i+1})$ and $x_i^{-1} x_{i+1} \in T$ for all $0 \leq i < m$.

A useful restatement of the Bruhat order is given in Exercise 1.64.

Remark 1.60 In the definition of the relation $\to$, and thus of the Bruhat order, we use right multiplication by reflections. However, since $xt = (xtx^{-1})x$ and xtx^{-1} is a reflection if t is a reflection, left multiplication defines the same relation.

Example 1.61 The following is the Hasse diagram of the Bruhat order on S_3:

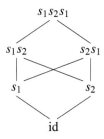

Exercise 1.62 Draw the Hasse diagram of the Bruhat order on the dihedral group $I_2(m)$ for m finite and $m = \infty$.

Exercise 1.63 Draw the Hasse diagram of the Bruhat order on S_4. (The answer is not pretty, so you should stop once you've convinced yourself you know how to crank the machinery.)

Exercise 1.64 Fix $y \in W$ and let $\underline{y} = (s_1, \ldots, s_d)$ denote a fixed reduced expression for y. Show that $x \leq y$ if and only if $x = s_{i_1} \cdots s_{i_k}$ for some $1 \leq i_1 < i_2 < \cdots < i_k \leq d$. (In this case, we say that x may be obtained as a *subexpression* of the expression $\underline{y}$.) Does the choice of reduced expression for y matter?

Exercise 1.65 The *weak right Bruhat graph* of a Coxeter group W is the directed graph whose vertices are the elements of W and which has an arrow from x to y if $x = ys$ for some $s \in S$ with $\ell(x) < \ell(y)$. Note the difference to the Bruhat graph which more generally has an arrow when $x = yt$ for $t \in T$. Also note that here, in contrast to the Bruhat order as explained in Remark 1.60, right and left multiplication by simple reflections define distinct relations. Draw the weak right Bruhat graph for $I_2(5)$, $I_2(\infty)$, and S_4.

1.2.7 Additional Exercises

We finish this chapter with some entirely optional (but enjoyable!) additional exercises to become familiar with Coxeter groups.

Exercise 1.66 To practice with Coxeter groups, we play with some embeddings.

1. Let $\{s, t, u\}$ be the simple reflections in the Coxeter group of type A_3, in the same order as given by the A_3-diagram. Show that the subgroup generated by su and t is a Coxeter group of type $B_2 = I_2(4)$, with simple reflections $\{su, t\}$, by checking the braid relation. This implies that B_2 embeds inside A_3 as the invariants under a certain automorphism σ, induced by a diagram automorphism of A_3.

2. Let $\{s, t, u, v\}$ be the simple reflections in the Coxeter group of type A_4, in the same order as given by the A_4-diagram. Show that the subgroup generated by su and tv is a Coxeter group of type $H_2 = I_2(5)$, with simple reflections $\{su, tv\}$. However, this subgroup is not the invariants of any diagram automorphism.
3. Embed the Coxeter group of type $I_2(m)$ inside the Coxeter group of type A_{m-1} for $m \geq 3$, using products of distinct simple reflections.

Exercise 1.67 Here are some more embedding exercises.

1. Embed H_3 inside D_6. Embed H_4 inside E_8. Generalize this.
2. Look at "star-shaped" Coxeter groups:

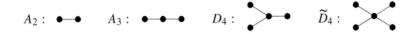

and so forth. Consider the subgroup generated by the "hub" (vertex incident to all edges) and by the product of the "spokes" (the remaining vertices). What subgroups do you get?

Exercise 1.68 Now we do the previous exercises "in reverse." Let (W, S) be a Coxeter system, and fix $s \in S$. Consider the set Γ_s of elements of W which have a unique reduced expression, and which have s in their right descent set. We make Γ_s into a labeled graph, where each element $w \in \Gamma_s$ is labeled by the (unique!) element $t \in S$ in its left descent set, and where w, x are connected by an edge if and only if $w = ux$ for some $u \in S$.

1. Let $\{s, t\}$ be the simple reflections in type B_2. Compute that Γ_s is the A_3 Coxeter graph, with the labelings corresponding to the embedding of B_2 inside A_3 from Exercise 1.66.
2. Do the same for $I_2(m)$ and A_{m-1}.
3. Let (W, S) be the Coxeter group of type H_4. For $s \in S$, compute the labeled graph Γ_s.
4. Repeat the exercise for $I_2(\infty)$. What labeled graph do you obtain?

(If you know about such things, Γ_s is the W-graph of the left cell containing s. See Lusztig's paper [128].)

Exercise 1.69 Here is a non-Coxeter presentation of S_4, which is still quite interesting. The generators are $s = (1, 2), t = (1, 3), u = (1, 4)$.

1. Which braid relations do these generators satisfy? If there were only braid relations, what Coxeter group would it be?
2. What additional relations are satisfied?
3. Count the number of elements of each length with respect to this presentation.

Exercise 1.70 The even-signed symmetric group ESS_n was defined as the subgroup of the signed symmetric group SS_n where the number of sign changes was a multiple of 2. Let $m \in \mathbb{Z}$, $m > 2$. Prove that (unless n is small) the subset of SS_n where the number of sign changes is a multiple of m is not a subgroup.

Exercise 1.71 Coxeter systems (W, S) are equipped with a standard length function ℓ, but can also be equipped with non-standard length functions, sometimes called *weights*. A weight function is a function $L : W \to \mathbb{Z}$ satisfying $L(uv) = L(u) + L(v)$ whenever $\ell(uv) = \ell(u) + \ell(v)$. Deduce the following elementary facts.

1. A weight function L is determined by the weights $L(s)$ of the simple reflections. Moreover, $L(s) = L(t)$ for $s, t \in S$ whenever m_{st} is odd.
2. Suppose one has an embedding of Coxeter groups $\iota : (W, S) \hookrightarrow (W', S')$ as in Exercise 1.66, where each simple reflection $s \in S$ is sent to a product $\prod t$ of commuting simple reflections $t \in S'$. This equips (W, S) with a weight L, given by $L(s) = \ell(\iota(s))$. For each possible value of m_{st}, what are the possible values of the ratio of $L(s)$ to $L(t)$? It will help to remind yourself of the classification of finite Coxeter groups (Theorem 1.34).

Chapter 2
Reflection Groups and Coxeter Groups

This chapter is based on expanded notes of a lecture given by the authors and taken by
Geordie Williamson

Abstract We explain why reflection groups are Coxeter groups. This motivates the general definition of Coxeter groups. We also explain the geometric meaning of reduced expressions and braid relations.

2.1 Reflections and Affine Reflections

This chapter concerns some fundamentals of reflection groups and Coxeter groups. Standard references for this material include [29, Chapter V] and [157, Chapter 2].

Let V denote a finite-dimensional Euclidean vector space over the real numbers. Recall that this means that V is equipped with an inner product, i.e. a positive definite symmetric bilinear form $(-, -)$. If one wishes, one can take $V = \mathbb{R}^n$ with its standard inner product.

Recall that a *reflection* is an orthogonal transformation (i.e. preserving angles and lengths) whose fixed subspace is a hyperplane (the *reflecting hyperplane*). Given any hyperplane $H \subset V$ there is a unique reflection which has H as reflecting hyperplane. Indeed, V is spanned by H and any nonzero vector normal to H, and a reflection has to fix the former and act by -1 on the latter. More precisely, if n denotes a vector normal to H of unit length, then the reflection in the hyperplane perpendicular to H is given by the formula

$$v \mapsto v - 2(v, n)n \,. \tag{2.1}$$

We now forget the origin of V. Recall that a *translation* $V \to V$ is a map of the form $v \mapsto v + b$ for some fixed $b \in V$ and that an *affine transformation* is the composition of a linear transformation and a translation, i.e. a map of the form $v \mapsto f(v) + b$ for some fixed $f \in GL(V)$ and $b \in V$. The set of all

---

G. Williamson
School of Mathematics and Statistics, University of Sydney, Sydney, Australia

affine transformations forms a group Aff(V) under composition. An *affine reflection* is obtained by conjugating a reflection by a translation. Alternatively, an affine reflection is the reflection in an affine hyperplane (i.e. a hyperplane which does not necessarily pass through the origin). We still refer to this fixed locus as the *reflecting hyperplane*. If n is as above then for any $\gamma \in \mathbb{R}$ the formula

$$v \mapsto v - 2(v, n)n + 2\gamma n \tag{2.2}$$

defines an affine reflection which fixes the affine hyperplane obtained by translating by γn the hyperplane of vectors perpendicular to n. Any affine reflection is of this form.

2.2 Affine Reflection Groups

Let us fix a subgroup W of the group of affine transformations of V. We assume:

- W is generated by affine reflections;
- W is *proper*, i.e. for any compact sets $K, L \subset V$ the set of $w \in W$ such that $K \cap wL \neq \emptyset$ is finite.

Such a subgroup is called an *affine reflection group*.

Remark 2.1 The above assumptions are satisfied if W is finite, in which case W is a *finite reflection group* (see Sects. 1.1.6 and 1.1.7). All finite reflection groups can be generated by ordinary (rather than affine) reflections. For this reason, the literature often uses the term "affine reflection group" to refer only to infinite affine reflection groups.

Lemma 2.2 *Every orbit of W is a discrete subset of V with its natural topology.*

Proof Let $v \in V$. Since V is a metric space, there is a compact neighborhood K of v. Choose an open ball U of v contained in K. Since w acts continuously on V, the set wU is open and contains wv. Taking $L = K$ in the assumption on W, we conclude that $U \cap wU \neq \emptyset$ for only finitely many $w \in W$. With only finitely many open sets to worry about, we can shrink U until $U \cap wU = \emptyset$ for all w. Then, since the sets wU for $w \in W$ cover the orbit Wv, we conclude that Wv is discrete. □

Example 2.3 Here are three examples of affine reflection groups.

1. Consider $V = \mathbb{R}$ with its standard Euclidean structure. Let W denote the group generated by reflections in the integral points $\mathbb{Z} \subset \mathbb{R}$:

2.3 Affine Reflection Groups are Coxeter Groups

2. Consider $V = \mathbb{R}^2$ with its standard Euclidean structure. Let W denote the affine reflection group generated by reflections in the following affine arrangement of hyperplanes:

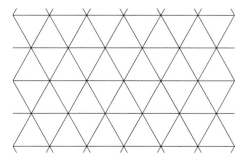

3. Consider $V = \mathbb{R}^2$ with its standard Euclidean structure. Consider the group generated by reflections in the following arrangement:

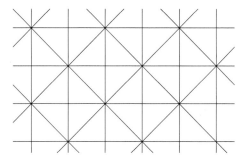

(This is the group of symmetries of the "square lattice.")

Exercise 2.4 Show that the group generated by the reflections in the first example of Example 2.3 is the infinite dihedral group $I_2(\infty)$. Thus $I_2(\infty)$ is an affine reflection group. In contrast, the action of $I_2(\infty)$ on the Euclidean plane using an irrational angle, as constructed in Remark 1.32, fails to make $I_2(\infty)$ into an affine reflection group; what part of the definition is violated?

2.3 Affine Reflection Groups are Coxeter Groups

We will show that any affine reflection group is naturally a Coxeter group. We first have to introduce some terminology to be able to study the combinatorics of the reflecting hyperplanes. Let

$$\Phi := \{H \mid H \text{ is a reflecting hyperplane for some reflection in } W\} . \qquad (2.3)$$

Given $H \in \Phi$ we denote by s_H the corresponding reflection in W. Choosing a vector n normal to H, we can write

$$V \setminus H = \{v \in V \mid (v, n) > 0\} \cup \{v \in V \mid (v, n) < 0\}. \tag{2.4}$$

These two subsets are the connected components of $V \setminus H$ and we call them the *half-spaces* defined by H. We say that $v, w \in V$ lie on the *same side* of H if they are contained in the same half-space; if v belongs to one of the half-spaces and w to the other, we say they lie on *opposite sides* and that they are *separated* by H.

We define an equivalence relation $\sim$ on V (relative to Φ) as follows: $v \sim w$ if for any $H \in \Phi$ either $v, w \in H$ or v, w lie on the same side of H. The equivalence classes of this relation are called the *facets* of V (relative to Φ). For a facet F the subset

$$\operatorname{supp} F := \bigcap_{\substack{F \subseteq H \\ H \in \Phi}} H \tag{2.5}$$

is an affine subspace, called the *support* of F. The *dimension* of F is defined as the dimension of $\operatorname{supp} F$.

Exercise 2.5 Show that the subset Φ of V is *locally finite*, i.e. any $v \in V$ has a neighborhood that intersects only finitely many of the $H \in \Phi$. (*Hint:* Use properness of W).

The exercise implies that the set $V \setminus \bigcup_{H \in \Phi} H$ is open in V. Let

$$\begin{aligned}\mathcal{A} &:= \text{connected components of } V \setminus \bigcup_{H \in \Phi} H, \\ \overline{\mathcal{A}} &:= \{\overline{A} \mid A \in \mathcal{A}\}.\end{aligned} \tag{2.6}$$

We refer to elements of $\mathcal{A}$ (resp. $\overline{\mathcal{A}}$) as *alcoves* (resp. *closed alcoves*). A *face* of an alcove A is a facet contained in the closure of A whose support is a hyperplane; and a *wall* of A is a hyperplane that is the support of a face of A. Note that

$$V = \bigcup_{\overline{A} \in \overline{\mathcal{A}}} \overline{A}. \tag{2.7}$$

Example 2.6 The picture below illustrates the two hyperplanes H and H' in $\mathbb{R}^2$ spanned by the vectors $(-1, 1)$ and $(1, 1)$, respectively. There are 4 alcoves which are denoted by $A_0, \ldots, A_3$ and are colored below. Each alcove borders two walls, namely H and H'. The half of H which is contained in the closure of A_0, but with the origin removed, is a face of A_0. The set of facets in this example consists of the four alcoves (dimension 2), their four faces (dimension 1), and the origin (dimension 0).

2.3 Affine Reflection Groups are Coxeter Groups

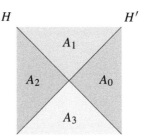

Let us make an arbitrary but fixed choice of $\Delta \in \mathcal{A}$ and call it the *fundamental alcove*. We denote by Φ_Δ the set of hyperplanes which contain the walls of Δ. The set of reflections in the walls of Δ will be denoted by S. Set

$$W_S := \langle S \rangle . \tag{2.8}$$

Note that W acts naturally on the sets $\mathcal{A}$ and $\overline{\mathcal{A}}$. In fact:

Lemma 2.7 *W_S acts transitively on $\mathcal{A}$ and on $\overline{\mathcal{A}}$, i.e. for any $A \in \mathcal{A}$ there is $w \in W_S$ such that $wA = \Delta$ and $w\overline{A} = w\overline{\Delta}$.*

Proof We first show that for any $v \in V$ there is $w \in W_S$ with $w(v) \in \overline{\Delta}$. If $v \in \overline{\Delta}$, this is obvious. Assume that $v \notin \overline{\Delta}$. Fix $\rho \in \Delta$. Then there is a hyperplane $H \in \Phi_\Delta$ which separates ρ and v. If s denotes the reflection in H then $||v-\rho|| > ||s(v)-\rho||$. Because W_S has discrete orbits by Lemma 2.2, there are finitely many points in the W_S-orbit of v which are of distance at most $||v - \rho||$ from ρ. Hence, using reflections from W_S we can keep reducing the distance from ρ to v until this is no longer possible, i.e. until $v \in \overline{\Delta}$.

Now, let $A \in \mathcal{A}$ and pick $v \in \overline{A}$. By the above we know that there is $w \in W_S$ with $w(v) \in \overline{\Delta}$. Since W acts on $\overline{\mathcal{A}}$, it follows that $w\overline{A}$ must be a closed alcove; as $w\overline{A} \cap \overline{\Delta} \neq \emptyset$, we must have $w\overline{A} = \overline{\Delta}$. Finally, since w acts continuously, it follows that $wA = \Delta$. □

Lemma 2.8 *$W = W_S$, i.e. W is generated by S.*

Proof Because W is generated by the reflections it contains, it is enough to show that any reflection in W belongs to W_S. To this end, fix $H \in \Phi$ and let r denote the corresponding affine reflection. Choose an alcove $A \in \mathcal{A}$ having H as wall and denote by F the corresponding face of A. By Lemma 2.7, there is $w \in W_S$ such that $wA = \Delta$. Let $s \in S$ be the reflection in the wall containing the face $wF \subset wA = \Delta$ of Δ. Then $w^{-1}sw = r$. Indeed, the left hand side is a reflection which fixes F and hence H, and hence must be r. □

Lemma 2.9 *Suppose that H and H' belong to Φ_Δ (i.e. H and H' constitute walls of Δ). If H and H' intersect, then they do so at an angle $\leq \pi/2$. Moreover, this angle is of the form π/m for some $m \in \mathbb{N}$.*

Proof Suppose for contradiction that H and H' intersect at an angle $> \pi/2$. Then reflecting H' in the hyperplane H would yield a hyperplane in the interior of Δ, which is a contradiction:

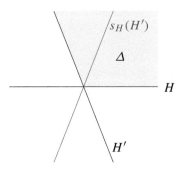

To see the second claim is a piece of cake since by properness, the cake is cut into finitely many pieces:

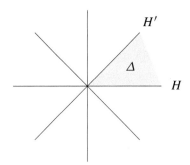

□

If s and t denote the reflections in the hyperplanes $H, H' \in \Phi_\Delta$ then we define

$$m_{st} := \begin{cases} m \text{ (of the previous lemma)} & \text{if } H \text{ and } H' \text{ meet,} \\ \infty & \text{if } H \text{ and } H' \text{ do not meet.} \end{cases} \quad (2.9)$$

The composition of two reflections in distinct parallel hyperplanes is a non-trivial translation, thus of infinite order. Meanwhile, the composition of two reflections in hyperplanes meeting at an angle of π/m is a rotation through $2\pi/m$, thus of order m. Hence:

Lemma 2.10 *For s, t as above, the order of $st \in W$ is m_{st}.*

We have established the easy part (i.e. that the relations are satisfied) of the following fundamental theorem:

2.4 Expressions and Strolls

Theorem 2.11 *W admits the following Coxeter presentation:*

$$W = \langle s \in S \mid s^2 = \mathrm{id} \text{ for all } s \in S, (st)^{m_{st}} = \mathrm{id} \text{ for all distinct } s, t \in S \rangle.$$

Exercise 2.12 Use the theorem to calculate presentations for all groups described in Example 2.3. Draw the corresponding Coxeter graph.

2.4 Expressions and Strolls

We continue with the notation from above. A *stroll* is a sequence $\underline{A} := (A_0, A_1, \ldots, A_k)$ of alcoves such that $A_0 = \Delta$ and A_{i-1} and A_i share a codimension 1 face F_i for all $1 \leq i \leq k$. We think of a stroll as a path in V beginning in Δ and only passing through codimension 1 parts of the hyperplane arrangement Φ (see the examples below). The *length* $\ell(\underline{A})$ is the number of hyperplane crossings of the path (i.e. if $\underline{A}$ is as above then $\ell(\underline{A}) = k$). A stroll is *reduced* if F_i and F_j are never contained in the same hyperplane for $i \neq j$, i.e. if our stroll "never crosses the same reflecting hyperplane twice".

Example 2.13 Two strolls ending in the same element; one is reduced, one is not:

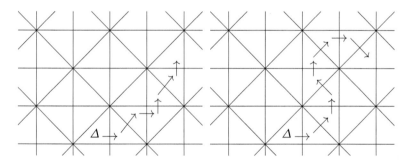

Remark 2.14 Starting in Sect. 3.3.4, we will redefine a stroll so that it also allows $A_i = A_{i-1}$. That is, a stroll is like a walk from alcove to alcove, where one might pause to admire the scenery. For the rest of this chapter, however, $A_i \neq A_{i-1}$.

Let $\underline{x} = (s_1, s_2, \ldots, s_k)$ be an expression. Note that Δ and $s_1 \Delta$ share a common face. Similarly, Δ and $s_2 \Delta$ share a common face. Multiplying by s_1, we deduce that $s_1 \Delta$ and $s_1 s_2 \Delta$ share a common face. Continuing like this we see that

$$\underline{A}(\underline{x}) := (A_0 = \Delta, A_1 = s_1 \Delta, A_2 = s_1 s_2 \Delta, \ldots, A_k = s_1 s_2 \cdots s_k \Delta) \quad (2.10)$$

is a stroll. We can thus associate to any expression a stroll. The following proposition tells us that (reduced) expressions and (reduced) stroll are essentially the same thing:

Proposition 2.15 *An expression $\underline{x}$ for $x \in W$ is reduced if and only if the corresponding stroll $\underline{A}(\underline{x})$ is reduced. Moreover, we have*

$$\ell(x) = \#\{H \in \Phi \mid H \text{ separates } \Delta \text{ and } x\Delta\}.$$

Example 2.16 In Example 2.13, if we assume that $x\Delta$ is the end point of the drawn path, then we see that there are six hyperplanes separating Δ and $x\Delta$, so $\ell(x) = 6$.

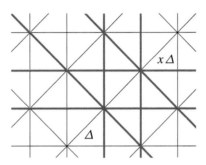

Proof Let us temporarily define

$$\ell'(x) := \#\{H \in \Phi \mid H \text{ separates } \Delta \text{ and } x\Delta\}.$$

We will argue by induction on $\ell(x)$ that $\ell(x) = \ell'(x)$ and that any reduced expression for x yields a reduced stroll. Let $\underline{x} = (s_1, \ldots, s_k)$ denote a reduced expression for x and let $\underline{y} = (s_1, \ldots, s_{k-1})$. Then $\underline{y}$ is a reduced expression for $y = s_1 \cdots s_{k-1}$ since an expression of length $< k-1$ for y would yield an expression of length $< k$ for x, contradicting $\ell(x) = k$. Thus we can apply induction to conclude that $\ell(y) = \ell'(y)$ and that $\underline{A}(\underline{y})$ crosses $k - 1$ distinct hyperplanes. Now consider $\underline{A}(\underline{x})$. Either $\ell(x) = \ell'(x)$ or the hyperplane H crossed from $y\Delta$ to $x\Delta$ has already been crossed in $\underline{A}(\underline{y})$:

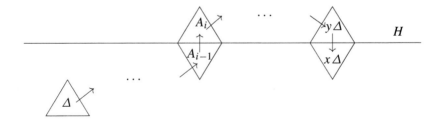

2.4 Expressions and Strolls

Let A_{i-1} and A_i with $i < k$ be two alcoves where this hyperplane is crossed earlier. Then $(s_1, \ldots, s_{i-1}, s_{i+1}, \ldots, s_{k-1})$ is an expression for x which is shorter than k. The corresponding stroll is obtained by reflecting the stroll between i and $k-1$ in the hyperplane H:

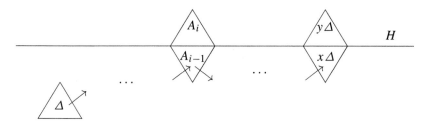

This contradicts the fact that $\ell(x) = k$. Hence $\ell(x) = \ell'(x)$ and we are done. □

Corollary 2.17 $x\Delta = \Delta$ *if and only if* $x = \mathrm{id}$.

Proof If $x\Delta = \Delta$ then x is of length zero in the generators, and hence $x = \mathrm{id}$. □

Combining this result with Lemma 2.7 yields:

Corollary 2.18 $\overline{\Delta}$ *is a* fundamental domain *for the W-action on V, i.e. every W-orbit in V meets $\overline{\Delta}$ in exactly one point.*

In particular the map $x \mapsto x\Delta$ is a bijection. We can use this bijection to identify W and $\mathcal{A}$. This is particularly useful as it allows us to deduce properties of W via the geometry of V and its decomposition into the sets $\mathcal{A}$.

Exercise 2.19 Modify the proof of Proposition 2.15 to prove the Exchange Condition and the Deletion Condition for W (see Sect. 1.2.3).

Recall the braid relations of a Coxeter group from (1.3). We now want to give an outline of why Matsumoto's theorem (Theorem 1.56) can be explained naturally in the context of affine reflection groups.

Theorem 2.20 (Matsumoto's Theorem) *Any two reduced expressions for $x \in W$ may be related by braid relations.*

Sketch of Proof We induct on the length $\ell(x)$ of a reduced expression. Let $\underline{x}_1 = (s_1, \ldots, s_\ell)$ and $\underline{x}_2 = (s'_1, \ldots, s'_\ell)$ denote two reduced expressions. If $s_\ell = s'_\ell$ then we can apply induction to deduce that there is a sequence of braid relations relating $(s_1, \ldots, s_{\ell-1})$ and $(s'_1, \ldots, s'_{\ell-1})$ and the result follows. Now assume that $s_\ell \neq s'_\ell$ and consider the hyperplane H (resp. H') separating $x\Delta$ and $xs_\ell\Delta$ (resp. $x\Delta$ and $xs'_\ell\Delta$):

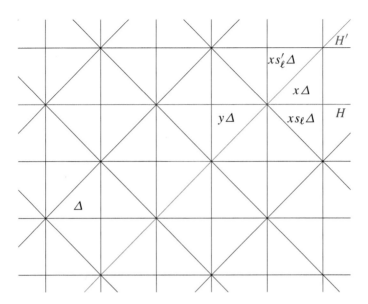

There is a unique alcove which contains $H \cap H'$ in its closure and lies on the same side of H and H' as the fundamental alcove. Suppose this alcove is labeled by $y \in W$. If we choose a reduced stroll from Δ to $y\Delta$ then we may extend it in two ways to a reduced stroll from Δ to $x\Delta$. Indeed, if our fixed reduced stroll from Δ to $y\Delta$ corresponds to the reduced expression $\underline{w} = (t_1, \ldots, t_k)$, then the two ways are

$$\underline{w}_1 := (t_1, \ldots, t_k, \underbrace{\ldots, s'_\ell, s_\ell}_{m}) \quad \text{and} \quad \underline{w}_2 := (t_1, \ldots, t_k, \underbrace{\ldots, s_\ell, s'_\ell}_{m}).$$

Note that $\underline{w}_1$ and $\underline{w}_2$ are related by a single braid relation.

Now we are done: by induction we have $\underline{x}_1 \sim \underline{w}_1 \sim \underline{w}_2 \sim \underline{x}_2$, where $\sim$ indicates that we can pass between the two expressions via a sequence of braid relations. □

Exercise 2.21 Use similar ideas to prove Proposition 1.54 for affine reflection groups.

2.5 Classification of Affine Reflection Groups

Let $\widetilde{W}$ be an affine reflection group, and $v \in V$ a 0-dimensional facet (think: the origin). If one considers the subgroup of $\widetilde{W}$ generated by those reflections which fix v, the result will be a finite reflection group $W \subset \widetilde{W}$ (by properness).

The affine reflection groups $\widetilde{W}$ in Example 2.3 all share a common feature: there is a finite reflection subgroup $W \subset \widetilde{W}$ and a lattice (a free abelian subgroup) $\Lambda \subset V$ such that every reflecting hyperplane is just a translation, by an element of Λ, of a reflecting hyperplane which fixes v. One can verify that Λ must be preserved by W

2.6 The Coxeter Complex

for this to make sense. Properness also implies that Λ must be a lattice, rather than a non-discrete subgroup.

We have no desire to go into the proof, but these considerations can be used to classify all affine reflection groups. In particular, they are all obtained by taking a finite reflection group W which preserves a lattice Λ and translating the hyperplanes accordingly. In the classification of Theorem 1.34, the Coxeter groups of type H or of type $I_2(m)$ for $m \neq 2, 3, 4, 6$ can not preserve a lattice, so they will not give rise to affine reflection groups. (Reflection groups preserving a lattice of full rank are called *crystallographic*.) The crystallographic finite reflection groups coincide with the so-called (finite) Weyl groups. Thus each Weyl group has its associated affine Weyl groups. If the original Weyl group is irreducible then the corresponding affine group is generated by the finite Weyl group and a single affine reflection. We encourage the reader to look up the list of affine Weyl groups and their Coxeter graphs; all affine reflection groups are affine Weyl groups.

2.6 The Coxeter Complex

In the previous sections we have seen that affine reflection groups are Coxeter groups in a natural way. Along the way we have seen that the action of W on the set of alcoves makes fundamental properties of W (for example the deletion and exchange conditions, and Matsumoto's theorem) geometrically transparent.

In the context of general Coxeter groups it is natural to ask whether there exists a space with a W-action similar to the collection of alcoves considered above. It turns out that there is such a space, called the Coxeter complex. Here we give a brief introduction, for more detail the reader is referred to [157, Chapter 2].

Let (W, S) be a Coxeter group of rank n. Our aim is to build an $(n-1)$-dimensional simplicial complex from (W, S). To this end, consider the standard $(n-1)$-simplex Δ embedded in affine $(n-1)$-space, and fix a coloring of the n faces of Δ by S. For any $s \in S$ we can reflect Δ along the face colored by s. This gives a way to glue two copies of Δ together, which we refer to as an *s-glueing*.

Example 2.22 An example of an s-glueing ($|S| = 3$ and s is red):

The *Coxeter complex* of (W, S) is constructed as follows:

- Take one copy of Δ as above for each $w \in W$ and call it Δ_w.
- For all $w \in W$ and $s \in S$, glue Δ_w to Δ_{ws} via an s-glueing.

The result is a simplicial complex of dimension $n - 1$ with faces (i.e. simplices of dimension $n - 2$) colored by S. The following exercises help to get used to the Coxeter complex.

Exercise 2.23 Let (W, S) be the type A_2 Coxeter system with $S = \{s, t\}$ (so that $W = \{\text{id}, s, t, st, ts, sts\}$). Show that its Coxeter complex is the following simplicial complex:

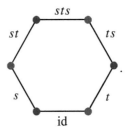

Remark 2.24 It is easy to see that W acts faithfully on its Coxeter complex "on the left" in a natural way. The action of w identifies Δ_x with Δ_{wx}. However, there is no action of W "on the right," as attempting to identify Δ_x and Δ_{xw} for each x simultaneously will not work (it can be defined on the interior of each simplex, but will not extend continuously to the boundary). This asymmetry between left and right multiplication is a common source of confusion. To help clarify this, consider a simple reflection s. Left multiplication by s (i.e. sending Δ_x to Δ_{sx}) is a global operation like a reflection, which can send a simplex to another simplex far, far away (e.g. t and st in Example 2.23). Meanwhile, "right multiplication" by s (i.e. crossing an s-colored face, sending Δ_x to Δ_{xs}) is a local operation, sending a simplex to an adjacent simplex.

Exercise 2.25 Describe the Coxeter complex for $I_2(m)$ for m finite, as well as for $I_2(\infty)$.

Exercise 2.26 Show that if W is of type A_3 then its Coxeter complex is isomorphic to the barycentric subdivision of a tetrahedron. Can you give a similar description for W of type B_3?

On the supplementary website to the book (see page ix for the link) the reader can find the Coxeter complex for types A_3, B_3, and $\tilde{A}_2$.

Exercise 2.27 Describe the automorphism group of the Coxeter complex, both with and without the coloring of its faces. (First, define what an automorphism is.)

The following exercise should convince you that the Coxeter complex provides the correct generalization of the alcove picture to arbitrary Coxeter groups.

Exercise 2.28 Describe the Coxeter complexes for all groups in Example 2.3.

The above description of the Coxeter complex is somewhat ad hoc. In the following exercise we give an alternative definition, which is often more useful for

2.6 The Coxeter Complex

theoretical considerations. (For this exercise you will need to know what an abstract simplicial complex is.)

Exercise 2.29 A *standard parabolic subgroup* of W is a subgroup $W_I \subset W$ generated by a subset $I \subset S$. Consider the set $X = \{xW_I \mid x \in W, I \subset S\}$ of all left cosets of standard parabolic subgroups of W. View X as a poset by reverse inclusion. Show that for any $xW_I \in X$, the set of elements less than xW_I is naturally in bijection with subsets of $S \setminus I$. In particular, our poset X describes an abstract simplicial complex. Check in a few examples that its realization agrees with your calculations of the Coxeter complex above.

We conclude by mentioning an interesting topological result, evidence for which we have seen in the exercises above.

Theorem 2.30 *When W is finite of rank n, the Coxeter complex is homeomorphic to the sphere S^{n-1}. When W is infinite, the Coxeter complex is contractible.*

For further discussion and a proof, see [37, §IV.5-6]. For a beautiful proof via the concept of a "shelling," see [27, Theorem 2.1, Corollary 2.2(iii)].

Chapter 3
The Hecke Algebra and Kazhdan–Lusztig Polynomials

This chapter is based on expanded notes of a lecture given by the authors and taken by
Joel Gibson and Alexander Kerschl

Abstract The Hecke algebra associated to a Coxeter system (W, S) is a $\mathbb{Z}[v, v^{-1}]$-algebra that recovers the group algebra $\mathbb{Z}[W]$ under the specialization $v \mapsto 1$. It has two important bases: the standard basis, and a certain "self-dual" basis called the Kazhdan–Lusztig basis, with the change of basis between the two given by the Kazhdan–Lusztig polynomials. In this chapter, we briefly review the definition and construction of the Hecke algebra and its Kazhdan–Lusztig basis.

3.1 The Hecke Algebra

This chapter covers fundamentals concerning the Hecke algebra and Kazhdan–Lusztig polynomials. We follow the streamlined presentation of Soergel [163, §2]. Some other references for this material are [85, Chapter 7] and [76].

Throughout, we fix a Coxeter system (W, S). Moreover, we fix an indeterminate v over $\mathbb{Z}$ and denote by $\mathbb{Z}[v, v^{-1}]$ the *Laurent polynomial ring*, i.e. the localization of the polynomial ring $\mathbb{Z}[v]$ in the multiplicative set generated by v.

Definition 3.1 The *Hecke algebra* $H := H(W)$ is the unital associative algebra over $\mathbb{Z}[v, v^{-1}]$ generated by the symbols $\{\delta_s \mid s \in S\}$, subject to the following relations:

- (*Quadratic relation*) For all simple reflections $s \in S$,

$$\delta_s^2 = (v^{-1} - v)\delta_s + 1. \qquad (3.1)$$

J. Gibson · A. Kerschl
School of Mathematics and Statistics, University of Sydney, Sydney, NSW, Australia

- (*Braid relation*) For all $s, t \in S$ with $m_{st} < \infty$,

$$\underbrace{\delta_s \delta_t \delta_s \cdots}_{m_{st}} = \underbrace{\delta_t \delta_s \delta_t \cdots}_{m_{st}}. \tag{3.2}$$

Remark 3.2 The definition above is the "one-parameter" or "equal parameter" Hecke algebra. The Hecke algebra and much of the theory of its Kazhdan–Lusztig basis can be developed in a more general setting ("multiparameter" or "unequal parameter"); see Lusztig [126].

The Hecke algebra is one of the fundamental objects of study in representation theory. More explicit motivation for the Hecke algebra can be found in the introduction to this book.

For any commutative ring A, a ring morphism $\varphi : \mathbb{Z}[v, v^{-1}] \to A$ gives A the structure of a $\mathbb{Z}[v, v^{-1}]$-algebra, and we may form the *specialization*

$$H_\varphi := A \otimes_{\mathbb{Z}[v,v^{-1}]} H \tag{3.3}$$

of H. A specialization of H is uniquely determined by the ring A and an invertible element $\varphi(v) \in A^\times$. When A is understood, we simply call H_φ the *specialization of* H *at* $v \mapsto \varphi(v)$. When $A = \mathbb{Z}$, the specialization of H at $v \mapsto 1$ is isomorphic to the group algebra $\mathbb{Z}[W]$, with δ_s being sent to s for each $s \in S$. For this reason, the Hecke algebra is called a *deformation* of the group algebra of W.

Remark 3.3 The quadratic relation is equivalent to

$$(\delta_s - v^{-1})(\delta_s + v) = 0. \tag{3.4}$$

Thus the eigenvalues of the action of δ_s on H are v^{-1} and $-v$. (This leads to a useful mnemonic for the quadratic relation. Both eigenvalues have a minus sign somewhere: one in the exponent, and one in the coefficient.) When v is specialized to 1, one recovers the usual statement that an involution s has eigenvalues $+1$ and -1.

3.1.1 The Standard Basis

Given $x \in W$, choose a reduced expression $\underline{x} = (s_1, s_2, \ldots, s_m)$ for x and define the element

$$\delta_x := \delta_{s_1} \delta_{s_2} \cdots \delta_{s_m} \tag{3.5}$$

3.1 The Hecke Algebra

of H. By Matsumoto's theorem (Theorem 1.56) and the braid relation (3.2), δ_x does not depend on the choice of reduced expression. Note that $\delta_{\text{id}} = 1$, and that this notation is consistent with our notation for the generators δ_s.

Definition 3.4 The *standard basis* of the Hecke algebra H is the set $\{\delta_x \mid x \in W\}$.

The word "basis" in Definition 3.4 is justified by Theorem 3.5 below. Before sketching a proof, we give some basic formulae for multiplying an arbitrary standard basis element δ_x and δ_s for some simple reflection s.

Let $x \in W$, $\underline{x} = (s_1, \ldots, s_n)$ a reduced expression for x, and $s \in S$. Recall the Bruhat order on W defined in Sect. 1.2.6. If $xs > x$, then $(s_1, \ldots, s_n, s)$ is reduced, and we have $\delta_x \delta_s = \delta_{xs}$. On the other hand, if $xs < x$, then by Corollary 1.49, we may replace $\underline{x}$ with a reduced expression ending in $s_n = s$. Applying the quadratic relation (3.1), we find

$$\delta_x \delta_s = \delta_{s_1} \cdots \delta_{s_{n-1}} \delta_s^2 = (v^{-1} - v)\delta_x + \delta_{xs}.$$

A similar analysis holds for left multiplication by δ_s, leading to the following two formulae:

$$\delta_x \delta_s = \begin{cases} \delta_{xs} & \text{if } x < xs, \\ (v^{-1} - v)\delta_x + \delta_{xs} & \text{if } x > xs, \end{cases} \tag{3.6}$$

and

$$\delta_s \delta_x = \begin{cases} \delta_{sx} & \text{if } x < sx, \\ (v^{-1} - v)\delta_x + \delta_{sx} & \text{if } x > sx. \end{cases} \tag{3.7}$$

Theorem 3.5 *The set $\{\delta_x \mid x \in W\}$ is a $\mathbb{Z}[v, v^{-1}]$-basis of the Hecke algebra H.*

Proof To show that the elements $\{\delta_x \mid x \in W\}$ span H, it is enough to show that for any $x \in W$ and $s \in S$, the products $\delta_x \delta_s$ and $\delta_s \delta_x$ can be expressed as linear combinations of the other δ_y. This is immediate from the multiplication rules (3.6) and (3.7).

Showing linear independence is considerably more difficult. A standard method to show that some elements of an algebra are linearly independent is to find a convenient representation on which that algebra acts, and to argue that a nontrivial linear dependence would give a contradiction. Here, we construct a model of the regular representation of H in our putative basis $\{\delta_x \mid x \in W\}$, and check by hand that the action is well-defined. We only sketch the proof, omitting several tedious verifications; for a complete proof see [85, Chapter 7, Sections 1–3].

Let E be the free $\mathbb{Z}[v, v^{-1}]$-module with basis $\{e_x \mid x \in W\}$: this will be the underlying module for our model of the regular representation. Firstly, we wish to construct for each $s \in S$, operators $L_s, R_s : E \to E$ mimicking left and right multiplication by the generators δ_s. In similarity to (3.6) and (3.7) we define

$$R_s(e_x) = \begin{cases} e_{xs} & \text{if } x < xs, \\ (v^{-1} - v)e_x + e_{xs} & \text{if } x > xs, \end{cases} \quad (3.8)$$

$$L_s(e_x) = \begin{cases} e_{sx} & \text{if } x < sx, \\ (v^{-1} - v)e_x + e_{sx} & \text{if } x > sx. \end{cases} \quad (3.9)$$

One can show by a case-by-case analysis that L_s and R_t commute for any $s, t \in S$.

Next, define $A = \langle L_s \mid s \in S \rangle$, the subalgebra of $\text{End}_{\mathbb{Z}[v,v^{-1}]}(E)$ generated by the operators L_s. Let $\phi : A \to E$ be the evaluation map $\phi(f) = f(e_{\text{id}})$; we wish to show that ϕ is an isomorphism of $\mathbb{Z}[v, v^{-1}]$-modules. If $(s_1, \ldots, s_n)$ is a reduced expression for $x \in W$, then by (3.9) we have $L_{s_1} \cdots L_{s_n}(e_{\text{id}}) = e_x$, and hence ϕ is surjective. To show injectivity of ϕ, suppose that $\phi(f) = 0$. We immediately have that $f(e_{\text{id}}) = 0$. If $f(e_w) = 0$ and $ws > w$ for some simple reflection $s \in S$, then $f(e_{ws}) = f(R_s(e_w)) = R_s(f(e_w)) = 0$, where we have used the fact that R_s commutes with the subalgebra A. By induction on the length of w, we conclude that $f = 0$, and hence ϕ is an isomorphism.

After checking that the operators L_s satisfy the quadratic relation (3.1) and the braid relation (3.2) (these are the tedious verifications alluded to above), we may invoke the universal property of H to get a surjection of algebras $\psi : \text{H} \to A$, defined on the generators by $\psi(\delta_s) = L_s$. Composing with ϕ gives a surjection of $\mathbb{Z}[v, v^{-1}]$-modules $\text{H} \to E$, taking δ_x to e_x, and hence the elements $\{\delta_x \mid x \in W\}$ form a basis of H. □

3.1.2 Inversion

For a simple reflection s, the quadratic relation implies

$$\delta_s^{-1} = \delta_s + (v - v^{-1}), \quad (3.10)$$

so each generator δ_s is invertible. Being products of the δ_s's, every δ_w is invertible. It will be useful to have an idea of what δ_w^{-1} looks like when expanded in the standard basis.

3.1 The Hecke Algebra

Lemma 3.6 *For all $w \in W$, the standard basis element δ_w is invertible in H. Moreover, we have that*

$$\delta_{w^{-1}}^{-1} = \delta_w + \sum_{x<w} a_x \delta_x \qquad (3.11)$$

for some coefficients $a_x \in \mathbb{Z}[v, v^{-1}]$.

In the proof, we will need the following important concept that we introduced already in Exercise 1.64.

Definition 3.7 Let $\underline{x} = (s_1, \ldots, s_m)$ be an arbitrary expression. A *subexpression* of $\underline{x}$ is a string $\underline{e} = e_1 \ldots e_m$ of length m, where each $e_i \in \{0, 1\}$. We write $\underline{e} \subset \underline{x}$ to mean that $\underline{e}$ is a subexpression of $\underline{x}$.

We think of a subexpression as an expression obtained by "crossing out" some of the terms in $\underline{x}$, where $e_i = 0$ (resp. 1) indicates that s_i is "crossed out" (resp. kept) in our subexpression.

Proof of Lemma 3.6 Suppose that $\underline{w} = (s_1, \ldots, s_m)$ is a reduced expression for w, and $\underline{e} \subset \underline{w}$ is a subexpression. We first show that

$$\delta_{s_1}^{e_1} \delta_{s_2}^{e_2} \cdots \delta_{s_m}^{e_m} \in \sum_{x \leq w} \mathbb{Z}[v, v^{-1}] \delta_x. \qquad (3.12)$$

Consider the expression $(r_1, \ldots, r_k)$ obtained by omitting the terms in $\underline{w}$ where $e_i = 0$. If this expression is already reduced, then $\delta_{r_1} \delta_{r_2} \cdots \delta_{r_k} = \delta_{r_1 r_2 \cdots r_k}$ and we are done. Otherwise, let $1 \leq i < k$ be maximal such that $(r_1, r_2, \ldots, r_i)$ is reduced. Then $(r_1, r_2, \ldots, r_i, r_{i+1})$ is not reduced and so by the deletion condition (Corollary 1.50) there exist $1 \leq a < b \leq i+1$ such that $(r_1, \ldots, \widehat{r_a}, \ldots, \widehat{r_b}, \ldots, r_{i+1})$ is reduced. So we obtain

$$\delta_{r_1} \delta_{r_2} \cdots \delta_{r_k} = (\delta_{r_1 r_2 \cdots r_i} \delta_{r_{i+1}}) \delta_{r_{i+2}} \cdots \delta_{r_k}$$
$$= ((v^{-1} - v)\delta_{r_1 r_2 \cdots r_i} + \delta_{r_1 r_2 \cdots r_i r_{i+1}}) \delta_{r_{i+2}} \cdots \delta_{r_k}$$
$$= (v^{-1} - v)\delta_{r_1}\delta_{r_2} \cdots \widehat{\delta_{r_{i+1}}} \cdots \delta_{r_k} + \delta_{r_1}\delta_{r_2} \cdots \widehat{\delta_{r_a}} \cdots \widehat{\delta_{r_b}} \cdots \delta_{r_k}.$$

From here, (3.12) follows by induction on k.

Now, let $(s_1, \ldots, s_m)$ be a reduced expression for w. Then

$$(\delta_{w^{-1}})^{-1} = \delta_{s_1}^{-1} \cdots \delta_{s_m}^{-1} = (\delta_{s_1} + (v - v^{-1})) \cdots (\delta_{s_m} + (v - v^{-1})). \qquad (3.13)$$

The last product expands to δ_w plus a sum of products of δ_{s_i}'s (with coefficients) which all give proper subexpressions of $\underline{w}$, so the lemma follows from (3.12). □

3.2 The Kazhdan–Lusztig Basis

The Hecke algebra admits another basis known as the *Kazhdan–Lusztig basis*. We first introduce an involution on H, which will be used to characterize this new basis.

Definition 3.8 The *Kazhdan–Lusztig involution* (or *bar involution*)

$$H \to H : h \mapsto \overline{h} \tag{3.14}$$

is the $\mathbb{Z}$-linear map defined on the generators as

$$\overline{\delta_s} = \delta_s^{-1} = \delta_s + (v - v^{-1}), \tag{3.15}$$

on Laurent polynomials by $\overline{v} = v^{-1}$, and extended to products as a ring automorphism, i.e. $\overline{ab} = \overline{a}\,\overline{b}$.

The *Kazhdan–Lusztig anti-involution*

$$\omega : H \to H \tag{3.16}$$

is defined in the same way on generators and Laurent polynomials, but extended to products as a ring anti-automorphism, i.e. $\omega(ab) = \omega(b)\omega(a)$.

For any $x \in W$, by choosing a reduced expression and writing δ_x as the corresponding product in the generators δ_s, we find that

$$\overline{\delta_x} = \delta_{x^{-1}}^{-1}, \tag{3.17}$$

$$\omega(\delta_x) = \delta_x^{-1}. \tag{3.18}$$

Definition 3.9 A *Kazhdan–Lusztig basis* is a set $\{b_x \mid x \in W\} \subseteq H$ with the following two properties: for any $x \in W$,

1. (*self-duality*): $\overline{b_x} = b_x$, and
2. (*degree bound*): b_x has the form

$$b_x = \delta_x + \sum_{y < x} h_{y,x} \delta_y \quad \text{for some } h_{y,x} \in v\mathbb{Z}[v], \tag{3.19}$$

where $<$ denotes the Bruhat order.

The coefficients $h_{y,x} \in v\mathbb{Z}[v]$ are called *Kazhdan–Lusztig polynomials*. We also set $h_{x,x} = 1$ and $h_{y,x} = 0$ if $y \not\leq x$.

Any set $\{b_x \mid x \in W\}$ satisfying the degree bound condition in Definition 3.9 is automatically a $\mathbb{Z}[v, v^{-1}]$-basis of H. This is readily seen by considering the $\mathbb{Z}[v, v^{-1}]$-linear map $\varphi : H \to H$, $\varphi(\delta_x) = b_x$. In the standard basis (with any total order on W refining the Bruhat order), the matrix of φ will be triangular

3.2 The Kazhdan–Lusztig Basis

with 1's along the diagonal, and hence φ is an isomorphism, and Kazhdan–Lusztig polynomials are the entries of the change of basis matrix from the standard basis to a Kazhdan–Lusztig basis. (It is also worth noting for this argument that only finitely many elements of W are less than a given x in the Bruhat order, even when W is infinite.)

We will show in Sect. 3.3 that a Kazhdan–Lusztig basis exists. For now, we show that such a basis must be unique.

Lemma 3.10 *A Kazhdan–Lusztig basis is unique.*

Proof Suppose there exist two sets of elements $\{b_x \mid x \in W\}$ and $\{c_x \mid x \in W\}$ satisfying the self-duality and degree bound conditions

$$b_x = \delta_x + \sum_{y<x} h_{y,x} \delta_y \quad \text{and} \quad c_x = \delta_x + \sum_{y<x} h'_{y,x} \delta_y$$

for some $h_{y,x}, h'_{y,x} \in v\mathbb{Z}[v]$. By self-duality, we have

$$\sum_{y<x}(h_{y,x} - h'_{y,x})\delta_y = b_x - c_x = \overline{b_x - c_x} = \sum_{y<x} \overline{(h_{y,x} - h'_{y,x})}\, \overline{\delta_y}. \tag{3.20}$$

If b_x and c_x are not equal, there is some $z < x$ maximal such that $h_{z,x} - h'_{z,x}$ is nonzero. By (3.17), $\overline{\delta_z} = \delta_{z^{-1}}^{-1}$, and by Lemma 3.6 we have that $\delta_{z^{-1}}^{-1}$ is equal to δ_z modulo terms lower in the Bruhat order. By the maximality of z, comparing coefficients of δ_z on both sides of (3.20) gives $h_{y,x} - h'_{y,x} = \overline{h_{y,x} - h'_{y,x}}$, and since this element lies in $v\mathbb{Z}[v] \cap v^{-1}\mathbb{Z}[v^{-1}] = \{0\}$, we have a contradiction. □

The Kazhdan–Lusztig basis element for a simple reflection has a particularly simple form:

Lemma 3.11 *For $s \in S$, the Kazhdan–Lusztig basis element is $b_s = \delta_s + v$.*

Proof The element $\delta_s + v$ satisfies the degree bound condition, and it is self-dual:

$$\overline{\delta_s + v} = \delta_s^{-1} + v^{-1} = \delta_s + (v - v^{-1}) + v^{-1} = \delta_s + v.$$

It follows by uniqueness that $b_s = \delta_s + v$. □

3.2.1 The Standard Form on H

In this section we will introduce a form and trace on the Hecke algebra, which is very important in the theory. The proofs in this section make good exercises, so the reader should attempt them before reading the solution.

Definition 3.12 The *standard trace* $\epsilon : \mathrm{H} \to \mathbb{Z}[v, v^{-1}]$ is the $\mathbb{Z}[v, v^{-1}]$-linear map that is given on the standard basis by extracting the coefficient of the identity: $\epsilon(\delta_{\mathrm{id}}) = 1$, and $\epsilon(\delta_x) = 0$ for $x \neq \mathrm{id}$.

We will say that a $\mathbb{Z}$-bilinear form $(-, -) : \mathrm{H} \times \mathrm{H} \to \mathbb{Z}[v, v^{-1}]$ is a *sesquilinear form* if it is $\mathbb{Z}[v, v^{-1}]$-linear in the second argument, and satisfies $(vx, y) = v^{-1}(x, y)$ in the first.

Definition 3.13 The *standard form* $(-, -) : \mathrm{H} \times \mathrm{H} \to \mathbb{Z}[v, v^{-1}]$ is the sesquilinear form $(a, b) := \epsilon(\omega(a)b)$ for $a, b \in \mathrm{H}$.

An important property of the standard form is *biadjointness* for elements fixed under ω: for any elements $a, x, y \in \mathrm{H}$ we have that

$$(ax, y) = (x, \omega(a)y) \tag{3.21}$$

If in addition $a = \omega(a)$, we have $(ax, y) = (x, ay)$. This holds in particular for the Kazhdan–Lusztig basis element b_s corresponding to a simple reflection $s \in S$, so

$$(b_s x, y) = (x, b_s y). \tag{3.22}$$

Remark 3.14 Eventually we will see that the Hecke algebra is categorified by the category of Soergel bimodules. The standard form is important to us because it will end up describing the graded ranks of homomorphisms between Soergel bimodules, see the Soergel Hom formula (Theorem 5.27). This is an instance of a general principle: a categorification of a mathematical structure (like a module or algebra) often leads to an interesting form which encodes the "sizes" of the morphism spaces in the categorification. Moreover, certain properties of Soergel bimodules will imply properties of this form. Sesquilinearity and biadjointness, which may seem mysterious at the moment, correspond to very natural properties in the categorification, see Remarks 5.28 and 8.28.

Lemma 3.15 *For all* $x, y \in W$,

$$\epsilon(\delta_x \delta_y) = \begin{cases} 1 & \text{if } x = y^{-1}, \\ 0 & \text{otherwise.} \end{cases}$$

Exercise 3.16 Prove Lemma 3.15. (*Hint:* First assume that $\ell(x) \geq \ell(y)$ and apply (3.6) repeatedly.)

Note that $x = y^{-1}$ if and only if $x^{-1} = y$. This fact, combined with Lemma 3.15 immediately gives the following corollary (justifying the name "trace"):

Corollary 3.17 *The standard trace ϵ satisfies $\epsilon(ab) = \epsilon(ba)$ for any $a, b \in \mathrm{H}$.*

3.2 The Kazhdan–Lusztig Basis

Furthermore, the standard basis has a simple dual basis under the standard form:

Corollary 3.18 *The bases* $\{\overline{\delta_x} \mid x \in W\}$ *and* $\{\delta_x \mid x \in W\}$ *are dual under the standard form:*

$$(\overline{\delta_x}, \delta_y) = \begin{cases} 1 & \text{if } x = y, \\ 0 & \text{otherwise.} \end{cases}$$

Proof Since (3.17) and (3.18) give that $\omega(\overline{\delta_x}) = \delta_{x^{-1}}$, we have

$$(\overline{\delta_x}, \delta_y) = \epsilon(\omega(\overline{\delta_x})\delta_y) = \epsilon(\delta_{x^{-1}}\delta_y) = \begin{cases} 1 & \text{if } x = y, \\ 0 & \text{otherwise,} \end{cases}$$

and the result follows. □

The final property we wish to show in this section is that the Kazhdan–Lusztig basis is "almost" orthonormal with respect to the standard form (Theorem 3.21). When we compute pairings with self-dual elements, we will often use the following lemma.

Lemma 3.19 *Let* $b = \sum_{x \in W} c_x \delta_x$ *be self-dual, i.e.* $\overline{b} = b$. *Then*

$$\left(b, \sum_{x \in W} d_x \delta_x\right) = \sum_{x \in W} c_x d_x. \tag{3.23}$$

Proof By linearity in the second argument, it is enough to show that $(b, \delta_x) = c_x$ for each $x \in W$. By Corollary 3.18 and the self-duality of b, we have

$$(b, \delta_x) = (\overline{b}, \delta_x) = (\overline{c_x \delta_x}, \delta_x) = c_x(\overline{\delta_x}, \delta_x) = c_x$$

which finishes the proof. □

Remark 3.20 Lemma 3.19 allows us to "pretend" that the pairing is bilinear and that the standard basis is orthonormal. We call this the *false orthonormality of the standard basis*. Do not be confused: the pairing is not bilinear, and the standard basis is not orthonormal, but by pretending one gets the right answer anyway if one restricts to self-dual elements.

Theorem 3.21 *The Kazhdan–Lusztig basis is* asymptotically orthonormal: *for all* $x, y \in W$ *we have*

$$(b_x, b_y) \in \begin{cases} 1 + v\mathbb{Z}[v] & \text{if } x = y, \\ v\mathbb{Z}[v] & \text{otherwise.} \end{cases}$$

Remark 3.22 In other words, the Kazhdan–Lusztig basis is "orthonormal at $v = 0$." The terminology "asymptotically orthonormal" comes from quantum groups. Quantum groups arose first in statistical mechanics, where v corresponds to temperature, thus a basis is asymptotically orthonormal if it approaches orthonormality as the temperature approaches absolute zero.

Proof Applying the degree bound condition of Definition 3.9 we have

$$(b_x, b_y) = \left(\delta_x + v \sum_{z<x} p_z \delta_z, \; \delta_y + v \sum_{z<y} q_z \delta_z\right)$$

for some polynomials $p_z, q_z \in \mathbb{Z}[v]$. The element b_x is self-dual, so by Lemma 3.19 we have

$$(b_x, b_y) = \begin{cases} 1 + v r_{x,y} & \text{if } x = y, \\ v r_{x,y} & \text{otherwise,} \end{cases}$$

for some polynomial $r_{x,y} \in \mathbb{Z}[v]$. □

Remark 3.23 In fact, the standard form also equips the Hecke algebra with the structure of a *Frobenius algebra*. This might not be surprising since the Hecke algebra is a deformation of a group algebra, and group algebras are Frobenius algebras. For references on Frobenius algebras and their properties see for example [43, Chapter IX] or Sect. 8.1.4.

3.3 Existence of the Kazhdan–Lusztig Basis

In this section we will show that the Kazhdan–Lusztig basis exists by describing an algorithm to construct it. In many ways, this algorithm is a more useful way to think about and compute with the Kazhdan–Lusztig basis than its defining properties.

3.3.1 A Motivating Example

We begin with an example. Let $W = S_3$ and $S = \{s, t\}$, so that

$$W = \langle s, t \mid s^2 = t^2 = \text{id}, \; sts = tst \rangle.$$

Recall the Coxeter complex from Sect. 2.6, especially Example 2.23. We will express an element $h = \sum f_x \delta_x \in H$ by writing the coefficients $f_x \in \mathbb{Z}[v, v^{-1}]$ on the Coxeter complex (see Sect. 2.6). Each wall is colored red for s or blue for t, so that crossing a wall corresponds to right multiplication by that simple reflection.

3.3 Existence of the Kazhdan–Lusztig Basis

For example, under the labeling

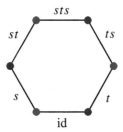

the element $v + (1 - v^{-1})\delta_{st} + v^2 \delta_{sts} - \delta_t$ may be written as follows:

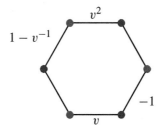

We begin with some known Kazhdan–Lusztig basis elements, namely $b_{\mathrm{id}} = 1$, $b_s = \delta_s + v$, and $b_t = \delta_t + v$, and would like to inductively construct the whole Kazhdan–Lusztig basis, thereby giving a constructive proof of its existence. Supposing we know the coefficients of $b_x = \delta_x + \sum_{y<x} h_{y,x} \delta_y$, we may find $b_x b_s$ termwise by the rule

$$\delta_x b_s = \begin{cases} \delta_{xs} + v\delta_x & \text{if } x < xs, \\ \delta_{xs} + v^{-1}\delta_x & \text{if } x > xs, \end{cases} \qquad (3.24)$$

which is an immediate consequence of (3.6). On the Coxeter complex, crossing a face between two cells corresponds to right multiplication by δ_s for some $s \in S$, such that the face separating the cells w and ws is labeled by s. So we may interpret the formula above in the following way: the coefficient of δ_x will cross the face labeled by s, and in doing so either move up or down in the Bruhat order. If it moved up, it leaves v times itself behind, and if it moved down, it leaves v^{-1} times itself behind.

An example of this is the following calculation of $b_t b_s$ by first expanding b_t in the standard basis, then right-multiplying by b_s using the above formula:

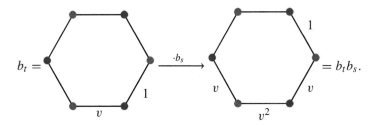

Both coefficients have moved across a red face, going up in the Bruhat order, and leaving v times itself behind. Being a product of self-dual elements, $b_t b_s$ is self-dual. Moreover, $b_t b_s$ satisfies the degree bound condition, hence $b_t b_s = b_{ts}$ by uniqueness.

Next, right multiplying b_{ts} by b_t, we find:

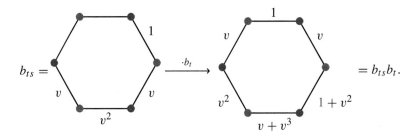

Being a product of self-dual elements, $b_{ts} b_t$ is again self-dual. However, it is not the Kazhdan–Lusztig basis element b_{tst} since the coefficient 1 (colored magenta in the picture for emphasis) violates the degree bound condition.

To obtain the Kazhdan–Lusztig basis element, we need to remove this extra 1. We cannot just subtract δ_t, since it is not self-dual. We *can* subtract b_t, however. Doing so yields

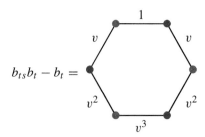

which must be b_{tst}. Thus $b_{tst} = b_t b_s b_t - b_t$.

3.3 Existence of the Kazhdan–Lusztig Basis

Exercise 3.24 Imitate this procedure to calculate all Kazhdan–Lusztig basis elements for dihedral groups (type $I_2(m)$). (Exercise 2.25 might be helpful.) Also, keep track of all the products $b_w b_s$ for $s \in S$, with $ws > w$. What do you observe?

3.3.2 Construction of the Kazhdan–Lusztig Basis

The examples in the previous two sections lead us to the following general method for constructing the Kazhdan–Lusztig basis element b_x using smaller basis elements $\{b_z \mid z < x\}$. Fix a y immediately below x in the Bruhat order, so $ys = x$ for some $s \in S$. The product $b_y b_s$ is self-dual, and almost satisfies the degree bound condition: the only thing that could go wrong is that the coefficients of the $\{\delta_z \mid z < x\}$ appearing in $b_y b_s$ could lie in $\mathbb{Z}[v]$ rather than $v\mathbb{Z}[v]$. We show that in this case, it is always possible to subtract a $\mathbb{Z}$-linear combination of the $\{b_z \mid z < x\}$ to correct this.

Theorem 3.25 *There exists a Kazhdan–Lusztig basis for* H.

Proof We have already seen in Lemma 3.11 that for simple reflections $s \in S$, the Kazhdan–Lusztig basis element is $b_s = \delta_s + v$, and also $b_{\mathrm{id}} = \delta_{\mathrm{id}} = 1$. Now proceed by induction, fixing $x \in W$ and assuming that b_z exists for all $z < x$.

Choose some simple reflection $s \in S$ such that $xs < x$. Define $y = xs$; by the inductive hypothesis, b_y exists and satisfies the degree bound condition

$$b_y = \delta_y + \sum_{z<y} h_{z,y} \delta_z$$

for some $h_{z,y} \in v\mathbb{Z}[v]$. Right multiplying by b_s gives

$$b_y b_s = \delta_y b_s + \sum_{z<y} h_{z,y} \delta_z b_s.$$

By (3.24), we have that $\delta_y b_s = \delta_x + v\delta_y$, and for each $z < y$ we see $\delta_z b_s = \delta_{zs} + v^{\pm 1} \delta_z$, where the sign depends on whether $zs > z$ or $zs < z$ in the Bruhat order. In particular, multiplication by b_s creates at most one factor of v^{-1}, and hence

$$b_y b_s = \delta_x + \sum_{z<x} h'_{z,x} \delta_z$$

for some polynomials $h'_{z,x} \in \mathbb{Z}[v]$. The product $b_y b_s$ is self-dual since each factor is, but we may have (as in the examples in the previous sections) some polynomials $h'_{z,x}$ with a nonzero constant term. Denote by $\mu(z, y; s)$ the constant term of the polynomial $h'_{z,x}$. For any $z < x$ such that $\mu(z, y; s) \neq 0$, we simply subtract the appropriate multiple of the basis element b_z, leaving

$$b_x := b_y b_s - \sum_{z<x} \mu(z, y; s) b_z \qquad (3.25)$$

$$= \delta_x + \sum_{z<x} \left(h'_{z,x} \delta_z - \mu(z, y; s) b_z \right). \qquad (3.26)$$

The element b_x is self-dual by (3.25) because each of b_y, b_s, and b_z (for every $z < x$) is, and the coefficients $\mu(z, y; s) \in \mathbb{Z}$ are fixed under the involution. It also satisfies the degree bound condition for x, since inside each summand for $z < x$ in (3.26), we have

$$h'_{z,x} \delta_z - \mu(z, y; s) b_z = \left(h'_{z,x} - \mu(z, y; s) \right) \delta_z - \mu(z, y; s) \sum_{w<z} h_{w,z} \delta_w$$

where both $h'_{z,x} - \mu(z, y; s)$ and $h_{w,z}$ belong to $v\mathbb{Z}[v]$. □

There are a few more useful observations to be obtained from the above proof, which we turn to now. The first is:

$$\mu(z, y; s) = 0 \quad \text{unless } zs < z. \qquad (3.27)$$

Indeed, if $zs > z$ then by (3.24) we have $h'_{z,x} = vh_{z,xs} + h_{zs,xs}$ which belongs to $v\mathbb{Z}[v]$ by induction.

Lemma 3.26

$$h_{ws,x} = vh_{w,x} \quad \text{if } ws < w \text{ and } xs < x. \qquad (3.28)$$

Proof Let $a \in H$ be an arbitrary element, and consider $ab_s \in H$. The Eq. (3.24) should be interpreted as saying that, whenever $ws < w$, the coefficient of δ_{ws} in ab_s is v times the coefficient of δ_w in ab_s.

We continue to use the notation $h'_{w,x}$ for the coefficient of δ_w in $b_y b_s$ when $y = xs < x$. Then by the above paragraph, $h'_{ws,x} = vh'_{w,x}$ when $ws < w$. By (3.26),

$$h_{w,x} = h'_{w,x} - \sum_{z<x} \mu(z, y; s) h_{w,z}, \qquad (3.29)$$

$$h_{ws,x} = h'_{ws,x} - \sum_{z<x} \mu(z, y; s) h_{ws,z}. \qquad (3.30)$$

3.3 Existence of the Kazhdan–Lusztig Basis

But by (3.27), we need only consider those z in the sum for which $zs < z$. For such z we can assume (3.28) by induction. Thus each individual term in (3.30) is v times the corresponding term in (3.29), as desired.[1] □

Let us introduce some important new[2] notation:

$$\mu(z, y) := \text{coefficient of } v \text{ in } h_{z,y}.$$

We have the following important multiplication formula:

Theorem 3.27 *For $y \in W$ and $s \in S$ we have:*

$$b_y b_s = \begin{cases} (v + v^{-1})b_y & \text{if } ys < y, \\ b_{ys} + \displaystyle\sum_{\substack{z < y \\ zs < z}} \mu(z, y)b_z & \text{if } ys > y. \end{cases}$$

Proof The first case is an immediate consequence of (3.28). Indeed, let us compute the coefficient of δ_z in $b_y b_s$. If $zs > z$, then using (3.24) and (3.28), this coefficient is

$$vh_{z,y} + h_{zs,y} = vh_{z,y} + v^{-1}h_{z,y} = (v + v^{-1})h_{z,y}.$$

Similarly, if $zs < z$ we get the coefficient

$$v^{-1}h_{z,y} + h_{zs,y} = v^{-1}h_{z,y} + vh_{z,y} = (v + v^{-1})h_{z,y}.$$

In both cases this agrees with the coefficient of δ_z in $(v + v^{-1})b_y$.

We now turn to the second case. Note that (again by (3.24))

$$\mu(z, y; s) = \begin{cases} \mu(z, y) & \text{if } zs < z, \\ 0 & \text{otherwise.} \end{cases}$$

The claim now follows from (3.25). □

Exercise 3.28 Show by induction that $v^{-(\ell(x)-\ell(y))}h_{y,x} \in \mathbb{Z}[v^{-2}]$. Deduce that the coefficients of $h_{y,x}$ in odd (resp. even) degrees are zero if $\ell(y) - \ell(x)$ is even (resp. odd).

Exercise 3.29 Let $W = S_4$, with $S = \{s, t, u\}$ as usual.

---

[1] We leave the reader to confirm that this argument still works when $w = x$, despite the superficial difference in (3.26).
[2] Or should that be "mu"?

1. Consider the reduced expression (s, t, s, u, t, s) for the longest element w_0. Use the inductive algorithm above to compute the Kazhdan–Lusztig basis element b_{w_0}. Along the way, you will compute the Kazhdan–Lusztig basis elements b_s, b_{st}, b_{sts}, etc. You will also compute several products and how they decompose into the Kazhdan–Lusztig basis: $b_s b_t$, $b_{st} b_s$, $b_{sts} b_u$, etc.
2. Repeat the calculation for the reduced expression (t, u, s, t, u, s). What is different this time? What non-trivial Kazhdan–Lusztig polynomials have you found?
3. Repeat the calculation for (s, u, t, s, u, t).
4. Which other Kazhdan–Lusztig basis elements can you deduce, using the symmetries of S_4? Which Kazhdan–Lusztig basis elements are missing? Which products $b_w b_x$ do you know, for $x \in S$ with $wx > w$, and which are missing?
5. If there are any missing Kazhdan–Lusztig basis elements or products, compute them. List all the non-trivial Kazhdan–Lusztig polynomials for S_4.

For more examples of Kazhdan–Lusztig basis elements in types A_3, B_3, and $\widetilde{A}_2$, see the book website mentioned on page ix. For the exercise above, you will find it helpful to run the inductive algorithm on the Coxeter complex, which you can find on the website as well.

3.3.3 The Kazhdan–Lusztig Presentation

Consider the Hecke algebra H in type A_2, with $S = \{s, t\}$ and $m_{st} = 3$. Since the standard generators δ_s and δ_t generate H, so do b_s and b_t. The quadratic relation on δ_s can be rewritten in terms of b_s as

$$b_s b_s = (v + v^{-1}) b_s. \quad (3.31)$$

What relation corresponds to the braid relation $\delta_s \delta_t \delta_s = \delta_t \delta_s \delta_t$? We have seen that $b_{sts} = b_s b_t b_s - b_s$. But also, $b_{sts} = b_t b_s b_t - b_t$. Therefore, one has

$$b_s b_t b_s - b_s = b_t b_s b_t - b_t. \quad (3.32)$$

Exercise 3.30 Verify that (3.31) and (3.32) give a presentation of H with generators b_s and b_t, in type A_2.

Exercise 3.31 Let W have type $I_2(m)$ with $S = \{s, t\}$. Suppose that $m < \infty$. The goal of this exercise is to generalize (3.32) from $m = 3$ to the general case.

1. When $m = 4$, verify that

$$b_{w_0} = b_s b_t b_s b_t - 2 b_s b_t = b_t b_s b_t b_s - 2 b_t b_s. \quad (3.33)$$

Verify that (3.31) and (3.33) give a presentation of H with generators b_s and b_t.

3.3 Existence of the Kazhdan–Lusztig Basis

2. Find a similar algebraic expression for b_{w_0} in terms of b_s and b_t, when $m = 2, 3, 4, 5, 6$.
3. Can you find a reasonable pattern for computing the coefficients which appear? (*Hint:* try to do this for all m at once, and arrange the coefficients into a triangle.)

3.3.4 Deodhar's Formula

Let $x \in W$, and fix a reduced expression $\underline{x} = (s_1, \ldots, s_m)$ for x. The product $b_{s_1} \cdots b_{s_m}$ is of the form $\delta_x + \sum_{z < x} c_z \delta_z$. In this section, we give an explicit combinatorial formula for the coefficients c_z in terms of *Deodhar defects*, which are certain combinatorial statistics on subexpressions of reduced words (recall the notion of a subexpression given in Definition 3.7).

Definition 3.32 Let $\underline{x} = (s_1, \ldots, s_m)$ be an expression and let $\underline{e} \subset \underline{x}$ be a subexpression. We may evaluate the subexpression $\underline{e}$ by defining $\underline{x}^{\underline{e}} := s_1^{e_1} \cdots s_m^{e_m}$; we say that $\underline{e}$ is a *subexpression with target* $\underline{x}^{\underline{e}}$. The *stroll*[3] associated to the subexpression $\underline{e}$ is the sequence $x_0, \ldots, x_m$ in W defined by $x_i := s_1^{e_1} \cdots s_i^{e_i}$.

The stroll x_i records how the element $\underline{x}^{\underline{e}}$ is built, from left to right, from the subexpression $\underline{e}$. The words defining the x_i do not need to be reduced, and so a step from x_i to x_{i+1} may move either up or down in the Bruhat order, or not change anything at all. We decorate the stroll $x_0, \ldots, x_m$ by a sequence $d_1, \ldots, d_m \in \{U0, U1, D0, D1\}$, defined as follows:

$$d_i = \begin{cases} Ue_i & \text{if } x_{i-1}s_i > x_{i-1}, \\ De_i & \text{if } x_{i-1}s_i < x_{i-1}. \end{cases} \quad (3.34)$$

The element d_i records how we *could* have moved in the Bruhat order from x_{i-1} to $x_{i-1}s_i$ (either U (up) or D (down)), along with whether we actually made that move (either $e_i = 0$ or 1).

Example 3.33 Let $W = S_3$ and $S = \{s, t\}$, so that

$$W = \langle s, t \mid s^2 = t^2 = \text{id}, \, sts = tst \rangle.$$

Take $\underline{e} = 001 \subset (s, t, s) = \underline{x}$. Then the stroll corresponding to $\underline{e} \subset \underline{x}$ is $x_0 = x_1 = x_2 = \text{id}$, and $x_3 = s$. In this case, at each step we have $x_{i-1}s_i > x_{i-1} = \text{id} \in W$, hence the directions are all up, and the decorated sequence is $U0, U0, U1$. Taking the same word $\underline{x} = (s, t, s)$, but a different subexpression $\underline{f} = 101$, the stroll instead is $x_0 = \text{id}$, $x_1 = x_2 = s$, $x_3 = \text{id}$, with the corresponding decorated sequence $U1, U0, D1$.

---

[3] In [59], a *stroll* is called a *Bruhat stroll*.

Definition 3.34 The *(Deodhar) defect* of a subexpression $\underline{e} \subset \underline{x}$, denoted by defect($\underline{e}$), is the number of $U0$'s minus the number of $D0$'s in the decorated sequence associated to $\underline{e}$:

$$\text{defect}(\underline{e}) = \#U0 - \#D0.$$

So the defect counts how often we could have gone up but did not, minus how often we could have gone down but did not.

Example 3.35 In the previous example, we have defect($\underline{e}$) = 2 and defect($\underline{f}$) = 1.

The defect is the correct combinatorial statistic we need to describe how a product $b_{s_1} \cdots b_{s_m}$ decomposes in terms of the standard basis elements. For an expression $\underline{x} = (s_1, \ldots, s_m)$, set

$$b_{\underline{x}} := b_{s_1} \cdots b_{s_m}. \tag{3.35}$$

Lemma 3.36 (Deodhar's Lemma) *For any expression $\underline{x}$, we have*

$$b_{\underline{x}} = \sum_{\underline{e} \subset \underline{x}} v^{\text{defect}(\underline{e})} \delta_{\underline{x}^{\underline{e}}}, \tag{3.36}$$

$$\varepsilon(b_{\underline{x}}) = \sum_{\substack{\underline{e} \subset \underline{x} \\ \underline{x}^{\underline{e}} = \text{id}}} v^{\text{defect}(\underline{e})}. \tag{3.37}$$

Proof Note that (3.37) follows immediately from (3.36) and the definition of the standard trace ε. To prove (3.36), we proceed by induction[4] on $\ell(\underline{x})$. If $\underline{x}'$ denotes the sequence $(s_1, \ldots, s_{m-1})$ leaving out the last simple reflection in $\underline{x}$ then we need to check that

$$\left(\sum_{\underline{e} \subset \underline{x}'} v^{\text{defect}(\underline{e})} \delta_{\underline{x}^{\underline{e}}} \right) b_{s_m} = \sum_{\underline{e} \subset \underline{x}} v^{\text{defect}(\underline{e})} \delta_{\underline{x}^{\underline{e}}}.$$

This equality follows by (3.24) from the fact that

$$v^{\text{defect}(\underline{e})} \delta_{\underline{x}^{\underline{e}}} b_{s_m} = v^{\text{defect}(\underline{e}_1)} \delta_{\underline{x}^{\underline{e}_1}} + v^{\text{defect}(\underline{e}_2)} \delta_{\underline{x}^{\underline{e}_2}},$$

where $\underline{e}_1$ and $\underline{e}_2$ are the two subexpressions of $\underline{x}$ whose first $m - 1$ terms agree with $\underline{e}'$. □

We end this chapter with a sample calculation of Kazhdan–Lusztig basis elements and polynomials using Deodhar's lemma.

---

[4] Note that $\ell(\underline{x})$ is the length m of the sequence $\underline{x}$, not the length of the element x it expresses.

3.3 Existence of the Kazhdan–Lusztig Basis

Example 3.37 As above, let $W = S_3$ and $S = \{s, t\}$. Consider the reduced expression $\underline{x} = (s, t, s)$ for the longest element. For each subexpression $\underline{e} \subset \underline{x}$, we compute its defect defect($\underline{e}$) and the evaluation $\underline{x}^{\underline{e}} \in W$:

$\underline{e}$	111	110	101	100	011	010	001	000
$\underline{x}^{\underline{e}}$	sts	st	id	s	ts	t	s	id
defect($\underline{e}$)	0	1	1	0	1	2	2	3

Applying (3.36), we get

$$b_s b_t b_s = \delta_{sts} + v\delta_{st} + v\delta_{ts} + (1 + v^2)\delta_s + v^2\delta_t + (v + v^3). \tag{3.38}$$

The only coefficient in the expansion which violates the degree bound condition is the $(1 + v^2)$ coefficient of δ_s. As before, subtracting b_s will cancel the problematic coefficient while preserving self-duality, and we arrive at

$$b_{sts} = b_s b_t b_s - b_s = \delta_{sts} + v\delta_{st} + v\delta_{ts} + v^2\delta_s + v^2\delta_t + v^3. \tag{3.39}$$

We have thus calculated the Kazhdan–Lusztig polynomials $h_{x,sts} = v^{3-\ell(x)}$ for all $x \in W$.

Exercise 3.38 In the following s, t, u denote distinct simple reflections.

1. Use the Deodhar defect formula to write $b_s b_s b_s$ in the standard basis.
2. Use the Deodhar defect formula to compute $b_s b_t b_u$. Is this product equal to b_{stu}?

Exercise 3.39 Let $\{s, t, u, v\}$ be the simple reflections in type D_4, where s, u, v all commute.

1. Compute the product $b_{\underline{w}}$ for the reduced expression $\underline{w} = (s, u, v, t, s, u, v)$. (*Hint:* there are 2^7 subexpressions, which is a lot. However, for each given element $x < w$, there are not many subexpressions for x. There is a lot of symmetry, so the number of x one must examine is relatively small.)
2. Compute the product $b_{\underline{w}}$ in terms of the Kazhdan–Lusztig basis. (*Hint:* One need not compute all the Kazhdan–Lusztig basis elements, as in the inductive construction. Instead, when writing $b_{\underline{w}}$ in the standard basis, try to explain where each coefficient with exponent v^k for $k \leq 0$ can come from.)
3. Compute the Kazhdan–Lusztig polynomial $h_{suv,suvtsuv}$.

Exercise 3.40 Compute the pairing $(b_s b_t b_s, b_s)$ in two different ways:

1. Use biadjunction and the quadratic relation to express this pairing in terms of $\epsilon(b_t b_s)$.
2. Use the Deodhar defect formula on both sides, and the false orthonormality of the standard basis.

Exercise 3.41 Let $\underline{w} = (s_1, \ldots, s_m)$ be an expression. We write $x \le \underline{w}$ if there exists a subexpression $\underline{e}$ of $\underline{w}$ with $x = \underline{w}^{\underline{e}}$. Consider two subexpressions $\underline{e}, \underline{e}'$ of $\underline{w}$, and let $x_0, x_1, \ldots$ and $x'_0, x'_1, \ldots$ be the associated strolls. We define the *path dominance order* on subexpressions of $\underline{w}$ by saying that $\underline{e} \le \underline{e}'$ if $x_i \le x'_i$ for $1 \le i \le \ell(\underline{w})$.

Fix $x \le \underline{w}$. Show that there is a unique subexpression $\underline{e}$ of $\underline{w}$ representing x, the *canonical subexpression* for x, which is characterized by the following equivalent conditions:

1. $\underline{e} \le \underline{e}'$ for any subexpression $\underline{e}'$ of $\underline{w}$ with $\underline{w}^{\underline{e}'} = x$ (i.e. $\underline{e}$ is the unique minimal element representing x in the path dominance order);
2. $\underline{e}$ has no D's in its UD-labeling;
3. $\underline{e}$ is of maximal defect among all subexpressions $\underline{e}'$ of $\underline{w}$ with $\underline{w}^{\underline{e}'} = x$.

See also Definition 12.21 in Chap. 12. (If you know about Bott–Samelson resolutions (see Example 16.16): What geometric fact does the existence of $\underline{e}$ correspond to? Do you think there is a unique maximal element in the path dominance order?)

Chapter 4
Soergel Bimodules

This chapter is based on expanded notes of a lecture given by the authors and taken by
Sean Rogers and Ziqing Xiang

Abstract Soergel bimodules provide an algebraic categorification of the Hecke algebra. This chapter provides a first introduction to these bimodules. First we consider the symmetric algebra on the geometric representation of the Coxeter group, and state the Chevalley theorem concerning its invariant subrings. We then discuss Demazure operators. Finally, we introduce Bott–Samelson bimodules and Soergel bimodules, and state Soergel's categorification theorem.

4.1 Gradings

Throughout this book, graded objects play a fundamental role, so here we fix our conventions. Unless otherwise specified, "graded" will mean $\mathbb{Z}$-graded. A *graded vector space* is a vector space M equipped with a decomposition

$$M := \bigoplus_{i \in \mathbb{Z}} M^i \tag{4.1}$$

into subspaces M^i. The M^i are the *graded pieces* of M. It is useful to imagine the graded pieces of M as arranged by height, with M^i occupying height i. An element m of M contained in some M^i will be called *homogeneous*, in which case its *degree* is i. The grading is *bounded above* (resp. *bounded below*) if $M^i = 0$ for $i \gg 0$ (resp. $i \ll 0$).

S. Rogers
Department of Mathematics, University of North Carolina, Chapel Hill, NC, USA

Z. Xiang
Department of Mathematics, University of Georgia, Athens, GA, USA

© The Editor(s) (if applicable) and The Author(s), under exclusive licence to Springer Nature Switzerland AG 2020
B. Elias et al., *Introduction to Soergel Bimodules*, RSME Springer Series 5, https://doi.org/10.1007/978-3-030-48826-0_4

More generally, one considers gradings for all sorts of objects like rings, modules, algebras, bimodules, etc.—usually with some additional compatibility conditions. For example, a *graded ring* is a ring R with a decomposition $R = \bigoplus_{i \in \mathbb{Z}} R^i$ into subgroups R^i of the additive group of R such that $R^i R^j \subseteq R^{i+j}$. A typical example is the polynomial ring $K[x_1, \ldots, x_n]$ with the usual grading. A *graded module* over a graded ring R is an R-module M with a decomposition $M = \bigoplus_{i \in \mathbb{Z}} M^i$ into subgroups M^i of M such that $R^i M^j \subseteq M^{i+j}$. Notice that a graded vector space over a field K is the same as a graded K-module when considering K with the trivial grading, i.e. $K^0 = K$.

Given a graded object M (vector space, module, ring, etc.) and $i \in \mathbb{Z}$ we can define a new object $M(i)$ with graded pieces

$$M(i)^j := M^{i+j}. \tag{4.2}$$

If we imagine the graded pieces of M arranged vertically as above, then $M(1)$ is obtained by shifting M *down*. We say that $M(i)$ is obtained from M via a shift in the grading.

A morphism $M \to N$ between graded objects is typically assumed to be *homogeneous of degree* 0, meaning that it sends M^i to N^i for each $i \in \mathbb{Z}$. One might also consider maps which are homogeneous of degree k for some $k \in \mathbb{Z}$, which send M^i to N^{i+k}. Note that a degree k map $M \to N$ is the same data as a degree 0 map $M \to N(k)$, or a degree 0 map $M(-k) \to N$. Two graded objects M and N are *isomorphic up to shift* if $M \simeq N(i)$ for some i. The *graded Hom* space between graded objects M and N is defined to be the direct sum of morphism spaces of all degrees:

$$\mathrm{Hom}^\bullet(M, N) := \bigoplus_{k \in \mathbb{Z}} \mathrm{Hom}(M, N(k)). \tag{4.3}$$

Given an object M in an additive category and $m \in \mathbb{Z}_{\geq 0}$, one can consider the iterated direct sum $M^{\oplus m}$. If M is graded, then one can take a number of copies of M, shift them, and add them together. Given a Laurent polynomial $p = \sum p_i v^i \in \mathbb{Z}_{\geq 0}[v^{\pm 1}]$ with positive integer coefficients we set

$$M^{\oplus p} := \bigoplus_{i \in \mathbb{Z}} M(i)^{\oplus p_i}. \tag{4.4}$$

Thus "multiplication by v" corresponds to a shift down; we return to this in Sect. 4.8.

When we consider a graded module M, we will use the following conventions.

By a *submodule* of M we always mean a *graded submodule*, i.e. a submodule of M generated by homogeneous elements.

By a *direct summand* of M we always mean a direct summand of M in the graded module category, i.e. a *graded* module N such that there is another graded module N' and an isomorphism $M \simeq N \oplus N'$ of graded modules. It is not hard to show that a *graded* submodule N of M is a direct summand if and only if it is a direct summand as ungraded modules, see e.g. [143, §2.3.4].

When we say that a graded R-module M is *free*, we always mean that it has an R-basis consisting of *homogeneous* elements of M. Suppose that $\{m_i\}_{i \in I}$ is such a basis. Then the map $r \mapsto rm_i$ is an isomorphism $R(-\deg m_i) \simeq Rm_i \subset M$ of graded modules. Hence, if the basis is finite, we have an isomorphism $M \simeq R^{\oplus p}$ for a uniquely defined $p \in \mathbb{Z}_{\geq 0}[v^{\pm 1}]$, and we call p the *graded rank* of M.

4.2 Polynomials

The data needed to define a Soergel bimodule will be a Coxeter system (W, S) together with its geometric representation V over $\mathbb{R}$ (see Definition 1.27). Recall that V is a real vector space of dimension $|S|$ with basis $\{\alpha_s \mid s \in S\}$ indexed by simple reflections. These basis elements α_s are known as *simple roots*. We defined an action of W on V where the simple reflections act by

$$s(\alpha_t) = \alpha_t - 2(\alpha_s, \alpha_t)\alpha_s. \tag{4.5}$$

Here $(-,-) : V \times V \to \mathbb{R}$ is the symmetric bilinear form where $(\alpha_s, \alpha_t) = -\cos(\frac{\pi}{m_{st}})$. When $m_{st} = \infty$, this is interpreted as $(\alpha_s, \alpha_t) = -1$.

Remark 4.1 With the bilinear form above, all simple roots have the same length 1. Readers familiar with Lie theory and root systems may expect roots to have different lengths. In Sect. 5.7 we describe the notion of a realization of W, which generalizes the geometric representation and allows for non-symmetric Cartan matrices, different base fields, simple roots which do not form a basis, and more. For now, we stick to the geometric representation.

Let

$$R = \operatorname{Sym}(V) = \bigoplus_{i \in \mathbb{Z}_{\geq 0}} \operatorname{Sym}^i(V) \tag{4.6}$$

be the *symmetric algebra* of V, and view it as a graded algebra in which $\deg(V) = 2$. In down-to-earth terms, R is the polynomial ring

$$R = \mathbb{R}[\alpha_s \mid s \in S] \tag{4.7}$$

with grading

$$\deg \alpha_s = 2 \tag{4.8}$$

for all $s \in S$. Note that $\text{Sym}^i(V)$ occupies degree $2i$. Soergel bimodules, to be defined below, will be certain graded bimodules over this graded algebra R. Note that the W-action on V induces a W-action on R via

$$w \cdot \prod_{s \in S} \alpha_s^{k_s} = \prod_{s \in S} (w(\alpha_s))^{k_s} \tag{4.9}$$

on monomials, and then extended linearly to polynomials.

4.2.1 Invariant Polynomials

For a subset $I \subset S$, define $W_I := \langle I \rangle \subset W$, the *(standard) parabolic subgroup* generated by I. We say that I is *finitary* if W_I is a finite group. Let R^I be the ring of W_I-invariants of R, i.e.

$$R^I = \{ f \in R \mid w \cdot f = f \text{ for all } w \in W_I \}. \tag{4.10}$$

We sometimes write R^W instead of R^S for the invariants under the entire Coxeter group. We also write R^s instead of $R^{\{s\}}$, $R^{s,t}$ instead of $R^{\{s,t\}}$, etc.

The following theorem is the algebraic foundation upon which the theory of Soergel bimodules is built.

Theorem 4.2 (Chevalley–Shephard–Todd [42, 160], See Also [85, Chapter 3]) *Let $I \subset S$ be finitary. Then R^I is a polynomial ring. Moreover, R is a graded free module of finite rank over R^I.*

A more refined version of the Chevalley–Shephard–Todd theorem also specifies the degrees of the generators of R^I in terms of the Poincaré polynomial of W_I.

Example 4.3 Let $W = S_5$ act on $\mathbb{R}[x_1, \ldots, x_5]$. Let $I = \{s_1, s_3, s_4\}$, so $W_I \simeq S_2 \times S_3$. Then we have that

$$R^I = \mathbb{R}[x_1 + x_2, \; x_1 x_2, \; x_3 + x_4 + x_5, \; x_3 x_4 + x_3 x_5 + x_4 x_5, \; x_3 x_4 x_5].$$

Here, the Chevalley–Shephard–Todd theorem implies that R^I has 5 algebraically independent generators (just like R), though in different degrees.

Remark 4.4 Note that $\mathbb{R}[x_1, \ldots, x_5]$ is not the ring R we associated to the Coxeter group S_5, since it comes from the permutation representation, not the geometric representation considered above. The reader is welcome throughout this book to use the permutation representation instead of the geometric representation for S_n (see Remark 4.1 above).

4.2 Polynomials

Exercise 4.5 When $W = S_3$ acts on its geometric representation (not the permutation representation), find a description of R^I for all parabolic subgroups I. That is, find expressions for generators of R^I in terms of the simple roots. (For a big hint, see Remark 4.10.)

Remark 4.6 In fact, the Chevalley–Shephard–Todd theorem includes an "if and only if" statement for finite groups of linear transformations of a complex vector space: the invariants are a polynomial ring if and only if the action is generated by *pseudoreflections*.[1] The next exercises explore how polynomiality of the invariants can fail if the group is not finite or not generated by pseudoreflections.

Exercise 4.7 Let $W = S_2$ and $S = \{s\}$. Let W act on $\mathbb{R}^2$, the span of $\{x, y\}$, by $s \cdot x = -x$ and $s \cdot y = -y$. Hence W acts on $\mathbb{R}[x, y]$. Prove that the ring of invariants is the subring of $\mathbb{R}[x, y]$ generated by the homogeneous elements x^2, xy, y^2. These three elements are not algebraically independent (why?), but all three are needed to generate the ring of invariants, and so the invariants are not a polynomial ring. In particular, R is not free over R^W. Why is this not in violation of the Chevalley–Shephard–Todd theorem?

Exercise 4.8 Let W have type $I_2(\infty)$ and $S = \{s, t\}$. Let $R = \mathbb{R}[\alpha_s, \alpha_t]$ be defined using the geometric representation. Prove that R^W is the subring $\mathbb{R}[\alpha_s + \alpha_t] \subset R$. In particular, R^W is a polynomial ring, but with a different number of generators. Thus, R is a graded free module over R^W, but is not of finite rank. Why is this not in violation of the Chevalley–Shephard–Todd theorem?

Exercise 4.9 Now let θ be an irrational angle, and consider the action of $I_2(\infty)$ on the Euclidean plane given in Remark 1.32. Let $R = \mathbb{R}[\alpha_s, \alpha_t]$ be the corresponding polynomial ring. Prove that R^W is the subring $\mathbb{R}[z] \subset R$ generated by a single polynomial of degree 2. (*Hint:* what is the quadratic form?) Once again, R^W is a polynomial ring over which R is graded free, but R is not of finite rank.

Remark 4.10 Let W have type $I_2(m)$ and $S = \{s, t\}$. The Chevalley–Shephard–Todd theorem implies that $R^{s,t}$ is generated by a polynomial of degree 2 and a polynomial of degree m. The degree 2 invariant polynomial is constructed in the same way as in the previous exercise, and the reader should try to find it explicitly. The degree m invariant polynomial is slightly trickier to construct. First find a linear polynomial which is fixed by s. Its orbit under W has size m, so the product of these m distinct linear factors will be W-invariant. For more details, see [47, Sect. 3.7].

In the next examples, we examine the crucial case when W_I is generated by a single simple reflection.

Example 4.11 Let $W = S_2$, $S = \{s\}$. Let $R = \mathbb{R}[\alpha]$, with the action of W given by $s(\alpha) = -\alpha$. Then it is clear that $R^s = \mathbb{R}[\alpha^2]$. Note that R now admits a direct sum decomposition $R = R^s \oplus (R^s \cdot \alpha)$ into s-invariants and s-antiinvariants. (By

---

[1] A *pseudoreflection* is a diagonalizable linear transformation of a complex vector space whose fixed space is a hyperplane, i.e. is of codimension 1.

definition an element is *s-antiinvariant* if $sf = -f$.) This is the classical fact (often learned in high school) that any function of one variable can be decomposed into an even and an odd function.

Example 4.12 Generalizing the above example, let (W, S) be an arbitrary Coxeter system. Then for every $s \in S$, R^s is generated by α_s^2 and the elements

$$\alpha_t + (\cos \frac{\pi}{m_{st}})\alpha_s \quad \text{for all} \quad t \in S \setminus \{s\}. \tag{4.11}$$

Moreover,

$$R = R^s \oplus R^s \alpha_s \tag{4.12}$$

is a splitting of R into s-invariants and s-antiinvariants. In other words, any polynomial $f \in R$ can be written uniquely in the form $f = g + h\alpha_s$, where $g, h \in R^s$. (We will find g and h explicitly in the next section.) Because α_s has degree 2 we have an isomorphism

$$R \simeq R^s \oplus R^s(-2) \tag{4.13}$$

as graded R^s-bimodules.

We introduce some more notation for later use.

Definition 4.13 If W is finite, let R_+^W denote the graded subspace of R^W consisting of everything in strictly positive degrees. Let I_W be the homogeneous ideal in R generated by R_+^W. The *coinvariant algebra* of W, denoted by C, is the graded algebra

$$C := R/I_W. \tag{4.14}$$

4.3 Demazure Operators

Definition 4.14 For $s \in S$, the *Demazure operator* ∂_s (also called a *BGG operator* or *divided difference operator*) is the graded map

$$\partial_s : R \to R^s(-2),$$

$$f \mapsto \frac{f - s(f)}{\alpha_s}. \tag{4.15}$$

As discussed in Example 4.12, the s-antiinvariants are generated (as an R^s-module) by α_s, so $f - s(f)$ is divisible by α_s. Thus ∂_s is well-defined. Note that the numerator and the denominator are both s-antiinvariant, so this fraction is s-invariant.

4.3 Demazure Operators

The Demazure operator can be used to make the R^s-module splitting $R \simeq R^s \oplus R^s \cdot \alpha_s$ from (4.13) explicit. One computes that

$$\partial_s(f\alpha_s) = f + s(f) \tag{4.16}$$

for any $f \in R$. Meanwhile,

$$\alpha_s \partial_s(f) = f - s(f). \tag{4.17}$$

Consequently,

$$f = \partial_s(f\frac{\alpha_s}{2}) + \frac{\alpha_s}{2}\partial_s(f) \tag{4.18}$$

is a formula for writing f as a sum of its invariant and antiinvariant parts. The isomorphism $R \to R^s \oplus R^s(-2)$ can be explicitly given by

$$f \mapsto (\partial_s(f\frac{\alpha_s}{2}), \partial_s(f)), \tag{4.19}$$

with inverse

$$(g, h) \mapsto g + h\frac{\alpha_s}{2}. \tag{4.20}$$

The Demazure operator satisfies a number of important properties.

Lemma 4.15 *Let $s \in S$. Then ∂_s satisfies the following properties.*

1. ∂_s *is an R^s-bimodule map.*
2. $s \circ \partial_s = \partial_s$ *and* $\partial_s \circ s = -\partial_s$.
3. $\partial_s^2 = 0$.
4. *There exists a short exact sequence*

$$0 \to R^s \to R \xrightarrow{\partial_s} R^s(-2) \to 0. \tag{4.21}$$

5. *The twisted Leibniz rule holds: for $f, g \in R$, we have*

$$\partial_s(fg) = \partial_s(f)g + s(f)\partial_s(g). \tag{4.22}$$

6. *We have*

$$\partial_s(I_W) \subset I_W. \tag{4.23}$$

Hence ∂_s induces a map on the coinvariant algebra.

7. *Let $(-, -)_s$ be the pairing $R \times R \to R^s$ given by*

$$(f, g)_s \mapsto \partial_s(fg). \tag{4.24}$$

This is a perfect pairing over R^s. More precisely, $\{1, \frac{\alpha_s}{2}\}$ is a basis for R over R^s, with dual basis $\{\frac{\alpha_s}{2}, 1\}$.

8. The Demazure operators satisfy the braid relations. That is, if $s, t \in S$ are distinct simple reflections with $m_{st} < \infty$, then

$$\partial_s \partial_t \partial_s \cdots = \partial_t \partial_s \partial_t \cdots, \quad (4.25)$$

where each side is a product of m_{st} Demazure operators.

Exercise 4.16 Prove Lemma 4.15. You should only check the braid relations for $m_{st} \in \{2, 3\}$.

Exercise 4.17 (This is a more advanced exercise, and requires understanding the theory of positive roots.) Prove the braid relations for $m_{st} \geq 4$, and find an explicit formula for the operator in (4.25), analogous to the formula for ∂_s in (4.15).

Exercise 4.18 Suppose that $m_{st} = 3$, and $f \in R^s$. Prove that $\partial_s \partial_t(f) \in R^{s,t}$.

Definition 4.19 The *Demazure operator* with respect to $w \in W$, denoted by ∂_w, is given by

$$\partial_w := \partial_{s_1} \cdots \partial_{s_n}. \quad (4.26)$$

for any reduced expression $\underline{w} = (s_1, \ldots, s_n)$ for w.

Because the Demazure operators satisfy the braid relations, Matsumoto's theorem (Theorem 1.56) implies that this definition of ∂_w does not depend on the choice of reduced expression.

Exercise 4.20 What happens when you take a product $\partial_{s_1} \cdots \partial_{s_n}$, when $(s_1, \ldots, s_n)$ is not a reduced expression?

Exercise 4.21 Let w_I be the longest element of a finite parabolic subgroup W_I. Prove that ∂_{w_I} sends R to R^I. (*Hint:* to show that a polynomial is s-invariant, one can either prove that it is in the image of ∂_s, or in the kernel of ∂_s.)

Remark 4.22 In fact, the operators $\{\partial_w\}_{w \in W}$ are linearly independent in $\mathrm{End}_\mathbb{R}(R)$, and a composition of two is either zero or another such operator. They form a basis for what is known as the *nilCoxeter algebra* associated with W, which can be viewed as a subalgebra of $\mathrm{End}_\mathbb{R}(R)$.

4.4 Bimodules and Tensor Products

Recall our conventions on gradings in Sect. 4.1. Let $R\text{-gbim}_{qc}$ denote the category of graded R-bimodules. The "qc" stands for quasi-coherent, and is intended to remind the reader that we make no assumptions on finite generation for bimodules in this category. It has a shift functor (n) for each integer n which sends $M \mapsto M(n)$. It

4.5 Bott–Samelson Bimodules

also has a tensor product $-\otimes_R -$ which makes $R\text{-gbim}_{qc}$ into a monoidal category (a category with tensor products, see e.g. [64, §2.1] for a precise definition). The tensor product is graded: $(M \otimes_R N)^k$ is the image of $\bigoplus_{i+j=k} M^i \otimes_{\mathbb{Z}} N^j$ in $M \otimes_R N$. The monoidal identity is the regular bimodule R. Tensor product and grading shift commute: given graded R-bimodules M, N and $n \in \mathbb{Z}$ we have canonical identifications

$$(M(n)) \otimes_R N = M \otimes_R (N(n)) = (M \otimes_R N)(n). \qquad (4.27)$$

For two modules M and N in $R\text{-gbim}_{qc}$, we usually abbreviate

$$MN := M \otimes_R N. \qquad (4.28)$$

We view $R\text{-gbim}_{qc}$ as a graded category, with morphisms those graded R-bimodule maps $M \to N$ which are homogeneous of degree 0. A degree k map $M \to N$ is the same data as a (degree 0) morphism $M \to N(k)$.

Let R-gbim denote the category of graded R-bimodules that are finitely generated as both left and right R-modules. It is a full monoidal subcategory of $R\text{-gbim}_{qc}$ thanks Exercise 4.24 below.

> **Assumption 4.23** In this book, all R-bimodules we consider will be finitely generated both as left and right R-modules, i.e. will belong to R-gbim.

Exercise 4.24 Consider the following full subcategories of $R\text{-gbim}_{qc}$:

$R\text{-gbim}_l := \{M \in R\text{-gbim}_{qc} \mid M \text{ is finitely generated as a left } R\text{-module}\}$;

$R\text{-gbim}_r := \{M \in R\text{-gbim}_{qc} \mid M \text{ is finitely generated as a right } R\text{-module}\}$;

$R\text{-gbim}_b := \{M \in R\text{-gbim}_{qc} \mid M \text{ is finitely generated as an } R\text{-bimodule}\}$.

Show that $R\text{-gbim}_l$ and $R\text{-gbim}_r$ are monoidal subcategories of $R\text{-gbim}_{qc}$, but that $R\text{-gbim}_b$ is not. Deduce that R-gbim is monoidal.

4.5 Bott–Samelson Bimodules

Definition 4.25 For $s \in S$, we let B_s denote the graded R-bimodule

$$B_s := R \otimes_{R^s} R(1). \qquad (4.29)$$

We will see momentarily that B_s belongs to R-gbim. An element in B_s can be represented as $\sum_i f_i \otimes g_i$ for some f_i and g_i in R. We encourage the reader to replace the tensor product symbol with a wall, as in

$$\sum_i f_i |_s g_i. \tag{4.30}$$

The symbol $|_s$ is meant to represent a porous wall which only appropriate polynomials can pass through. In this case, $f |_s 1 = 1 |_s f$ if and only if f is s-invariant. Recall that, from our grading convention, the element $1|_s 1$ lives in degree -1.

Definition 4.26 The *Bott–Samelson bimodule* corresponding to an expression $\underline{w} = (s_1, \ldots, s_n)$, denoted by $\mathrm{BS}(\underline{w})$, is the graded R-bimodule given by

$$\mathrm{BS}(\underline{w}) := B_{s_1} B_{s_2} \cdots B_{s_n}. \tag{4.31}$$

That is, it is an iterated tensor product of the bimodules B_{s_i} defined above. By convention, if $\underline{w}$ is the empty expression then $\mathrm{BS}(\underline{w}) = R$.

It is easy to see that we have a canonical isomorphism

$$\mathrm{BS}(\underline{w}) = R \otimes_{R^{s_1}} R \otimes_{R^{s_2}} \cdots \otimes_{R^{s_n}} R(\ell(\underline{w})). \tag{4.32}$$

Thus, an element of $\mathrm{BS}(\underline{w})$ is of the form $\sum_i f_i \otimes g_i \otimes \cdots \otimes h_i$ for some $f_i, g_i, \ldots, h_i \in R$, which we may similarly denote by

$$\sum_i f_i |_{s_1} g_i |_{s_2} \cdots |_{s_n} h_i. \tag{4.33}$$

The element $1 |_{s_1} 1 |_{s_2} \cdots |_{s_n} 1$ lives in degree $-\ell(\underline{w})$, and will be referred to as the *1-tensor*.

Given two expressions $\underline{u}$ and $\underline{v}$ we have

$$\mathrm{BS}(\underline{u}) \mathrm{BS}(\underline{v}) = \mathrm{BS}(\underline{uv}), \tag{4.34}$$

where $\underline{uv}$ denotes the concatenation of $\underline{u}$ and $\underline{v}$. Thus Bott–Samelson bimodules are closed under tensor product. Note that Bott–Samelson bimodules are not closed under taking grading shifts or direct sums.

We have seen in (4.13) above that R is graded free as an R^s-module. By (4.13) we deduce that, as graded left R-modules,

$$B_s \simeq R \otimes_{R^s} (R^s \oplus R^s(-2))(1) \simeq R(1) \oplus R(-1). \tag{4.35}$$

In particular, B_s is graded free as a left R-module. An identical argument shows that the same is true when B_s is regarded as a right R-module. Because tensor products of bimodules which are free of finite rank as left (resp. right) modules are free of finite rank as left (resp. right) modules, we conclude:

Lemma 4.27 *Any Bott–Samelson bimodule is graded free of finite rank as a left (resp. right) R-module.*

Inside the bimodule B_s, consider the elements

$$c_{\mathrm{id}} := 1 \otimes 1, \qquad c_s := \frac{1}{2}(\alpha_s \otimes 1 + 1 \otimes \alpha_s) \qquad (4.36)$$

of degree -1 and 1, respectively.

Exercise 4.28 Prove the following, for any $f \in R$:

$$f \cdot c_{\mathrm{id}} = c_{\mathrm{id}} \cdot s(f) + c_s \cdot \partial_s(f), \qquad f \cdot c_s = c_s \cdot f. \qquad (4.37)$$

(*Hint:* prove this formula for $f \in R^s$, and prove it for $f = \alpha_s$. Now try to deduce it for all $f \in R$.)

Write down a similar formula for $c_{\mathrm{id}} \cdot f$, in terms of $g \cdot c_{\mathrm{id}}$ and $h \cdot c_s$ for some polynomials g, h. Conclude that c_{id} and c_s yield a basis of B_s, when regarded as either a left or right R-module.

As we will discuss in the next chapter, $R \cdot c_s$ is the submodule of B_s consisting of elements x such that $fx = xf$ for all $f \in R$. In other words, c_s generates a copy of the bimodule $R(-1)$, as a sub-bimodule of B_s.

The crucial relations (4.37) allow us to relate the left action of R on B_s with the right action, and vice versa. For future reference, the first equation is called the *polynomial forcing relation*. We discuss bases of Bott–Samelson bimodules in Chap. 12.

4.6 Soergel Bimodules

Recall our conventions about direct summands of graded modules in Sect. 4.1. We can now define Soergel bimodules.

Definition 4.29 A *Soergel bimodule* is a direct summand of a finite direct sum of grading shifts of Bott–Samelson bimodules. The *category of Soergel bimodules*, denoted by $\mathbb{S}\mathrm{Bim}$, is the strictly full subcategory of R-gbim consisting of Soergel bimodules.[2]

Exercise 4.30 Show that $\mathbb{S}\mathrm{Bim}$ is closed under tensor products.

---

[2] *Strictly full* means that $\mathbb{S}\mathrm{Bim}$ is closed under isomorphisms, i.e. $\mathbb{S}\mathrm{Bim}$ is the full subcategory of R-gbim whose objects are isomorphic to Soergel bimodules. Being closed under isomorphism is not of essential importance, and is a matter of categorical taste.

Equivalently, the category of Soergel bimodules is the smallest strictly full subcategory of R-gbim containing R and B_s for all $s \in S$ that is closed under tensor products, direct sums, direct summands and shifts. The reader should think that adding direct sums and shifts to a subcategory is a relatively boring and formal process, while adding direct summands is quite exciting (see Chap. 11 for more details).

Lemma 4.31 *Any Soergel bimodule is graded free as a left or right R-module.*

This is a consequence of Lemma 4.27 and the following exercise.

Exercise 4.32 Let M denote a graded left R-module which is free of finite rank. Show that any graded summand N of M is also graded free.[3]

Note that morphisms between Soergel bimodules are assumed to be homogeneous of degree 0, that is

$$\mathrm{Hom}_{\mathbb{S}\mathrm{Bim}}(B, B') = \mathrm{Hom}_{R\text{-gbim}}(B, B').$$

The space of homogeneous morphisms $B \to B'$ of degree k can still be studied within $\mathbb{S}\mathrm{Bim}$, being isomorphic to the space $\mathrm{Hom}_{\mathbb{S}\mathrm{Bim}}(B, B'(k))$. However, there is a notational difference between Soergel bimodules and Bott–Samelson bimodules, since the latter are not closed under grading shifts. We still want to remember the space of homogeneous morphisms of all degrees between Bott–Samelson bimodules. For this reason, we make the following definition.

Definition 4.33 The *category of Bott–Samelson bimodules*, denoted $\mathbb{BS}\mathrm{Bim}$, is the monoidal category defined as follows. Its objects are Bott–Samelson bimodules. The morphism space between two objects B and B' is the graded vector space

$$\mathrm{Hom}_{\mathbb{BS}\mathrm{Bim}}(B, B') = \bigoplus_{k \in \mathbb{Z}} \mathrm{Hom}_{R\text{-gbim}}(B, B'(k)), \tag{4.38}$$

whose degree k piece is the space of homogeneous bimodule maps of degree k. The monoidal structure is the tensor product of R-bimodules.

4.7 Examples of Soergel Bimodules

Recall that an object M of an additive category (like bimodules) is called *indecomposable* if it cannot be expressed as a direct sum $M' \oplus M''$ for nonzero subobjects M', M''. In this book, an extremely important role is played by the indecomposable

---

[3]If we drop the assumption that M and N are graded, this exercise becomes a difficult theorem of Quillen and Suslin, which was known for many years as Serre's conjecture.

4.7 Examples of Soergel Bimodules

Soergel bimodules. The following lemma is useful for identifying the simplest indecomposable Soergel bimodules.

Lemma 4.34 *Suppose that M is a graded R-bimodule which is generated as an R-bimodule by a homogeneous element $m \in M$. Then M is indecomposable.*

Proof Let d denote the degree of m. Because M is generated by m and $R^0 = \mathbb{R}$, we have $M^d = \mathbb{R}m$. Suppose $M = L \oplus N$. Then $M^d = L^d \oplus N^d$, and we may assume that $m \in L$. Then $M = R \cdot m \cdot R \subset L$ and hence $N = 0$. □

It follows from the lemma that the bimodules R and B_s are indecomposable.

Example 4.35 (Soergel Bimodules of Type A_1) Let $W = S_2$, generated by the simple reflection s. The bimodules R and B_s are indecomposable Soergel bimodules. The square of B_s is

$$B_s B_s \simeq R \otimes_{R^s} R \otimes_{R^s} R(2) \tag{4.32}$$

$$\simeq R \otimes_{R^s} (R^s \oplus R^s(-2)) \otimes_{R^s} R(2) \quad \text{(see Example 4.12)}$$

$$\simeq R \otimes_{R^s} R(2) \oplus R \otimes_{R^s} R$$

$$\simeq B_s(1) \oplus B_s(-1).$$

It thus follows that R and B_s give representatives for all indecomposable Soergel bimodules in type A_1, up to shift and isomorphism.

Remark 4.36 The isomorphism in the above example is very similar to the equation

$$b_s b_s = (v + v^{-1}) b_s = b_s v + b_s v^{-1} \tag{4.39}$$

in the Hecke algebra H of type A_1, where b_s is the Kazhdan–Lusztig basis element in H. We will have more to say about this below.

Example 4.37 One may use the calculations in Example 4.12 to conclude that R is generated by the subrings R^s and R^t, whenever $s \neq t$ and $m_{st} \neq \infty$. In particular, the bimodules

$$B_s B_t = R \otimes_{R^s} R \otimes_{R^t} R(2) \quad \text{and} \quad B_t B_s = R \otimes_{R^t} R \otimes_{R^s} R(2)$$

are indecomposable, because they are generated by the 1-tensor $1 \otimes 1 \otimes 1$. This implies that B_s and B_t are not isomorphic: we have seen that $B_s B_s$ is decomposable, whereas $B_s B_t$ is not.

Henceforth we write

$$B_{st} := B_s B_t \quad \text{and} \quad D_{ts} := B_t B_s. \tag{4.40}$$

Remark 4.38 For a practical reason why R is generated by R^s and R^t when $m_{st} < \infty$, let us return to the setting of the geometric representation (see Sect. 1.1.7).

Recall that $R = \mathrm{Sym}(V)$, where V is the geometric representation. As a polynomial ring, R is generated by its linear terms, the vectors in V. Meanwhile, the linear terms in R^s are the vectors fixed by s, which is precisely the "reflecting" hyperplane H_s of vectors perpendicular to α_s. (Even though our form on V is not Euclidean, the space of vectors perpendicular to α_s is still a hyperplane.) The linear terms in R^t are the vectors in H_t. Since the hyperplanes H_s and H_t are distinct, they span the entire representation V.

In the case when $S = \{s, t\}$ and $m_{st} = \infty$, the hyperplanes H_s and H_t are actually the same! Both are spanned by $\alpha_s + \alpha_t$. Thus R is not generated by R^s and R^t, because they only generate a one-dimensional space of linear terms. One can still prove that $B_s B_t$ is indecomposable, though it is not cyclic. We leave this as a weird (and entirely optional) exercise to the reader. This motivates why one might want to work with a different representation than the geometric representation, something we will allow in subsequent chapters.

Example 4.39 (Soergel Bimodules of Type A_2) Here, let $W = A_2$, with $S = \{s, t\}$ and $m_{st} = 3$. By the considerations in Examples 4.35 and 4.37, the indecomposable modules R, B_s, B_t are distinct, and the products $B_s B_s$, $B_t B_t$ offer us no new indecomposables. In addition, B_{st} and B_{ts} are also indecomposable. Let us write $B_{sts} := R \otimes_{R^{s,t}} R(3)$ (the notation will be justified shortly). Observe that B_{sts} is generated by $1 \otimes 1$ in degree -3, so we could add it to our list of indecomposables in $\mathbb{S}\mathrm{Bim}$ by Lemma 4.34 if we knew B_{sts} actually is in $\mathbb{S}\mathrm{Bim}$. Indeed it is, as follows from either of the following isomorphisms:

$$B_s B_t B_s \simeq B_{sts} \oplus B_s, \qquad B_t B_s B_t \simeq B_{sts} \oplus B_t. \qquad (4.41)$$

For more on these isomorphisms, see Exercise 4.41 below.

Notice that

$$B_s B_t B_s \simeq B_{sts} \oplus B_s \not\simeq B_s B_t(-1) \oplus B_s B_t(1) \simeq B_s B_s B_t,$$

since all grading shifts of B_s, B_{st}, B_{ts}, B_{sts} are indecomposable. (In this argument and elsewhere in these examples, we have implicitly used the uniqueness of direct summand decompositions; see Remark 4.45 for more on this.) This shows that $B_{ts} \not\simeq B_{st}$.

The final computations necessary to finalize the classification in this example are

$$B_{sts} B_s \simeq B_s B_{sts} \simeq B_{sts}(1) \oplus B_{sts}(-1) \simeq B_{sts} B_t \simeq B_t B_{sts}.$$

(See Exercise 4.42 below.) The way B_{sts} interacts with B_s and B_t under the monoidal product ensure that B_{sts} is not isomorphic to any grading shift of our previously found indecomposables, and also that we cannot produce new indecomposables if we keep multiplying the bimodules with each other. Thus, our list of distinct indecomposables up to grading shift is $\{R, B_s, B_t, B_{st}, B_{ts}, B_{sts}\}$. Notice how this list corresponds bijectively with W.

4.7 Examples of Soergel Bimodules

Example 4.40 Again, the above calculations have echoes in the Hecke algebra. The reader is invited to match the following equalities in the Hecke algebra with isomorphisms considered in the previous example:

$$b_s b_t = b_{st}, \qquad b_t b_s = b_{ts},$$

$$b_{st} b_t = (v + v^{-1}) b_{st}, \qquad b_{ts} b_s = (v + v^{-1}) b_{ts},$$

$$b_s b_t b_s = b_{sts} + b_s, \qquad b_t b_s b_t = b_{sts} + b_t,$$

$$b_{sts} b_s = b_s b_{sts} = (v + v^{-1}) b_{sts}, \quad \text{and} \quad b_{sts} b_t = b_t b_{sts} = (v + v^{-1}) b_{sts}.$$

Exercise 4.41 This exercise explores the Bott–Samelson bimodule $B_s B_t B_s$ when $m_{st} = 3$.

1. Show that there is an injective map $B_{sts} \to B_s B_t B_s$ which sends

$$1 \otimes 1 \mapsto 1|1|1|1.$$
$$s\,t\,s$$

2. Show that there is an injective map $B_s \to B_s B_t B_s$ which sends

$$1|1 \mapsto \frac{1}{2}(1|\alpha_t|1|1 + 1|1|\alpha_t|1).$$
$$\phantom{1|1 \mapsto \frac{1}{2}(1|}s\,t\,s s\,t\,s$$

3. Show that the images of these two maps are disjoint, and span $B_s B_t B_s$. This is sufficient to prove (4.41).
4. Find formulas for the projection maps $B_s B_t B_s \to B_s$ and $B_s B_t B_s \to B_{sts}$. In other words, say explicitly where $1|f|g|1$ goes under each projection, for $f, g \in R$.
$$s\,t\,s$$

The last part of this exercise is surprisingly difficult! Even with the hints below, it is still difficult. In fact, the point of this exercise is not necessarily to succeed, but it is to struggle and to demonstrate just how difficult the simplest problems can be when one is forced to work directly with polynomials. This is one of the major motivations to use diagrammatic techniques instead.

Here are some hints (but give it a go first!).

- The projection map $B_s B_t B_s \to B_s$ sends $1|f|g|1 \mapsto -\partial_s(fg)|1$. Check that this is a left inverse to the inclusion map from B_s, and check that it annihilates this image of B_{sts}. (But how would you come up with this formula? It becomes clear once one knows about Jones–Wenzl projectors in Chap. 9. See in particular (9.21).)
- Whatever the projection map to B_{sts} is, it sends $1|f|g|1$ to $\sum_i h_i \otimes k_i$ for some $h_i, k_i \in R$. Composing this with the inclusion map $B_{sts} \to B_s B_t B_s$, we get an idempotent e_{sts} which sends $1|f|g|1$ to $\sum_i h_i|1|1|k_i$. If only you could compute a formula for e_{sts}....

- Let e_s be the idempotent which composes the maps $B_s B_t B_s \to B_s \to B_s B_t B_s$. You have explicit formulas for both the projection and inclusion, so an explicit formula for e_s. Also, $\mathrm{id}_{B_s B_t B_s} = e_s + e_{sts}$. So this gives an explicit formula for e_{sts}!
- Unfortunately, the explicit formula just computed for e_{sts} will not obviously send $1|f|g|1\atop s\;t\;s$ to a sum of tensors of the form $h_i|1|1|k_i\atop\;\;s\;t\;s$. So now you must do some work to rewrite your answer in this form.

Exercise 4.42 Continue to assume that $m_{st} = 3$. In Example 4.35 we used the fact that $R \simeq R^s \oplus R^s(-2)$ as (R^s, R^s)-bimodules to prove that $B_s B_s \simeq B_s(1) \oplus B_s(-1)$. Decompose R as an $(R^{s,t}, R^s)$-bimodule. Use this to deduce that $B_{sts} B_s \simeq B_{sts}(1) \oplus B_{sts}(-1)$.

Example 4.43 (Soergel Bimodules of Type $A_1 \times A_1$) Suppose that $m_{st} = 2$. One can prove that

$$B_s B_t \simeq R \otimes_{R^{s,t}} R(2) \simeq B_t B_s. \tag{4.42}$$

In particular, $B_{st} B_s$ and $B_{st} B_t$ give us no new indecomposable objects, so that the list of distinct indecomposables up to grading shift is $\{R, B_s, B_t, B_{st}\}$.

Once again, the isomorphisms have echoes in the Hecke algebra, namely the equalities $b_s b_t = b_t b_s = b_{st}$.

Exercise 4.44 For any $m_{st} < \infty$, there is a bimodule map $R \otimes_{R^{s,t}} R(2) \to B_s B_t$, sending

$$1 \otimes 1 \mapsto 1|1|1.\atop\;\;\;\;\;\;\;\;\;\;\;\;s\;t$$

This map is surjective because $B_s B_t$ is cyclic. Thus it is an isomorphism if and only if it is injective. Prove that the map is not injective when $m_{st} > 2$, by a dimension count. When $m_{st} = 2$ write down the inverse map explicitly, by saying where $1|f|1\atop\;\;s\;t$ goes.

Remark 4.45 In the examples above, we often took a decomposable Bott–Samelson bimodule (like $B_s B_s$ or $B_s B_t B_s$ when $m_{st} = 3$), and found a way to write it as a direct sum of known indecomposable bimodules. Then, we asserted that we had found all the indecomposable Soergel bimodules. What if there were more than one way to decompose the bimodule into indecomposable summands, and some other exotic decompositions lead to new indecomposable bimodules? Then our assertions above that we had found all indecomposable bimodules would be false. Thankfully the *Krull–Schmidt property* comes to the rescue, as it implies the uniqueness of decompositions into indecomposable objects. For more on the Krull–Schmidt property, see Sect. 11.3.4.

4.8 A First Glimpse of Categorification

In the previous section we saw striking parallels between the behavior of indecomposable Soergel bimodules and the Kazhdan–Lusztig basis of the Hecke algebra. This book is involved to a large extent with explaining this connection. We are now in a position to see the first concrete connection between Soergel bimodules and the Hecke algebra.

Let us consider the *split Grothendieck group* $[\mathbb{S}\text{Bim}]_\oplus$ of the category of Soergel bimodules. By definition, this is an abelian group generated by symbols $[B]$ for each object B in $\mathbb{S}\text{Bim}$, subject to the relations

$$[B] = [B'] + [B''] \quad \text{whenever } B \simeq B' \oplus B''. \tag{4.43}$$

Because $\mathbb{S}\text{Bim}$ is a monoidal category, $[\mathbb{S}\text{Bim}]_\oplus$ is a ring, via

$$[B][B'] := [BB']. \tag{4.44}$$

We may also make $[\mathbb{S}\text{Bim}]_\oplus$ into a $\mathbb{Z}[v^{\pm 1}]$-algebra via

$$v[B] := [B(1)]. \tag{4.45}$$

The following is the first portion of the *Soergel categorification theorem*, which we state more precisely in Chap. 5.

Theorem 4.46 *The assignment $b_s \mapsto [B_s]$ for $s \in S$ yields a $\mathbb{Z}[v^{\pm 1}]$-algebra homomorphism*

$$\mathrm{H} \to [\mathbb{S}\text{Bim}]_\oplus.$$

Let us give an outline of the proof when W is simply laced, i.e. when $m_{st} \in \{2, 3\}$ for all s, t. We have seen in Sect. 3.3.3 that H is generated as a $\mathbb{Z}[v^{\pm 1}]$-algebra by the elements b_s for all $s \in S$ subject to the relations

$$b_s^2 = (v + v^{-1})b_s,$$
$$b_s b_t = b_t b_s \quad \text{if } s \neq t \text{ and } st = ts,$$
$$b_s b_t b_s - b_s = b_t b_s b_t - b_t \quad \text{if } s \neq t \text{ and } sts = tst.$$

We have also seen that these relations hold in $[\mathbb{S}\text{Bim}]_\oplus$, as illustrated in Examples 4.35, 4.39 and 4.43. Thus the homomorphism is well-defined in this case.

To prove Theorem 4.46 in general involves checking one relation for each pair $s, t \in S$ with $m_{st} < \infty$. The direct computations involving polynomials are extremely difficult (try Exercise 4.41 for the flavor). Instead, Soergel [165] proves the result with a clever appeal to coherent geometry, and by developing the theory of standard modules. Some of these techniques will be seen in the next chapter. Meanwhile Chap. 9 gives the background to understand Elias's alternate proof in [47], which is more explicit and computational.

Chapter 5
The "Classical" Theory of Soergel Bimodules

This chapter is based on expanded notes of a lecture given by the authors and taken by
Cihan Bahran and **Ethan Kowalenko**

Abstract We define standard bimodules and standard filtrations of Soergel bimodules, and explore the technique of localization. The theory of standard filtrations allows one to define the character map, which is an algebra homomorphism from the split Grothendieck group of Soergel bimodules to the Hecke algebra, and is the inverse to the homomorphism in Theorem 4.46. This allows us to state the Soergel categorification theorem. We also introduce realizations of a Coxeter system, which generalize the construction of Soergel bimodules beyond the geometric representation.

5.1 Twisted Actions

Let A be a graded algebra over a base ring $\Bbbk$. Assume in addition that A is commutative. In this case an A-bimodule structure on a $\Bbbk$-module M is equivalent to a *left $A \otimes_\Bbbk A$-module structure on M*.

If we have a $\Bbbk$-algebra automorphism $\eta : A \to A$, we can "twist" an A-module structure by precomposing with η. If M is an A-bimodule, and the bimodule structure on M is encoded by $\rho : A \otimes_\Bbbk A \to \text{End}_\Bbbk(M)$, then the composite map $\rho \circ (\text{id} \otimes \eta)$ defines a new A-bimodule structure on the same underlying $\Bbbk$-module M, which leaves the left action alone but changes the right action. We denote this new bimodule by M_η. If η and ψ are two automorphisms, then

$$\text{id} \otimes (\eta \circ \psi) = (\text{id} \otimes \eta) \circ (\text{id} \otimes \psi).$$

C. Bahran
School of Mathematics, University of Minnesota, Minneapolis, MN, USA

E. Kowalenko
University of California, Riverside, Riverside, CA, USA

Consequently we have

$$M_{\eta \circ \psi} = (M_\eta)_\psi. \tag{5.1}$$

The category of A-bimodules has a natural monoidal structure given by tensoring over A (*not* $\Bbbk$!). With this monoidal structure, we see that the bimodule M_η can be naturally identified with $M \otimes_A A_\eta$. We deduce the elementary but important identification

$$A_{\eta \circ \psi} \simeq (A_\eta)_\psi \simeq A_\eta \otimes_A A_\psi. \tag{5.2}$$

In formulas, we have

$$m \cdot_\eta r := m \cdot \eta(r), \tag{5.3}$$

for $m \in M$ and $r \in A$. Here $(-) \cdot_\eta r$ denotes the right action of r on M_η, while $\cdot$ without the subscript is the usual right action of r on M.

5.2 Standard Bimodules

We return to the setting of Chap. 4, letting R be the polynomial ring of the geometric representation, acted on by W. We will specifically consider automorphisms of R of the form

$$\eta_x : R \to R$$
$$a \mapsto xa$$

for $x \in W$. Of course η_{id} with the identity $\mathrm{id} \in W$ is just the identity map, but other η_x are more interesting. Twisting the regular R-bimodule R by $\mathrm{id} \otimes \eta_x$, we obtain the notion of a standard bimodule.

Definition 5.1 The *standard bimodules* are the R-bimodules of the form

$$R_x := R_{\eta_x} \tag{5.4}$$

obtained by twisting the regular bimodule R on the right by η_x, for some $x \in W$. Let $Std\mathrm{Bim}$ be the smallest strictly full subcategory of R-gbim which contains R_x for each $x \in W$, and is closed under finite direct sums and grading shifts.

Using (5.2) we see that

$$R_x \otimes R_y \simeq R_{xy}. \tag{5.5}$$

Thus $Std\mathrm{Bim}$ is monoidal.

5.2 Standard Bimodules

For $M, N \in R$-gbim, recall the graded Hom space from (4.3):

$$\operatorname{Hom}^\bullet(M, N) := \bigoplus_{i \in \mathbb{Z}} \operatorname{Hom}(M, N(i)). \tag{5.6}$$

Lemma 5.2 *For any $x, y \in W$ we have*

$$\operatorname{Hom}^\bullet(R_x, R_y) = \begin{cases} R & \text{if } x = y, \\ 0 & \text{otherwise,} \end{cases} \tag{5.7}$$

as graded vector spaces.

Proof Notice that R_x is a cyclic R-bimodule for any $x \in W$, so $\varphi \in \operatorname{Hom}(R_x, R_y)$ is uniquely determined by $\varphi(1) = c$. Then we have

$$yr \cdot c = c \cdot_y r = \varphi(1 \cdot_x r) = \varphi(xr \cdot 1) = xr \cdot \varphi(1) = xr \cdot c \tag{5.8}$$

for every $r \in R$. Since the left action of R on R_y is untwisted, this simply reads $(yr)c = (xr)c$ as elements in R. Since R is a domain, this forces $xr = yr$ for all $r \in R$, or $c = 0$. Because W acts faithfully on its geometric representation (Proposition 1.28), we deduce that $xr = yr$ for all $r \in R$ if and only if $x = y$. □

Exercise 5.3 Use the above lemma to show that StdBim is closed under direct summands. (That is, unlike Soergel bimodules, there are no "interesting" direct summands in this category.)

We see that StdBim is an easily understood category. The indecomposable objects in StdBim are exactly the standard bimodules R_x and their grading shifts. They tensor in the same way that elements of W multiply.

Remark 5.4 In the terminology of [60], StdBim is isomorphic to the *graded 2-groupoid for W over R*. This is a formal construction one can make for any group, where group elements become indecomposable objects which do not interact (i.e. they satisfy Lemma 5.2), and the group multiplication is transformed into a monoidal structure.

In particular, the split Grothendieck group $[Std\text{Bim}]_\oplus$ (defined in Sect. 4.8) is isomorphic to the group algebra $\mathbb{Z}[v^{\pm 1}][W]$, via an isomorphism sending the (symbol of the) standard bimodule $[R_x]$ to the standard basis element x. We say that StdBim is a *categorification* of this group algebra.

Remark 5.5 Meanwhile, Soergel bimodules categorify the Hecke algebra of W, and the natural basis of indecomposables corresponds to the Kazhdan–Lusztig basis. This fact is several orders of magnitude harder to prove! The Hecke algebra is a deformation of the group algebra, and morally speaking, Soergel bimodules are like a deformation of the category StdBim.

5.3 Soergel Bimodules and Standard Filtrations

For the next several sections, we will be making one tacit assumption of some importance, which would be distracting to introduce now. We return to this in Sect. 5.6.

We continue the study of Soergel bimodules, as introduced in Chap. 4. Recall that

$$B_s = R \otimes_{R^s} R(1),$$

and that the Bott–Samelson bimodule $\mathrm{BS}(\underline{x})$ associated with an expression $\underline{x} = (s_1, s_2, \ldots, s_d)$ is the bimodule

$$\mathrm{BS}(\underline{x}) = B_{s_1} B_{s_2} \cdots B_{s_d} = R \otimes_{R^{s_1}} R \otimes_{R^{s_2}} \cdots \otimes_{R^{s_d}} R(d).$$

We consider the homogeneous elements $c_s, d_s \in B_s$ of degree 1, defined by

$$2c_s := \alpha_s \otimes 1 + 1 \otimes \alpha_s, \tag{5.9}$$

$$2d_s := \alpha_s \otimes 1 - 1 \otimes \alpha_s. \tag{5.10}$$

The element c_s was already defined in (4.36), and its properties were explored in Exercise 4.28, which the reader should revisit. Here is the corresponding exercise for d_s.

Exercise 5.6 Prove the following, for any $f \in R$:

$$f \cdot c_{\mathrm{id}} = c_{\mathrm{id}} \cdot f + d_s \cdot \partial_s(f), \qquad f \cdot d_s = d_s \cdot s(f). \tag{5.11}$$

Deduce that $\{c_{\mathrm{id}}, d_s\}$ is also a basis for B_s as a left or right R-module.

The fact that $fc_s = c_s f$ and $fd_s = d_s s(f)$ implies that c_s generates a copy of $R(-1)$ inside B_s, and d_s generates a copy of $R_s(-1)$ inside B_s. (The shifts come from the fact that c_s and d_s are of degree 1 in B_s.) In fact, these sub-bimodules fit into nice short exact sequences in R-gbim

$$0 \longrightarrow R_s(-1) \xrightarrow{1 \mapsto d_s} B_s \xrightarrow{\mu_{\mathrm{id}}} R(1) \longrightarrow 0 \tag{Δ}$$

and

$$0 \longrightarrow R(-1) \xrightarrow{1 \mapsto c_s} B_s \xrightarrow{\mu_s} R_s(1) \longrightarrow 0, \tag{∇}$$

with $\mu_{\mathrm{id}}(f \otimes g) = fg$, and $\mu_s(f \otimes g) = f \cdot s(g)$.

5.3 Soergel Bimodules and Standard Filtrations

Remark 5.7 Notice that R_s is both a quotient and a submodule, up to shift, of B_s, but is *not a summand* of B_s. In fact, R_s is not a Soergel bimodule. Since R_s is not in $\mathbb{S}$Bim, we see that $\mathbb{S}$Bim is *not* closed under taking submodules! Hence $\mathbb{S}$Bim is not an abelian category, but is merely an additive category. When we discuss short exact sequences, we must work inside the ambient abelian category R-gbim.

Given an expression $\underline{w} = (s_1, \ldots, s_d)$ of w, we can tensor the sequences (Δ) together to get a filtration of the Bott–Samelson bimodule BS($\underline{w}$). Let's spell this out, for $B_s B_s$. Note that all modules in question are free as right or left R-modules, so taking the tensor product is exact. Tensoring (Δ) on the right with B_s, we get the short exact sequence

$$0 \to R_s B_s(-1) \to B_s B_s \to B_s(1) \to 0.$$

Meanwhile, tensoring (Δ) on the left with $R_s(-1)$ gives a filtration on the submodule $R_s B_s(-1)$. Tensoring (Δ) on the left with $R(1)$ gives a filtration on the quotient $B_s(1)$. Putting these together in the usual way, one gets a filtration on $B_s B_s$.

Again, by tensoring the sequences (Δ) together, we get a filtration on any Bott–Samelson bimodule. A subquotient in this filtration will be a tensor product of some copies of $R(1)$ and some copies of $R_{s_i}(-1)$, and hence will be a standard bimodule (up to shift). Using the notation for subexpressions developed in Sect. 3.3.4, we note that a choice of subquotient in the filtration corresponds to a choice of subexpression $\underline{e} \subset \underline{x}$, where $e_i = 1$ if $R_{s_i}(-1)$ is chosen, and $e_i = 0$ if $R(1)$ is chosen. If $x = \underline{w}^{\underline{e}}$ is the target of the subexpression, then $R_x(k)$ appears as a subquotient, where k is the number of zeroes minus the number of ones in $\underline{e}$ (this is stated for sake of clarity, it will not end up being important).

Thus any Bott–Samelson bimodule has a filtration by shifted standard bimodules. Nor is the previous paragraph the only way to construct such a filtration. One could also tensor the sequences (∇) together, or tensor together a mix of (Δ) and (∇). A Bott–Samelson bimodule has many filtrations by shifted standard bimodules, and it is unclear at first which are to be preferred. One issue with these filtrations is that the grading shifts which appear in the subquotients depend on the choice of filtration, which can already be seen by comparing (Δ) and (∇). Should R_s appear as a subquotient of B_s with grading shift $+1$ or -1? Another issue is that there is no apparent compatibility between the order of the various R_x in the filtration, and the Bruhat order on $x \in W$.

To correct these issues, we fix an enumeration $x_0, x_1, \ldots$ of W such that $x_i \leq x_j$ in Bruhat order implies $i \leq j$. For example, we might choose id $< s < t < st < ts < sts$ for A_2. In general one could order the elements by length, and then choose an arbitrary order between elements of the same length. Using this order we may make the following definition. Recall our notation for direct sums of grading shifts introduced in (4.4).

Definition 5.8 For an enumeration of W as above, a Δ-*filtration* of a Soergel bimodule B is a filtration $0 = B^k \subset B^{k-1} \subset \cdots \subset B^0 = B$ with subquotients

$$B^i/B^{i+1} \simeq R_{x_i}^{\oplus h_{x_i}}, \qquad (5.12)$$

where $h_{x_i} \in \mathbb{Z}_{\geq 0}[v^{\pm 1}]$. Note that, even if W is infinite, this filtration is required to be of finite length.

Example 5.9 Of the two filtrations given above on B_s, the Δ-filtration of B_s is $0 \subset R_s(-1) \subset B_s$, given by ($\Delta$), with successive subquotients $R(1) = R^{\oplus v}$ and $R_s(-1) = R_s^{\oplus v^{-1}}$.

The natural question is how many Δ-filtrations exist on a given Soergel bimodule B. The following is a theorem of Soergel.

Theorem 5.10 (Soergel [165, Lemma 6.3]) *For a fixed enumeration of W, any Soergel bimodule B has a unique Δ-filtration. Moreover, for any $x \in W$ the graded multiplicity h_x of R_x in the Δ-filtration (5.12) depends only on B and x, and not on the choice of enumeration of W.*

Notice that we only demand that our ordering on W respects the Bruhat order, and this is enough to determine uniquely defined grading shifts on the standard bimodules which appear in the filtration! Henceforth we denote these multiplicities by $h_x(B) \in \mathbb{Z}_{\geq 0}[v^{\pm 1}]$ to indicate that they depend on B, but not on any other data. We can therefore make the following definition.

Definition 5.11 The Δ-*character* of a Soergel bimodule B is the element

$$\operatorname{ch}_\Delta(B) := \sum_{x \in W} v^{\ell(x)} h_x(B) \delta_x, \qquad (5.13)$$

of H, where δ_x are the standard basis elements (see Sect. 3.1.1).

Example 5.12 As computed in the previous example, we have $h_{\mathrm{id}}(B_s) = v^1$ and $h_s(B_s) = v^{-1}$, giving

$$\operatorname{ch}_\Delta(B_s) = v\delta_{\mathrm{id}} + v \cdot v^{-1}\delta_s = v + \delta_s, \qquad (5.14)$$

so $\operatorname{ch}_\Delta(B_s) = b_s$ for any $s \in S$.

Each of the above definitions and results has an analogue which uses (∇) instead of (Δ).

Definition 5.13 For an enumeration of W as above, a ∇-*filtration* of a Soergel bimodule B is a filtration $0 = B^0 \subset B^1 \subset \cdots \subset B^k = B$ with subquotients

$$B^{i+1}/B^i \simeq R_{x_i}^{\oplus h'_{x_i}}, \qquad (5.15)$$

where $h'_{x_i} \in \mathbb{Z}_{\geq 0}[v^{\pm 1}]$.

5.3 Soergel Bimodules and Standard Filtrations

Soergel also proves that ∇-filtrations on a Soergel bimodule are unique, and that their multiplicities are independent of the choice of enumeration.

Example 5.14 Of the two filtrations given above on B_s, the ∇-filtration of B_s is $0 \subset R(-1) \subset B_s$, given by (∇), with successive subquotients $R^{\oplus v^{-1}}$ and $R_s^{\oplus v}$.

Definition 5.15 The ∇-*character* of a Soergel bimodule B is

$$\operatorname{ch}_\nabla(B) := \sum_{x \in W} v^{\ell(x)} \overline{h'_x(B)} \delta_x \in H. \tag{5.16}$$

Recall that $\overline{(-)} : \mathbb{Z}[v, v^{-1}] \to \mathbb{Z}[v, v^{-1}]$ is the ring automorphism determined by $\overline{v} = v^{-1}$.

Example 5.16 As computed in the previous example, we have $h'_{\mathrm{id}}(B_s) = v^{-1}$ and $h'_s(B_s) = v$, giving

$$\operatorname{ch}_\nabla(B_s) = \overline{v^{-1}} \delta_{\mathrm{id}} + v \cdot \overline{v} \delta_s = v + \delta_s = b_s. \tag{5.17}$$

Thus $\operatorname{ch}_\Delta(B_s) = \operatorname{ch}_\nabla(B_s)$.

It is easy to see from the definitions that

$$\operatorname{ch}_\Delta(B \oplus B') = \operatorname{ch}_\Delta(B) + \operatorname{ch}_\Delta(B'), \qquad \operatorname{ch}_\nabla(B \oplus B') = \operatorname{ch}_\nabla(B) + \operatorname{ch}_\nabla(B') \tag{5.18}$$

and

$$\operatorname{ch}_\Delta(B(1)) = v \operatorname{ch}_\Delta(B), \qquad \operatorname{ch}_\nabla(B(1)) = v^{-1} \operatorname{ch}_\nabla(B) \tag{5.19}$$

for all Soergel bimodules B and B'. It follows from (5.18) that both characters induce $\mathbb{Z}$-linear maps

$$\operatorname{ch}_\Delta, \operatorname{ch}_\nabla : [\mathbb{S}\mathrm{Bim}]_\oplus \to H \tag{5.20}$$

from the split Grothendieck group of $\mathbb{S}\mathrm{Bim}$. However, with our convention that $v[B] = [B(1)]$ in the split Grothendieck group, it follows from (5.19) that only ch_Δ is $\mathbb{Z}[v, v^{-1}]$-linear. This explains our preference for

$$\operatorname{ch} := \operatorname{ch}_\Delta \tag{5.21}$$

in what follows. In Sect. 5.5 below we will see that $\operatorname{ch} : [\mathbb{S}\mathrm{Bim}]_\oplus \to H$ is a $\mathbb{Z}[v, v^{-1}]$-algebra isomorphism.

We saw in the examples above that $\operatorname{ch}_\Delta(B_s) = \operatorname{ch}_\nabla(B_s)$. More generally, there is a certain contravariant duality on bimodules that fixes every R_x, hence exchanges Δ-filtrations with ∇-filtrations. This duality fixes every B_x and intertwines ch_Δ and ch_∇, so that $\operatorname{ch}_\Delta(B_x) = \operatorname{ch}_\nabla(B_x)$ for all $x \in W$. Given (5.18) and (5.19), it follows that $\operatorname{ch}_\Delta(B) = \operatorname{ch}_\nabla(B)$ for B a direct sum of B_x without grading shift, but not for an arbitrary Soergel bimodule. Some more discussion of this duality can be found in Definition 18.8 and the subsequent discussion.

Exercise 5.17 Explicitly find Δ- and ∇-filtrations on the bimodule $B_s B_t$, and compute the corresponding characters.

Remark 5.18 In [165], Soergel views R-bimodules as quasi-coherent sheaves on $V \times V$. He constructs Δ-filtrations and ∇-filtrations using support filtrations, and studies the category of objects with both Δ- and ∇-filtrations. This also gives a geometric perspective to the next section on localization. These ideas are quite interesting, but they will not be pursued further in this book outside some isolated exercises (e.g. Exercise 6.38).

Remark 5.19 One consequence of Soergel's approach is that the Δ- and ∇-filtrations are functorial. In particular, they commute with direct sums. One can use this to prove that a direct summand of a Δ-filtered bimodule is Δ-filtered, a fact which is not as obvious as it might seem.

5.4 Localization

Since R is a domain, we may consider its field of fractions Q. We will be considering tensoring our Bott–Samelson bimodules with Q on the right; this will do away with grading considerations, since Q is ungraded. Furthermore, recall that Q is a flat R-module, meaning that $- \otimes_R Q$ is an exact functor on R-modules.

As a first computation, let $s \in S$ and consider $B_s \otimes_R Q$. The natural inclusion $R \hookrightarrow Q$ induces an injection $B_s \otimes_R Q \simeq R \otimes_{R^s} Q \hookrightarrow Q \otimes_{Q^s} Q$, where Q^s denotes the ring of s-invariants of Q. This map is also surjective since for any nonzero $f \in R$ we have

$$\frac{1}{f} \otimes 1 = \frac{s(f)}{fs(f)} \otimes 1 = s(f) \otimes \frac{1}{fs(f)}, \tag{5.22}$$

the element on the right being in the image of our map. Thus, we have that $B_s \otimes_R Q \simeq Q \otimes_{Q^s} Q$ as (R, Q)-bimodules, i.e. left R-module and right Q-module. Inductively, we obtain the following lemma.

Lemma 5.20 *For any expression* $\underline{x} = (s, t, \ldots, u)$, *we have*

$$\mathrm{BS}(\underline{x}) \otimes_R Q = R \otimes_{R^s} R \otimes_{R^t} \cdots \otimes_{R^u} R \otimes_R Q \simeq Q \otimes_{Q^s} \cdots \otimes_{Q^u} Q \tag{5.23}$$

as (R, Q)-*bimodules.*

5.4 Localization

In particular, this shows that $BS(\underline{x}) \otimes_R Q$, which was a priori only an (R, Q)-bimodule, is in fact a Q-bimodule.[1]

We have a similar story for standard bimodules. For any $x \in W$, we let Q_x denote the standard Q-bimodule obtained by twisting the regular bimodule Q on the right by x (i.e. defined analogously to R_x). Then we have an isomorphism $R_x \otimes_R Q \simeq Q_x$ of (R, Q)-bimodules, which shows that $R_x \otimes_R Q$ is in fact a Q-bimodule.

The following observation is what makes the passage to Q-bimodules so useful for studying Soergel bimodules.

Lemma 5.21 *For $s \in S$, we have $B_s \otimes_R Q \simeq Q_s \oplus Q_{\mathrm{id}}$ as Q-bimodules.*

Proof Notice that (Δ) and (∇) "split each other" after localization, since the composition $R \xrightarrow{1 \mapsto c_s} B_s \xrightarrow{\mu_{\mathrm{id}}} R$ is multiplication by α_s, which is an invertible element of Q. □

Remark 5.22 Soergel in [165, Lemma 6.10] also proves that, after localization, any Δ-filtration or ∇-filtration splits.

We let $\mathbb{BSBim}_Q$ be the smallest full subcategory of Q-bimodules containing $B_{s,Q}$ for all $s \in S$, and closed under tensor product. We define $\mathbb{SBim}_Q$ to be the smallest full subcategory of Q-bimodules containing $\mathbb{BSBim}_Q$ and closed under finite direct sums and direct summands. We let $Std\mathrm{Bim}_Q$ denote the full subcategory consisting of finite direct sums of shifts of various standard bimodules Q_x. Lemma 5.20 implies that the functor $(-) \otimes_R Q$ (and the functor of forgetting the grading) induces a monoidal functor

$$\mathbb{BSBim} \to \mathbb{BSBim}_Q.$$

Taking direct summands (or more formally, the idempotent completion), we obtain a functor

$$\mathrm{Loc} : \mathbb{SBim} \to \mathbb{SBim}_Q$$

which we call the *localization functor*.

Lemma 5.21 tells us that any Bott–Samelson bimodule over Q, i.e. any object of $\mathbb{BSBim}_Q$, splits as the direct sum of standard bimodules Q_x with shifts. The same is true of any Soergel bimodule. Note that this splitting is unique by the Krull–Schmidt property (see Sect. 11.3.4). That is, there is an equivalence of categories

$$\mathbb{SBim}_Q \simeq Std\mathrm{Bim}_Q \qquad (5.24)$$

which is truly wonderful! The category $Std\mathrm{Bim}_Q$ is a very simple one, so this allows one to view objects and morphisms in $\mathbb{SBim}$, after localization, in a simple way.

---

[1] Note that a torsion free left R-module admits at most one Q-module structure extending the given R-module structure.

In particular, this implies that $[\mathbb{S}\text{Bim}_Q]_\oplus \simeq \mathbb{Z}[W]$ as rings, where $[Q_x]$ is sent to the element x. Note the conspicuous absence of $\mathbb{Z}[v^{\pm 1}]$, as we no longer have any gradings. This gives a character function on Soergel bimodules over Q, which is equal to the formal sum of the elements $x \in W$ weighted by the number of summands of Q_x appearing in its decomposition into standard bimodules.

A very nice trait of localization is that it can be seen as a categorification of specialization at $v = 1$. Specifically, we obtain the following commutative diagram:

$$\begin{array}{ccc} \mathbb{S}\text{Bim} & \xrightarrow{\text{ch}} & \text{H} \\ \downarrow{\text{Loc}} & & \downarrow{v \mapsto 1} \\ \mathbb{S}\text{Bim}_Q & \xrightarrow{\text{ch}} & \mathbb{Z}[W] \end{array}$$

(5.25)

The squiggly arrows are technically maps from the Grothendieck group, not from the category itself.

Remark 5.23 This entire discussion of localization could be adapted to the homogeneous fraction field Q' of R, obtained by inverting all nonzero homogeneous polynomials. The ring Q' is still graded, so one can define localization as a functor between graded bimodule categories. However, $Q' \simeq Q'(2)$, since any nonzero degree 2 polynomial (like α_s) is invertible. Thus the action of $\mathbb{Z}[v^{\pm 1}]$ on $[\mathbb{S}\text{Bim}_{Q'}]_\oplus$ factors through the quotient $\mathbb{Z}[v^{\pm 1}]/(v^2-1)$, and we have $[\mathbb{S}\text{Bim}_{Q'}]_\oplus \simeq \mathbb{Z}[v^{\pm 1}]/(v^2-1)[W]$. This is very similar to the specialization at $v = 1$, but it keeps track of some simple information about parity.

5.5 Soergel's Categorification Theorem

Now we make more precise some of the ideas stated in Sect. 4.8. For $M, N \in R$-gbim, note that $\text{Hom}^\bullet(M, N)$ is itself a graded R-bimodule via

$$(f \cdot \varphi \cdot g)(m) = f \cdot \varphi(m) \cdot g \qquad (5.26)$$

for $f, g \in R$, $\varphi \in \text{Hom}^\bullet(M, N)$, and $m \in M$.

Theorem 5.24 (Soergel's Categorification Theorem [165]) *Under the technical assumptions to be discussed in Sect. 5.6, we have the following results.*

1. *There is a $\mathbb{Z}[v^{\pm 1}]$-algebra homomorphism*

$$c : \text{H} \to [\mathbb{S}\text{Bim}]_\oplus \qquad (5.27)$$

sending $b_s \mapsto [B_s]$ for all $s \in S$.

5.5 Soergel's Categorification Theorem

2. *There is a bijection between W and the set of indecomposable objects of $\mathbb{S}\mathrm{Bim}$ up to shift and isomorphism:*

$$W \xleftrightarrow{1:1} \{ \text{ indec. objects in } \mathbb{S}\mathrm{Bim}\}/\simeq, (1) \tag{5.28}$$
$$w \longleftrightarrow B_w.$$

The indecomposable object B_w appears as a direct summand of the Bott–Samelson bimodule $\mathrm{BS}(\underline{w})$ for a reduced expression of w. Moreover, all other summands of $\mathrm{BS}(\underline{w})$ are shifts of B_x for $x < w$ in the Bruhat order.

3. *The character function $\mathrm{ch} = \mathrm{ch}_\Delta$ defined above descends to a $\mathbb{Z}[v^{\pm 1}]$-module homomorphism*

$$\mathrm{ch}\colon [\mathbb{S}\mathrm{Bim}]_\oplus \to H \tag{5.29}$$

which is the inverse to c. Thus, both are isomorphisms.

The characterization of B_w in part 2 of the theorem suggests an algorithm for describing them more explicitly. Suppose one has already found all the indecomposables B_x for $x < w$. Choose a reduced expression $\underline{w}$ for w, and look at the Bott–Samelson bimodule $\mathrm{BS}(\underline{w})$. Then find all the direct summands that you already know, and eliminate them (i.e. take a complement). What remains must be the indecomposable bimodule B_w.

Remark 5.25 The algorithm just described should seem vague at the moment. In practice, how does one find direct summands, and how does one take their complements? Later in this book we will discuss how to perform these computations explicitly using morphisms between Bott–Samelson bimodules.

Let us also note an equivalent characterization of the indecomposable Soergel bimodules.

Lemma 5.26 *In Theorem 5.24, the last sentence of part 2 could be replaced with the following sentence: Moreover, B_w has $R_w(-\ell(w))$ exactly once in its Δ-filtration, and all other standard modules in its Δ-filtration are shifts of R_x for $x < w$ in the Bruhat order.*

Proof By part 1 and part 3 of Theorem 5.24, $\mathrm{ch}(\mathrm{BS}(\underline{w})) = b_{\underline{w}}$ for any expression $\underline{w}$. Thus the standard modules appearing as subquotients in a Δ-filtration of $\mathrm{BS}(\underline{w})$ are controlled by the Deodhar defect formula (3.36). If $\underline{w}$ is a reduced expression, then the standard module R_w appears exactly once (with shift) inside $\mathrm{BS}(\underline{w})$, and this shift is precisely $R_w(-\ell(w))$. In particular, exactly one direct summand of $\mathrm{BS}(\underline{w})$ will have $R_w(-\ell(w))$ in its Δ-filtration, and all other direct summands will only contain shifts of R_x for $x < w$ in the Bruhat order.

Now we compare the two potential characterizations of B_w, given that B_w is a direct summand of $\mathrm{BS}(\underline{w})$, and show that they are equivalent. The proof is in a footnote;[2] we want the reader to try it as an exercise. □

One further part of the Soergel categorification theorem has its own name.

Theorem 5.27 (Soergel Hom formula [165]) *For any two Soergel bimodules B, B', the graded Hom space $\mathrm{Hom}^{\bullet}_{\mathbb{S}\mathrm{Bim}}(B, B')$ is free as a left graded R-module with graded rank $(\mathrm{ch}(B), \mathrm{ch}(B'))$:*

$$\underline{\mathrm{rk}}\, \mathrm{Hom}^{\bullet}_{\mathbb{S}\mathrm{Bim}}(B, B') = (\mathrm{ch}(B), \mathrm{ch}(B')). \tag{5.30}$$

Here, $(-, -)$ denotes the standard form on $\mathbf{H}$ (see Definition 3.13), and $\underline{\mathrm{rk}}$ denotes the graded rank (see the end of Sect. 4.1). It is also free as a right graded R-module with the same graded rank.

Remark 5.28 Recall that the standard from is $\mathbb{Z}[v, v^{-1}]$-sesquilinear (antilinear in the first variable, linear in the second variable). Together with (5.19), this ensures that the right hand side of (5.30) interacts with grading shifts in the way one expects from the graded rank of a Hom form.

Later in this book, we will see why the Soergel Hom formula motivates the following conjecture:

Conjecture 5.29 (Soergel's Conjecture) For any $x \in W$, $\mathrm{ch}(B_x) = b_x$. In other words, the Kazhdan–Lusztig polynomial $h_{x,y}$ is equal to $h_x(B_y)$.

It will turn out that this conjecture is equivalent to the statement that certain degree zero morphisms are isomorphisms. A proof of this conjecture (now a theorem) is the focus of Part IV of this book.

Remark 5.30 In particular, the truth of Soergel's conjecture implies that the Kazhdan–Lusztig polynomials $h_{x,y}$ have non-negative coefficients, since $h_x(B)$ is manifestly in $\mathbb{Z}_{\geq 0}[v^{\pm 1}]$ for any Soergel bimodule B. The non-negativity of $h_{x,y}$ is often called the Kazhdan–Lusztig Positivity Conjecture, and it has no known proof which does not use categorification.

Exercise 5.31 We have already classified the indecomposable Soergel bimodules for type A_2 in Example 4.39. Did we name them correctly? Verify Soergel's Categorification Theorem and Soergel's Conjecture for type A_2.

---

[2]If all other summands are shifts of B_x for $x < w$, then only B_w could contain $R_w(-\ell(w))$, so it must. Conversely, if each B_w has $R_x(-\ell(x))$ in its Δ-filtration, then it is impossible for a shift of B_x to appear as a summand of $\mathrm{BS}(\underline{w})$ unless $x \leq w$, because it would contradict the Deodhar formula. Moreover, B_w can appear exactly once as a summand.

5.6 A Technical Wrinkle

Exercise 5.32 Use the Soergel Hom formula to compute the graded rank (as a free left R-module) of the following graded Hom spaces.

$$\text{Hom}^\bullet(B_s, B_t), \qquad \text{Hom}^\bullet(B_s B_s, B_s), \qquad \text{Hom}^\bullet(B_s, B_s B_t B_s)$$

Does this computation depend on the value of m_{st}?

5.6 A Technical Wrinkle

Soergel proves all of the results above under one technical assumption, which we address now. Henceforth, we write V_{geom} for the geometric representation defined in Sect. 1.1.7, a vector space over $\mathbb{R}$. We will let V denote a more general representation of W, possibly over some other base field $\Bbbk$. Our construction of Soergel bimodules did not depend on any of the particular features of the geometric representation. From any representation V one can construct the graded ring $R = \text{Sym}(V)$, and could attempt to define bimodules $B_s = R \otimes_{R^s} R(1)$ for $s \in S$, and hence the categories $\mathbb{B}\mathbb{S}\text{Bim}(V)$ and $\mathbb{S}\text{Bim}(V)$. One can also define bimodules R_x for $x \in W$ and the category $\mathit{Std}\text{Bim}(V)$. What assumptions on V make this theory worth studying?

If $\Bbbk$ is a field of characteristic 2, the very notion of a "reflection" is not entirely clear. The action of a simple reflection is diagonalizable if and only if it is trivial. So let us assume that $\Bbbk$ does not have characteristic 2.

If the representation V is not faithful, all sorts of problems can arise in the discussions above. To begin with Lemma 5.2 will fail, as $R_x \simeq R_y$ whenever x and y act in the same way on V. Hence Δ- and ∇-filtrations will not be well-defined.

Thankfully, the geometric representation is always faithful. However, there is a stronger condition which the geometric representation will not possess when W is for example an affine Weyl group.

Definition 5.33 A representation V of W is *reflection faithful* if it is faithful, and an element $x \in W$ has a fixed point set of codimension one if and only if x is a reflection in W.

In other words, we ask that $x \in W$ acts as a reflection if and only if it is a reflection in W. Let $H_x \subset V$ denote the fixed hyperplane of a reflection $x \in W$. One can deduce from this definition that $H_x = H_y$ if and only if $x = y$, for two reflections $x, y \in W$.

Example 5.34 In Remark 4.38 we discussed how, for the geometric representation V_{geom} of the infinite dihedral group, $H_s = H_t$. As a consequence, V_{geom} is not reflection faithful. This led to some technical problems, such as B_{st} not being cyclic.

Soergel's proofs work under the assumption that V is reflection faithful, and $\Bbbk$ is an infinite field of characteristic not equal to 2. He proved that any Coxeter system (W, S) admits a reflection faithful representation where $\Bbbk = \mathbb{R}$; let us refer to the reflection faithful representation that Soergel constructs (see [165, §2]) as the *Kac–Moody representation* V_{KM}. It was later proved by Libedinsky [120] that various theorems (e.g. the Soergel categorification theorem, Soergel's conjecture) hold for $\mathbb{S}\text{Bim}(V_{\text{geom}})$ if and only if they hold for $\mathbb{S}\text{Bim}(V_{KM})$.

Exercise 5.35 Construct a reflection faithful action of the infinite dihedral group on a three-dimensional space V, where V_{geom} is a subrepresentation. (*Hint:* add a new basis vector, and have s and t act in such a way that $H_s \neq H_t$.) Now H_s and H_t are distinct planes, meeting in the line spanned by $\alpha_s + \alpha_t$, so they will span V.

Unfortunately, reflection faithfulness is a fairly restrictive condition on a representation, because even faithfulness is a restrictive condition!

Exercise 5.36 Prove that there is no faithful finite-dimensional representation of an infinite Coxeter group W over a finite field, or even over its algebraic closure.

Interesting categorifications of the Hecke algebra arise in finite characteristic, so it is extremely worthwhile to have a theory which works well in finite characteristic as well. Unfortunately, Soergel bimodules will not fit the bill.

All this discussion is to motivate the diagrammatic replacement for Soergel bimodules, which is the focus of the next part of this book. We will construct a monoidal category by generators and relations, which is equivalent to $\mathbb{S}\text{Bim}(V)$ when the bimodules are well-behaved, but has the desired properties even when the bimodules do not (e.g. for non-faithful representations). Even when the diagrammatic category is equivalent to $\mathbb{S}\text{Bim}(V)$, it will give a new tool for computing with this category, complementary to the algebraic methods above.

Finally, let us note that while the category $\mathbb{S}\text{Bim}(V)$ was defined purely from a representation V, in our study of it above we made frequent use of the simple roots $\alpha_s \in V_{\text{geom}}$ which come as part of the construction of the geometric representation. The simple roots were also used in the definition of the Demazure operators ∂_s. Our diagrammatic replacement for Soergel bimodules will be defined by generators and relations, and these generators (being very specific maps) will depend on a choice of simple roots α_s and a choice of Demazure operators ∂_s. So, we conclude this chapter with the definition of a realization, which will be the data needed as input to our diagrammatic machinery.

5.7 Realizations of a Coxeter System

We now define a realization of a Coxeter system, following [59, Definition 3.1].

Definition 5.37 Let $\Bbbk$ be a commutative integral domain. A *realization* of (W, S) over $\Bbbk$ is a triple $(\mathfrak{h}, \{\alpha_s\}_{s \in S}, \{\alpha_s^\vee\}_{s \in S})$, where $\mathfrak{h}$ is a free, finite rank $\Bbbk$-module, together with subsets

$$\{\alpha_s^\vee\}_{s \in S} \subseteq \mathfrak{h}, \qquad \{\alpha_s\}_{s \in S} \subseteq \mathfrak{h}^* = \text{Hom}_\Bbbk(\mathfrak{h}, \Bbbk).$$

5.7 Realizations of a Coxeter System

We call the α_s *simple roots* and the $\alpha_s^\vee$ *simple coroots* of the realization. This data is required to satisfy the following conditions:

1. $\alpha_s(\alpha_s^\vee) = 2$ for all $s \in S$;
2. the assignment

$$S \times \mathfrak{h} \to \mathfrak{h}$$
$$(s, v) \mapsto v - \alpha_s(v)\alpha_s^\vee$$

extends to a W-action on $\mathfrak{h}$;
3. Some very minor technical conditions, see (5.40).

Remark 5.38 The technical conditions only exist to rule out some unusual possibilities which can arise: in characteristic two, or when two simple roots and coroots are colinear, or if one wishes to weaken the condition that $\Bbbk$ is a domain.

A realization $(\mathfrak{h}, \{\alpha_s\}_{s\in S}, \{\alpha_s^\vee\}_{s\in S})$ of (W, S) is in particular a W-representation $\mathfrak{h}$. Starting from this underlying representation, we can study Bott–Samelson and Soergel bimodules as before. We will sometimes speak of (the category of) Bott–Samelson and Soergel bimodules associated to the realization $(W, \mathfrak{h})$.

Notation 5.39 In this book, whenever a realization $(\mathfrak{h}, \{\alpha_s\}, \{\alpha_s^\vee\})$ over $\Bbbk$ is fixed, we write R for the symmetric algebra

$$R = \operatorname{Sym}(\mathfrak{h}^*) = \bigoplus_{i \in \mathbb{Z}} \operatorname{Sym}^i(\mathfrak{h}^*),$$

viewed as a graded $\Bbbk$-algebra with $\deg \mathfrak{h}^* = 2$. As an example, if $\{\alpha_s\}_{s\in S}$ forms a $\Bbbk$-basis for $\mathfrak{h}^*$, then R is simply the polynomial algebra $\Bbbk[\alpha_s : s \in S]$ where each α_s has degree 2.

Note that since $\Bbbk$ is a domain, so is R. This is relevant in arguments involving localization.

Let us recast the geometric representation in this new framework. First, note that given a realization $(\mathfrak{h}, \{\alpha_s\}, \{\alpha_s^\vee\})$ of (W, S), the contragredient action of W on $\mathfrak{h}^*$ satisfies a very similar identity to its action on $\mathfrak{h}$. Given $s \in S$ and $\gamma \in \mathfrak{h}^*$, for every $v \in \mathfrak{h}$ we have

$$(s\gamma)(v) = \gamma(s^{-1}v) = \gamma(sv) = \gamma(v - \alpha_s(v)\alpha_s^\vee)$$
$$= \gamma(v) - \alpha_s(v)\gamma(\alpha_s^\vee)$$
$$= (\gamma - \gamma(\alpha_s^\vee)\alpha_s)(v),$$

or in other words,

$$s\gamma = \gamma - \gamma(\alpha_s^\vee)\alpha_s \in \mathfrak{h}^*. \tag{5.31}$$

We are interested in the action of W on R, so it is the action of W on $\mathfrak{h}^*$ which really interests us, not its action on $\mathfrak{h}$. In particular, $s(\alpha_s) = -\alpha_s$ for all $s \in S$.

Example 5.40 Let V_{geom} be the geometric representation of (W, S). Let $\Bbbk = \mathbb{R}$ and $\mathfrak{h} = V_{\text{geom}}^*$. Let $\alpha_s \in \mathfrak{h}^* = V_{\text{geom}}$ be the usual simple root, and let $\alpha_s^\vee$ be the linear functional

$$\alpha_s^\vee = \frac{2(\alpha_s, -)}{(\alpha_s, \alpha_s)}. \tag{5.32}$$

Then this data[3] is a realization, called[4] the *geometric realization*. Note that the simple roots α_s are linearly independent by construction. Meanwhile, the simple coroots $\alpha_s^\vee$ need not be linearly independent. For example, in the geometric realization of $I_2(\infty)$, $\alpha_s^\vee = -\alpha_t^\vee$. When W is finite so that $(-, -)$ is positive definite, the simple coroots will be linearly independent.

Note that R is the symmetric algebra of $\mathfrak{h}^*$, not of $\mathfrak{h}$. Since $\mathfrak{h} = V_{\text{geom}}^*$ for the geometric realization, we recover $R = \text{Sym}(V_{\text{geom}}) = \Bbbk[\alpha_s : s \in S]$.

Example 5.41 (Root Datum Realization) (For those who know some Lie theory.) The following important class of realizations over $\mathbb{Z}$ play an important role in modular representation theory. Let G be a connected reductive group over $\mathbb{C}$, with Borel subgroup and maximal torus $B \supset T$, and let (W, S) be the corresponding Weyl group and simple reflections, which is a (finite crystallographic) Coxeter system. By the structure theory of split reductive groups, such $G \supset B \supset T$ are classified by a combinatorial data called a *based root datum*, which may be viewed as a realization of (W, S) over $\mathbb{Z}$.

We will have more to say about these realizations in Chap. 27. For examples, see Example 27.5 and Exercise 27.6.

Example 5.42 Consider a realization $(\mathfrak{h}, \{\alpha_s\}, \{\alpha_s^\vee\})$ of (W, S), and let $\mathfrak{h}' = \mathfrak{h} \oplus \Bbbk \cdot e$. Setting $\alpha_s(e) = 0$ for all $s \in S$, we get elements $\alpha_s \in (\mathfrak{h}')^*$. Then $(\mathfrak{h}', \{\alpha_s\}, \{\alpha_s^\vee\})$ is a realization, where e is fixed by the action of W.

Example 5.43 Consider a realization $(\mathfrak{h}, \{\alpha_s\}, \{\alpha_s^\vee\})$ of (W, S), and pick some subset $I \subset S$. Then $(\mathfrak{h}, \{\alpha_s\}_{s \in I}, \{\alpha_s^\vee\}_{s \in I})$ is a realization of the parabolic subgroup (W_I, I), which is called the *restriction* to I.

---

[3] As we originally defined the geometric representation, $(\alpha_s, \alpha_s) = 1$, which makes the denominator in (5.32) look silly. However, this definition also constructs a realization whenever we rescale α_s by an invertible scalar in $\mathbb{R}$. In this way, one can get the more familiar representations in non-simply-laced type, where roots can have different lengths.

[4] In [59] it is the dual of this realization (see Example 5.44 below) which is called the geometric realization instead.

5.7 Realizations of a Coxeter System

Note that restriction does not change the size of the polynomial ring R, only of the group that acts on it. In particular, the definition of the R-bimodule B_s for $s \in I$ is identical, whether one uses the original realization or the restricted realization. In this way, $\mathbb{S}$Bim for a restricted realization is a full subcategory of $\mathbb{S}$Bim for the original.

Example 5.44 Consider a realization $(\mathfrak{h}, \{\alpha_s\}, \{\alpha_s^\vee\})$. Then the *(Langlands) dual realization* is $(\mathfrak{h}^*, \{\alpha_s^\vee\}, \{\alpha_s\})$ (i.e. exchange $\mathfrak{h}$ with $\mathfrak{h}^*$, and simple roots with simple coroots).

From these examples, one can see that the simple roots need not span $\mathfrak{h}^*$, and they need not be linearly independent, and similarly for the simple coroots.

Exercise 5.45 Show that the permutation representation of S_n can be equipped with the structure of a realization, constructed by extending the geometric realization using the method of Example 5.42.

Exercise 5.46 Verify that $\partial_s(\lambda) = \lambda(\alpha_s^\vee)$ for any $\lambda \in \mathfrak{h}^*$, so that the coroots are really determining the action of the Demazure operator on linear terms of R. Verify that the operator ∂_s is uniquely determined by its action on linear terms and by the twisted Leibniz rule.

The following condition on a realization plays a crucial role everywhere in the theory.

Definition 5.47 A realization $(\mathfrak{h}, \{\alpha_s\}, \{\alpha_s^\vee\})$ over $\Bbbk$ is said to satisfy *Demazure surjectivity* if the maps

$$\alpha_s : \mathfrak{h} \to \Bbbk, \qquad \alpha_s^\vee : \mathfrak{h}^* \to \Bbbk$$

are surjective for all $s \in S$.

We will need Demazure surjectivity in order to assert that $R^s \subset R$ is a Frobenius extension for each $s \in S$, a fact we exploit in Chap. 8. Note that Demazure surjectivity is automatically satisfied if 2 is invertible in $\Bbbk$.

Assumption 5.48 In this book, we always assume that our realization satisfies Demazure surjectivity.

Given a realization $(\mathfrak{h}, \{\alpha_s\}, \{\alpha_s^\vee\})$ of (W, S), its *Cartan matrix* is the square matrix

$$(a_{st})_{s,t \in S} = (\alpha_t(\alpha_s^\vee))_{s,t \in S}$$

whose rows and columns are indexed by S. Conversely, given a $\Bbbk$-valued matrix $(a_{st})_{s,t \in S}$ with $a_{ss} = 2$ for all $s \in S$, set

$$\mathfrak{h} = \bigoplus_{s \in S} \Bbbk \alpha_s^\vee$$

and define $\alpha_s \in \mathfrak{h}^*$ by $\langle \alpha_s^\vee, \alpha_t \rangle = a_{st}$. We can define an operator s on $\mathfrak{h}$ by the formula from Definition 5.37. If this produces a representation of W on $\mathfrak{h}$, we say that $(a_{st})_{s,t \in S}$ is a *Cartan matrix for* (W, S), and this data is a realization. Note that if $\{\alpha_s^\vee\}$ form a basis of $\mathfrak{h}$, then the realization is determined by its Cartan matrix.

Remark 5.49 Unlike for Cartan matrices of semisimple Lie algebras or generalized Cartan matrices (see Definition 27.1), there are no integrality conditions on the entries of Cartan matrices in the sense above. This leads to some exotic realizations for which the theory of the diagrammatic Hecke category is still well-behaved.

For example, suppose that W is dihedral with $S = \{s, t\}$, and suppose that $\mathfrak{h}^* = \Bbbk \alpha_s \oplus \Bbbk \alpha_t$. The Cartan matrix is

$$\begin{bmatrix} 2 & a_{st} \\ a_{ts} & 2 \end{bmatrix}$$

for some scalars a_{st} and a_{ts}. Then the action of s and t on $\mathfrak{h}^*$ is given by

$$s \mapsto \begin{bmatrix} -1 & -a_{st} \\ 0 & 1 \end{bmatrix}, \quad t \mapsto \begin{bmatrix} 1 & 0 \\ -a_{ts} & -1 \end{bmatrix}. \tag{5.33}$$

The exercises below will use this setup.

Exercise 5.50 Show that

$$\begin{bmatrix} 2 & 1 \\ 1 & 2 \end{bmatrix}$$

is a Cartan matrix of type A_2. That is, check (by directly calculating the action of sts and tst) that this assignment defines a representation of W on $\mathfrak{h}^*$. Note that this differs from the usual Cartan matrix of type A_2. For the usual Cartan matrix, the vectors α_s and α_t form an 120° angle. What about for this unusual Cartan matrix?

The following exercise is particularly important for understanding dihedral groups, and hence for understanding Soergel bimodules outside of type A.

Exercise 5.51 In this exercise we make two assumptions: α_s and α_t are linearly independent, and $a_{st} = a_{ts}$.

In the geometric representation we had $a_{st} = -2\cos(\frac{\pi}{m_{st}})$. Let us instead write $a_{st} = -(q + q^{-1})$ for some parameter $q \in \Bbbk$; after all, when $\Bbbk = \mathbb{C}$ and $q = e^{\frac{\pi i}{m_{st}}}$, the two formulae agree. Writing $a_{st} = -(q + q^{-1})$ will allow us to derive formulae

5.7 Realizations of a Coxeter System

which work simultaneously for all dihedral groups, using what are called quantum numbers.

1. Define the *quantum number* $[n]$, $n \in \mathbb{N}$, by

$$[n] = \frac{q^n - q^{-n}}{q - q^{-1}} = q^{n-1} + q^{n-3} + \cdots + q^{3-n} + q^{1-n}. \tag{5.34}$$

For example, $a_{st} = -[2]$. One has $[1] = 1$ and $[0] = 0$. Prove that

$$[2][n] = [n+1] + [n-1] \tag{5.35}$$

for all $n \geq 1$. We set $[-n] = -[n]$ by convention, so that (5.35) holds for all integers n.

2. Show that the action of $(st)^k$ on the two-dimensional subspace of $\mathfrak{h}^*$ with basis $\{\alpha_s, \alpha_t\}$ is given by the matrix

$$\begin{bmatrix} [2k+1] & -[2k] \\ [2k] & -[2k-1] \end{bmatrix}. \tag{5.36}$$

Deduce that $(st)^m$ acts as the identity on this subspace if and only if

$$[2m] = 0 \quad \text{and} \quad [2m-1] = -1. \tag{5.37}$$

3. For certain values of q it is possible that $[m] = 0$ for some m. If $[m] = 0$ then (5.35) implies that $[m+1] = -[m-1]$. Prove that $[m+k] = -[m-k]$ for all k. Deduce that if $[m] = 0$ then $[2m] = 0$ and $[2m-1] = -1$.

4. The statement that q^2 is a primitive m-th root of unity is equivalent to what statement about quantum numbers? The statement that q is a primitive $2m$-th root of unity is equivalent to what statement about quantum numbers? What about when q is a primitive m-th root of unity for m odd? In all these cases, compare $[m-k]$ and $[k]$.

5. Assume that q is a primitive $2m$-th root of unity. The *positive roots* for the dihedral group are the elements in the W-orbit of the simple roots $\{\alpha_s, \alpha_t\}$, which have the form $a\alpha_s + b\alpha_t$ for $a, b \geq 0$. Find a simple enumeration of these roots as linear combinations of α_s and α_t.

Exercise 5.52 Start with the rank 2 realization above for a dihedral group W, and extend it almost as in Example 5.42, but instead set $\alpha_s(e) = \kappa_s$ for some arbitrary scalars $\kappa_s \in \Bbbk$. Is this a realization of W? For simplicity, you should assume that $a_{st} = a_{ts} = -(q + q^{-1})$ as in the previous exercise.

The computations of Exercise 5.51 can all be done algebraically, letting q or even $(q + q^{-1}) = [2]$ be a formal variable. Observe that (5.35) implies that $[n]$ can be described as $p_n([2])$ for some unique polynomial $p_n \in \mathbb{Z}[x]$ of degree $n - 1$. So if we have fixed an element $[2] \in \Bbbk$, we can define

$$[n] := p_n([2]), \tag{5.38}$$

and these abstract quantum numbers will satisfy (5.35), which is all one needs to reproduce Exercise 5.51 in this abstract context. We record this in the following lemma.

Lemma 5.53 *Fix $s, t \in S$ and a realization where α_s and α_t are linearly independent, or where $\alpha_s^\vee$ and $\alpha_t^\vee$ are linearly independent. Suppose that $a_{st} = a_{ts}$, and denote these entries by $-[2] \in \Bbbk$. If $m_{st} < \infty$ then*

$$[2m_{st}] = 0 \quad \text{and} \quad [2m_{st} - 1] = -1. \tag{5.39}$$

Here, quantum numbers are defined as in (5.38).

Proof Since $(st)^{m_{st}} = \mathrm{id}$, this follows immediately from Exercise 5.51. Note that this exercise can be easily adapted to study the span of $\alpha_s^\vee$ and $\alpha_t^\vee$ in $\mathfrak{h}$, instead of the span of α_s and α_t in $\mathfrak{h}^*$. □

As noted in the exercise, (5.39) would follow from $[m_{st}] = 0$. However, the two statements are not equivalent. It turns out that $[m_{st}] = 0$ is a highly desirable feature in a realization, so much so that we require it! Our final condition on a realization, stated in the special case when $a_{st} = a_{ts}$, is that

$$p_{m_{st}}(-a_{st}) = 0. \tag{5.40}$$

For non-symmetric Cartan matrices where $a_{st} \neq a_{ts}$, the technology of *two-colored quantum numbers*, developed in the appendix of [47], is the appropriate replacement for quantum numbers, and there is an analogue of (5.40) in this setting.

5.8 More Technicalities

We conclude with some technical and distracting remarks on quantum numbers. Please, skip to the next chapter!

As noted, (5.37) does not imply that $[m] = 0$. It does imply the weaker condition

$$2[m] = [2][m] = 0, \tag{5.41}$$

but even under our standing assumption that $\Bbbk$ is a domain, it still might be the case that $\Bbbk$ has characteristic 2, and also that $[2] = 0$, which could permit $[m] \neq 0$.

The following lemma gives a simplification of the technical condition (5.40) in the case of a symmetric Cartan matrix over a domain.

5.8 More Technicalities

Lemma 5.54 *Suppose that $\Bbbk$ is a domain, $a_{st} = a_{ts} = -[2] \in \Bbbk$, and (5.37) holds for some fixed $m < \infty$. Then $[m] = 0$ if and only if the following condition holds: if $[2] = 0$ then m is even.*

Proof First, observe that if $[2] = 0$, then by Exercise 5.51 we have $[2k] = 0$ and $[2k+1] = \pm 1$ for any $k \in \mathbb{Z}$.

Now, if $[m] = 0$, then the condition holds by the previous paragraph: $[2] = 0$ implies that m is even.

Conversely, suppose the condition holds. The equation $[2][m] = 0$ implies that either $[m] = 0$ or $[2] = 0$. If $[2] = 0$ then m is even so $[m] = 0$. Either way, $[m] = 0$. □

Remark 5.55 In the case of a non-symmetric Cartan matrix where $a_{st} \ne a_{ts}$, the lemma should be modified as follows. Suppose that $\Bbbk$ is a domain and (5.37) holds for some fixed $m < \infty$, for both colors of quantum numbers. Then $[m] = 0$ for both colors of quantum numbers if and only if the following conditions hold:

1. If $a_{st} = a_{ts} = 0$ then m is even.
2. If $m = 2$ then $a_{st} = a_{ts} = 0$.

The main difference between this version and the previous is that when $m = 2$ one replaces $[2]^2 = 0$ with $a_{st}a_{ts} = 0$. In a domain, the former implies $[2] = 0$, while the latter only implies that either $a_{st} = 0$ or $a_{ts} = 0$. What we want is for both to be zero.

Let us quickly discuss the degenerate situation where α_s and α_t are colinear, spanning a line. Then both s and t act by -1 on this line, and $(st)^m = 1$ on the span of $\{\alpha_s, \alpha_t\}$, for any m. This can give rise to representations of (W, S) whose Cartan matrices could never give rise to a representation of (W, S) had the roots been independent.

Exercise 5.56 Suppose α_s and α_t are linearly dependent. Deduce that $a_{st}a_{ts} = 4$ in $\Bbbk$. If $a_{st} = a_{ts}$ and the characteristic of $\Bbbk$ is not 2, then deduce that $\alpha_s = \pm \alpha_t$. In this context, what is $[m]$? When is it also true that $[2m] = 0$ and $[2m - 1] = -1$?

The upshot of the exercise above is that our requirement that $[m] = 0$ in a realization rules out many of the degenerate situations which can arise when α_s and α_t are colinear.

Finally, this book assumes one other technical property of the Cartan matrix, called *balancedness* in [59, Definition 3.6]. When $a_{st} = a_{ts} = -[2]$ as in Exercise 5.51, balancedness is the condition that $[m_{st} - 1] = 1$. This condition leads to a major simplification of the diagrammatic Hecke category when m_{st} is odd, and prevents utter disaster when m_{st} is even. We refer the reader to the appendix of [47] for more details.

Assumption 5.57 In this book, we always assume that our realization is balanced.

Chapter 6
Sheaves on Moment Graphs

This chapter is based on expanded notes of a lecture given by the authors and taken by

Geordie Williamson

Abstract We give a brief introduction to the theory of sheaves on moment graphs. This theory is due to Braden–MacPherson and Fiebig. It provides a very useful "local" perspective on Soergel bimodules. Our aim is a brief introduction, as well as the discussion of a few interesting examples.

6.1 Roots in the Geometric Representation

Let (W, S) denote a Coxeter system. Let V_{geom} denote its geometric representation and let $\alpha_s \in V_{\text{geom}}$ denote the simple root associated to $s \in S$. Recall that the set

$$T := \bigcup_{x \subset W} x S x^{-1} \tag{6.1}$$

of conjugates of S in W are called reflections. For any $t \in T$, write $t = xsx^{-1}$ for some $s \in S$ and $x \in W$ such that $xs > x$ and define

$$\alpha_t := x(\alpha_s). \tag{6.2}$$

We call α_t the *positive root associated to t*.

Exercise 6.1 Compute the positive roots associated to all reflections in the symmetric group S_n (see Exercises 1.24 and 1.29).

G. Williamson
School of Mathematics and Statistics, University of Sydney, Sydney, Australia

The positive roots associated to arbitrary reflections are well defined thanks to the following exercise:

Exercise 6.2 Show that α_t for $t \in T$ is well defined as follows:

1. Show that s fixes a hyperplane in V_{geom} and sends α_s to $-\alpha_s$. Make a similar conclusion for xsx^{-1} and thus deduce that α_t is well defined up to scalar.
2. By exploiting a suitable form on V_{geom} show that $(\alpha_t, \alpha_t) = 1$, and thus deduce that α_t is well defined up to ± 1.
3. Show that (under the above assumption that $xs > x$), we have

$$x(\alpha_s) \in \bigoplus_{s \in S} \mathbb{R}_{\geq 0} \alpha_s.$$

Deduce that α_t is well defined. (*Hint:* This part of the exercises requires more work. The interested reader is referred to [85, §5.7].)

Remark 6.3 More generally, consider a realization $(\mathfrak{h}, \{\alpha_s\}_{s \in S}, \{\alpha_s^\vee\}_{s \in S})$ (in the sense of Definition 5.37) of (W, S) over $\Bbbk$. One can proceed as in the previous exercise (and use the fact that $\Bbbk$ is always assumed to be an integral domain) to deduce that α_t is well-defined up to a scalar. This is the most one can hope for in a general realization. What happens for the unusual Cartan matrix of Example 5.50?

6.2 Bruhat Graphs

Let (W, S) be a Coxeter system.

Definition 6.4 The *Bruhat graph* of W is the directed, edge-labeled graph defined as follows. The vertices are the elements $x \in W$ and there is a directed edge from x to tx if $x < tx$ with $t \in T$, in which case this edge is labeled by $\alpha_t \in V_{\text{geom}}$. More generally, for $x \in W$ the *Bruhat graph of* $\{\leq x\}$ is the full subgraph with vertices $\{y \mid y \leq x\}$.

Recall from Sect. 1.2.6 that $x \leq y$ in Bruhat order if and only if there exists a directed path from x to y in the Bruhat graph.

Example 6.5 Let (W, S) be a Coxeter system of type A_1. Its Bruhat graph is:

$$\begin{array}{c} s \\ \uparrow \alpha_s \\ \text{id} \end{array} \tag{6.3}$$

6.2 Bruhat Graphs

Exercise 6.6 Let (W, S) be a Coxeter system of type A_2, with simple reflections s and t, and simple roots α_s and α_t. Show that the Bruhat graph is as follows:

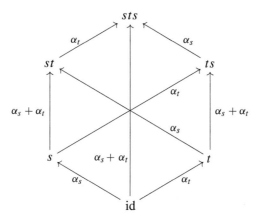

Exercise 6.7 Let (W, S) denote a Coxeter group of type $I_2(5)$, with simple reflections s and t. Show that

$$\alpha_{sts} = \varphi\alpha_s + \alpha_t, \quad \alpha_{ststs} = \varphi\alpha_s + \varphi\alpha_t \quad \text{and} \quad \alpha_{tst} = \alpha_s + \varphi\alpha_t.$$

(Here $\varphi = \frac{1+\sqrt{5}}{2} = 2\cos(\pi/5)$ is the golden ratio.) Fill in the missing edge labels on the following graph to obtain the Bruhat graph of W:

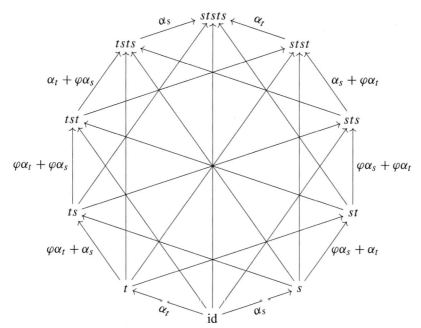

Exercise 6.8 Let (W, S) be a Coxeter group of type A_3, with generators s, t, u with $su = us$. Draw the Bruhat graph of $\{\leq tsut\}$.

As one can see from the examples above, even relatively simple groups can produce rather complicated Bruhat graphs. The following "singular" version is useful in practice, usually because the graphs one obtains are more manageable. Recall our Coxeter system from above, and fix a subset $I \subset S$. We denote by $W_I \subset W$ the corresponding standard parabolic subgroup.

Definition 6.9 The *I-singular Bruhat graph* (or simply *singular Bruhat graph*) is the directed, edge-labeled graph defined as follows. The vertices are the cosets W/W_I and there is a directed edge from xW_I to yW_I if there exist representatives $x' \in xW_I$, $y' \in yW_I$ and $t \in T$ such that $tx' = y'$ and $tx' < y'$, in which case this edge is labeled by $\alpha_t \in V_{\text{geom}}$. More generally, for $x \in W/W_I$ the *singular Bruhat graph of* $\{\le x\}$ is the full subgraph with vertices $\{yW_I \mid y \le x\}$.

Remark 6.10 Two things are not obvious from the above definition. The first is that the Bruhat order on W induces a partial order on W/W_I with good properties (for example, there are no loops in the singular Bruhat graph). The second is that, given xW_I and yW_I there is at most one t such that $txW_I = yW_I$ (hence the edge labels are well defined). Both statements are true. We will see examples below. For further discussion the reader is referred to [85, §5.13] and [177, §1.2].

Exercise 6.11 Suppose that (W, S) is of type A_n with simple reflections $s_1, s_2, \ldots, s_n$. Let the corresponding simple roots be denoted by $\alpha_1, \ldots, \alpha_n$. Fix $1 \le i \le n$ and let $I = S \setminus \{s_i\}$. In this exercise we describe the I-singular Bruhat graph explicitly.

1. By identifying W with the symmetric group S_{n+1}, show that W/W_I may be identified with subsets of $\{1, \ldots, n+1\}$ of cardinality i.
2. Under this identification, show that two distinct subsets $J, J' \subset \{1, \ldots, n+1\}$ are connected by an edge in the singular Bruhat graph if and only if $|J \cap J'| = i - 1$.
3. Describe explicitly the direction and label of the edges in terms of the vertices J and J'.
4. Check that your recipe produces the following graph, for $n = 3$ and $i = 2$:

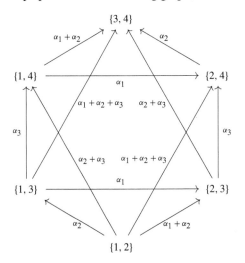

6.2 Bruhat Graphs

Exercise 6.12 Consider W of type D_4 with generators

and simple roots $\alpha, \beta, \gamma, \zeta$ such that $\alpha = \alpha_s$, $\beta = \alpha_t$, $\gamma = \alpha_u$ and $\zeta = \alpha_z$. Let $I = \{s, t, u\}$. Show that the I-singular Bruhat graph for $\{\leq x = stuz\}$ is given as follows:

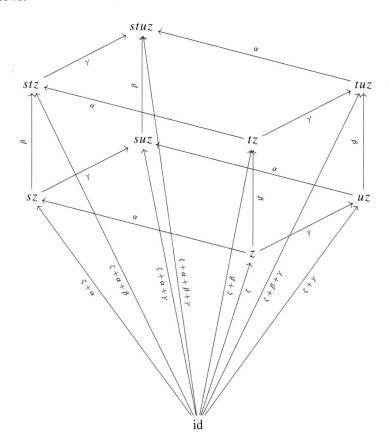

Exercise 6.13 Let (W, S) be a Coxeter system of type $I_2(m)$ for $m \geq 2$, with generators s and t, and let $I = \{s\}$. Describe the I-singular Bruhat graph explicitly (*Hint:* it might be helpful to revisit Exercise 6.7).

6.3 Moment Graphs

Let $\Bbbk$ denote a commutative ring (usually we will take $\Bbbk = \mathbb{R}$). Let V denote a finitely-generated and free $\Bbbk$-module.

Definition 6.14 A *V-moment graph* (or simply *moment graph*) consists of

- a graph $\mathcal{G}$,
- a partial order $\leq$ on the vertices of $\mathcal{G}$,
- a labeling $E \mapsto \alpha_E \in V$ of the edges of $\mathcal{G}$ by elements of V.

We assume that the graph structure and partial order are compatible. More precisely, we assume that any two incident vertices are comparable, and that the relations $v \leq v'$ for incident vertices v and v' generate the partial order $\leq$ on Γ.

Remark 6.15 It is useful to extend the partial order to the edges of $\mathcal{G}$ as follows. Given an edge $E : x \to y$ in $\mathcal{G}$ we declare $x \leq E \leq y$, and extend this to the vertices and edges via transitivity. By abuse of notation we continue to use $\leq$ for the resulting partial order on the vertices and edges of $\mathcal{G}$.

The notion of a V-moment graph is far too loose to be useful in general. In the following, we will only consider moment graphs satisfying the following additional condition.

Definition 6.16 A V-moment graph is said to be *Goresky–Kottwitz–MacPherson* (or simply *GKM*) if it satisfies the following condition:

$$\text{For any two edges } E, E' \text{ incident to the same vertex } v, \text{ the labels } \alpha_E \text{ and } \alpha_{E'} \text{ are linearly independent.} \tag{6.4}$$

Example 6.17 The most important example of a moment graph (and the only example we will consider in this book) is the Bruhat graph or singular Bruhat graph of $\{\leq x\}$ for some x in a Coxeter system. This is a moment graph where $V = V_{\text{geom}}$. In the previous section we saw a number of examples of such graphs.

Exercise 6.18 Show that the Bruhat graph or singular Bruhat graph is always GKM.

Remark 6.19 (For readers with some background in algebraic geometry.) Another important source of examples of moment graphs arises from equivariant algebraic geometry. Suppose that an algebraic torus T acts on an algebraic variety with finitely many fixed points, and finitely many one-dimensional orbits. Then from this action one obtains a moment graph in which the vertices (resp. edges) encode the fixed points (resp. one-dimensional orbits). This is the origin of the term "moment graph." (In this level of generality the directions on the edges are missing. They encode the additional data of a stratification on our variety.) For Weyl groups, the example of (partial) flag varieties gives rise to (singular) Bruhat graphs. For more details, the reader is referred to [31, 68].

Remark 6.20 For all the important applications, it turns out that rescaling the edge labels α_E makes no difference. For example, the GKM condition is preserved by rescaling the edge labels, and the definition below of a sheaf on a moment graph is unchanged. For this reason, one can define Bruhat graphs associated to arbitrary realizations of (W, S), see Remark 6.3.

6.4 Sheaves on Moment Graphs

Let $\Bbbk$ and V be as above, and fix a V-moment graph $\mathcal{G}$. Let R denote the symmetric algebra of V, graded so that V is in degree 2.

Definition 6.21 A *sheaf* $\mathcal{B}$ on the moment graph $\mathcal{G}$ consists of the following data:

- graded R-modules $\mathcal{B}_x$ for each vertex x in $\mathcal{G}$,
- graded $R/(\alpha_E)$-modules $\mathcal{B}_E$ for each edge E in $\mathcal{G}$,
- for each edge E incident to x, a map

$$\rho_{x,E} : \mathcal{B}_x \to \mathcal{B}_E$$

of graded R-modules.

Sheaves on the moment graph $\mathcal{G}$ form a category: morphisms from $\mathcal{B}$ to $\mathcal{B}'$ are families of degree zero maps $\mathcal{B}_x \to \mathcal{B}'_x$, $\mathcal{B}_E \to \mathcal{B}'_E$ commuting with the $\rho_{x,E}$. The shift functor on graded bimodules induces a shift functor $\mathcal{B} \mapsto \mathcal{B}(1)$ in a natural way. We denote the resulting category by $\mathrm{Sh}(\mathcal{G})$.

Example 6.22 The most important example of a sheaf on the moment graph is given by the *structure sheaf* $\mathcal{A}$. We have $\mathcal{A}_x = R$ for all vertices x, $\mathcal{A}_E = R/(\alpha_E)$ for all edges E, and $\rho_{x,E} : R \to R/(\alpha_E)$ is the canonical quotient map.

Example 6.23 Another important example is the *skyscraper sheaf* $\mathcal{I}(x)$, for x a vertex of $\mathcal{G}$. We have $\mathcal{I}(x)_x = R$, $\mathcal{I}(x)_y = 0 = \mathcal{I}(x)_E$ for all vertices $y \neq x$ and all edges E.

One can depict a sheaf $\mathcal{M}$ on the moment graph by drawing $\mathcal{M}_x$ at each vertex, and $\mathcal{M}_E$ along each edge. Here is such a picture of a sheaf on the moment graph of type A_1 (see Example 6.5):

(One usually omits the direction of the arrows in the moment graph, as they lead to confusion. We will also often omit the maps $\rho_{x,E}$, however it should not be forgotten that these are important pieces of data.)

Example 6.24 Let $\mathcal{G}$ be the Bruhat graph of type A_1, as above. Here we depict the structure sheaf $\mathcal{A}$, as well as the skyscraper sheaves $\mathcal{I}(\mathrm{id})$ and $\mathcal{I}(s)$:

$$\begin{array}{c} R \\ | \\ R/(\alpha_s) \\ | \\ R \end{array} \quad , \quad \begin{array}{c} 0 \\ | \\ 0 \\ | \\ R \end{array} \quad , \quad \begin{array}{c} R \\ | \\ 0 \\ | \\ 0 \end{array} \quad .$$

Definition 6.25 Let $\mathcal{G}$ be a moment graph and $\mathcal{M}$ a sheaf on $\mathcal{G}$. Given a subset U of the vertices and edges of $\mathcal{G}$ we define

$$\Gamma(U, \mathcal{M}) = \left\{ (f_u) \in \bigoplus_{u \in U} \mathcal{M}_u \;\middle|\; \begin{array}{c} \text{for all pairs } (x, E) \text{ of incident} \\ \text{vertices and edges of } U \\ \text{we have } \rho_{x,E}(f_x) = f_E. \end{array} \right\}. \tag{6.5}$$

We call $\Gamma(U, \mathcal{M})$ the *sections of $\mathcal{M}$ over U*. If U consists of all vertices and edges we abbreviate

$$\Gamma(\mathcal{M}) := \Gamma(U, \mathcal{M}) \tag{6.6}$$

and call $\Gamma(\mathcal{M})$ the *global sections of $\mathcal{M}$*.

Remark 6.26 For a general subset U the sections $\Gamma(U, \mathcal{M})$ are not very meaningful. In the theory a particularly important role is played by $\Gamma(U, \mathcal{M})$ where U is *upwardly closed*; that is, if $x \in U$, then so are any vertices x' or edges E with $x \leq x'$ or $x \leq E$ (see Remark 6.15). The most important upwardly closed sets that we will encounter are

$$\{> x\} := \{\text{vertices } y \mid y > x\} \cup \{\text{edges } E \mid E > x\} \tag{6.7}$$

for a vertex x.

Exercise 6.27 Consider the Bruhat graph of type A_1 (where $V = \mathbb{R}\alpha_1$):

$$\begin{array}{c} s \\ \uparrow \\ \alpha_1 \\ | \\ \mathrm{id} \end{array} \tag{6.8}$$

Show that the global sections of the structure sheaf are free over R of graded rank $1 + v^2$.

6.5 The Braden–MacPherson Algorithm

Exercise 6.28 Consider the following moment graph (where $V = \mathbb{R}\alpha_1 \oplus \mathbb{R}\alpha_2$):

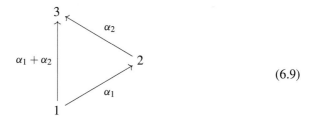
(6.9)

Show that the global sections of the structure sheaf is free of graded rank $1+v^2+v^4$. Do you recognize this moment graph? (*Hint:* See Exercise 6.11.)

Remark 6.29 We have been a little hasty in our treatment of sheaves on moment graphs. For more detail of this theory, the reader is referred to [31, 65, 68, 89].

6.5 The Braden–MacPherson Algorithm

In this section we assume for simplicity that $\mathbb{k}$ is a field.

Let $\mathcal{G}$ denote a moment graph and fix a vertex $x \in \mathcal{G}$. We wish to describe an algorithm (the *Braden–MacPherson algorithm*) which produces a sheaf $\mathcal{B}(x)$ on the moment graph. The algorithm is via downward induction on the partial order on the vertices of $\mathcal{G}$.

Step 0 For all vertices y and edges E which are not less than x (see Remark 6.15), we set $\mathcal{B}(x)_y = \mathcal{B}(x)_E = 0$.

Step 1 Set $\mathcal{B}(x)_x := R$.

Step 2 Suppose that $\mathcal{B}(x)_y$ is defined, but $\mathcal{B}(x)_E$ is not, where $E : y' \to y$ is an edge. Define

$$\mathcal{B}(x)_E := \mathcal{B}(x)_y / \alpha_E \mathcal{B}(x)_y = R/(\alpha_E) \otimes_R \mathcal{B}(x)_y \quad (6.10)$$

and let $\rho_{y,E} : \mathcal{B}(x)_y \to \mathcal{B}(x)_y / \alpha_E \mathcal{B}(x)_y$ be the canonical quotient map.

Step 3 Suppose that $\mathcal{B}(x)_y$ is not defined, but that $\mathcal{B}(x)$ is defined on all vertices and edges above y. Consider the natural map

$$\Gamma(\mathcal{B}(x), \{> y\}) \to \bigoplus_{E:y \to y'} \mathcal{B}(x)_E, \quad (6.11)$$

and let $M_{y,x}$ denote its image (a graded R-module). Define $\mathcal{B}(x)_y$ to be the projective cover of $M_{y,x}$ in the category of graded R-modules. For all edges

$E' : y \to y'$ let $\rho_{y,E'}$ denote the composition

$$\mathcal{B}(x)_y \to M_{y,x} \to \mathcal{B}(x)_{E'} \qquad (6.12)$$

where the second map is induced by the projection $\bigoplus_{E:y\to y'} \mathcal{B}(x)_E \to \mathcal{B}(x)_{E'}$.

Remark 6.30 All steps in the algorithm computing the Braden–MacPherson sheaf are essentially trivial, with the exception of Step 3.

Definition 6.31 The sheaf on the moment graph $\mathcal{B}(x)$ constructed via the above algorithm is called the *Braden–MacPherson sheaf* associated to the vertex $x \in \mathcal{G}$. We denote by $\mathrm{BM}(\mathcal{G})$ the full subcategory of $\mathrm{Sh}(\mathcal{G})$ consisting of direct sums of shifts of Braden–MacPherson sheaves (for any vertex). If the poset underlying $\mathcal{G}$ is graded with length function ℓ, then we define the *normalized Braden–MacPherson sheaf* to be

$$\widetilde{\mathcal{B}}(x) := \mathcal{B}(x)(\ell(x)). \qquad (6.13)$$

Remark 6.32 In all useful examples of the theory the poset underlying $\mathcal{G}$ is graded, and hence normalized Braden–MacPherson sheaves are defined. It is easy to see that the normalized Braden–MacPherson sheaf $\widetilde{\mathcal{B}}(x)$ is obtained via the above algorithm, after instead setting $\widetilde{\mathcal{B}}(x)_x := R(\ell(x))$ at Step 1.

Example 6.33 Let $\mathcal{G}$ denote the Bruhat graph of type A_1:

$$\begin{array}{c} s \\ | \\ \alpha_E \\ | \\ \mathrm{id} \end{array} \qquad (6.14)$$

We denote by E the unique edge in $\mathcal{G}$, and set $\alpha := \alpha_E$. Thus $R = \mathbb{R}[\alpha]$ is a polynomial ring in one variable. There are two vertices in $\mathcal{G}$ and hence two Braden–MacPherson sheaves. When $x = \mathrm{id}$ we apply Steps 0, 1 and 2 above to obtain the following Braden–MacPherson sheaf:

$$\mathcal{B}(\mathrm{id}) = \begin{array}{c} 0 \\ | \\ 0 \\ | \\ R \end{array} \qquad (6.15)$$

When $x = s$ we need to work a little bit harder. Steps 1 and 2 allow us to conclude that $\mathcal{B}(s)_s = R$ and $\mathcal{B}(s)_E = R/(\alpha) = \mathbb{R}$, with $\rho_{s,E}$ the canonical quotient map.

6.5 The Braden–MacPherson Algorithm

The only unknown is $\mathcal{B}(s)_{\mathrm{id}}$. We now carry out Step 3. We first compute

$$\Gamma(\mathcal{B}(s), \{> \mathrm{id}\}) = \{(f, g) \in \mathcal{B}(s)_s \times \mathcal{B}(s)_E \mid \rho_{s,E}(f) = g\} = R$$

because $\rho_{s,E}$ is surjective, and hence g is determined by f. Hence

$$M_{\mathrm{id},x} := \mathrm{Im}(\Gamma(\mathcal{B}(s), \{> \mathrm{id}\}) \to \mathcal{B}(s)_E) = R/(\alpha) = \mathbb{R}.$$

Hence $\mathcal{B}(s)_{\mathrm{id}} = R$, the projective cover of $\mathbb{R}$, and $\rho_{\mathrm{id},E}$ is the canonical quotient map:

$$\mathcal{B}(s) \quad = \quad \begin{array}{c} R \\ | \\ R/(\alpha) \\ | \\ R \end{array} \tag{6.16}$$

We conclude that $\mathcal{B}(s)$ is isomorphic to the structure sheaf of $\mathcal{G}$.

Exercise 6.34 Let $\mathcal{G}$ denote the Bruhat graph of type A_2.

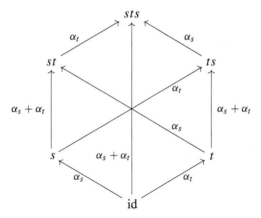

Here $R = \mathbb{R}[\alpha_s, \alpha_t]$ is a polynomial ring in two variables. This a guided exercise in which we compute the Braden–MacPherson sheaf $\mathcal{B}(sts)$ associated to the longest element of W. For ease of notation we abbreviate $\mathcal{B} := \mathcal{B}(sts)$.

1. Argue as in the previous exercise to show that $\mathcal{B}$ has the following form

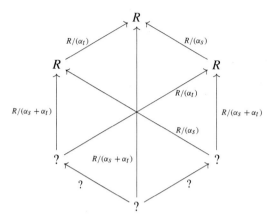

with all restriction maps the canonical quotient maps.

2. We now compute the space of sections $\Gamma(\mathcal{B}, \{> s\})$. Show that any such section is determined by its value in $\mathcal{B}_{st}$, $\mathcal{B}_{ts}$ and $\mathcal{B}_{sts}$ and the space of sections is spanned (as an R-module) by the following 3 sections:

$$\begin{array}{ccc} 1 & 0 & 0 \\ 1 \quad 1 \ , & \alpha_t \quad 0 \ , & 0 \quad \alpha_s \ . \end{array}$$

3. Deduce that

$$M_{s,sts} = \langle (1,1), (\alpha_t, 0), (0, \alpha_s) \rangle \subset R/(\alpha_s + \alpha_t) \oplus R/(\alpha_t).$$

and hence that $\mathcal{B}_s = R$. (Hint: In $R/(\alpha_s + \alpha_t) \oplus R/(\alpha_t)$ one has $(\alpha_s + \alpha_t) \cdot (1, 1) = (0, \alpha_s)$ and $\alpha_t \cdot (1, 1) = (\alpha_t, 0)$.)

4. Via a symmetry argument, conclude that $\mathcal{B}$ has the following form

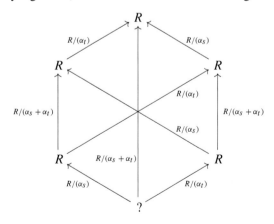

with all restriction maps the canonical quotient maps.

6.5 The Braden–MacPherson Algorithm

5. We now compute the space of sections $\Gamma(\mathcal{B}, \{> \text{id}\})$. Show that any such section is determined by its value in $\mathcal{B}_x$ for $x \in \{s, t, st, ts, sts\}$ and that the space of sections is spanned (as an R-module) by the following 5 sections:

$$\begin{matrix} & 1 & & \\ 1 & & 1 & \\ 1 & & 1 & \end{matrix}, \quad \begin{matrix} & 0 & & \\ \alpha_t & & 0 & \\ \alpha_t & & \alpha_s + \alpha_t & \end{matrix}, \quad \begin{matrix} & 0 & & \\ 0 & & \alpha_s & \\ \alpha_s + \alpha_t & & \alpha_s & \end{matrix},$$

$$\begin{matrix} & 0 & & \\ 0 & & 0 & \\ 0 & & \alpha_s(\alpha_s + \alpha_t) & \end{matrix}, \quad \begin{matrix} & 0 & & \\ 0 & & 0 & \\ \alpha_t(\alpha_s + \alpha_t) & & 0 & \end{matrix}.$$

Deduce that

$$M_{\text{id},sts} = \langle (1, 1, 1), (\alpha_t, 0, \alpha_s + \alpha_t), (\alpha_s + \alpha_t, 0, \alpha_s),$$
$$(0, 0, \alpha_s(\alpha_s + \alpha_t)), (\alpha_t(\alpha_s + \alpha_t), 0, 0) \rangle$$
$$\subset R/(\alpha_s) \oplus R/(\alpha_s + \alpha_t) \oplus R/(\alpha_t).$$

Finally, show that $M_{\text{id}} = R \cdot (1, 1, 1)$; deduce that $\mathcal{B}_{\text{id}} = R$, and that $\mathcal{B}$ is isomorphic to the structure sheaf.

In these two examples the Braden–MacPherson algorithm did not provide any new examples of sheaves on moment graphs. We now consider two examples where the Braden–MacPherson algorithm produces something genuinely new.

Exercise 6.35 We consider the moment graph obtained by considering the full subgraph of elements less than $\{2, 4\}$ in Example 6.11:

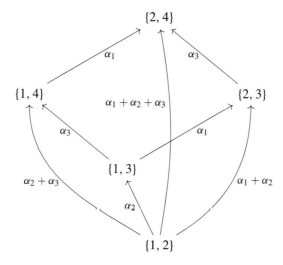

Fig. 6.1 The Braden–MacPherson sheaf $\mathcal{B}$ from Exercise 6.35

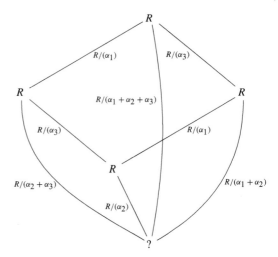

(We have rearranged the organization of the vertices slightly, to make the order more apparent.) Recall that $R = \mathbb{R}[\alpha_1, \alpha_2, \alpha_3]$. We abbreviate $\mathcal{B} := \mathcal{B}(\{2, 4\})$ for the Braden–MacPherson sheaf corresponding to the maximal vertex.

1. Show that "nothing interesting happens" in the first few steps of the Braden–MacPherson algorithm; that is, that $\mathcal{B}$ has the form shown in Fig. 6.1.
2. Deduce that

$$M_{\{1,2\},\{2,4\}} = \langle (1, 1, 1, 1), (\alpha_1, \alpha_1, 0, 0), (0, \alpha_3, 0, \alpha_3), (0, \alpha_1\alpha_3, 0, 0) \rangle$$
$$\subset R/(\alpha_2 + \alpha_3) \oplus R/(\alpha_2) \oplus R/(\alpha_1 + \alpha_2 + \alpha_3) \oplus R/(\alpha_1 + \alpha_2).$$

3. Conclude that $\mathcal{B}_{\{1,2\}}$ is of graded rank $1 + v^2$.

Exercise 6.36 Calculate the Braden–MacPherson sheaf for the moment graph in Example 6.12 (a singular Bruhat graph of type D_4). In particular, show that $\mathcal{B}(suvz)_{\mathrm{id}}$ is of graded rank $1 + 2v^2$. (*Hint:* Imitate the strategy of the previous exercise.)

6.6 Stalks and Standard Bimodules

In this section we lay the groundwork for a functor from Soergel bimodules to sheaves on moment graphs. The results of the rest of this chapter are due to Fiebig, see [65] for more detail.

For the rest of this chapter we will break with tradition and allow us to view graded R-bimodules (and in particular Soergel bimodules) as graded modules over $R \otimes R := R \otimes_{\mathbb{R}} R$. This is allowed because R is commutative. Recall the standard

6.6 Stalks and Standard Bimodules

bimodule R_y associated to $y \in W$ from Sect. 5.2. As an $R \otimes R$-module we have

$$R_y = R \otimes R / \langle yr \otimes 1 - 1 \otimes r \mid r \in R \rangle. \tag{6.17}$$

Given a bimodule (or equivalently $R \otimes R$-module) M we define

$$\Gamma^y M := R_y \otimes_{R \otimes R} M. \tag{6.18}$$

Exercise 6.37 Show that $\Gamma^y M$ may be characterized as a bimodule as follows: $\Gamma^y M$ is the largest quotient Q of M on which one has the relation $q \cdot r = y(r) \cdot q$ for all $q \in Q$.

Exercise 6.38 (For readers with some background in algebraic geometry.) Identify $R \otimes R$ with the regular functions on $V^* \times V^*$ and consider the subvariety given by the "twisted graph"

$$\mathrm{Gr}_y := \{(y\lambda, \lambda) \mid \lambda \in V^*\} \subset V^* \times V^*. \tag{6.19}$$

Identify R_y with the regular functions on Gr_y, and show that if we view M as a quasi-coherent sheaf on $V^* \times V^*$, then $\Gamma^y M$ is its stalk along Gr_y.[1]

As well as the twisted bimodules R_y another set of bimodules will be important below. Given $x \in W$ and $t \in T$, define:

$$R_{x,tx} := R \otimes_{R^t} R_x. \tag{6.20}$$

Exercise 6.39 Show that $R_{x,tx} = R_{tx,x}$ canonically.

Exercise 6.40 The following exercise justifies the notation $R_{x,tx}$ somewhat and will be important below:

1. Show that $R_{x,tx}$ has filtrations of the form

$$0 \to R_x(-2) \to R_{x,tx} \to R_{tx} \to 0 \tag{6.21}$$

and

$$0 \to R_{tx}(-2) \to R_{x,tx} \to R_x \to 0. \tag{6.22}$$

(If you are stuck, reread Sect. 5.3.)

---

[1] This exercise explains part of the title of this subsection. At least for the author it is very useful to think of $\Gamma^y M$ as a "stalk" of M.

2. Deduce that

$$\Gamma^y R_{x,tx} = \begin{cases} R_y & \text{if } y \in \{tx, x\}, \\ 0 & \text{otherwise.} \end{cases} \tag{6.23}$$

Given any bimodule M, define

$$\Gamma^{x,tx} M := R_{x,tx} \otimes_{R \otimes R} M. \tag{6.24}$$

The canonical maps $R_{x,tx} \to R_x$ and $R_{x,tx} \to R_t$ give us canonical maps:

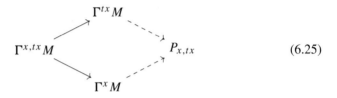

(6.25)

We define $P_{x,tx}$ to be the pushout (in the category of graded $R \otimes R$-modules) of this diagram.

Exercise 6.41 Show that α_t acts as zero on $P_{x,tx}$. (*Hint:* Show that any element $p \in P_{x,tx}$ satisfies both $p \cdot r = x(r) \cdot p$ and $p \cdot r = tx(r) \cdot p$ for all $r \in R$.)

6.7 A Functor to Sheaves on Moment Graphs

Let $\mathcal{G}$ denote the Bruhat graph of W, which we view as a V-moment graph, where V is the geometric representation. Let B denote an R-bimodule. We define a sheaf on the moment graph $\mathcal{B}$ as follows:

1. $\mathcal{B}_x := \Gamma^x B$ for all vertices x of $\mathcal{G}$;
2. $\mathcal{B}_E := P_{x,tx}$ for all edges $E : x \to tx$ of $\mathcal{G}$;
3. Given an edge $E : x \to tx$ the maps

$$\rho_{x,E} : \mathcal{B}_x \to \mathcal{B}_E \quad \text{and} \quad \rho_{tx,E} : \mathcal{B}_{tx} \to \mathcal{B}_E$$

are the same maps in the diagram (6.25).

It is easy to check that $\mathcal{B}$ defines a sheaf on the moment graph. (In fact, if you did Exercise 6.41 you have already checked this.) We obtain in this way a functor

$$\Lambda : R\text{-gbim} \to \text{Sh}(\mathcal{G}) \tag{6.26}$$

from bimodules to sheaves on the moment graph $\mathcal{G}$. Following Fiebig, we call Λ the *localization functor*. For an arbitrary bimodule, Λ is not very interesting, however remarkably one has the following:

Theorem 6.42 *The localization functor induces an equivalence between the additive categories of Soergel bimodules and Braden–MacPherson sheaves:*

$$\Lambda : \mathbb{SBim} \xrightarrow{\sim} \mathrm{BM}(\mathcal{G}). \tag{6.27}$$

Moreover, Λ sends indecomposable Soergel bimodules to indecomposable normalized Braden–MacPherson sheaves:

$$\Lambda(B_x) \simeq \widetilde{\mathcal{B}}(x). \tag{6.28}$$

Exercise 6.43 Let W denote the symmetric group S_3 with simple reflections s and t. Compute the images of the bimodules B_{id}, B_s, B_{st} under Λ. Verify that one does indeed obtain normalized Braden–MacPherson sheaves.

As the reader might guess, the proof of the above theorem is quite involved, and we refer the reader to [65]. A key ingredient is the definition of functors on $\mathrm{Sh}(\mathcal{G})$ which correspond under Λ to tensoring on the right with B_s. Fiebig also shows that the functor of global sections may be upgraded to land in R-bimodules, and this upgraded global sections functor provides an inverse to Λ.

Remark 6.44 Recall that Soergel bimodules are constructed inductively, starting from B_{id} and working up the Bruhat order. On the other hand, Braden–MacPherson sheaves are constructed by working down the Bruhat order, and (in contrast to Soergel bimodules) one does not need any knowledge of previous Braden–MacPherson sheaves to carry out the algorithm.

Remark 6.45 Another interesting aspect of Fiebig's theory is that one may characterize the Braden–MacPherson sheaves as the projective objects in an exact structure on certain subcategories of $\mathrm{Sh}(\mathcal{G})$. This gives a remarkable intrinsic characterization of Braden–MacPherson sheaves.

Remark 6.46 The attentive reader might ask: where have sheaves on singular Bruhat graphs gone? The answer is singular Soergel bimodules, which will be discussed in Chap. 24.

6.8 Soergel's Conjecture and the Braden–MacPherson Algorithm

In Sect. 5.5 we discussed Soergel's conjecture. Soergel's conjecture takes a particularly appealing[2] form when seen from the perspective of sheaves on the moment graph.

---

[2] And perhaps deceptively simple...

Let $\mathcal{G}$ denote the Bruhat graph, viewed as a V_{geom}-moment graph as above. Let $\mathcal{B}$ denote a Braden–MacPherson sheaf (i.e. an object of $\text{BM}(\mathcal{G})$, not necessarily indecomposable). Each $\mathcal{B}_x$ is a free graded R-module, and hence we have

$$\mathcal{B}_x \simeq R^{\oplus h_x(\mathcal{B})} \tag{6.29}$$

for some $h_x(\mathcal{B}) \in \mathbb{Z}[v^{\pm 1}]$. We define:

$$\text{ch}(\mathcal{B}) := \sum_{x \in W} v^{-\ell(x)} h_x(\mathcal{B}) \delta_x \in \text{H}. \tag{6.30}$$

This character is related to the ∇-character ch_∇ (see Sect. 5.3) via

$$\text{ch}(\Lambda(B)) = \text{ch}_\nabla(B) \tag{6.31}$$

(see [65, §8.2]). Thus, via Theorem 6.42, Soergel's conjecture (Conjecture 5.29) translates into the following statement:

Conjecture 6.47 For all $x \in W$ we have

$$\text{ch}(\widetilde{\mathcal{B}}(x)) = b_x. \tag{6.32}$$

The following exercise is straightforward and worthwhile:

Exercise 6.48 Show that $\text{ch}(\widetilde{\mathcal{B}}(x)) = b_x$ is equivalent to the fact that $M_{y,x}$ is generated in degrees $< \ell(x) - \ell(y)$ for all $y < x$, where $M_{y,x}$ is the R-module occurring in Step 3 of the Braden–MacPherson algorithm.

Remark 6.49 The condition that $M_{y,x}$ is generated in degrees $< \ell(x) - \ell(y)$ is a categorification of the condition that $h_{y,x} \in v\mathbb{Z}[v]$. In [31], Braden and MacPherson deduce it from the hard Lefschetz theorem in intersection cohomology. The Hodge theory of Soergel bimodules (the topic of Part IV of this book) provides another proof. The paper [179] provides a proof of the analogue of the relevant hard Lefschetz statement used by Braden and MacPherson, in the context of Soergel bimodules. It would be wonderful to have a proof entirely within the context of sheaves on moment graphs.

Part II
Diagrammatic Hecke Category

Chapter 7
How to Draw Monoidal Categories

This chapter is based on expanded notes of a lecture given by the authors and taken by
Anna Cepek and **Andrew Stephens**

Abstract In this chapter we take a short break from Soergel bimodules to learn about diagrammatics. We illustrate the method of string diagrams for drawing morphisms in 2-categories and contrast it with the way that they are typically drawn. By viewing a monoidal category as a 2-category with a single object, we are also able to draw morphisms in monoidal categories. With these diagrams in hand, we then define the Temperley–Lieb category. In subsequent chapters we will use string diagrams to understand the morphisms in the monoidal category of Soergel bimodules.

7.1 Linear Diagrams for Categories

For a category $\mathcal{C}$ with objects M, N, a morphism $f \in \mathrm{Hom}_{\mathcal{C}}(M, N)$ is typically drawn as an arrow $M \xrightarrow{f} N$. If $g \in \mathrm{Hom}_{\mathcal{C}}(N, P)$, then the composition of f and g is usually depicted as

$$M \xrightarrow{f} N \xrightarrow{g} P. \qquad (7.1)$$

In this approach an object is depicted by something zero-dimensional and a morphism by something one-dimensional. We can dualize this to obtain another way to draw categories, which we will call *linear diagrams*, where now objects are represented by something one-dimensional and morphisms by something zero-dimensional. For example, the following linear diagram (read from right to left)

A. Cepek
Department of Mathematical Sciences, Montana State University, Bozeman, MT, USA

A. Stephens
Department of Mathematics, University of Oregon, Eugene, OR, USA

© The Editor(s) (if applicable) and The Author(s), under exclusive licence to Springer Nature Switzerland AG 2020
B. Elias et al., *Introduction to Soergel Bimodules*, RSME Springer Series 5, https://doi.org/10.1007/978-3-030-48826-0_7

represents the composition gf:

$$\underset{P}{\bullet} \overset{g}{} \underset{N}{\bullet} \overset{f}{} \underset{M}{\bullet}$$

A generic point on the line has an associated object, and a generic interval is a morphism from the right endpoint to the left. Composition of morphisms is given by concatenation of intervals. An interval that only contains an object is the identity morphism on that object. Note that in a linear diagram there are many distinct points that represent the same object. Likewise there are many ways to represent a morphism by choosing where we place the dot:

$$\underset{N}{} \overset{f}{\underset{M}{\bullet}} \qquad \underset{N}{\bullet} \overset{f}{} \underset{M}{}$$

(Compare the diagrams above with the equality of $N \xleftarrow{f} M$ and $N \xleftarrow{f} M$.)

Remark 7.1 One can view repositioning the dot, as in the pictures above, as composition with the appropriate identity morphisms. More formally, we can label the above diagrams as

$$\overset{\mathbb{1}_N}{\vdash\!\!-\!\!\dashv} \overset{f}{\vdash\!\bullet\!\dashv} \qquad \overset{f}{\vdash\!\bullet\!\dashv} \overset{\mathbb{1}_M}{\vdash\!\!-\!\!\dashv}$$

(for clarity we have omitted the labels for the objects M, N). The identity axiom of a category then says that the following diagrams are equal:

$$\overset{\mathbb{1}_N}{\vdash\!\!-\!\!\dashv} \overset{f}{\vdash\!\bullet\!\dashv} = \overset{f}{\vdash\!\bullet\!\dashv} = \overset{f}{\vdash\!\bullet\!\dashv} \overset{\mathbb{1}_M}{\vdash\!\!-\!\!\dashv} \qquad (7.2)$$

Remark 7.2 Associativity of composition means we do not have to worry about parenthesizing morphisms.

In light of the remarks above we have the following proposition:

Proposition 7.3 *The identity and composition axioms of a category imply that a diagram, up to linear isotopy, unambiguously represents a morphism.*

Here, *linear isotopy* means isotopy constrained in a line. Informally, it means we can stretch intervals and slide dots along a line, but not slide them past other dots.

Remark 7.4 Note that we have chosen to read linear diagrams from right to left, rather than from left to right as in (7.1). Our choice is compatible with the standard notation for composition: $g \circ f$ is drawn as g to the left of f.

7.2 Planar Diagrams for 2-Categories

The procedure for (strict) 2-categories is analogous to that for categories. For the definition of a 2-category, see e.g. [119]. We start with some examples of 2-categories and then describe how to draw planar diagrams to represent objects, 1-morphisms, and 2-morphisms (morphisms between morphisms).

Example 7.5 $\mathcal{C}at$ is the 2-category where the objects are categories themselves. The 1-morphisms of $\mathcal{C}at$ are functors, and the 2-morphisms are natural transformations.

Example 7.6 The 2-category $\mathcal{B}im$ has rings as its objects. For two rings R, S, a 1-morphism from R to S is an (S, R)-bimodule, $_S M_R$. Composition of 1-morphisms is given by tensor product. That is, the (T, R)-bimodule $_T N \otimes_S M_R$ is the composition of $_S M_R$ with $_T N_S$. The 2-morphisms of $\mathcal{B}im$ are bimodule homomorphisms.

In a 2-category, a 2-morphism α is traditionally drawn as follows:

(7.3)

The objects $\mathcal{C}$ and $\mathcal{D}$ are depicted by something zero-dimensional, the 1-morphisms F and G by something one-dimensional, and the 2-morphism α by something two-dimensional, a 2-cell filling in the diagram.

As before, we will dualize this method to obtain our preferred way to draw (strict) 2-categories. Objects will label regions in the plane, morphisms will label lines, and 2-morphisms will label dots. We will call these diagrams *planar diagrams* (also called *string diagrams*) and read them from right to left and bottom to top. For example, the following planar diagram describes the same information as (7.3):

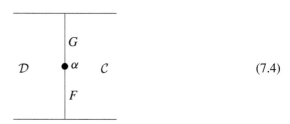
(7.4)

Meanwhile, the planar diagram

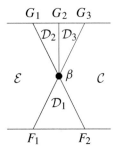

represents a natural transformation $\beta : F_1 \circ F_2 \to G_1 \circ G_2 \circ G_3$.

In a planar diagram a generic point is an object, a generic horizontal interval is a 1-morphism from the right end-point to the left end-point, and a generic rectangle is a 2-morphism from the bottom edge to the top edge. Composition of 1-morphisms will be given by horizontal concatenation. Vertical (resp. horizontal) composition of 2-morphisms will be given by vertical (resp. horizontal) concatenation. By convention, "emptiness is identity." So the following diagram (which has no dots) represents the identity 2-morphism on the 1-morphism F:

$$\boxed{}\, F \;:=\; \mathrm{id}_F.$$

The following diagram (which has no strings or dots) represents the identity 2-morphism from the identity morphism of $\mathcal{C}$ to itself:

$$\overline{}\, \mathcal{C} \;:=\; \mathrm{id}_{\mathbb{1}_\mathcal{C}}.$$

Remark 7.7 The diagrammatics used here are only for strict 2-categories, where 1-morphism composition is associative. Our composition of 1-morphisms is horizontal concatenation, which is automatically associative.

The literature is also full of weak 2-categories, where 1-morphism composition is associative up to isomorphism. Given a weak 2-category, one can get a strict 2-category by a process called *strictification*. The 1-morphisms in a strictified 2-category are words in the 1-morphisms of the weak 2-category.

In a 2-category, 2-morphisms are required to satisfy axioms which can become tedious to keep track of when performing calculations. In the language of planar diagrams these axioms have a nice interpretation. For example, the identity axiom of vertical composition says that for a 2-morphism $\alpha : F \to G$, there are equalities $\alpha \circ \mathrm{id}_F = \alpha = \mathrm{id}_G \circ \alpha$. Diagrammatically this translates to the equality of the following pictures.

7.2 Planar Diagrams for 2-Categories

$$\begin{array}{c} G \\ \bullet\,\alpha \\ F \end{array} \;=\; \begin{array}{c} G \\ \bullet\,\alpha \\ F \end{array} \;=\; \begin{array}{c} G \\ \bullet\,\alpha \\ F \end{array} \tag{7.5}$$

The identity axiom of horizontal composition implies that the following diagrams are equal.

$$\begin{array}{c} F \\ \\ F \end{array} \;=\; \begin{array}{c} F \\ \\ F \end{array} \;=\; \begin{array}{c} F \\ \\ F \end{array} \tag{7.6}$$

Exercise 7.8 Show that the axioms of a 2-category imply the following equalities.

$$\mathcal{E} \;\; \mathcal{D}_{\bullet\alpha} \;\; {}^{\bullet\beta}\mathcal{C} \;=\; \mathcal{E} \;\; {}^{\bullet\alpha}\mathcal{D}^{\bullet\beta} \;\; \mathcal{C} \;=\; \mathcal{E} \;\; {}^{\bullet\alpha}\mathcal{D} \;\; {}_{\bullet\beta}\mathcal{C} \tag{7.7}$$

(with strands labeled F_2, G_2 on top and F_1, G_1 on bottom)

$$\mathcal{C}\,{}^{\bullet\alpha}_{\bullet\beta} \;\; \mathcal{C} \;=\; \mathcal{C} \;\; {}^{\bullet\beta}_{\bullet\alpha}\mathcal{C} \;=\; \mathcal{C}\,{}_{\bullet\beta} \;\; {}^{\bullet\alpha}\mathcal{C} \tag{7.8}$$

(with strand labeled G_2 on top and F_1 on bottom)

Similar to before, when we considered linear diagrams up to linear isotopy, we consider planar diagrams up to *rectilinear isotopy*, which is isotopy defined by the equalities in (7.5)–(7.8).

Proposition 7.9 *The axioms of a (strict) 2-category imply that a planar diagram, up to rectilinear isotopy, unambiguously represents a 2-morphism.*

Remark 7.10 Many 2-categories also have an additive structure, so that the 2-morphisms between two given 1-morphisms form an abelian group. In this case, a linear combination of diagrams also represents a 2-morphism.

7.3 Drawing Monoidal Categories

A (strict) monoidal category can be thought of as a (strict) 2-category with a single object. The objects of the monoidal category become 1-morphisms when viewed in this context. Similarly, the morphisms of the monoidal category become 2-morphisms. The tensor product of objects gives composition of 1-morphisms, etc. See [119, Example 2.2] for more information.

By viewing a monoidal category as a 2-category, we can use our diagrammatics to draw monoidal categories as illustrated in the following examples.

Example 7.11 The *symmetric category* is a $\mathbb{C}$-linear monoidal category where the objects are the natural numbers. The morphisms are defined by $\text{Hom}(m, n) = \delta_{m,n}\mathbb{C}[S_n]$. The tensor product of objects is addition, $m \otimes n = m + n$. The tensor product of morphisms is induced from the inclusion of group algebras $\mathbb{C}[S_n] \times \mathbb{C}[S_m] \hookrightarrow \mathbb{C}[S_{n+m}]$. We will draw the symmetric category as follows:

- Objects: The object 1 will be drawn as a single dot on a line ──•──, and we will draw the tensor product as horizontal concatenation. For example,

    ──•──•──•── is the object $1 \otimes 1 \otimes 1 = 3$.

    With this convention, the monoidal identity is an interval with no dots on it.
- Morphisms: We will use the diagrams introduced in Sect. 1.1.2 for elements of the symmetric group. In particular,

    ⌶ is the identity morphism in $\text{End}(1) = \mathbb{C}[S_1] = \mathbb{C}$,

    ⤫ represents the transposition $s \in \mathbb{C}[S_2]$.

With these two diagrams in hand, we can actually construct any morphism by taking linear combinations of tensor products and compositions. For example,

⤫| $= s \otimes 1$, |⤫ $= 1 \otimes s$,

$$\text{(diagram)} = (1 \otimes s) \circ (s \otimes 1).$$

7.4 The Temperley–Lieb Category

The horizontal line in the middle was added for clarity, to show how two morphisms are being concatenated vertically. We will not usually draw these internal horizontal lines when composing morphisms.

As a monoidal category, the symmetric category has the following presentation:

Generating object:

$$\text{———} \tag{7.9}$$

Generating morphism:

$$\diagup\!\!\!\diagdown \tag{7.10}$$

Relations:

$$\diagup\!\!\!\diagdown\diagup\!\!\!\diagdown = || \qquad \diagup\!\!\!\diagdown\!\!\!\diagup\!\!\!\diagdown = \diagdown\!\!\!\diagup\!\!\!\diagdown\!\!\!\diagup \tag{7.11}$$

These relations can be applied "locally" to a subdiagram of another diagram, just as relations in an ordinary monoid can be applied "locally" to contiguous subwords of a word. Note that the equality

$$\diagup\!\!\!\diagdown\,| = |\,\diagup\!\!\!\diagdown \tag{7.12}$$

is not needed as a relation since it already holds by rectilinear isotopy.

For the reader seeking a rigorous treatment of presentations of monoidal categories, see [102, §XII.1].

7.4 The Temperley–Lieb Category

Our next example is important enough to merit its own section. We consider $\text{Vect}_\mathbb{C}$, the monoidal category of finite-dimensional complex vector spaces with the usual tensor product. We will use diagrammatics only for certain objects and certain morphisms. Let V denote $\mathbb{C}^2 = \mathbb{C}\langle e_1, e_2 \rangle$. We will draw V as ———, so that $V \otimes V$ is ———. As before, the monoidal identity $\mathbb{C}$ is an interval with no dots. Consider the following map $n : V \otimes V \to \mathbb{C}$, defined by

$$\begin{cases} e_1 \otimes e_1 \mapsto 0, \\ e_2 \otimes e_2 \mapsto 0, \\ e_1 \otimes e_2 \mapsto -1, \\ -e_2 \otimes e_1 \mapsto -1. \end{cases}$$

We will draw this map as a "cap":

$$n :=\ \cap$$

We also want a diagram for the morphism $u : \mathbb{C} \to V \otimes V$ defined by $1 \mapsto e_1 \otimes e_2 - e_2 \otimes e_1$. We will draw it as a "cup":

$$u :=\ \cup$$

(The motivation for this particular choice of maps comes from identifying $\mathbb{C}$ as a vector space with the exterior product $\Lambda^2 V$.)

Now, any planar diagram which is built from the above pieces can be interpreted using composition. For example,

represents the morphism $n \circ u : \mathbb{C} \to \mathbb{C}$ factoring through $V \otimes V$. We compute $n \circ u(1) = n(e_1 \otimes e_2 - e_2 \otimes e_1) = -2$, so that $n \circ u$ is multiplication by -2. We can encode this computation by the following diagrammatic equation:

$$\bigcirc = -2 \qquad (7.13)$$

This is our first example involving a linear combination of diagrams, and it says that $n \circ u$ is equal to -2 times the identity morphism of $\mathbb{C}$.

Similarly, we can check that

$$\rotatebox{90}{$\cup$}\!\cap = \ | \ = \cup\!\rotatebox{90}{$\cap$} \qquad (7.14)$$

7.4 The Temperley–Lieb Category

We check the first equality and leave the second to the reader. The first thing we do is decompose the leftmost diagram into pieces so that we know how to interpret it. We can factor it as

where the bottom strip represents the morphism $\mathrm{id}_V \otimes u : V \otimes \mathbb{C} \to V \otimes V \otimes V$ and the top strip the morphism $n \otimes \mathrm{id}_V$. To check that this composition is the identity on V we verify that it is the identity on e_1 and e_2. Indeed,

$$(n \otimes \mathrm{id}_V) \circ (\mathrm{id}_V \otimes u)(e_1 \otimes 1) = n \otimes \mathrm{id}_V(e_1 \otimes e_1 \otimes e_2 - e_1 \otimes e_2 \otimes e_1) = 1 \otimes e_1,$$

and similarly $e_2 \otimes 1 \mapsto 1 \otimes e_2$.

Now that we have verified (7.13) and (7.14) we can apply the relations diagrammatically, without needing to check the computations on elements. For example,

The computations above show that the choice of notation, for n and u, caps and cups, which a priori didn't mean anything diagrammatically, is indeed a good choice. By our choice of notation actual isotopy appeared in (7.14), not just rectilinear isotopy. Now that isotopy has appeared, we can use topological arguments to show that all diagrams which are made from caps, cups, and identity morphisms are spanned by what are called *crossingless matchings*. A crossingless matching is a way of matching up dots in pairs so that strands don't cross each other. For example,

(7.15)

is a crossingless matching, while

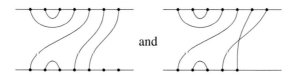

are not.

Remark 7.12 Note that the diagram in (7.15) represents a morphism from $V^{\otimes 5}$ to $V^{\otimes 7}$. That is, it represents a $2^7 \times 2^5$ matrix. Composing these diagrams is something that is easily done by hand, while the corresponding matrix multiplication is much more time consuming! This showcases the incredible efficiency of diagrammatic computation.

This discussion motivates the following definition.

Definition 7.13 The *Temperley–Lieb category*, $\mathcal{TL}$, is the $\mathbb{C}$-linear monoidal category given by the following presentation:

Generating object:

$$\text{―•―} \tag{7.16}$$

Generating morphisms:

$$\bigcap \quad \text{and} \quad \bigcup \tag{7.17}$$

Relations:

$$\bigcirc = -2 \tag{7.18}$$

$$\rotatebox{0}{$\cap\!\!\cup$} = \, | \, = \rotatebox{0}{$\cup\!\!\cap$} \tag{7.19}$$

The discussion above gives the following proposition.

Proposition 7.14 *There is a $\mathbb{C}$-linear monoidal functor $\mathcal{F} : \mathcal{TL} \to \text{Vect}_{\mathbb{C}}$ which sends*

$$\text{―•―} \mapsto V,$$

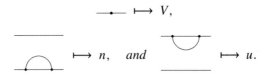

$$\bigcap \mapsto n, \quad \text{and} \quad \bigcup \mapsto u.$$

Remark 7.15 In fact, this functor $\mathcal{F}$ gives an equivalence of categories between $\mathcal{TL}$ and the full subcategory of representations of $\mathfrak{sl}_2$ consisting of tensor products of the standard representation $V = \mathbb{C}^2$. Thus $\mathcal{TL}$ can be said to control the representation theory of $\mathfrak{sl}_2$.

Exercise 7.16 We can view the algebra $A = \mathbb{R}[x]/(x^2)$ as an object in the monoidal category of $\mathbb{R}$-vector spaces. Let $\cap : A \otimes A \to \mathbb{R}$ denote the linear map which sends

7.5 More About Isotopy

$f \otimes g$ to the coefficient of x in fg. Let $\cup : \mathbb{R} \to A \otimes A$ denote the map which sends 1 to $x \otimes 1 + 1 \otimes x$.

1. We wish to encode these maps diagrammatically, drawing $\cap$ as a cap and $\cup$ as a cup. Justify this diagrammatic notation, by checking the isotopy relations.
2. Draw a sequence of nested circles, as in an archery target. Evaluate this morphism.

Exercise 7.17 This question is about the Temperley–Lieb category.

1. Finish the proof that the isotopy relation holds in vector spaces.
2. There is a map $V \otimes V \to V \otimes V$ which sends $x \otimes y \mapsto y \otimes x$. Draw this as an element of the Temperley–Lieb category (a linear combination of diagrams).
3. Find an endomorphism of 2 strands which is killed by placing a cap on top. Can you find one which is an idempotent? Also find an endomorphism killed by putting a cup on bottom.
4. (Harder) Find an idempotent endomorphism of 3 strands which is killed by a cap on top (for either of the two placements of the cap).

7.5 More About Isotopy

We conclude this chapter with a discussion about isotopy, using the example of the 2-category $\mathcal{C}at$. The ideas are motivated by the Temperley–Lieb category, where we observed true isotopy instead of rectilinear isotopy. Suppose we have functors $E : \mathcal{C} \to \mathcal{D}$, and $E^\vee : \mathcal{D} \to \mathcal{C}$ and an adjunction $E \dashv E^\vee$. Then we have the counit $\epsilon : EE^\vee \to 1_\mathcal{D}$ and unit $\eta : 1_\mathcal{C} \to E^\vee E$ of adjunction. Diagrammatically, if we choose to draw the identity natural transformation of E as an upward oriented line, and the identity map of $E^\vee$ as the same symbol but with a downward orientation, that is

$$E := \quad \bigg\vert \, \mathcal{C} \qquad E^\vee := \quad \bigg\vert \, \mathcal{D} \, ,$$

and we also choose to draw the counit and unit as

$$\stackrel{\frown}{} \mathcal{D} \qquad \stackrel{\smile}{} \mathcal{C} \, ,$$

then the axioms of adjunction are equivalent to

$$\bigg\vert \!\!\smile\!\!\bigg\vert \, \mathcal{C} \;=\; \bigg\vert \, \mathcal{C} \quad \text{and} \quad \bigg\vert \!\!\frown\!\!\bigg\vert \, \mathcal{D} \;=\; \bigg\vert \, \mathcal{D} \, . \qquad (7.20)$$

Let $\mathcal{B}$ be a category, and let $X : \mathcal{B} \to \mathcal{C}$ and $Y : \mathcal{B} \to \mathcal{D}$ be functors. Then the adjunction $E \dashv E^\vee$ can equivalently be viewed as a bijection of 2-morphism spaces

$$\operatorname{Hom}(EX, Y) \simeq \operatorname{Hom}(X, E^\vee Y) \qquad (7.21)$$

given using the unit and counit by

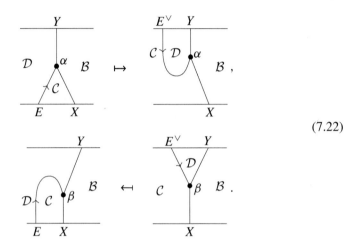

$$(7.22)$$

These maps are mutual inverses because of (7.20). Moreover, (7.21) is natural in X and Y in an obvious sense. If $\mathcal{B}$ is the category consisting of a single object with only the identity endomorphism, functors $X : \mathcal{B} \to \mathcal{C}$ and $Y : \mathcal{B} \to \mathcal{D}$ can be viewed as objects $X \in \operatorname{Ob}(\mathcal{C})$ and $Y \in \operatorname{Ob}(\mathcal{D})$, and (7.21) reduces to the familiar Hom adjunction.

Suppose in addition that $E^\vee \dashv E$, that is E and $E^\vee$ are *biadjoint*. Then we also have the natural transformations

and we have the diagrammatic relations

7.5 More About Isotopy

By choosing appropriate notation, we get actual isotopy for biadjoint 1-morphisms.

Now suppose that $E, F : \mathcal{C} \to \mathcal{D}$ are two such 1-morphisms with biadjoints $E^\vee, F^\vee$. We will draw E as above and we will draw F, $F^\vee$ and the units and counits similarly but using dashed lines instead. Given a 2-morphism $\alpha : E \to F$,

we can form the following 2-morphisms $^\vee\alpha, \alpha^\vee : F^\vee \to E^\vee$ which are called (respectively) the *left* and *right mates* of α with respect to the chosen biadjunction:

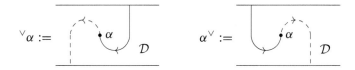

In general $^\vee\alpha$ and $\alpha^\vee$ are not equal. When they are equal, we say that α is *cyclic*. In this case, we can draw the mates unambiguously using a rotation of the original picture for α, as below.

$$^\vee\alpha = \alpha^\vee =$$

Proposition 7.18 *If all 1-morphisms have biadjoints and all 2-morphisms are cyclic, then diagrams up to true isotopy unambiguously represent a 2-morphism (when using the above conventions for representing biadjoints, units and counits of adjunction, and mateship).*

Exercise 7.19 One can think about the right mate and the left mate as "twisting" or "rotating" α by 180° to the right or to the left. Visualize what it would mean to twist α by 360° to the right, yielding another 2-morphism $\alpha^{\vee\vee} : E \to F$. Verify that $^\vee\alpha = \alpha^\vee$, if and only if $\alpha = \alpha^{\vee\vee}$. Thus cyclicity is the same as "360 degree rotation invariance," which one might expect from any planar picture.

Exercise 7.20 Suppose that B is an object in a monoidal category with biadjoints, and $\phi : B \otimes B \otimes B \to \mathbb{1}$ is a cyclic morphism. What should it mean to "rotate" ϕ by 120°? Suppose that $\operatorname{Hom}(B \otimes B \otimes B, \mathbb{1})$ is one-dimensional over $\mathbb{C}$. What can you say about the 120° rotation of ϕ, vis a vis ϕ? What if $\operatorname{Hom}(B \otimes B \otimes B, \mathbb{1})$ is one-dimensional over $\mathbb{R}$?

Chapter 8
Frobenius Extensions and the One-Color Calculus

This chapter is based on expanded notes of a lecture given by the authors and taken by

Nicolle E. S. González

Abstract We use the planar diagrammatics of the previous chapter to describe a Frobenius algebra object in a monoidal category. We then explain how the graded Frobenius extension $R^s \subset R$ gives a Frobenius algebra object structure to the generating Bott–Samelson bimodule B_s. The result is the one-color calculus: diagrammatics for Bott–Samelson bimodules involving a single simple reflection.

8.1 Frobenius Structures

We now head towards a diagrammatic presentation for the monoidal category of Bott–Samelson bimodules, starting in this chapter with Bott–Samelson bimodules involving a single simple reflection. What allows a particularly nice presentation is the fact that the generating Bott–Samelson bimodule $B_s = R \otimes_{R^s} R(1)$ is a (graded) Frobenius algebra object. We start by explaining this notion.

8.1.1 Frobenius Algebra Objects

Let $\mathcal{C}$ be a monoidal category. A *monoid object* in $\mathcal{C}$ is a triple (A, μ, η), where $A \in \mathrm{Ob}(\mathcal{C})$, and $\mu : A \otimes A \to A$ and $\eta : \mathbb{1} \to A$ are morphisms making the

---

N. E. S. González
Department of Mathematics, University of Southern California, Los Angeles, CA, USA

© The Editor(s) (if applicable) and The Author(s), under exclusive licence to Springer Nature Switzerland AG 2020
B. Elias et al., *Introduction to Soergel Bimodules*, RSME Springer Series 5, https://doi.org/10.1007/978-3-030-48826-0_8

following diagrams commute:

Associativity

$$(A \otimes A) \otimes A \xrightarrow{\alpha} A \otimes (A \otimes A) \xrightarrow{\mathrm{id}_A \otimes \mu} A \otimes A$$
$$\downarrow \mu \otimes \mathrm{id}_A \qquad\qquad\qquad\qquad\qquad\qquad \downarrow \mu$$
$$A \otimes A \xrightarrow{\qquad\qquad \mu \qquad\qquad} A \tag{8.1}$$

Unit

$$\mathbb{1} \otimes A \xrightarrow{\eta \otimes \mathrm{id}_A} A \otimes A \xleftarrow{\mathrm{id}_A \otimes \eta} A \otimes \mathbb{1}$$
$$\searrow_\lambda \quad \downarrow \mu \quad \swarrow_\rho$$
$$A \tag{8.2}$$

Here, α is the associator, and λ, ρ are the unitors. (Recall that these natural isomorphisms are part of the structure of a monoidal category.)

A *comonoid object* in $\mathcal{C}$ is the dual notion: a triple (A, δ, ϵ), with $\delta : A \longrightarrow A \otimes A$ and $\epsilon : A \longrightarrow \mathbb{1}$ satisfying the coassociativity and counit axioms.

A *Frobenius algebra object* in $\mathcal{C}$ is a quintuple $(A, \mu, \eta, \delta, \epsilon)$ such that (A, μ, η) is a monoid object, (A, δ, ϵ) is a comonoid object, and the following diagrams commute:

Frobenius Associativity

$$A \otimes A \xrightarrow{\delta \otimes \mathrm{id}_A} A \otimes A \otimes A \qquad\qquad A \otimes A \xrightarrow{\mathrm{id}_A \otimes \delta} A \otimes A \otimes A$$
$$\downarrow \mu \qquad\qquad \downarrow \mathrm{id}_A \otimes \mu \quad \text{and} \quad \downarrow \mu \qquad\qquad \downarrow \mu \otimes \mathrm{id}_A$$
$$A \xrightarrow{\delta} A \otimes A \qquad\qquad\qquad A \xrightarrow{\delta} A \otimes A. \tag{8.3}$$

That is, we demand a compatibility between the monoid object and comonoid object structures, which we call *Frobenius associativity*. Note that $\delta \circ \mu$ appears twice in (8.3), so that Frobenius associativity is the equality of three morphisms.

Remark 8.1 The reader may be familiar with the definition of a Hopf algebra object, which is also a monoid object and a comonoid object with some compatibility between these structures. However, these two definitions are not the same, as the compatibility requirement is different! A motivation for Frobenius algebra objects will be given in Sect. 8.1.4.

8.1.2 Diagrammatics for Frobenius Algebra Objects

In Chap. 7 we described diagrammatics for morphisms in (strict) monoidal categories. Let us put this into practice to encode the structure of Frobenius algebra objects.

Remark 8.2 Because we work with strict monoidal categories in order to employ diagrammatic techniques, the associator and unitor isomorphisms are the identity maps, and will be "invisible" below.

Let $(A, \mu, \eta, \delta, \epsilon)$ be a Frobenius algebra object in a strict monoidal category $\mathcal{C}$. We draw A using the color red, so that the identity morphism of A is drawn as a vertical red line:

$$\mathrm{id}_A = \quad \bigg| \quad .$$

We depict the structure maps using the following planar diagrams:

$$\eta := \quad , \quad \mu := \quad , \quad \epsilon := \quad , \quad \delta := \quad . \tag{8.4}$$

The diagrams representing η and ϵ are called *dots* or *univalent vertices*, while the diagrams representing μ and δ are called *trivalent vertices*.

With this convention, the axioms of a Frobenius algebra object can be drawn diagrammatically as follows.

Unit

$$\quad = \quad = \quad \tag{8.5a}$$

Counit

$$\quad = \quad = \quad \tag{8.5b}$$

Associativity

$$\includegraphics{} = \includegraphics{} \qquad (8.5c)$$

Coassociativity

$$\includegraphics{} = \includegraphics{} \qquad (8.5d)$$

Frobenius Associativity

$$\includegraphics{} = \includegraphics{} = \includegraphics{} \qquad (8.5e)$$

Exercise 8.3 Verify that these relations correspond to the relations given in the standard categorical language from Sect. 8.1.1.

It turns out that for any Frobenius algebra object A, the functor $A \otimes (-)$ is adjoint to itself. This adjunction structure is determined by the units and counits of adjunction, which are typically drawn as cups and caps.

Cap

$$\includegraphics{} := \includegraphics{} \qquad (8.6)$$

Cup

$$\includegraphics{} := \includegraphics{} \qquad (8.7)$$

Following the discussion of biadjunction in Sect. 7.5, we can confirm that these maps represent the unit and counit of the adjunction $A \dashv A$ by checking the following isotopy relations.

8.1 Frobenius Structures

Isotopy

$$\text{(8.8)}$$

Exercise 8.4 Deduce the isotopy relations (8.8) from the (diagrammatic) axioms of a Frobenius algebra object. (*Hint:* After replacing the cup and cap using their definitions, apply Frobenius associativity.)

Not only is A adjoint to itself, but the structure maps are all cyclic with respect to this adjunction. Even more is true: the mate of η is ϵ, justifying the fact that we have drawn ϵ as the 180 degree rotation of η.

Rotation of Univalent Vertices

$$\text{(8.9)}$$

Together, these imply the cyclicity of both η and ϵ.

Rotation of Trivalent Vertices

$$\text{(8.10)}$$

Similarly, the mate of δ is μ. Further still, rotating μ by 60 degrees gives the morphism δ, and vice versa (this is one way of stating (8.10) aloud). Both μ and

δ are invariant under 120 degree rotation, which justifies having drawn them as trivalent vertices which have an inborn rotational symmetry.

Exercise 8.5 Deduce the rotation relations (8.9) and (8.10) from the axioms of a Frobenius algebra object. Use (8.10) to deduce that δ and μ are mates.

Together, these isotopy and rotational relations have the following consequence.

Proposition 8.6 *Planar diagrams (built as concatenations of trivalent vertices, univalent vertices, caps, and cups as above) which are isotopic represent equal morphisms in $\mathcal{C}$.*

In particular, we see that planar diagrams are the perfect tool for studying Frobenius algebra objects, as one can use isotopy to visually simplify many computations. Let us summarize the results in a similar way we did with the Temperley–Lieb category in Chap. 7.

Definition 8.7 Let Frob be the monoidal category given by the following presentation. It has one generating object, drawn in red. Its generating morphisms are given by the diagrams in (8.4). The relations are those in (8.5).

Proposition 8.8 *For any Frobenius algebra object A in any monoidal category $\mathcal{C}$, there is a functor $\mathcal{F}$: Frob $\to \mathcal{C}$, sending the generating object to A, and the generating morphisms to the structure maps of the Frobenius structure on A.*

We have thus given a diagrammatic presentation of a category Frob which possesses a universal Frobenius algebra object. This category encompasses everything which is true about an arbitrary Frobenius algebra object. However, nothing states that the functor $\mathcal{F}$ is full or faithful! For any given Frobenius algebra object A, there may be additional relations between these diagrams.

Remark 8.9 As an example, if A is a Frobenius algebra over a field $\Bbbk$ (i.e. a Frobenius algebra object in the category $\text{Vect}_{\Bbbk}$ of vector spaces over $\Bbbk$), then the composition $\epsilon \circ \eta$ is an endomorphism of $\Bbbk$ in $\text{Vect}_{\Bbbk}$, hence equal to multiplication by a scalar $a \in \Bbbk$:

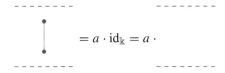

However, such a relation does not hold in Frob. After all, the element a depends on the Frobenius algebra A, and is not universal.

Exercise 8.10 Prove that the functor which sends each diagram to its vertical flip (i.e. its reflection across a horizontal axis) is an autoequivalence of Frob. This autoequivalence is monoidal and contravariant. Prove that the functor which sends each diagram to its horizontal flip (i.e. its reflection across a vertical axis) is an autoequivalence of Frob. This autoequivalence is covariant and contravariantly-

8.1 Frobenius Structures

monoidal, meaning that $M \otimes N$ is sent to $N \otimes M$. (*Hint:* To construct each functor, you need only check that the relations are satisfied after applying the desired flip.)

8.1.3 Playing with Isotopy

We now introduce some notational shorthand. A priori, there is no intrinsic meaning to a horizontal red line, such as might occur if the diagram for a cap were stretched.

This diagram has no meaning, because it is not built as a concatenation of pictures that we understand. However, under the understanding that diagrams are isotopy invariant, there is no harm in accepting such a diagram as being equal to the cap, and allowing a more general notion of isotopy which allows for such "nonsense" pictures. For example, we may use the diagram

to represent either the right hand side or the left hand side of (8.5e), since these diagrams are all isotopic to each other. Meanwhile, the middle diagram in (8.5e) is not related to the other two by isotopy, so that Frobenius associativity can be rewritten as follows.

Frobenius Associativity (Allowing for Isotopy)

$$\vcenter{\hbox{⋈}} \quad = \quad \vcenter{\hbox{⋈}} \qquad (8.11)$$

Remark 8.11 To reiterate, the vertical line in the middle of the left hand side of (8.11) has meaning, and it implies that this morphism $A \otimes A \to A \otimes A$ factors through A. Meanwhile, the horizontal line in the middle of the right hand side of (8.11) has no meaning, and to interpret this diagram as a composition of more elementary morphisms, one must replace it with an isotopic diagram like either side of (8.5e).

Similarly, one has

Unit (Allowing for Isotopy)

 (8.12)

With these conventions in place, we can make sense out of any graph embedded in the planar strip (possibly with edges running to the boundary) which has only univalent and trivalent vertices. That is, instead of rigidly viewing the graph as being built from rectangular glueings of the 4 generators in (8.4), we can work directly with isotopy classes of graphs. When working with isotopy classes of graphs, the presentation of Frob is dramatically simplified. For example, (8.12) implies the unit and counit axioms, as well as the definition of the cups and caps, while (8.11) implies associativity, coassociativity, and Frobenius associativity. Thus (8.11) and (8.12) are the only relations one needs!

Remark 8.12 One could give a definition of Frob where morphisms are (linear combinations of) isotopy classes of uni- and trivalent graphs, modulo the two relations (8.12) and (8.11). One might call this an *isotopy presentation* of Frob. We do not bother to give this definition precisely here, but we will practice it in subsequent chapters; also see [59, §5.1] for more examples.

8.1.4 Frobenius Extensions

Here is a definition which gives rise to many Frobenius algebra objects in algebraic settings.

Definition 8.13 Let $\iota \colon A \to B$ be an inclusion of commutative rings. A *Frobenius extension* is the additional structure of an A-linear map $\partial \colon B \to A$, for which there exist dual bases $\{b_i\}$ and $\{b_i^*\}$ of B (as a finite rank free A-module), i.e. such that

$$\partial(b_i b_j^*) = \begin{cases} 1 & \text{if } i = j, \\ 0 & \text{otherwise.} \end{cases}$$

The map ∂ is often called the *Frobenius trace*. Associated to this Frobenius extension are four bimodule maps: the A-bimodule maps

$$\iota \colon A \to B, \qquad (8.13a)$$

$$\partial \colon B \to A, \qquad (8.13b)$$

8.1 Frobenius Structures

and the B-bimodule maps

$$m \colon B \otimes_A B \to B, \tag{8.13c}$$

$$\Delta \colon B \to B \otimes_A B. \tag{8.13d}$$

Here, m is the multiplication map, and the coproduct map Δ sends 1 to the finite sum $\sum b_i \otimes b_i^*$.

Exercise 8.14 Prove the following facts.

1. For any $f \in B$,

$$\sum_i b_i \partial(f b_i^*) = f. \tag{8.14}$$

2. For any $f \in B$,

$$f \Delta(1) = \Delta(1) f \tag{8.15}$$

inside $B \otimes_A B$. (*Hint:* write $f b_i = \sum_j c_{ij} b_j$, and prove that $f b_i^* = \sum_j c_{ji} b_j^*$.) In particular, Δ is a B-bimodule map.

3. The element $\sum b_i \otimes b_i^*$ in $B \otimes_A B$ does not depend on the choice of dual bases.

Exercise 8.15 Returning to the setting of Chaps. 4 and 5, prove that R is a Frobenius extension over R^s. More specifically, prove that ∂_s is a Frobenius trace, with dual bases $\{1, \frac{\alpha_s}{2}\}$ and $\{\frac{\alpha_s}{2}, 1\}$. Suppose that $m_{st} = 3$. Prove that $\{1, -\alpha_t\}$ and $\{\alpha_s + \alpha_t, 1\}$ are also dual bases for R over R^s. This gives two descriptions for $\Delta(1) \in R \otimes_{R^s} R$; show explicitly that they are equal.

Exercise 8.16 For any Frobenius extension $A \subset B$, show that $B \otimes_A B$ is a Frobenius algebra object in the monoidal category of B-bimodules. Ironically, when one does this, the multiplication map m becomes the counit map ϵ, and the coproduct Δ becomes the unit η. Where do the other two maps (μ and δ) of the Frobenius algebra object structure come from?

Combining the last two exercises (with some remarks on gradings later in this section) we see that $B_s = R \otimes_{R^s} R(1)$ is a Frobenius algebra object inside R-gbim, which will allow us to study it diagrammatically. This we do soon, after further discussion of Frobenius extensions.

Remark 8.17 In Chap. 24 we study diagrammatics for Frobenius extensions, not just for Frobenius algebra objects. The reader can head to Chap. 24 after finishing this chapter and the next.

Perhaps the most significant feature of Frobenius extensions is that the restriction and induction functors

$$\text{Res}_A^B : B\text{-mod} \to A\text{-mod}, \qquad \text{Ind}_A^B : A\text{-mod} \to B\text{-mod}$$

are biadjoint. This feature is called *Frobenius reciprocity*.

Exercise 8.18 For any ring extension $A \subset B$, show that Ind_A^B is left adjoint to Res_A^B. More precisely, show that the inclusion map $\iota \colon A \to B$ and the multiplication map $m \colon B \otimes_A B \to B$ satisfy the requirements to be the unit and counit of this adjunction.

Exercise 8.19 For any Frobenius extension, show that Res_A^B is left adjoint to Ind_A^B. More precisely, show that the trace map $\partial \colon B \to A$ and the coproduct map $\Delta \colon B \to B \otimes_A B$ satisfy the requirements to be the unit and counit of this adjunction.

Exercise 8.20 Let us attempt to prove a converse. Let $A \subset B$ be any extension of commutative rings and suppose that Res_A^B is left adjoint to Ind_A^B. Let $\partial \colon B \to A$ and $\Delta \colon B \to B \otimes_A B$ be the unit and counit of this adjunction. Write $\Delta(1) = \sum_i b_i \otimes b_i^*$ for some $b_i, b_i^* \in B$.

1. Prove that (8.14) holds, for any $f \in B$.
2. Deduce that $\{b_i\}$ and $\{b_i^*\}$ are spanning sets for B as an A-module, so B is finitely generated as an A-module.
3. Suppose that B is free as an A module. We may as well assume that $\{b_i\}$ is a basis for B over A (why?). Then prove that $\{b_i^*\}$ is a dual basis. In particular, ∂ is a Frobenius trace, making $A \subset B$ into a Frobenius extension.

Remark 8.21 As the previous exercise showed, Frobenius reciprocity is roughly equivalent to the extension $A \subset B$ being Frobenius. In our definition of a Frobenius extension we required that B is free over A, while Frobenius reciprocity can also occur when B is finitely-generated projective over A. (A missing step in the exercise above: prove that B is projective over A.) See [96, §1.3] for more details.

The biadjunction between induction and restriction implies that the composition of induction with restriction is self-biadjoint. This gives another reason why the Frobenius algebra object $B \otimes_A B$ is a self-biadjoint B-bimodule, because the functor $(B \otimes_A B) \otimes_B (-)$ is the composition of induction with restriction.

Let us define the graded analogue of a Frobenius extension.

Definition 8.22 Let $\iota \colon A \to B$ be an inclusion of graded commutative rings. A *Frobenius extension of degree* ℓ is the additional structure of an A-linear map $\partial \colon B \to A$, homogeneous of degree -2ℓ, for which there exist homogeneous dual bases $\{b_i\}$ and $\{b_i^*\}$ of B (as a finite rank free graded A-module) such that

$$\partial(b_i b_j^*) = \begin{cases} 1 & \text{if } i = j, \\ 0 & \text{otherwise.} \end{cases}$$

Note that $\deg b_i + \deg b_i^* = 2\ell$ for each i.

For a graded Frobenius extension, as for any ring extension, Ind_A^B is left adjoint to Res_A^B. However, Ind_A^B is right adjoint to $\text{Res}_A^B(2\ell)$. The functors of induction and restriction are only biadjoint up to shift. There is a more satisfactory convention: henceforth, whenever $A \subset B$ is a graded Frobenius extension of degree ℓ, Res_A^B will denote restriction with a built-in grading shift by ℓ, namely

$$\text{Res}_A^B = {}_A B(\ell) \otimes_B (-) : B\text{-gmod} \to A\text{-gmod}. \tag{8.16}$$

There is no grading shift on induction. Now Ind_A^B is left adjoint to $\text{Res}_A^B(-\ell)$ and right adjoint to $\text{Res}_A^B(\ell)$.

The upshot of this grading convention is that $B \otimes_A B(\ell)$, the bimodule corresponding to the composition of Ind_A^B and Res_A^B, is a Frobenius algebra object where the "dot" maps η and ϵ are homogeneous of degree $+\ell$, and the "trivalent vertices" μ and Δ are homogeneous of degree $-\ell$.

We conclude with some familiar examples of Frobenius extensions. When $A = \Bbbk$ is a field, B is called a *Frobenius algebra* over $\Bbbk$. Frobenius algebras are quite ubiquitous in nature, although this book will be focused on Frobenius extensions. One should also mention that Frobenius extensions can also be defined for non-commutative rings.

Example 8.23 Let G be a finite group. The group algebra $\Bbbk[G]$ is a (possibly non-commutative) Frobenius algebra over $\Bbbk$, with trace map $\partial(g) := 1$ if $g = 1$ and $\partial(g) := 0$ otherwise. Thus $\Delta(1) = \sum_{g \in G} g \otimes g^{-1}$.

Exercise 8.24 Let $H \subset G$ be an inclusion of finite groups. Show that $\Bbbk[H] \subset \Bbbk[G]$ is a Frobenius extension. This leads to the most famous example of Frobenius reciprocity, which is originally due to Frobenius.

Example 8.25 Let X be a compact oriented n-dimensional manifold. Then the cohomology ring $H^*(X; \Bbbk)$ is a Frobenius algebra over $\Bbbk$. The trace map is given by integration over X, and the nondegeneracy of this trace map follows from Poincaré duality.

Example 8.26 Any finite-dimensional Hopf algebra is a Frobenius algebra. Note that the trace map does not agree with the familiar counit, but is instead related to an element of the Hopf algebra known as an integral. See [118] for details. (Note that the previous example gives many examples of Frobenius algebras which are not Hopf algebras.)

8.2 A Tale of One Color

Fix a Coxeter system (W, S). From the geometric representation (or the Kac–Moody representation, or any other realization, see Chap. 5) we have a polynomial ring R, a collection of simple roots α_s for $s \in S$ living in R, and a collection of Demazure operators ∂_s acting on R. Recall the monoidal category $\mathbb{BSB}$im of Bott–Samelson bimodules (a subcategory of R-bimodules), generated under tensor product by the bimodules $B_s = R \otimes_{R^s} R(1)$ for each simple reflection s. In this chapter we fix one simple reflection $s \in S$, and describe diagrammatically the morphisms in $\mathbb{BSB}$im between tensor products of B_s.

8.2.1 Frobenius Structure

Lemma 8.27 *The object B_s in $\mathbb{BSB}$im is a graded Frobenius algebra object of degree 1.*

This lemma follows directly from Exercises 8.15 and 8.16. In particular, the functor $B_s \otimes (-)$ is self-biadjoint (up to grading shift), as is the functor $(-) \otimes B_s$.

Remark 8.28 Thus for any Soergel bimodules M and N,

$$\mathrm{Hom}^{\bullet}(B_s \otimes M, N) \simeq \mathrm{Hom}^{\bullet}(M, B_s \otimes N), \tag{8.17}$$

and in particular they have the same graded rank as a free left R-module. Via the Soergel Hom formula (Theorem 5.27), this categorifies the fact that b_s is self-adjoint with respect to the standard pairing on the Hecke algebra, i.e.

$$(b_s m, n) = (m, b_s n). \tag{8.18}$$

Said another way, the self-adjointness of b_s gives rise to the expectation that its categorical analogue B_s is self-biadjoint, and motivates our exploration of Frobenius structures.

It is useful to write down the four structure maps for the Frobenius algebra structure on B_s explicitly, as R-bimodule maps.

$$\begin{array}{c} B_s \otimes B_s \\ \delta \uparrow \\ B_s \end{array} \quad = \quad \begin{array}{c} R \otimes_s R \otimes_s R(2) \\ \uparrow \\ R \otimes_s R(1) \end{array} \quad \text{where} \quad \begin{array}{c} f \otimes 1 \otimes g \\ \uparrow \\ f \otimes g \end{array} \tag{8.19}$$

$$\begin{array}{c} R \\ \epsilon \uparrow \\ B_s \end{array} \quad = \quad \begin{array}{c} R \\ \uparrow \\ R \otimes_s R(1) \end{array} \quad \text{where} \quad \begin{array}{c} fg \\ \uparrow \\ f \otimes g \end{array} \tag{8.20}$$

8.2 A Tale of One Color

$$\begin{array}{ccccc} B_s & & R \otimes_s R(1) & & \partial_s(g)f \otimes h \\ \mu \uparrow & = & \uparrow & \text{where} & \uparrow \\ B_s \otimes B_s & & R \otimes_s R \otimes_s R(2) & & f \otimes g \otimes h \end{array} \quad (8.21)$$

$$\begin{array}{ccccc} B_s & & R \otimes_s R(1) & & \Delta(1) = \frac{1}{2}(1 \otimes \alpha_s + \alpha_s \otimes 1) \\ \eta \uparrow & = & \uparrow & \text{where} & \uparrow \\ R & & R & & 1 \end{array}$$
(8.22)

Note that $\Delta(1)$ was defined using the dual bases $\{1, \frac{\alpha_s}{2}\}$ and $\{\frac{\alpha_s}{2}, 1\}$, but other dual bases would give rise to the same element, see Exercises 8.14 and 8.15.

We draw these structure maps in the usual way:

$$\delta := \quad \begin{array}{c}\text{Y}\end{array} \quad , \quad \eta := \quad \begin{array}{c}\bullet\\|\end{array} \quad , \quad \mu := \quad \begin{array}{c}\text{A}\end{array} \quad , \quad \epsilon := \quad \begin{array}{c}|\\\bullet\end{array} \quad . \tag{8.23}$$

As homogeneous bimodule maps, η and ϵ have degree $+1$, while δ and μ have degree -1.

The following exercise is redundant by now, but quite useful to anyone new to diagrammatics.

Exercise 8.29 Check directly that the Frobenius relations (8.5) hold between these bimodule maps. That is, show that both sides of the equation act in the same way on a tensor of polynomials. From easiest to hardest, the relations are: coassociativity, counit, Frobenius associativity, associativity, unit.

Consequently, by Proposition 8.6, we deduce that isotopic diagrams yield equal bimodule morphisms.

8.2.2 Additional Generators and Relations

The monoidal identity of R-bimodules is R, and for any $f \in R$, multiplication by f is an endomorphism of R. We denote this endomorphism as follows, and call it a box:

$$\boxed{f} \quad . \tag{8.24}$$

It turns out that the univalent and trivalent vertices above, together with boxes, will generate all morphisms between tensor products of B_s. Let us consider some additional relations which these maps satisfy.

Multiplication

$$\boxed{f}\,\boxed{g} \;=\; \boxed{fg} \tag{8.25a}$$

Keyhole

$$\left(\boxed{f}\right) \;=\; \boxed{\partial_s f} \tag{8.25b}$$

Barbell

$$\big\vert \;=\; \boxed{\alpha_s} \tag{8.25c}$$

The diagram on the left hand side of (8.25c) is called a *barbell*. The literature sometimes calls it a "double dot."

Fusion

$$\;=\; \frac{1}{2}\left(\boxed{\alpha_s} \;+\; \boxed{\alpha_s}\right) \tag{8.25d}$$

The diagram on the left hand side of (8.25d) is called a "broken line," which has been "fused" on the right hand side at the cost of placing polynomials in various regions. This relation used the dual bases $\{\frac{\alpha_s}{2}, 1\}$ and $\{1, \frac{\alpha_s}{2}\}$, but other dual bases would give rise to an equivalent relation.

8.2 A Tale of One Color

Polynomial Slide

$$\boxed{f} \quad = \quad \boxed{f} \qquad \text{when } f \in R^s \qquad (8.25\text{e})$$

Exercise 8.30 Confirm that these relations hold, as maps between bimodules.

Remark 8.31 Each of these additional relations has an analogue for a general Frobenius extension. For example, the barbell is always equal to the *product-coproduct element* $m(\Delta(1)) = \sum b_i b_i^*$. We leave the reader to ponder the other relations.

Exercise 8.32 Deduce the following relations, which describe an empty keyhole or an empty "needle":

$$\bigcirc = 0 \quad \text{and} \quad \bigcirc = 0. \qquad (8.26)$$

Exercise 8.33 Evaluate the triangle. What about an empty n-cycle? What about a circle?

Exercise 8.34 Deduce the following relation.

$$\Bigg\Vert \quad = \quad \frac{1}{2}\Bigg(\; \underset{\boxed{\alpha_s}}{\bigtimes} \; + \; \overset{\boxed{\alpha_s}}{\bigtimes} \;\Bigg). \qquad (8.27)$$

(*Hint:* From the left hand side to the right, create dots in the middle using (8.12), apply fusion (8.25d), apply Frobenius associativity (8.11).)

Exercise 8.35 Derive the general *polynomial forcing relation*:

$$\boxed{f} \quad = \quad \boxed{sf} \quad + \quad \boxed{\partial_s f} \qquad (8.28)$$

There are two good proofs, and you should try both.

1. Show that (8.28) holds if and only if it holds for $f \in R^s$ and holds for $f = \alpha_s$. Why does it hold in these special cases?
2. Show that if (8.28) holds for f and holds for g, then it holds for fg. Show that it holds for any linear term.

8.2.3 The Moral of the Tale

For sake of simplicity, let us assume that we define $\mathcal{H}_{\text{BS}}$ and $\mathbb{BSBim}$ using the geometric representation. In Chap. 10 we will discuss, in the general case, what happens for other realizations.

Definition 8.36 Let (W, S) be the Coxeter system of type A_1, with $S = \{s\}$. Let R be the polynomial ring associated to the geometric representation. The *one-color diagrammatic Hecke category*

$$\mathcal{H}_{\text{BS}} = \mathcal{H}_{\text{BS}}(s)$$

is the $\mathbb{R}$-linear monoidal category given by the following presentation. It has one generating object s, drawn in red. Its generating morphisms are given by the diagrams in (8.4) together with the box (8.24) labeled by any homogeneous $f \in R$. The relations are those in (8.5) and (8.25).

Theorem 8.37 *Let (W, S) be a Coxeter system equipped with its geometric representation, and pick some $s \in S$. Let $\mathbb{BSBim}$ be the corresponding category of Bott–Samelson bimodules, see Definition 4.33. Then there exists a ($\mathbb{R}$-linear monoidal) functor*

$$\mathcal{F} : \mathcal{H}_{\text{BS}}(s) \to \mathbb{BSBim},$$

sending $s \mapsto B_s$, and sending the generators of $\mathcal{H}_{\text{BS}}(s)$ to the corresponding bimodule maps, as in Sect. 8.2.1.

Theorem 8.38 *Moreover, $\mathcal{F}$ is fully faithful.*

The well-definedness of the functor $\mathcal{F}$ has been completed in the exercises. Proving fullness and faithfulness can be done through a variety of means, see [47] or [59] for details. In Chaps. 10 and 11 we will explain more of the ideas behind this result.

8.2.4 A Direct Sum Decomposition, Diagrammatically

The point of this section is to show how one can study direct sum decompositions diagrammatically, or morphism-theoretically.

Exercise 8.39 Let X, M, N be objects in an additive category. Prove that $X \simeq M \oplus N$ if and only if there are morphisms

$$i_M \colon M \to X, \quad i_N \colon N \to X, \quad p_M \colon X \to M, \quad p_N \colon X \to N, \qquad (8.29)$$

which satisfy

$$p_M \circ i_N = 0, \quad p_N \circ i_M = 0, \qquad (8.30a)$$

$$p_M \circ i_M = \mathrm{id}_M, \quad p_N \circ i_N = \mathrm{id}_N, \qquad (8.30b)$$

$$i_M \circ p_M + i_N \circ p_N = \mathrm{id}_X. \qquad (8.30c)$$

Let us apply this exercise to the direct sum decomposition

$$B_s \otimes B_s \simeq B_s(1) \oplus B_s(-1) \qquad (8.31)$$

in $\mathbb{S}\mathrm{Bim}$. We refer to the projection maps as p_{+1} and p_{-1} respectively, and similarly for the inclusion maps. We aim to describe the morphisms in $\mathcal{H}_{\mathrm{BS}}(s)$ which are sent to these morphisms by $\mathcal{F}$.

Let

$$i_{+1} = \;\;\raisebox{-0.5em}{\scalebox{1}[1]{$\bigvee$}}\;\;, \quad i_{-1} = \;\;\raisebox{-0.5em}{\scalebox{1}[1]{$\bigvee\!\boxed{\tfrac{\alpha_s}{2}}$}}\;\;, \qquad (8.32)$$

$$p_{+1} = \;\;\raisebox{0.2em}{\scalebox{1}[1]{$\bigwedge\!\boxed{\tfrac{\alpha_s}{2}}$}}\;\;, \quad p_{-1} = \;\;\raisebox{0.2em}{\scalebox{1}[1]{$\bigwedge$}}\;\;.$$

Let us consider once again the equality (8.27):

$$\Big\Vert\Big\Vert = \tfrac{1}{2}\Big(\;\raisebox{-0.3em}{$\bigvee\!\!\bigwedge\;\boxed{\alpha_s}$}\; + \;\raisebox{-0.3em}{$\boxed{\alpha_s}\;\bigvee\!\!\bigwedge$}\;\Big). \qquad (8.33)$$

This equality clearly corresponds to (8.30c). We have added an extra dashed line to indicate how the morphisms factor.

Exercise 8.40 Verify the other equalities in (8.30).

Thus we have proven the direct sum decomposition (8.31) by concretely exhibiting the projections and inclusions. In Chap. 11 we formally add direct sums, grading shifts, and direct summands to $\mathcal{H}_{BS}$ to obtain the category $\mathrm{Kar}(\mathcal{H}_{BS})$, where the same computation will give rise to the direct sum decomposition

$$s \otimes s \simeq s(1) \oplus s(-1). \tag{8.34}$$

Remark 8.41 The inclusion map i_{+1} and the projection map p_{-1} each live in a one-dimensional piece of the graded Hom space, so they are unique up to scalar. However, there is more flexibility in the choice of i_{-1} and p_{+1}; see the next exercise. Note that neither idempotent $e_{+1} = i_{+1} \circ p_{+1}$ nor $e_{-1} = i_{-1} \circ p_{-1}$ is canonically defined. This is typical behavior in a graded category, where the projections and inclusions of minimal degree are canonical (up to scalar), but the remaining inclusions and projections are not. From experience with semisimple categories the reader may have the intuition that idempotents projecting to isotypic components are canonical, but this is no longer true in the graded setting.

Exercise 8.42 Redefine i_{-1} and p_{+1} as follows:

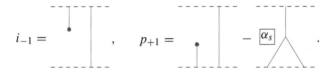

Prove that the four maps $i_{\pm 1}$, $p_{\pm 1}$ also imply the idempotent decomposition (8.31).

Chapter 9
The Dihedral Cathedral

This chapter is based on expanded notes of a lecture given by the authors and taken by
Ben Elias and Nicolle E. S. González

Abstract We introduce the diagrammatic category $\mathcal{H}_{\mathrm{BS}}$ for the infinite and finite dihedral groups.

9.1 A Tale of Two Colors

We now consider a dihedral Coxeter system (W, S), for $S = \{s, t\}$. We refer to the elements of S as colors. For each color s and t we have already defined the category $\mathcal{H}_{\mathrm{BS}}$ in the previous chapter. Meanwhile, a Bott–Samelson bimodule for (W, S) will have the form $B_s B_t B_s B_s \cdots$, indexed by sequences of s's and t's. Between such bimodules we can already produce a host of morphisms by combining the one-color diagrams already known. The following is an example of such a diagram:

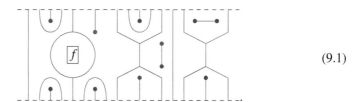 (9.1)

Diagrams made exclusively out of concatenations of one-color diagrams (i.e. built from univalent vertices, trivalent vertices, and boxes) are known as *universal diagrams*.

B. Elias
Department of Mathematics, University of Oregon, Fenton Hall, Eugene, OR, USA

N. E. S. González
Department of Mathematics, University of Southern California, Los Angeles, CA, USA

© The Editor(s) (if applicable) and The Author(s), under exclusive licence to Springer Nature Switzerland AG 2020
B. Elias et al., *Introduction to Soergel Bimodules*, RSME Springer Series 5, https://doi.org/10.1007/978-3-030-48826-0_9

Exercise 9.1

1. Consider a universal diagram without boxes, and view it as a colored graph (embedded in $\mathbb{R} \times [0, 1]$, with boundary on $\mathbb{R} \times \{0, 1\}$). Prove that any two trees with the same boundary are equal. Prove that any connected graph which is not a tree is equal to zero.
2. Consider a universal diagram with boxes. Prove that it is in the span of universal diagrams where all boxes appear in the leftmost region. (*Hint:* Use polynomial forcing.)
3. Prove that any universal diagram with empty boundary (i.e. a closed diagram) is equal to a polynomial. (*Hint:* induct on the number of connected components.)

Exercise 9.2 In Exercise 5.32 you computed the graded rank of the morphism spaces

$$\mathrm{Hom}^\bullet(B_s, B_t), \qquad \mathrm{Hom}^\bullet(B_s B_s, B_s), \qquad \mathrm{Hom}^\bullet(B_s, B_s B_t B_s).$$

For each of these spaces, universal diagrams can be used to construct a basis as a left R-module. Find a set of diagrams of the appropriate degrees which look like a basis. For a challenge, try to write all other universal diagrams as linear combinations of your diagrams. (*Hint:* By the previous exercise, you can ignore diagrams whose graphical structure is too complicated. You can assume there are no boxes, no closed components, and that all connected components are trees. Why?)

The observant reader will notice that so far no distinction has been made between the finite and infinite dihedral groups. It turns out that universal diagrams are sufficient to produce all morphisms between Bott–Samelson bimodules when $m_{st} = \infty$! Moreover, no new relations appear when red diagrams and blue diagrams are combined.

Definition 9.3 Let (W, S) be an infinite dihedral Coxeter system, equipped with its Kac–Moody representation. The associated *universal two-color diagrammatic Hecke category* $\mathcal{H}_{\mathrm{BS}}$ is the $\Bbbk$-linear monoidal category whose objects are expressions $\underline{w}$ in $S = \{s, t\}$, and whose morphisms are the $\Bbbk$-span of universal diagrams, modulo the one-color relations from Chap. 8.

We can also define $\mathcal{H}_{\mathrm{BS}}$ using a presentation, just as we defined the one-color diagrammatic Hecke category in Definition 8.36. The morphism generators would be univalent and trivalent vertices of both colors s and t, together with boxes for the polynomials $f \in R$.

Remark 9.4 When defining $\mathcal{H}_{\mathrm{BS}}$ one of our inputs is the polynomial ring R, so we know what kinds of polynomials we can put in the boxes. In this case R is the polynomial ring of the Kac–Moody representation rather than the geometric representation, see Sect. 5.6. As discussed in Remark 4.38, there are reasons why, for an affine Weyl group like the infinite dihedral group, the geometric representation is not quite satisfactory. Again, in Chap. 10 we will discuss, in the general case, what happens for other realizations.

9.1 A Tale of Two Colors

Theorem 9.5 *The functors $\mathcal{F}$ from $\mathcal{H}_{BS}(s)$ and $\mathcal{H}_{BS}(t)$ to $\mathbb{BSBim}$, defined by the formulas in Sect. 8.2.1, glue together to give a functor $\mathcal{F}\colon \mathcal{H}_{BS} \to \mathbb{BSBim}$.*

Theorem 9.6 *This functor $\mathcal{F}$ is fully faithful.*

At this point, one could attempt to use diagrammatic arguments to find a spanning set for universal diagrams in $\mathcal{H}_{BS}$, and algebraic arguments to find the graded rank of all morphism spaces in $\mathbb{BSBim}$ (via the Soergel Hom formula). Then one could try to match these two sides and prove that $\mathcal{F}$ is fully faithful. This sounds daunting now, but it is the proof we will eventually use, and it becomes easier in Chap. 10 when we introduce combinatorial bases for morphism spaces in $\mathcal{H}_{BS}$. For a different proof, see [47].

Remark 9.7 Suppose that (W, S) is a *universal Coxeter system*, where $m_{st} = \infty$ for all $s \neq t \in S$. Then one can define universal diagrams as above, but with more than two colors. It turns out that, again, no additional modification is needed to produce a diagrammatic category $\mathcal{H}_{BS}$ which is equivalent to $\mathbb{BSBim}$. This explains why the diagrams above are called universal diagrams.

When $m_{st} = m < \infty$, we will need a new two-color generator to produce all the morphisms in $\mathbb{BSBim}$, and with it we will need new two-color relations. Our new generator is drawn as a $2m$-valent vertex, and has degree zero:

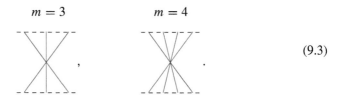

$$: \underbrace{(s, t, s, t, \ldots)}_{m} \to \underbrace{(t, s, t, s, \ldots)}_{m}. \quad (9.2)$$

The colors on the edges alternate. One of the purple edges is red and the other is blue, but which is which depends on the parity of m.

For example, for $m \in \{3, 4\}$ the new generator has the following form:

$$\begin{array}{cc} m = 3 & m = 4 \end{array} \quad (9.3)$$

Note that the vertex with the colors swapped is also included as a generator.

Later in this chapter we will give a convincing motivation for why we need a new generator of this form, but for now the reader should take it on faith. The next remark begins this discussion. Before describing the relations, it is important to study the case $m_{st} = \infty$ in more depth, and to introduce another 2-category that will play a crucial role in this story.

Remark 9.8 What effect does the passage from $m_{st} = \infty$ to $m_{st} = m < \infty$ have on the size of Hom spaces in $\mathbb{BSBim}$? This can be computed in the Hecke algebra, using the Soergel Hom formula. For many morphism spaces this change appears to make no difference! In the Exercise 5.32 the size of certain Hom spaces did not depend on the value of m_{st}, and in Exercise 9.2 it was argued that universal diagrams should suffice to describe those Hom spaces. However, eventually the braid relation does have an effect on the size of Hom spaces, and the "first place" it matters is in the size of the degree zero morphism space where the $2m$-valent vertex lives.

Exercise 9.9 Compute the following standard traces in the Hecke algebra, both when $m_{st} = 3$ and when $m_{st} = \infty$.

$$\epsilon(b_s b_t b_s), \quad \epsilon(b_s b_t b_s b_t), \quad \epsilon(b_s b_t b_s b_t b_s), \quad \epsilon(b_s b_t b_s b_t b_s b_t). \tag{9.4}$$

Where can you find a difference between the computation for $m_{st} = 3$ and for $m_{st} = \infty$? Does this explain the existence of the 6-valent vertex?

9.2 The Temperley–Lieb 2-Category

Recall the Temperley–Lieb category $\mathcal{TL}$ from Definition 7.13. This is a monoidal category with objects $n \in \mathbb{N}$, where $\mathrm{Hom}(m,n)$ is spanned by crossingless matchings from m dots on bottom to n dots on top. One of the relations in this category was

$$\bigcirc = -2 \tag{9.5}$$

The *(generic) Temperley–Lieb category* $\mathcal{TL}_\delta$ is defined in the same way, except that the base ring is $\mathbb{Z}[\delta]$ for a formal variable δ, and

$$\bigcirc = \delta \tag{9.6}$$

After specializing to $\delta = -2$ we obtain the category $\mathcal{TL}$ from before.

Remark 9.10 The specialization $\delta = -2$ controls the representation theory of $\mathfrak{sl}_2$, as discussed in Remark 7.15. The specialization $\delta = -(q + q^{-1})$ controls the representation theory of the quantum group of $\mathfrak{sl}_2$, and other specializations are important in the theory of tilting modules.

We now enhance the Temperley–Lieb category as follows.

9.2 The Temperley–Lieb 2-Category

Definition 9.11 The *two-colored Temperley–Lieb 2-category* $2\mathcal{TL}_\delta$ is the 2-category defined as follows. Its objects are colored regions (blue and red). The 1-morphisms are tensor generated by a 1-morphism from red to blue and a 1-morphism from blue to red. That is, 1-morphisms alternate between red and blue. Thus for any $n \in \mathbb{Z}$ there are two associated 1-morphisms, the one starting at red (and ending at red if n is even, blue if n is odd), and the one starting at blue. The 2-morphisms are $\mathbb{Z}[\delta]$-linear combinations of crossingless matchings with appropriately colored regions (the colors on regions must alternate). Composition is as in the Temperley–Lieb category, where the composition of the cup and cap (aka the bubble) evaluates to δ.

For example, here is the 1-morphism corresponding to the number 4, starting and ending at red.

$$\tag{9.7}$$

Here is a colored crossingless matching which represents a 2-morphism in $2\mathcal{TL}_\delta$.

$$\tag{9.8}$$

The reason we introduce the two-colored Temperley–Lieb 2-category is that there is a strong relationship between crossingless matchings and universal diagrams. This relationship can be expressed as a (slightly awkward) functor. Because of this functor we can use the representation theory of $\mathfrak{sl}_2$ to study the diagrammatic Hecke category for the dihedral group!

Definition 9.12 Let (W, S) be a (finite or infinite) dihedral group with its geometric representation, and set $a_{st} = \partial_s(\alpha_t) = -2\cos(\pi/m_{st})$. Note that $a_{st} = a_{ts}$. We let $2\mathcal{TL}_{a_{st}}$ denote the specialization of the two-colored Temperley–Lieb 2-category at $\delta = a_{st}$. Then there is a functor

$$\Sigma: 2\mathcal{TL}_{a_{st}} \to \mathbb{BSBim} \tag{9.9}$$

defined as follows. It acts on identity 1-morphisms as

$$\tag{9.10}$$

and on a general 1-morphism as

$$\longmapsto B_s B_t B_s B_t B_s \cdots \tag{9.11}$$

On 2-morphisms, Σ applies a deformation retract to all crossingless matchings. On generators it acts as

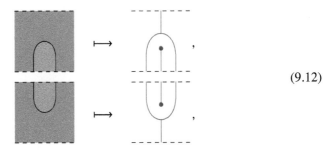

(9.12)

and similarly for blue and red switched. Note that the generators are mapped to diagrams of degree 0.

Proposition 9.13 *The functor Σ is well defined.*

Proof Checking the isotopy relation in $2\mathcal{TL}_{a_{st}}$ we leave as an exercise. The interesting relation is

$$\text{(diagram)} \longmapsto \text{(diagram)} = \partial_s(\alpha_t) \qquad \text{(diagram)} \longleftarrow a_{st} \qquad \text{(diagram)}.$$

(9.13)

This agrees with our specialization of the bubble to $a_{st} = a_{ts}$. □

Remark 9.14 Note that we have not called Σ a 2-functor, because it is not! The identity 1-morphism (of red, or blue) is not sent to the monoidal identity, so this functor is not monoidal. It is a functor in that it respects vertical concatenation of diagrams, but it does not respect horizontal concatenation, which raises eyebrows. The reason for the awkwardness of Σ is that the monoidal category $\mathbb{BSBim}$ is not the appropriate target; rather, there is a 2-category of singular Soergel bimodules, defined and explored in Chap. 24, which admits a genuine 2-functor from $2\mathcal{TL}_{a_{st}}$. One can recover information about Soergel bimodules from singular Soergel bimodules by applying a non-monoidal functor, and this explains the issue. We return to this story in Sect. 24.2.3.

Here is a little more detail on the notion of a deformation retract. It makes sense to discuss crossingless matchings on the (planar) disk, although these will not be arranged into a category or 2-category.

9.2 The Temperley–Lieb 2-Category

(9.14)

Given a crossingless matching on the disk, each colored region deformation retracts to a tree, producing a universal diagram on the disk.

(9.15)

Note that this universal diagram has degree $+2$, not degree zero. When one transforms a diagram from the planar disk to the planar strip there is some duplication, as the rightmost region (red in the picture above) becomes a region in both the top and bottom boundary, and similarly for the leftmost region. In universal diagrams, this duplication is accomplished with two trivalent vertices.

(9.16)

This lowers the degree by 2, yielding the degree zero diagram in the image of Σ.

Not every universal diagram is a deformation retract of a crossingless matching. In particular, the image of Σ only has morphisms of degree zero, so the map $B_s \to B_s$ given by the broken line is not in the image of the functor. The key idea is that Σ is fully faithful onto universal diagrams of minimal degree, which is the essence of the following exercise.

Exercise 9.15 Consider a universal diagram on the disk, whose boundary alternates $B_s B_t B_s B_t \cdots B_s B_t$ when reading clockwise. Prove that the minimal degree of any nonzero universal diagram is 2, and that all such universal diagrams are deformation retracts of colored Temperley–Lieb diagrams. (*Hint:* what characterizes universal diagrams in the image of the functor from Temperley–Lieb? What happens when you try to connect two different connected components of the same color?)

Remark 9.16 The specialization $\delta = a_{st}$ is closely related to the specialization $\delta = -(q + q^{-1})$ which relates to quantum $\mathfrak{sl}_2$. As noted in Exercise 5.51,

$a_{st} = -2\cos(\pi/m_{st})$ is equal to $-(q + q^{-1})$ where q is a primitive $2m$-th root of unity, or $q = 1$ when $m_{st} = \infty$.

Remark 9.17 This section can be adapted to more general realizations without much difficulty. If the Cartan matrix is symmetric, one simply specializes δ to $a_{st} = a_{ts}$. If the Cartan matrix is non-symmetric, one must redefine $2\mathcal{TL}$ so that there are two formal parameters δ_{st} and δ_{ts}, which describe the evaluation of a bubble with red and blue interiors respectively. Then these can be specialized to a_{st} and a_{ts}. For simplicity, we stick with symmetric Cartan matrices in this chapter.

9.3 Jones–Wenzl Projectors

Let $\mathrm{TL}_{n,\delta}$ denote the endomorphism ring of the object n in $\mathcal{TL}_\delta$. Then $\mathrm{TL}_{n,\delta}$ is a $\mathbb{Z}[\delta]$-algebra, often called the *Temperley–Lieb algebra*. In this section we extend scalars and consider $\mathcal{TL}_\delta$ and $\mathrm{TL}_{n,\delta}$ over $\mathbb{Q}(\delta)$.

Theorem 9.18 *There is a unique element* $\mathrm{JW}_n \in \mathrm{TL}_{n,\delta}$ *satisfying:*

- JW_n *sent to zero by postcomposition with any cap;*
- JW_n *is sent to zero by precomposition with any cup;*
- JW_n *has 1 as the coefficient of the identity matching* id_n*, when it is expressed as a linear combination of crossingless matchings.*

Moreover, JW_n *is an idempotent:* $\mathrm{JW}_n^2 = \mathrm{JW}_n$.

Definition 9.19 The *Jones–Wenzl projector* JW_n is the unique element of $\mathrm{TL}_{n,\delta}$ satisfying the conditions of Theorem 9.18.

We delay the proof of Theorem 9.18 to the exercises. For the uniqueness of JW_n, and for a reason it is an idempotent, see Exercise 9.25. Jones–Wenzl projectors are hard to describe explicitly as linear combinations of crossingless matchings. However they can be shown to exist via recursive formulas, see Exercises 9.26 and 9.27.

Remark 9.20 Jones–Wenzl projectors play a key role in the representation theory of $\mathfrak{sl}_2$, as JW_n projects to the top irreducible summand (of highest weight n) inside the tensor product $V^{\otimes n}$. They were defined by Jones and Wenzl, and a recursive formula was given by Wenzl [175].

Clearly $\mathrm{JW}_1 = \mathrm{id}_1$. The Jones–Wenzl projectors JW_2 and JW_3 were explored in Exercise 7.17. To give away the answer, here is JW_2.

$$\boxed{\mathrm{JW}_2} \;=\; \Big|\Big| \;-\; \frac{1}{\delta}\; \cup_\cap \qquad\qquad (9.17)$$

9.3 Jones–Wenzl Projectors

Let us write this another way:

$$\text{(diagram)} = \boxed{JW_2} + \frac{1}{\delta} \text{(cup-cap diagram)}. \tag{9.18}$$

We see that (9.18) writes the identity of the object $2 \in \mathcal{TL}$ as a sum of two orthogonal idempotents. Let us interpret this in the light of Sect. 8.2.4. The second idempotent factors as $i \circ p$, factoring through the object $0 \in \mathcal{TL}$, so the image of this idempotent is isomorphic to the object 0. The first idempotent JW_2 does not (necessarily) factor, so it represents some new direct summand that we may not have seen. We might write

$$2 \simeq 0 \oplus T_2, \tag{9.19}$$

where $T_2 = \text{Im}(JW_2)$ is the image of the idempotent JW_2.

Remark 9.21 This is a slight lie, since T_2 is not an object in $\mathcal{TL}$. However, there is a formal process which takes an additive category and adds all images of idempotents (i.e. direct summands) as new objects. This is called taking the Karoubi envelope, and it is described in Chap. 11. In the Karoubi envelope, T_2 is an object and (9.19) holds.

Now, let us color JW_2 with red on the right, and apply the functor Σ to obtain an endomorphism of $B_s B_t B_s$. The result, which we denote by $JW_{(s,t,s)}$, is still an idempotent. Then (9.18) is transformed into

$$\text{(diagram)} = \boxed{JW_{(s,t,s)}} + \frac{1}{a_{st}} \text{(dotted cup-cap diagram)} \tag{9.20}$$

In other words,

$$B_s B_t B_s \simeq B_s \oplus \text{Im}(JW_{(s,t,s)}). \tag{9.21}$$

We already know that $B_s B_t B_s \simeq B_{sts} \oplus B_s$, so we conclude that $\text{Im}(JW_{(s,t,s)}) \simeq B_{sts}$!

Jones–Wenzl projectors play an extremely important role in the Hecke category, as can be seen in the following theorem.

Theorem 9.22 *Suppose that $n < m_{st}$. Take the Jones–Wenzl projector JW_n, color it with red on the far right, and apply the functor Σ. The image inside $\mathbb{BS}\text{Bim}$, denoted by $JW_{\underline{w}}$, is an idempotent inside $\text{End}(\cdots B_t B_s)$, where this sequence $\underline{w} = (\ldots, t, s)$*

is a reduced expression for an element $w \in W$ of length $n + 1$. In fact, the image of this idempotent is precisely the indecomposable summand B_w.

Of course, by switching the role of red and blue, we obtain idempotents projecting to B_w where $w = \cdots st$, and hence we can pick out B_w for all w in the dihedral group. So, under the functor Σ, the Jones–Wenzl projectors pick out exactly the indecomposable objects inside $\mathbb{S}\text{Bim}$!

Remark 9.23 For the generalization of this result to all universal Coxeter groups, see [55].

The observant reader might have one quibble. In our formula for JW_2 above there was a denominator, δ, which became a_{st}. If $a_{ts} = a_{st} = 0$ then this Jones–Wenzl projector cannot be defined. Thankfully, this happens precisely when $m_{st} = 2$, which is exactly when we no longer need JW_2 because (s, t, s) is not a reduced expression.

More generally, Jones–Wenzl projectors have denominators, so they are not defined generically (i.e. in $\mathbb{Z}[\delta]$) but only after specialization to a ring where the denominators are invertible. Typical formulas for these projectors involve quantum numbers $[n]$, which are certain polynomials in δ, with the convention that $\delta = -[2]$. For example, $[3] = [2]^2 - 1$. For more on quantum numbers see Exercise 5.51. It turns out that when $\delta = a_{st} = -2\cos(\pi/m)$ then $[m] = 0$ and $[m - 1] = 1$. As a consequence, JW_n is defined for all $n < m$, and has particularly nice coefficients when $n = m - 1$.

In particular, all the Jones–Wenzl projectors we need for Theorem 9.22 are well-defined. When $m_{st} = \infty$, JW_n is defined for all n, and we are capable of recovering all the indecomposable objects inside $I_2(\infty)$ that our hearts could possibly desire.

Remark 9.24 The reader may be intrigued at what seems to be a deep connection between representations of $\mathfrak{sl}_2$ and the Hecke category for the dihedral group. This is actually the first example of the *quantum geometric Satake equivalence*. See [47] and [49] for more details.

Exercise 9.25 Let our base ring be some specialization of $\mathbb{Z}[\delta]$. Inside $\text{TL}_{n,\delta}$ let T be the vector space of elements which are killed by all the $(n-1)$ caps on top, and let B be the space killed by cups on the bottom. For an element $x \in \text{TL}_{n,\delta}$ let $\overline{x}$ denote the same element with each diagram flipped upside down. Thus, for example, $x \in T$ if and only if $\overline{x} \in B$.

1. Show that any crossingless matching is either the identity diagram, or has both a cap on bottom and a cup on top.
2. We now make the following assumption:

$$\text{There exist some } f \in T \text{ for which the coefficients of the identity diagram is invertible.} \tag{9.22}$$

Why is this equivalent to the analogous assumption for B?

9.3 Jones–Wenzl Projectors

3. Let $f \in T$, with invertible coefficient c for the identity diagram. Let $g \in B$, with invertible coefficient d for the identity diagram. Compute the composition fg in two ways and deduce that f and g are colinear.
4. Assuming (9.22) deduce that $T = B$, that this space is one-dimensional, and that $f = \overline{f}$ for $f \in T$.
5. Thus, assuming (9.22), there is a unique element $\text{JW}_n \in T$ whose identity coefficient is 1. Prove that JW_n is idempotent. (If we construct JW_n in some other way, this proves (9.22).)

Exercise 9.26 Let TL_n be the Temperley–Lieb algebra with n-strands where the bubble evaluates to $-[2] = q + q^{-1} \in \mathbb{Q}(q)$. Clearly JW_1 is just the identity element, where the condition of being killed by caps and cups is vacuous. Verify the following recursive formula:

$$\text{JW}_{n+1} = \text{JW}_n \Big| + \sum_{a=1}^{n} \frac{[a]}{[n+1]} \;\; \text{JW}_n \tag{9.23}$$

In this last diagram, the cup on top matches the a-th and $(a+1)$-st boundary points, counting from the left.

The next two exercises should be done together. Both are proven by induction, and the inductive step of one exercise uses the result of the other exercise.

Exercise 9.27 Prove the following recursive formula.

$$\text{JW}_{n+1} = \text{JW}_n \Big| + \frac{[n]}{[n+1]} \begin{array}{c} \text{JW}_n \\ \text{JW}_n \end{array} \tag{9.24}$$

Exercise 9.28 The *trace* of an element $a \in \text{TL}_n$ is the evaluation in $\mathbb{Z}[q, q^{-1}]$ of the following closed diagram:

$$\tag{9.25}$$

1. Calculate the trace of JW_n.
2. Suppose that q is a primitive $2m$-th root of unity. What is the trace of JW_{m-1}? What do you get when you rotate JW_{m-1} by one strand?

Remark 9.29 The particularly nice coefficients of JW_{m-1} translate to this idempotent being rotation-invariant in $\mathrm{TL}_{m-1,\delta}$. This rotation invariance will lead to the cyclicity of the $2m$-valent vertex. More generally, this is the reason for the balanced condition, see Assumption 5.57.

9.4 Two-color Relations

When $m_{st} = m < \infty$, the Kazhdan–Lusztig basis element $b_{st\dots}$ and $b_{ts\dots}$ of length m are identified, because both $st\cdots$ and $ts\cdots$ are equal to the longest element w_0. Lifting this fact, there should be an isomorphism between the image of $\mathrm{JW}_{(s,t,\dots)}$ inside $B_s B_t \cdots$ and the image of $\mathrm{JW}_{(t,s,\dots)}$ inside $B_t B_s \cdots$, because both should give B_{w_0}. (Both Jones–Wenzl projectors come from the rotation-invariant idempotent JW_{m-1} in TL_{m-1}.) This motivates the $2m$-valent vertex, which is supposed to represent the composition

$$B_s B_t \cdots \to B_{w_0} \to B_t B_s \cdots. \tag{9.26}$$

We can now, as foretold, state the new relations needed to accommodate the interaction between the two colors. We impose the following relations (and their color swaps):

Cyclicity

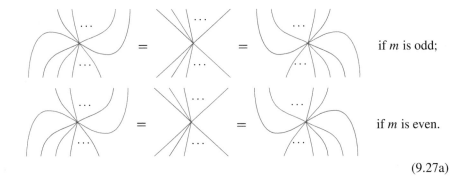

$$\tag{9.27a}$$

9.4 Two-color Relations

Elias–Jones–Wenzl

$$\begin{array}{c}\includegraphics{fig}\end{array} = \boxed{\text{JW}_{(s,t,\ldots)}} \qquad (9.27\text{b})$$

Two-color Associativity

$$\begin{aligned} &\text{diagram} = \text{diagram} && \text{if } m \text{ is odd;} \\ &\text{diagram} = \text{diagram} && \text{if } m \text{ is even.} \end{aligned} \qquad (9.27\text{c})$$

In particular, (9.27b) confirms that the $2m$-valent vertex factors through the image of the idempotent $\text{JW}_{(s,t,\ldots)}$. Also, (9.27a) implies that two isotopic diagrams represent the same morphism.

Here are some other useful relations.

Two-color dot Contraction (Aka the Jones–Wenzl Relation)

$$\begin{aligned} &\text{diagram} = \boxed{\text{JW}_{(s,t,\ldots)}} && \text{if } m \text{ is odd;} \\ &\text{diagram} = \boxed{\text{JW}_{(s,t,\ldots)}} && \text{if } m \text{ is even.} \end{aligned} \qquad (9.28)$$

More Two-color Associativity

$$\text{[diagram]} = \text{[diagram]} \quad \text{if } m \text{ is odd;}$$

$$\text{[diagram]} = \text{[diagram]} \quad \text{if } m \text{ is even.} \tag{9.29}$$

$$\text{[diagram]} = \text{[diagram]} \quad \text{if } m \text{ is odd;}$$

$$\text{[diagram]} = \text{[diagram]} \quad \text{if } m \text{ is even.} \tag{9.30}$$

The culmination of our tale should no longer be surprising.

Definition 9.30 Let (W, S) be a finite dihedral group with $S = \{s, t\}$ and $m = m_{st}$, equipped with its geometric representation.[1] The *two-color diagrammatic Hecke category* is the $\mathbb{R}$-linear monoidal category

$$\mathcal{H}_{BS}$$

generated over $\mathcal{H}_{BS}(s)$ and $\mathcal{H}_{BS}(t)$ by the $2m$-valent vertices, modulo the relations in (9.27).

---

[1] For finite dihedral groups, the geometric representation and the Kac–Moody representation coincide.

9.4 Two-color Relations

Theorem 9.31 *There is an $\mathbb{R}$-linear monoidal functor*

$$\mathcal{F}: \mathcal{H}_{\mathrm{BS}} \longrightarrow \mathbb{BSBim}.$$

Theorem 9.32 *This functor $\mathcal{F}$ is an equivalence of categories.*

For a detailed proof, and an implicit description of the image of the $2m$-valent vertex under $\mathcal{F}$, see [47]. Again, the ideas behind this proof will be explored in Chaps. 10 and 11.

Remark 9.33 We have been working with the geometric realization. In Sect. 10.2.6 we discuss the status of these results over more general realizations.

Exercise 9.34

1. Write down the two-color relations when $m = 2$. Prove that $B_s B_t \simeq B_t B_s$ by constructing inverse isomorphisms.
2. Write down the two-color relations when $m = 3$. Prove that $B_s B_t B_s \simeq X \oplus B_s$, where X is the image of an idempotent constructed using two 6-valent vertices, by following the rubric of Exercise 8.39.
3. (Still for $m = 3$) Similarly, one has $B_t B_s B_t \simeq Y \oplus B_t$. Prove that X is isomorphic to Y. Extend the rubric of Exercise 8.39 to a rubric which describes when two summands of different objects are isomorphic.

Exercise 9.35 Prove that there is an autoequivalence of $\mathcal{H}_{\mathrm{BS}}$ which flips each diagram vertically (resp. horizontally). See Exercise 8.10 for inspiration.

Exercise 9.36 Show that the diagram obtained by attaching a "handle" to the left or right of a Jones–Wenzl projector equals 0. For example,

$$\boxed{\mathrm{JW}_{(s,t,\ldots)}} = 0. \qquad (9.31)$$

(*Hint:* use (9.16).)

Exercise 9.37

1. A *pitchfork* is a diagram of the form

$$\qquad (9.32)$$

(or its color swap). The *death by pitchfork* relation states that the diagram obtained by placing a pitchfork anywhere on top or bottom of a Jones–Wenzl projector equals 0. For example:

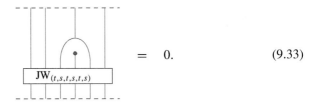

$$\quad (9.33)$$

Why is death by pitchfork implied by the defining property of the Jones–Wenzl projector?

2. Use (9.28) and (9.31) to prove that the diagram obtained by placing a pitchfork anywhere on top or bottom of the $2m$-valent vertex equals 0. We also call this *death by pitchfork*.

Exercise 9.38

1. Prove (9.29) and (9.30) using the relations in (9.27). (*Hint:* each relation follows from two careful applications of (9.27c). Alternatively, (9.29) can be proved by repeatedly applying (9.30).
2. Prove (9.27b) using (9.28) and the other relations in (9.27). (*Hint:* first use (8.12) to create a dot and a trivalent vertex on the left hand side, and then dispose of the trivalent vertex with two-color associativity.)

The following exercise is harder, but very worthwhile.

Exercise 9.39

1. Prove (9.28) using the relations in (9.27).
2. Prove that the Elias–Jones–Wenzl relation (9.27b) follows from two-color associativity (9.27c) and two-color dot contraction (9.28).

Chapter 10
Generators and Relations for Bott–Samelson Bimodules and the Double Leaves Basis

This chapter is based on expanded notes of a lecture given by the authors and taken by
Alexander Sistko

Abstract In the previous chapters we described a diagrammatic category $\mathcal{H}_{BS}$ associated to dihedral Coxeter systems, and a functor to the category $\mathbb{BS}\text{Bim}$ of Bott–Samelson bimodules (which is meant to be an equivalence). To extend this definition to an arbitrary Coxeter system, it turns out that the only new relations needed are associated with finite rank 3 parabolic subgroups. In this chapter we describe the category $\mathcal{H}_{BS}$ for an arbitrary Coxeter system, discussing the new details and giving a different set of motivations. We also briefly discuss what happens for other realizations, where the diagrammatic category may behave better than the algebraic category $\mathbb{BS}\text{Bim}$.

Using morphisms called light leaves, Libedinsky constructed a right R-basis for the Hom-spaces of $\mathbb{BS}\text{Bim}$, called the double leaves basis. We describe Libedinsky's light leaves and double leaves diagrammatically.

10.1 Why Present $\mathbb{BS}\text{Bim}$?

The goal of this chapter is to give the presentation of a diagrammatic category which models $\mathbb{BS}\text{Bim}$ for any Coxeter system. Before we begin the presentation, we feel it is worthwhile to discuss some of the philosophy which is operating behind the scenes.

The category we are truly interested in is $\mathbb{S}\text{Bim}$, the category of Soergel bimodules, which categorifies the Hecke algebra. The category of Bott–Samelson bimodules $\mathbb{BS}\text{Bim}$ was seemingly only a stepping stone to constructing $\mathbb{S}\text{Bim}$. So why do we focus on modeling $\mathbb{BS}\text{Bim}$ in particular, rather than finding a presentation

A. Sistko
Department of Mathematics, University of Iowa, Iowa, IA, USA

© The Editor(s) (if applicable) and The Author(s), under exclusive licence to Springer Nature Switzerland AG 2020
B. Elias et al., *Introduction to Soergel Bimodules*, RSME Springer Series 5, https://doi.org/10.1007/978-3-030-48826-0_10

for $\mathbb{S}$Bim? Why do we look for a presentation at all, instead of studying $\mathbb{BS}$Bim and $\mathbb{S}$Bim algebraically as we did in Chap. 4?

The first point we wish to make is that algebra is hard! If we were forced to think about Bott–Samelson bimodules as linear combinations of tensors of polynomials, we would soon reach a dead end. Suppose we want to understand or construct a particular indecomposable Soergel bimodule B_w. It appears as a direct summand of a Bott–Samelson bimodule, but identifying which sums of tensors of polynomials live in this summand is extremely difficult. The reader who attempted Exercise 4.41 has already seen how hard it is to construct the projection map from $B_s B_t B_s$ to its summand B_{sts} in terms of operations on polynomials. This is the easiest (nontrivial) case, and it only gets harder from there on. Meanwhile, the great advantage of diagrammatics is its efficiency. One diagram (for example, the $2m$-valent vertex) can encode an incredibly complicated operation on polynomials, and so long as the relations involving that diagram are easy enough to work with, it didn't matter how complicated this operation was. An explicit formula for the $2m$-valent vertex as an operation on linear combinations of tensors of polynomials cannot be found in the literature (except for small m), because it is extremely annoying to derive—but working with the $2m$-valent vertex is not so challenging.

Remark 10.1 There are other algebraic approaches to computing with indecomposable Soergel bimodules, such as the technology of moment graphs introduced in Chap. 6. Instead of thinking of a Bott–Samelson or Soergel bimodule as a collection of tensors of polynomials, they are viewed as a collection of polynomials smeared across the Bruhat graph. This is considerably easier, it turns out, but still has its technical limitations. Diagrammatics, it seems, can take us much further.

Now to address the question: why present $\mathbb{BS}$Bim instead of $\mathbb{S}$Bim? A first reason is that $\mathbb{BS}$Bim is simpler so the problem is tractable. Objects in $\mathbb{BS}$Bim are completely combinatorial in nature, being described by sequences of simple reflections. Thankfully, the combinatorial nature of these objects is reflected in the combinatorial nature of morphisms between these objects, which we are able to describe using decorated planar graphs. While the indecomposable Soergel bimodules B_w are parameterized by the Coxeter group W (a combinatorial object), the bimodule B_w itself is some mysterious direct summand, only implicitly defined. With no explicit description of B_w, it is not surprising that there is no known description (diagrammatic or otherwise) of the Hom spaces between Soergel bimodules, except in small cases.

A second reason, or perhaps a prerequisite for the whole topic, is that presenting $\mathbb{BS}$Bim does give one enough information to reconstruct $\mathbb{S}$Bim. This is because $\mathbb{S}$Bim can be realized as Kar($\mathbb{BS}$Bim), the *Karoubi envelope* or *idempotent completion* of $\mathbb{BS}$Bim (see Chap. 11 for details). Roughly speaking, the Karoubi envelope of an additive category $\mathcal{C}$ is obtained by adding (as new objects) the direct summands of objects in $\mathcal{C}$. Taking the Karoubi envelope is a formal process which only involves the morphisms in $\mathcal{C}$: it can be applied to diagrammatic categories (where the objects are abstract, intangible things) just as well as to algebraic categories like R-bimodules. The data of a direct summand is the same as the data of the idempotent

10.1 Why Present $\mathbb{BS}$Bim?

which projects to it (the morphism-based approach to direct summands was also discussed in Sect. 8.2.4). Now, one can extend a presentation of $\mathcal{C}$ to a presentation of its Karoubi envelope by computing all the idempotent endomorphisms in $\mathcal{C}$. This reduces the difficult problem of describing morphisms in $\mathbb{S}$Bim to the easier combinatorial problem of describing morphisms in $\mathbb{BS}$Bim, along with the (still formidable) problem of computing idempotents within $\mathbb{BS}$Bim.

Remark 10.2 This philosophy, of studying an additive category $\mathcal{C}'$ by means of a simpler (Morita equivalent) full subcategory $\mathcal{C}$ with $\text{Kar}(\mathcal{C}) \simeq \mathcal{C}'$, has found application to many other problems in representation theory. For a brief survey see [48, §1.2].

A third reason to focus on $\mathbb{BS}$Bim is that it has the potential of a characteristic-independent presentation. This topic deserves some elaboration, and is explored in more depth in Chap. 27.

Suppose (W, S) is a crystallographic Coxeter group (see Definition 1.36). There is a variant $\mathfrak{h}$ of the geometric representation which is defined over $\mathbb{Z}$. We can then specialize it to any ring $\Bbbk$, with main examples being fields of positive characteristic, or p-adic integers. One can still define the categories of Bott–Samelson bimodules and Soergel bimodules as categories of bimodules over $\text{Sym}(\mathfrak{h} \otimes \Bbbk)$. For reasons to be discussed (see Chap. 27 or [91]), we are particularly interested in positive-characteristic realizations of finite and affine Weyl groups, for their applications to modular representation theory.

Remark 10.3 For non-crystallographic Coxeter groups, we can define $\mathfrak{h}$ over any extension of $\mathbb{Z}$ which contains the coefficients in the Cartan matrix. For example, $I_2(5)$ has a realization defined over $\mathbb{Z}[\phi]$, where ϕ is the golden ratio.

Let us assume temporarily that $\Bbbk$ is an infinite field of characteristic not equal to 2, and that $\mathfrak{h} \otimes \Bbbk$ is reflection faithful, see Sect. 5.6. Then the Soergel categorification theorem (Theorem 5.24) applies, so that $\mathbb{S}$Bim categorifies the Hecke algebra. We know exactly how big the Bott–Samelson bimodules are (e.g. what their graded rank is as a free right R-module), as they always categorify the same element of the Hecke algebra. Moreover, by the Soergel Hom formula, we know how big the Hom spaces are between Bott–Samelson bimodules. All this is independent of the characteristic.

However, Soergel's conjecture 5.29 may fail (it was not even conjectured in finite characteristic). To briefly explain why, remember that for $w \in W$ the indecomposable bimodule B_w is defined implicitly as the unique new summand of a Bott–Samelson bimodule. It may be the case that the idempotent projecting to the indecomposable Soergel bimodule B_w over $\mathbb{R}$ has a coefficient $\frac{1}{p}$. Then this idempotent cannot be defined in characteristic p, and several generically-indecomposable objects can "stick together" into one larger indecomposable object. In contrast to

Bott–Samelson bimodules, we do not know how big[1] the indecomposable Soergel bimodules are, and their size depends on the characteristic (as does the size of Hom spaces between Soergel bimodules).

When one presents a category by generators and relations, one always produces an A-linear *integral form* of this category, where A is the smallest extension of $\mathbb{Z}$ containing all the coefficients in the relations. One is particularly interested in *flat* presentations, where the size of the category (i.e. the graded rank of all the Hom spaces) does not change after specialization to any other A-algebra $\Bbbk$. Then one might construct an A-basis of morphism spaces for the integral form which stays a basis after any specialization, which is a useful computational tool; in contrast, given a non-flat presentation one would have to devise ad hoc methods to study each specialization. So, tractable presentations should be flat. When $A = \mathbb{Z}$, one has the ability to specialize the category to any field of finite characteristic. On the other hand, it might be the case that a presentation is so bloated that $A = \mathbb{Q}$, in which case there are no positive-characteristic specializations, and the A-linear category hardly deserves to be called an integral form at all. Such a presentation would have to have infinitely many relations, with infinitely many annoying coefficients, which is computationally problematic! Put succinctly, tractable presentations should produce flat integral forms over small base rings.

For example, the only coefficients in the relations from Chaps. 8 and 9 were the coefficients a_{st} in the Cartan matrix, and the coefficients in the negligible Jones–Wenzl projector when $m_{st} < \infty$. These generate a very small extension of $\mathbb{Z}$ (just $\mathbb{Z}$ itself, when W is crystallographic). Later in this chapter we construct Libedinsky's double leaves basis of $\mathcal{H}_{\mathrm{BS}}$, and since this construction is independent of the base ring, it proves that $\mathcal{H}_{\mathrm{BS}}$ (a model for $\mathbb{BS}\mathrm{Bim}$) is flat over this small extension of $\mathbb{Z}$. Meanwhile, any flat presentation of $\mathbb{S}\mathrm{Bim}$ must involve the numbers $\frac{1}{p}$ for all primes p where something interesting happens (i.e. an idempotent needs this coefficient), or else some specialization will change the size of the Hom spaces. For infinite Coxeter groups like affine Weyl groups, every prime should eventually appear, and one expects no tractable presentation. This explains why finding a presentation for $\mathbb{S}\mathrm{Bim}$ is a more difficult problem than for $\mathbb{BS}\mathrm{Bim}$.

Finally, we can discuss one more reason why studying a diagrammatically-presented category like $\mathcal{H}_{\mathrm{BS}}$ is better than studying $\mathbb{BS}\mathrm{Bim}$ algebraically. Suppose we consider a specialization $\mathfrak{h} \otimes \Bbbk$ which is not faithful (or not reflection faithful). Then the Soergel categorification theorem will not apply. As noted in Exercise 5.36 of Sect. 5.6, there is no faithful realization of an infinite Coxeter groups over a finite field or its algebraic closure. In particular, this is true for affine Weyl groups, of major importance in modular representation theory. When the Soergel categorification theorem does not apply, the size of the Bott–Samelson bimodules may not change, but the size of the morphism spaces between Bott–Samelson bimodules will change! It may no longer be the case that $\mathbb{S}\mathrm{Bim}$ categorifies the

---

[1] Again, "bigness" or "size" in this section will commonly refer to the graded rank of a free left or right R-module.

Hecke algebra, and the entire algebraic story is thrown out the window. However, the diagrammatic category $\mathcal{H}_{BS}$ is defined by a flat presentation over $\mathbb{Z}$, and it will continue to categorify the Hecke algebra (as we prove in Chap. 11) regardless of the specialization. If Kazhdan–Lusztig theory in characteristic p is to be studied anywhere, it will be with the diagrammatic category $\mathcal{H}_{BS}$ rather than the (inequivalent) algebraic category $\mathbb{S}$Bim.

Remark 10.4 Consider the case when $\Bbbk$ has characteristic 2, and suppose we want to prove that $B_s B_s \simeq B_s(1) \oplus B_s(-1)$. The idempotent decomposition of (8.33) is not defined, because it requires the coefficient $\frac{1}{2}$. In $\mathcal{H}_{BS}$ there is another idempotent decomposition which is defined over $\mathbb{Z}$ (see Exercise 8.42). One might not be able to transfer this idempotent to $\mathbb{BS}$Bim, because the functor $\mathcal{H}_{BS} \to \mathbb{BS}$Bim might not be defined; the formula for the image of one of the univalent vertices in (8.22) also requires the coefficient $\frac{1}{2}$. Thus, in type A_1, one can already see a difference between $\mathcal{H}_{BS}$ and $\mathbb{BS}$Bim in characteristic 2.

Remark 10.5 For crystallographic Coxeter systems (W, S) (which includes finite and affine Weyl groups), one can also study Kazhdan–Lusztig theory in characteristic p geometrically using *parity complexes* with characteristic p coefficients on the flag variety of a (complex) Kac–Moody group with Weyl group W. Moreover, geometry and diagrammatics agree: Kac–Moody groups are also defined from a kind of realization over $\mathbb{Z}$, and there is a canonical monoidal equivalence from the associated diagrammatic Hecke category to the category of parity complexes. We will say more about this in Chap. 27 (see Sect. 27.4), but let us emphasize the following right away: although parity complexes on affine flag varieties are enough to *define* the *p-canonical basis* appearing in modular representation theory, diagrammatics are still extremely important for *computing* many things related to this basis.

10.2 Generators and Relations

The goal of this section is to give a presentation (by generators and relations) of a monoidal category $\mathcal{H}_{BS}$ which models the category of Bott–Samelson bimodules associated to a Coxeter system. The notion of a diagrammatic presentation was discussed in Chap. 7. Important aspects of our presentation (Frobenius structures, interaction with the Temperley–Lieb algebra) have been discussed at length in Chaps. 8 and 9. Now we plan to consolidate our presentation in one place, so the reader doesn't have to flip between different chapters for the whole picture. Consequently, there will be some repetition of material in this chapter, for ease of reference later. When possible, we try to give a complementary set of motivations and ideas to what has been done in previous chapters.

10.2.1 A Diagrammatic Reminder

Let's give a rough description of $\mathcal{H}_{BS}$, reminding the reader of the diagrammatic technology we will use. Its objects will be generated by the simple reflections $s \in S$, with tensor product given by juxtaposition. In other words, an arbitrary object will be an expression $\underline{w} = (s_1, \ldots, s_n)$, where $s_i \in S$ for each i. The empty expression, denoted by $\varnothing$, acts as the monoidal unit of this category. The morphisms will be represented by $\mathbb{R}$-linear combinations of certain decorated planar graphs, embedded in the planar strip $\mathbb{R} \times [0, 1]$.

As an example, consider the decorated graph in Fig. 10.1, which appears in [82, §2]. Here, s, t, and u are distinct elements of S, and we have colored s red, t blue, and u green. This graph has 13 vertices, represented by crossings, dots, and bifurcation points (or arguably more, depending on how you choose to interpret the red cap on the right hand side). Note that if we think of this graph as being embedded in $\mathbb{R} \times [0, 1]$, we do *not* count the 16 endpoints lying on the boundary as vertices in the graph. To distinguish them, we call them *boundary points*: these give us the source and target of the morphism. In the graph above, the source is (s, t, s, t, t, s, u, t), and the target is (t, s, t, u, s, t, u, u). In general, we follow the convention that graphs are read bottom-to-top. Note that each edge is colored by one of the elements s, t, or u. Also note that there are two boxes in the diagram, labeled f_1 and f_2: these are meant to represent elements of a polynomial ring R. We also impose a $\mathbb{Z}$-grading on the Hom-spaces of $\mathcal{H}_{BS}$, which will be described in the next section.

By a *region*, we mean a connected component of the complement of the graph in the planar strip. For general morphisms in $\mathcal{H}_{BS}$, one allowable decoration is that elements of R can be placed in the regions of the graph. This endows the Hom-spaces in $\mathcal{H}_{BS}$ with a natural right (resp. left) R-module structure, given by inserting elements of R on the unique right-most (resp. left-most) region.

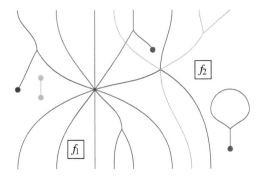

Fig. 10.1 A morphism $(s, t, s, t, t, s, u, t) \to (t, s, t, u, s, t, u, u)$ in $\mathcal{H}_{BS}$

10.2 Generators and Relations

We do not allow arbitrary graphs, but only those with certain types of vertices. For instance, the graph below *cannot* represent a morphism in $\mathcal{H}_{\mathrm{BS}}$.

The allowable vertices will be univalent vertices, trivalent vertices, and $2m_{st}$-valent vertices, which we describe in Sect. 10.2.2.

Composition of morphisms is performed by stacking. For instance, the morphism $\varnothing \to (s, s)$ represented by the diagram

$$\tag{10.1}$$

can be realized as the composition of a "dot" map $\varnothing \to (s)$ and a "trivalent vertex" map $(s) \to (s, s)$:

$$\tag{10.2}$$

In (10.1) and (10.2), we drew the boundary of the planar strip $\mathbb{R} \times [0, 1]$ as dashed lines for the convenience of the reader. We recommend that the novice diagrammatician always draw the boundary, though often we will omit it, as in (10.1).

Implicit in the statement that morphisms will be represented by graphs is the technology of biadjunction, cyclicity, and isotopy. This idea was discussed at greater length in Sect. 8.1.3. When giving a monoidal presentation, one produces certain generators, and a general morphism is obtained by stacking these generators horizontally and vertically. However, if the presentation contains caps and cups which give a biadjunction structure, and all morphisms are cyclic with respect to the biadjunction, then two diagrams which are related by an isotopy of $\mathbb{R} \times [0, 1]$ fixing the boundary will actually represent equal morphisms. For instance, we have the following equality of morphisms in $\mathcal{H}_{\mathrm{BS}}$:

$$\tag{10.3}$$

These diagrams are built by stacking certain trivalent vertices and dots which are generators of $\mathcal{H}_{\mathrm{BS}}$. Meanwhile, the symbol

$$\tag{10.4}$$

is not built from our generators, and it contains a horizontal colored strand, which does not have an intrinsic meaning in a monoidal category. Nevertheless, once isotopy invariance is established we will use (10.4) to refer to either side of (10.3), with the understanding that we can perturb the dot slightly in any direction without changing the morphism.

There are two good ways of defining $\mathcal{H}_{\mathrm{BS}}$. In the first, the isotopy is built-in: morphisms are defined to be linear combinations of isotopy classes of embedded graphs, modulo local relations. In the second, morphisms are built from generators, and isotopy invariance is a consequence of the relations. Thus (10.4) is a genuine morphism in the first description, and is automatically equal to (10.3). In the second description the two sides of (10.3) are morphisms which are only equal because of some relations, while (10.4) only represents a morphism by convention. Only the second description is a genuine presentation of $\mathcal{H}_{\mathrm{BS}}$ by generators and relations! By contrast, we call the first one an *isotopy presentation*. In Chap. 8 we defined $\mathcal{H}_{\mathrm{BS}}$ for the one-color case using a presentation, while in Chap. 9 we defined it in the two-color case using an isotopy presentation, in order that the reader might have practice with both.

Both presentations and isotopy presentations are useful. Presentations are useful when one wants to check that a functor is well-defined, because it gives a precise list of generators to define the functor on, and relations to check. At the moment, they are more useful for computerized calculations. Once one is used to diagrammatics and what it means to consider a diagram up to isotopy, it becomes entirely natural to work and compute by hand with isotopy presentations.

For the reader's benefit, we give both a presentation and an isotopy presentation of $\mathcal{H}_{\mathrm{BS}}$.

10.2.2 An Isotopy Presentation of $\mathcal{H}_{\mathrm{BS}}$

Throughout, we fix a Coxeter system (W, S). We let R be the graded polynomial ring associated to the geometric representation, see Chap. 4. Alternatively, we can choose any realization[2] of (W, S) and let R be its polynomial ring. The important

---

[2] As in Sect. 5.7, we assume the realization is balanced and satisfies Demazure surjectivity.

10.2 Generators and Relations

thing is that R contains elements α_s of degree 2, comes equipped with Demazure operators ∂_s of degree -2, and possesses a compatible action of W.

Definition 10.6 The *diagrammatic Hecke category* or *diagrammatic Bott–Samelson category* $\mathcal{H}_{\mathrm{BS}}$ is the $\mathbb{R}$-linear strict monoidal category defined by the following isotopy presentation:

- The objects are generated by S, so that an arbitrary object will be an expression $\underline{w} = (s_1, \ldots, s_n)$ in S.
- The morphism space from $\underline{w}$ to $\underline{x}$ will be a graded vector space, given by the $\mathbb{R}$-span of the S-graphs (defined next) with bottom boundary $\underline{w}$ and top boundary $\underline{x}$, modulo the local relations in (10.7) below.

Definition 10.7 An *S-graph* is an isotopy class of decorated graphs embedded in the planar strip, built from the vertices in (10.5). Each S-graph is homogeneous, with a degree given by the sums of the degrees of each vertex, and the degrees of each polynomial.

Here are the vertices and decorations allowed in an S-graph, and along with their name and degrees:

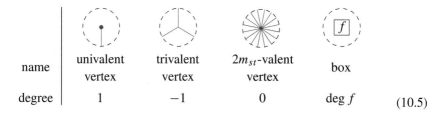

name	univalent vertex	trivalent vertex	$2m_{st}$-valent vertex	box
degree	1	-1	0	deg f

(10.5)

Here, univalent and trivalent vertices are allowed for any simple reflection $s \in S$, and $2m_{st}$-valent vertices are allowed for each pair $s \neq t \in S$ with $m_{st} < \infty$. For boxes, we assume that $f \in R$ is homogeneous.

The reader who followed Chaps. 8 and 9 will have already seen and explored all these generating vertices, and will have seen their motivation.

Example 10.8 An example of an S-graph representing a morphism $(u, s, t, s) \to (u)$ (here $m_{su} = m_{tu} = 2$ and $m_{st} = 3$):

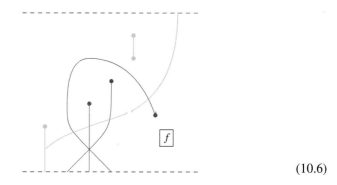

(10.6)

Remark 10.9 An s-colored strand is allowed to cross a t-colored strand if and only if $m_{st} = 2$. Of course, this crossing is really a $2m_{st}$-valent vertex.

Now we impose several relations on the span of S-graphs. The reader who followed Chaps. 8 and 9 will have already seen most of these relations, but new three-color relations appear at the end.

First, we have the "1-color relations":

Frobenius Unit

$$\left(\;\vcenter{\hbox{$\dashv$}}\;\right) = \left(\;\vcenter{\hbox{$|$}}\;\right) \tag{10.7a}$$

Frobenius Associativity

$$\left(\;\vcenter{\hbox{$\bowtie$}}\;\right) = \left(\;\vcenter{\hbox{$\bowtie$}}\;\right) \tag{10.7b}$$

Barbell Relation

$$\left(\;\vcenter{\hbox{$\mathrel{\raise1pt\hbox{$\scriptstyle\bullet$}}\!\!-\!\!\mathrel{\raise-1pt\hbox{$\scriptstyle\bullet$}}$}}\;\right) = \left(\;\boxed{\alpha_s}\;\right) \tag{10.7c}$$

Polynomial Forcing

$$\left(\;\boxed{f}\;\Big|\;\right) = \left(\;\Big|\;\boxed{sf}\;\right) + \left(\;\boxed{\partial_s f}\;\dashv\;\right) \tag{10.7d}$$

Needle Relation

$$\left(\;\vcenter{\hbox{$\bigcirc\!\!-\!\!$}}\;\right) = 0 \tag{10.7e}$$

10.2 Generators and Relations

For each rank 2 finitary subset $\{s, t\} \subset S$, we have the "2-color relations." The first collection is called "two-color associativity." We provide examples for m_{st} even and odd. From these, one can guess the general form of the relation.

Two-color Associativity

$m_{st} = 2$ (type $A_1 \times A_1$):

$$\tag{10.7f}$$

$m_{st} = 3$ (type A_2):

$$\tag{10.7g}$$

$m_{st} = 4$ (type B_2):

$$\tag{10.7h}$$

The next relation is easiest to state in terms of the Jones–Wenzl projector. This subject of Jones–Wenzl projectors and their relation to Soergel diagrammatics was discussed in the previous chapter. For more detail, the reader may consult Chap. 9 or [59, §5.2], or [47, §4.1].

Two-color dot Contraction (Aka the Jones–Wenzl Relation)

The general form is as follows:

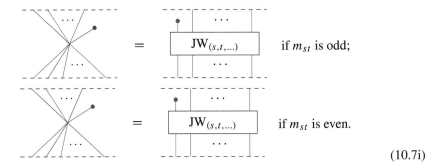

$$\text{if } m_{st} \text{ is odd;}$$

$$\text{if } m_{st} \text{ is even.} \tag{10.7i}$$

Here, $\text{JW}_{(s,t,\ldots)}$ is the idempotent arising from $\text{JW}_{m_{st}-1}$. A few specific instances of this relation follow:

$m_{st} = 2$ (type $A_1 \times A_1$):

$$\tag{10.7j}$$

$m_{st} = 3$ (type A_2):

$$\tag{10.7k}$$

$m_{st} = 4$ (type B_2).

$$\tag{10.7l}$$

Remark 10.10 In the previous chapter, the above relation was deduced from the Elias–Jones–Wenzl relation (9.27b). However we have seen in Exercise 9.39 that (assuming cyclicity and two-color associativity), the two-color dot contraction and the Elias–Jones–Wenzl relation are equivalent.

Finally, we have the "3-color relations," also called the *Zamolodchikov equations* in [59]. There is a 3-color relation for each finitary rank 3 subset of S, and the specific form of the equation depends on the type of subgroup. In each case, the

10.2 Generators and Relations

relation can be viewed as occuring in the Hom space between two different reduced expressions for the longest element in the rank 3 subgroup. By Theorem 1.34, all finite rank 3 Coxeter groups are isomorphic to one of A_3, $A_1 \times I_2(m)$ for some m, B_3, or H_3. We show the relations for A_3 and $A_1 \times I_2(m)$, and refer the reader to [59] for the remaining relations.

Type A_3:

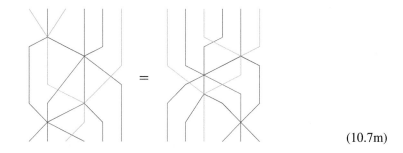

(10.7m)

Type $A_1 \times I_2(m)$:

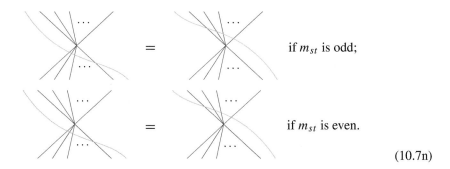

if m_{st} is odd;

if m_{st} is even.

(10.7n)

We will give some motivation for the form of the 3-color relations in Sect. 10.3.

This ends the definition of $\mathcal{H}_{BS}$. Before we continue, here are some remarks on notation and nomenclature.

Remark 10.11 The A_3 Zamolodchikov relation was also called *3-color associativity* in [53]. If you look at (10.7m) and ignore everything but the red (resp. green) strands, what is left looks like the associativity relation in one color. The 2-color associativity relation when $m_{st} = 3$ has the same feature.

Remark 10.12 We will denote the objects of $\mathcal{H}_{BS}$ by expressions. So the generating object is (s) or s. We do this in contrast to the generating object of $\mathbb{BSBim}$, which is B_s. Because the empty expression $\emptyset$ can be confusing, we may denote the monoidal identity in $\mathcal{H}_{BS}$ by $\mathbb{1}$ or (abusively) R.

Remark 10.13 The category $\mathcal{H}_{BS}$ was first defined in type A by Elias–Khovanov [53], then for dihedral groups by Elias [47], and finally for all Coxeter systems by Elias–Williamson [59]. More recently, one encounters the term *Elias–Williamson diagrammatic category* in the literature. In type A, one should say *Elias–Khovanov diagrammatic category*.

Remark 10.14 In the literature the term "diagrammatic Hecke category" can refer to several things: $\mathcal{H}_{BS}$ itself, or its additive envelope, or its Karoubi envelope, or even the homotopy category of this Karoubi envelope. To distinguish $\mathcal{H}_{BS}$ from these other choices, one might call it the "diagrammatic Bott–Samelson category."

10.2.3 Examples and Exercises

Example 10.15 Let us use the above relations to simplify a 6-valent vertex with a dot on all 6 incoming strands:

$$= \quad + \quad$$

$$= \quad + \quad = \quad \boxed{\alpha_t \alpha_s (\alpha_t + \alpha_s)}.$$

The first equality follows from the Jones–Wenzl relation, the second from the Frobenius unit, and the third from the barbell relations. Note that $\alpha_s \alpha_t (\alpha_s + \alpha_t)$ is the product of the positive roots of A_2. In fact, for a general $2m_{st}$-valent vertex, we have

$$= \quad \boxed{\mu_{st}}, \qquad (10.8)$$

where μ_{st} is the product of all positive roots in the root subsystem corresponding to s and t, see [47], [82, §2.6] and Chap. 24.

Exercise 10.16 The Zamolodchikov relations are essential, but if you put a dot on them then they already follow from the previous relations.

1. Put a green dot on the upper left strand on both sides of (10.7n). Resolve both diagrams and prove that they are equal, without using (10.7n). (Diagrams below are when m is odd.)

2. Repeat this exercise for a dot on a red or blue strand instead.
3. Now put a green dot on the upper left strand on both sides of (10.7m). Resolve both diagrams and prove that they are equal, without using (10.7m).

Exercise 10.17 Here are some exercises in multi-colored associativity and the Zamolodchikov relations. Prove the equality of the following two diagrams. Here, s (in red), t (in blue), u (in green) are simple reflections in an A_3 configuration.[3]

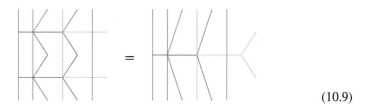

(10.9)

Now prove the equality of the following two diagrams. Here, s (in red), t (in blue), u (in green), v (in olive) are simple reflections in an A_4 configuration.[4]

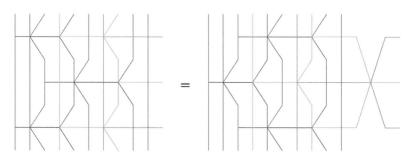

(10.10)

Sanity check for the reader: the top and bottom of both equations is a reduced expression for the longest element.

---

[3] They satisfy $m_{st} = m_{tu} = 3$ and $m_{su} = 2$.
[4] They satisfy $m_{st} = m_{tu} = m_{uv} = 3$ and $m_{su} = m_{tv} = m_{sv} = 2$.

Once we introduce light leaves at the end of this chapter, there will be many more exercises to play with.

10.2.4 A Presentation of $\mathcal{H}_{BS}$

We now define $\mathcal{H}_{BS}$ again, this time using a genuine presentation instead of an isotopy presentation.

Definition 10.18 The *diagrammatic Hecke category* $\mathcal{H}_{BS}$ is the $\mathbb{R}$-linear strict monoidal category defined by the following presentation:

- The objects are generated by S, so that an arbitrary object will be an expression $\underline{w} = (s_1, \ldots, s_n)$ in S.
- The morphism generators are listed in (10.11), and each is homogeneous of the listed degree. The relations between morphisms are discussed below.

Here are the generating morphisms with their names and degrees.

Univalent Vertices Aka dots (Degree 1)

$$: (s) \to \varnothing, \qquad : \varnothing \to (s),$$

"startdot"   "enddot"   (10.11a)

Trivalent Vertices (Degree -1)

$$: (s, s) \to (s), \qquad : (s) \to (s, s).$$

"merge"   "split"   (10.11b)

$2m_{st}$-valent Vertices (Degree 0)

$$: \underbrace{(s, t, s, t, \ldots)}_{m_{st}} \to \underbrace{(t, s, t, s, \ldots)}_{m_{st}}$$

(10.11c)

10.2 Generators and Relations

for each ordered pair $(s, t) \subset S \times S$ which generates a finite subgroup of W. Here, the colors on the edges alternate. One of the purple edges is red and the other is blue, but which is which depends on the parity of m_{st}.

Polynomials (Degree $\deg(f)$)

$$\boxed{f} \quad : \emptyset \to \emptyset,$$

"box" (10.11d)

for each homogeneous $f \in R$.

This ends the list of generators. We now define some shorthands for additional morphisms, the *cap* and *cup*, which have degree zero.

Cap

$$\cap \; := \; \curlyvee \qquad (10.12)$$

Cup

$$\cup \; := \; \curlywedge \qquad (10.13)$$

First we impose the following relations, which make up the *isotopy relations*.

Biadjunction

$$\text{(diagram)} \quad = \quad \text{(diagram)} \quad = \quad \text{(diagram)}$$
$$\text{(diagram)} \quad = \quad \text{(diagram)} \quad = \quad \text{(diagram)} \qquad (10.14a)$$

Rotation of Univalent Vertices

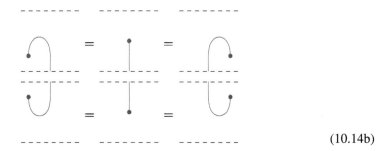

(10.14b)

Rotation of Trivalent Vertices

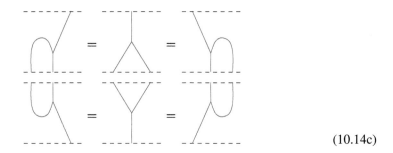

(10.14c)

Rotation of $2m_{st}$-valent Vertices

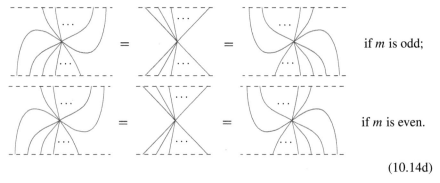

if m is odd;

if m is even.

(10.14d)

Once the isotopy relations are known, Proposition 7.18 implies that an isotopy class of diagrams unambiguously represents a morphism in $\mathcal{H}_{BS}$. Now we can freely use isotopy classes of graphs to represent morphisms, so the following sentence makes sense.

Now we impose the relations of (10.7). This ends the presentation of $\mathcal{H}_{BS}$. It is clear that this definition is equivalent to the definition given in Sect. 10.2.2.

10.2 Generators and Relations

Remark 10.19 We chose to impose first the isotopy relations, and then the Frobenius unit and associativity relations, which were phrased already using isotopy invariance. We could have instead imposed the Frobenius relations in Sect. 8.1.2, and the result would be the same.

10.2.5 The Functor to Bimodules

Theorem 10.20 *When the categories are defined using the geometric representation, there is an $\mathbb{R}$-linear monoidal functor $\mathcal{F} : \mathcal{H}_{BS} \to \mathbb{BSBim}$, defined on the generating objects by $(s) \mapsto B_s$ and on generating morphisms as in Fig. 10.2.*

It is noted in [59] that the explicit form for the map on the $2m_{st}$-valent vertex is not particularly enlightening, and is rather tedious to write out. Nevertheless, it has a nice conceptual interpretation: if $w_{s,t}$ is the longest element of $\langle s, t \rangle \subset W$ then $B_s B_t \cdots$ and $B_t B_s \cdots$ each contain a summand isomorphic to $B_{w_{s,t}}$ with multiplicity one. The map corresponding to the $2m_{st}$-valent vertex is then the composition of projection and inclusion of this common summand. The projections and inclusions are unique up to a scalar. The overall scalar is fixed so as to send c_{bot} to c_{bot} (see Chap. 12 for the definition of c_{bot}).

The content of Theorem 10.20 is that the functor is well-defined, i.e. the relations hold in $\mathbb{BSBim}$ after applying the functor. Later in this chapter we construct a basis for morphism spaces in $\mathcal{H}_{BS}$ (as left or right R-modules), and argue that $\mathcal{F}$ is faithful. Given that the Soergel Hom formula (Theorem 5.27) describes the size of Hom spaces in $\mathbb{BSBim}$, we can use this to prove that $\mathcal{F}$ is full as well. Thus we can prove that $\mathcal{F}$ is an equivalence of categories between $\mathcal{H}_{BS}$ and $\mathbb{BSBim}$. Using this,

$$\mathcal{F}\left(\boxed{f}\right) : R \to R, \qquad 1 \mapsto f.$$

$$\mathcal{F}\left(\bullet\right) : B_s \to R, \qquad f \otimes g \mapsto fg.$$

$$\mathcal{F}\left(\bullet\right) : R \to B_s, \qquad 1 \mapsto \tfrac{1}{2}(1 \otimes \alpha_s + \alpha_s \otimes 1).$$

$$\mathcal{F}\left(\bigwedge\right) : B_s B_s \to B_s, \qquad 1 \otimes g \otimes 1 \mapsto \partial_s g \otimes 1.$$

$$\mathcal{F}\left(\bigvee\right) : B_s \to B_s B_s, \qquad f \otimes g \mapsto f \otimes 1 \otimes g.$$

$$\mathcal{F}\left(\bowtie\right) : \underbrace{B_s B_t \cdots}_{m_{st}} \to \underbrace{B_t B_s \cdots}_{m_{st}}, \qquad \underbrace{B_s B_t \cdots}_{m_{st}} \twoheadrightarrow B_{w_{s,t}} \hookrightarrow \underbrace{B_t B_s \cdots}_{m_{st}}.$$

Fig. 10.2 The functor $\mathcal{F} : \mathcal{H}_{BS} \to \mathbb{BSBim}$

one can transfer the results of the Soergel categorification theorem (Theorem 5.24) to the Karoubi envelope Kar($\mathcal{H}_{BS}$). Alternatively, in Chap. 11 we prove directly that Kar($\mathcal{H}_{BS}$) satisfies its own version of the Soergel categorification theorem, without needing to use the functor $\mathcal{F}$.

Example 10.21 Let's use our relations and the functor described above to compute the map illustrated in (10.6) (recall that $m_{su} = m_{tu} = 2$ and $m_{st} = 3$). We begin by simplifying the diagram slightly, and dividing it into two horizontal parts:

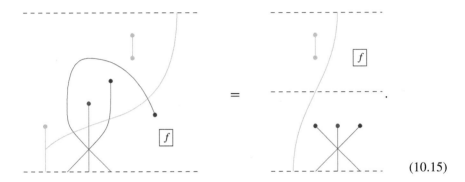

(10.15)

Here we have used the Jones–Wenzl relation (for $m_{su} = 2$ and $m_{tu} = 2$) and isotopy invariance to simplify our diagram. Using the relations from the previous section, we simplify the bottom part as follows:

The functor sends this bottom part to a map $B_u B_t B_s B_t \mapsto B_u$. From the last expression, this map sends

$$a \otimes b \otimes c \otimes d \otimes e \mapsto a \otimes bcde + a \otimes (\partial_t(cd)be\alpha_t)$$
$$= a \otimes be[\alpha_t \partial_t(cd) + cd] = a \otimes be[2(cd) - t \cdot (cd)]$$

The top part can be simplified as follows:

10.2 Generators and Relations

The functor sends this part to the map $B_u \to B_u$ sending

$$x \otimes y \mapsto (1 \otimes \alpha_u + \alpha_u \otimes 1) \cdot (xyf) - (x \otimes y) \cdot (\alpha_u f).$$

Therefore, the map represented by Diagram 10.6 is the composition

$$a \otimes b \otimes c \otimes d \otimes e$$
$$\mapsto a \otimes be\,[2(cd) - t \cdot (cd)]$$
$$\mapsto (1 \otimes \alpha_u + \alpha_u \otimes 1) \cdot (abef\,[2(cd) - t \cdot (cd)]) - a \otimes bef\alpha_u\,[2(cd) - t \cdot (cd)].$$

At this point, the masochistic reader may be disheartened by the beautiful simplicity of diagrammatics. The rest of us, however, will be impressed by its ability to keep track of what would otherwise be excruciating polynomial expressions.

10.2.6 General Realizations

Let us conclude with some remarks about general realizations.

The category of Soergel bimodules only depends on the ring R and its action of W. However, the diagrammatic Hecke category $\mathcal{H}_{\text{BS}}$ depends more finely on a choice of simple roots and Demazure operators, as they are used in the barbell and the polynomial forcing relations. The functor $\mathcal{F}$ also depends on this choice. Changing the simple roots (say, by rescaling them) will change the Frobenius structure on the ring extension $R^s \hookrightarrow R$, which changes what your dot maps and trivalent vertices represent. So the category $\mathcal{H}_{\text{BS}}$ really depends[5] on a choice of realization, as in Sect. 5.7. When we wish to emphasize the choice of W or of the realization $\mathfrak{h}$ we might write $\mathcal{H}_{\text{BS}}(\mathfrak{h}, W)$.

For an arbitrary realization, the definition of $\mathcal{H}_{\text{BS}}$ makes sense, so the first important question is whether an analogue of Theorem 10.20 holds. To define the functor on dots we need $R^s \subset R$ to be a Frobenius extension, which is equivalent to our assumption of Demazure surjectivity, see Definition 5.47. The startdot is sent to a sum over dual bases. Thus with Demazure surjectivity assumed, the functor is well-defined on the one-color calculus.

In [47] the image of the $2m_{st}$-valent vertex under $\mathcal{F}$ is defined explicitly (in some sense of the word) under two assumptions.

- Letting $R^{s,t}$ denote the invariants under s and t, $R^{s,t} \subset R$ is also a graded Frobenius extension, with Frobenius trace $\partial_{w_{s,t}}$.
- The ring R is generated by R^s and R^t, and R^t contains an element ϖ_s which is sent by ∂_s to 1.

---

[5] We do not discuss here when two versions of $\mathcal{H}_{\text{BS}}$ defined for different realizations are actually equivalent as monoidal categories.

This is explained in more detail in Chap. 24. The two-color relations are checked under these assumptions in [47], and the 3-color relations in [59].

Remark 10.22 Simon Riche kindly pointed out to us (Elias and Williamson) that we have asserted in [59] that $\mathcal{F}$ is defined in more generality, even though we only define it under the above assumptions. The functor should exist in slightly more generality, and this is the subject of [61].

The next question is what additional assumptions are needed for $\mathcal{F}$ to be an equivalence of categories. The rest of this chapter is devoted to describing a basis for morphism spaces in $\mathcal{H}_{\mathrm{BS}}$, which will work under the above assumptions. This basis has the appropriate size to obey the Soergel Hom formula. Thus $\mathcal{F}$ cannot be an equivalence of categories unless $\mathbb{BSBim}$ also satisfies the Soergel Hom formula. At the moment Soergel's categorification theorem and Hom formula are only proven (for $\mathbb{BSBim}$) for reflection faithful realizations in characteristic not equal to 2, and it is unknown whether they hold in any greater generality.

10.3 Rex Moves and the 3-color Relations

In the presentation of the Hecke algebra given in Chap. 3, we saw that the generators come from (finite) rank 1 parabolic subgroups (i.e. simple reflections), while the relations come from finite rank ≤ 2 parabolic subgroups. Now, in the categorification $\mathcal{H}_{\mathrm{BS}}$, the generating objects come from (finite) rank 1 parabolic subgroups, the generating morphisms come from finite rank ≤ 2 parabolic subgroups, and the relations come from finite rank ≤ 3 parabolic subgroups. It is certainly an interesting phenomenon that only 3-color relations are required. Why are there no 4-color relations?[6] We wish to motivate this, and to motivate the form of the 3-color relations; more details can be found in [59]. To do this we must first introduce rex moves, which will also play an important role in the definition of light leaves.

Given an element $w \in W$, its *rex graph* is a graph whose vertices are reduced expressions for w. There is an edge between two reduced expressions if they differ by a single application of a braid relation. By Matsumoto's theorem (Theorem 1.56), this graph is connected. We previously defined the rex graph in Exercise 1.58, where the reader was to compute the rex graphs for the longest element of $A_1 \times A_1 \times A_1$, $A_1 \times A_2$, and A_3, and observe that each graph had a cycle. In fact, each finite rank 3 Coxeter group has a cycle in the rex graph for the longest element, which we call a *Zamolodchikov cycle*.

It is a general result in Coxeter theory that all cycles in any rex graph can be built from boring *disjoint cycles* (which come from performing two braid relations on disjoint parts of an expression in either order; some of these can be observed in

---

[6]One might like to ponder this miracle one level down the categorical ladder: why are Coxeter groups and their Hecke algebras defined by relations which come from rank ≤ 2?

10.3 Rex Moves and the 3-color Relations

the rex graph for the longest element in type A_3) and Zamolodchikov cycles coming from rank 3 finite parabolic subgroups. This result seems to have been first noticed in the study of buildings (see e.g. [157]). This result can also be proven using the topology of the Coxeter complex, see [60] for more details.

To each vertex in the rex graph one has a corresponding object in $\mathcal{H}_{\mathrm{BS}}$. A path in the rex graph is a sequence of reduced expressions which differ by one braid relation at a time. To such a path one can associate a morphism in $\mathcal{H}_{\mathrm{BS}}$, which applies the corresponding sequence of $2m_{st}$-valent vertices. We call this morphism the *rex move* associated to the path.

Example 10.23 Suppose $W = \langle s, u, t \rangle$ with Coxeter graph

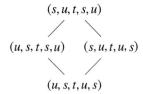

Here we give the rex graph for the element $w = sutsu$:

$$(s, u, t, s, u)$$
$$\diagup \qquad \diagdown$$
$$(u, s, t, s, u) \qquad (s, u, t, u, s)$$
$$\diagdown \qquad \diagup$$
$$(u, s, t, u, s)$$

This is an example of a disjoint cycle, emerging from the application of two disjoint braid relations. Note that $(s, u, t, u, s) \to (s, u, t, s, u) \to (u, s, t, s, u)$ and $(s, u, t, u, s) \to (u, s, t, u, s) \to (u, s, t, s, u)$ are two distinct paths in the rex graph from (s, u, t, u, s) to (u, s, t, s, u).

Exercise 10.24 Draw the rex graph of the longest element of $A_1 \times I_2(m)$. Verify that (10.7n) is an equality between two rex moves, corresponding to two paths which go halfway around the Zamolodchikov cycle.

Exercise 10.25 Draw the rex graph of the longest element of A_3. Verify that (10.7m) is an equality between two rex moves, corresponding to two paths which go halfway around the Zamolodchikov cycle.

The Zamolodchikov relations above appear to say (see the caveat below) that the two different paths between antipodal reduced expressions in the rex graph actually give rise to the same morphism in $\mathcal{H}_{\mathrm{BS}}$! The two paths around a disjoint cycle also agree in $\mathcal{H}_{\mathrm{BS}}$, a consequence of the axioms of a monoidal category. A false but motivational argument says: if one wished to prove that two paths in the rex graph give rise to the same morphism in $\mathcal{H}_{\mathrm{BS}}$, one need only check this for disjoint cycles and Zamolodchikov cycles; the former would hold by isotopy invariance, and the latter by the Zamolodchikov equations. Thus we do not need any 4-color relations to handle the complexity of the rex graph.

It is false, however, that two paths in the rex graph (with the same source and target) give rise to the same morphism in $\mathcal{H}_{BS}$. This is false even for the *edge loop* which goes back and forth along a single edge; the result is not the identity map, but is instead some idempotent (see the Elias–Jones–Wenzl relation (9.27b)). Moreover, despite appearances, the Zamolodchikov equations do *not* say that two different paths between antipodal reduced expressions in the rex graph actually give rise to the same morphism in $\mathcal{H}_{BS}$. One must choose the correct antipodal reduced expressions for this to be true, even in type A_3! See Exercise 10.27 below for an example.

Instead, it will be the case that two paths in the rex graph (with the same source and target) have rex moves which are equivalent modulo lower terms. The notion of lower terms will be explained later in this chapter, see Remark 10.30. This is guaranteed for the Zamolodchikov cycle by the Zamolodchikov relation, for disjoint cycles by monoidality, and for the edge loop by (9.27b), so it will be true for all cycles in the rex graph.

Remark 10.26 In type H_3 there is no choice of vertices such that the two rex moves around the Zamolodchikov cycle will be equal on the nose, instead of modulo lower terms. Unfortunately, the lower terms in the H_3 Zamolodchikov relation are not known, meaning that our presentation is not complete for Coxeter groups with parabolic subgroups of the form H_3.

Exercise 10.27 The relation (10.7m) is an equality of two diagrams in

$$\mathrm{Hom}((s,t,u,s,t,s),(u,t,u,s,t,u)).$$

1. Produce instead two diagrams in

$$\mathrm{Hom}((s,t,u,t,s,t),(t,u,t,s,t,u)),$$

   corresponding to the two rex moves around the Zamolodchikov cycle. (*Hint:* You can chop a 6-valent vertex off the top of (10.7m) and add it to the bottom, or vice versa.)
2. Now prove that these two diagrams are *not* equal in $\mathcal{H}_{BS}$. (*Hint:* Put a dot on some strand and resolve using the other relations, as in Exercise 10.16. You may not yet have the technology to say that a linear combination of diagrams is nonzero, but you will in the next section.)

10.4 Light Leaves and Double Leaves

In [120] Libedinsky introduced certain morphisms in $\mathbb{BSBim}$ that he called "light leaves." A diagrammatic interpretation of light leaves was given in [59]. Using light leaves, one can construct a basis for all Hom spaces in $\mathcal{H}_{BS}$, called a "double leaves basis."

10.4 Light Leaves and Double Leaves

10.4.1 Overview

Let $\underline{w}$ be an arbitrary expression, and $\underline{e} \subset \underline{w}$ be a subexpression (see Definition 3.7) with target $x = \underline{w}^{\underline{e}}$. Then a light leaf $LL_{\underline{w},\underline{e}}$ will be a particular morphism in $\mathcal{H}_{BS}$ from $\underline{w}$ to x, for some reduced expression $\underline{x}$ for x. The degree of $LL_{\underline{w},\underline{e}}$ will equal the defect of $\underline{e}$, see Definition 3.34. Here are some examples to give a quick flavor.

Example 10.28 Let $W = S_4 = \langle s, t, u \mid (st)^3 = (tu)^3 = (su)^2 = 1 \rangle$. Here are two possible light leaves associated to the expression $\underline{w} = (s, t, s, u, t, u, t)$, for the subexpressions $\underline{f} = 0101001$ and $\underline{e} = 1111111$:

$$LL_{\underline{w},\underline{f}} = \quad \ldots \quad , \quad LL_{\underline{w},\underline{e}} = \quad \ldots$$

$$U0, U1, U0, U1, U0, D0, U1 \qquad U1, U1, U1, U1, U1, D1, D1$$

(10.16)

Light leaves are defined by a recursive algorithm (in Sect. 10.4.2 below) using the decorations $\{U0, U1, D0, D1\}$ on a subexpression coming from its stroll, see Sect. 3.3.4. For example, a $U0$ will always give rise to an enddot, and a $D0$ to a merging trivalent vertex. We depict light leaves as trapezoids.

$$\boxed{LL_{\underline{w},\underline{e}}}^{\underline{x}}_{\underline{w}}$$

(10.17)

The trapezoidal shape reflects the fact that light leaves always get "narrower." The length of $\underline{x}$ is at most equal to the length of $\underline{w}$, and in fact, the length will weakly decrease at each step in the recursive construction.

For each light leaf $LL_{\underline{w},\underline{e}}$ we have a dual map $\overline{LL}_{\underline{w},\underline{e}}$ defined by flipping the light leaf map upside-down. For example, for $\underline{w}, \underline{f}$ as in the previous example,

$$\overline{LL}_{\underline{w},\underline{f}} = \quad \ldots \quad .$$

(10.18)

Now let $\underline{w}$ and $\underline{y}$ be arbitrary expressions, and let $\underline{e} \subset \underline{w}$ and $\underline{f} \subset \underline{y}$ be subexpressions such that $\underline{w}^{\underline{e}} = \underline{y}^{\underline{f}} = x$. A *double leaf map* is a composition of the form

$$LL^x_{\underline{f},\underline{e}} := \overline{LL_{\underline{y},\underline{f}}} \circ LL_{\underline{w},\underline{e}} : \underline{w} \to \underline{y}, \tag{10.19}$$

factoring through a reduced expression $\underline{x}$ for x.

$$LL^x_{\underline{f},\underline{e}} = \begin{array}{c}\overline{LL_{\underline{y},\underline{f}}} \\ LL_{\underline{w},\underline{e}}\end{array} \begin{array}{c}\underline{y} \\ \underline{x} \\ \underline{w}\end{array} \tag{10.20}$$

The goal is to construct a basis of $\mathrm{Hom}(\underline{w}, \underline{y})$ as a right R-module (or a left R-module) consisting of double leaves. The Soergel Hom formula implies (see Exercise 10.29) that a basis for $\mathrm{Hom}^\bullet(\mathrm{BS}(\underline{w}), \mathrm{BS}(\underline{y}))$ should have the same size as the set of all subexpressions $\underline{e} \subset \underline{w}$ and $\underline{f} \subset \underline{y}$ such that $\underline{w}^{\underline{e}} = \underline{y}^{\underline{f}}$. This is the indexing set for double leaves, so it is realistic to expect a basis of this form.

Exercise 10.29 Use the Deodhar defect formula (3.36), together with the false orthonormality of the standard basis proven in Lemma 3.19, to confirm the following fact. If $\underline{w}$ and $\underline{y}$ are two expressions, then

$$\left(b_{\underline{w}}, b_{\underline{y}}\right) = \sum_{\substack{\underline{e} \subset \underline{w} \\ \underline{f} \subset \underline{y} \\ \underline{w}^{\underline{e}} = \underline{y}^{\underline{f}}}} v^{\mathrm{defect}(\underline{e}) + \mathrm{defect}(\underline{f})}. \tag{10.21}$$

Now apply the Soergel Hom formula (Theorem 5.27) to deduce that (10.21) agrees with the graded rank, as a right R-module, of $\mathrm{Hom}^\bullet(\mathrm{BS}(\underline{w}), \mathrm{BS}(\underline{y}))$.

Remark 10.30 One key feature is that, for each $w \in W$, the double leaves which factor through (reduced expressions for) elements $x \in W$ satisfying $x < w$ span an ideal $(\mathcal{H}_{\mathrm{BS}})_{<w}$. We can now explain the lower terms which appeared in Sect. 10.3. When measuring the differences between rex moves in the rex graph of w, lower terms refer to morphisms in the ideal $(\mathcal{H}_{\mathrm{BS}})_{<w}$.

Remark 10.31 The hourglass shape of double leaves symbolizes the idea that "this morphism factors through some object which is smaller than the source and the target." In Chap. 11 we explore this idea further, defining the notion of an object-adapted cellular category, which has a basis consisting of hourglass morphisms.

An important subtlety is that the recursive algorithm we give below to construct a light leaf is not deterministic. Light leaves may have a number of $2m$-valent vertices, and there is often no canonical choice for which $2m$-valent vertices appear.

10.4 Light Leaves and Double Leaves

For example, adding a blue-green 6-valent vertex to the top of the first light leaf in (10.16) would give another valid choice for a light leaf $LL_{\underline{w},f}$ (whose target is a different reduced expression for $tut = \underline{w}^f$). Adding yet another blue-green 6-valent vertex on top would give yet another valid choice (whose target now agrees with the original one pictured in (10.16)).

If double leaves are to form a basis, to avoid redundancy we will need to choose a single morphism to be "the" light leaf associated to the subexpression $\underline{e} \subset \underline{w}$. In order to make our algorithm deterministic, we would need to fix the following data:

1. for each $x \in W$, a reduced expression $\underline{x}$ for x;
2. for each $s \in S$ in the right descent set of x, a reduced expression $\underline{x}_s$ for x ending in s;
3. for any two reduced expressions $\underline{x}$ and $\underline{x}'$ for x, a path $p_{\underline{x},\underline{x}'}$ from $\underline{x}$ to $\underline{x}'$ in the rex graph of x.

This will give us, for any subexpression $\underline{e} \subset \underline{w}$ expressing x, a uniquely-defined morphism $LL_{\underline{w},\underline{e}} \colon \underline{w} \to \underline{x}$. However, the notation involved in implementing this process quickly becomes oppressive. Even then, we are not clearly rewarded for our rigor. If we change any of the data above, we end up with a different basis of double leaves which is just as good. We could even choose different data for each double leaf $\mathbb{LL}^x_{\underline{f},\underline{e}}$, and still get a basis.

It seems that the way out of this conundrum is to maintain a level of flexibility for as long as possible. What we really do is provide a non-deterministic recursive algorithm which takes a subexpression $\underline{e} \subset \underline{w}$ and produces a class of morphisms from $\underline{w}$ to various reduced expressions $\underline{x}$ for x. When it comes time to choose a basis for $\mathrm{Hom}(\underline{w},\underline{y})$, we construct "the" double leaf $\mathbb{LL}^x_{\underline{f},\underline{e}}$ by choosing a single morphism from this class to be "the" light leaf $LL_{\underline{w},\underline{e}}$, and another to give "the" upside-down light leaf $\overline{LL}_{\underline{y},f}$, ensuring that these morphisms share the same reduced expression for x so that they are composable.

Theorem 10.32 *Fix expressions $\underline{w}, \underline{y}$ in (W, S). Let $\mathbb{LL}_{\underline{w},\underline{y}}$ denote a collection containing one map $\mathbb{LL}^x_{\underline{f},\underline{e}}$ for each $x \in W$ and pair $(\underline{w}, \underline{e})$, $(\underline{y}, \underline{f})$ of subexpressions with $\underline{w}^{\underline{e}} = x = \underline{y}^{\underline{f}}$. Then $\mathbb{LL}_{\underline{w},\underline{y}}$ is a basis for $\mathrm{Hom}_{\mathcal{H}_{\mathrm{BS}}}(\underline{w}, \underline{y})$ as a right (or left) R-module.*

This is Theorem 6.11 of [59]. We give some indication of its proof below.

10.4.2 The Algorithm

First, define $LL_{\varnothing,\varnothing} \colon R \to R$ to be the identity map, where $\varnothing$ denotes both the length 0 expression and its unique subexpression. Next, take a deep breath: if you wear glasses (resp. contacts), now is the time to put them on (resp. in).

The rubric for building LL inductively is this picture:

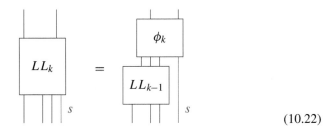

(10.22)

In this picture, LL_k is the morphism built at the k-th step of the algorithm (and similarly for LL_{k-1}), while s is the k-th simple reflection in $\underline{w}$. Defining the morphism ϕ_k will be the interesting part of the algorithm.

To make this more precise, let us introduce some notation. Let $\underline{w} = (s_1, \ldots, s_m)$ be an expression, and $\underline{e} \subset \underline{w}$ a subexpression. Following [59, §6.1], for each $k \leq m$, we define $\underline{w}_{\leq k}$ to be the expression $(s_1, \ldots, s_k)$, and $\underline{e}_{\leq k}$ to be the subexpression $(e_1, \ldots, e_k)$, expressing an element $x_k = s_1^{e_1} \cdots s_k^{e_k}$. At the k-th step, we will construct a morphism

$$LL_k \colon \underline{w}_{\leq k} \to \underline{x_k} \qquad (10.23)$$

for some reduced expression $\underline{x_k}$ for x_k. This will be done by defining a morphism

$$\phi_k \colon \underline{x_{k-1}} \otimes s_k \to \underline{x_k} \qquad (10.24)$$

which depends on e_k and its decoration in $\{U0, U1, D0, D1\}$. Assuming inductively that LL_{k-1} has been defined, we then set

$$LL_k = \phi_k \circ (LL_{k-1} \otimes \mathrm{id}_{s_k}), \qquad (10.25)$$

which is the algebraic version of (10.22).

Remark 10.33 It may be the case that $x_k = x_\ell$ for $k \neq \ell$, but we do not assume that the reduced expressions $\underline{x_k}$ and $\underline{x_\ell}$ are the same.

We now describe ϕ_k for each of the possible decorations $\{U0, U1, D0, D1\}$ for e_k. Set $s = s_k$. We use red for s in the diagrams below.

U0:

$$\phi_k = \quad \text{[diagram with } \alpha\text{]} \qquad (10.26)$$

10.4 Light Leaves and Double Leaves

The decoration $U0$ says that $x_{k-1}s > x_{k-1}$ but $e_k = 0$, so $x_k = x_{k-1}$. The interesting part of ϕ_k is an s-colored startdot. However, the reduced expressions $\underline{x}_{k-1}$ and $\underline{x}_k$ need not be the same, so we choose (arbitrarily) a rex move α which goes between them. In practice, one can choose $\underline{x}_{k-1}$ and $\underline{x}_k$ to be the same reduced expression, and α to be the identity map.

U1:

$$\phi_k = \quad \boxed{\alpha} \tag{10.27}$$

The decoration $U1$ says that $x_{k-1}s > x_{k-1}$ and $e_k = 1$, so $x_k = x_{k-1}s$. Consequently $\underline{x}_{k-1}s$ is a reduced expression for x_k, and the interesting part of ϕ_k is just id_s. Again, $\underline{x}_{k-1}s$ need not agree with $\underline{x}_k$, so we choose any rex move α which goes between them.

D0:

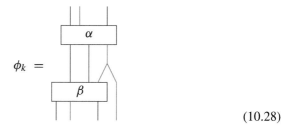

(10.28)

The decoration $D0$ says that $x_{k-1}s < x_{k-1}$ but $e_k = 0$, so that $x_k = x_{k-1}$. The interesting part of ϕ_k is an s-colored trivalent vertex, but it might not make sense initially since the reduced expression $\underline{x}_{k-1}$ might not end in s. By Corollary 1.49, x_{k-1} does have some reduced expression ending in s. We choose some such reduced expression $\underline{x}_{k-1,s}$, and some rex move β from $\underline{x}_{k-1}$ to $\underline{x}_{k-1,s}$. Then we can apply the s-colored trivalent vertex. Finally, we choose a rex move from $\underline{x}_{k-1,s}$ to our destination $\underline{x}_k$.

D1:

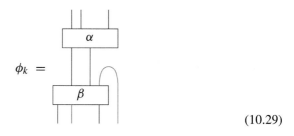

(10.29)

The decoration $D1$ says that $x_{k-1}s < x_{k-1}$ and $e_k = 1$, so that $x_k = x_{k-1}s$. The interesting part of ϕ_k is an s-colored cap, but as in the previous case, we must first apply a rex move β to bring s to the right of the reduced expression of x_{k-1}.

That concludes the recursive algorithm. We let $LL_{\underline{w},f}$ be the morphism LL_m constructed in the final step, whose target is the reduced expression $\underline{x}_m$ for $x = x_m$.

Remark 10.34 The rex moves β are a necessary part of the algorithm. The rex moves α are all "optional." By choosing reduced expressions as desired, one can always assume α is the identity map. We include them as they might be necessary to rigidify the algorithm, see the discussion after Remark 10.31. A final rex move α might be necessary in the last step, to ensure that the target of $LL_{\underline{w},e}$ is the desired reduced expression $\underline{x}$.

Exercise 10.35 Verify that the morphisms in Example 10.28 can be constructed by this algorithm. One can always choose α to be the identity map. Where does β appear?

Exercise 10.36 Describe all light leaves maps from $\underbrace{(s, s, \ldots, s)}_{m \text{ times}}$.

Exercise 10.37 The following diagram $(s, s) \to (s)$ of degree $+1$, which is a horizontal reflection of the light leaf for $01 \subset (s, s)$, is not a light leaf.

Rewrite this morphism as an R-linear combination of double leaves.

Exercise 10.38 Is the left hand side of (10.7m) a light leaf? For which subexpression?

Exercise 10.39 When $m_{st} = 3$, write down light leaves for all 8 subexpressions of (s, t, s).

Exercise 10.40 There is a unique subexpression of $(s, t, s, t, s, t, s, t, s, t)$ which expresses sts and has minimal defect. Write down a light leaf for this subexpression.

Exercise 10.41 Consider a Coxeter system of type D_4 with generators labeled as follows:

10.4 Light Leaves and Double Leaves

Consider the expression $\underline{w} = (s_1, s_3, u, s_2, s_1, s_3, u)$. Find all subexpressions expressing $w_I = s_1 s_3 u$ and draw the corresponding light leaf morphisms.

Exercise 10.42 Let W be of type A_7, with generators $s_1, s_2, \ldots, s_7$ and let $\underline{w}$ be the reduced expression

$$\underline{w} = (s_1, s_3, s_5, s_7, s_2, s_4, s_6, s_3, s_5, s_2, s_4, s_6, s_1, s_3, s_5, s_7).$$

1. What braid relations of the form $sts = tst$ can be applied to $\underline{w}$?
2. Show that

$$\underline{e} = 1111010110100000$$

is the unique subexpression expressing

$$w_I = s_1 s_3 s_4 s_3 s_5 s_4 s_3 s_7.$$

with defect zero. (Note that w_I is the longest element of the parabolic subgroup for $I = \{1, 3, 4, 5, 7\}$.)
3. Draw the corresponding light leaf.
4. Take this light leaf, and precompose it with the upside-down version of itself, to obtain a morphism $\underline{w_I} \to \underline{w} \to \underline{w_I}$. Compute this morphism, modulo terms lower than w_I. (A lengthy calculation, this is a supplementary exercise.)

10.4.3 Diagrammatics and Bimodules

Let us quickly outline a proof of Theorem 10.32 that double leaves form a basis for morphisms in $\mathcal{H}_{BS}$, assuming that we are in a setting where Soergel bimodules behave well. The first statement is that double leaves span. The proof of this in [59] was entirely diagrammatic and extremely complicated, and we will say no more.

The second statement is that double leaves are linearly independent. Linear independence can be checked after applying a functor, so we can apply the functor $\mathcal{F}$ to $\mathbb{BSBim}$ given in Theorem 10.20, and can check linear independence there. In fact, we can go one step further and apply the functor $\mathbb{BSBim} \to \mathbb{BSBim}_Q$ given by localization at the fraction field Q of R, which was the focus of Sect. 5.4. In the next section we outline the proof that light leaves are linearly independent after localization. As a consequence, we deduce that they are linearly independent in $\mathcal{H}_{BS}$, and hence form a basis of $\mathcal{H}_{BS}$. Moreover, we deduce that $\mathcal{F}$ is faithful, since it sends a basis to a linearly independent set.

Since double leaves are a basis, by Exercise 10.29, the Hom spaces in $\mathcal{H}_{\mathrm{BS}}$ are free right R-modules whose graded rank agrees with that of the Hom spaces in $\mathbb{BSBim}$. Consequently, since $\mathcal{F}$ is faithful, it is also full (see Exercise 10.44), so it is an equivalence of categories.

Corollary 10.43 *For the Kac–Moody realization (or another realization where the Soergel categorification theorem applies), the functor $\mathcal{F}$ is an equivalence of categories between $\mathcal{H}_{\mathrm{BS}}$ and $\mathbb{BSBim}$.*

Exercise 10.44 Let M and N be free graded right R-modules of the same finite graded rank. If $f: M \to N$ is an injective graded (degree 0) right R-module map, prove that f is also surjective.

Remark 10.45 The proof in [59] that double leaves are linearly independent makes use of diagrammatic localization techniques, whose rigorous treatment lies beyond the scope of this book. The advantage is that one need not use the functor $\mathcal{F}$ or any properties of bimodules, which allows it to function in additional generality. Thus the analogue of the Soergel Hom formula holds for $\mathcal{H}_{\mathrm{BS}}$. In Chap. 11 we give a proof of the rest of the Soergel categorification theorem for $\mathrm{Kar}(\mathcal{H}_{\mathrm{BS}})$ directly, rather than via an equivalence with $\mathbb{SBim}$.

10.4.4 Light Leaves and Localization

The reader is encouraged to review Sect. 5.4. Let Q denote the fraction field of R, and let $\mathbb{BSBim}_Q$ denote the localization of $\mathbb{BSBim}$ at Q. We ignore gradings after we localize. For $w \in W$, let R_w denote the standard bimodule associated to w, which is isomorphic to R as a left R-module, and whose right R-action is given by $x \cdot r = w(r)x$. Let Q_w denote the localization of R_w, i.e. $Q_w = R_w \otimes_R Q$. Recall that $Q_w \otimes_Q Q_x \simeq Q_{wx}$ and that $\mathrm{Hom}(Q_w, Q_x)$ is zero if $w \neq x$, and is isomorphic to Q if $w = x$.

When $\underline{w} = (s_1, \ldots, s_m)$ is an expression and $\underline{e} \subset \underline{w}$ is a subexpression, we write $Q_{\underline{e}}$ for the following standard bimodule:

$$Q_{\underline{e}} \simeq Q_{s_1^{e_1}} \otimes_Q Q_{s_2^{e_2}} \otimes_Q \cdots \otimes_Q Q_{s_m^{e_m}}. \tag{10.30}$$

Of course, if $\underline{w}^{\underline{e}} = x$ then $Q_{\underline{e}} \simeq Q_x$.

In Lemma 5.21 it was proven that

$$B_s \otimes_R Q \simeq Q_s \oplus Q_{\mathrm{id}}. \tag{10.31}$$

(If desired, one can fix inclusion and projection maps for this direct sum decomposition.) From (10.31) it is easy to prove the following fact about Bott–Samelson bimodules after localization.

10.4 Light Leaves and Double Leaves

Lemma 10.46 *Let $\underline{w} = (s_1, \ldots, s_m)$ be an expression. Then $\mathrm{BS}(\underline{w}) \otimes_R Q \simeq Q \otimes_{Q^{s_1}} Q \otimes_{Q^{s_2}} \cdots \otimes_{Q^{s_m}} Q$. Furthermore,*

$$\mathrm{BS}(\underline{w}) \otimes_R Q \simeq \bigoplus_{\underline{e} \subset \underline{w}} Q_{\underline{e}}. \tag{10.32}$$

Definition 10.47 For each pair $\underline{e}, \underline{f} \subset \underline{w}$ we define an element $q_{\underline{e},\underline{f}} \in Q$ as follows. Let $x = \underline{w}^{\underline{e}}$, and choose a reduced expression $\underline{x}$ for x. Then $\mathrm{B\overline{S}}(\underline{x}) \otimes_R Q$ has a unique standard summand isomorphic to Q_x, namely $Q_{\underline{x}}$. Then a light leaf $LL_{\underline{w},\underline{e}}$ (after applying $\mathcal{F}$ and localization) induces a map $\tilde{q}_{\underline{e},\underline{f}}$ from the summand $Q_{\underline{f}}$ of $\mathrm{BS}(\underline{w})$ to $Q_{\underline{x}}$. When $\underline{w}^{\underline{f}} \neq x$ then $\mathrm{Hom}(Q_{\underline{f}}, Q_{\underline{x}}) = 0$ and we set $q_{\underline{e},\underline{f}} = 0$. When $\underline{w}^{\underline{f}} = x$ then $\tilde{q}_{\underline{e},\underline{f}}$ is some map $Q_x \to Q_x$, so it is multiplication by an element $q_{\underline{e},\underline{f}} \in Q$.

Remark 10.48 Note that, a priori, $\tilde{q}_{\underline{e},\underline{f}}$ depends on the choice of light leaf $LL_{\underline{w},\underline{e}}$, although one can prove that it does not actually depend on this choice (two different light leaves for the same subexpression will differ by lower terms, which are orthogonal to Q_x).

Recall the path dominance order on subexpressions of $\underline{w}$ defined in Exercise 3.41.

Theorem 10.49 (Path Dominance Upper-Triangularity) *If $q_{\underline{e},\underline{f}} \neq 0$, then $\underline{f} \leq \underline{e}$ in the path dominance order. Moreover, $q_{\underline{e},\underline{e}} \neq 0$, and is a product of simple roots independent of the light leaf.*

Proof See [59, Proposition 6.6] or Exercise 10.52. □

Corollary 10.50 *Fix $x \leq w$. For each $\underline{e} \subset \underline{w}$ expressing x choose a light leaf $LL_{\underline{w},\underline{e}}$, and let $E(\underline{w}, x)$ be the set of these chosen light leaves. Let p be the projection map from $\mathrm{BS}(\underline{x}) \otimes_R Q$ to $Q_x \simeq Q_{\underline{x}}$. Then the set of maps*

$$\{p \circ LL \mid LL \in E(\underline{w}, x)\} \tag{10.33}$$

forms a basis for $\mathrm{Hom}(\mathrm{BS}(\underline{w}) \otimes_R Q, Q_x)$ in the category of Q-bimodules.

Proof Left as an exercise. □

Corollary 10.51 *Let $\underline{w}$ be an expression in (W, S). Then $\{LL_{\underline{w},\underline{e}} \mid \underline{w}^{\underline{e}} = \mathrm{id}\}$ is R-linearly independent in $\mathrm{Hom}_{\mathcal{H}_{\mathrm{BS}}}(\underline{w}, \varnothing)$.*

Exercise 10.52 If $\underline{f} \not\leq \underline{e}$ in the path dominance order, prove that $q_{\underline{e},\underline{f}} = 0$. (*Hint:* Consider how the map LL_k, constructed in the k-th step of the light leaf algorithm for $LL_{\underline{w},\underline{e}}$, acts upon the summand $Q_{\underline{f}_{\leq k}}$ of $\mathrm{BS}(\underline{w}_{\leq k})$.)

Exercise 10.53 Using Theorem 10.49, prove Corollary 10.50.

Chapter 11
The Soergel Categorification Theorem

This chapter is based on expanded notes of a lecture given by the authors and taken by

Ulrich Thiel

Abstract From the diagrammatic Bott–Samelson category $\mathcal{H}_{\mathrm{BS}}$ defined in Chap. 10 we can obtain the diagrammatic Hecke category $\mathcal{H}$ by a formal analogue of the process that turned $\mathbb{BSBim}$ into $\mathbb{SBim}$: the Karoubi envelope. We have decided to include a comprehensive discussion of Karoubi envelopes and the Krull–Schmidt property in this chapter.

The Soergel categorification theorem—due in this context to Elias–Williamson—states that $\mathcal{H}$ is a categorification of the Hecke algebra. The central ingredient in the proof is the classification of the indecomposable objects. Elias and Williamson noticed that this is a consequence of formal properties of Libedinsky's light leaves, namely the fact that they form a cellular basis of the morphism spaces. This argument was generalized to certain kinds of cellular categories by Elias and Lauda, and we explain that generalization here.

11.1 Introduction

In Chap. 10 we defined the diagrammatic Bott–Samelson category $\mathcal{H}_{\mathrm{BS}}$ for any Coxeter system (W, S), starting with the geometric representation or with any[1] realization. Let us recall a few basic properties of this category. First, it is a monoidal category whose objects are monoidally generated by the set S. An arbitrary object is thus a tensor product of elements of S, which can be viewed as an expression $\underline{w} = (s_1, \ldots, s_n)$. The morphism spaces are R-bimodules generated by certain local diagrams modulo certain local relations, where $R = \mathbb{R}[\mathfrak{h}] = \mathrm{Sym}(\mathfrak{h}^*)$ is the polynomial ring of the W-representation $\mathfrak{h}$. Each morphism space is naturally

---

[1] To avoid making annoying clarifications, we continue to assume that the realization is balanced and satisfies Demazure surjectivity.

U. Thiel
Department of Mathematics, University of Kaiserslautern, Kaiserslautern, Germany

© The Editor(s) (if applicable) and The Author(s), under exclusive licence to Springer Nature Switzerland AG 2020
B. Elias et al., *Introduction to Soergel Bimodules*, RSME Springer Series 5, https://doi.org/10.1007/978-3-030-48826-0_11

a $\mathbb{Z}$-graded R-bimodule with respect to the degree 2 grading on R. The grading is compatible with composition and with the monoidal structure. Summarizing, $\mathcal{H}_{BS}$ is a graded R-bilinear monoidal category. The theory of Libedinsky's light leaves shows that each Hom-space of $\mathcal{H}_{BS}$ is free of finite rank as a right R-module, whose graded rank (via Soergel's Hom formula) is compatible with Deodhar's defect formula.

In Sect. 10.2.5 we defined a functor $\mathcal{H}_{BS} \to \mathbb{BS}\text{Bim}$ and in Corollary 10.43 we have seen that this is an equivalence of monoidal categories, under certain assumptions on the realization. We should thus be able to get from $\mathcal{H}_{BS}$ a diagrammatic variant $\mathcal{H}$ of the category $\mathbb{S}\text{Bim}$ of Soergel bimodules by running it through the same process that turned $\mathbb{BS}\text{Bim}$ into $\mathbb{S}\text{Bim}$: we take direct summands of finite direct sums of grading shifts of objects. We refer to this process as taking the *(graded) Karoubi envelope*, and use the symbol Kar, so that $\mathcal{H} = \text{Kar}(\mathcal{H}_{BS})$. A priori, it makes sense to talk about a direct summand of a Bott–Samelson bimodule, but not a direct summand of an object in $\mathcal{H}_{BS}$ since there is no enclosing additive context (e.g. the category of R-bimodules). However, we already know how to deal with direct summands morphism-theoretically (see Sect. 8.2.4), and there is a straightforward formal analogue of a direct summand. We review the Karoubi envelope in Sect. 11.2. What we get is an additive monoidal $\mathbb{R}$-bilinear category $\mathcal{H}$ equipped with a shift functor—the *diagrammatic Hecke category*.

The first crucial property of $\mathcal{H}$ is that it is a Krull–Schmidt category, which we prove more generally for Karoubi envelopes in Theorem 11.26. This means that every object admits a unique decomposition into indecomposable objects (with local endomorphism rings). One consequence is that its split Grothendieck group $[\mathcal{H}]_\oplus$ is a free $\mathbb{Z}[v, v^{-1}]$-module, with v acting by grading shift. Another is that to parameterize isomorphism classes of indecomposable objects it is sufficient to do this only up to shift. For convenience, we have added a detailed review of Krull–Schmidt categories in Appendix 1.

After having established this basic fact, the main theorem we want to prove is:

Theorem 11.1 (Soergel Categorification Theorem)

1. *For each reduced expression $\underline{w}$ the object $\underline{w} \in \mathcal{H}$ has a unique indecomposable direct summand $B_{\underline{w}}$ which does not occur as a direct summand in any shorter expression.*
2. *Let $w \in W$. If $\underline{w}$ and $\underline{w}'$ are reduced expressions for w, then $B_{\underline{w}}$ and $B_{\underline{w}'}$ are isomorphic.[2] We denote the isomorphism class of $B_{\underline{w}}$ by B_w.*
3. *Up to shift, any indecomposable object of $\mathcal{H}$ is isomorphic to some B_w.*
4. *The map $b_s \mapsto [s]$ for $s \in S$ induces a $\mathbb{Z}[v^{\pm 1}]$-algebra isomorphism*

$$c: \text{H} \to [\mathcal{H}]_\oplus. \qquad (11.1)$$

---

[2]They are isomorphic via any choice of rex move. There is a canonical isomorphism if Soergel's conjecture holds.

11.1 Introduction

5. *(Soergel Hom formula) For any two objects X, Y of $\mathcal{H}$, let $x, y \in H$ be the elements for which $c(x) = [X]$ and $c(y) = [Y]$. Then the graded Hom space $\mathrm{Hom}^\bullet_{\mathcal{H}}(X, Y)$ is free as a left graded R-module with graded rank (x, y).*

Remark 11.2 We have already stated a nearly identical version of the Soergel categorification theorem in Sect. 5.5, but that theorem was about the category $\mathbb{S}\mathrm{Bim}$ of R-bimodules, rather than the diagrammatically-defined category $\mathcal{H}$. Moreover, the theorem about $\mathbb{S}\mathrm{Bim}$ required $\Bbbk$ to be an infinite field of characteristic not equal to 2, and required $\mathfrak{h}$ to be a reflection faithful representation, while this theorem only requires the assumptions of Demazure surjectivity and balancedness of the realization (and balancedness can be weakened).

Remark 11.3 Recall that we denote the generating objects of $\mathcal{H}_{\mathrm{BS}}$ by s, while the generating bimodules of $\mathbb{BSBim}$ were denoted B_s. Despite this difference, following the literature, we continue to denote the indecomposable objects in $\mathcal{H}$ as B_w, $w \in W$, just as they were denoted in $\mathbb{S}\mathrm{Bim}$ (and as a consequence, the object s can also be denoted B_s). This abuse of notation is not terrible, as we will only be interested in $\mathbb{S}\mathrm{Bim}$ when it is equivalent to $\mathcal{H}$, and this equivalence matches the isomorphism classes named B_w.

The aim of this chapter is to argue that Theorem 11.1 is actually a consequence of formal properties of Libedinsky's light leaves, namely the fact that they form a "cellular basis" of the morphisms spaces of $\mathcal{H}_{\mathrm{BS}}$. What is so special about this basis? Recall from Chap. 10 that a basis element looks like this:

$$\mathbb{LL}^w_{\underline{e},\underline{f}} = \overline{LL_{\underline{x},\underline{e}}} \circ LL_{\underline{y},\underline{f}} = \begin{array}{c}\overline{LL_{\underline{x},\underline{e}}} \\ LL_{\underline{y},\underline{f}}\end{array} \begin{array}{c} \underline{x} \\ w \\ \underline{y}\end{array} \quad (11.2)$$

It has an "hourglass shape," meaning that it is composed of two morphisms, factoring in the middle through a reduced expression for w. Moreover, there is a "flipping operation" on the diagrams which gives an involution on the category:

$$\begin{array}{c} LL_{\underline{x},\underline{e}} \end{array}\begin{array}{c}w \\ \underline{x}\end{array} \quad \rightsquigarrow \quad \begin{array}{c} \overline{LL_{\underline{x},\underline{e}}} \end{array}\begin{array}{c}\underline{x} \\ w\end{array} \quad (11.3)$$

These two properties (factoring and flip) are the characteristic features of a so-called "object-adapted cellular category"—a special kind of a cellular category, which in turn is the "multi-object" analogue of a cellular algebra. We give the precise definition in Sect. 11.3. Another key example is the Temperley–Lieb category $\mathcal{TL}$, see Exercise 11.31.

In an object-adapted cellular category, there is a set Λ of objects through which a basis of morphisms will factor. For $\mathcal{H}_{\mathrm{BS}}$ this set Λ is in bijection with the elements of the Coxeter group W. In any such category $\mathcal{C}$ (satisfying some additional assumptions) it is possible to show that the indecomposable objects of Kar($\mathcal{C}$) are enumerated by Λ, in a way resembling the Soergel categorification theorem. This is proven in Theorem 11.39, which is the main result of this chapter. This approach is due to Elias–Williamson [59] and Elias–Lauda [54].

11.2 Prelude: From $\mathcal{H}_{BS}$ to $\mathcal{H}$

We start by reviewing the formal process that turns $\mathcal{H}_{BS}$ into $\mathcal{H}$. This involves converting a graded category to a category with shift functor (Sect. 11.2.1), then taking its additive closure (Sect. 11.2.2) and finally taking its Karoubian closure (Sect. 11.2.3). In Sect. 11.2.4 we show that Karoubi envelopes (typically) satisfy the Krull–Schmidt property. In Sect. 11.2.5 we discuss the interaction between diagrammatics and the Karoubi envelope. The reader is welcome to skip any parts that they are familiar with.

11.2.1 Graded Categories

In the literature, two distinct concepts go under the name "graded category": one involves gradings on the Hom-spaces and the other involves shift functors. After reviewing the definitions, we will show that both concepts are related via an adjunction of 2-categories which allows us to convert between the two.

Definition 11.4 A *graded category* is a category $\mathcal{C}$ which is enriched over the category of graded abelian groups, i.e. $\text{Hom}_\mathcal{C}(X, Y)$ is a $\mathbb{Z}$-graded abelian group for all $X, Y \in \mathcal{C}$ and the compositions are graded bilinear maps. So

$$\text{Hom}^i_\mathcal{C}(Y, Z) \circ \text{Hom}^j_\mathcal{C}(X, Y) \subseteq \text{Hom}^{i+j}_\mathcal{C}(X, Z) , \tag{11.4}$$

where the superscripts denote the homogeneous components. A functor between such categories is a pre-additive functor preserving degrees of morphisms.

Example 11.5 $\mathcal{H}_{BS}$ is a graded category, as is $\mathbb{BSBim}$.

Graded categories naturally arise from the following competing concept.

Definition 11.6 A *shift functor* (also called *translation functor*) on a pre-additive category $\mathcal{C}$ is an additive automorphism $(1) \colon \mathcal{C} \to \mathcal{C}$. A functor between categories with shift is a pre-additive functor commuting with the shift functors.

Example 11.7 Let R be a graded commutative ring. In Sect. 4.5 we have introduced the category R-gbim of graded R-bimodules (finitely generated on both sides), where the morphisms are graded R-bimodule morphisms of degree zero. This is a category with shift functor (1) as defined in Sect. 4.1.

11.2 Prelude: From $\mathcal{H}_{BS}$ to $\mathcal{H}$

Thus a graded category has graded morphism spaces, but no built-in shift functor (note that $\mathbb{BSBim}$ is not closed under grading shifts). Meanwhile, a category with a shift functor has ungraded morphism spaces (think: just the degree zero part).

A category $\mathcal{C}$ with shift functor has an *associated graded category* $\mathcal{C}^{gr}$ as follows. For $n \in \mathbb{Z}$ we denote by $(n) := (1)^n$ the n-fold composition of the functor (1), where $(-1) = (1)^{-1}$. For $X, Y \in \mathcal{C}$ we define

$$\mathrm{Hom}_{\mathcal{C}}^n(X, Y) := \mathrm{Hom}_{\mathcal{C}}(X, Y(n)) \tag{11.5}$$

and

$$\mathrm{Hom}_{\mathcal{C}}^{\bullet}(X, Y) := \bigoplus_{n \in \mathbb{Z}} \mathrm{Hom}_{\mathcal{C}}^n(X, Y). \tag{11.6}$$

If $f \in \mathrm{Hom}_{\mathcal{C}}^n(X, Y)$, we say f is of *degree n*. We call $\mathrm{Hom}_{\mathcal{C}}^{\bullet}(X, Y)$ the space of *graded morphisms*. We denote by $\mathcal{C}^{gr}$ the category with the same objects as $\mathcal{C}$ but with morphisms

$$\mathrm{Hom}_{\mathcal{C}^{gr}}(X, Y) := \mathrm{Hom}_{\mathcal{C}}^{\bullet}(X, Y). \tag{11.7}$$

It is clear that $\mathcal{C}^{gr}$ is a graded category. By definition we have

$$\mathrm{Hom}_{\mathcal{C}^{gr}}(X, Y(n)) = \mathrm{Hom}_{\mathcal{C}^{gr}}(X, Y)(n) \tag{11.8}$$

as graded abelian groups for all $n \in \mathbb{Z}$. Here, the grading shift on the right hand side is the grading shift on graded abelian groups as introduced in Sect. 4.1. Moreover, the shift functor (1) on $\mathcal{C}$ induces a shift functor on $\mathcal{C}^{gr}$ which is also a graded functor.

Example 11.8 Let R be a graded commutative ring. The graded category R-gbim^{gr} associated to R-gbim has as objects graded R-bimodules and as morphisms graded R-bimodule morphisms of *arbitrary* degree.

The construction $(-)^{gr}$ is in fact a 2-functor from the 2-category of categories with shift to the 2-category of graded categories. Namely, if $F: \mathcal{C} \to \mathcal{D}$ is a functor of categories with shift, define $F^{gr}: \mathcal{C}^{gr} \to \mathcal{D}^{gr}$ as follows. On objects X, we define $F^{gr}(X) := F(X)$. A morphism $f \in \mathrm{Hom}_{\mathcal{C}^{gr}}(X, Y) = \bigoplus_{n \in \mathbb{Z}} \mathrm{Hom}_{\mathcal{C}}(X, Y(n))$ is a family $f = (f_n)_{n \in \mathbb{Z}}$ of morphisms $f_n: X \to Y(n)$. We have

$$F(f_n) \in \mathrm{Hom}_{\mathcal{D}}(F(X), F(Y(n))) = \mathrm{Hom}_{\mathcal{D}}(F(X), F(Y)(n)) = \mathrm{Hom}_{\mathcal{D}^{gr}}^n(F(X), F(Y)),$$

where we used the fact that F commutes with the shift functors. Now, define

$$F^{gr}(f) := F(f) = (F(f_n))_{n \in \mathbb{Z}} \in \mathrm{Hom}_{\mathcal{D}^{gr}}(F^{gr}(X), F^{gr}(Y)), \tag{11.9}$$

using the fact that F is pre-additive.

Conversely, every graded category $\mathcal{C}$ has an *associated category* $\mathcal{C}^{sh}$ *with shift* as follows. We define $\mathcal{C}^{sh}$ to be the category with formal objects $X(n)$ for $X \in \mathcal{C}$ and $n \in \mathbb{Z}$, and with morphisms

$$\mathrm{Hom}_{\mathcal{C}^{sh}}(X(n), Y(m)) = \mathrm{Hom}_{\mathcal{C}}^{m-n}(X, Y) . \tag{11.10}$$

We have a shift functor (1) on $\mathcal{C}^{sh}$ sending $X(n)$ to $X(n+1)$ and being the identity on morphisms. The construction $(-)^{sh}$ is in fact a 2-functor from the 2-category of graded categories to the 2-category of categories with shift. Namely, if $F \colon \mathcal{C} \to \mathcal{D}$ is a functor of graded categories, we define $F^{sh} \colon \mathcal{C}^{sh} \to \mathcal{D}^{sh}$ by $F^{sh}(X(n)) := F(X)(n)$ on objects and if $f \in \mathrm{Hom}_{\mathcal{C}^{sh}}(X(n), Y(m)) = \mathrm{Hom}_{\mathcal{C}}^{m-n}(X, Y)$, we define

$$\begin{aligned} F^{sh}(f) &:= F(f) \in \mathrm{Hom}_{\mathcal{D}}^{m-n}(F(X), F(Y)) = \mathrm{Hom}_{\mathcal{D}^{sh}}(F(X)(n), F(Y)(m)) \\ &= \mathrm{Hom}_{\mathcal{D}^{sh}}(F(X(n)), F(Y)(m)) , \end{aligned} \tag{11.11}$$

where we used the fact that F is graded.

We now have a pair of 2-functors,

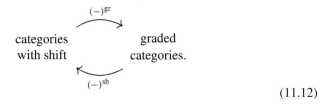

$$\tag{11.12}$$

Proposition 11.9 *The 2-functor $(-)^{sh}$ is left adjoint to the 2-functor $(-)^{gr}$.*

Proof We need to show that for a graded category $\mathcal{C}$ and a category $\mathcal{D}$ with shift we have a functorial isomorphism

$$\mathrm{Fun}(\mathcal{C}^{sh}, \mathcal{D}) \simeq \mathrm{Fun}(\mathcal{C}, \mathcal{D}^{gr})$$

of functor categories (the left hand side being functors of categories with shift, the right hand side being functors of graded categories). Let $F \colon \mathcal{C}^{sh} \to \mathcal{D}$ be a functor of categories with shift. Define the functor $F^{\flat} \colon \mathcal{C} \to \mathcal{D}^{gr}$ as follows. On objects we set $F^{\flat}(X) := F(X(0))$ and on morphisms we define $F^{\flat}(f) := F(f)$. To see that the latter is well-defined, note that if

$$f = (f_n)_{n \in \mathbb{Z}} \in \mathrm{Hom}_{\mathcal{C}}(X, Y) = \bigoplus_{n \in \mathbb{Z}} \mathrm{Hom}_{\mathcal{C}}^n(X, Y) = \bigoplus_{n \in \mathbb{Z}} \mathrm{Hom}_{\mathcal{C}^{sh}}(X(0), Y(n)) ,$$

then

$$\begin{aligned} F(f_n) &\in \mathrm{Hom}_{\mathcal{D}}(F(X(0)), F(Y(n))) = \mathrm{Hom}_{\mathcal{D}}(F(X)(0), F(Y)(n)) \\ &= \mathrm{Hom}_{\mathcal{D}^{gr}}^n(F(X)(0), F(Y)(0)) = \mathrm{Hom}_{\mathcal{D}^{gr}}^n(F^{\flat}(X), F^{\flat}(Y)) , \end{aligned}$$

11.2 Prelude: From $\mathcal{H}_{BS}$ to $\mathcal{H}$

using the fact that F commutes with the shift functors, so

$$F^\flat(f) = F(f) = (F(f_n))_{n\in\mathbb{Z}} \in \mathrm{Hom}_{\mathcal{D}^{\mathrm{gr}}}(F^\flat(X), F^\flat(Y)),$$

using the fact that F is pre-additive. It is clear that $F^\flat$ is a functor of graded categories. Conversely, let $F \colon \mathcal{C} \to \mathcal{D}^{\mathrm{gr}}$ be a functor of graded categories. We define the functor $F^\sharp \colon \mathcal{C}^{\mathrm{sh}} \to \mathcal{D}$ as follows. On objects we set $F^\sharp(X(n)) := F(X)(n)$ and on morphisms we define $F^\sharp(f) := F(f)$. To see that the latter is well-defined, note that if

$$f \in \mathrm{Hom}_{\mathcal{C}^{\mathrm{sh}}}(X(n), Y(m)) = \mathrm{Hom}_{\mathcal{C}}^{m-n}(X, Y),$$

then

$$F^\sharp(f) = F(f) \in \mathrm{Hom}_{\mathcal{D}^{\mathrm{gr}}}^{m-n}(F(X), F(Y)) = \mathrm{Hom}_{\mathcal{D}}(F(X)(n), F(Y)(m))$$
$$= \mathrm{Hom}_{\mathcal{D}}(F^\sharp(X(n)), F^\sharp(Y(m))),$$

using (11.8). It is clear that $F^\sharp$ is a functor of categories with shift. From the constructions it follows immediately that $(-)^\flat$ and $(-)^\sharp$ are mutually inverse. □

Lemma 11.10 *Suppose that $\mathcal{C}$ is a full subcategory of $\mathcal{D}^{\mathrm{gr}}$ for a category $\mathcal{D}$ with shift. Then $\mathcal{C}^{\mathrm{sh}}$ is equivalent to the full subcategory of $\mathcal{D}$ consisting of the shifts of objects of $\mathcal{C}$.*

Proof Let [1] be the shift functor on $\mathcal{D}$ and let $\mathcal{C}'$ be the full subcategory of $\mathcal{D}$ consisting of the shifts of objects of $\mathcal{C}$. We have

$$\mathrm{Hom}_{\mathcal{C}^{\mathrm{sh}}}(X(n), Y(m)) = \mathrm{Hom}_{\mathcal{C}}^{m-n}(X, Y) = \mathrm{Hom}_{\mathcal{D}^{\mathrm{gr}}}^{m-n}(X, Y)$$
$$= \mathrm{Hom}_{\mathcal{D}}(X[n], Y[m]) = \mathrm{Hom}_{\mathcal{C}'}(X[n], Y[m]).$$

Hence, the functor $\mathcal{C}^{\mathrm{sh}} \to \mathcal{C}'$ mapping $X(n)$ to $X[n]$ is fully faithful. Clearly, it is essentially surjective, thus an equivalence. □

Example 11.11 The category $\mathbb{BS}\mathrm{Bim}$ of Bott–Samelson bimodules is a full subcategory of $R\text{-gbim}^{\mathrm{gr}}$, see Definition 4.33. Hence, $\mathbb{BS}\mathrm{Bim}^{\mathrm{sh}}$ is the category of shifts of Bott–Samelson bimodules and with morphisms of degree zero (i.e. morphisms in $R\text{-gbim}$).

Exercise 11.12 Show that if $\mathcal{C}$ is *graded monoidal* (spell out what this means) with tensor product $\otimes$, then $\mathcal{C}^{\mathrm{sh}}$ is monoidal with tensor product defined by

$$X(n) \otimes Y(m) := (X \otimes Y)(n + m) \tag{11.13}$$

and the same tensor product on morphisms.

11.2.2 Additive Closure

Let $\mathcal{C}$ be a pre-additive category, i.e. enriched over the category of abelian groups. There is a straightforward minimal way to embed $\mathcal{C}$ into an additive category: the *additive closure* $\mathcal{C}^\oplus$ of $\mathcal{C}$ is the category whose objects are formal finite direct sums $\bigoplus_{i \in I} X_i$ of objects X_i of $\mathcal{C}$, and whose morphisms are given by

$$\mathrm{Hom}_{\mathcal{C}^\oplus}(\bigoplus_{i \in I} X_i, \bigoplus_{j \in J} Y_j) := \bigoplus_{i \in I, j \in J} \mathrm{Hom}_{\mathcal{C}}(X_i, Y_j) . \tag{11.14}$$

Said another way, the morphisms in $\mathcal{C}^\oplus$ are matrices of morphisms in $\mathcal{C}$. Clearly, $\mathcal{C}$ embeds into $\mathcal{C}^\oplus$ and $\mathcal{C}^\oplus$ is an additive category.

The construction $(-)^\oplus$ is in fact a 2-functor from the 2-category of pre-additive categories to the 2-category of additive categories. Namely, if $F \colon \mathcal{C} \to \mathcal{D}$ is a functor of pre-additive categories, then we define $F^\oplus \colon \mathcal{C}^\oplus \to \mathcal{D}^\oplus$ to be the functor with $F^\oplus(\bigoplus_{i \in I} X_i) := \bigoplus_{i \in I} F(X_i)$ on objects and $F^\oplus(f) := F(f)$ on morphisms. The following lemma is immediate.

Lemma 11.13 *The additive closure of $\mathcal{C}$ satisfies the following universal property: for every pre-additive functor $F \colon \mathcal{C} \to \mathcal{D}$ into an additive category $\mathcal{D}$ there is a unique additive functor $F^\oplus \colon \mathcal{C}^\oplus \to \mathcal{D}$ such that*

$$\begin{array}{ccc} \mathcal{C} & \xrightarrow{F} & \mathcal{D} \\ \downarrow & \nearrow_{F^\oplus} & \\ \mathcal{C}^\oplus & & \end{array}$$

commutes. In other words, the 2-functor $(-)^\oplus$ is left adjoint to the forgetful 2-functor from additive categories to pre-additive categories.

Lemma 11.14 *If $\mathcal{C}$ is a full subcategory of an additive category $\mathcal{D}$, then $\mathcal{C}^\oplus$ is equivalent to the full subcategory of $\mathcal{D}$ consisting of finite direct sums of objects of $\mathcal{C}$.*

Proof Let $\mathcal{C}'$ be the full subcategory of $\mathcal{D}$ consisting of finite direct sums of objects of $\mathcal{C}$. For clarity, we denote an object of $\mathcal{C}^\oplus$ by $(X_i)_{i \in I}$. We can define a functor $\mathcal{C}^\oplus \to \mathcal{C}'$ mapping $(X_i)_{i \in I}$ to $\bigoplus_{i \in I} X_i$, the direct sum taken in $\mathcal{D}$. We have

$$\mathrm{Hom}_{\mathcal{C}^\oplus}((X_i)_{i \in I}, (Y_j)_{j \in J}) = \bigoplus_{i \in I, j \in J} \mathrm{Hom}_{\mathcal{D}}(X_i, Y_j) = \mathrm{Hom}_{\mathcal{C}'}(\bigoplus_{i \in I} X_i, \bigoplus_{j \in J} Y_j) ,$$

so the functor is fully faithful. It is moreover essentially surjective, thus an equivalence. □

Example 11.15 The additive closure $\mathbb{BSBim}^{\mathrm{sh}, \oplus}$ of $\mathbb{BSBim}^{\mathrm{sh}}$ is the full subcategory of R-gbim consisting of direct sums of shifts of Bott–Samelson bimodules.

Exercise 11.16 Show that if $\mathcal{C}$ is R-linear (resp. monoidal, resp. a category with shift functor), then so is $\mathcal{C}^\oplus$.

11.2.3 Karoubian Closure

Let $\mathcal{C}$ be an additive category. We call $\mathcal{C}$ *Karoubian* if every idempotent splits. That is, if X is an object and $e \in \mathrm{End}(X)$ is an idempotent endomorphism, then there are objects Y and Z such that $X \simeq Y \oplus Z$ and e is projection to Y. Since every direct summand should be the image of an idempotent, being Karoubian is a more precise way to say that a category should contain all direct summands of objects.

There is a straightforward way to embed $\mathcal{C}$ into a Karoubian category. The *Karoubian closure* $\mathrm{Kar}\,\mathcal{C}$ of $\mathcal{C}$ is the category consisting of pairs (X, e), where $X \in \mathcal{C}$ is an object and $e \in \mathrm{End}(X)$ is an idempotent. One should think of (X, e) as the "image of e." Morphisms $(X, e) \to (Y, f)$ are morphisms $\phi\colon X \to Y$ with $f\phi e = \phi$ (equivalently, $f\phi = \phi = \phi e$). Equivalently, every morphism $(X, e) \to (Y, f)$ has the form $f\psi e$ for some (not necessarily unique) morphism $\psi\colon X \to Y$:

$$\mathrm{Hom}_{\mathrm{Kar}(\mathcal{C})}((X, e), (Y, f)) := \{\phi \in \mathrm{Hom}_\mathcal{C}(X, Y) \mid f\phi e = \phi\} \tag{11.15}$$

$$= \{f\psi e \in \mathrm{Hom}_\mathcal{C}(X, Y) \mid \psi \in \mathrm{Hom}_\mathcal{C}(X, Y)\}. \tag{11.16}$$

Composition of morphisms is the same as that in $\mathcal{C}$. The category $\mathrm{Kar}\,\mathcal{C}$ is additive with direct sum defined by $(X, e) \oplus (Y, f) = (X \oplus Y, e \oplus f)$. We have a full additive embedding of $\mathcal{C}$ into $\mathrm{Kar}\,\mathcal{C}$ via $X \mapsto (X, \mathrm{id}_X)$. The lemma below will imply that $(X, \mathrm{id}_X) \simeq (X, e) \oplus (X, 1 - e)$ for any idempotent $e \in \mathrm{End}(X)$.

Lemma 11.17 *The category $\mathrm{Kar}\,\mathcal{C}$ is Karoubian.*

Proof Let f be an idempotent in $\mathrm{End}((X, e))$. Then $ef = f = fe$ by definition of the morphisms in $\mathrm{Kar}\,\mathcal{C}$, so

$$(e - f)^2 = e^2 - ef - fe + f^2 = e - f - f + f = e - f,$$

i.e. $e - f$ is an idempotent of X. It is now easy to see that (X, e) is isomorphic to $(X, e - f) \oplus (X, f)$ via the maps $\begin{pmatrix} e - f \\ f \end{pmatrix}$ and $(e - f \;\; f)$. Hence, f splits. $\square$

Like the additive closure, the Karoubian closure is in fact a 2-functor from the 2-category of additive categories to the 2-category of Karoubian categories and it is left adjoint to the forgetful functor in the other direction. We leave it to the reader to show this and to prove the following lemma.

Lemma 11.18 *If $\mathcal{C}$ is a full subcategory of a Karoubian category $\mathcal{D}$, then $\operatorname{Kar}\mathcal{C}$ is equivalent to the full subcategory of $\mathcal{D}$ consisting of all direct summands of objects of $\mathcal{C}$.*

Exercise 11.19 Prove Lemma 11.18.

Example 11.20 The Karoubian closure $\operatorname{Kar}\mathbb{BS}\operatorname{Bim}^{\text{sh},\oplus}$ is the full subcategory of R-gbim consisting of direct summands of direct sums of shifts of Bott–Samelson bimodules. This is precisely the category $\mathbb{S}\operatorname{Bim}$ of Soergel bimodules.

Example 11.21 The Karoubian closure of the category of finitely generated free modules over a ring is the category of finitely generated projective modules.

Exercise 11.22 Show that if $\mathcal{C}$ is R-linear (resp. monoidal, resp. a category with shift functor), then so is $\operatorname{Kar}\mathcal{C}^{\oplus}$. The monoidal structure is induced via

$$(X, e) \otimes (Y, f) := (X \otimes Y, e \otimes f). \tag{11.17}$$

11.2.4 Karoubi Envelopes Are Krull–Schmidt

We fix a graded category $\mathcal{C}$. We denote by

$$\operatorname{Kar}\mathcal{C} := \operatorname{Kar}(\mathcal{C}^{\text{sh},\oplus}) \tag{11.18}$$

the Karoubian closure of the additive closure of the associated category with shifts, see Sect. 11.2. The main examples we have in mind are

$$\mathbb{S}\operatorname{Bim} = \operatorname{Kar}\mathbb{BS}\operatorname{Bim} \quad \text{and} \quad \mathcal{H} := \operatorname{Kar}\mathcal{H}_{\text{BS}}. \tag{11.19}$$

Remark 11.23 Any pre-additive (not necessarily graded) category $\mathcal{C}$ can be viewed as a graded category by concentrating all morphisms in degree 0. The associated category with shift is just a coproduct of $\mathbb{Z}$ copies of $\mathcal{C}$, so it is not worth considering shifts and we set

$$\operatorname{Kar}\mathcal{C} := \operatorname{Kar}\mathcal{C}^{\oplus} \tag{11.20}$$

in this case. This whole section thus has a non-graded formulation, and the reader is invited to spell out the assumptions and results in the non-graded setting. For convenience, we have added the non-graded version of the main Theorem 11.39 as Corollary 11.42. The reason we consider graded categories throughout is that our main examples $\mathbb{BS}\operatorname{Bim}$ and $\mathcal{H}_{\text{BS}}$ are graded.

Exercise 11.24 Using the previous section, spell out how the objects and morphisms of $\operatorname{Kar}\mathcal{C}$ look like in terms of those of $\mathcal{C}$.

11.2 Prelude: From $\mathcal{H}_{\mathrm{BS}}$ to $\mathcal{H}$

Our aim is to parameterize the indecomposable objects of $\operatorname{Kar} \mathcal{C}$. But to be able to say anything useful about this, we will need some additional assumptions on $\mathcal{C}$ which are natural in our context.

> **Assumption 11.25** We assume that $\Bbbk$ is a henselian local ring (e.g. a field or a complete local ring) and that R is a non-negatively graded commutative $\Bbbk$-algebra of finite type with $R^0 = \Bbbk$, such that all homogeneous components of R are finitely generated $\Bbbk$-modules. We assume that $\mathcal{C}$ is enriched over graded R-modules and that all endomorphism algebras of $\mathcal{C}$ are finitely generated R-modules.

Our main examples $\mathbb{BSBim}$ and $\mathcal{H}_{\mathrm{BS}}$ satisfy all the assumptions, with $\Bbbk = \mathbb{R}$ and $R = \mathbb{R}[\mathfrak{h}]$. We can now show the following important preliminary fact about $\operatorname{Kar} \mathcal{C}$. We outsource some basics to Appendix 1 where we have a detailed review of Krull–Schmidt categories.

Theorem 11.26 $\operatorname{Kar} \mathcal{C}$ *is Krull–Schmidt and the split Grothendieck group* $[\operatorname{Kar} \mathcal{C}]_\oplus$ *is a free* $\mathbb{Z}[v, v^{-1}]$-*module, with* v *acting by shift.*

Proof We start with a general observation. Let M be a finitely generated R-module. Then M is a quotient of some finite direct sum $\bigoplus_{i=1}^n R(n_i)$ by a graded submodule. Since R is non-negatively graded, it follows that the degrees of M are bounded below. Moreover, as all homogeneous components of R are finitely generated $\Bbbk$-modules, every homogenous component of M is a finitely generated $\Bbbk$-module.

By our assumptions, each endomorphism ring $\operatorname{End}_\mathcal{C}(X)$ is a finitely generated R-module. Hence, by the above, every homogenous component is a finitely generated $\Bbbk$-module. Now, Hom-spaces of $\mathcal{C}^{\mathrm{sh}}$ are by definition just homogeneous components of Hom-spaces of $\mathcal{C}$. Hence, all endomorphism rings in $\mathcal{C}^{\mathrm{sh}}$ are finitely generated $\Bbbk$-modules. From the definition of the additive closure and of the Karoubian closure, it is clear that also in $\operatorname{Kar} \mathcal{C} = \operatorname{Kar} \mathcal{C}^{\mathrm{sh}, \oplus}$ all endomorphism rings are finitely generated $\Bbbk$-modules. Since $\Bbbk$ is henselian, Proposition 11.60 thus implies that all endomorphism rings of $\operatorname{Kar} \mathcal{C}$ are semiperfect. Hence, $\operatorname{Kar} \mathcal{C}$ is a Krull–Schmidt category by Theorem 11.53. In particular, it has unique decompositions by Theorem 11.50.

To prove the second claim we will employ Proposition 11.62 and to this end we need to know that the degrees of the graded endomorphism rings of $\operatorname{Kar} \mathcal{C}$, see (11.6), are bounded below. First, since each endomorphism ring $\operatorname{End}_\mathcal{C}(X)$ is a finitely generated R-module, we know from our general observations at the beginning that its degrees are bounded below. Graded endomorphism rings in $\mathcal{C}^{\mathrm{sh}}$ are just endomorphism rings in $\mathcal{C}$ since

$$\operatorname{End}^\bullet_{\mathcal{C}^{\mathrm{sh}}}(X(n)) = \bigoplus_{m \in \mathbb{Z}} \operatorname{End}_{\mathcal{C}^{\mathrm{sh}}}(X(n), X(n)(m)) = \bigoplus_{m \in \mathbb{Z}} \operatorname{End}^m_\mathcal{C}(X) = \operatorname{End}_\mathcal{C}(X).$$
(11.21)

Hence, their degrees are bounded below. The same is then still true for $\text{Kar}\,\mathcal{C}$ by definition of the additive and of the Karoubian closure. □

11.2.5 Diagrammatics and Karoubi Envelopes

We have defined a diagrammatic category which encodes morphisms between Bott–Samelson bimodules. Can we use this category to understand morphisms between arbitrary Soergel bimodules? More generally, if we have a diagrammatic description of a category, can we also understand its Karoubi envelope? The answer is yes, assuming that one can compute the relevant idempotents, using the description (11.16) of morphisms in the Karoubi envelope.

For example, suppose $s, t \in S$ with $m_{st} = 3$ and one is interested in morphisms from B_{sts} to some Bott–Samelson bimodule Y. We know that B_{sts} is a direct summand of (s, t, s), the image of an idempotent $e \in \text{End}((s, t, s))$. The idempotent $e = \text{JW}_{(s,t,s)}$ is explicitly given as follows:

(11.22)

Then we can identify $\text{Hom}^\bullet(B_{sts}, Y)$ with the set of morphisms $(s, t, s) \to Y$ of the form ψe for some $\psi : (s, t, s) \to Y$. Here is an example:

(11.23)

Note that two different morphisms $\psi, \psi' : (s, t, s) \to Y$ can give rise to the same morphism $\psi e = \psi' e : B_{sts} \to Y$.

11.3 Grothendieck Groups of Object-Adapted Cellular Categories

Similarly, we can draw morphisms in $\mathrm{Hom}^\bullet(B_{sts} \otimes (u,s), t \otimes B_{sutus} \otimes u)$ by pre- and post-composing a morphism $\psi \in \mathrm{Hom}((s,t,s,u,s),(t,s,u,t,u,s,u))$ with the appropriate idempotents e and f:

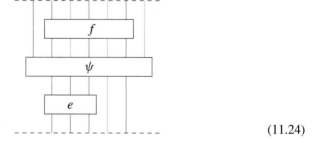

(11.24)

Of course, finding the idempotent f projecting from (s,u,t,u,s) to B_{sutus} is probably not an easy task for the reader at this point. Finding the idempotent which projects from $\underline{w}$ to B_w for all w is an incredibly difficult open problem.

Exercise 11.27 We assume that $m_{st} = 3$. Construct a basis for $\mathrm{Hom}^\bullet(B_{sts}, B_s)$ as a right R-module diagrammatically. (*Hint:* Use the double leaves basis for $\mathrm{Hom}((s,t,s),s)$. What happens when you precompose with e?)

Exercise 11.28 Construct a basis for $\mathrm{End}^\bullet(B_{sts})$ diagrammatically. For a challenge, multiply two elements of your basis and express it in terms of the basis.

11.3 Grothendieck Groups of Object-Adapted Cellular Categories

Let $\mathcal{C}$ be some graded category. Describing a basis of $[\mathrm{Kar}\,\mathcal{C}]_\oplus$ is hopeless in this generality because we have nothing to start with. This is where we make use of the cellular structure defined by the double leaves basis.

11.3.1 Object-Adapted Cellular Categories

Here is a definition due to Elias–Lauda [54]. It is less terrible than it looks (see Remark 11.32 for the context and motivation of the definition).

Definition 11.29 Consider a (graded) R-linear category $\mathcal{C}$ over a (graded) commutative ring.[3] A *(graded) object-adapted cellular category* is[4] such a category $\mathcal{C}$ equipped with the following data:

- there is a set Λ of objects of $\mathcal{C}$ and a partial order $\leq$ on Λ.
- for each object $X \in \mathcal{C}$ and each $\lambda \in \Lambda$ there are two finite families

$$(LL_T)_{T \in E(X,\lambda)} \subset \mathrm{Hom}_{\mathcal{C}}(X, \lambda) \tag{11.25}$$

and

$$(\overline{LL}_S)_{S \in \overline{E}(\lambda, X)} \subset \mathrm{Hom}_{\mathcal{C}}(\lambda, X) \tag{11.26}$$

of (homogeneous) morphisms, indexed by *finite* sets $E(X, \lambda)$ and $\overline{E}(\lambda, X)$,
- for each object $X \in \mathcal{C}$ and each $\lambda \in \Lambda$ there are pairwise inverse maps between the indexing sets $E(X, \lambda)$ and $\overline{E}(\lambda, X)$ that we denote by $\overline{\cdot}$ in either direction.

This data is subject to the following requirements:

CC1. Set

$$\mathbb{LL}^{\lambda}_{S,T} := \overline{LL}_S \circ LL_T \in \mathrm{Hom}_{\mathcal{C}}(X, Y) \tag{11.27}$$

for $T \in E(X, \lambda)$ and $S \in \overline{E}(\lambda, Y)$, and

$$\mathbb{LL}^{\lambda}(X, Y) := (\mathbb{LL}^{\lambda}_{S,T})_{\substack{T \in E(X,\lambda) \\ S \in \overline{E}(\lambda, Y)}} \subset \mathrm{Hom}_{\mathcal{C}}(X, Y) \,. \tag{11.28}$$

Then the family

$$\mathbb{LL}(X, Y) := \bigcup_{\lambda \in \Lambda} \mathbb{LL}^{\lambda}(X, Y) \tag{11.29}$$

is an R-basis of $\mathrm{Hom}_{\mathcal{C}}(X, Y)$.

CC2. There is an R-linear contravariant involution $\overline{(\cdot)}$ on $\mathcal{C}$ with the property that

$$\overline{LL_T} = \overline{LL}_{\overline{T}} \tag{11.30}$$

for all $T \in E(X, \lambda)$.

CC3. For each $\lambda \in \Lambda$ the sets $E(\lambda, \lambda)$ and $\overline{E}(\lambda, \lambda)$ are just a singleton $\{*\}$ and

$$LL_* = \mathrm{id}_{\lambda} = \overline{LL}_* \,. \tag{11.31}$$

---

[3] The definition has both a graded and an ungraded version; we leave it to the reader to make the adaptions.

[4] In [54] this was called a *strictly* object-adapted cellular category.

11.3 Grothendieck Groups of Object-Adapted Cellular Categories

CC4. For each $\lambda \in \Lambda$ the collection of morphisms

$$\text{Hom}_{\leq \lambda}(X, Y) := \text{span}_R \bigcup_{\mu \leq \lambda} \mathbb{LL}^\mu(X, Y) \subset \text{Hom}_{\mathcal{C}}(X, Y) \qquad (11.32)$$

forms a *two-sided ideal* $\mathcal{C}_{\leq \lambda}$ in $\mathcal{C}$, meaning that it is stable under pre- and post-composition with arbitrary morphisms in $\mathcal{C}$.

If it is important to emphasize the object λ, we may write LL_T^λ instead of just LL_T. Similarly, to distinguish between $* \in E(\lambda, \lambda)$ and $* \in E(\mu, \mu)$ we might write $*_\lambda$ or $*_\mu$.

Property CC4 is really the crucial one. The ideals $\mathcal{C}_{\leq \lambda}$ are called *cellular ideals*. Note that $\mathcal{C}_{\leq \lambda}$ is not a subcategory, since it does not contain the identity maps of all objects. However, it does contain the identity maps of all objects $\mu \in \Lambda$ with $\mu \leq \lambda$.

Exercise 11.30 A subset $I \subset \Lambda$ is called an *ideal* if $\lambda \in I$ and $\mu < \lambda$ implies $\mu \in I$. Prove that $\mathcal{C}_I$, defined to be the R-span of the morphisms in $\mathbb{LL}^\mu(X, Y)$ for $\mu \in I$, is a (homogeneous) two-sided ideal in $\mathcal{C}$.

Thus any filtration of Λ by ideals gives rise to a filtration of $\mathcal{C}$ by two-sided ideals.

Exercise 11.31 This exercise will prove that the Temperley–Lieb category $\mathcal{TL}$, see Definition 7.13, is a (non-graded) object-adapted cellular category. First we need some terminology. An (m, n)-*crossingless matching* is a crossingless matching between m points on the bottom boundary and n points on the top. Inside this diagram, a *cap* is a strand which pairs two bottom boundary points, a *cup* is a strand which pairs two top boundary points, and a *thru-strand* is a strand which pairs a top boundary point with a bottom boundary point.

1. For an (m, n)-crossingless matching, let k be the number of thru-strands. Prove that $k \leq \min(m, n)$, $m + n$ is even, and $k + m$ is even. Find a formula for the number of caps and the number of cups. Conversely, for each such triple m, n, k, construct at least one crossingless matching.
2. Prove that the identity is the unique (n, n)-crossingless matching with exactly n thru-strands.
3. A *cap diagram* is a crossingless matching with no cups, and a *cup diagram* is a crossingless matching with no caps. Prove that any (m, n)-crossingless matching with k through strands is the composition of an (m, k)-cap diagram with a (k, n) cup diagram. (The picture of the proof is the important part, don't worry about making a rigorous topological argument.)
4. Finally, let us equip $\mathcal{TL}$ with the following data:

   - $\Lambda = \mathbb{N}$ with the usual order $\leq$.
   - $E(n, \lambda)$ is the set of (n, λ)-cap diagrams, and for $T \in E(n, \lambda)$, LL_T is the corresponding morphism in $\mathcal{TL}$. Similarly, $\overline{E}(\lambda, n)$ is the set of (λ, n)-cup diagrams.
   - The involution $\overline{(\cdot)}$ flips a diagram upside-down.

Prove that the conditions (CC1) through (CC4) are satisfied. Two of these conditions match previous parts of this exercise. What do the other two conditions say in this context?

Remark 11.32 Graham and Lehrer first defined cellular algebras [80], which are algebras with special bases of the form $\{c_{S,T}^\lambda\}$. As in property CC4, this basis gives a filtration of the algebra by ideals, and the subquotients have a particularly nice form (see Lemma 11.36 below). However, there is no factorization property for the elements $c_{S,T}^\lambda$ (after all, this is an algebra and not a category, so there are no other objects to factor through). Westbury [176] introduced a generalization, the cellular category, but did not assert any factorization properties either. The prefix "object-adapted" indicates that our cellular basis $\mathbb{LL}_{S,T}^\lambda$ factors through certain particular objects, as a composition $\overline{LL}_S \circ LL_T$. This factorization leads to easy explanations for many of the more subtle features of cellular algebras. For example, the definitions in [80] and [176] do not state (CC4) as we have, but instead have a fairly complicated formula (the *cellular formula*, not reprinted here) for resolving a product $a \cdot c_{S,T}^\lambda$ in the cellular basis. The cellular formula implies the existence of a filtration by ideals. However, in the object-adapted context, our axiom (CC4) implies the cellular formula, and is much more intuitive to state.

Remark 11.33 Often in the literature it is assumed that the poset Λ is finite, or at least satisfies the DCC. Since this is not necessary for our arguments, we omit this assumption.

11.3.2 First Properties

The first lemma reinterprets the cellular ideals in terms of their factorization properties. The ideal $\mathcal{C}_{<\lambda}$ is defined as in Exercise 11.30.

Lemma 11.34 *The following holds for all $\lambda, \mu \in \Lambda$:*

1. *The sets $E(\lambda, \mu)$ and $\overline{E}(\mu, \lambda)$ are empty unless $\mu \leq \lambda$.*
2. *The ideal $\mathcal{C}_{\leq \lambda}$ is generated by $\{\mathrm{id}_\mu\}_{\mu \leq \lambda}$. In other words, it is the ideal spanned by morphisms which factor through some object less than λ.*
3. $\mathrm{End}_\mathcal{C}(\lambda)/\mathrm{End}_{<\lambda}(\lambda) \simeq R.$

Proof

1. Suppose $E(\lambda, \mu)$ is nonempty. Let $T \in E(\lambda, \mu)$. Then for $* = *_\mu \in \overline{E}(\mu, \mu)$ we have

$$\mathbb{LL}_{*,T}^\mu = \overline{LL}_*^\mu \circ LL_T = \mathrm{id}_\mu \circ LL_T = LL_T.$$

Thus LL_T is one of our basis elements, associated to cell μ. However,

$$\mathrm{id}_\lambda = \mathbb{LL}_{*,*}^\lambda,$$

11.3 Grothendieck Groups of Object-Adapted Cellular Categories

so $\text{id}_\lambda \in \mathcal{C}_{\leq\lambda}$. Since $\mathcal{C}_{\leq\lambda}$ is an ideal, $LL_T = LL_T \circ \text{id}_\lambda$ is also in $\mathcal{C}_{\leq\lambda}$. But by definition of $\mathcal{C}_{\leq\lambda}$, this is possible if and only if $\mu \leq \lambda$.

2. By definition, $\mathcal{C}_{\leq\lambda}$ is spanned by elements which factor through objects μ for $\mu \leq \lambda$, which shows one inclusion. Since $\text{id}_\mu \in \mathcal{C}_{\leq\lambda}$ by the above argument, the reverse inclusion holds as well.

3. By part 2, all of $\text{End}(\lambda)$ lies in $\mathcal{C}_{\leq\lambda}$. By part 1, the unique basis element of $\text{End}(\lambda)$ which is in $\mathcal{C}_{\leq\lambda}$ but not in $\mathcal{C}_{<\lambda}$ is $\mathbb{LL}^\lambda_{*,*} = \text{id}_\lambda$, since $E(\lambda,\lambda) = \{*\} = \overline{E}(\lambda,\lambda)$ by CC3.

$\square$

Definition 11.35 For $X \in \mathcal{C}$ and $\lambda \in \Lambda$ we can now define the *cellular pairing* as the map

$$\varphi_X^\lambda : E(X,\lambda) \times \overline{E}(\lambda,X) \to R$$
$$(T, S) \mapsto LL_T \circ \overline{LL}_S + \mathcal{C}_{<\lambda} \in \text{End}(\lambda)/\text{End}_{<\lambda}(\lambda) \simeq R \ . \tag{11.33}$$

In other words, $\varphi_X^\lambda(T, S)$ is the coefficient of id_λ in $LL_T \circ \overline{LL}_S \in \text{End}(\lambda)$ with respect to the cellular basis.

The next lemma indicates how the cellular pairing controls multiplication in the associated graded (of the filtration by cellular ideals).

Lemma 11.36 *Let $T \in E(Y, \lambda)$, $S \in \overline{E}(\lambda, Z)$, $V \in E(X, \lambda)$, and $U \in \overline{E}(\lambda, Y)$. Then modulo $\mathcal{C}_{<\lambda}$ we have*

$$\mathbb{LL}^\lambda_{S,T} \circ \mathbb{LL}^\lambda_{U,V} = \varphi_Y^\lambda(T, U)\mathbb{LL}^\lambda_{S,V} \ . \tag{11.34}$$

Proof This is a straightforward computation:

$$\mathbb{LL}^\lambda_{S,T} \circ \mathbb{LL}^\lambda_{U,V} = (\overline{LL}_S \circ LL_T) \circ (\overline{LL}_U \circ LL_V)$$
$$= \overline{LL}_S \circ (LL_T \circ \overline{LL}_U) \circ LL_V$$
$$\equiv \overline{LL}_S \circ (\varphi_Y^\lambda(T,U)\text{id}_\lambda) \circ LL_V$$
$$= \varphi_Y^\lambda(T,U)\overline{LL}_S \circ LL_V$$
$$= \varphi_Y^\lambda(T,U)\mathbb{LL}^\lambda_{S,V} \ .$$

The third line has an equivalence modulo $\mathcal{C}_{<\lambda}$.

$\square$

11.3.3 The Main Example

Our main example of a strictly object-adapted cellular category is the category $\mathcal{H}_{\text{BS}}$. This follows from the theorem below.

Theorem 11.37 (Elias–Williamson) *The diagrammatic Bott–Samelson category $\mathcal{H}_{BS}$ is a strictly object-adapted cellular category with the following data:*

- *We identify W with a subset Λ of $\mathcal{H}_{BS}$ by choosing a reduced expression $\underline{w}$ for each element w, and equip Λ with the Bruhat order under this identification.*
- *For any $\underline{x} \in \mathcal{H}_{BS}$ and $w \in \Lambda$, the sets $E(\underline{x}, w)$ and $\overline{E}(w, \underline{x})$ are both equal to the set of subexpressions of $\underline{x}$ that express w, i.e.,*

$$E(\underline{x}, w) = \{\underline{e} \subset \underline{x} \mid \underline{x}^{\underline{e}} = w\} = \overline{E}(w, \underline{x}) . \quad (11.35)$$

- *For $w \in \Lambda$ and $\underline{e} \in E(\underline{x}, w) = \overline{E}(w, \underline{x})$ set*

$$LL_{\underline{x},\underline{e}} := \begin{array}{c} \underline{w} \\ \diagup \overline{LL_{\underline{x},\underline{e}}} \diagdown \\ \underline{x} \end{array} \quad (11.36)$$

and

$$\overline{LL}_{\underline{x},\underline{e}} := \begin{array}{c} \underline{x} \\ \diagdown \overline{LL_{\underline{x},\underline{e}}} \diagup \\ \underline{w} \end{array} \quad (11.37)$$

- *The contravariant autoequivalence $\overline{(-)}$ flips a diagram upside-down. The bijection $\overline{(-)}$ between $E(\underline{x}, w)$ and $\overline{E}(w, \underline{x})$ is the identity.*

Above, $LL_{\underline{x},\underline{e}}$ is a light leaf constructed so that $\underline{w}$ is the target. Moreover, when $\underline{w}$ is one of these chosen reduced expressions and $\underline{e} \subset \underline{w}$ is all ones, then we choose our light leaf $LL_{\underline{w},\underline{e}}$ to be the identity map of $\underline{w}$ (rather than some other rex move).

The two difficulties in the proof are to show CC1 and CC4, i.e. that $\mathbb{LL}(\underline{x}, \underline{y})$ forms a basis of $\mathrm{Hom}_{\mathcal{H}_{BS}}(\underline{x}, \underline{y})$ and that $\mathcal{H}_{BS \leq w}$ is an ideal. Both were done by Elias–Williamson [59]. The statement that light leaves are a basis is the diagrammatic analogue of Libedinsky's work [121]. The remaining properties are then straightforward.

11.3.4 Classifying Indecomposables in Object-Adapted Cellular Categories

Assumption 11.38 For the rest of this section we assume that $\mathcal{C}$ is a (graded) R-linear object-adapted cellular category satisfying Assumption 11.25. We moreover assume that R is an integral domain and that $\Bbbk$ is Noetherian.

11.3 Grothendieck Groups of Object-Adapted Cellular Categories

In this setting we want to prove:

Theorem 11.39 (Elias–Williamson, Elias–Lauda)

1. For each $\lambda \in \Lambda$ the object $\lambda \in \operatorname{Kar} \mathcal{C}$ has a unique direct summand B_λ which does not occur as a direct summand of any $\mu \in \Lambda$ with $\mu < \lambda$.
2. $(B_\lambda)_{\lambda \in \Lambda}$ is a $\mathbb{Z}[v, v^{-1}]$-basis of $[\operatorname{Kar} \mathcal{C}]_\oplus$.

The proof follows Elias–Lauda [54], we just added some details every now and then. We need to prepare the proof with a few elementary results.

For any $\lambda \in \Lambda$, the set of elements μ for which $\nu \not\geq \lambda$ is an ideal in Λ. Thus the collection of morphisms $\mathcal{C}_{\not\geq \lambda}$, defined as in Exercise 11.30, is an ideal in $\mathcal{C}$. We define the quotient category $\mathcal{C}^{\geq \lambda} := \mathcal{C}/\mathcal{C}_{\not\geq \lambda}$. By definition, we have

$$\operatorname{Hom}_{\mathcal{C}^{\geq \lambda}}(X, Y) = \operatorname{Hom}_\mathcal{C}(X, Y)/\operatorname{Hom}_{\not\geq \lambda}(X, Y). \tag{11.38}$$

Since $\mathcal{C}_{\not\geq \lambda}$ is a homogeneous ideal, $\mathcal{C}^{\geq \lambda}$ is still a graded category.

Remark 11.40 It is worth noting that, inside Λ, the ideal $\not\geq \lambda$ and the ideal $< \lambda$ agree after intersecting with the ideal $\leq \lambda$. Thus when examining $\operatorname{End}_\mathcal{C}(\lambda)$ or any other Hom space determined to live entirely within $\mathcal{C}_{\leq \lambda}$, the quotient by $\mathcal{C}_{\not\geq \lambda}$ is the same as the quotient by $\mathcal{C}_{\leq \lambda}$.

Recall that R is assumed to be positively graded. Hence, the subset $R_{>0}$ of elements of positive degree is an ideal in R. Pick some object X of $\mathcal{C}$, and look at $\operatorname{End}_{\mathcal{C}_{\not\geq \lambda}}(X)$. We define

$$J_\lambda(X) := R_{>0} \cdot \{\mathbb{LL}^\lambda_{S,T} \mid S \in \overline{E}(\lambda, X), T \in E(X, \lambda)\}, \tag{11.39}$$

which we consider as a graded R-submodule of $\operatorname{End}_{\mathcal{C}^{\geq \lambda}}(X)$. By Lemma 11.36 we have

$$\mathbb{LL}^\lambda_{S,T} \circ \mathbb{LL}^\lambda_{U,V} = \varphi^\lambda_Y(T, U)\mathbb{LL}^\lambda_{S,V} \tag{11.40}$$

modulo $\mathcal{C}_{<\lambda}$ for $S, U \in \overline{E}(\lambda, X)$ and $T, V \in E(X, \lambda)$, so this is going to be an actual equality in $\operatorname{End}_{\mathcal{C}^{\geq \lambda}}(X)$. In particular, $J_\lambda(X)$ is an ideal in $\operatorname{End}_{\mathcal{C}^{\geq \lambda}}(X)$.

Lemma 11.41 *Any element $y \in J_\lambda(X)$ of degree ≤ 0 is nilpotent. In particular, the non-positive part of $J_\lambda(X)$ is contained in the graded Jacobson radical of $\operatorname{End}_{\mathcal{C}^{\geq \lambda}}(X)$.*

Proof Let $y \in J_\lambda(X)$ be of degree $k \leq 0$. By definition of $J_\lambda(X)$ we can write

$$y = \sum_{S \in \overline{E}(\lambda, X), T \in E(X, \lambda)} r_{S,T} \mathbb{LL}^\lambda_{S,T}$$

with $r_{S,T}$ homogeneous of positive degree. Since y is in the (image of the) ideal $C_{\leq \lambda}$, so y^n will be as well, so we may write

$$y^n = \sum_{S \in \overline{E}(\lambda, X), T \in E(X, \lambda)} r_{S,T}^{(n)} \mathrm{LL}_{S,T}^{\lambda}$$

for some coefficients $r_{S,T}^{(n)} \in R$.

We have

$$y^2 = \sum_{\substack{S, U \in \overline{E}(\lambda, X) \\ T, V \in E(X, \lambda)}} r_{S,T} r_{U,V} \mathrm{LL}_{S,T}^{\lambda} \mathrm{LL}_{U,V}^{\lambda} = \sum_{\substack{S, U \in \overline{E}(\lambda, X) \\ T, V \in E(X, \lambda)}} r_{S,T} r_{U,V} \varphi_X^{\lambda}(T, U) \mathrm{LL}_{S,V}^{\lambda}$$

$$= \sum_{\substack{S \in \overline{E}(\lambda, X) \\ V \in E(X, \lambda)}} \left(\sum_{\substack{U \in \overline{E}(\lambda, X) \\ T \in E(X, \lambda)}} r_{S,T} r_{U,V} \varphi_X^{\lambda}(T, U) \right) \mathrm{LL}_{S,V}^{\lambda},$$

from which we compute that

$$r_{S,V}^{(2)} = \sum_{\substack{U \in \overline{E}(\lambda, X) \\ T \in E(X, \lambda)}} r_{S,T} r_{U,V} \varphi_X^{\lambda}(T, U). \tag{11.41}$$

Similarly we compute that

$$r_{S,V}^{(n)} = \sum_{\substack{U \in \overline{E}(\lambda, X) \\ T \in E(X, \lambda)}} r_{S,T}^{(n-1)} r_{U,V} \varphi_X^{\lambda}(T, U). \tag{11.42}$$

Since $r_{U,V}$ is in the ideal $R_{>0}$ for all U and V, then so is $r_{S,V}^{(n)}$ for all n, S, V.

Note that $E(X, \lambda)$ is a finite set, so there is an integer d such that $\deg \mathrm{LL}_T \geq d$ for all $T \in E(X, \lambda)$. Now, for each $S \in \overline{E}(\lambda, X)$ let $\mathcal{V}_S^{(n)}$ be the set of all $V \in E(X, \lambda)$ such that $r_{S,V}^{(n)} \neq 0$. If this set is nonempty, then

$$d_S^{(n)} := \max_{V \in \mathcal{V}_S^{(n)}} \deg \mathrm{LL}_V \tag{11.43}$$

is a well-defined integer. Note that by (11.42), if $r_{S,T}^{(n-1)} = 0$ for all T, then $r_{S,V}^{(n)} = 0$ for all V, so once $\mathcal{V}_S^{(n)}$ is empty it stays empty henceforth. Our approach will be to show that

$$d_S^{(n)} < d_S^{(n-1)} \tag{11.44}$$

11.3 Grothendieck Groups of Object-Adapted Cellular Categories

so long as the set $\mathcal{V}_S^{(n)}$ is nonempty. But because of the global lower bound $d_S^{(n)} \geq d$, this means that $\mathcal{V}_S^{(n)}$ must eventually be empty. Since $\overline{E}(\lambda, X)$ is a finite set, we can find an $N > 0$ for which $\mathcal{V}_S^{(N)}$ is empty for all S, meaning that $y^N = 0$.

Suppose $r_{U,V} \neq 0$. Since y has degree ≤ 0 and $r_{U,V}$ has degree > 0, it must be the case that $\overline{LL_U} \circ LL_V$ has degree < 0. Thus

$$\deg \overline{LL_U} + \deg LL_V < 0. \tag{11.45}$$

Let $V \in \mathcal{V}_S^{(n)}$ so that $r_{S,V}^{(n)} \neq 0$. Then by (11.42) this means there exist T, U with $r_{S,T}^{(n-1)} r_{U,V} \varphi_X^\lambda(T, U) \neq 0$. Hence all factors are nonzero. For $\varphi_X^\lambda(T, U)$ to be nonzero it must have degree ≥ 0, since it is an element of R. But the $\varphi_X^\lambda(T, U)$ is the coefficient of the identity in the composition $LL_T \circ \overline{LL_U}$, so

$$\deg LL_T + \deg \overline{LL_U} \geq 0. \tag{11.46}$$

Combining this with (11.45) we get

$$\deg LL_T > \deg LL_V. \tag{11.47}$$

In particular, the degree of LL_V is strictly smaller than the degrees of all the LL_T which contribute to it, for $T \in \mathcal{V}_S^{(n-1)}$. Consequently, $d_S^{(n)} < d_S^{(n-1)}$, and the result is proven. □

Now, we can come to the proof of Theorem 11.39.

Proof Let $\lambda \in \Lambda$. By our assumptions, the degree zero part $\text{End}_\mathcal{C}^0(\lambda)$ of the endomorphism ring of λ in $\mathcal{C}$ is a finitely generated $\Bbbk$-module, thus a noetherian ring. We can therefore write the identity $\text{id}_\lambda \in \text{End}_\mathcal{C}^0(\lambda)$ as a finite sum of mutually orthogonal primitive idempotents:

$$\text{id}_\lambda = e_1 + \cdots + e_n.$$

We have

$$\text{End}_\mathcal{C}^0(\lambda) / \text{End}_{<\lambda}^0(\lambda) = (\text{End}_\mathcal{C}(\lambda) / \text{End}_{<\lambda}(\lambda))^0 = R^0 = \Bbbk$$

by Lemma 11.34 and by our assumptions. The image of the idempotents above in this quotient remain mutually orthogonal idempotents. But $\Bbbk$ is assumed to be local and as a local ring has no non-trivial idempotents. So there must be a unique one of the idempotents, denote it e_λ, whose image is nonzero. Let us denote the object (λ, e_λ) of $\text{Kar}\,\mathcal{C}$ by B_λ; it is a direct summand of λ in the Karoubi envelope. This object cannot be a direct summand of some μ with $\mu < \lambda$, since the idempotent would factor through the object μ, and thus live inside $\mathcal{C}_{\leq \mu} \subset \mathcal{C}_{<\lambda}$, contradicting the assumption that the image of e_λ is nonzero.

It remains to show that any indecomposable object Y of $\operatorname{Kar}\mathcal{C}$ is isomorphic to a shift of some B_λ. Since $\operatorname{Kar}\mathcal{C} = \operatorname{Kar}(\mathcal{C}^{\operatorname{sh},\oplus})$ and Y is indecomposable, it is up to shift of the form $Y \simeq (X, e)$ with $X \in \mathcal{C}$ and $e \in \operatorname{End}_\mathcal{C}^0(X)$ an idempotent of degree zero. Let

$$e = \sum_{\substack{\mu \in \Lambda \\ T \in E(X,\mu), S \in \overline{E}(\mu,X)}} r_{S,T}^\mu \operatorname{LL}_{S,T}^\mu$$

be the basis representation of e, with homogeneous $r_{S,T}^\mu \in R$. This is a finite expression, hence there is a λ which is maximal among the μ with $r_{S,T}^\mu \neq 0$ for some S, T. We will prove that $Y \simeq B_\lambda$ up to shift.

The only element $\mu \in \Lambda$ for which $\mu \geq \lambda$ and $r_{S,T}^\mu \neq 0$ is $\mu = \lambda$. Thus the image of e in the quotient $\operatorname{End}_{\mathcal{C} \geq \lambda}(X) = \operatorname{End}(X)/\operatorname{End}_{\not\geq \lambda}(X)$ is just the image of those terms in the sum with $\mu = \lambda$, and is nonzero. (Note that λ is also maximal with respect to the property that the image of e in $\operatorname{End}_{\mathcal{C} \geq \lambda}(X)$ is nonzero.)

The image of e in $\operatorname{End}_{\mathcal{C} \geq \lambda}(X)$ (which we abusively denote by e) can thus be written as

$$e = \sum_{T \in E(X,\lambda), S \in \overline{E}(\lambda,X)} r_{S,T}^\lambda \operatorname{LL}_{S,T}^\lambda \in \operatorname{End}_{\mathcal{C} \geq \lambda}(X) .$$

From this we obtain

$$e^3 = \left(\sum_{\substack{S \in \overline{E}(\lambda,X) \\ V \in E(X,\lambda)}} r_{S,V}^\lambda \operatorname{LL}_{S,V}^\lambda \right) e \left(\sum_{\substack{U \in \overline{E}(\lambda,X) \\ T \in E(X,\lambda)}} r_{U,T}^\lambda \operatorname{LL}_{U,T}^\lambda \right)$$

$$= \sum_{\substack{S,U \in \overline{E}(\lambda,X) \\ V,T \in E(X,\lambda)}} r_{S,V}^\lambda r_{U,T}^\lambda (\overline{LL}_S LL_V) e (\overline{LL}_U LL_T)$$

$$= \sum_{\substack{S,U \in \overline{E}(\lambda,X) \\ V,T \in E(X,\lambda)}} r_{S,V}^\lambda r_{U,T}^\lambda \overline{LL}_S (LL_V e \overline{LL}_U) LL_T$$

Note that $LL_V \circ e \circ \overline{LL}_U \in \operatorname{End}_{\mathcal{C} \geq \lambda}(\lambda) \simeq R$ is a homogeneous element, which commutes with the other maps since the category is R-linear. We thus get

$$e^3 = \sum_{\substack{S,U \in \overline{E}(\lambda,X) \\ V,T \in E(X,\lambda)}} r_{S,V}^\lambda r_{U,T}^\lambda (LL_V e \overline{LL}_U) \overline{LL}_S LL_T$$

$$= \sum_{\substack{S,U \in \overline{E}(\lambda,X) \\ V,T \in E(X,\lambda)}} r_{S,V}^\lambda r_{U,T}^\lambda (LL_V e \overline{LL}_U) \operatorname{LL}_{S,T}^\lambda .$$

11.3 Grothendieck Groups of Object-Adapted Cellular Categories

Since $e^3 = e$, comparing coefficients shows that

$$r^\lambda_{S,T} = \sum_{\substack{U \in \overline{E}(\lambda, X) \\ V \in E(X, \lambda)}} r^\lambda_{S,V} r^\lambda_{U,T} (LL_V e \overline{LL}_U) . \tag{11.48}$$

Now suppose that $LL_{V,\lambda} e \overline{LL}_{U,\lambda}$ is either zero or contained in $R_{>0}$ for all $U \in \overline{E}(X, \lambda)$ and $V \in E(X, \lambda)$. Then the equation above shows us that

$$e \in R_{>0} \cdot \{\mathbb{LL}^\lambda_{S,T} \mid S, T\} = J_\lambda(X) .$$

But e is homogeneous of degree zero, and any degree 0 element of $J_\lambda(X)$ is nilpotent by Lemma 11.41, which is a contradiction to the fact that e is a nonzero idempotent. Hence, there must exist $U \in \overline{E}(\lambda, X)$ and $V \in E(X, \lambda)$ such that

$$LL_V e \overline{LL}_U \in \Bbbk^\times \subset \Bbbk = \mathrm{End}^0_{\mathcal{C}_{\geq \lambda}}(\lambda) . \tag{11.49}$$

We then have $\deg LL_V = k$ and $\deg \overline{LL}_U = -k$ for some $k \in \mathbb{Z}$.

Let $e_\lambda \in \mathrm{End}_\mathcal{C}(\lambda)$ be the idempotent from above defining the indecomposable object B_λ. It descends to the identity of $\mathrm{End}_{\mathcal{C}_{\geq\lambda}}(X)$. Then in $\mathrm{End}^0_{\mathcal{C}_{\geq\lambda}}(\lambda) = \Bbbk$ the image of

$$e_\lambda (LL_V e \overline{LL}_U) e_\lambda$$

is still invertible. The morphism $p := e_\lambda LL_V e \in \mathrm{Hom}_\mathcal{C}(X, \lambda)$ is of degree $-k$ and thus defines a morphism $Y \to B_\lambda(-k)$ in $\mathrm{Kar}\,\mathcal{C}$. Similarly, the morphism $i := e \overline{LL}_U e_\lambda \in \mathrm{Hom}_\mathcal{C}(\lambda, X)$ is of degree k and thus defines a morphism $B_\lambda(-k) \to Y$ in $\mathrm{Kar}\,\mathcal{C}$. In particular, $pi \in \mathrm{End}_{\mathrm{Kar}\,\mathcal{C}}(B_\lambda(-k))$. Since $\mathrm{Kar}\,\mathcal{C}$ is Krull–Schmidt and $B_\lambda(-k)$ is indecomposable, we know that $\mathrm{End}_{\mathrm{Kar}\,\mathcal{C}}(B_\lambda(-k))$ is a local ring. In the quotient category we work in all the summands of λ other than B_λ are zero, so that $\mathrm{End}^0_{\mathcal{C}_{\geq\lambda}}(\lambda) = \mathrm{End}^0_{\mathcal{C}_{\geq\lambda}}(B_\lambda)$ is a quotient of $\mathrm{End}_{\mathrm{Kar}\,\mathcal{C}}(B_\lambda(-k))$. We know that pi is invertible in the quotient by the above. Consequently, pi cannot be contained in the maximal ideal of $\mathrm{End}_{\mathrm{Kar}\,\mathcal{C}}(B_\lambda(-k))$, so pi is invertible. Moreover, in the quotient ring $\mathrm{End}_{\mathcal{C}_{\geq\lambda}}(X)$, pi is an invertible scalar multiple of the identity map.

Meanwhile, $ip \in \mathrm{End}_{\mathrm{Kar}\,\mathcal{C}}(Y)$, and this ring is also local since Y is indecomposable. But ip can not be in the maximal ideal, or it would descend to the maximal ideal modulo $\mathcal{C}_{\not\geq \lambda}$ (or rather, the corresponding ideal of $\mathrm{Kar}\,\mathcal{C}$). In this quotient, since pi is a multiple of the identity, ip squares to a multiple of itself, so it is an invertible scalar multiple of an idempotent. Thus it can not live in the maximal ideal.

Since both $ip \in \mathrm{End}_{\mathrm{Kar}\,\mathcal{C}}(Y)$ and $pi \in \mathrm{End}_{\mathrm{Kar}\,\mathcal{C}}(B_\lambda(-k))$ are invertible, living outside the maximal ideal of their respective local rings, we see that i and p give isomorphisms between Y and $B_\lambda(-k)$. □

For reference, let us state the following non-graded version of Theorem 11.39.

Corollary 11.42 *Let $\Bbbk$ be a henselian local ring and let $\mathcal{C}$ be a $\Bbbk$-linear category such that all Hom-spaces are finitely generated over $\Bbbk$. Assume that $\mathcal{C}$ is a strictly object-adapted cellular category (as before, ignoring the homogeneity assumption). Then:*

1. *For each $\lambda \in \Lambda$ the object $\lambda \in \operatorname{Kar}\mathcal{C}$ has a unique direct summand B_λ which does not occur as a direct summand of any $\mu \in \Lambda$ with $\mu < \lambda$.*
2. *$(B_\lambda)_{\lambda \in \Lambda}$ is a $\mathbb{Z}$-basis of $[\operatorname{Kar}\mathcal{C}]_\oplus$.*

Exercise 11.43 Extend the contravariant autoequivalence $\overline{(\cdot)}$ in CC2 from $\mathcal{C}$ to its Karoubi envelope. Prove that the object B_λ is isomorphic to $\overline{B}_\lambda$.

Appendix 1: Krull–Schmidt Categories

This section provides a detailed review of the Krull–Schmidt property for categories. The survey paper by Krause [114] is a valuable reference.

Categories with Unique Decompositions

Let $\mathcal{C}$ be an additive category. Recall that an object $X \in \mathcal{C}$ is called *indecomposable* if it is nonzero and cannot be decomposed into a non-trivial direct sum, i.e. $X = X_1 \oplus X_2$ implies $X_1 = 0$ or $X_2 = 0$. We say that $X \in \mathcal{C}$ has a *unique decomposition* if it has a finite decomposition $X = \bigoplus_{i=1}^n X_i$ into indecomposable objects X_i of $\mathcal{C}$, and if $X = \bigoplus_{j=1}^m X'_j$ is another such decomposition, then $n = m$ and there is a permutation $\sigma \in S_n$ such that X_i is isomorphic to $X'_{\sigma(i)}$ for each i. We say that $\mathcal{C}$ itself has *unique decompositions* if every nonzero object of $\mathcal{C}$ has a unique decomposition.

Example 11.44 The category of free modules of finite rank over a commutative ring R obviously has unique decompositions (with R being the unique indecomposable object). In particular, the category R-proj of finitely generated projective modules over any principal ideal domain or any noetherian local ring has unique decompositions (since such modules are already free). Moreover, for any principal ideal domain R the whole category R-mod of finitely generated modules has unique decompositions by the structure theorem of finitely generated modules over principal ideal domains.

Exercise 11.45 Here is an example of the failure of unique decompositions in the category R-proj of finitely generated projective modules for more general (but still nice!) rings: let $R = \mathbb{Z}[\sqrt{-5}]$, $I = (3, 1 + \sqrt{-5})$, and $J = (3, 1 - \sqrt{-5})$.

1. Show that R, I, and J are indecomposable R-modules. (*Hint*: decomposability implies non-trivial idempotents in R.)

11.3 Grothendieck Groups of Object-Adapted Cellular Categories

2. Show that I and J are not isomorphic to R. (*Hint:* are the ideals principal?)
3. Show that $I \oplus J$ is isomorphic to $R \oplus R$. (*Hint:* $I + J = R$, $IJ = 3R$, and there is a split exact sequence $0 \to I \cap J \to I \oplus J \to I + J \to 0$.)
4. Conclude that R-proj does *not* have unique decompositions. (A similar argument works for any Dedekind domain which is not a principal ideal domain.)

Having unique decompositions is equivalent to the property that the isomorphism classes of the indecomposable objects yield a basis of the split Grothendieck group of $\mathcal{C}$. Let us recall this.

Definition 11.46 The *split Grothendieck group* $[\mathcal{C}]_\oplus$ of $\mathcal{C}$ is the quotient of the free abelian group on the set of isomorphism classes $[X]$ of objects X of $\mathcal{C}$ modulo the subgroup generated by the relations $[X \oplus Y] = [X] + [Y]$.

Remark 11.47 The reader may have noticed that we should assume that our category $\mathcal{C}$ is *essentially small*, i.e. that the class of isomorphism classes of objects is a *set*. We will silently assume this when working with Grothendieck groups. The important example of the category of finitely generated modules over a ring is small.

Exercise 11.48 Show the following properties:

1. Every element of $[\mathcal{C}]_\oplus$ can be written in the form $[X] - [Y]$ for $X, Y \in \mathcal{C}$.
2. $[X] = [Y]$ in $[\mathcal{C}]_\oplus$ if and only if $X \oplus Z \simeq Y \oplus Z$ for some $Z \in \mathcal{C}$.
3. $\mathcal{C}$ has unique decompositions if and only if $[\mathcal{C}]_\oplus$ is the free abelian group with basis the isomorphism classes of indecomposable objects of $\mathcal{C}$.
4. If $\mathcal{C}$ is monoidal, then $[\mathcal{C}]_\oplus$ is a ring (we always assume that $\otimes$ is additive).

Krull–Schmidt Categories

How can we ensure that an additive category $\mathcal{C}$ has unique decompositions? We approach this question for an object $X \in \mathcal{C}$ by translating it into a ring-theoretic question about its endomorphism ring $\Gamma := \mathrm{End}_\mathcal{C}(X)$. Krause [114] calls this process "projectivization." For any $Y \in \mathcal{C}$ the abelian group $\mathrm{Hom}_\mathcal{C}(X, Y)$ is naturally a right Γ-module with action $f\varphi = f \circ \varphi$ for $f \in \mathrm{Hom}_\mathcal{C}(X, Y)$ and $\varphi \in \Gamma$. We thus have a functor

$$\mathrm{Hom}_\mathcal{C}(X, -) \colon \mathcal{C} \to \mathrm{Mod}\text{-}\Gamma \qquad (11.50)$$

into the category of right Γ-modules. Now, assume that Y is a direct summand of X. Since the Hom-functor is additive, it is obvious that $\mathrm{Hom}_\mathcal{C}(Y, X)$ is a direct summand of Γ. Hence, $\mathrm{Hom}_\mathcal{C}(X, -)$ maps the full subcategory add X of $\mathcal{C}$ consisting of all direct summands of finite direct sums of X to the category of projective Γ-modules.

There is a more concrete description of this: a direct summand Y of an object X defines via the composition of the projection $\pi \colon X \twoheadrightarrow Y$ followed by the inclusion

$\iota: Y \hookrightarrow X$ an *idempotent* e in the ring Γ. The Γ-module $\text{Hom}_\mathcal{C}(X, Y)$ is then naturally isomorphic to $e\Gamma$; in particular, it is a finitely generated projective module. Moreover, this gives a canonical isomorphism

$$\text{End}_\mathcal{C}(Y) \simeq \text{End}_\Gamma(e\Gamma) \simeq e\Gamma e \ . \tag{11.51}$$

Exercise 11.49 Show that $\text{Hom}_\mathcal{C}(X, -)$ induces a full embedding of add X into the category proj-Γ of finitely generated projective Γ-modules.

Using this observation we can transfer the classical Krull–Schmidt theorem for modules to objects in general categories. The following elementary proof for modules is from [117].

Theorem 11.50 (Krull–Schmidt) *Suppose that X has a decomposition $X = \bigoplus_{i=1}^n X_i$ with local endomorphism rings. Then the X_i are indecomposable and X has a unique decomposition.*

Proof The full embedding add $X \subset$ proj-Γ allows us to assume that $\mathcal{C}$ is a module category over a ring. Let $e_i \in \Gamma$ be the idempotent corresponding to X_i and set $\Gamma_i := e_i \Gamma e_i$, a local ring by assumption. Let $X = \bigoplus_{i=1}^m X'_i$ be another decomposition into indecomposable modules with corresponding idempotents e'_i. Then $\text{id}_X = e_1 + \cdots + e_n = e'_1 + \cdots + e'_m$, hence $e_1 = e_1 e_1 = e_1 e'_1 e_1 + \cdots + e_1 e'_m e_1$. Note that $e_1 e'_i e_1 \in \Gamma_1$. Since the sum of the $e_1 e'_i e_1$ is the unit e_1 in Γ_1, the sum is invertible and as Γ_1 is local, one summand, say $e_1 e'_j e_1$, must be invertible (in a local ring a sum of non-units is a non-unit). Renumbering the X'_i if needed, we may assume that $e_1 e'_1 e_1$ is invertible. Hence, there is $g \in \Gamma$ with $e_1 = (e_1 g e_1)(e_1 e'_1 e_1) = e_1 g e_1 e'_1 e_1$. Let us rewrite this equation using projections and inclusions:

$$\text{id}_{X_1} = \text{id}_{X_1} \text{id}_{X_1} = (\pi_1 \iota_1)(\pi_1 \iota_1) = \pi_1(\iota_1 \pi_1)\iota_1 = \pi_1 e_1 \iota_1 = \pi_1(e_1 x e_1 e'_1 e_1)\iota_1$$
$$= \pi_1 e_1 x e_1 e'_1 \iota_1 \pi_1 \iota_1 = \pi_1 e_1 x e_1 e'_1 \iota_1 = \pi_1 e_1 x e_1 \iota'_1 \pi'_1 \iota_1 = (\pi_1 e_1 x e_1 \iota'_1)(\pi'_1 \iota_1) \ .$$

This shows that the map $\pi'_1 \iota_1 : X_1 \to X'_1$ is a section of the map $\pi_1 e_1 x e_1 \iota'_1 : X'_1 \to X_1$. It thus identifies X_1 with a direct summand of X'_1. But X'_1 is indecomposable by assumption, so the two maps are actually isomorphisms $X_1 \simeq X'_1$. If $x \in X_1 \cap X'_i$ for $i \geq 2$, then $\pi'_1 \iota_1(x) = 0$, so $x = 0$ since $\pi'_1 \iota_1$ is an isomorphism. Hence, we can consider the direct sum $X' = X_1 \oplus X'_2 \oplus \cdots \oplus X'_m$. If we can show that $X'_1 \subset X'$, then $X = X'$. Let $x' \in X'_1$. Then $x' = \pi'_1 \iota_1(x)$ for some $x \in X_1$. Hence, $\pi'_1(\iota'_1(x') - \iota_1(x)) = x' - \pi'_1 \iota_1(x) = x' - x' = 0$, so $\iota'_1(x') - \iota_1(x) \in \text{Ker}\, \pi'_1 = X'_2 + \cdots + X'_m$, consequently, $x' \in X_1 + X'_2 + \cdots + X'_m$. Now, modding out by X_1 yields $X_2 \oplus \cdots \oplus X_n \simeq X/X_1 \simeq X'_2 \oplus \cdots \oplus X'_m$. We proceed inductively with the same argument above and deduce that the two decompositions of X are isomorphic. □

How can we ensure that X admits a *local decomposition* as in the theorem? Using projectivization (11.50) we can translate this into a question about Γ: we want $\text{id}_X \in$

11.3 Grothendieck Groups of Object-Adapted Cellular Categories

Γ to decompose into pairwise orthogonal idempotents with local endomorphism rings. Rings with this property have a special name:

Definition 11.51 A ring A is called *semiperfect* if there is a decomposition $1 = e_1 + \cdots + e_n$ into pairwise orthogonal idempotents e_i with local endomorphism rings $\text{End}_A(Ae_i) = e_i A e_i = \text{End}_A(e_i A)$.

But we need to be careful! We just have a full embedding add $X \hookrightarrow$ proj-Γ and not an equivalence. So, even if Γ is semiperfect, this does not necessarily induce a local decomposition of X. We can enforce this with the following property:

Definition 11.52 The category $\mathcal{C}$ is *Karoubian* if $\text{Hom}_\mathcal{C}(X, -)$ induces an equivalence add $X \simeq$ proj-Γ for all $X \in \mathcal{C}$.

This coincides with the definition given at the beginning of Sect. 11.2.3, see also Exercise 11.56 below. From Theorem 11.50 and the definitions we then immediately get:

Theorem 11.53 (Krull–Schmidt) *The following are equivalent:*

1. *Every object of $\mathcal{C}$ has a local decomposition.*
2. *$\mathcal{C}$ is Karoubian and all endomorphism rings are semiperfect.*

Such a category has unique decompositions, and an object is indecomposable if and only if its endomorphism ring is local.

Definition 11.54 A category as in the theorem is called a *Krull–Schmidt* category.

> **Caution** Depending on the literature, *Krull–Schmidt* may also just mean having unique decompositions—and this is a weaker property (see Example 11.59 below). For the development of the theory here we really need the stronger property.

The Karoubian Property

How do we know that a category is Karoubian? This property is a special condition on *idempotents* of $\mathcal{C}$ (i.e. idempotent endomorphisms of objects). Namely, note that idempotents coming from direct summands are of the following special kind:

Definition 11.55 An idempotent $e \in \text{End}_\mathcal{C}(X)$ is called *split* if there is a factorization $X \xrightarrow{\pi} Y \xrightarrow{\iota} X$ of e with $\pi \circ \iota = \text{id}_Y$.

Exercise 11.56 Show the following:

1. $\mathcal{C}$ is Karoubian if and only if all idempotents in $\mathcal{C}$ split.

2. An idempotent e of X splits if and only if $\mathrm{id}_X - e$ has a cokernel, if and only if $\mathrm{id}_X - e$ has a kernel. (Cokernel and kernel are then given by π and ι, respectively.)
3. If an idempotent e splits and also $\mathrm{id}_X - e$ splits, then Y is a direct summand of X. In particular, in a Karoubian category, idempotents of X are in bijection with direct summands of X.
4. If $\mathcal{C}$ has kernels (or cokernels), it is Karoubian. In particular, any abelian category is Karoubian.
5. A full subcategory of a Karoubian category is Karoubian if and only if it is closed under direct summands.

Example 11.57 Let A be a ring. The category A-Mod of all (not necessarily finitely generated) A-modules is Karoubian since it is abelian. The categories A-mod and A-proj of finitely generated A-modules and finitely generated projective A-modules, respectively, are Karoubian since they are full subcategories which are closed under direct summands.

Example 11.58 For a non-Karoubian example, consider $R = \mathbb{Z}/6\mathbb{Z}$ and the category $\mathcal{C}$ of free modules of finite rank. We have $R \simeq \mathbb{Z}/3\mathbb{Z} \oplus \mathbb{Z}/2\mathbb{Z}$, so $P = \mathbb{Z}/2\mathbb{Z}$ is a direct summand of a free module. But it is clearly not free. So, $\mathcal{C}$ is not closed under direct summands, thus not Karoubian.

Example 11.59 Since A-proj is Karoubian for any ring A, it follows that A-proj is Krull–Schmidt if and only if A is semiperfect. The category $\mathbb{Z}$-proj has unique decompositions by Example 11.44 and it is Karoubian, but it is not Krull–Schmidt: $\mathbb{Z}$ is an indecomposable object but $\mathrm{End}_\mathcal{C}(\mathbb{Z}) = \mathbb{Z}$ is not local.

Semiperfect Rings

How do we know that endomorphism rings of $\mathcal{C}$ are semiperfect? The following proposition due to Azumaya [12] provides a large class of examples of semiperfect rings. The proof is elementary and relies on the fact that the property of being semiperfect is related to an idempotent lifting property.

Proposition 11.60 *For a local commutative ring $\Bbbk$ the following are equivalent:*

1. *Every $\Bbbk$-algebra which is finitely generated as a $\Bbbk$-module is semiperfect.*
2. *Hensel's lemma holds: if $P \in \Bbbk[X]$ is a monic polynomial, then any factorization of its reduction to the residue field of $\Bbbk$ into a product of coprime monic polynomials can be lifted to a factorization of P in $\Bbbk[X]$.*

In this case, $\Bbbk$ is called henselian.

Now, it is not too difficult to prove:

Lemma 11.61 *The following rings are all henselian: fields, complete local rings, local artinian rings.*

The Split Grothendieck Group of a Category with Shift Functor

Let $\mathcal{C}$ be a category with shift functor (1). The Laurent polynomial ring $\mathbb{Z}[v, v^{-1}]$ naturally acts on the split Grothendieck group $[\mathcal{C}]_\oplus$ with v acting by the shift (1). If $X \in \mathcal{C}$ is indecomposable, then so is any shift $X(n)$ since a decomposition of $X(n)$ would yield, after shifting with $(-n)$, a decomposition of X. Recall from Exercise 11.48 that if $\mathcal{C}$ has unique decompositions (e.g. if it is Krull–Schmidt), then $[\mathcal{C}]_\oplus$ is a free $\mathbb{Z}$-module. In the presence of a shift functor we would actually like $[\mathcal{C}]_\oplus$ to be a free $\mathbb{Z}[v, v^{-1}]$-module. Here is a common situation where this holds:

Proposition 11.62 *If $\mathcal{C}$ has unique decompositions and for each $X \in \mathcal{C}$ the degrees of $\mathrm{End}^\bullet_\mathcal{C}(X)$ are bounded above or below, then $[\mathcal{C}]_\oplus$ is a free $\mathbb{Z}[v, v^{-1}]$-module with basis the orbits of shifts of isomorphism classes of indecomposable objects of $\mathcal{C}$.*

This proposition follows directly from the following lemma.

Lemma 11.63 *If the degrees of $\mathrm{End}^\bullet_\mathcal{C}(X)$ are bounded below (or above), then $X \simeq X(n)$ implies $n = 0$ or $X = 0$.*

Proof We assume the degrees are bounded below (the other case follows analogously). Assume that $n \neq 0$. If $X \simeq X(n)$, then also $X \simeq X(n) \simeq (X(n))(n) = X(2n)$, so, inductively we obtain $X \simeq X(kn)$ for any $k \in \mathbb{Z}$. Choose k such that kn is less than the minimal degree of $\mathrm{End}^\bullet_\mathcal{C}(X)$. An isomorphism $X \to X(kn)$ is an element of $\mathrm{End}^{kn}_\mathcal{C}(X)$, but this homogeneous component is zero by assumption. Hence such an isomorphism must be zero, and hence $X = 0$. □

Appendix 2: Composition Forms, Cellular Forms, and Local Intersection Forms

This chapter has focused on two main themes. The first is the classification of the indecomposable objects for the Hecke category: they are all grading shifts of objects B_w for $w \in W$. The second is the Krull–Schmidt property, stating that any object in a sufficiently nice category (like the Hecke category) splits as a finite direct sum of indecomposable objects with (graded) local endomorphism rings. Putting these together, we see that any object X in the Hecke category has a splitting

$$X \simeq \bigoplus_{w \in W} B_w^{\oplus m_w}$$

where $m_w \in \mathbb{Z}[v, v^{-1}]$ is the graded multiplicity of B_w in X.

Now a practical question arises: given an arbitrary object X (e.g. a Bott–Samelson object), how does one compute these multiplicities m_w? The main tool here is the idea of the *local intersection form*, which we spend a little time motivating.

In Sect. 8.2.4 and specifically Exercise 8.39, we explained a morphism-theoretic interpretation of direct sums. An object X is isomorphic to a direct sum $M \oplus N$ if and only if there are inclusion and projection maps satisfying some standard relations. Let us extrapolate from this exercise to the case when there are many isomorphic direct summands. Suppose that B is an indecomposable object, and we wish to prove that $B \oplus B \oplus B$ is a direct summand of X. We will need to construct three inclusion maps $i_1, i_2, i_3 \colon B \to X$, and three projection maps $p_1, p_2, p_3 \colon X \to B$ which pair against each other to be orthonormal in the sense that

$$p_j \circ i_k = \delta_{j,k} \mathrm{id}_B. \tag{11.52}$$

If we can construct these maps, then the compositions $e_j = i_j \circ p_j \in \mathrm{End}(X)$ are mutually orthogonal idempotents, each picking out a copy of B as a direct summand of X. There may be other direct summands too, but this would be sufficient to prove that $B \oplus B \oplus B$ is a direct summand.

Let us rephrase this idea in the following definition and lemma.

Definition 11.64 The *composition pairing of X at B* is the map

$$\mathrm{Hom}(X, B) \times \mathrm{Hom}(B, X) \to \mathrm{End}(B), \qquad (f, g) \mapsto f \circ g \tag{11.53}$$

for any objects B and X. We call subsets $\{i_j\}_{j=1}^m \subset \mathrm{Hom}(B, X)$ and $\{p_j\}_{j=1}^m \subset \mathrm{Hom}(X, B)$ which satisfy (11.52) *dual sets*. (They are not dual bases since they need not span, though they are necessarily linearly independent). The *rank* of the composition pairing is the maximal size m of a pair of dual sets.

The above discussion implies the following:

Lemma 11.65 *In a Karoubian category, the rank of the composition pairing of X at B agrees with the maximal number m such that $B^{\oplus m} \stackrel{\oplus}{\subset} X$, which is usually called the* multiplicity *of B in X.*

One is usually interested in the case when B is indecomposable, although this lemma works for arbitrary objects B and X.

In the presence of a contravariant duality functor $\overline{(\cdot)}$ on the category (such as the involution coming from a cellular structure) which fixes the objects X and B, there is an isomorphism between $\mathrm{Hom}(B, X)$ and $\mathrm{Hom}(X, B)$, so that the composition pairing can be transferred to a *composition form* on $\mathrm{Hom}(X, B)$:

$$\mathrm{Hom}(X, B) \times \mathrm{Hom}(X, B) \to \mathrm{End}(B), \qquad (f, g) \mapsto f \circ \overline{g}. \tag{11.54}$$

The composition form is symmetric whenever $\overline{(\cdot)}$ acts trivially on $\mathrm{End}(B)$.

11.3 Grothendieck Groups of Object-Adapted Cellular Categories

Exercise 11.66 In the Temperley–Lieb category (Definition 7.13) where the circle evaluates to -2, use the basis of crossingless matchings to compute the following composition forms.

- The composition form of 2 at 0 (this is a 1×1 matrix).
- The composition form of 4 at 0 (this is a 2×2 matrix).
- (Extra credit) The composition form of 6 at 0 (this is a 5×5 matrix).
- The composition form of 3 at 1.
- The composition form of 5 at 1.

In each case, observe that the form is actually positive or negative definite! Make a conjecture as to the signature of the composition forms of $2n$ at 0, and $2n + 1$ at 1.

The next exercise uses this idea to compute explicit idempotents giving the direct sum decomposition of a particular Bott–Samelson bimodule.

Exercise 11.67 In this exercise, we work with Soergel bimodules in type B_2. We let $S = \{s, t\}$, so that $m_{st} = 4$, and use a realization over $\mathbb{R}$ with the non-symmetric Cartan matrix with $a_{st} = -1$ and $a_{ts} = -2$.

1. Write $b_s b_t b_s b_t$ as a sum of Kazhdan–Lusztig basis elements. How do you expect $B_s B_t B_s B_t$ to decompose? Can there be any summands in $B_s B_t B_s B_t$ besides B_{st} and B_{stst}? Why or why not?
2. Calculate $\underline{\text{rk}} \, \text{Hom}^\bullet(B_{st}, B_s B_t B_s B_t)$ and find a diagrammatic basis (over $\mathbb{R}$) of maps in degree 0. (It should be 2-dimensional.)
3. Calculate $\underline{\text{rk}} \, \text{Hom}^\bullet(B_s B_t B_s B_t, B_{st})$ and find a diagrammatic basis (over $\mathbb{R}$) of maps in degree 0. Given part (2), why is this really easy?
4. Calculate $\underline{\text{rk}} \, \text{End}^\bullet(B_{st})$ and deduce that the only degree-0 endomorphisms of B_{st} are $\mathbb{R}$-multiples of the identity.
5. Compute the matrix of the composition form on $B_s B_t B_s B_t$ at B_{st}, viewed as a bilinear form on $\text{Hom}^0(B_s B_t B_s B_t, B_{st})$ valued in $\text{End}^0(B_{st}) \simeq \mathbb{R}$.
6. Whenever two maps $B_s B_t B_s B_t \to B_{st}$ pair under the form in part (5) to give 1, one can construct an idempotent in $\text{End}^0(B_s B_t B_s B_t)$ that factors through B_{st}. Find a complete set of primitive orthogonal idempotents for $\text{End}^0(B_s B_t B_s B_t)$. Is your solution unique? (*Hint:* given an idempotent e, its complementary idempotent is $1 - e$. This is how one can find the idempotent projecting to the direct summand B_{stst} inside $B_s B_t B_s B_t$.)

In this exercise we used the grading to focus our attention on degree 0 maps in the category, which helped because the form became valued in $\mathbb{R}$ rather than in the infinite-dimensional graded ring $\text{End}^\bullet(B_{st})$. However, one must be more careful in general.

Suppose that we are in a Karoubian category with a grading shift, and that we are interested in computing the graded multiplicity of B in X: how many orthogonal summands of *shifts* of B appear inside X. In the corresponding graded category (whose Hom spaces are the graded Hom spaces of the original category), we now have a graded composition form or pairing of X at B, and we can speak of *homogeneous dual sets*, i.e. dual sets consisting of homogeneous inclusions $\{i_j\}$

and homogeneous projections $\{p_j\}$. The desired graded multiplicity agrees with the *graded rank* of this form, defined in an obvious way from homogeneous dual sets of maximal size.

It is true that, in order for homogeneous elements i_j and p_j to satisfy $p_j \circ i_j = \mathrm{id}_B$, we need $\deg i_j + \deg p_j = 0$. However, since B may have nonzero endomorphisms of nonzero degree, it is possible for $j \neq k$ that $\deg i_j + \deg p_k \neq 0$ and yet $p_j \circ i_k \neq 0$. In this generality, we therefore cannot restrict our attention to the part of the composition form or pairing that involve morphisms whose degrees add to zero. What allowed us to restrict our attention to degree 0 maps in Exercise 11.67 was the absence of any negative degree maps in $\mathrm{Hom}^\bullet(B_s B_t B_s B_t, B_{st})$, which implies that all inclusion and projection maps must have degree 0.

Exercise 11.68 In this exercise, we work with Soergel bimodules in type A_1 with $S = \{s\}$. Assume that $R = \mathbb{R}[\alpha_s]$.

1. Write $b_s b_s$ as a sum of Kazhdan–Lusztig basis elements. How do you expect $B_s B_s$ to decompose?
2. Calculate $\underline{\mathrm{rk}}\,\mathrm{Hom}^\bullet(B_s, B_s B_s)$. Find a diagrammatic basis (over $\mathbb{R}$) of maps in degree -1. (It should be 1-dimensional.) Find a diagrammatic basis (over $\mathbb{R}$) of maps in degree $+1$. (It should be 3-dimensional. Your basis should include: one light leaf of degree $+1$, one double leaf of degree $+1$ factoring through R, and a light leaf of degree -1 together with α_s on the left.)
3. Calculate $\underline{\mathrm{rk}}\,\mathrm{Hom}^\bullet(B_s B_s, B_s)$ and find a diagrammatic basis (over $\mathbb{R}$) of maps in degrees -1 and $+1$. Given part (2), why is this really easy?
4. Calculate $\underline{\mathrm{rk}}\,\mathrm{End}^\bullet(B_s)$. Deduce that the identity morphism and the broken line (the unique double leaf factoring through R, with degree 2) form a (left) R-basis for $\mathrm{End}^\bullet(B_s)$.
5. Compute the 4×4 matrix of the composition form on $B_s B_s$ at B_s, viewed as an $\mathrm{End}^\bullet(B_s)$-valued bilinear form on $\mathrm{Hom}^{-1}(B_s B_s, B_s) \oplus \mathrm{Hom}^{+1}(B_s B_s, B_s)$. Your answer should agree (up to relabeling the rows and columns) with the following matrix.

$$\begin{pmatrix} 0 & | & 0 & 0 \\ | & \boxed{\alpha_s} & \vdots & \boxed{\alpha_s} \\ 0 & \vdots & 0 & 0 \\ 0 & \boxed{\alpha_s} & 0 & 0 \end{pmatrix} \quad (11.55)$$

6. Find a complete set of primitive orthogonal idempotents for $\mathrm{End}^\bullet(B_s B_s)$. Is your solution unique? (*Hint:* see Sect. 8.2.4.)

Let's delve in to the computation of Exercise 11.68 more closely. The graded local ring $\mathrm{End}^\bullet(B_s)$ has a maximal ideal $\mathfrak{m}$ consisting of all positive degree endomorphisms. Taking the quotient by $\mathfrak{m}$, the composition form simplifies to the

11.3 Grothendieck Groups of Object-Adapted Cellular Categories

$\mathbb{R}$-valued form with matrix

$$\begin{pmatrix} 0 & 1 & 0 & 0 \\ 1 & 0 & 0 & 0 \\ 0 & 0 & 0 & 0 \\ 0 & 0 & 0 & 0 \end{pmatrix}.$$

The first two morphisms (the light leaves of degrees -1 and $+1$ respectively) can be used to construct dual sets of maximal size for this (rank 2) $\mathbb{R}$-valued form. However, they are not true projection and inclusion maps since they satisfy (11.52) only modulo $\mathfrak{m}$, and the true projection and inclusion maps will involve the other basis maps which are in the kernel of the $\mathbb{R}$-valued form. Thankfully, the rank of the true composition form agrees with the rank of this $\mathbb{R}$-valued form, and this is no accident.

Let us prove the foundational algebraic result which says that, when computing the multiplicity of an indecomposable object B, one can work modulo the maximal ideal of $\mathrm{End}(B)$ and still compute the rank of the composition pairing. This is a common idea in ring theory called idempotent lifting, and we have adapted the proof of [117, Proposition 21.22] in our proof below.

Proposition 11.69 *Let B and X be two objects in a (graded) additive category, and assume the ring $\mathrm{End}(B)$ is (graded) local, having (graded) Jacobson radical[5] $\mathfrak{m}_B$. Let $\{i_j\}_{j=1}^m \subset \mathrm{Hom}(B, X)$ and $\{p_j\}_{j=1}^m \subset \mathrm{Hom}(X, B)$ be (homogeneous) almost dual sets in that*

$$p_j \circ i_k \equiv \delta_{j,k} \mathrm{id}_B \pmod{\mathfrak{m}_B} \tag{11.56}$$

for all $1 \le j, k \le m$. Then one can modify the almost-inclusion maps i_j to find (homogeneous) dual sets $\{i'_j\}_{j=1}^m$ and $\{p_j\}_{j=1}^m$ satisfying $p_j \circ i'_k = \delta_{j,k} \mathrm{id}_B$. Moreover, $i'_j - i_j \in \mathrm{Hom}(B, X) \cdot \mathfrak{m}_B$, so that the new inclusion maps agree with the old almost-inclusion maps modulo the action of $\mathfrak{m}_B$.

---

[5] We do not assume that $\mathrm{End}(B)$ is commutative! In a local ring, there is a unique maximal left ideal, which coincides with the unique maximal right ideal, and with the Jacobson radical. It consists of all non-unit elements of the local ring. The quotient of $\mathrm{End}(B)$ by $\mathfrak{m}_B$ is a division algebra. (See [117, Theorem 19.1] for these facts about (noncommutative) local rings.) For any element x of the Jacobson radical, $1 + x$ is invertible, and using locality one can prove that the right inverse is actually a two-sided inverse.

In a graded local ring, there is a unique maximal homogeneous left ideal, coinciding with the unique maximal homogeneous right ideal and with the graded Jacobson radical. It contains all homogeneous non-unit elements of the graded local ring (as follows from [143, Proposition 2.9.1(vi)]). It is not true that $1 + x$ is invertible for any x in the graded Jacobson radical; consider for example the polynomial ring $\Bbbk[x]$. However, the degree zero part of a graded local ring is an ordinary local ring, and the degree zero part of the graded Jacobson radical is the ordinary Jacobson radical of the degree zero part of the ring [143, Corollary 2.9.3]. Hence, if x is in the graded Jacobson radical and x has degree zero, then $1 + x$ is invertible.

Proof We will argue in several steps. In each step we replace the entire suite of almost-inclusion maps $\{i_j\}$ by successively better almost-inclusion maps, until we finally reach $\{i'_j\}$. For sanity's sake, we call the input to each step i_j, and the output of each step i'_j, even though it is only the final product which will satisfy the requirements of the proposition.

Let us first argue that we can modify the inclusion maps to assume that $p_j \circ i'_j = \mathrm{id}_B$ on the nose. Note that $p_j \circ i_j = \mathrm{id}_B + r_j$ for some $r_j \in \mathfrak{m}_B$, and that $\mathrm{id}_B + r_j$ is invertible in $\mathrm{End}(B)$ (this is also true in the graded setting, since r_j must have degree zero). Now set $i'_j = i_j(\mathrm{id}_B + r_j)^{-1}$. Then $p_j \circ i'_j = \mathrm{id}_B$ as desired. Note also that $(\mathrm{id}_B + r_j)^{-1} \equiv \mathrm{id}_B \pmod{\mathfrak{m}_B}$, as is easily computed in the ring $\mathrm{End}(B)/\mathfrak{m}_B$. So $i'_j - i_j \in \mathrm{Hom}(B, X) \cdot \mathfrak{m}_B$.

For the rest of the proof we can assume that $p_j \circ i_j = \mathrm{id}_B$ for all $1 \leq j \leq m$. For $j \neq k$ we write $r_{jk} = p_j \circ i_k \in \mathfrak{m}_B$. We will prove the result by induction on m, but for pedagogical reasons, let us first examine the case $m = 2$. Note that $r_{12}r_{21} \in \mathfrak{m}_B$ so $\mathrm{id}_B - r_{12}r_{21}$ is invertible (again, this is also true in the graded setting, since $r_{12}r_{21}$ must have degree zero). Let us set

$$i'_1 = (\mathrm{id}_X - i_2 p_2) \circ i_1 \circ (\mathrm{id}_B - r_{12}r_{21})^{-1}, \tag{11.57}$$

$$i'_2 = i_2 - i'_1 \circ r_{12}. \tag{11.58}$$

Note that we have modified i_1 and i_2 by adding elements of $\mathrm{Hom}(B, X) \cdot \mathfrak{m}_B$. Then one can compute that $p_2 i'_1 = 0$ since $p_2(1 - i_2 p_2) = 0$. We can compute that

$$p_1 i'_1 = (\mathrm{id}_B - r_{12}r_{21}) \circ (\mathrm{id}_B - r_{12}r_{21})^{-1} = \mathrm{id}_B.$$

From this we see that

$$p_1 i'_2 = p_1 i_2 - r_{12} = 0$$

and

$$p_2 i'_2 = \mathrm{id}_B - 0 = \mathrm{id}_B.$$

Hence $\{i'_1, i'_2\}$ and $\{p_1, p_2\}$ are dual sets.

Now we do the general step of the induction. Suppose that $\{i_2, \ldots, i_m\}$ and $\{p_2, \ldots, p_m\}$ are dual sets, and that $p_1 i_1 = \mathrm{id}_B$, but that $p_1 i_j = r_{1j}$ and $p_j i_1 = r_{j1}$ might be nonzero elements of $\mathfrak{m}_B$. Set

$$i'_1 = (\mathrm{id}_X - i_2 p_2 - i_3 p_3 - \ldots - i_m p_m) \circ i_1 \circ (\mathrm{id}_B - \sum_{j \neq 1} r_{1j} r_{j1})^{-1}, \tag{11.59}$$

and for all $j \neq 1$ set

$$i'_j = i_j - i'_1 \circ r_{1j}. \tag{11.60}$$

11.3 Grothendieck Groups of Object-Adapted Cellular Categories

We leave the reader to confirm that these elements $\{i'_j\}$ form a dual set to $\{p_j\}$. □

So to restate the result, let us make a definition.

Definition 11.70 Let B and X be two objects in a (graded) additive category, and assume the ring $\mathrm{End}(B)$ is (graded) local, having (graded) Jacobson radical $\mathfrak{m}_B$. The *local composition pairing* of X at B is the composition

$$\mathrm{Hom}(X, B) \times \mathrm{Hom}(B, X) \to \mathrm{End}(B) \to \mathrm{End}(B)/\mathfrak{m}_B \tag{11.61}$$

which sends (f, g) to the image of $f \circ g$ in this quotient.

Being a bilinear pairing valued in a division algebra, it already makes sense to discuss the rank of this pairing. The following corollary is an immediate implication of Proposition 11.69.

Corollary 11.71 *The (graded) rank of the composition pairing is equal to the (graded) rank of the local composition pairing.*

In theory the local composition pairing is easier to compute than the composition pairing, because while $\mathrm{End}(B)$ can be rather complicated, often $\mathrm{End}(B)/\mathfrak{m}_B$ is just a field, spanned by the identity map. If one is only interested in the rank of these forms (i.e. the multiplicity of B in X) then the local composition pairing is a useful simplification. However, the local composition pairing does not make it any easier to explicitly compute the idempotents themselves.

Let us focus on the diagrammatic Hecke category $\mathcal{H}$, which we will view as a graded category (rather than a category with a grading shift). What is the Jacobson radical $\mathfrak{m}_w$ of $\mathrm{End}(B_w)$? We claim that it is actually the sum of certain 2-sided ideals of the category $\mathcal{H}$ itself. First, we have the cellular ideal $\mathcal{H}_{<w}$ consisting of all morphisms which factor through indecomposables B_x for $x < w$ (or which factor through the Bott–Samelson bimodules for their reduced expressions). Using the double leaves basis, we see that $\mathrm{End}(B_w)/\mathrm{End}_{<w}(B_w) = R \cdot \mathrm{id}_{B_w}$ is a free R-module of rank 1. Now this has a maximal proper graded ideal $R_+ \cdot \mathrm{id}_{B_w}$ consisting of positive degree elements. In fact, $R_+ \cdot \mathcal{H}$ is a 2-sided ideal in the category, spanned by all double leaves with positive degree polynomials on the left (there is also a version where R_+ acts on the right). Let $\mathcal{H}_{<w,+}$ denote the sum of these two ideals. Then $\mathrm{End}_{<w,+}(B_w)$ is a proper two-sided ideal inside $\mathrm{End}(B_w)$ and must be contained in the Jacobson radical. Moreover, $\mathrm{End}(B_w)/\mathrm{End}_{<w,+}(B_w) \simeq \Bbbk \cdot \mathrm{id}_{B_w}$. If $\Bbbk$ is a field, then we have clearly identified the Jacobson radical as $\mathfrak{m}_w = \mathrm{End}_{<w,+}(B_w)$. When $\Bbbk$ is a commutative local ring, we should also throw in the maximal ideal of $\Bbbk$. For ease of discussion, let us assume $\Bbbk$ is a field below.

Remark 11.72 One can also say that $\mathfrak{m}_w$ is spanned by $\mathrm{End}_{<w}(B_w)$ and by all endomorphisms of positive degree. This is because, modulo $\mathrm{End}_{<w}(B_w)$, all endomorphisms of positive degree lie within $R_+ \cdot \mathrm{id}_{B_w}$.

The fact that $\mathfrak{m}_w$ is the span inside $\mathrm{End}(B_w)$ of an ideal in the category makes the computation of the local composition pairing even easier, thanks to the following exercise.

Exercise 11.73 Continue the setup of Definition 11.70. Suppose that I is an ideal in the category such that $\mathrm{End}_I(B) \subset \mathfrak{m}_B$. Then any elements of $\mathrm{Hom}_I(X, B)$ or $\mathrm{Hom}_I(B, X)$ are in the kernel of the local composition pairing.

This leads us to the following definition. For $X, Y \in \mathcal{H}$, define the graded R-module

$$\mathrm{Hom}_{\not< w}(X, Y) := \mathrm{Hom}(X, Y) / \mathrm{Hom}_{<w}(X, Y)$$

and the graded $\Bbbk$-module

$$\mathrm{Hom}_{\not< w, \Bbbk}(X, Y) := \mathrm{Hom}(X, Y) / \mathrm{Hom}_{<w, +}(X, Y).$$

We also write $\mathrm{End}_{\not< w}(X)$ for $\mathrm{Hom}_{\not< w}(X, X)$ and $\mathrm{End}_{\not< w, \Bbbk}(X)$ for $\mathrm{Hom}_{\not< w, \Bbbk}(X, X)$.

Definition 11.74 Let X be an object in $\mathcal{H}$, and $w \in W$. The *local intersection pairing* of X at w is the $\Bbbk$-valued pairing

$$I_{X,w} : \mathrm{Hom}_{\not< w, \Bbbk}(X, B_w) \times \mathrm{Hom}_{\not< w, \Bbbk}(B_w, X) \to \mathrm{End}_{\not< w, \Bbbk}(B_w) \simeq \Bbbk \qquad (11.62)$$

induced by composition: it takes the classes of two map $f \in \mathrm{Hom}(X, B_w)$ and $g \in \mathrm{Hom}(B_w, X)$ and returns the degree zero coefficient of the identity map when $f \circ g \in \mathrm{End}(B_w)$ is expanded in the double leaves basis. Using duality, we can view this as a form on the space $\mathrm{Hom}_{\not< w, \Bbbk}(X, B_w)$, which we call the *local intersection form*. Sometimes we abuse terminology and also refer to the R-valued pairing

$$\mathrm{Hom}_{\not< w}(X, B_w) \times \mathrm{Hom}_{\not< w}(B_w, X) \to \mathrm{End}_{\not< w}(B_w) \simeq R \qquad (11.63)$$

as the *local intersection pairing*.

We can refine this to take degree shifts into account, as follows. For any $d \in \mathbb{Z}$, morphisms in $\mathrm{Hom}^d_{\not< w, \Bbbk}(B_{\underline{x}}, B_w)$ can only pair nontrivially with morphisms in $\mathrm{Hom}^{-d}_{\not< w, \Bbbk}(B_w, B_{\underline{x}})$. We therefore consider the restriction

$$I^d_{X,w} : \mathrm{Hom}^d_{\not< w, \Bbbk}(X, B_w) \times \mathrm{Hom}^{-d}_{\not< w, \Bbbk}(B_w, X) \to \Bbbk \qquad (11.64)$$

of $I_{X,w}$, which we call the *d-th graded piece* of the local intersection pairing. The *graded rank* of the local intersection pairing is defined to be

$$\sum_{d \in \mathbb{Z}} \mathrm{rk}(I^d_{X,w}) v^d \in \mathbb{Z}_{\geq 0}[v, v^{-1}]. \qquad (11.65)$$

11.3 Grothendieck Groups of Object-Adapted Cellular Categories

Corollary 11.75 *The graded rank of the local intersection pairing $I_{X,w}$ agrees with the graded multiplicity of B_w as a summand in X.*

Proof By Exercise 11.73, the graded rank of the local intersection pairing agrees with the graded rank of the local composition pairing. By Corollary 11.71, this agrees with the graded rank of the composition pairing and the graded multiplicity of B_w in X. □

Now, consider the case when $X = \mathrm{BS}(\underline{x})$ is a Bott–Samelson object in $\mathcal{H}$. If we choose a reduced expression $\underline{w}$ for w, then

$$\mathrm{Hom}_{\not< w}(\mathrm{BS}(\underline{x}), B_w) \simeq \mathrm{Hom}_{\not< w}(\mathrm{BS}(\underline{x}), \mathrm{BS}(\underline{w}))$$

is a graded free R-module with basis given by the (images of) light leaves maps with target w (not the double leaves maps), so $\mathrm{Hom}_{\not< w, \Bbbk}(\mathrm{BS}(\underline{x}), B_w)$ is a graded free $\Bbbk$-module with the same basis. Consequently, the local intersection form on $\mathrm{BS}(\underline{x})$ at w can be viewed as an R-valued or $\Bbbk$-valued form on the space of light leaves with target w.

To reiterate, the local intersection form on $\mathrm{BS}(\underline{x})$ at w is an R-valued or $\Bbbk$-valued matrix with rows and columns labeled by subexpressions of $\underline{x}$ which express w. One computes the entries of the matrix by composing one light leaf with the other light leaf turned upside-down to get an endomorphism of $\mathrm{BS}(\underline{w})$, and then computing the coefficient of the identity map $\mathrm{id}_{\mathrm{BS}(\underline{w})}$ in the double leaves basis (i.e. killing all diagrams which factor through lower terms).

Let us record the following observation.

Proposition 11.76 *For any expression $\underline{x}$ and $w \in W$, the local intersection pairing of $\mathrm{BS}(\underline{x})$ at w is precisely the cellular pairing (Definition 11.35) between $E(\underline{x}, w)$ and $\overline{E}(w, \underline{x})$.*

Exercise 11.77 Prove Proposition 11.76.

Exercise 11.78 More generally, let $\mathcal{C}$ be any object-adapted cellular category over a ring R which is either local or graded local. Fix an object X and a cell $\lambda \in \Lambda$. Prove that the rank of the cellular pairing between $E(X, \lambda)$ and $\overline{E}(\lambda, X)$ agrees with the multiplicity of B_λ in X.

Exercise 11.79 Continuing Exercise 11.66, compute the cellular form of 4 at 2 and the cellular form of 5 at 3. What are the ranks of these pairings? What multiplicity did you just compute? Make observations on the definiteness of these forms.

Exercise 11.80 Compute the following local intersection forms. (As usual, s, t, u, v denote distinct simple reflections.)

1. $\mathrm{BS}(s, t, s, t)$ at st in type B_2 (cf. Exercise 11.67).
2. $\mathrm{BS}(s, t, s, t, s)$ at sts and s in type H_2.
3. $\mathrm{BS}(s, u, v, t, s, u, v)$ at suv in type D_4, where s, u, v all commute (cf. Exercise 3.39).

In each case, make observations about definiteness and signature in degree 0.

Remark 11.81 All these observations about the signature of various local intersection forms will be explained in Chap. 18.

Chapter 12
How to Draw Soergel Bimodules

This chapter is based on expanded notes of a lecture given by
Leonardo Patimo
and taken by
Seth Shelley-Abrahamson and **Siddharth Venkatesh**

Abstract In this chapter, we use the diagrammatic descriptions of morphisms between Bott–Samelson bimodules to give diagrammatic descriptions of two different bases for these bimodules. First, we describe the 01-basis, which is constructed by taking an iterated tensor product of a natural basis of B_s. The 01-basis is depicted diagrammatically as a sequence of straight vertical broken or unbroken lines. We define a commutative multiplication structure on Bott–Samelson bimodules, relate this multiplication to the diagrammatics of the 01-basis, and use these tools to define and prove the nondegeneracy of the global intersection form, a certain invariant R-bilinear form on Bott–Samelson bimodules. We then describe and give diagrammatics for another basis using light leaves. We show that this latter basis is compatible with the standard filtration on Bott–Samelson bimodules in a natural way, and we give a concrete example in the case of $\mathrm{BS}(s, t, s)$ with $m_{st} = 3$.

12.1 The 01-Basis

Let (W, S) be a Coxeter system. For any simple reflection $s \in S$, recall from Chap. 4 that $B_s := R \otimes_{R^s} R(1)$ is free over R as a left and right R-module with a basis given by $c_{\mathrm{id}} = 1 \otimes 1$ and $c_s = \frac{\alpha_s}{2} \otimes 1 + 1 \otimes \frac{\alpha_s}{2}$. Fix an expression $\underline{w} = (s_1, \ldots, s_m)$. Then we can obtain a right (or left) R-module basis for

$$\mathrm{BS}(\underline{w}) = B_{s_1} \otimes_R B_{s_2} \otimes \cdots \otimes_R B_{s_m}$$

L. Patimo
Max Planck Institute for Mathematics, Bonn, Germany

S. Shelley-Abrahamson · S. Venkatesh
Department of Mathematics, Massachusetts Institute of Technology, Cambridge, MA, USA

© The Editor(s) (if applicable) and The Author(s), under exclusive licence
to Springer Nature Switzerland AG 2020
B. Elias et al., *Introduction to Soergel Bimodules*, RSME Springer Series 5,
https://doi.org/10.1007/978-3-030-48826-0_12

simply by tensoring together the bases for each B_{s_i}. That is, for each subexpression $\underline{e} \subset \underline{w}$, viewed as a 01-sequence $e_1, \ldots, e_m \in \{0, 1\}$, set

$$c_{\underline{e}} := c_{s_1}^{e_1} \otimes \cdots \otimes c_{s_m}^{e_m}. \tag{12.1}$$

Then $\{c_{\underline{e}} : \underline{e} \subset \underline{w}\}$ forms a right (or left) R-module basis for $\mathrm{BS}(\underline{w})$.

Remark 12.1 Keep in mind that $c_{\underline{e}}$ depends on the expression $\underline{w}$. Usually, $\underline{w}$ will be fixed or will be obvious from context, and we suppress it from the notation.

Remark 12.2 While the set $\{c_{\underline{e}}\}$ is a basis for $\mathrm{BS}(\underline{w})$ as both a right and a left R-module, these right and left R-module structures are in general distinct. In particular, the coefficients of an element $b \in \mathrm{BS}(\underline{w})$ with respect to the 01-basis is, in general, different for the right and left R-module structures. For the rest of this chapter, we will only be concerned with the 01-basis as a basis for the right R-module structure.

We identify two special elements in this basis.

Definition 12.3 Define

$$c_{\mathrm{bot}} := c_{0\cdots 0} = 1 \otimes \cdots \otimes 1 \quad \text{and} \quad c_{\mathrm{top}} := c_{1\cdots 1} = c_{s_1} \otimes \cdots \otimes c_{s_m}. \tag{12.2}$$

Note that the degree of the element $c_{\underline{e}}$ is equal to the number of 1s appearing in $\underline{e}$ minus the number of 0s:

$$\deg c_{\underline{e}} = \#\{i \mid e_i = 1\} - \#\{i \mid e_i = 0\}.$$

Hence, c_{bot} is the unique basis element of minimum degree $-\ell(\underline{w})$, and c_{top} is the unique basis element of maximum degree $\ell(\underline{w})$.

We can represent the elements of the 01-basis diagrammatically by viewing them as images of c_{bot} under certain endomorphisms of $\mathrm{BS}(\underline{w})$.

Definition 12.4 Given $\underline{e} \subset \underline{w}$, let $\phi_{\underline{e}} : \mathrm{BS}(\underline{w}) \to \mathrm{BS}(\underline{w})$ be the tensor product $\phi_1 \otimes \cdots \otimes \phi_m$ of morphisms $\phi_i : B_{s_i} \to B_{s_i}$, $1 \leq i \leq m$, where each ϕ_i is defined as follows:

1. If $e_i = 0$, then ϕ_i is the identity map. Diagrammatically, this map is depicted by a straight vertical line:

2. If $e_i = 1$, then ϕ_i is the map sending $c_0 \mapsto c_{s_i}$. Diagrammatically, this map is depicted by a broken vertical line:

12.2 Commutative Ring Structure on a Bott–Samelson Bimodule

We write $\phi_{\text{bot}} = \phi_{0\cdots 0}$ and $\phi_{\text{top}} = \phi_{1\cdots 1}$ in analogy with c_{bot} and c_{top}.

Example 12.5 If $\underline{w} = (s, t)$, the endomorphisms $\phi_{\underline{e}}$ of $\text{BS}(\underline{w})$ are depicted diagrammatically as follows:

$\underline{e} \subset (s, t)$	00	01	10	11
$\phi_{\underline{e}}$				

It is clear from the definitions that $\phi_{\underline{e}}(c_{\text{bot}}) = c_{\underline{e}}$. As an abuse of notation, we will often depict $c_{\underline{e}}$ by the diagram depicting $\phi_{\underline{e}}$.

We end this section by noting a useful property of c_{top}.

Lemma 12.6 *For any $f \in R$,*

$$f c_{\text{top}} = c_{\text{top}} f. \tag{12.3}$$

Proof Diagrammatics makes this apparent; since every line in ϕ_{top} is broken, one can slide f from left to right:

□

12.2 Commutative Ring Structure on a Bott–Samelson Bimodule

Let $\underline{w} = (s_1, \ldots, s_m)$ be an expression with length $m = \ell(\underline{w})$. Since R is a commutative ring, so is

$$R \otimes_{R^{s_1}} R \otimes_{R^{s_2}} \cdots \otimes_{R^{s_m}} R,$$

where the multiplication is component-wise. The identity of this ring is $1 \otimes 1 \otimes \cdots \otimes 1$. For example, when $m = 1$, we have

$$(f_1 \otimes f_2) \cdot (g_1 \otimes g_2) = (f_1 g_1) \otimes (f_2 g_2) \in R \otimes_{R^{s_1}} R.$$

Hence, for any expression $\underline{w}$, the Bott–Samelson bimodule

$$\text{BS}(\underline{w}) = R \otimes_{R^{s_1}} R \otimes_{R^{s_2}} \cdots \otimes_{R^{s_m}} R(m)$$

is a commutative ring, but with an unusual grading convention, since the identity element c_{bot} lives in degree $-m$. One has $\deg(fg) = \deg(f) + \deg(g) + m$. Of course, the grading shift $\text{BS}(\underline{w})(-m)$ is a graded ring in the ordinary sense.

Remark 12.7 One potential confusion is that

$$\text{BS}(\underline{w}) = B_{s_1} \otimes B_{s_2} \otimes \cdots \otimes B_{s_m}$$

and that each B_s is a Frobenius algebra object, so one might attempt to define a ring structure using the multiplication map from this Frobenius structure, componentwise. This is not well-defined, nor what we are discussing here.

For any element $f = f_1 \otimes \cdots \otimes f_{m+1} \in \text{BS}(\underline{w})$, the operator of multiplication by f commutes with the R-bimodule action on $\text{BS}(\underline{w})$. Thus there is a graded algebra homomorphism

$$\text{BS}(\underline{w})(-m) \to \text{End}^\bullet(\text{BS}(\underline{w})). \tag{12.4}$$

Let P denote the image of this map, a commutative subalgebra of $\text{End}^\bullet(\text{BS}(\underline{w}))$. It is clear from the definition of multiplication in $\text{BS}(\underline{w})$ that we have

$$f(c_{\text{bot}})g(c_{\text{bot}}) = (f \circ g)(c_{\text{bot}}) \tag{12.5}$$

for all $f, g \in P$.

Remark 12.8 We are phrasing this discussion in the algebraic context of R-bimodules. However, it is equally easy to define a graded algebra homomorphism

$$R \otimes_{R^{s_1}} R \otimes_{R^{s_2}} \cdots \otimes_{R^{s_m}} R \to \text{End}_{\mathcal{H}_{\text{BS}}}(\underline{w}) \tag{12.6}$$

in the diagrammatic context, sending $f_1 \otimes \cdots \otimes f_{m+1}$ to

$$\boxed{f_1}\,\boxed{f_2} \quad \cdots \quad \boxed{f_m}\,\boxed{f_{m+1}}.$$

Because of (8.25d), each $\phi_{\underline{e}}$ is in P. Using (12.5), we have the following description of multiplication in the 01-basis.

Proposition 12.9 For all $\underline{e}, \underline{f} \subset \underline{w}$, we have:

$$c_{\underline{e}} \cdot c_{\underline{f}} = (\phi_{\underline{e}} \circ \phi_{\underline{f}})(c_{\text{bot}}).$$

Example 12.10 Consider $\underline{w} = (s, t, s)$ with $m_{st} = 3$. Let us compute $c_{101}c_{100}$ diagrammatically. We have

$$\phi_{101} \circ \phi_{100} = \quad \cdots \quad = \boxed{\alpha_s} \quad \cdots \quad = \boxed{t(\alpha_s)} \quad - \quad \cdots$$

by the polynomial forcing relation (10.7), since $\partial_t(\alpha_s) = -1$ when $m_{st} = 3$. Hence,

$$c_{101}c_{100} = -c_{\text{top}} + c_{101}t(\alpha_s).$$

12.3 Trace and the Global Intersection Form

Let $\underline{w}$ be any expression. Using the 01-basis for $\text{BS}(\underline{w})$ as a right R-module, we can define a trace function on $\text{BS}(\underline{w})$ as follows.

Definition 12.11 The *trace* on $\text{BS}(\underline{w})$ is the map

$$\text{Tr}: \text{BS}(\underline{w}) \to R \qquad (12.7)$$

sending any element b to the coefficient of c_{top} when b is expressed in the 01-basis.

Remark 12.12 It is very important to remember that we have fixed the 01-basis to be a *right* R-module basis for $\text{BS}(\underline{w})$. The trace function that would be defined by taking coefficients as a left R-module basis is different from the one defined above.

Using this trace and multiplication, we can define a symmetric bilinear form on $\text{BS}(\underline{w})$.

Definition 12.13 The *global intersection form* on $\text{BS}(\underline{w})$ is the R-valued pairing

$$\langle -, - \rangle : \text{BS}(\underline{w}) \times \text{BS}(\underline{w}) \to R \qquad (12.8)$$

defined by

$$\langle a, b \rangle = \text{Tr}(ab). \qquad (12.9)$$

Example 12.14

1. If $\underline{w} = (s, t, s)$ with $m_{st} = 3$, then the computation in Example 12.10 shows that $\langle c_{101}, c_{100} \rangle = -1$.
2. Let $\underline{w} = (s_1, \ldots, s_m)$. Since $c_{s_i}c_{s_i} = \alpha_{s_i}c_{s_i}$, we have

$$C_{\text{top}} C_{\text{top}} = \prod_{i=1}^{m} \alpha_{s_i} C_{\text{top}} = C_{\text{top}} \prod_{i=1}^{m} \alpha_{s_i}$$

as polynomials commute with c_{top}. Hence,

$$\langle c_{\text{top}}, c_{\text{top}} \rangle = \prod_{i=1}^{m} \alpha_{s_i}.$$

The global intersection form is symmetric (i.e. $\langle a, b \rangle = \langle b, a \rangle$) because multiplication is commutative. Here are some other properties of this form.

Proposition 12.15 *Let $\underline{w}$ be an expression. Let $a, b \in \text{BS}(\underline{w})$ and $f \in R$. Then,*

1. $\deg(\langle a, b \rangle) = \deg(a) + \deg(b)$.
2. $\langle fa, b \rangle = \langle a, fb \rangle$.
3. $\langle af, b \rangle = \langle a, bf \rangle = \langle a, b \rangle f$.

The proofs of these properties are obvious from the description of multiplication as vertical stacking of diagrams in the 01-basis and from the interaction of the multiplication and grading. Note the difference between left and right multiplication: the form is bilinear for right multiplication by $f \in R$, but is only self-adjoint for left multiplication.

Example 12.16 Let $\underline{w} = (s, t)$ with $m_{st} = 3$. Using the diagrams for the morphisms $\phi_{\underline{e}}$ from Example 12.5, we can compute $c_{\underline{e}} c_{\underline{f}}$ and compute their intersection pairing. Here is the resulting Gram matrix of the global intersection form on $\text{BS}(\underline{w})$ in the 01-basis.

	c_{00}	c_{01}	c_{10}	c_{11}
c_{00}	0	0	0	1
c_{01}	0	0	1	α_t
c_{10}	0	1	-1	α_s
c_{11}	1	α_t	α_s	$\alpha_s \alpha_t$

The entry in the row corresponding to $c_{\underline{e}}$ and the column corresponding to $c_{\underline{f}}$ is $\langle c_{\underline{e}}, c_{\underline{f}} \rangle$. Reordering the rows, we get

	c_{00}	c_{01}	c_{10}	c_{11}
c_{11}	1	α_t	α_s	$\alpha_s \alpha_t$
c_{10}	0	1	-1	α_s
c_{01}	0	0	1	α_t
c_{00}	0	0	0	1

12.3 Trace and the Global Intersection Form

which is upper triangular with 1s on the diagonal. Hence, the Gram matrix has determinant ± 1, and the global intersection form is nondegenerate.

This example generalizes, and the global intersection form on Bott–Samelson bimodules is nondegenerate in general. To show this, we need to use the lexicographic order on subexpressions.

Definition 12.17 Fix an expression $\underline{w}$. The *lexicographic order* on the set of subexpressions $\underline{e} \subset \underline{w}$ is the total order defined by setting $\underline{e} < \underline{f}$ if and only there exists some index i, $1 \leq i \leq \ell$, such that $e_i = 0$, $f_i = 1$ and $e_j = f_j$ for $j < i$. In plain English, this means that $\underline{e} \neq \underline{f}$ and, when reading from left to right, at the first index i where $\underline{e}$ and $\underline{f}$ differ we have $e_i = 0$ and $f_i = 1$.

Proposition 12.18 *Fix an expression $\underline{w}$. For all $\underline{e}, \underline{f} \subset \underline{w}$,*

$$\underline{e} < \underline{f} \quad \Rightarrow \quad \langle c_{\underline{e}}, c_{\underline{f}^\circ} \rangle = 0. \tag{12.10}$$

Here, $\underline{f}^\circ$ is the subexpression of $\underline{w}$ obtained by replacing every 0 in $\underline{f}$ by 1 and vice versa, i.e. $(\underline{f}^\circ)_i = 1 - f_i$. Additionally,

$$\langle c_{\underline{e}}, c_{\underline{e}^\circ} \rangle = 1. \tag{12.11}$$

Hence, the global intersection form on $\mathrm{BS}(\underline{w})$ *is nondegenerate.*

Proof Suppose $\underline{f} > \underline{e}$ in lexicographic order. Let i be the smallest number such that $e_i \neq f_i$, so that $e_i = 0$, $f_i = 1$. The first $i-1$ entries of $\underline{e}$ and $\underline{f}$ are the same, so the first $i-1$ entries of $\underline{e}$ and $\underline{f}^\circ$ are different and the i-th entry of both $\underline{e}$ and $\underline{f}^\circ$ is 0. We proved earlier that multiplication in the 01-basis is composition, i.e. vertical stacking, of the corresponding morphisms. So, if we stack $\phi_{\underline{e}}$ with $\phi_{\underline{f}^\circ}$, the first $k-1$ strands are broken vertical strands ϕ_1 and the k-th strand is an unbroken vertical strand ϕ_0. In particular, barbells only appear to the right of the unbroken strand in position k. In order to determine the coefficient in $c_{\underline{e}} c_{\underline{f}^\circ}$ of each basis element $c_{\underline{e}}$, we need to push all polynomials to the right using the polynomial forcing relation (10.7). This may break some strands along the way but, critically, since polynomials only appear to the right of the unbroken strand in position k, the k-th strand will never break. In particular, all basis elements $c_{\underline{g}}$ appearing with nonzero coefficient in this manner must have $g_i = 0$, and in particular $c_{\underline{g}} \neq c_{\mathrm{top}}$. This proves the first statement.

The second statement follows from the fact that

$$\phi_{\underline{e}} \circ \phi_{\underline{e}^\circ} = \phi_{\mathrm{top}}. \tag{12.12}$$

Finally, to see the nondegeneracy, note that if we order the columns lexicographically and the rows reverse lexicographically (so that $\underline{e}^\circ$ is ordered lexicographically along the rows), then the Gram matrix of the intersection form becomes upper triangular with 1s on the diagonal. □

12.4 Bott–Samelson Bimodules and the Light Leaves Basis

The 01-basis for Bott–Samelson bimodules is very well-suited to computing multiplication and the global intersection form. We now describe a different basis that is more compatible with the standard filtration on Bott–Samelson bimodules.

Fix an expression $\underline{w}$, and let $b \in \mathrm{BS}(\underline{w})$. The existence of the 01-basis shows that we can find some endomorphism ϕ of $\mathrm{BS}(\underline{w})$ such that $\phi(c_{\mathrm{bot}}) = b$. Recall the double leaves basis from Sect. 10.4. In particular, the graded endomorphism space $\mathrm{End}^\bullet(\mathrm{BS}(\underline{w}))$ has a right R-basis

$$\{\overline{LL}_{\underline{w},\underline{f}} \circ LL_{\underline{w},\underline{e}} : \text{subexpressions } \underline{e}, \underline{f} \subset \underline{w} \text{ such that } \underline{w}^{\underline{e}} = \underline{w}^{\underline{f}}\},$$

where $LL_{\underline{w},\underline{e}}$ denotes the light leaf corresponding to $\underline{w}, \underline{e}$ and $\overline{LL}_{\underline{w},\underline{f}}$ the upside down light leaf corresponding to $\underline{w}, \underline{f}$. Diagrammatically, $\overline{LL}_{\underline{w},\underline{f}} \circ LL_{\underline{w},\underline{e}}$ was depicted as an hourglass

where the middle horizontal line separating the trapezoids represents the Bott–Samelson bimodule corresponding to a chosen reduced expression for $\underline{w}^{\underline{e}} = \underline{w}^{\underline{f}}$.

We can therefore write

$$\phi = \sum_{\substack{\underline{e},\underline{f}\subset\underline{w}:\\ \underline{w}^{\underline{e}}=\underline{w}^{\underline{f}}}} \left(\overline{LL}_{\underline{w},\underline{f}} \circ LL_{\underline{w},\underline{e}}\right) g_{\underline{e},\underline{f}}$$

for uniquely defined coefficients $g_{\underline{e},\underline{f}} \in R$. To compute $\phi(c_{\mathrm{bot}})$, let us analyze what light leaves do to c_{bot}.

Let $\underline{w}$ be an expression. For each subexpression $\underline{e} \subset \underline{w}$, we defined in Sect. 3.3.4 its associated stroll and decorated sequence $d_1, \ldots, d_{\ell(\underline{w})} \in \{U0, U1, D0, D1\}$. If $d_i \in \{U0, U1\}$ for all $1 \leq i \leq \ell(\underline{w})$, we will abusively write $\underline{e} = U \ldots U$ and say that the stroll for $\underline{e}$ consists solely of ups.

Proposition 12.19 *For any expression $\underline{w}$ and subexpression $\underline{e} \subset \underline{w}$, we have*

$$LL_{\underline{w},\underline{e}}(c_{\mathrm{bot}}) = \begin{cases} c_{\mathrm{bot}} \in \mathrm{BS}(\underline{w}^{\underline{e}}) & \text{if the stroll for } \underline{e} \text{ consists solely of ups,} \\ 0 & \text{otherwise.} \end{cases}$$

(12.13)

12.4 Bott–Samelson Bimodules and the Light Leaves Basis

Proof Fix an expression $\underline{w} = (s_1, \ldots, s_m)$. This proposition relies on the inductive construction of light leaves (see Sect. 10.4.2). If $\underline{e} = U \ldots U$, then the light leaf corresponding to $\underline{e}$ consists of enddots and identity maps (unbroken strands) in each tensor component composed in some order, potentially with some rex moves thrown in. As a direct consequence of Soergel's categorification theorem and grading considerations, rex moves preserve c_{bot} up to nonzero scalar, and by convention the rex moves send c_{bot} to c_{bot}. Enddot sends $1 \otimes 1 \in B_s$ to $1 \in R$. The identity map obviously sends $1 \otimes 1 \in B_{s_i}$ to itself. Hence, the elements $c_{\text{bot}} = 1 \otimes \cdots \otimes 1$ are preserved by enddots, identity maps, and rex moves. We conclude that if $\underline{e} = U \ldots U$, then $LL_{\underline{w},\underline{e}}(c_{\text{bot}}) = c_{\text{bot}} \in \text{BS}(\underline{w}^{\underline{e}})$.

Now, suppose that the decoration associated to $\underline{e}$ contains a down somewhere, say $\underline{e} = U \ldots UD \ldots$. As above, the light leaf corresponding to the initial string of ups sends c_{bot} to c_{bot}. At the first down, the inductive construction of light leaves says to apply an upward trivalent vertex to the end of c_{bot}, i.e. we apply a Demazure operator in a tensor component. As every tensor component of c_{bot} is 1, any Demazure operator kills c_{bot}. This proves the second case of the proposition. □

This proposition shows that we only need to consider double leaves $\overline{LL}_{\underline{w},\underline{f}} \circ LL_{\underline{w},\underline{e}}$ where $\underline{e} = U \ldots U$. How many of these are there?

Proposition 12.20 *Let $\underline{w}$ be an expression. Suppose $x \leq \underline{w}$, i.e. there exists some subexpression $\underline{e}' \subset \underline{w}$ such that $x = \underline{w}^{\underline{e}'}$. Then, there exists a unique subexpression $\underline{e} \subset \underline{w}$ such that $\underline{w}^{\underline{e}} = x$ and the stroll for $\underline{e}$ consists solely of ups.*

Proof The proof follows by induction on the length of the expression $\underline{w}$. Before we begin the induction, note that a subexpression has a stroll without any $D1$s if and only if it is reduced.

Let us now induct on $\ell(\underline{w})$. The base case for length 0 is trivial. Now let $\underline{w}'$ be an expression of nonzero length. We have $\underline{w}' = (\underline{w}, s)$ for some simple reflection s and we may assume that the proposition holds for all expressions of length at most $\ell(\underline{w})$. Suppose $x' \leq \underline{w}'$, and let $\underline{e}'$ be a subexpression of $\underline{w}'$ such that $x' = (\underline{w}')^{\underline{e}'}$. By the deletion condition, we may assume that $\underline{e}'$ gives a reduced expression for x'. We now split into two cases.

A. $x's > x'$: In this case no reduced expression for x' can end in an s, so $\underline{e}'$ ends in a 0 and so must any expression with the desired property. In particular, we have $\underline{e}' = (\underline{e}, 0)$ for some unique subexpression $\underline{e} \leq \underline{w}$. By the inductive hypothesis we may replace $\underline{e}$ so that still $\underline{w}^{\underline{e}} = x'$ but that also the stroll associated to $\underline{e}$ consists only of ups. The expression $\underline{e}' \leq \underline{w}'$ then also consists only of ups and satisfies $(\underline{w}')^{\underline{e}'} = x'$, proving existence. As the choice of $\underline{e}$ was unique and as the desired sequence had to end in a 0, uniqueness follows.

B. $x's < x'$: It follows from the deletion condition that there is a subexpression $\underline{e}' \leq \underline{w}'$ such that $(\underline{w}')^{\underline{e}'} = x'$ and the last term of e' is 1, and furthermore any expression of the desired form must end in a 1. Let $\underline{e}' = (\underline{e}, 1)$. Then, $\underline{w}^{\underline{e}} \leq w$

and hence, by induction, we may uniquely replace $\underline{e}$ so that both $(\underline{w}')^{\underline{e}'} = x'$ and also the stroll associated to $\underline{e}'$ has only ups. Then, as $x's < x'$, we see that $\underline{e}'$ has only ups as well and evaluates to x', showing existence. As the choice of $\underline{e}$ was unique and as the desired sequence had to end in a 1, uniqueness follows.
□

Definition 12.21 For any expression $\underline{w}$ and any $x \leq w$, we call the unique subexpression $\underline{e} \subset \underline{w}$ from the previous proposition the *canonical subexpression* for x and denote it by can_x (the expression $\underline{w}$ will be clear from context).

Exercise 12.22 Consider the expression (s, s, s, s, s, s). Find the canonical subexpressions can_{id} and can_s.

Exercise 12.23 Pick a reduced expression $\underline{w}$ for the longest element of S_4. For every other element of S_4, find the canonical subexpression.

Combining Propositions 12.19 and 12.20, we conclude:

Corollary 12.24 *Recall that any endomorphism ϕ of $\mathrm{BS}(\underline{w})$ can be written in the form*

$$\phi = \sum_{\substack{\underline{e}, \underline{f} \subset \underline{w}: \\ \underline{w}^{\underline{e}} = \underline{w}^{\underline{f}}}} \left(\overline{LL_{\underline{w}, \underline{f}}} \circ LL_{\underline{w}, \underline{e}} \right) g_{\underline{e}, \underline{f}}$$

for uniquely defined coefficients $g_{\underline{e}, \underline{f}} \in R$. For ϕ as above, we have

$$\phi(c_{\text{bot}}) = \sum_{\underline{f} \subset \underline{w}} \overline{LL_{\underline{w}, \underline{f}}}(c_{\text{bot}, \underline{w}^{\underline{f}}}) \, g_{\text{can}_{\underline{w}^{\underline{f}}}, \underline{f}}. \tag{12.14}$$

Here, the first c_{bot} is in $\mathrm{BS}(\underline{w})$ while $c_{\text{bot}, \underline{w}^{\underline{f}}}$ is the c_{bot} for $\mathrm{BS}(\text{can}_{\underline{w}^{\underline{f}}})$, where $\text{can}_{\underline{w}^{\underline{f}}}$ is the canonical (and hence reduced) expression for $\underline{w}^{\underline{f}} \leq w$.

Since every element of $\mathrm{BS}(\underline{w})$ can be obtained as the image of c_{bot} under an endomorphism, this corollary shows that

$$\left\{ \overline{LL_{\underline{w}, \underline{f}}}(c_{\text{bot}, \underline{w}^{\underline{f}}}) : \underline{f} \subset \underline{w} \right\} \tag{12.15}$$

spans $\mathrm{BS}(\underline{w})$ as a right R-module. Since this set has the same size as the 01-basis, we conclude:

Theorem 12.25 *The set (12.15) is a basis for $\mathrm{BS}(\underline{w})$ as a right R-module.*

12.5 Light Leaves Basis and the Standard Filtration on Bott–Samelson Bimodules

Recall from Sect. 5.2 that for all $x \in W$ we have the associated standard R-bimodule R_x defined by giving the free rank 1 left R-module R the structure of a right R-module via $m.r := x(r)m$. Recall also from Sect. 5.3 that, fixing a total ordering $x_1 < x_2 < \cdots$ of W refining the Bruhat ordering on W, a standard filtration on a Soergel bimodule B is an exhaustive increasing bimodule filtration

$$0 = B_0 \subset B_1 \subset \cdots$$

of B such that for each i we have an isomorphism of graded R-bimodules

$$B_i/B_{i-1} \simeq R_{x_i}^{h_{x_i}(B)}$$

for some Laurent polynomial $h_{x_i}(B) \in \mathbb{Z}_{\geq 0}[v^{\pm 1}]$. The following proposition gives a concrete construction of such standard filtrations for Bott–Samelson bimodules $B = \mathrm{BS}(\underline{w})$:

Proposition 12.26 *Fix an expression $\underline{w}$ and a total ordering $x_1 < x_2 < \cdots$ of W refining the Bruhat order. For each i let $B_i \subset B$ be the right R-submodule of $\mathrm{BS}(\underline{w})$ with basis*

$$\left\{ \overline{LL}_{\underline{w}, \underline{f}}(c_{\mathrm{bot}}) : \underline{w}^{\underline{f}} = x_j, j \leq i \right\}. \tag{12.16}$$

(Here, in the expression $\overline{LL}_{\underline{w}, \underline{f}}(c_{\mathrm{bot}})$, we have $c_{\mathrm{bot}} \in \mathrm{BS}(\mathrm{can}_{\underline{w}^{\underline{f}}})$ as in Theorem 12.25). Then $0 = B_0 \subset B_1 \subset \cdots$ is a standard filtration on $\mathrm{BS}(\underline{w})$ with respect to the total ordering $<$.

Proof Let $\underline{f} \subset \underline{w}$ be a subexpression with $\underline{w}^{\underline{f}} = x_i \in W$. Let $\underline{x_i}$ be the reduced expression for x_i such that the reversed light leaf $\overline{LL}_{\underline{w}, \underline{f}}$ has source $\mathrm{BS}(\underline{x_i})$. For any $g \in R$, using the polynomial forcing relation (10.7) to pass g from right to left through the bottom strands of the diagram representing $\overline{LL}_{\underline{w}, \underline{f}}$, it follows that

$$x_i(g)\overline{LL}_{\underline{w}, \underline{f}}(c_{\mathrm{bot}}) = \overline{LL}_{\underline{w}, \underline{f}}(c_{\mathrm{bot}})g + (\text{lower order terms})$$

where (lower order terms) lies in the right R-span of the elements $\overline{LL}_{\underline{w}, \underline{f}}(c_{\underline{e}})$ for various $\underline{e} \subset \underline{x_i}$ with $\underline{e} \neq \underline{0}$. In particular, the proposition will follow as soon as we show that every $\overline{LL}_{\underline{w}, \underline{f}}(c_{\underline{e}})$, $\underline{e} \neq \underline{0}$, lies in the right R-span of elements $\overline{LL}_{\underline{w}, \underline{f'}}(c_{\mathrm{bot}})$ satisfying $\underline{w}^{\underline{f'}} < x_i$.

As above, take $\underline{e} \subset \underline{x_i}$ with $\underline{e} \neq \underline{0}$. We have $c_{\underline{e}} = \phi_{\underline{e}}(c_{\text{bot}})$, so $\overline{LL}_{\underline{w},\underline{f}}(c_{\underline{e}})$ is the image of $c_{\text{bot}} \in BS(\underline{x_i})$ under $\overline{LL}_{\underline{w},\underline{f}} \circ \phi_{\underline{e}}$. Here is an example:

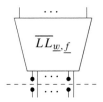

Consider the factorization of this morphism along the dashed line shown above, breaking $\phi_{\underline{e}}$ into two parts. The bottom factor sends $c_{\text{bot}} \in BS(\underline{x_i})$ to $c_{\text{bot}} \in BS(\underline{y_i})$ for some proper subexpression $\underline{y_i}$ of $\underline{x_i}$ with evaluation $y_i < x_i$, so $\overline{LL}_{\underline{w},\underline{f}}(c_{\underline{e}}) = \psi(c_{\text{bot}})$, where $\psi : BS(\underline{y_i}) \to BS(\underline{w})$ is the top factor. Writing ψ in the double leaves basis and applying Proposition 12.19, it follows that $\overline{LL}_{\underline{w},\underline{f}}(c_{\underline{e}})$ lies in the right R-span of elements $\overline{LL}_{\underline{w},\underline{f'}}(c_{\text{bot}})$ for subexpressions $\underline{f'} \subset \underline{w}$ such that $\underline{w}^{\underline{f'}} \leq y_i < x_i$, and the result follows. □

Example 12.27 Let $W = \langle s, t \rangle$ with $m_{st} = 3$. Fix the total ordering id $< s < t < st < ts < sts = tst$ extending the Bruhat order on W, and let $\underline{w} = (s, t, s)$. The basis elements $\overline{LL}_{\underline{w},\underline{f}}(c_{\text{bot}})$ are represented diagrammatically as follows (application to c_{bot} is implicit):

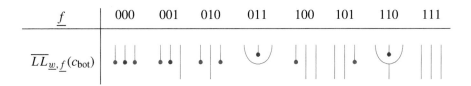

In particular, recalling that a dot (resp. a trivalent vertex) has degree $+1$ (resp. -1) and that $c_{\text{bot}} \in BS(\underline{x})$ lies in degree $-\ell(\underline{x})$, we see that this basis is homogeneous with

- $\overline{LL}_{\underline{w},000}(c_{\text{bot}})$ lying in degree 3,
- $\overline{LL}_{\underline{w},001}(c_{\text{bot}})$, $\overline{LL}_{\underline{w},010}(c_{\text{bot}})$, and $\overline{LL}_{\underline{w},101}(c_{\text{bot}})$ lying in degree 1,
- $\overline{LL}_{\underline{w},011}(c_{\text{bot}})$, $\overline{LL}_{\underline{w},110}(c_{\text{bot}})$, and $\overline{LL}_{\underline{w},100}(c_{\text{bot}})$ lying in degree -1, and
- $\overline{LL}_{\underline{w},111}(c_{\text{bot}})$ lying in degree -3.

By Proposition 12.26, $BS(\underline{w})$ has a standard filtration $B_\bullet$ with respect to the chosen total order, such that

- the elements $\overline{LL}_{\underline{w},\underline{f}}(c_{\text{bot}})$ with $\underline{f} \in \{000, 101\}$ give a basis for B_1,
- those with $\underline{f} \in \{001, 100\}$ give a basis for B_2/B_1,
- $\overline{LL}_{\underline{w},010}(c_{\text{bot}})$ spans B_3/B_2,

12.5 Light Leaves Basis and the Standard Filtration on Bott–Samelson Bimodules

- $\overline{LL}_{\underline{w},110}(c_{\text{bot}})$ spans B_4/B_3,
- $\overline{LL}_{\underline{w},011}(c_{\text{bot}})$ spans B_5/B_4, and
- $\overline{LL}_{\underline{w},111}(c_{\text{bot}})$ spans B_6/B_5.

As a final application of the basis of Theorem 12.25, let us now compute a basis for the indecomposable summand $B_s \stackrel{\oplus}{\subset} \text{BS}(\underline{w}) = B_{sts} \oplus B_s$. Recall from Sect. 9.3 the projector $\text{JW}_{\underline{w}} \in \text{End}(\text{BS}(\underline{w}))$ to B_{sts} arising from the Jones–Wenzl idempotent:

$$\text{JW}_{\underline{w}} = \Big|\Big|\Big| + \text{\Large $\curlyvee$}.$$

We have

$$\text{JW}_{\underline{w}} \circ \overline{LL}_{\underline{w},101} = \bigcup \Big| + \text{\large $\curlyvee$} = \bigcup \Big| + \boxed{\alpha_t}$$

$$= \bigcup \Big| + \partial_s(\alpha_t) \text{\large $\frown$} + s(\alpha_t) \text{\large $\frown$}$$

$$= \bigcup - \bigcup + 0 = 0,$$

so $\text{JW}_{\underline{w}}(\overline{LL}_{\underline{w},101}(c_{\text{bot}})) = 0$. A similar calculation shows that $\text{JW}_{\underline{w}}$ kills $\overline{LL}_{\underline{w},100}(c_{\text{bot}})$ as well. As $B_s = \text{Ker}(\text{JW}_{\underline{w}})$ is a free right R-module of rank 2, it follows that

$$\left\{ \overline{LL}_{\underline{w},101}(c_{\text{bot}}), \overline{LL}_{\underline{w},100}(c_{\text{bot}}) \right\} \tag{12.17}$$

is a basis for B_s.

Appendix 1: A Crucial Positivity Result

The main result of this appendix, Proposition 12.29, is a nice application of the results of this chapter, combined with some elementary Coxeter group theory. It can be appreciated as is, but it plays a major role in the Hodge theory section of this book, especially in Chap. 18. It is difficult to explain its importance now, so we will not try.

Definition 12.28 Let $\underline{x}$ be any expression, and let c_{bot} and c_{top} be the corresponding elements of $\text{BS}(\underline{x})$. Let $f \in \mathfrak{h}^*$, so that $f^{\ell(\underline{x})} \in R$ has degree $2\ell(\underline{x})$. We define

$$N_{\underline{x}}(f) := \text{Tr}(f^{\ell(\underline{x})} c_{\text{bot}}), \tag{12.18}$$

i.e. the coefficient of c_{top} when $f^{\ell(\underline{x})} c_{\text{bot}}$ is written in the 01-basis of $\text{BS}(\underline{x})$ (as a right R-module). For degree reasons, $N_{\underline{x}}(f) \in \mathbb{R}$ is a scalar.

Said another way,

$$N_{\underline{x}}(f) := \langle f^{\ell(\underline{x})} c_{\text{bot}}, c_{\text{bot}} \rangle, \tag{12.19}$$

where $\langle -, - \rangle$ is the global intersection form on $\text{BS}(\underline{x})$.

Proposition 12.29 *Assume that our realization is the Kac–Moody realization, with base field $\mathbb{R}$. Suppose that $\underline{x}$ is a reduced expression, and $\rho \in \mathfrak{h}^*$ is regular dominant, meaning that $\partial_s(\rho) > 0$ for all $s \in S$. Then $N_{\underline{x}}(\rho) > 0$.*

Remark 12.30 In fact, this proposition also holds for the dual geometric realization. Note that a given realization may not have any regular dominant elements. In any realization over $\mathbb{R}$ where the simple coroots $\alpha_s^\vee$ are linearly independent, there will be a regular dominant element. However, Proposition 12.29 requires more "positivity properties" of the realization than merely the existence of a regular dominant element.

We begin with some preliminaries that only make use of the polynomial forcing relation. We are ultimately interested in $f^{\ell(\underline{x})} c_{\text{bot}}$, but we begin by studying $f c_{\text{bot}}$.

Lemma 12.31 *Let $\underline{x}$ be any expression, $d = \ell(\underline{x})$, and $f \in \mathfrak{h}^*$. For $1 \le i \le d$ define a scalar $\lambda_i \in \mathbb{R}$ as*

$$\lambda_i = \partial_{\underline{x}_i}(x_{i-1}^{-1} f), \tag{12.20}$$

where $\underline{x}_i$ is the i-th simple reflection in the expression, and x_{i-1} is the element expressed by the first $(i-1)$ simple reflections of the expression $\underline{x}$.

Then $f c_{\text{bot}}$ expands as follows in the 01-basis:

$$f c_{\text{bot}} = \sum_{i=1}^{d} c_{0\cdots010\cdots0} \cdot \lambda_i + c_{\text{bot}} \cdot x^{-1}(f). \tag{12.21}$$

Here, x is the element expressed by $\underline{x}$, and the basis element $c_{0\cdots010\cdots0}$ in the i-th term of the sum is indexed by the 01-sequence with 1 in the i-th position and 0 elsewhere.

Proof This is a straightforward consequence of the polynomial forcing relation (10.7). □

12.5 Light Leaves Basis and the Standard Filtration on Bott–Samelson Bimodules

Remark 12.32 We can also view (12.21) as an equality of endomorphisms of $\mathrm{BS}(\underline{x})$:

$$f \cdot \mathrm{id}_{\mathrm{BS}(\underline{x})} = \sum_{i=1}^{d} \lambda_i \left| \cdots \left| \begin{smallmatrix} \bullet \\ \bullet \end{smallmatrix} \right| \cdots \right|_{\underline{x}_i} + \mathrm{id}_{\mathrm{BS}(\underline{x})} \cdot x^{-1}(f), \tag{12.22}$$

where the diagrams appearing on the right hand side are obtained by "breaking" the i-th line. This point of view will be the more important one in Hodge theory.

Example 12.33 Consider $W = S_4$ with $S = \{s, t, u\}$ as usual, with s red, t blue, and u black. Let $\mathfrak{h}$ be the geometric realization, so the simple roots $\alpha_s, \alpha_t, \alpha_u$ form a basis for $\mathfrak{h}^*$, and let $\varpi_s, \varpi_t, \varpi_u$ be the dual basis for $\mathfrak{h}$ (the fundamental weights). Let $\rho = \varpi_s + \varpi_t + \varpi_u$, so $\partial_s(\rho) = \partial_t(\rho) = \partial_u(\rho) = 1 > 0$, and let $\underline{x} = (s, t, u, s)$. Then by applying the polynomial forcing relation repeatedly, left multiplication by ρ on $\mathrm{BS}(\underline{x})$ can be rewritten as

$$\boxed{\rho}\,|||| = \,\overset{\bullet}{\underset{\bullet}{|}}\,||| + \boxed{\rho - \alpha_s}\,|||$$

$$= \,\overset{\bullet}{\underset{\bullet}{|}}\,||| + 2\,\overset{\bullet}{\underset{\bullet}{|}}\,|| + \boxed{\rho - \alpha_s - 2\alpha_t}\,||$$

$$= \,\overset{\bullet}{\underset{\bullet}{|}}\,||| + 2\,\overset{\bullet}{\underset{\bullet}{|}}\,|| + 3\,\overset{\bullet}{\underset{\bullet}{|}}\,| + \boxed{\rho - \alpha_s - 2\alpha_t - 3\alpha_u}$$

$$= \,\overset{\bullet}{\underset{\bullet}{|}}\,||| + 2\,\overset{\bullet}{\underset{\bullet}{|}}\,|| + 3\,\overset{\bullet}{\underset{\bullet}{|}}\,| + \,\overset{\bullet}{\underset{\bullet}{|}}\, + \boxed{\rho - 2\alpha_s - 2\alpha_t - 3\alpha_u}\,.$$

The first equality arises because $\partial_s(\rho) = 1$ and $s(\rho) = \rho - \alpha_s$, the next equality arises because $\partial_t(s(\rho)) = 2$ and $t(s(\rho)) = \rho - \alpha_s - 2\alpha_t$, and so forth.

In particular, $\lambda_1 = 1, \lambda_2 = 2, \lambda_3 = 3, \lambda_4 = 1$ in the notation of the lemma.

The reason that regular dominant elements are special is the following lemma.

Lemma 12.34 (Positive Breaking Lemma) *Assume that our realization is the Kac–Moody realization, with base field $\mathbb{R}$. Let $\rho \in \mathfrak{h}^*$ be regular dominant. Then $\underline{x}$ is a reduced expression if and only if each of the scalars λ_i of the previous lemma is (strictly) positive, $\lambda_i > 0$.*

The proof uses some standard Coxeter group theory (see [85, §5]) which we have not yet covered: the theory of positive roots, positive coroots, etcetera. This theory works for the geometric realization and similar realizations (like the Kac–Moody realization), but not for a general realization.

Proof For each $s \in S$, recall that the action of ∂_s on linear polynomials is given by pairing with the simple coroot $\alpha_s^\vee \in \mathfrak{h}$. Thus

$$\lambda_i = \partial_{\underline{x}_i}(x_{i-1}^{-1}(\rho)) = \langle \alpha_{\underline{x}_i}^\vee, x_{i-1}^{-1}(\rho) \rangle. \tag{12.23}$$

Using the W-invariance of this pairing between $\mathfrak{h}$ and $\mathfrak{h}^*$, we get that

$$\lambda_i = \langle x_{i-1}(\alpha_{\underline{x}_i}^\vee), \rho \rangle. \tag{12.24}$$

Now, a coroot pairs against ρ to be positive if and only if it is a positive coroot; after all, each simple coroot pairs against ρ to be positive, and a general coroot is either a purely positive or purely negative linear combination of simple coroots. Finally, one basic property of the Bruhat order is that $ws > w$ if and only if $w(\alpha_s^\vee)$ is a positive coroot. Thus λ_i is positive if and only if $x_{i-1}\underline{x}_i > x_{i-1}$, and this is true for all i if and only if $\underline{x}$ is a reduced expression. □

Exercise 12.35 Make sure you understand the assertions in the previous proof.

Exercise 12.36 (Easy) Conversely, suppose that $\rho \in \mathfrak{h}^*$ is such that $\lambda_i > 0$ for any reduced expression $\underline{x}$ and for all i. Deduce that ρ is regular dominant.

Exercise 12.37 Consider W the dihedral group of type $I_2(m)$, with $S = \{s, t\}$. Let $\mathfrak{h}$ be the geometric realization. As in Example 12.33, let $\rho = \varpi_s + \varpi_t$ be the sum of the fundamental weights, so $\partial_s(\rho) = \partial_t(\rho) = 1 > 0$, and let $\underline{x} = (s, t, s, t, \ldots)$ be a reduced expression for the longest element, having length m.

Recall from Exercise 5.51 that $\partial_s(\alpha_t) = \partial_t(\alpha_s) = -(q + q^{-1}) = -[2]_q$, where q is the complex number $e^{\frac{\pi i}{m}}$. Recall also the definitions of the other quantum numbers $[n]_q$. Clearly $[2]_q > 0$, since $[2]_q$ is $2\operatorname{Re}(q)$. The fact that q is a primitive $2m$-th root of unity implies that $[m] = 0$ and $[m-1] = 1$. Which other quantum numbers are positive real numbers?

For each $1 \le i \le m$ compute that

$$\lambda_i = [i-1]_q + [i]_q = [2i-1]_{\sqrt{q}}. \tag{12.25}$$

Deduce that $\lambda_i > 0$ for all i.

Remark 12.38 The previous exercise demonstrates a special positivity property of the geometric representation (and related realizations). One can attempt to construct a realization of the dihedral group where $\partial_s(\alpha_t) = -[2]_q$ for some other primitive $2m$-th root of unity $q \in \mathbb{C}$. This Cartan matrix will satisfy all the requisite algebraic properties to give a realization of the dihedral group. However, unless q is the root of unity with the largest real part (as in the geometric realization), there will exist $i < m$ such that $[i] < 0$. Consequently, one can deduce a failure of the Positive Breaking Lemma 12.34, and of the statement in its proof that $ws > w$ if and only if $w(\alpha_s^\vee)$ has non-negative coefficients.

12.5 Light Leaves Basis and the Standard Filtration on Bott–Samelson Bimodules

Now we return to the general element $f \in \mathfrak{h}^*$ and unravel a simple implication of (12.21), to be used for an inductive proof of Proposition 12.29.

Lemma 12.39 *We have*

$$N_{\underline{x}}(f) = \sum_{i=1}^{d} N_{\underline{x}_{\hat{i}}}(f)\lambda_i, \tag{12.26}$$

where $\underline{x}_{\hat{i}}$ is the expression $\underline{x}$ with the i-th simple reflection removed.

Proof We are interested in the coefficient of c_{top} when $f^d c_{\text{bot}}$ is expanded in the 01-basis. By (12.21), we have

$$f^d c_{\text{bot}} = \sum_{i=1}^{d} f^{d-1} c_{0\cdots 010\cdots 0} \cdot \lambda_i + c_{\text{bot}} \cdot f^{d-1} x^{-1}(f).$$

The final term contributes nothing to the coefficient of c_{top}. For each term in the sum, factor the diagrammatics for $f^{d-1} c_{0\cdots 010\cdots 0}$ vertically as follows:

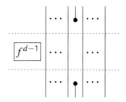

The middle part is $f^{\ell(\underline{x}_{\hat{i}})} c_{\text{bot}}$ inside $\text{BS}(\underline{x}_{\hat{i}})$. Expanding this part in the 01-basis of $\text{BS}(\underline{x}_{\hat{i}})$, we see that only the coefficient of c_{top}, which by definition is $N_{\underline{x}_{\hat{i}}}(f)$, contributes to the coefficient of c_{top} for $f^{d-1} c_{0\cdots 010\cdots 0}$ in $\text{BS}(\underline{x})$. Adding these contributions together, we get the desired formula. □

Next comes a crucial lemma which simplifies the inductive step enormously.

Lemma 12.40 *If the expression $\underline{x}$ is not reduced, then $N_{\underline{x}}(f) = 0$ for any $f \in \mathfrak{h}^*$.*

Proof Consider the submodule $\text{BS}(\underline{x}) \cdot R_+$ of $\text{BS}(\underline{x})$, i.e. the submodule generated by the ideal R_+ of elements of R in strictly positive degree, under right multiplication. Using the 01-basis, it is clear that any element of $\text{BS}(\underline{x}) \cdot R_+$ has coefficient of c_{top} living in R_+. If we can show that $f^{\ell(\underline{x})} c_{\text{bot}} \in \text{BS}(\underline{x}) \cdot R_+$, then $N_{\underline{x}}(f) = 0$ for degree reasons.

Let $\underline{w}$ be any expression. If $m > \ell(\underline{w})$, then $f^m c_{\text{bot}}$ lies in $\text{BS}(\underline{w}) \cdot R_+$. This is a straightforward pigeon-hole principle argument using polynomial forcing, which we leave to the reader.

We will show that $f^{\ell(\underline{x})} c_{\text{bot}} \in \text{BS}(\underline{x}) \cdot R_+$ when $\underline{x}$ is not a reduced expression, using the light leaves basis. We can write any element of $\text{BS}(\underline{x})$, including the identity, as a linear combination of images of $c_{\text{bot}} \in \text{BS}(\underline{w})$ under light leaves maps, for various reduced expressions $\underline{w}$ with $w \leq x$. But since $\underline{x}$ is not reduced, $\ell(\underline{w}) = \ell(w) < \ell(\underline{x})$, implying that $f^{\ell(\underline{x})}c_{\text{bot}} \in \text{BS}(\underline{w}) \cdot R_+$. Postcomposing with a bimodule map sends $\text{BS}(\underline{w}) \cdot R_+$ to $\text{BS}(\underline{x}) \cdot R_+$. □

Exercise 12.41 Time for a sanity check. The above lemma implies that $N_{(s,s)}(\alpha_s) = 0$. Verify this directly.

Exercise 12.42 Suppose that $m_{st} = 3$, and $f \in \mathfrak{h}^*$ satisfies $\partial_s(f) = a$ and $\partial_t(f) = b$. Compute $N_{(s,t,s)}(f)$. Directly verify that $N_{(s,t,s,t)}(f) = 0$.

Now we can put these lemmas together to prove the proposition.

Proof (Proof of Proposition 12.29) The result is true for the empty sequence. Let us assume inductively that $N_{\underline{w}}(\rho) > 0$ for any reduced expression $\underline{w}$ with length $\ell(\underline{w}) < \ell(\underline{x})$.

By (12.26) we may rewrite $N_{\underline{x}}(\rho)$ as a sum of various $N_{\underline{x}_{\hat{i}}}(\rho)$ rescaled by various λ_i. By Lemma 12.40, we can ignore those terms in the sum for which $\underline{x}_{\hat{i}}$ is not reduced. For the subsequences $\underline{x}_{\hat{i}}$ which are reduced, we know that $N_{\underline{x}_{\hat{i}}}(\rho) > 0$ by the inductive hypothesis. We also know that $\lambda_i > 0$ by Lemma 12.34. Thus $N_{\underline{x}}(\rho) > 0$ as well. □

Exercise 12.43 Continue the setup of Example 12.33. Compute $N_{\underline{x}}(\rho)$.

Part III
Historical Context: Category $\mathcal{O}$ and the Kazhdan–Lusztig Conjectures

Chapter 13
Category $\mathcal{O}$ and the Kazhdan–Lusztig Conjectures

This chapter is based on expanded notes of a lecture given by the authors and taken by

Joshua Jeishing Wen

Abstract We give a historical introduction to the Kazhdan–Lusztig conjectures. Our account begins with Verma's thesis and introduces the problems in category $\mathcal{O}$ these conjectures are meant to solve. After meandering through connections with the flag variety, we consider Soergel's proof of these conjectures, where Soergel bimodules played their first major role.

13.1 Introduction

To a complex semisimple Lie algebra one can associate two very different kinds of objects (in two different settings): Schubert varieties (in the setting of geometry), and highest weight representations (in the setting of infinite-dimensional representation theory). The Kazhdan–Lusztig conjectures [108] posited a combinatorial structure that can detect pathologies in both of these settings simultaneously. These conjectures are especially striking when situated in their historical context. Even more striking are the relationships that were unveiled in the process of proving them. We hope to give a sense of this by presenting a historical introduction to the conjectures, and a summary of the machinery going into their proofs.

Since this story spans Lie theory, Coxeter groups, and algebraic geometry, there are many definitions involved and we present only some of them. More background can be found in Chaps. 3, 14, and 15.

J. J. Wen
Department of Mathematics, University of Illinois at Urbana-Champaign, Urbana, IL, USA

> For the reader who is new to representation theory or to category $\mathcal{O}$, the exercises in this chapter may be too difficult. We recommend working through Chaps. 14–15 first, and then returning to these exercises.

13.2 The Verma Problem

13.2.1 Verma Modules

Let $\mathfrak{g}$ be a complex semisimple Lie algebra, e.g. $\mathfrak{sl}_2(\mathbb{C})$. Given a symmetry object like $\mathfrak{g}$, a fundamental problem is to study its category of representations. A naive solution to this problem is quite hopeless, but if we focus on *finite-dimensional* representations, the classical theory of such Lie algebras tells us to do the following. First, take a triangular decomposition $\mathfrak{g} = \mathfrak{n}_- \oplus \mathfrak{h} \oplus \mathfrak{n}_+$. In the case of $\mathfrak{g} = \mathfrak{sl}_n(\mathbb{C})$, such a decomposition amounts to a choice of ordered basis of n-dimensional space: $\mathfrak{n}_\pm$ are the strictly upper- and lower-triangular matrices respectively, and $\mathfrak{h}$ are the diagonal matrices. For each $\lambda \in \mathfrak{h}^*$, there is a one-dimensional representation $\mathbb{C}_\lambda$ of $\mathfrak{b} := \mathfrak{h} \oplus \mathfrak{n}_+$ where $\mathfrak{h}$ acts via λ and $\mathfrak{n}_+$ acts by zero. Our next step is to consider the *Verma module* $\Delta(\lambda)$ constructed using this data:

$$\Delta(\lambda) := U(\mathfrak{g}) \otimes_{U(\mathfrak{b})} \mathbb{C}_\lambda \tag{13.1}$$

where $U(\mathfrak{g})$ and $U(\mathfrak{b})$ are the universal enveloping algebras.

Each $\Delta(\lambda)$ has a unique simple quotient $L(\lambda)$. Note that $\Delta(\lambda)$ is infinite dimensional, and in general, so is $L(\lambda)$. If λ is *dominant* and *integral*, then $L(\lambda)$ is finite dimensional, and in fact one can construct all simple finite-dimensional representations of $\mathfrak{g}$ in this manner. Thus, to construct and classify finite-dimensional representations, one encounters the infinite-dimensional Verma modules as an intermediate step.

To ground ourselves in reality, here is the simplest case. In Chap. 14 more detail will be given.

Example 13.1 Let $\mathfrak{g} = \mathfrak{sl}_2(\mathbb{C})$. If we let

$$f = \begin{pmatrix} 0 & 0 \\ 1 & 0 \end{pmatrix}, \quad h = \begin{pmatrix} 1 & 0 \\ 0 & -1 \end{pmatrix}, \quad e = \begin{pmatrix} 0 & 1 \\ 0 & 0 \end{pmatrix}, \tag{13.2}$$

then a triangular decomposition is given by $\mathbb{C}f \oplus \mathbb{C}h \oplus \mathbb{C}e$. For $\lambda \in \mathbb{C} \simeq \mathfrak{h}^*$, let v_λ be a nonzero vector in $\mathbb{C}_\lambda$. By definition, $ev_\lambda = 0$, $hv_\lambda = \lambda v_\lambda$, and our Verma module $\Delta(\lambda)$ has a basis given by $\{\frac{1}{n!} f^n v_\lambda \mid n \in \mathbb{N}\}$. One can deduce from $[h, f] = -2f$ that this is in fact an eigenbasis of $\Delta(\lambda)$ for the h-action. The rest of the structure of $\Delta(\lambda)$ is illustrated by the following diagram.

13.2 The Verma Problem

$$\cdots \xleftarrow{5} \tfrac{1}{5!}f^5 v_\lambda \xleftarrow{4} \tfrac{1}{4!}f^4 v_\lambda \xleftarrow{3} \tfrac{1}{3!}f^3 v_\lambda \xleftarrow{2} \tfrac{1}{2!}f^2 v_\lambda \xleftarrow{1} f v_\lambda \; v_\lambda$$

with bottom arrows labeled $\lambda-4, \lambda-3, \lambda-2, \lambda-1, \lambda$ and eigenvalues $\cdots\ \lambda-10,\ \lambda-8,\ \lambda-6,\ \lambda-4,\ \lambda-2,\ \lambda$. (13.3)

Here, the bottom row gives the eigenvalues for h, the top arrows give the action of f, and the bottom arrows give the action of e. The interpretation of an arrow is "take the source vector and send it to the target vector rescaled by the label on the arrow." For example, $e(\tfrac{1}{2!}f^2 v_\lambda) = (\lambda-1)f v_\lambda$.

When an arrow has a zero label, it becomes possible to find a proper submodule. Notice that if $\lambda \notin \mathbb{Z}_{\geq 0}$, then $\Delta(\lambda) = L(\lambda)$ is simple because there are no arrows with zero label. If $\lambda \in \mathbb{Z}_{\geq 0}$, then $\Delta(-\lambda-2)$ is a submodule of $\Delta(\lambda)$, and we have a short exact sequence

$$0 \longrightarrow \Delta(-\lambda-2) \longrightarrow \Delta(\lambda) \longrightarrow L(\lambda) \longrightarrow 0. \tag{13.4}$$

In (13.4) we saw that each finite-dimensional irreducible module for $\mathfrak{sl}_2(\mathbb{C})$ has a resolution by Verma modules, which further motivates their study. A result of Bernstein–Gelfand–Gelfand (BGG) states that this is a general feature. Letting W be the Weyl group of $\mathfrak{g}$ and ρ the half-sum of the positive roots, we denote by $w \cdot \lambda$ the ρ-shifted action of $w \in W$ on $\lambda \in \mathfrak{h}^*$ (called the *dot action*), i.e.

$$w \cdot \lambda := w(\lambda + \rho) - \rho. \tag{13.5}$$

Theorem 13.2 (BGG Resolution [23]) *For λ dominant and integral, there is a resolution*

$$M^\bullet \longrightarrow L(\lambda) \tag{13.6}$$

where

$$M^{-k} = \bigoplus_{\substack{w \in W \\ \ell(w)=k}} \Delta(w \cdot \lambda) \tag{13.7}$$

and $\ell(-)$ is the length function.

Exercise 13.3 By the PBW theorem, $\Delta(\lambda)$ has the same weight space decomposition as the polynomial ring in $\mathfrak{n}_-$, but with an overall shift by λ. Use this and the BGG resolution to derive the Weyl character formula

$$\operatorname{ch}(L(\lambda)) = \frac{\sum_{w \in W}(-1)^{\ell(w)} e^{w \cdot \lambda}}{\prod_{\alpha \in \Phi^+}(1 - e^{-\alpha})}, \tag{13.8}$$

where Φ^+ is the set of positive roots.

13.2.2 Category $\mathcal{O}$ and Its Mysteries

Verma modules seem relatively straightforward: they have an explicit definition, and have a well-understood character. The BGG resolution and the Weyl character formula seem to say all there is to say about a finite-dimensional simple module $L(\lambda)$, by relating it to Verma modules. One can ask whether the nice results for finite-dimensional simple modules extend to the other (possibly infinite-dimensional) simple modules $L(\lambda)$. Arbitrary infinite-dimensional representations can be quite nasty, so let us minimally expand our scope beyond finite-dimensional representations to find an appropriate setting for the study of Verma modules and their simple quotients. This is the gist of category $\mathcal{O}$.

Definition 13.4 For a $\mathfrak{g}$-module M and $\lambda \in \mathfrak{h}^*$, we set

$$M_\lambda := \{m \in M : hm = \lambda(h)m \ \forall h \in \mathfrak{h}\},$$

a subspace (but not a submodule) of M. We say M is a *weight module* if $M = \bigoplus_\lambda M_\lambda$ as vector spaces. The *category* $\mathcal{O}$ is the full subcategory of all $\mathfrak{g}$-modules M such that:

- M is finitely generated,
- M is a weight module,
- and the action of $\mathfrak{n}_+$ on M is locally finite.

We call $v \in M$ a *highest weight vector* if $v \in M_\lambda$ for some $\lambda \in \mathfrak{h}^*$ and $\mathfrak{n}_+ v = 0$.

For a more detailed discussion of category $\mathcal{O}$, see Chap. 14.

Note that $\{\Delta(\lambda)\}_{\lambda \in \mathfrak{h}^*} \subset \mathcal{O}$. It turns out that $\{L(\lambda)\}_{\lambda \in \mathfrak{h}^*}$ gives a complete set of simple modules (up to isomorphism) in $\mathcal{O}$.

Our category now has two families of modules $\{\Delta(\lambda)\}$ and $\{L(\lambda)\}$ parametrized by $\mathfrak{h}^*$, and we wish to understand the relationship between them. We can focus our pursuit on the following two problems:

1. Mirroring the finite-dimensional case, we have by construction a surjection $\Delta(\lambda) \to L(\lambda)$, so one can try to extend this to a BGG-type resolution whose terms are indexed by the Weyl group.
2. Category $\mathcal{O}$ is a finite-length abelian category with simple objects $\{L(\lambda)\}$, so a Verma module has a Jordan–Hölder filtration whose subquotients are some $L(\lambda)$. A natural problem is to compute the multiplicity with which $L(\mu)$ appears in the filtration of $\Delta(\lambda)$, i.e. to find the decomposition numbers $[\Delta(\lambda) : L(\mu)]$.

In terms of the (abelian) Grothendieck group (see the first paragraph of Sect. 19.6) of $\mathcal{O}$, a solution to Problem 1 gives the transition matrix from the basis $\{[L(\lambda)]\}$ to the basis $\{[\Delta(\lambda)]\}$, while that of Problem 2 gives its inverse.

In his 1968 thesis Verma [174] showed that

$$\dim \mathrm{Hom}(\Delta(\mu), \Delta(\nu)) \le 1 \qquad (13.9)$$

13.3 The Kazhdan–Lusztig Conjectures

for all weights μ and ν. He also claimed to prove that

all submodules of $\Delta(\lambda)$ are generated by highest weight vectors. (13.10)

Exercise 13.5 Show that $\Delta(\lambda)$ satisfies a universal property concerning highest weight vectors. Use this to show that (13.9) and (13.10) above imply that any simple module occurs with multiplicity at most 1 in $\Delta(\lambda)$.

The upshot of (13.9) is that we do not have to think too much about what differentials can appear in a BGG-type resolution. Although it should not be clear to the reader why, one could use Exercise 13.5 and the (anachronistic) technology of Koszul duality (see Chap. 25) to deduce that the multiplicity of each Verma module in a BGG-resolution of any simple $L(\lambda)$ must also be at most 1. Thus Verma's results make BGG resolutions of arbitrary simples seem relatively straightforward.

In a remarkable paper from 1971, Bernstein–Gelfand–Gelfand [22] proved a series of results about category $\mathcal{O}$, two consequences of which were the introduction of the Weyl group in (13.9) and a dampening of our optimism.

Theorem 13.6 *The following holds:*

1. *If* $\text{Hom}(\Delta(\mu), \Delta(\nu)) = 1$, *then there exists a dominant* $\lambda \in \mathfrak{h}^*$ *and* $x, y \in W$ *such that* $\mu = x \cdot \lambda$, $\nu = y \cdot \lambda$, *and* $x \geq y$ *in the Bruhat order.*
2. *Point* (13.10) *is incorrect.*

By Exercise 13.5, we can contradict Verma's results by finding examples of Verma modules with higher multiplicities. The following chart gives the multiplicities that were known circa 1980 as computed by Bernstein–Gelfand–Gelfand and Jantzen:

Lie algebra	$\mathfrak{sl}_2$	$\mathfrak{sl}_3$	$\mathfrak{sl}_4$	$\mathfrak{sp}_4$	$\mathfrak{g}_2$	$\mathfrak{sp}_6$
Multiplicities	0, 1	0, 1	0, 1, 2	0, 1	0, 1	0, 1, 2, 3

13.3 The Kazhdan–Lusztig Conjectures

As we have seen above, Verma modules have turned out to be more complicated than previously thought. One would like a systematic way to detect higher multiplicities as well as a conceptual explanation for this phenomenon.

13.3.1 The Multiplicity Conjecture

In their 1979 paper [108], Kazhdan and Lusztig constructed a basis $\{b_w\}_{w \in W}$ of the Hecke algebra $H(W)$ of W satisfying some conditions with respect to a bar involution. Recall that $H(W)$ also has a standard basis $\{\delta_w\}_{w \in W}$, and by construction, the transition matrix from $\{\delta_w\}$ to $\{b_w\}$ is upper triangular with respect to the Bruhat order:

$$b_x = \sum_{y \leq x} h_{y,x}(v)\delta_y.$$

The *Kazhdan–Lusztig polynomials* are defined to be the coefficients $h_{y,x}(v)$ above. The basis $\{b_w\}$ and hence these polynomials are constructed recursively and only rely on Weyl group combinatorics. More background on the Hecke algebra was contained in Chap. 3.

The first statement of Theorem 13.6 implies that $[\Delta(\nu) : L(\mu)] \neq 0$ only if $\mu = x \cdot \lambda$ and $\nu = y \cdot \lambda$ with λ dominant and $x \geq y$. This begins to connect Verma modules to the Weyl group. The most interesting case of the multiplicity problem is when λ is not just dominant but integral as well, and via translation functors (see Chap. 15), it suffices to consider the case $\lambda = 0$. We will denote by $\mathcal{O}_0$ the *principal block*, the thick subcategory of $\mathcal{O}$ generated by $\{L(w \cdot 0)\}_{w \in W}$. Given the limited numerical data at the time, the following conjecture made in [108] is remarkable:

Conjecture 13.7 (Kazhdan–Lusztig) For all $x, y \in W$, we have

$$[\Delta(y \cdot 0) : L(x \cdot 0)] = h_{y,x}(1). \tag{13.11}$$

Using the recursive definition of the $h_{y,x}$s, Conjecture 13.7 promises a much easier way to compute the multiplicities of Verma modules. Addressing our question of BGG resolutions, Kazhdan and Lusztig also proved the following inversion formula:

Theorem 13.8 (Kazhdan–Lusztig Inversion Formula) *Let $w_0 \in W$ be the longest element. For all $x, y \in W$, we have*

$$\sum_{z \in W}(-1)^{\ell(z)+\ell(x)}h_{z,x}(v)h_{zw_0,yw_0}(v) = \begin{cases} 1 & \text{if } x = y, \\ 0 & \text{otherwise.} \end{cases} \tag{13.12}$$

Using this, Conjecture 13.7 then implies[1] the following identity in the Grothendieck group of $\mathcal{O}$:

$$[L(y \cdot 0)] = \sum_{y \leq x}(-1)^{\ell(y)+\ell(x)}h_{xw_0,yw_0}(1)[\Delta(x \cdot 0)]. \tag{13.13}$$

This is weaker than a resolution, but it still gives the character of $L(y \cdot 0)$ in terms of the characters of Verma modules. Therefore, many of our questions about Verma

---

[1]Actually, this is somewhat ahistorical. Kazhdan–Lusztig first stated their conjecture in the form (13.13); it was later realized that the inversion formula implied the equivalent form in Conjecture 13.7.

modules can be answered by the validity of this conjecture. Conjecture 13.7 raises yet more questions (which will be addressed in this book):

1. Why is the multiplicity given by the evaluation at 1 of a polynomial? What data about category $\mathcal{O}$ is encoded by the coefficients of $h_{y,x}(v)$?
2. The Hecke algebra has a Kazhdan–Lusztig basis for arbitrary Coxeter groups (rather than only Weyl groups). Is there a structure like $\mathcal{O}_0$ even without the presence of a Lie algebra?
3. What is the meaning of the inversion formula?

Just how bad can the multiplicity problem be? What kinds of polynomials can appear as Kazhdan–Lusztig polynomials? This is answered by the following theorem of Polo [146].

Theorem 13.9 *For any monic polynomial $p \in \mathbb{Z}_{\geq 0}[v^2]$, there exist m and N such that $v^m p = h_{y,x}(v)$ for some $x, y \in S_N$.*

Roughly speaking, the theorem says that "any polynomial with positive coefficients is a Kazhdan-Lusztig polynomial in some symmetric group". In particular, we can see that the multiplicities $[\Delta(y \cdot 0) : L(x \cdot 0)]$ can be arbitrarily large, even in type A.

13.3.2 Positivity and Schubert Varieties

Switching gears, we address another conjecture made by Kazhdan and Lusztig in [108] about their polynomials $\{h_{y,x}\}$, sometimes called the *Kazhdan–Lusztig positivity conjecture*.

Conjecture 13.10 The coefficients of $h_{y,x}(v)$ are non-negative integers.

This positivity conjecture (for Weyl groups) was proved the subsequent year by Kazhdan and Lusztig in [109] by relating $\{h_{y,x}\}$ to the geometry of the *flag variety*. For more background for what follows, see Chap. 15. Recall our triangular decomposition $\mathfrak{g} = \mathfrak{n}_- \oplus \mathfrak{h} \oplus \mathfrak{n}_+$. Let G be the simply-connected (complex semisimple) Lie group corresponding to $\mathfrak{g}$, then let B and U be the subgroups corresponding to $\mathfrak{b} = \mathfrak{h} \oplus \mathfrak{n}_+$ and $\mathfrak{n}_+$, respectively. The flag variety can be defined as G/B. Looking at left B-orbits (or equivalently U-orbits) gives the *Bruhat decomposition*:

$$G/B = \bigsqcup_{w \in W} BwB/B.$$

The *Schubert variety* X_w is the closure of BwB/B. We can define an order on W by $y \leq w$ if and only if $X_y \subset X_w$, and this order coincides with the Bruhat order. Note that $X_{w_0} = G/B$.

An interesting question is to determine whether X_w is singular, and if so, to understand its singularities. It was noticed first by Kazhdan and Lusztig that there appears to be a connection between their polynomials and (rational) smoothness of Schubert varieties. In the appendix to [108], Kazhdan and Lusztig proved the following result.

Theorem 13.11 *The Kazhdan–Lusztig polynomial $h_{y,w}(v) = v^{\ell(w)-\ell(y)}$ if and only if X_w is rationally smooth at a generic point of X_y.*

Remark 13.12 A point of $x \in X$ of a complex variety is *rationally smooth* if the rational cohomology of a punctured contractible neighborhood agrees with that of a smooth point. In simply laced type (i.e. types A, D, and E) the locus of smooth and rationally smooth points agree (Deodhar [45] in type A, extended to all simply laced types by Peterson in unpublished work).

In this geometric context, the powers of v have a natural meaning. A category with objects that can detect singularities of Schubert varieties is the category $\text{Perv}_{(B)}(G/B)$ of B-constructible *perverse sheaves* on G/B. The simple objects in this category are given by the *intersection cohomology* sheaves[2] of X_w, which we denote by IC_w. We have that X_w is smooth if and only if IC_w is isomorphic to the constant sheaf on X_w. We give an introduction to perverse sheaves in Chap. 16.

Kazhdan and Lusztig proved their positivity conjecture by proving the following theorem. The powers of v, which were mysterious in their relation to $\mathcal{O}$, here correspond to a cohomological grading.

Theorem 13.13 *Letting $IH^i_{X_y}(X_w)$ denote the stalk of the i-th hypercohomology of IC_w at a generic point of X_y, we have*

$$v^{\ell(w)-\ell(y)} h_{y,w}(v^{-1}) = \sum_i v^i \dim IH^{2i}_{X_y}(X_w). \tag{13.14}$$

13.4 Two Proofs of the Kazhdan–Lusztig Conjecture

The first proof of Conjecture 13.7 was given in 1981 by two teams of mathematicians, Beilinson–Bernstein and Brylinski–Kashiwara, using similar methods. A key result is the *localization theorem*, a consequence of which is an equivalence between $\mathcal{O}_0$ and the category of sheaves of U-equivariant modules over the ring of differential operators on G/B. We denote the latter category by $D^U_{G/B}\text{-mod}$. The Riemann–Hilbert correspondence can be applied to relate $D^U_{G/B}$-modules

---

[2] Intersection cohomology sheaves are not sheaves. They are (a special kind of) complexes of sheaves.

13.4 Two Proofs of the Kazhdan–Lusztig Conjecture

with perverse sheaves. Finally, perverse sheaves are related to Kazhdan–Lusztig polynomials as above. Here is a schematic summary:

$$\mathcal{O}_0 \xleftrightarrow{(1)} D^U_{G/B}\text{-mod} \xleftrightarrow{(2)} \text{Perv}_{(B)}(G/B) \xrightarrow{(3)} \{h_{y,x}\} \tag{13.15}$$

where

(1) is the localization theorem,
(2) is the Riemann–Hilbert correspondence, and
(3) is Theorem 13.13.

The real meat is an equivalence of abelian categories between $\mathcal{O}_0$ and $\text{Perv}_{(B)}(G/B)$, which sends simple objects to simple objects. This book will not go into the details of this proof.

The second proof was given by Soergel in 1990 [161]. Soergel's approach uses two functors: a $\mathbb{V}$ functor relating $\mathcal{O}_0$ to what are called Soergel modules, and an $\mathbb{H}$ functor relating perverse sheaves to Soergel modules. We will say more about these two functors in Chaps. 15 and 16, respectively. In the rest of this chapter, let us give an overview of Soergel's proof.

We first need to define Soergel modules; more details can be found in Sect. 15.3. The Weyl group W of $\mathfrak{g}$ has the natural structure of a (finite crystallographic, see Definition 1.36) Coxeter system, and the subalgebra $\mathfrak{h} \subset \mathfrak{g}$ is naturally a realization of this Coxeter system over $\mathbb{C}$. Consider Soergel bimodules associated to the action of W on $R = \text{Sym}(\mathfrak{h})$.[3] A *(right) Soergel module* is a graded right R-module of the form $\mathbb{C} \otimes_R B$, where B is a Soergel bimodule and $R \to \mathbb{C}$ sends $\mathfrak{h}$ to 0. That is, Soergel modules are obtained from Soergel bimodules by killing the action of higher degree polynomials on the left. Soergel showed that the modules

$$\overline{B}_x := \mathbb{C} \otimes_R B_x$$

remain indecomposable.

Remark 13.14 For the connection to $\mathcal{O}_0$, right Soergel modules are more convenient than left Soergel modules. Elsewhere in the book, we primarily consider left modules.

Any W-invariant polynomial $f \in R^W$ will pass freely through every tensor product $\otimes_{R^s}$ in a Bott–Samelson bimodule. Thus $f \in R^W$ will act the same on the right or the left of any Soergel bimodule. Thus if $f \in R^W$ has strictly positive degree, then f will act as zero on any Soergel module. Now, consider the coinvariant algebra (cf. Definition 4.13) $C = R/I_W$, where I_W denotes the homogeneous ideal generated by the set of W-invariant polynomials in R of strictly positive degree.

---

[3]Note that this is different from the usual $R = \text{Sym}(\mathfrak{h}^*)$. This is a first hint that the Langlands dual will play a role.

As just noted, I_W acts as zero on any Soergel module, so Soergel modules can be viewed as graded right C-modules.

To define $\mathbb{V}$, we need an additional class of modules in category $\mathcal{O}$ that we have yet to mention: each simple $L(\lambda)$ has an indecomposable projective cover denoted by $P(\lambda)$. We define $\mathbb{V}$ to be $\mathrm{Hom}(P(w_0 \cdot 0), -)$.

Soergel proved the following results about this functor.

Theorem 13.15 *There exist a canonical isomorphism*

$$\mathrm{End}(P(w_0 \cdot 0)) = \mathbb{V}(P(w_0 \cdot 0)) \simeq C. \qquad (13.16)$$

Using this isomorphism, we may view $\mathbb{V}$ as a functor

$$\mathbb{V}: \mathcal{O}_0 \to \mathrm{mod\text{-}}C$$

to ungraded right C modules.

Theorem 13.16 *Let* $\mathrm{Proj}\,\mathcal{O}_0 \subset \mathcal{O}_0$ *be the full subcategory of projective modules.*
1. $\mathbb{V}$ *is fully faithful on* $\mathrm{Proj}\,\mathcal{O}_0$.
2. $\mathbb{V}(P(x \cdot 0)) \simeq \overline{B}_x$ *(as ungraded modules) for all* $x \in W$.

We will discuss these results in some more detail in Chap. 15, but let us immediately make some remarks.

Remark 13.17 A key component of Soergel's approach is to study $\mathcal{O}_0$ using certain endofunctors called *translation functors* Θ_s, one for each simple reflection of W, which define a W-action on $\mathcal{O}_0$.

Remark 13.18 Let $P = \bigoplus_{w \in W} P(w \cdot 0)$ and $\overline{B} = \bigoplus_{w \in W} \overline{B}_w$. By general abelian category nonsense and the theorem above, we have

$$\mathrm{Hom}(P, -): \mathcal{O}_0 \xrightarrow{\sim} \mathrm{mod\text{-}}\mathrm{End}(P) \simeq \mathrm{mod\text{-}}\mathrm{End}_{\mathrm{mod\text{-}}C}(\overline{B}).$$

Since C admits a *grading*, so does $\mathrm{End}_{\mathrm{mod\text{-}}C}(\overline{B})$, which explains the appearance of Kazhdan–Lusztig *polynomials* in the combinatorics of $\mathcal{O}_0$.

By BGG reciprocity (see Theorem 14.27), the Kazhdan–Lusztig conjecture 13.7 can equivalently be phrased in terms of the character of the indecomposable projective modules $P(x \cdot 0)$. This in turn can be phrased, via the $\mathbb{V}$ functor, in terms of the size of the indecomposable Soergel modules $\overline{B}_x$.

It is at this point that Soergel states ominously, "Der Sündenfall naht heran." In this final step, Soergel needed geometry, and in particular the forbidden fruit called the decomposition theorem. Let $G^\vee$ be the complex reductive group that is Langlands dual to G. Let $B^\vee$ be a Borel subgroup, and let $D^b_{(B^\vee)}(G^\vee/B^\vee)$ be the bounded $B^\vee$-constructible derived category of sheaves (of $\mathbb{C}$-vector spaces) on the Langlands dual flag variety $G^\vee/B^\vee$. This is a triangulated category that contains the abelian category $\mathrm{Perv}_{(B^\vee)}(G^\vee/B^\vee)$ of perverse sheaves as a full subcategory. Using

13.4 Two Proofs of the Kazhdan–Lusztig Conjecture

the Borel isomorphism $C \simeq H^*(G^\vee/B^\vee; \mathbb{C})$, we may view total hypercohomology as a functor

$$\mathbb{H} : D^b_{(B^\vee)}(G^\vee/B^\vee) \to \text{gmod-}C.$$

Soergel proved the following, which is the analogue of Theorem 13.16 for $\mathbb{H}$.

Theorem 13.19 ([161]) *Let* $\text{SemiSimple}_{(B^\vee)}(G^\vee/B^\vee) \subset D^b_{(B^\vee)}(G^\vee/B^\vee)$ *be the full subcategory of semisimple complexes, i.e. finite direct sums of cohomological shifts of intersection cohomology sheaves.*

1. $\mathbb{H}$ *is fully faithful on* $\text{SemiSimple}_{(B^\vee)}(G^\vee/B^\vee)$.
2. $\mathbb{H}(\text{IC}_x) \simeq \overline{B}_x$ *for all* $x \in W$.

Combining Theorems 13.16 and 13.19, Soergel deduced that the characters of indecomposable projective modules in $\mathcal{O}_0$ are given in terms of the Kazhdan–Lusztig basis, hence deduced the Kazhdan–Lusztig conjecture. However, to prove Theorem 13.19(2), he needed the decomposition theorem.

This clear demarcation of where the algebra ends and the geometry begins suggests where a purely algebraic proof of both the Kazhdan–Lusztig positivity and multiplicity conjectures may lie. As we will see in Sect. 15.5, the Kazhdan–Lusztig conjecture follows from *Soergel's conjecture*, which says that the size of indecomposable Soergel bimodules is given by the Kazhdan–Lusztig basis, and which can be stated entirely in the algebraic language of Soergel bimodules. Soergel's conjecture was proved for all Coxeter groups by Elias and Williamson [57]; their proof is the subject of Part IV of this book.

Remark 13.20 The original proof of the Kazhdan–Lusztig conjectures relates the simple modules in $\mathcal{O}_0$ to simple perverse sheaves on the flag variety. In contrast, Soergel's proof via $\mathbb{V}$ and $\mathbb{H}$ relates a *projective* module to a simple perverse sheaf, and on the Langlands dual flag variety. Combining the two proofs, we obtain a relation between Langlands dual $\mathcal{O}_0$ (or flag varieties) exchanging simple objects with projective objects. This is *Koszul duality*, a topic we return to in Chaps. 25–26.

Chapter 14
Lightning Introduction to Category $\mathcal{O}$

This chapter is based on expanded notes of a lecture given by the authors and taken by
Anna Romanov and Sean Taylor

Abstract In this chapter we review the structure of Bernstein, Gelfand and Gelfand's category $\mathcal{O}$. We start with some basic results about complex semisimple Lie algebras, then define the category of modules in question. We state and prove some fundamental structural results about this category, such as the decomposition of objects into finite-dimensional weight spaces, the existence of Verma modules, and the parametrization of simple modules. We define a notion of duality in $\mathcal{O}$, and state the BGG reciprocity theorem. We introduce central characters and blocks of category $\mathcal{O}$, then complete the chapter with a careful examination of the example $\mathfrak{g} = \mathfrak{sl}_3(\mathbb{C})$. We prove a selection of the stated results to give the reader a flavor of the theory, and encourage the curious reader to examine the references for the remaining proofs.

14.1 Lie Algebra Basics

We begin this chapter with a brief review of the structure and representation theory of complex semisimple Lie algebras, taking $\mathfrak{sl}_2(\mathbb{C})$ as a running example. These running examples can be taken as running exercises for the novice, as we mostly give the answer without the computation, which the reader should verify. For a detailed treatment of the subject we refer the reader to [84].

Let $\mathfrak{g}$ be a finite-dimensional semisimple complex Lie algebra. The Lie algebra $\mathfrak{g}$ acts on itself via the adjoint action $\mathrm{ad} : \mathfrak{g} \to \mathrm{End}\,\mathfrak{g}$, where for $x, y \in \mathfrak{g}$, $\mathrm{ad}_x(y) = [x, y]$.

A. Romanov
Department of Mathematics, University of Utah, Salt Lake City, UT, USA

S. Taylor
Department of Mathematics, Louisiana State University, Baton Rouge, LA, USA

Example 14.1 Let $\mathfrak{sl}_2(\mathbb{C})$ denote the Lie algebra of traceless 2×2 matrices with entries in $\mathbb{C}$. The bracket operation is the commutator: $[x, y] = xy - yx$ for $x, y \in \mathfrak{sl}_2(\mathbb{C})$. A basis for $\mathfrak{sl}_2(\mathbb{C})$ is

$$f := \begin{pmatrix} 0 & 0 \\ 1 & 0 \end{pmatrix}, \quad h := \begin{pmatrix} 1 & 0 \\ 0 & -1 \end{pmatrix}, \quad e := \begin{pmatrix} 0 & 1 \\ 0 & 0 \end{pmatrix}.$$

The adjoint action is given with respect to the above basis by

$$\mathrm{ad}_f = \begin{pmatrix} 0 & 2 & 0 \\ 0 & 0 & -1 \\ 0 & 0 & 0 \end{pmatrix}, \quad \mathrm{ad}_h = \begin{pmatrix} -2 & 0 & 0 \\ 0 & 0 & 0 \\ 0 & 0 & 2 \end{pmatrix}, \quad \mathrm{ad}_e = \begin{pmatrix} 0 & 0 & 0 \\ 1 & 0 & 0 \\ 0 & -2 & 0 \end{pmatrix}.$$

Fix a Cartan subalgebra $\mathfrak{h} \subset \mathfrak{g}$. For each $h \in \mathfrak{h}$, the operator ad_h is diagonalizable. Since any two elements of $\mathfrak{h}$ commute, the operators ad_h for $h \in \mathfrak{h}$ are simultaneously diagonalizable.

Example 14.2 For $\mathfrak{g} = \mathfrak{sl}_2(\mathbb{C})$, $\mathfrak{h} = \mathbb{C}h$ is a Cartan subalgebra. We fix h as a basis for $\mathfrak{h}$. Then the basis $\{e, h, f\}$ of $\mathfrak{sl}_2(\mathbb{C})$ is an eigenbasis for ad_h.

Let $\mathfrak{g}$-Mod be the category of $\mathfrak{g}$-modules. A module $V \in \mathfrak{g}$-Mod is called a *weight module* if $V = \bigoplus_{\lambda \in \mathfrak{h}^*} V_\lambda$, where $V_\lambda = \{v \in V \mid hv = \lambda(h)v \text{ for all } h \in \mathfrak{h}\}$. One should think that a *weight* $\lambda \in \mathfrak{h}^*$ is a way of encoding, altogether, the eigenvalues for a family $\mathfrak{h}$ of commuting operators as they act on a simultaneous eigenvector. We call V_λ the *weight space* of V corresponding to the weight λ. Nonzero vectors $v \in V_\lambda$ are called *weight vectors*. One can prove that any finite-dimensional representation of $\mathfrak{g}$ is a weight module.

Example 14.3 For $\mathfrak{g} = \mathfrak{sl}_2(\mathbb{C})$, we may identify $\mathfrak{h}^*$ with $\mathbb{C}$ by sending $\lambda \in \mathfrak{h}^*$ to the scalar $\lambda(h)$. Alternatively, we may view $\mathfrak{h}^*$ as the span of the linear functional h^* which satisfies $h^*(h) = 1$. The standard representation of $\mathfrak{sl}_2(\mathbb{C})$ is a weight module: $\mathfrak{sl}_2(\mathbb{C})$ acts on $\mathbb{C}^2$ by matrix multiplication, and we have

$$\begin{pmatrix} 1 & 0 \\ 0 & -1 \end{pmatrix} \begin{pmatrix} x \\ 0 \end{pmatrix} = 1 \begin{pmatrix} x \\ 0 \end{pmatrix} \quad \text{and} \quad \begin{pmatrix} 1 & 0 \\ 0 & -1 \end{pmatrix} \begin{pmatrix} 0 \\ y \end{pmatrix} = -1 \begin{pmatrix} 0 \\ y \end{pmatrix}$$

for any $x, y \in \mathbb{C}$. Therefore, writing $e_1 = \begin{pmatrix} 1 \\ 0 \end{pmatrix}$ and $e_2 = \begin{pmatrix} 0 \\ 1 \end{pmatrix}$, this representation has two weight spaces $\mathbb{C}e_1$ and $\mathbb{C}e_2$, corresponding to weights $\lambda_1 = 1$ and $\lambda_2 = -1$, respectively. This gives us a weight space decomposition $\mathbb{C}^2 = \mathbb{C}e_1 \oplus \mathbb{C}e_2$.

The Lie algebra $\mathfrak{g}$ with its adjoint action is a weight module. We call the set $\Phi \subset \mathfrak{h}^*$ of nonzero weights of the representation $\mathfrak{g}$ the *roots* of $\mathfrak{g}$. The weight spaces $\mathfrak{g}_\alpha$ under this action are called *root spaces*. Nonzero vectors $x \in \mathfrak{g}_\alpha$ are called *root vectors*.

14.1 Lie Algebra Basics

Example 14.4 For $\mathfrak{g} = \mathfrak{sl}_2(\mathbb{C})$, $\Phi = \{\pm\alpha\}$, where $\alpha : \mathfrak{h} \to \mathbb{C}$ is given by $\alpha(h) = 2$, i.e. $\alpha = 2h^*$.

Define the bilinear form $(-|-) : \mathfrak{g} \times \mathfrak{g} \to \mathbb{C}$ by $(x|y) = \operatorname{tr}(\operatorname{ad}_x \operatorname{ad}_y)$ for $x, y \in \mathfrak{g}$. This is the *Killing form* on $\mathfrak{g}$. It is clearly symmetric. Since $\mathfrak{g}$ is semisimple, it is nondegenerate.

Example 14.5 In $\mathfrak{g} = \mathfrak{sl}_2(\mathbb{C})$,

$$(f|e) = \operatorname{tr}\begin{pmatrix} 2 & 0 & 0 \\ 0 & 2 & 0 \\ 0 & 0 & 0 \end{pmatrix} = 4, \quad (f|h) = \operatorname{tr}\begin{pmatrix} 0 & 0 & 0 \\ 0 & 0 & -2 \\ 0 & 0 & 0 \end{pmatrix} = 0,$$

$$(e|h) = \operatorname{tr}\begin{pmatrix} 0 & 0 & 0 \\ -2 & 0 & 0 \\ 0 & 0 & 0 \end{pmatrix} = 0, \quad (h|h) = \operatorname{tr}\begin{pmatrix} 4 & 0 & 0 \\ 0 & 0 & 0 \\ 0 & 0 & 4 \end{pmatrix} = 8,$$

$$(f|f) = 0 = (e|e).$$

The Gram matrix of the Killing form is thus

$$\begin{pmatrix} 0 & 0 & 4 \\ 0 & 8 & 0 \\ 4 & 0 & 0 \end{pmatrix}.$$

The restriction of the Killing form to $\mathfrak{h}$ is positive definite, hence it is an inner product, and the map $\mathfrak{h} \to \mathfrak{h}^* : x \mapsto (x|-)$ is an isomorphism. In particular, for any $\lambda \in \mathfrak{h}^*$ there is a unique element $h_\lambda \in \mathfrak{h}$ with $(h_\lambda|-) = \lambda$. We use this map to transfer the inner product from $\mathfrak{h}$ to $\mathfrak{h}^*$ by defining $(\lambda|\mu) := (h_\lambda|h_\mu)$. If $K_\mathfrak{h}$ is the Gram matrix of $(-|-)$ on $\mathfrak{h}$ with respect to a fixed basis, then $K_\mathfrak{h}^{-1}$ is the Gram matrix of $(-|-)$ on $\mathfrak{h}^*$ with respect to the dual basis.

Example 14.6 For $\mathfrak{g} = \mathfrak{sl}_2(\mathbb{C})$, the pairing $(-|-)$ on $\mathfrak{h}^*$ satisfies $(h^*|h^*) = \frac{1}{8}$. The length of $\alpha = 2h^*$ is thus $1/\sqrt{2}$.

The collection $\Phi \subset \mathfrak{h}^*$ of roots spans $\mathfrak{h}^*$. When fixing a basis of roots of $\mathfrak{h}^*$, every root is a *rational* linear combination of the basis roots. Hence, denoting by $E_\mathbb{Q} := \operatorname{Span}_\mathbb{Q}(\Phi) \subset \mathfrak{h}^*$ the rational span of the roots, we have $\dim_\mathbb{Q} E_\mathbb{Q} = \dim_\mathbb{C} \mathfrak{h}^*$. The inner product $(-|-)$ attains rational values on the roots, thus restricts to an inner product on $E_\mathbb{Q}$. Let $E := \mathbb{R} \otimes_\mathbb{Q} E_\mathbb{Q} = \operatorname{Span}_\mathbb{R}(\Phi) \subset \mathfrak{h}^*$ be the real span of the roots. This is a Euclidean space with inner product $(-|-)$ and Φ is a *root system* in E. There is a *base* Δ, a subset of Φ such that $\Phi = \Phi^+ \cup (-\Phi^+)$, where $\Phi^+ = \mathbb{N}\Delta \cap \Psi$. We call Δ a collection of *simple roots* and call the elements of Φ^+ the *positive roots* with respect to Δ.

For $\alpha \in \Phi$, let $s_\alpha \in O(\mathfrak{h}^*, (-|-))$ be the orthogonal reflection fixing $\langle\alpha\rangle^\perp$. The group $W = \langle s_\alpha \mid \alpha \in \Phi \rangle \subset GL(\mathfrak{h}^*)$ is the *Weyl group*. Let $S = \{s_\alpha \mid \alpha \in \Delta\}$. The pair (W, S) forms a finite crystallographic (see Definition 1.36) Coxeter system of the same type as $\mathfrak{g}$.

Example 14.7 For $\mathfrak{g} = \mathfrak{sl}_2(\mathbb{C})$, $\Delta = \{\alpha\}$ is a collection of simple roots. Hence, $\Phi^+ = \{\alpha\}$. The reflection s_α acts by -1 on both $\mathfrak{h}$ and $\mathfrak{h}^*$. The Weyl group $W \simeq S_2$ has type A_1.

For a root α, we define $\alpha^\vee \in \mathfrak{h}$ to be the unique element such that $\alpha = 2\frac{(\alpha^\vee|-)}{(\alpha^\vee|\alpha^\vee)}$. We call the $\alpha^\vee$ *coroots*. For $\lambda \in \mathfrak{h}^*$, we set $\langle\lambda, \alpha^\vee\rangle := \lambda(\alpha^\vee)$. We can then express reflections through hyperplanes orthogonal to roots in the following way: $s_\alpha(\lambda) = \lambda - \langle\lambda, \alpha^\vee\rangle\alpha$ for $\lambda \in \mathfrak{h}^*$.

Example 14.8 For $\mathfrak{g} = \mathfrak{sl}_2(\mathbb{C})$, $\alpha^\vee = h$.

Let $Q := \mathbb{Z}\Phi \subset \mathfrak{h}^*$ be the *root lattice* of Φ. Let $Q^+ := \mathbb{N}\Phi^+$. We define a partial order on $\mathfrak{h}^*$ in the following way: for $\mu, \lambda \in \mathfrak{h}^*$, say $\mu \leq \lambda$ if $\lambda - \mu \in Q^+$.

Example 14.9 For $\mathfrak{g} = \mathfrak{sl}_2(\mathbb{C})$, recall the identification $\mathfrak{h}^* \xrightarrow{\sim} \mathbb{C}$ given by $\lambda \mapsto \lambda(h)$. Under this identification we have $Q = \mathbb{Z}\alpha = 2\mathbb{Z}h^* \simeq 2\mathbb{Z}$, and $Q^+ = 2\mathbb{N}h^* \simeq 2\mathbb{N}$. For two complex numbers z, z' we have $z \leq z'$ if and only if $z' - z \in 2\mathbb{N}$. For example, $3 + 2i \leq 5 + 2i$, but $3 + 2i$ and $5 + 6i$ are unrelated since their difference is not a multiple of 2.

A weight $\lambda \in \mathfrak{h}^*$ is called *integral* $\langle\lambda, \alpha^\vee\rangle \in \mathbb{Z}$ for all $\alpha \in \Phi$ and *dominant integral* if $\langle\lambda, \alpha^\vee\rangle \in \mathbb{N}$ for all $\alpha \in \Phi$. Let X (resp. X^+) denote the set of integral (resp. dominant integral) weights. Then X is a lattice in $\mathfrak{h}^*$, called the *weight lattice* of $\mathfrak{g}$. We have $Q \subseteq X$ and $Q^+ \subseteq X^+$.

Example 14.10 For $\mathfrak{g} = \mathfrak{sl}_2(\mathbb{C})$, $X = \mathbb{Z}h^* \simeq \mathbb{Z}$, and $X^+ = \mathbb{N}h^* \simeq \mathbb{N}$.

The subspace $\mathfrak{n}^\pm := \bigoplus_{\alpha \in \pm\Phi^+} \mathfrak{g}_\alpha$ is a subalgebra of $\mathfrak{g}$. The *Borel subalgebra* of $\mathfrak{g}$ is the subalgebra $\mathfrak{b} := \mathfrak{h} \oplus \mathfrak{n}^+$.

Example 14.11 For $\mathfrak{g} = \mathfrak{sl}_2(\mathbb{C})$, we have $\mathfrak{n}^- = \mathbb{C}f$ and $\mathfrak{n}^+ = \mathbb{C}e$, so the Borel subalgebra is $\mathfrak{b} = \mathbb{C}h \oplus \mathbb{C}e$, the subalgebra of upper triangular matrices in $\mathfrak{sl}_2(\mathbb{C})$.

The *universal enveloping algebra* of $\mathfrak{g}$ is the associative algebra $U(\mathfrak{g}) := T(\mathfrak{g})/(x \otimes y - y \otimes x - [x, y])$. Any $\mathfrak{g}$-module can also be viewed as a $U(\mathfrak{g})$-module where the generators (the image of $\mathfrak{g} = T^1(\mathfrak{g}) \subset T(\mathfrak{g})$) act as expected. This gives an isomorphism between the category of $\mathfrak{g}$-modules (in the Lie algebra sense) and the category of $U(\mathfrak{g})$-modules (in the associative algebra sense). If $\mathfrak{b} \subset \mathfrak{g}$ is a Lie subalgebra, then $U(\mathfrak{b}) \subset U(\mathfrak{g})$ is a subalgebra.

For $\lambda \in \mathfrak{h}^*$, the corresponding *Verma module* is defined by

$$\Delta(\lambda) := U(\mathfrak{g}) \otimes_{U(\mathfrak{b})} \mathbb{C}_\lambda . \tag{14.1}$$

Here $\mathbb{C}_\lambda$ is the one-dimensional $\mathfrak{b}$-module where $\mathfrak{h}$ acts by λ and $\mathfrak{n}^+$ acts trivially. This is a *highest weight module* for λ. What we mean by this is the following. A nonzero vector v^+ in a $U(\mathfrak{g})$-module V is called a *maximal vector* of weight $\lambda \in \mathfrak{h}^*$

14.1 Lie Algebra Basics

if $v^+ \in V_\lambda$ and $\mathfrak{n}^+ \cdot v^+ = 0$. We say that a $U(\mathfrak{g})$-module V is a *highest weight module* of weight λ if there is a maximal vector $v^+ \in V_\lambda$ such that $V = U(\mathfrak{g}) \cdot v^+$.

By the Poincaré–Birkhoff–Witt (PBW) theorem, $U(\mathfrak{g}) = U(\mathfrak{n}^-)U(\mathfrak{b})$, so as a left $U(\mathfrak{n}^-)$-module, $\Delta(\lambda) = U(\mathfrak{n}^-) \otimes_\mathbb{C} \mathbb{C}_\lambda$. Therefore, $\Delta(\lambda)$ is a free $U(\mathfrak{n}^-)$-module of rank 1, and the vector $v^+ = 1 \otimes 1 \in \Delta(\lambda)$ has the property that $\mathfrak{n}^+ \cdot v^+ = 0$ and $h \cdot v^+ = \lambda(h)v^+$ for all $h \in \mathfrak{h}$. Thus v^+ is a maximal vector which generates the $U(\mathfrak{g})$-module $\Delta(\lambda)$.

Example 14.12 Let $\mathfrak{g} = \mathfrak{sl}_2(\mathbb{C})$, and $\lambda \in \mathfrak{h}^* \simeq \mathbb{C}$. Since $\mathfrak{n}^-$ is one-dimensional, $U(\mathfrak{n}^-) \simeq \mathbb{C}[f]$ is a polynomial ring in one variable. Our preferred basis for $U(\mathfrak{n}^-)$ is $\{\frac{1}{i!}f^i\}_{i \in \mathbb{N}}$. Hence, $\Delta(\lambda) = U(\mathfrak{n}^-) \otimes_\mathbb{C} \mathbb{C}_\lambda$ has $\mathbb{C}$-basis $\{v_i\}_{i \in \mathbb{N}}$, where $v_i = \frac{1}{i!}f^i(v^+)$. By direct computation,

$$h \cdot v_i = (\lambda - 2i)v_i, \quad e \cdot v_i = (\lambda - i + 1)v_{i-1}, \quad f \cdot v_i = (i+1)v_{i+1}. \quad (14.2)$$

Hence $\{v_i\}$ is a weight basis for $\Delta(\lambda)$ with weights $\lambda, \lambda - 2, \lambda - 4, \ldots$, and each weight occurs with multiplicity one. We capture this information in the following picture.

$$\cdots \bullet \rightleftarrows \bullet \rightleftarrows \bullet \cdots \bullet \rightleftarrows \bullet \rightleftarrows \bullet \rightleftarrows \bullet \quad (14.3)$$

Here, the $\bullet$ represent weight spaces, all of which are one-dimensional and of the form $\mathbb{C}v_i$. Since the weight spaces are one-dimensional, action by $\{f, e, h\}$ can be expressed by a complex number (i.e. a 1×1 matrix) and an arrow indicating where a weight vector is sent under the action. The rightward arrows (which should really be thought of as upward arrows, since they increase the weight) represent action by e, the leftward (i.e. downward) arrows represent action by f, and the loops represent action by h.

Exercise 14.13 This exercise is about the Verma module $\Delta(\lambda)$ for $\mathfrak{sl}_2(\mathbb{C})$, and uses the notation of the previous example.

1. Verify the formulas (14.2).
2. More generally, why should e send V_k to V_{k+2} for any weight module V?
3. Suppose instead that we used the basis $\{f^k v^+\}$ for $\Delta(\lambda)$. Compute the formulas analogous to (14.2), and draw the diagram analogous to (14.3) for this basis.

Exercise 14.14 Continue to let $\mathfrak{g} = \mathfrak{sl}_2(\mathbb{C})$.

1. Prove that there is a nonzero map $\Delta(\mu) \to \Delta(\lambda)$ if and only if $\Delta(\lambda)$ has a weight vector x of weight μ such that $ex = 0$, which is to say, a maximal vector of weight μ. (*Hint:* realize the cyclic module $\Delta(\mu)$ as the quotient of $U(\mathfrak{g})$ by the left ideal generated by $h - \mu$ and e.)
2. For which k is v_k a maximal vector of some weight?

3. Conclude that there is a nonzero map $\Delta(\mu) \to \Delta(\lambda)$ if and only if either $\mu = \lambda$ or $\lambda \in \mathbb{N}$ and $\mu = -\lambda - 2$.

Spoilers for this exercise are in Example 14.16 below.

Verma modules have a rich structure. We list some of their basic properties.

Theorem 14.15 *The following statements hold:*

1. $\Delta(\lambda)$ is a semisimple $\mathfrak{h}$-module.
2. $\Delta(\lambda)$ is an indecomposable $\mathfrak{g}$-module.
3. If μ is a weight of $\Delta(\lambda)$, then $\mu \leq \lambda$.
4. All weight spaces of $\Delta(\lambda)$ are finite dimensional. The λ-weight space is one-dimensional.
5. Each nonzero quotient of $\Delta(\lambda)$ is also a highest weight module with highest weight λ.
6. We can realize $\Delta(\lambda)$ as the quotient of $U(\mathfrak{g})$ by the left ideal $I \subset U(\mathfrak{g})$ generated by $\mathfrak{n}^+$ and all $h - \lambda(h) \cdot 1$ for $h \in \mathfrak{h}$. This ideal I is the annihilator of the maximal vector $v \in \Delta(\lambda)$. The ideal I also annihilates any maximal vector of weight λ which generates an arbitrary highest weight module V, so $\Delta(\lambda)$ maps naturally onto V. Because of this property, we say that $\Delta(\lambda)$ is the universal highest weight module *of weight* λ.

The Verma module $\Delta(\lambda)$ has a unique maximal submodule, and thus a unique irreducible quotient $L(\lambda)$. This gives an *injection* $L : \mathfrak{h}^* \to \text{Irr } \mathfrak{g}$. The module $L(\lambda)$ is finite dimensional if and only if $\lambda \in X^+$. The map L restricts to a *bijection* between X^+ and the set of isomorphism classes of finite-dimensional irreducible $\mathfrak{g}$-modules.

Example 14.16 For $\mathfrak{g} = \mathfrak{sl}_2(\mathbb{C})$, if $\lambda \in X^+ \simeq \mathbb{N}$, then the maximal submodule of $\Delta(\lambda)$ has basis $\{v_i\}_{i > \lambda}$. Hence, the maximal submodule is isomorphic to $\Delta(-\lambda - 2)$ and the irreducible quotient $L(\lambda)$ is of dimension $\lambda + 1$. We can see this in the following picture.

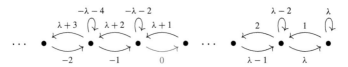

As before, • represent one-dimensional weight spaces $\mathbb{C}v_i$. If λ is a positive integer, then for $i = \lambda + 1$, the upward action by e annihilates v_i:

$$e \cdot v_{\lambda+1} = (\lambda - (\lambda + 1) + 1)v_\lambda = 0 \cdot v_\lambda.$$

14.2 Category $\mathcal{O}$

As explained above, we have a complete classification of the finite-dimensional irreducible modules of a complex semisimple Lie algebra $\mathfrak{g}$: they are in bijection with the set X^+ of dominant integral weights. Weyl's complete reducibility theorem states that the category $\mathfrak{g}$-mod^{fd} of finite-dimensional $\mathfrak{g}$-modules is semisimple, so every finite-dimensional $\mathfrak{g}$-module is a direct sum of some $L(\lambda)$, $\lambda \in X^+$. We thus have a very good understanding of the category $\mathfrak{g}$-mod^{fd}.

However, we are far away from understanding the whole category $\mathfrak{g}$-Mod of all $\mathfrak{g}$-modules. Recall that the map $L \colon \mathfrak{h}^* \to \operatorname{Irr}\mathfrak{g}$ is only an injection. In fact, the image (the highest weight irreducible modules) form only a *tiny* portion of the irreducible $\mathfrak{g}$-modules, and only for $\mathfrak{sl}_2(\mathbb{C})$ a classification of $\operatorname{Irr}\mathfrak{g}$ is known, see [28].

Nonetheless, certain subcategories of $\mathfrak{g}$-Mod containing $\mathfrak{g}$-mod^{fd} have been extensively studied. In this section, we list structural results about such a category introduced in the late 1970s by Joseph Bernstein, Israel Gelfand, and Sergei Gelfand, inspired by the work in Verma's thesis from 1966. This is the *BGG category $\mathcal{O}$*.[1]

Definition 14.17 *Category $\mathcal{O}$ is the full subcategory of $\mathfrak{g}$-Mod consisting of modules V such that:*

1. V is a finitely generated $U(\mathfrak{g})$-module;
2. V is a weight module; and
3. V is *locally* $\mathfrak{n}^+$-finite, i.e. $\dim U(\mathfrak{n}^+)v < \infty$ for all $v \in V$.

To discuss weight modules and $\mathfrak{n}^+$, we need to already have fixed a Cartan subalgebra $\mathfrak{h}$ and a choice of positive roots Φ^+ with base Δ. This data will be fixed for the rest of the chapter. We list some of the major structural results about category $\mathcal{O}$ in the following theorem.

Theorem 14.18 *Let $V \in \mathcal{O}$. The following holds:*

1. *All weight spaces of V are finite dimensional.*
2. *The weights of V are contained in the set $\bigcup_{i=1}^{r}(\lambda_i - \mathbb{Z}\Phi^+)$ for some $\lambda_i \in \mathfrak{h}^*$ and some $r \in \mathbb{N}$.*
3. *$\mathcal{O}$ is an abelian category, and is closed under taking submodules, quotients, and finite direct sums.*
4. *$\mathcal{O}$ contains the Verma module $\Delta(\lambda)$ for all $\lambda \in \mathfrak{h}^*$, and contains all finite-dimensional $\mathfrak{g}$-modules.*
5. *The map $L \colon \mathfrak{h}^* \to \operatorname{Irr}\mathfrak{g}$ is a bijection between $\mathfrak{h}^*$ and the irreducible objects up to isomorphism in $\mathcal{O}$.*

---

[1] Why is this category referred to as "category $\mathcal{O}$"? It could be that when first learning about it, one finds oneself exclaiming "Oh! What a nice category!" The more serious explanation is that "O" is the first letter of a Russian word meaning "basic."

6. $\mathcal{O}$ has finite length, so every object $V \in \mathcal{O}$ has a composition series and the Jordan–Hölder multiplicities $[V : L(\lambda)]$ are well-defined. Furthermore, if $[\Delta(\lambda) : L(\mu)] \neq 0$, for some $\mu, \lambda \in \mathfrak{h}^*$, then $\mu \leq \lambda$.
7. $\mathcal{O}$ has enough projectives, i.e. for any object V, there exists an epimorphism from a projective object to V.

Proof We prove 1, 2, and 3, and refer the reader to [86] for the remaining proofs. Let V be a module in $\mathcal{O}$. Conditions 2 and 3 in Definition 14.17 imply that V is a locally finite $\mathfrak{b}$-module. Therefore, by condition 1 in Definition 14.17, there exists a finite-dimensional $\mathfrak{b}$-invariant subspace U that generates V as a $U(\mathfrak{g})$-module. By the PBW theorem, $U(\mathfrak{g}) = U(\mathfrak{n}^-)U(\mathfrak{b})$, so $V = U(\mathfrak{g})U = U(\mathfrak{n}^-)U$. This gives a natural surjection $U(\mathfrak{n}^-) \otimes_{\mathbb{C}} U \to V$ by sending $Y \otimes u \mapsto Y \cdot u$. The vector space $U(\mathfrak{n}^-) \otimes_{\mathbb{C}} U$ is an $\mathfrak{h}$-module by the tensor product of the adjoint action and the restricted action of $\mathfrak{h}$ on U, and this map is an $\mathfrak{h}$-module morphism. This implies that the weights of V are a subset of the weights of $U(\mathfrak{n}^-) \otimes_{\mathbb{C}} U$. Since U is a finite-dimensional $\mathfrak{b}$-invariant subspace, the collection S_0 of the weights of U is finite. The weights of the adjoint action on $U(\mathfrak{n}^-)$ are non-positive sums of positive roots (i.e. of the form $-\sum_{\alpha \in \Phi^+} m_\alpha \alpha$ for $m_\alpha \in \mathbb{N}$). Since the weights of tensor product representations are sums of weights of the original representations (see also Exercise 14.40), we conclude that weights of $U(\mathfrak{n}^-) \otimes_{\mathbb{C}} U$, and thus weights of V, are of the form $\lambda_i - \sum_{\alpha \in \Phi^+} m_\alpha \alpha$ for $\lambda_i \in S_0$ and $m_\alpha \in \mathbb{N}$. Moreover, since the dimension of a weight space $U(\mathfrak{n}^-)_\nu$ is equal to the number of ways $\nu \in \mathfrak{h}^*$ can be expressed as a non-negative sum of positive roots (this quantity is often referred to as *Kostant's partition function*) and the weight spaces of U are finite dimensional, we conclude that the weight spaces of V must be finite dimensional as well. This proves 1 and 2.

Since $U(\mathfrak{g})$ is Noetherian, any submodule of a finitely generated $U(\mathfrak{g})$-module is finitely generated. (The fact that $U(\mathfrak{g})$ is Noetherian follows from the fact that a filtered ring is Noetherian if its associated graded ring is Noetherian. Since $U(\mathfrak{g})$ has a PBW filtration so that the associated graded algebra gr $U(\mathfrak{g})$ is $S(\mathfrak{g})$, this implies that $U(\mathfrak{g})$ is Noetherian.) This implies that condition 1 of Definition 14.17 holds for submodules, and conditions 2 and 3 are immediate. Both quotients and finite direct sums clearly inherit conditions 1, 2, and 3 of Definition 14.17, so the category is closed under quotients and finite direct sums. The fact that $\mathcal{O}$ is closed under quotients and submodules implies that it is closed under kernels and cokernels of homomorphisms, which in turn implies that it is an abelian category, since it is a full subcategory of the abelian category $\mathfrak{g}$-Mod. This proves 3. □

14.3 Duality in $\mathcal{O}$

In this section, we explore a useful notion of duality in category $\mathcal{O}$. For any $U(\mathfrak{g})$-module V, there is a natural action of $U(\mathfrak{g})$ on the dual vector space V^*: for $x \in$

14.3 Duality in $\mathcal{O}$

$\mathfrak{g}$, $v \in V$, and $f \in V^*$,

$$(x \cdot f)(v) = -f(x \cdot v). \tag{14.4}$$

With this action, V^* is a $U(\mathfrak{g})$-module. However, if V is an infinite-dimensional module in category $\mathcal{O}$, the dual module V^* will not be in $\mathcal{O}$. Firstly, if λ is a weight appearing in V, then any linear functional $\phi \in V_\lambda^*$ can be extended by zero to a linear functional in V^*. Then ϕ will be a weight vector, but having weight $-\lambda$ rather than λ. Thus if the weights of V are "bounded above" in the sense of Theorem 14.18 part 2, then the weights of V^* will instead be "bounded below." Secondly, taking duals of vector spaces only commutes with finite direct sums, so that V^* will not be the direct sum of its weight spaces V_λ^*.

To remedy these problems, we can define a different duality functor that preserves category $\mathcal{O}$. First we modify the action of $U(\mathfrak{g})$ on the dual vector space so that it preserves weights, rather than negating them. Then we restrict our attention to the direct sum of the weight spaces.

For each simple root $\alpha \in \Delta$ we can choose root vectors $x_\alpha \in \mathfrak{g}_\alpha$ and $y_\alpha \in \mathfrak{g}_{-\alpha}$, and set $h_\alpha = [x_\alpha, y_\alpha] \in \mathfrak{h}$. This can be accomplished in such a way that $\alpha(h_\alpha) = 2$ for all $\alpha \in \Delta$, so that $(x_\alpha, h_\alpha, y_\alpha)$ behaves just like the triple (e, h, f) in $\mathfrak{sl}_2(\mathbb{C})$.

Definition 14.19 There is an anti-involution $\tau : \mathfrak{g} \to \mathfrak{g}$ sending

$$x_\alpha \mapsto y_\alpha, \qquad y_\alpha \mapsto x_\alpha, \qquad h_\alpha \mapsto h_\alpha. \tag{14.5}$$

In the case of $\mathfrak{sl}_n(\mathbb{C})$, this is the usual matrix transpose map. There is a canonical extension of τ to an anti-automorphism of $U(\mathfrak{g})$. (If you add a minus sign, $-\tau$ is an involution on $\mathfrak{g}$ which is sometimes referred to as the *Chevalley involution*.)

We use τ to define a new action of $\mathfrak{g}$ on V^*. For $x \in \mathfrak{g}$, $f \in V^*$, and $v \in V$, define the *twisted action* of $\mathfrak{g}$ on V^* by

$$(x \cdot f)(v) := f(\tau(x) \cdot v). \tag{14.6}$$

From now on, when we refer to the action of $\mathfrak{g}$ on V^* we will be referring to this twisted action. Note that since τ fixes $\mathfrak{h}$ pointwise, $(h \cdot f)(v) = f(h \cdot v)$.

Let $\mathcal{C}$ be the subcategory of $U(\mathfrak{g})$-Mod consisting of weight modules with finite-dimensional weight spaces. For a module V in $\mathcal{C}$, a weight $\lambda \in \mathfrak{h}^*$, and a weight space V_λ, we can identify $(V_\lambda)^*$ with the set of $f \in V^*$ which vanish on V_μ for $\mu \neq \lambda$. Under this identification, $(V^*)_\lambda = (V_\lambda)^*$. Indeed, if $f \in (V^*)_\lambda$, $h \in \mathfrak{h}$, $v \in V_\mu$, and $\mu(h) \neq \lambda(h)$,

$$\mu(h) f(v) = f(\mu(h)v) = f(h \cdot v) = (h \cdot f)(v) = \lambda(h) f(v),$$

so f vanishes outside of V_λ. This allows us to drop parentheses and refer unambiguously to V_λ^*.

Definition 14.20 For a module V in $\mathcal{C}$, the *dual module* $\mathbb{D}(V)$ is

$$\mathbb{D}(V) := \bigoplus_{\lambda \in \mathfrak{h}^*} V_\lambda^*. \tag{14.7}$$

The contravariant functor $\mathbb{D} : \mathcal{C} \to \mathcal{C}$ is exact. We claim that this functor preserves category $\mathcal{O}$. Indeed, by construction, the dual $\mathbb{D}(V)$ of a module V in $\mathcal{O}$ decomposes into the direct sum of weight spaces, and the exactness of $\mathbb{D}$ implies that $\mathbb{D}(V)$ is finite length, and thus is finitely generated. Furthermore, the twisted action of $\mathfrak{g}$ on $\mathbb{D}(V)$ ensures that $x_\alpha \cdot \mathbb{D}(V)_\mu \subset \mathbb{D}(V)_{\alpha+\mu}$ for $\mu \in \mathfrak{h}^*$, so the locally $\mathfrak{n}^+$-finite condition is also preserved.

For $\lambda \in \mathfrak{h}^*$, we denote

$$\nabla(\lambda) := \mathbb{D}(\Delta(\lambda)). \tag{14.8}$$

This is the *dual Verma module*.

Example 14.21 Let $\mathfrak{g} = \mathfrak{sl}_2(\mathbb{C})$. The dual Verma $\nabla(\lambda)$ is a module in $\mathcal{O}$ with the same set of weights as $\Delta(\lambda)$. If $\{v_i\}_{i\geq 0}$ is the basis for $\Delta(\lambda)$ described in Example 14.12, then a dual basis for $\nabla(\lambda)$ is given by $\{\varphi_i\}_{i\geq 0}$, where $\varphi_i \in \Delta(\lambda)^*_{\lambda-2i}$ is defined by $\varphi_i(v_i) = 1$. The twisted action of $\mathfrak{g}$ on $\nabla(\lambda)$ is given by

$$h \cdot \varphi_i = (\lambda - 2i)\varphi_i, \qquad e \cdot \varphi_i = i\varphi_{i-1}, \qquad f \cdot \varphi_i = (\lambda - i)\varphi_{i+1}.$$

If we draw Verma modules with dots and arrows as we did in (14.3), taking the dual of a Verma module amounts to switching the numbers on the upward and downward arrows. Crucially, for λ dominant integral, this will switch which direction the arrow labeled 0 is pointing.

As a consequence, the finite-dimensional module $L(\lambda)$ is a submodule of $\nabla(\lambda)$, whereas it was a quotient of $\Delta(\lambda)$.

The vector $\varphi_0 \in \nabla(\lambda)_\lambda$ is a maximal vector since it is annihilated by e. However, this vector does not generate $\nabla(\lambda)$ as a $U(\mathfrak{g})$-module, since $f^{\lambda+1} \cdot \varphi_0 = 0$, so $\nabla(\lambda)$ is not a highest weight module. Instead, the maximal vector φ_0 generates an irreducible submodule of $\nabla(\lambda)$ which is isomorphic to $L(\lambda)$.

Exercise 14.22 Let λ be a dominant integral weight for $\mathfrak{sl}_2(\mathbb{C})$. Prove that the quotient of $\nabla(\lambda)$ by its irreducible submodule $L(\lambda)$ is isomorphic to the dual Verma module $\nabla(-\lambda - 2)$. Also prove that $\nabla(-\lambda - 2) \simeq \Delta(-\lambda - 2)$.

14.3 Duality in $\mathcal{O}$

We list some properties of dual modules in the following theorem, and refer the reader to [86] for proofs.

Theorem 14.23 *Let $\lambda, \mu \in \mathfrak{h}^*$. The following holds:*

1. $\mathbb{D}(L(\lambda)) \simeq L(\lambda)$.
2. $\nabla(\lambda)$ *has $L(\lambda)$ as its unique irreducible submodule. The other composition factors $L(\mu)$ of $\nabla(\lambda)$ satisfy $\mu < \lambda$.*
3. *("Fundamental vanishing")* $\mathrm{Ext}^1(\Delta(\lambda), \nabla(\mu)) = 0$.

14.3.1 Standard Filtrations and BGG Reciprocity

The following notion plays an important role in the study of $\mathcal{O}$.

Definition 14.24 Let $M \in \mathcal{O}$. A *standard filtration* (or a *Verma flag*) on M is a finite filtration

$$0 = M_0 \subset M_1 \subset \cdots \subset M_n = M$$

of M by objects in $\mathcal{O}$, such that each subquotient M_i/M_{i-1} is isomorphic to a Verma module. For $\lambda \in \mathfrak{h}^*$, the *standard multiplicity* of $\Delta(\lambda)$ in M, denoted by $(M : \Delta(\lambda))$, is the number of times that $\Delta(\lambda)$ occurs as a subquotient.

By character consideration, the standard multiplicity $(M : \Delta(\lambda))$ is independent of the choice of standard filtration on M. We emphasize that, unlike the Jordan–Hölder multiplicity $[M : L(\lambda)]$, standard multiplicity is only defined for those objects of $\mathcal{O}$ with a standard filtration. For instance, irreducible modules do not admit a standard filtration.

If $M, N \in \mathcal{O}$ admit a standard filtration, then clearly so does $M \oplus N$, and $(M \oplus N : \Delta(\lambda)) = (M : \Delta(\lambda)) + (N : \Delta(\lambda))$. Thus modules that admit a standard filtration form an additive category, but not an abelian category.

Remark 14.25 Standard filtrations and their multiplicities should be viewed as an analogue in category $\mathcal{O}$ of Δ-filtrations and their graded multiplicities for Soergel bimodules[2] (Definition 5.8). We will see in Chap. 15 that this is more than just an analogy.

Here are some other important facts about standard filtrations. Again, we refer the reader to [86] for proofs.

Proposition 14.26 *Let $M \in \mathcal{O}$ admit a standard filtration. The following holds:*

---

[2] One could also study *costandard filtrations* (where Verma modules are replaced by dual Verma modules), which are analogues of ∇-filtrations on Soergel bimodules. The two notions are related by the duality functor $\mathbb{D}$, just as Δ- and ∇-filtrations are related by a duality functor on bimodules.

1. For all $\lambda \in \mathfrak{h}^*$,

$$(M : \Delta(\lambda)) = \dim \operatorname{Hom}_{\mathcal{O}}(M, \nabla(\lambda)).$$

2. Any direct summand of M also admits a standard filtration.
3. Every projective object in $\mathcal{O}$ admits a standard filtration.

We finish this subsection with a fundamental "reciprocity" result of Bernstein–Gelfand–Gelfand. Before stating the theorem, we need to discuss projective objects in $\mathcal{O}$ in more detail.

Since $\mathcal{O}$ has enough projectives and is finite length, each object $V \in \mathcal{O}$ has a *projective cover*

$$\pi : P_V \to V. \tag{14.9}$$

That is, π is an essential epimorphism (meaning that no proper submodule of P_V is mapped onto V), and up to (non-unique) isomorphism, P_V is the unique projective object in $\mathcal{O}$ with this property. In the case $V = L(\lambda)$ for $\lambda \in \mathfrak{h}^*$, we denote the projective module obtained in this way by $P(\lambda)$. Since $\pi_\lambda : P(\lambda) \to L(\lambda)$ is an essential epimorphism, it must have a unique maximal submodule $\ker \pi_\lambda$, and thus is indecomposable. Moreover, we actually have epimorphisms

$$P(\lambda) \to \Delta(\lambda) \to L(\lambda), \tag{14.10}$$

and $P(\lambda)$ is also a projective cover of $\Delta(\lambda)$. We can now state the Bernstein–Gelfand–Gelfand reciprocity, which relates the multiplicities of irreducible modules in Verma modules to the multiplicities of Verma modules in projective modules.

Theorem 14.27 (BGG Reciprocity) *For all $\lambda, \mu \in \mathfrak{h}^*$,*

$$(P(\lambda) : \Delta(\mu)) = [\nabla(\mu) : L(\lambda)] = [\Delta(\mu) : L(\lambda)]. \tag{14.11}$$

14.4 Blocks of Category $\mathcal{O}$

In this section, we describe how a careful examination of the action of the center of the universal enveloping algebra of $\mathfrak{g}$ leads us to a decomposition of $\mathcal{O}$ into subcategories. Recall that a category $\mathcal{C}$ is *small* if its class of objects forms a set; it is *essentially small* if it is equivalent to a small category.

Lemma 14.28 *Category $\mathcal{O}$, in addition to being a finite length category and abelian, is essentially small.*

This follows from the fact that categories of finitely generated modules over a ring are essentially small, and $\mathcal{O}$ is a full subcategory of the category of finitely generated $U(\mathfrak{g})$-modules. This is important, because it implies something very strong about category $\mathcal{O}$. Namely,

Theorem 14.29 *Category $\mathcal{O}$ admits a unique decomposition*

$$\mathcal{O} = \bigoplus_{i \in I} \mathcal{O}_i \tag{14.12}$$

into indecomposable full subcategories $\mathcal{O}_i$.

An abelian category is called *indecomposable* if it is not equivalent to the direct sum of two nonzero categories. The $\mathcal{O}_i$ in the previous theorem are often referred to as the *blocks of category $\mathcal{O}$*. The direct sum decomposition means in particular that objects in different blocks do not extend: $\text{Ext}^1_{\mathcal{O}}(M, N) = \text{Ext}^1_{\mathcal{O}}(N, M) = 0$ for all $M \in \mathcal{O}_i$ and $N \in \mathcal{O}_j$ with $i \neq j$. For more information on block decompositions of categories, see [63, Section 1]. Note that $\Delta(\lambda)$ being indecomposable implies that it lies in a unique block of $\mathcal{O}$.

To describe the block decomposition of category $\mathcal{O}$ in more detail, we will analyze the action of the center of the universal enveloping algebra on modules in $\mathcal{O}$. We denote by $Z(\mathfrak{g}) = Z(U(\mathfrak{g}))$ the center of the universal enveloping algebra.

Remark 14.30 In the study of finite-dimensional $U(\mathfrak{g})$-modules, it is enough to examine only the action of a distinguished element c of $Z(\mathfrak{g})$ called the *Casimir element*. With just information about the action of c, one can obtain strong results such as Weyl's complete reducibility theorem. See Exercise 14.39 for a calculation with the Casimir element of $\mathfrak{sl}_2(\mathbb{C})$. In contrast to this, to understand the block decomposition of category $\mathcal{O}$, we must understand the whole center $Z(\mathfrak{g})$.

Note that the axioms of category $\mathcal{O}$ imply that any module $V \in \mathcal{O}$ is locally $Z(\mathfrak{g})$-finite, see [86, Theorem 1.1]. We start by introducing the notion of a central character. Let V be a highest weight module generated by a maximal vector $v^+ \in V$. For $z \in Z(\mathfrak{g})$ and $h \in \mathfrak{h}$, we have

$$h \cdot (z \cdot v^+) = z \cdot (h \cdot v^+) = z \cdot (\lambda(h)v^+) = \lambda(h)z \cdot v^+,$$

since the center commutes with $\mathfrak{h}$. Since $\dim V_\lambda = 1$ (a consequence of the definition of highest weight modules is that all highest weight modules have a one-dimensional highest weight space), this implies that z must act by a scalar, i.e. $z \cdot v^+ = \chi_\lambda(z)v^+$ for some $\chi_\lambda(z) \in \mathbb{C}$. This defines an algebra homomorphism

$$\chi_\lambda : Z(\mathfrak{g}) \to \mathbb{C} \tag{14.13}$$

such that for $z \in Z(\mathfrak{g})$,

$$z \cdot v^+ = \chi_\lambda(z)v^+.$$

Note that this homomorphism depends only on λ, not on V. The kernel of χ_λ is a maximal ideal in $Z(\mathfrak{g})$. Furthermore, since v^+ generates V, arbitrary elements of V are of the form uv^+ for $u \in U(\mathfrak{n}^-)$. This implies that for an arbitrary $uv^+ \in V$ and $z \in Z(\mathfrak{g})$,

$$z \cdot uv^+ = u \cdot (z \cdot v^+) = u \cdot (\chi_\lambda(z)v^+) = \chi_\lambda(z)uv^+,$$

so z acts on all vectors in V by the same scalar. We call the algebra homomorphism $\chi_\lambda : Z(\mathfrak{g}) \to \mathbb{C}$ obtained in this way the *central character* associated to λ, and more generally, we call any algebra homomorphism $\chi : Z(\mathfrak{g}) \to \mathbb{C}$ a central character. We say that a $U(\mathfrak{g})$-module V has central character χ if $z \cdot v = \chi(z)v$ for all $z \in Z(\mathfrak{g})$ and $v \in V$. The set of central characters is naturally in bijection with the set of maximal ideals of $Z(\mathfrak{g})$. Central characters play a key role in our analysis of category $\mathcal{O}$, and in the study of representation theory of Lie algebras in general. Next we will describe a useful homomorphism between $Z(\mathfrak{g})$ and $U(\mathfrak{h})$ that provides key information about the structure of $Z(\mathfrak{g})$.

The PBW theorem allows us to choose an ordered basis

$$y_1 < \cdots < y_m < h_1 < \cdots < h_\ell < x_1 < \cdots < x_m$$

of $\mathfrak{g}$ such that the monomials

$$y_1^{r_1} \cdots y_m^{r_m} h_1^{s_1} \cdots h_\ell^{s_\ell} x_1^{t_1} \cdots x_m^{t_m} \tag{14.14}$$

form a basis for $U(\mathfrak{g})$. We can choose our basis of $\mathfrak{g}$ such that $y_i \in \mathfrak{g}_{-\alpha_i}$, $x_i \in \mathfrak{g}_{\alpha_i}$ for $\alpha_i \in \Phi^+$, and $h_i = [x_i, y_i]$ for simple roots α_i. Since all elements of $U(\mathfrak{g})$ can be expressed as sums of these monomials, there is a decomposition

$$U(\mathfrak{g}) = U(\mathfrak{h}) + \mathfrak{n}^- U(\mathfrak{g}) + U(\mathfrak{g})\mathfrak{n}^+.$$

In this decomposition, $U(\mathfrak{h}) \cap (\mathfrak{n}^- U(\mathfrak{g}) + U(\mathfrak{g})\mathfrak{n}^+) = \{0\}$, so the first sum is direct:

$$U(\mathfrak{g}) = U(\mathfrak{h}) \oplus (\mathfrak{n}^- U(\mathfrak{g}) + U(\mathfrak{g})\mathfrak{n}^+).$$

The centralizer $U(\mathfrak{g})_0 = \{x \in U(\mathfrak{g}) \mid [h, x] = 0 \text{ for all } h \in \mathfrak{h}\}$ of $\mathfrak{h}$ in $U(\mathfrak{g})$ is generated by monomials of the form (14.14) such that

$$\sum_{i=1}^m (t_i - r_i)\alpha_i = 0.$$

This implies that if an element $x \in U(\mathfrak{g})_0$ is in $\mathfrak{n}^- U(\mathfrak{g})$, it is also in $U(\mathfrak{g})\mathfrak{n}^+$, so $x \in \mathfrak{n}^- U(\mathfrak{g}) \cap U(\mathfrak{g})\mathfrak{n}^+$. Since $Z(\mathfrak{g})$ is contained in $U(\mathfrak{g})_0$, we conclude that there

14.4 Blocks of Category $\mathcal{O}$

are inclusions

$$Z(\mathfrak{g}) \subset U(\mathfrak{g})_0 \subset U(\mathfrak{h}) \oplus U(\mathfrak{g})\mathfrak{n}^+.$$

Thus, any element $z \in Z(\mathfrak{g})$ can be written uniquely as $z = h + n$ for $h \in U(\mathfrak{h})$ and $n \in U(\mathfrak{g})\mathfrak{n}^+$. The utility of this decomposition comes when z acts on a highest weight vector v^+, since $(h + n)v^+ = hv^+$, and the action of h on v^+ is determined easily from the weight space that v^+ lives in.

Definition 14.31 The *Harish-Chandra homomorphism* is the algebra homomorphism

$$\gamma : Z(\mathfrak{g}) \to U(\mathfrak{h}) \tag{14.15}$$

sending $z = h + n \in Z(\mathfrak{g})$ to $h \in U(\mathfrak{h})$.

Note that for the central character χ_λ associated to λ and $z \in Z(\mathfrak{g})$ we have

$$\chi_\lambda(z) = \lambda(\gamma(z)) \in \mathbb{C}, \tag{14.16}$$

where we extend $\lambda \in \mathfrak{h}^*$ in the obvious way to an algebra morphism $\lambda : U(\mathfrak{h}) \to \mathbb{C}$. The Harish-Chandra homomorphism provides the first step to understanding the structure of $Z(\mathfrak{g})$, but for the full story we need to find the kernel and image of γ. To do this, we start by defining a shifted action of the Weyl group W on $\mathfrak{h}^*$. We define the following special integral weight:

$$\rho := \frac{1}{2} \sum_{\alpha \in \Phi^+} \alpha \in X. \tag{14.17}$$

Definition 14.32 Let $\lambda \in \mathfrak{h}^*$ and $w \in W$. We define the *dot action* of W on $\mathfrak{h}^*$ by

$$w \cdot \lambda = w(\lambda + \rho) - \rho. \tag{14.18}$$

For $\lambda, \mu \in \mathfrak{h}^*$, we say that λ and μ are *linked* if they are in the same dot-orbit of W. Linkage is an equivalence relation on $\mathfrak{h}^*$, and we call the dot-orbit $\{w \cdot \lambda \mid w \in W\}$ of λ the *linkage class* of λ.

Normally, linear operators preserve the origin (it lies in its own orbit), and reflections fix hyperplanes which contain the origin. The dot action effectively shifts the origin to lie at $-\rho$ instead, so that reflections fix a hyperplane through $-\rho$, and $-\rho$ is fixed by the dot action ($-\rho$ lies in a linkage class by itself). Linkage classes give us important information about central characters.

Exercise 14.33 Let $\mathfrak{g} = \mathfrak{sl}_2(\mathbb{C})$. For any $\lambda \in \mathfrak{h}^*$, show that λ is linked to $-\lambda - 2$. When λ is dominant integral, use the fact that there exists an embedding of $\Delta(-\lambda - 2)$ into $\Delta(\lambda)$ to deduce that $\chi_{-\lambda-2} = \chi_\lambda$.

Theorem 14.34 *If λ and μ are linked, then $\chi_\lambda = \chi_\mu$.*

Sketch of Proof Suppose first that λ and μ lie in the weight lattice $X \subset \mathfrak{h}^*$, that $\mu \leq \lambda$, and that $t \cdot \lambda = \mu$ for some reflection $t \in W$. Then there exists an embedding of $\Delta(\mu)$ into $\Delta(\lambda)$, which implies that $\chi_\lambda = \chi_\mu$. From this, one can deduce that $\chi_\mu = \chi_\lambda$ if $\lambda, \mu \in X$ are linked. Since the weight lattice X is Zariski dense in $\mathfrak{h}^*$, the claim for general $\lambda, \mu \in \mathfrak{h}^*$ follows from a density argument. $\square$

Now we can say more about the image and kernel of γ. To account for the ρ-shift in the dot action, we compose γ with the algebra endomorphism of $U(\mathfrak{h}) \simeq S(\mathfrak{h}) \simeq \mathbb{C}[\mathfrak{h}^*]$ sending $p(\lambda) \mapsto p(\lambda - \rho)$. Call this composite $\psi : Z(\mathfrak{g}) \to \mathbb{C}[\mathfrak{h}^*]$ the *twisted Harish-Chandra homomorphism*. Then, for a central character χ_λ and $z \in Z(\mathfrak{g})$, we have

$$\chi_\lambda(z) = (\lambda + \rho)(\psi(z)) . \tag{14.19}$$

In the proof of Theorem 14.34, we showed that polynomial functions on $\mathfrak{h}^*$ are constant on linkage classes. This implies that the image of ψ is the set of polynomials in $\mathbb{C}[\mathfrak{h}^*]$ which are invariant under the action of W. In addition to this, it can be shown that ψ is injective, which brings us to the following fundamental result of Harish-Chandra.

Theorem 14.35 (Harish-Chandra)

1. ψ *gives an isomorphism $Z(\mathfrak{g}) \simeq \mathbb{C}[\mathfrak{h}^*]^W$.*
2. *For $\lambda, \mu \in \mathfrak{h}^*$, $\chi_\lambda = \chi_\mu$ if and only if λ and μ are linked.*
3. *Every central character $\chi : Z(\mathfrak{g}) \to \mathbb{C}$ is of the form χ_λ for some $\lambda \in \mathfrak{h}^*$.*

This theorem gives us the critical information that we need to understand the block decomposition of $\mathcal{O}$. For $V \in \mathcal{O}$, we set

$$V^{\chi_\lambda} = \{v \in V \mid (z - \chi_\lambda(z))^{n_z} v = 0 \text{ for some } n_z > 0\} . \tag{14.20}$$

We can then define $\mathcal{O}_\lambda = \mathcal{O}_{\chi_\lambda}$ to be the full subcategory of $\mathcal{O}$ consisting of all modules $V \in \mathcal{O}$ such that $V = V^{\chi_\lambda}$. (See Exercise 14.40 for a description of such a module on which $Z(\mathfrak{g})$ acts by generalized eigenvalues.) Since Verma modules are highest weight modules, our earlier arguments imply that $\Delta(\lambda) \in \mathcal{O}_\lambda$. Therefore, we can describe the block decomposition of $\mathcal{O}$ in the following way.

Theorem 14.36 *Category $\mathcal{O}$ decomposes as*

$$\mathcal{O} = \bigoplus_{\lambda \in \mathfrak{h}^*/(W, \cdot)} \mathcal{O}_\lambda . \tag{14.21}$$

If λ is integral, then $\mathcal{O}_\lambda$ is a block of $\mathcal{O}$.

Remark 14.37 *If λ is not integral, then $\mathcal{O}_\lambda$ may decompose further.*

14.4 Blocks of Category $\mathcal{O}$

Definition 14.38 We call $\mathcal{O}_0 = \mathcal{O}_{\chi_0}$ the *principal block* of category $\mathcal{O}$. These are the $U(\mathfrak{g})$-modules with the same annihilator as the trivial module.

Since all irreducible modules in $\mathcal{O}_0$ have central character χ_0, Theorem 14.35 implies that $\{L(w \cdot 0) \mid w \in W\}$ is the set of irreducible objects in $\mathcal{O}_0$ up to isomorphism. In other words, we have a bijection

$$W \xleftrightarrow{1:1} \mathrm{Irr}\,\mathcal{O}_0, \tag{14.22}$$
$$w \longleftrightarrow L(w \cdot 0).$$

The dot action on $0 \in \mathfrak{h}^*$ is

$$w \cdot 0 = w(\rho) - \rho, \tag{14.23}$$

so an element $w \in W$ fixes $0 \in \mathfrak{h}^*$ under the dot action if and only if w fixes ρ under the regular action of W on $\mathfrak{h}^*$. Since ρ is in the dominant Weyl chamber and W acts transitively on Weyl chambers, no $w \in W$ fixes ρ. From this we see that $\mathrm{Irr}\,\mathcal{O}_0$ is parameterized by W.

Exercise 14.39 Consider $\mathfrak{g} = \mathfrak{sl}_2(\mathbb{C})$ with our usual notations. The element

$$c = 2ef + h^2 + 2fe \in U(\mathfrak{g}) \tag{14.24}$$

is called the *Casimir element*.

1. Show that $c = h^2 + 2h + 4fe$ and that $c \in Z(\mathfrak{g})$.
2. Compute cv^+ inside $\Delta(\lambda)$.
3. When $\mu = -\lambda - 2$, deduce from this computation that the central characters of $\Delta(\lambda)$ and $\Delta(\mu)$ agree on cv^+.
4. (Challenge) Using only the knowledge that $\chi_\lambda(z) = \chi_\mu(z)$ when $\mu = -\lambda - 2$ and $\lambda \in \mathbb{N}$, can you deduce that $\chi_\lambda(z) = \chi_\mu(z)$ when $\mu = -\lambda - 2$ for all $\lambda \in \mathbb{C}$?

Exercise 14.40 The goal of this exercise is to study an object of category $\mathcal{O}$ on which the center only acts by generalized eigenvalues. Consider $\mathfrak{g} = \mathfrak{sl}_2(\mathbb{C})$. Let $V = \Delta(-1) \otimes \mathbb{C}^2$. More precisely, let V denote the vector space with basis $\{(f^k v^+) \otimes \uparrow\} \cup \{(f^k v^+) \otimes \downarrow\}$. On the first tensor factor, $\mathfrak{sl}_2(\mathbb{C})$ acts as on $\Delta(-1)$. On the second tensor factor, $\uparrow$ has weight $+1$, $\downarrow$ has weight -1, and $e(\downarrow) = \uparrow$, $e(\uparrow) = 0$, $f(\uparrow) = \downarrow$, and $f(\downarrow) = 0$. Finally, on a pure tensor we have $e(a \otimes b) = ea \otimes b + a \otimes eb$, and similarly for h and f.

1. Prove that if a and b are weight vectors, then so is $a \otimes b$. What is its weight?
2. Compute how e, f, and h act on the basis of V.
3. Find a vector x of weight 0 with $ex = 0$, giving a map $\Delta(0) \to V$. Show that this map is injective.

4. Find a vector $\bar{y}$ in the quotient $V/\Delta(0)$, of weight -2, with $e\bar{y} = 0$. Show that this gives an isomorphism from $\Delta(-2)$ to the quotient. Thus V has a filtration by Verma modules.
5. Does $\bar{y}$ lift to an element $y \in V$ with $ey = 0$?
6. Consider the action of c on the -1 weight space of V. Prove that c acts by a nontrivial Jordan block. (Why did you know it had only one eigenvalue in the first place?)

14.5 Example: $\mathfrak{g} = \mathfrak{sl}_3(\mathbb{C})$

In this section, we illustrate the structure theory of the previous sections for the Lie algebra $\mathfrak{g} = \mathfrak{sl}_3(\mathbb{C})$.

Definition 14.41 We denote by $\mathfrak{sl}_3(\mathbb{C})$ the complex vector space of traceless 3×3 matrices. It is a Lie subalgebra of $\mathfrak{gl}_3(\mathbb{C})$ when given the Lie bracket

$$[x, y] = xy - yx.$$

The subspace of diagonal matrices in $\mathfrak{sl}_3(\mathbb{C})$ forms a Cartan subalgebra

$$\mathfrak{h} = \left\{ \begin{pmatrix} x_1 & 0 & 0 \\ 0 & x_2 & 0 \\ 0 & 0 & x_3 \end{pmatrix} \;\middle|\; x_1 + x_2 + x_3 = 0 \right\}$$

with basis

$$\left\{ h_1 := \begin{pmatrix} 1 & 0 & 0 \\ 0 & -1 & 0 \\ 0 & 0 & 0 \end{pmatrix}, \quad h_2 := \begin{pmatrix} 0 & 0 & 0 \\ 0 & 1 & 0 \\ 0 & 0 & -1 \end{pmatrix} \right\}.$$

We want to describe the root system of $\mathfrak{sl}_3(\mathbb{C})$, which is a root system of type A_2 in the terminology of abstract root systems. We fix some notation that will be helpful. We denote by $e_i \in \mathfrak{h}^*$, for $i \in \{1, 2, 3\}$, the linear functional

$$e_i \left(\begin{pmatrix} x_1 & 0 & 0 \\ 0 & x_2 & 0 \\ 0 & 0 & x_3 \end{pmatrix} \right) = x_i.$$

Then the root system is

$$\Phi = \{e_1 - e_2, e_2 - e_3, e_1 - e_3, e_2 - e_1, e_3 - e_2, e_3 - e_1\}.$$

14.5 Example: $\mathfrak{g} = \mathfrak{sl}_3(\mathbb{C})$

We can choose the following base Δ of Φ, with the corresponding set Φ^+ of positive roots:

$$\Delta = \{e_1 - e_2, e_2 - e_3\}, \qquad \Phi^+ = \{e_1 - e_2, e_2 - e_3, e_1 - e_3\},$$

For convenience, we write $\alpha_i = e_i - e_{i+1}$ for $i \in \{1, 2\}$.

Under the identification $\mathfrak{h} \simeq \mathfrak{h}^* : h \mapsto (h|-)$ induced by the Killing form, the dual basis to $\{h_1, h_2\}$ is $\{f_1, f_2\}$, given for $x = \begin{pmatrix} x_1 & 0 & 0 \\ 0 & x_2 & 0 \\ 0 & 0 & x_3 \end{pmatrix} \in \mathfrak{h}$ by

$$f_1(x) = (h_1|x) = c\,\mathrm{tr}(h_1 x) = c(x_1 - x_2), \qquad f_2(x) = (h_2|x) = c\,\mathrm{tr}(h_2 x) = c(x_2 - x_3),$$

for some scalar $c \in \mathbb{C}$. (Here we have used the fact that the Killing form for $\mathfrak{g} = \mathfrak{sl}_n(\mathbb{C})$ is a scalar multiple of the trace form, which is the bilinear form $(-, -) : \mathfrak{g} \times \mathfrak{g} \to \mathbb{C}$ defined by $(X, Y) = \mathrm{tr}(XY)$, see e.g. [72, §15.1].) Therefore, up to a scalar, the dual basis of $\mathfrak{h}^*$ given by the Killing form consists of the simple roots $\{\alpha_1, \alpha_2\}$.

Let us draw this root system in the Euclidean plane $E := \mathrm{Span}_{\mathbb{R}}(\Phi) \subset \mathfrak{h}^*$. For this, we use the formula $(\alpha_1|\alpha_2) = \cos\theta \|\alpha_1\| \|\alpha_2\|$ to determine the angle between α_1 and α_2. Using the identification $\mathfrak{h} \simeq \mathfrak{h}^*$ described above, we calculate that

$$\frac{(\alpha_1|\alpha_2)}{\|\alpha_1\|\|\alpha_2\|} = \frac{(h_1|h_2)}{\|h_1\|\|h_2\|} = -\frac{1}{2},$$

so $\theta = 120°$. We therefore obtain the following picture of Φ:

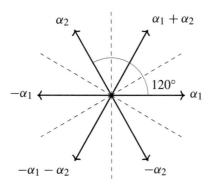

The dotted lines are the reflecting hyperplanes, and the solid lines are the root vectors. Let $s_i = s_{\alpha_i}$ for $i \in \{1, 2\}$. Then we can also see from this picture that the Weyl group W is the type A_2 Coxeter system with $S = \{s_1, s_2\}$. The longest element $s_1 s_2 s_1 = s_2 s_1 s_2$ is the reflection orthogonal to $\alpha_1 + \alpha_2$.

We have the following basis of $\mathfrak{sl}_3(\mathbb{C})$ corresponding to our choice of positive roots:

- A basis for the Cartan subalgebra $\mathfrak{h}$ is $\{h_1, h_2\}$ as above.
- A basis for $\mathfrak{n}^+$, the positive root spaces, can be given by

$$\left\{ \begin{pmatrix} 0 & 1 & 0 \\ 0 & 0 & 0 \\ 0 & 0 & 0 \end{pmatrix}, \begin{pmatrix} 0 & 0 & 1 \\ 0 & 0 & 0 \\ 0 & 0 & 0 \end{pmatrix}, \begin{pmatrix} 0 & 0 & 0 \\ 0 & 0 & 1 \\ 0 & 0 & 0 \end{pmatrix} \right\}.$$

- A basis for $\mathfrak{n}^-$, the negative roots spaces, can be given by

$$\left\{ \begin{pmatrix} 0 & 0 & 0 \\ 0 & 0 & 0 \\ 0 & 1 & 0 \end{pmatrix}, \begin{pmatrix} 0 & 0 & 0 \\ 0 & 0 & 0 \\ 1 & 0 & 0 \end{pmatrix}, \begin{pmatrix} 0 & 0 & 0 \\ 1 & 0 & 0 \\ 0 & 0 & 0 \end{pmatrix} \right\}.$$

The root lattice Q and its positive part Q^+ are

$$Q = \operatorname{Span}_{\mathbb{Z}}\{\alpha_1, \alpha_2\}, \qquad Q^+ = \operatorname{Span}_{\mathbb{N}}\{\alpha_1, \alpha_2\}.$$

To determine the weight lattice, we must find all $\lambda \in \mathfrak{h}^*$ satisfying $\lambda(\alpha_1^\vee), \lambda(\alpha_2^\vee) \in \mathbb{Z}$. By construction, all elements of the root lattice satisfy this condition. (The root lattice is always a sublattice of the weight lattice.) To see if any additional elements of $\mathfrak{h}^*$ lie in the weight lattice, we calculate the coroots $\alpha_i^\vee$ for $i \in \{1, 2\}$, which by definition is the unique element of $\mathfrak{h}$ such that

$$\alpha_i = 2 \frac{(\alpha_i^\vee | -)}{(\alpha_i^\vee | \alpha_i^\vee)}.$$

Since $\alpha_i(h_i) = 2$, we conclude that $\alpha_i^\vee = h_i$. Now, the elements $e_i \in \mathfrak{h}^*$ for $i \in \{1, 2, 3\}$ also lie in the weight lattice, since

$$e_1(\alpha_1^\vee) = e_1(h_1) = 1, \quad e_2(\alpha_1^\vee) = e_2(h_1) = -1, \quad e_3(\alpha_1^\vee) = e_3(h_1) = 0,$$
$$e_1(\alpha_2^\vee) = e_1(h_2) = 0, \quad e_2(\alpha_2^\vee) = e_2(h_2) = 1, \quad e_3(\alpha_2^\vee) = e_3(h_2) = -1.$$

Therefore, the weight lattice X and the set of dominant integral weights X^+ are

$$X = \operatorname{Span}_{\mathbb{Z}}\{e_1, e_2, e_3\}, \qquad X^+ = \operatorname{Span}_{\mathbb{N}}\{e_1, e_2, e_3\}.$$

Figure 14.1 shows the root lattice sitting inside the weight lattice, again drawn in the Euclidean plane E.

14.5 Example: $\mathfrak{g} = \mathfrak{sl}_3(\mathbb{C})$

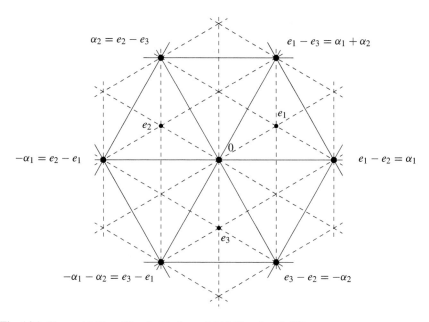

Fig. 14.1 The root lattice sitting inside the weight lattice, for $\mathfrak{sl}_3(\mathbb{C})$

The half-sum of positive roots is

$$\rho = \frac{1}{2}\left(\alpha_1 + (\alpha_1 + \alpha_2) + \alpha_2\right) = \alpha_1 + \alpha_2.$$

The standard action of the Weyl group is given by reflections: for $\alpha \in \Phi$ (hence $s_\alpha \in W$) and $\lambda \in \mathfrak{h}^*$,

$$s_\alpha(\lambda) = \lambda - \frac{2(\lambda|\alpha)}{(\alpha|\alpha)}\alpha,$$

where $(-|-)$ is the Killing form on $\mathfrak{h}^*$. (Geometrically, this is of course just reflection along the hyperplane orthogonal to α.) The dot action is therefore given by

$$s_\alpha \cdot \lambda = s_\alpha(\lambda + \rho) - \rho = (\lambda + \rho) - \frac{2(\lambda + \rho|\alpha)}{(\alpha|\alpha)}\alpha - \rho$$

$$= \lambda - \frac{2(\lambda + \rho|\alpha)}{(\alpha|\alpha)}\alpha = s_\alpha(\lambda) - \frac{2(\rho|\alpha)}{(\alpha|\alpha)}.$$

Highest weights of irreducible modules in the principal block $\mathcal{O}_0$ are those in the dot-orbit of the weight 0. To determine this orbit, we start with 0 and reflect

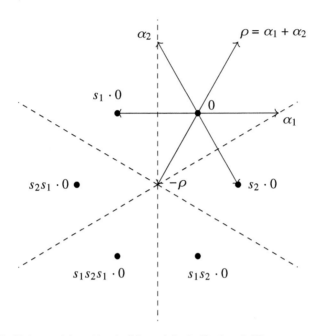

Fig. 14.2 The highest weights of irreducible modules in $\mathcal{O}_0$, for $\mathfrak{sl}_3(\mathbb{C})$

repeatedly along the shifted root hyperplanes (going through $-\rho$):

$$\mathrm{id}\cdot 0 = 0, \qquad s_1\cdot 0 = -\alpha_1, \qquad s_2\cdot 0 = -\alpha_2,$$
$$s_1 s_2\cdot 0 = -2\alpha_1 - \alpha_2, \qquad s_2 s_1\cdot 0 = -\alpha_1 - 2\alpha_2,$$
$$s_1 s_2 s_1\cdot 0 = s_2 s_1 s_2\cdot 0 = -2\rho = -2\alpha_1 - 2\alpha_2.$$

This is pictured in Fig. 14.2, where the shifted root hyperplanes are denoted by dashed lines. Hence $\mathcal{O}_0$ contains the following six irreducible highest weight modules up to isomorphism:

$$L(0),\ L(-\alpha_1),\ L(-\alpha_2),\ L(-2\alpha_1-\alpha_2),\ L(-\alpha_1-2\alpha_2),\ L(-2\alpha_1-2\alpha_2).$$

Chapter 15
Soergel's $\mathbb{V}$ Functor and the Kazhdan–Lusztig Conjecture

This chapter is based on expanded notes of a lecture given by the authors and taken by
Dmytro Matvieievskyi and **Boris Tsvelikhovsky**

Abstract We recall basic facts about category $\mathcal{O}$ and introduce the translation functors between its blocks. We then explain Soergel's approach to the Kazhdan–Lusztig conjecture via his $\mathbb{V}$ functor, which relates the principal block $\mathcal{O}_0$ to Soergel (bi)modules. In particular, we explain how Soergel's conjecture implies the Kazhdan–Lusztig conjecture.

15.1 Brief Reminder on Category $\mathcal{O}$

We begin with a brief reminder on category $\mathcal{O}$. We follow the notation of Chap. 14, where more details can be found.

Let $\mathfrak{g}$ be a complex semisimple Lie algebra. Fix a triangular decomposition $\mathfrak{g} = \mathfrak{n}^- \oplus \mathfrak{h} \oplus \mathfrak{n}^+$. This determines the weight lattice $X \subset \mathfrak{h}^*$, roots $\Phi \subset \mathfrak{h}^*$, positive roots $\Phi^+ \subset \Phi$, simple roots, and the Weyl group W. Moreover, W comes with the natural structure of a Coxeter system (W, S), and the simple roots α_s are indexed by $s \in S$. The dot action of W on $\mathfrak{h}^*$ is given by

$$w \cdot \lambda = w(\lambda + \rho) - \rho \qquad (w \in W,\ \lambda \in \mathfrak{h}^*),$$

where $\rho = \frac{1}{2} \sum_{\alpha \in \Phi^+} \alpha$.

D. Matvieievskyi · B. Tsvelikhovsky
Department of Mathematics, Northeastern University, Boston, MA, USA

The BGG category $\mathcal{O}$ consists of all finitely-generated $U(\mathfrak{g})$-modules that are $\mathfrak{h}$-semisimple (i.e. weight modules) and locally $\mathfrak{n}^+$-finite. Category $\mathcal{O}$ has a block[1] decomposition

$$\mathcal{O} = \bigoplus_{\lambda \in \mathfrak{h}^*/(W,\cdot)} \mathcal{O}_\lambda. \tag{15.1}$$

Thus the study of $\mathcal{O}$ is reduced to the study of each $\mathcal{O}_\lambda$.

It is helpful to visualize each $\mathcal{O}_\lambda$ as living over the point $\lambda \in \mathfrak{h}^*$. To help draw pictures, we focus on the real span $E = \mathbb{R} \otimes_\mathbb{Z} X \subset \mathfrak{h}^*$, which is equipped with an inner product $(-,-)$ induced from the Killing form. For the next few definitions, the reader should follow along using the picture in Example 15.1. For each root α, define the *ρ-shifted root hyperplane* H_α orthogonal to α:

$$H_\alpha = \{\lambda \in E \mid (\lambda + \rho, \alpha) = 0\} = \{\lambda \in E \mid (\lambda + \rho)(\alpha^\vee) = 0\} \subset E.$$

Note that $-\rho$ lies on every such hyperplane. Under the dot action, each $s \in S$ acts as orthogonal reflection across H_{α_s}.

The *ρ-shifted dominant Weyl chamber* is the set of weights $\lambda \in E$ satisfying $(\lambda + \rho)(\alpha_s^\vee) > 0$ for all $s \in S$. It is bounded by the hyperplanes H_{α_s} for $s \in S$, and its closure is a fundamental domain for the dot action on E. We will find it convenient to use these weights to label the blocks $\mathcal{O}_\lambda$. The points in the closure of the ρ-shifted dominant Weyl chamber are partitioned into $2^{|S|}$ *facets*, depending on whether or not they lie on each of the $|S|$ hyperplanes H_{α_s}. For instance, 0 lies on none of these hyperplanes, whereas $-\rho$ lies on all of them. The stabilizer of each facet of the ρ-shifted dominant Weyl chamber is a parabolic subgroup of W.

Example 15.1 Let $\mathfrak{g} = \mathfrak{sl}_3(\mathbb{C})$. We saw in Sect. 14.5 that $\Phi^+ = \{\alpha_s, \alpha_t, \alpha_s + \alpha_t\}$ and that E looks as follows:

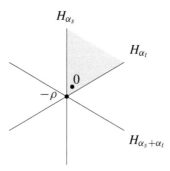

We have shaded the ρ-shifted dominant Weyl chamber.

---

[1] Recall that $\mathcal{O}_\lambda$ is indecomposable if λ is integral, but otherwise may decompose further, thus this is a "block decomposition" only in a loose sense.

15.2 Translation Functors

Thus we have an abelian category $\mathcal{O}_\lambda$ for each weight λ in the closure of the ρ-shifted dominant Weyl chamber. How does $\mathcal{O}_\lambda$ depend on λ? For instance, we saw in Chap. 14 that the simple objects in $\mathcal{O}_\lambda$ up to isomorphism are $\{L(w \cdot \lambda) \mid w \in W\}$, so the number of simple objects in $\mathcal{O}_\lambda$ up to isomorphism equals the size of the orbit $W \cdot \lambda$. This is a first hint that the structure of $\mathcal{O}_\lambda$ is related to which facet λ lies in; the bigger the stabilizer of the facet, the smaller the orbit.

15.2 Translation Functors

The goal of this section is to construct and study functors between different blocks $\mathcal{O}_\lambda$. To simplify the exposition, we will focus below mostly on integral weights λ. For a thorough treatment of the general case, see [86].

We begin by discussing the tensor product of $\mathfrak{g}$-modules, which will be used to construct these functors.

15.2.1 Tensor Products

Given two $\mathfrak{g}$-modules M, N, recall that the tensor product representation $M \otimes N$ is determined by

$$X(m \otimes n) = Xm \otimes n + m \otimes Xn \tag{15.2}$$

for $X \in \mathfrak{g}$, $m \in M$, $n \in N$. Tensor product gives a monoidal structure on $\mathfrak{g}$-modules, but it does not preserve $\mathcal{O}$; see Example 15.3 below. However, if V is a finite-dimensional $\mathfrak{g}$-module, then it is easy to show that $V \otimes (-)$ preserves $\mathcal{O}$:

$$V \otimes (-) : \mathcal{O} \to \mathcal{O}.$$

Let us see how this works in a small example.

Example 15.2 Let $\mathfrak{g} = \mathfrak{sl}_2(\mathbb{C})$, and let λ be a weight (identified with a complex number as in Chap. 14). Let $V = \mathbb{C}^2 = \mathbb{C}e_1 \oplus \mathbb{C}e_2$ be the standard representation (see Example 14.3). Then $V \otimes \Delta(\lambda)$ can be visualized as follows:

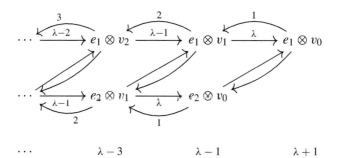

Here, as in Example 14.12, we have chosen the basis of weight vectors $\{v_j\}_{j\in\mathbb{N}}$ for $\Delta(\lambda)$, where $v_j = \frac{f^j(v^+)}{j!}$ and v^+ is a highest weight vector for $\Delta(\lambda)$. Then $\{e_i \otimes v_j\}_{i\in\{1,2\}, j\in\mathbb{N}}$ forms a basis of weight vectors for $V \otimes \Delta(\lambda)$. In the diagram above, these basis vectors are listed according to their weight, given in the bottom row. Arrows that decrease (resp. increase) the weight indicate the action of f (resp. e) on this basis.

It is not hard to see that $\Delta(\lambda + 1)$ is a submodule of $V \otimes \Delta(\lambda)$, generated by the highest weight vector $e_1 \otimes v_0$. Moreover, the quotient by this submodule will be $\Delta(\lambda - 1)$, so that there is a short exact sequence

$$0 \to \Delta(\lambda + 1) \to V \otimes \Delta(\lambda) \to \Delta(\lambda - 1) \to 0. \tag{15.3}$$

The question at hand is whether this sequence splits.

A direct computation shows that the vector $e_1 \otimes v_1 - e_2 \otimes \lambda v_0$ spans the space of all highest weight vectors of weight $\lambda - 1$. When $\lambda \neq -1$, this highest weight vector is not proportional to

$$f(e_1 \otimes v_0) = e_1 \otimes v_1 + e_2 \otimes v_0 \in \Delta(\lambda + 1) \subset V \otimes \Delta(\lambda),$$

so that it induces a map $\Delta(\lambda - 1) \to V \otimes \Delta(\lambda)$ which splits the sequence above. Thus if $\lambda \neq -1$, then

$$V \otimes \Delta(\lambda) \simeq \Delta(\lambda + 1) \oplus \Delta(\lambda - 1). \tag{15.4}$$

However, in the case $\lambda = -1$, we have

$$f(e_1 \otimes v_0) = e_1 \otimes v_1 + e_2 \otimes v_0 = e_1 \otimes v_1 - e_2 \otimes \lambda v_0$$

and any map $\Delta(\lambda - 1) \to V \otimes \Delta(\lambda)$ will have image contained inside $\Delta(\lambda + 1)$. Thus there is no way to split the sequence (15.3), so that we have a nonsplit short exact sequence

$$0 \to \Delta(0) \to V \otimes \Delta(0) \to \Delta(-2) \to 0. \tag{15.5}$$

Example 15.3 If $M \in \mathcal{O}$ is infinite dimensional, then the functor $M \otimes (-)$ does not preserve $\mathcal{O}$. For example, consider $\mathfrak{g} = \mathfrak{sl}_2(\mathbb{C})$ and the module $\Delta(0) \otimes \Delta(0)$. The dimension of its weight space with weight $-2n$ equals the number of decompositions $2n = a + b$, where $a, b \in \mathbb{N}$. There is no common upper bound for such decompositions for all n, so $\Delta(0) \otimes \Delta(0)$ cannot be finitely generated over $U(\mathfrak{n}^-)$.

The following proposition generalizes the phenomenon observed in Example 15.2.

Proposition 15.4 *Let V be a finite-dimensional irreducible representation of $\mathfrak{g}$, and let $\lambda \in \mathfrak{h}^*$. Then $V \otimes \Delta(\lambda)$ admits a standard filtration, i.e. a finite filtration*

$$V \otimes \Delta(\lambda) = M_0 \supset M_1 \supset \cdots \supset M_{\ell+1} = 0$$

by $\mathfrak{g}$-modules such that each successive subquotient M_i/M_{i+1} is isomorphic to some Verma module $\Delta(\lambda + \nu)$. Moreover, ν runs over the weights of V with their multiplicities.

Proof We first check that $V \otimes \Delta(\lambda) \simeq U(\mathfrak{g}) \otimes_{U(\mathfrak{b})} (V \otimes \mathbb{C}_\lambda)$. Indeed, this follows from the Yoneda lemma and the following chain of isomorphisms, functorial in $M \in \mathcal{O}$:

$$\mathrm{Hom}_\mathfrak{g}(V \otimes \Delta(\lambda), M) \simeq \mathrm{Hom}_\mathfrak{g}(\Delta(\lambda), V^* \otimes M) \simeq \mathrm{Hom}_\mathfrak{b}(\mathbb{C}_\lambda, V^* \otimes M)$$
$$\simeq \mathrm{Hom}_\mathfrak{b}(V \otimes \mathbb{C}_\lambda, M) = \mathrm{Hom}_\mathfrak{g}(U(\mathfrak{g}) \otimes_{U(\mathfrak{b})} (V \otimes \mathbb{C}_\lambda), M).$$

Here, the second isomorphism is Frobenius reciprocity.

As a $\mathfrak{b}$-module, $V \otimes \mathbb{C}_\lambda$ admits a filtration

$$V \otimes \mathbb{C}_\lambda = K_0 \supset K_1 \supset \cdots \supset K_\ell \supset K_{\ell+1} = 0$$

with $K_i = \mathbb{C}\langle v_i, \ldots, v_\ell \rangle$, where the indices of v_i's are chosen in such a way that v_ℓ is the highest weight vector and $v_i \geq v_j$ (the weights of v_i and v_j) whenever $i \geq j$. As $U(\mathfrak{g})$ is a free right $U(\mathfrak{b})$-module, the functor $U(\mathfrak{g}) \otimes_{U(\mathfrak{b})} (-)$ is exact. We define $M_i := U(\mathfrak{g}) \otimes_{U(\mathfrak{b})} K_i$. It follows that $M_i/M_{i+1} \simeq U(\mathfrak{g}) \otimes_{U(\mathfrak{b})} K_i/K_{i+1} \simeq \Delta(\lambda + v_i)$. □

One can now ask to what extent the filtration in Proposition 15.4 splits. There are no extensions between $\Delta(\lambda)$ and $\Delta(\mu)$ when λ and μ are not linked (i.e. in the same dot action orbit), forcing many parts of the filtration to split. This explains the splitting (15.4) from Example 15.2 when $\lambda \neq -1$. Conversely, when λ and μ are linked, they (heuristically) tend to stick together as much as possible, as in (15.5) from Example 15.2 when $\lambda = -1$.

To make these ideas precise, we introduce translation functors.

15.2.2 Definition of Translation Functors and First Properties

For any $\lambda \in \mathfrak{h}^*$, let

$$i_\lambda : \mathcal{O}_\lambda \hookrightarrow \mathcal{O}, \qquad \mathrm{pr}_\lambda : \mathcal{O} \twoheadrightarrow \mathcal{O}_\lambda$$

be the inclusion and projection along the block decomposition (15.1).

Definition 15.5 We say that $\lambda, \mu \in \mathfrak{h}^*$ are *compatible* if $\mu - \lambda$ is integral.

Definition 15.6 Let $\lambda, \mu \in \mathfrak{h}^*$ be compatible. Let ν be the unique dominant integral weight in $W(\mu - \lambda)$ (linear action, not the dot action), and let $V = L(\nu)$, the (finite-dimensional) irreducible representation of highest weight ν.

The *translation functor* T_λ^μ is the composition

$$T_\lambda^\mu := \mathrm{pr}_\mu \circ (V \otimes (-)) \circ i_\lambda : \mathcal{O}_\lambda \to \mathcal{O}_\mu.$$

It is exact, being a composition of exact functors.

We will soon focus on the case of dominant integral weights λ, μ. Let us discuss some first properties of translation functors that hold for any weights.

Lemma 15.7 *Let λ, μ be compatible. The functors*

$$T_\lambda^\mu : \mathcal{O}_\lambda \rightleftarrows \mathcal{O}_\mu : T_\mu^\lambda$$

are biadjoint.

Proof (Sketch of Proof) If $V = L(\nu)$ is as in the definition of T_λ^μ, then $V^* = L(-w_0\nu)$, where w_0 is the longest element of W, and $T_\mu^\lambda := \mathrm{pr}_\lambda \circ (V^* \otimes (-)) \circ i_\mu$. The lemma follows from the biadjunction between $V \otimes (-)$ and $V^* \otimes (-)$. □

Exercise 15.8 Verify the statements made in the proof sketch above.

Lemma 15.9 *Translation functors send projective modules to projective modules.*

Proof This follows from Lemma 15.7 and the following exercise. □

Exercise 15.10 Let $\mathcal{A}, \mathcal{B}$ be abelian categories, and consider functors $F : \mathcal{A} \to \mathcal{B}$ and $G : \mathcal{B} \to \mathcal{A}$. If F is left adjoint to G and G is exact, then show that F sends projective objects to projective objects. (*Hint:* an object P is projective if and only if the functor $\mathrm{Hom}(P, -)$ is exact.)

15.2.3 Effect on Verma Modules

Recall from Chap. 14 that $\Delta(\lambda)$ lies in $\mathcal{O}_\mu$ if and only if $\lambda \in W \cdot \mu$. Let M be a module with a standard filtration (Definition 14.24). If we apply the (exact) projection functor pr_μ to this filtration, we get a filtration of $\mathrm{pr}_\mu(M)$ whose subquotients are either zero or a Verma module $\Delta(\lambda)$ for $\lambda \in W \cdot \mu$. Applying pr_μ to the filtration in Proposition 15.4 we get the following result.

Corollary 15.11 *Let $\lambda, \mu \in \mathfrak{h}^*$ be compatible, and let ν be the unique dominant integral weight in $W(\mu - \lambda)$. The module $T_\lambda^\mu(\Delta(\lambda))$ admits a filtration with successive subquotients isomorphic to $\Delta(\lambda + \nu')$, where ν' runs over the weights of $L(\nu)$ such that $\lambda + \nu' \in W \cdot \mu$, together with their multiplicities.*

15.2 Translation Functors

For special choices of weights λ and μ, we can say more. Recall that there is a partial order $\leq$ on $\mathfrak{h}^*$ defined by $\mu \leq \lambda$ if $\lambda - \mu$ lies in Q^+, i.e. if $\lambda - \mu$ equals an $\mathbb{N}$-linear combination of simple roots. We will work with weights μ that are *maximal in its dot orbit* with respect to this partial order: for all $w \in W$, $\mu \leq w \cdot \mu$ implies $\mu = w \cdot \mu$.

Remark 15.12 A dominant integral weight is maximal in its dot orbit, and the reader can consider this a primary application. However, a completely generic weight (with no integrality properties) will be incomparable with anything else in its dot orbit, so it will also be maximal in its dot orbit, regardless of which chamber it lies in.

The following proposition plays a crucial role in what follows.

Proposition 15.13 *Let $\lambda, \mu \in \mathfrak{h}^*$ be compatible. Assume moreover that μ is maximal in its dot orbit, and that $\lambda - \mu$ is dominant integral. Then for any $w \in W$,*

1. $T_\lambda^\mu(\Delta(w \cdot \lambda)) \simeq \Delta(w \cdot \mu)$;
2. $T_\mu^\lambda(\Delta(w \cdot \mu))$ *admits a filtration with successive subquotients isomorphic to $\Delta(ww' \cdot \mu)$, where w' runs over the stabilizer of μ under the dot action, with each such w' appearing exactly once.*

Let us illustrate the special case addressed by Proposition 15.13 in the case that λ, μ both lie in the Euclidean space E. Then λ lies in the ρ-shifted dominant Weyl chamber, while μ may lie in the ρ-shifted dominant Weyl chamber or in one of the facets in its closure. In the second case, T_λ^μ is called *translation onto a wall*, and T_μ^λ is called *translation out of a wall*. The following picture illustrates the two possibilities for $\mathfrak{g} = \mathfrak{sl}_3(\mathbb{C})$:

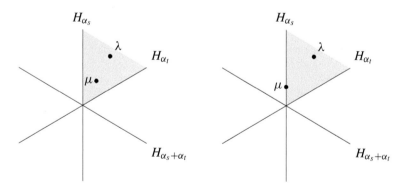

In the second picture, T_λ^μ (resp. T_μ^λ) is translation onto (resp. out of) the s-wall, where μ has stabilizer $\{\mathrm{id}, s\}$. We can also translate onto and out of the most singular point $-\rho$, which has stabilizer W.

Proof For both claims, we use Corollary 15.11. Let $V = L(\lambda - \mu)$, so we have $V^* = L(w_0(\mu - \lambda))$.

1. One needs to find weights ν' of V^* such that $w \cdot \lambda + \nu' = x \cdot \mu$ for some $x \in W$, or equivalently,

$$\lambda + \rho + w^{-1}\nu' = w^{-1}x(\mu + \rho) \quad \text{for some } x \in W. \tag{15.6}$$

Since $\mu - \lambda$ is the lowest weight of V^*, we have $\mu - \lambda \leq w^{-1}\nu'$. Combining this with (15.6), we get

$$\mu + \rho = \lambda + \rho + (\mu - \lambda) \leq \lambda + \rho + w^{-1}\nu' = w^{-1}x(\mu + \rho),$$

so $\mu \leq w^{-1}x \cdot \mu$. Since μ is maximal in its dot orbit, this implies $\mu = w^{-1}x \cdot \mu$, or equivalently $w \cdot \mu = x \cdot \mu$. Moreover, this forces the inequality $\mu - \lambda \leq w^{-1}\nu'$ to be an equality, so $\nu' = w(\mu - \lambda)$. This is an extremal weight of V^*, so it has multiplicity one.

2. One needs to find weights ν' of V such that $w \cdot \mu + \nu' = x \cdot \lambda$ for some $x \in W$, or equivalently,

$$x^{-1}w(\mu + \rho) + x^{-1}\nu' = \lambda + \rho \quad \text{for some } x \in W.$$

Combine this with $\lambda - \mu \geq x^{-1}\nu'$ to get

$$x^{-1}w(\mu + \rho) + (\lambda - \mu) \geq \lambda + \rho,$$

or $x^{-1}w \cdot \mu \geq \mu$. Arguing as before, this implies $x^{-1}w \cdot \mu = \mu$ and $\lambda - \mu = x^{-1}\nu'$, so $\nu' = x(\lambda - \mu)$, an extremal weight of V, where x has the form $x = ww'$ for w' in the dot action stabilizer of μ. □

As a first consequence of Proposition 15.13, we obtain an equivalence between dominant integral blocks.

Theorem 15.14 *If λ, μ are dominant integral weights, then there is an equivalence $\mathcal{O}_\lambda \simeq \mathcal{O}_\mu$ sending $\Delta(w \cdot \lambda)$ to $\Delta(w \cdot \mu)$ for all $w \in W$.*

Proof (Sketch of Proof) First we find functors between these blocks which send Verma modules to Verma modules. When $\mu - \lambda$ is dominant integral, the first statement in Proposition 15.13 implies that the translation functor T_λ^μ sends $\Delta(w \cdot \lambda)$ to $\Delta(w \cdot \mu)$ for all $w \in W$, and vice versa for T_μ^λ. Otherwise, let $\nu = \lambda + \mu$ (or any integral weight ν such that both $\nu - \lambda$ and $\nu - \mu$ are dominant integral), and consider the compositions $T_\nu^\mu \circ T_\lambda^\nu$ and $T_\nu^\lambda \circ T_\mu^\nu$. These compositions have the desired effect on Verma modules, again by Proposition 15.13.

Now, recall that the Grothendieck group $[\mathcal{O}_\lambda]$ has a basis given by $\{[\Delta(w \cdot \lambda)] \mid w \in W\}$. The theorem is an application of Exercise 15.15 below. □

15.2 Translation Functors

Exercise 15.15 Let $\mathcal{A}, \mathcal{B}$ be finite-length abelian categories, and consider exact functors $F : \mathcal{A} \to \mathcal{B}$ and $G : \mathcal{B} \to \mathcal{A}$ such that F is left adjoint to G. Then F and G are mutually inverse quasi-equivalences if and only if they induce mutually inverse isomorphisms on Grothendieck groups. (*Hint:* show that the unit of the adjunction is a natural isomorphism. If you need help, see [73, Lemma 4.27].)

Remark 15.16 In the situation of Theorem 15.14, T_λ^μ and T_μ^λ are mutually inverse quasi-equivalences (sending Verma modules to Verma modules) even when $\mu - \lambda$ is not dominant integral. Thus there was no need to choose an auxiliary weight ν in the proof above. The key lemma for this more general statement is [86, §7.5], which relies on Weyl chamber geometry in E. We have chosen to stick to the easier case where $\mu - \lambda$ is dominant integral, and the price we pay is having to introduce some intermediate dominant integral weight ν in the proof of Theorem 15.14 and in Definition 15.17 below.

Recall that a fundamental question about $\mathcal{O}$ is the multiplicity of simple modules in Verma modules. The block decomposition reduces this to the same question for each block $\mathcal{O}_\lambda$. For integral weights λ, Theorem 15.14 and a further study of translation onto and out of the wall further reduces this question to the case of the principal block $\mathcal{O}_0$.

Since 0 has trivial dot action stabilizer, $\mathcal{O}_0$ contains $|W|$ non-isomorphic Verma (resp. simple) modules, which we label as follows using the weight 0:

$$\Delta_w := \Delta(w \cdot 0), \qquad L_w := L(w \cdot 0).$$

From now on, we restrict our attention to $\mathcal{O}_0$.

15.2.4 Wall-Crossing Functors

We will study the principal block $\mathcal{O}_0$ using a collection of endofunctors Θ_s indexed by $s \in S$, called the *wall-crossing functors*.

Definition 15.17 Let $s \in S$. Choose an integral weight μ in the closure of the ρ-shifted dominant Weyl chamber such that ρ lies on H_{α_s}, but not on any other root hyperplane (i.e. the dot action stabilizer of μ is $\{\text{id}, s\}$). Next, choose a dominant integral weight ν such that $\nu - \mu$ is also dominant integral. Then define

$$\Theta_s := T_\nu^0 \circ T_\mu^\nu \circ T_\nu^\mu \circ T_0^\nu.$$

Here is a picture of the situation for $\mathfrak{g} = \mathfrak{sl}_3(\mathbb{C})$:

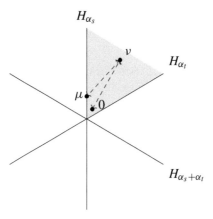

By the proof of Theorem 15.14, T_0^ν and T_ν^0 are mutually quasi-inverse equivalences. Note that Θ_s is exact and self-biadjoint by Lemma 15.7. It can be shown that, up to natural isomorphism, Θ_s does not depend on the choice of μ and ν; see Remark 15.18 and [86, Example 10.8].

Remark 15.18 The intermediate weight ν is not necessary; defining $\Theta_s := T_\mu^0 \circ T_0^\mu$ yields a naturally isomorphic functor. Again see Remark 15.16. Thus Θ_s can be defined as translation onto the s-wall followed by translation out of the s-wall, which explains the term "wall-crossing" (though perhaps "wall-bouncing" is more accurate).

The following is another consequence of Proposition 15.13.

Lemma 15.19 *For any $w \in W$ and $s \in S$, we have a nonsplit short exact sequence*

$$0 \to \Delta_w \to \Theta_s \Delta_w \to \Delta_{ws} \to 0 \quad \text{if } ws > w,$$
$$0 \to \Delta_{ws} \to \Theta_s \Delta_w \to \Delta_w \to 0 \quad \text{if } ws < w.$$

Exercise 15.20 Prove Lemma 15.19.

Remark 15.21 Since Lemma 15.19 involves Δ_{ws} rather than Δ_{sw}, one should think of Θ_s as acting on the right of $\mathcal{O}_0$.

Since $[\Theta_s \Delta_w] = [\Delta_w] + [\Delta_{ws}]$ in the Grothendieck group, we immediately obtain the following.

Corollary 15.22 *The group isomorphism*

$$[\mathcal{O}_0] \to \mathbb{Z}[W] : [\Delta_w] \mapsto w$$

intertwines the map induced by Θ_s with right multiplication by $1 + s$.

15.2 Translation Functors

Just like $\mathcal{O}$, the principal block $\mathcal{O}_0$ is not monoidal, so its Grothendieck group is $\mathbb{Z}[W]$ as a $\mathbb{Z}$-module, not as a ring. Corollary 15.22 remedies this somewhat: $\mathcal{O}_0$ equipped with the wall-crossing functors $\{\Theta_s\}_{s \in S}$ categorifies the abelian group $\mathbb{Z}[W]$ equipped with right multiplication by $\{1+s\}_{s \in S}$. We will further upgrade this below.

15.2.5 Effect on Projective Modules

By Lemma 15.9, wall-crossing functors preserve the class of projectives. Let us first discuss the structure of projectives in $\mathcal{O}_0$.

It was already mentioned in Chap. 14 that every simple $L(\lambda)$ in $\mathcal{O}$ has an indecomposable projective cover $P(\lambda)$ (in fact, this can be proved using translation functors). For each $w \in W$, let

$$P_w := P(w \cdot 0).$$

Lemma 15.23 *The set* $\{P_w\}_{w \in W}$ *is a complete set of non-isomorphic indecomposable projective objects in $\mathcal{O}_0$. Moreover, every projective object $Q \in \mathcal{O}_0$ is isomorphic to a finite direct sum*

$$Q \simeq \bigoplus_{w \in W} P_w^{\oplus \mathrm{mult}(P_w, Q)}$$

for uniquely defined multiplicities $\mathrm{mult}(P_w, Q) \geq 0$. *In fact,*

$$\mathrm{mult}(P_w, Q) = \dim \mathrm{Hom}(Q, L_w).$$

This lemma is a fairly general statement about finite length abelian categories with enough projectives.

Proof Let $Q \in \mathcal{O}_0$ be projective. Choose a filtration $0 = M_0 \subset M_1 \subset \cdots \subset M_n = Q$ with simple successive subquotients. Then we have a natural projection $Q \to M_n/M_{n-1} \simeq L_x$ for some $x \in W$. As both Q and P_x are projective we have maps $f : P_x \to Q$ and $g : Q \to P_x$. Then $gf(P_x) \subset P_x$ is a submodule of P_x. As P_x is indecomposable and g and f commute with the surjective maps to L_x, we have $gf(P_x) = P_x$, so $P_x \stackrel{\oplus}{\subset} Q$ (i.e. P_x is a direct summand of Q). By induction on the length of the filtration, we obtain a decomposition

$$Q \simeq \bigoplus_{w \in W} P_w^{\oplus m_w}, \qquad m_w \geq 0.$$

Let $y \in W$. We have $\operatorname{Hom}(P_w, L_y) = 0$ for all $w \neq y$. Since L_y is the unique simple quotient of P_y and $\operatorname{End}(L_y) = \mathbb{C}$, we have $\dim \operatorname{Hom}(P_y, L_y) = 1$. So

$$\dim \operatorname{Hom}(Q, L_y) = \sum_w m_w \cdot \operatorname{Hom}(P_w, L_y) = m_y,$$

which proves the claim about the multiplicities. □

For an expression $\underline{x} = (s_1, s_2, \ldots, s_m)$, define

$$P_{\underline{x}} := \Theta_{s_m} \cdots \Theta_{s_2} \Theta_{s_1}(P_{\mathrm{id}}). \tag{15.7}$$

Note the (reverse) order in which we apply the translation functors, which agrees with Remark 15.21. For the empty expression, $P_\emptyset = P_{\mathrm{id}} \simeq \Delta_{\mathrm{id}}$.

Proposition 15.24 *For any $x \in W$, the indecomposable projective P_x admits a standard filtration such that Δ_y appears as a subquotient in this filtration only if $y \leq x$, and Δ_x appears exactly once. Moreover, for any reduced expression $\underline{x}$ for x, we have*

$$P_{\underline{x}} \simeq P_x \oplus \bigoplus_{y < x} P_y^{\oplus m_y} \tag{15.8}$$

for some multiplicities m_y, where $<$ is the Bruhat order.

Proof Let $\underline{x}$ be a reduced expression for x. By repeatedly using Lemma 15.19, we see that $P_{\underline{x}}$ admits a standard filtration with multiplicities as in the first statement. The same lemma also shows that $P_{\underline{x}}$ surjects to Δ_x and hence to L_x, so $P_x \stackrel{\oplus}{\subset} P_{\underline{x}}$ by the last statement of Lemma 15.23. As a direct summand of a module with a standard filtration, P_x again has a standard filtration (see Proposition 14.26), with the same properties.

The last statement follows by comparing the multiplicities in the standard filtration on $P_{\underline{x}}$ with those in the standard filtration of the various P_y. □

Corollary 15.25 *Let $x \in W$, and choose a reduced expression $\underline{x}$ for x. Then P_x is the unique indecomposable direct summand of $P_{\underline{x}}$ that does not appear as a direct summand of $P_{\underline{w}}$ for any expression $\underline{w}$ with $\ell(\underline{w}) < \ell(\underline{x})$.*

Observe the similarity between the description of indecomposable projectives in Corollary 15.25 and the Bott–Samelson description of indecomposable Soergel bimodules from Theorem 5.24. In this analogy, the wall-crossing functor Θ_s correspond to the endofunctor $(-) \otimes B_s$. Then (15.7) defines the "Bott–Samelson projectives" $P_{\underline{x}}$, and Proposition 15.24 and Corollary 15.25 characterize the indecomposable projective P_x as its "biggest" direct summand.

Soergel showed that this is more than an analogy: he defined the $\mathbb{V}$ *functor* (also called Soergel's "functor to combinatorics") relating the projectives in $\mathcal{O}_0$ to *Soergel modules*. We will describe the $\mathbb{V}$ functor and Soergel's approach to the Kazhdan–

15.3 Soergel Modules

Lusztig conjecture in Sect. 15.5 below. But first, we take a brief break from category $\mathcal{O}$ to discuss Soergel modules in the general setting.

15.3 Soergel Modules

In this section, we fix a Coxeter system (W, S) (not necessarily a Weyl group) and a W-representation $\mathfrak{h}$ over some field $\Bbbk$. Consider the associated Bott–Samelson and Soergel bimodules. In particular, $R = \mathrm{Sym}(\mathfrak{h}^*)$ with $\deg \mathfrak{h}^* = 2$, and $\mathbb{BS}\mathrm{Bim}(\mathfrak{h}, W)$ and $\mathbb{S}\mathrm{Bim}(\mathfrak{h}, W)$ are certain categories of graded R-bimodules.

We now define (right) Bott–Samelson and Soergel modules, which are certain graded right R-modules. In what follows, $\Bbbk$ is viewed as a graded R-module concentrated in degree 0 (in the only way possible, i.e. all positive degree polynomials act by 0).

Definition 15.26 Let $\underline{w} = (s_1, \ldots, s_m)$ be an expression. The *Bott–Samelson module* $\overline{\mathrm{BS}}(\underline{w})$ is the graded right R-module

$$\overline{\mathrm{BS}}(\underline{w}) := \Bbbk \otimes_R \mathrm{BS}(\underline{w})$$
$$= \Bbbk \otimes_R B_{s_1} \otimes_R \cdots \otimes_R B_{s_m}$$
$$\simeq \Bbbk \otimes_R R \otimes_{R^{s_1}} R \otimes_{R^{s_2}} \cdots \otimes_{R^{s_m}} R(m).$$

A *Soergel module* is any graded right R-module that is isomorphic to a finite direct sum of shifts of direct summands of Bott–Samelson modules. Let $\overline{\mathbb{S}\mathrm{Bim}}(\mathfrak{h}, W)$ denote the full subcategory of gmod-R consisting of Soergel modules.

Consider the functor $\Bbbk \otimes_R (-) : R\text{-gmod-}R \to \text{gmod-}R$, i.e. quotient on the left by the action of positive degree polynomials. This functor is additive and sends $\mathrm{BS}(\underline{w})$ to $\overline{\mathrm{BS}}(\underline{w})$, hence restricts to a functor

$$\Bbbk \otimes_R (-) : \mathbb{S}\mathrm{Bim}(\mathfrak{h}, W) \to \overline{\mathbb{S}\mathrm{Bim}}(\mathfrak{h}, W). \tag{15.9}$$

Soergel modules may equivalently be defined as graded R-modules that arise as direct summands of $\Bbbk \otimes_R B$, where B is a Soergel bimodule. In other words, $\overline{\mathbb{S}\mathrm{Bim}}(\mathfrak{h}, W)$ is naturally identified with the Karoubian envelope of the image of $\mathbb{S}\mathrm{Bim}(\mathfrak{h}, W)$ under (15.9).

What is not clear is if the functor (15.9) is essentially surjective; a priori, this functor may send an indecomposable Soergel bimodule to a decomposable Soergel module. For this, we need the following result due to Soergel ([162] for W a finite Weyl group, [150, Proposition 1.13] for W any finite Coxeter group).

Proposition 15.27 *If W is finite and $\mathfrak{h}$ is reflection faithful, then the natural map*

$$\Bbbk \otimes_R \mathrm{Hom}^\bullet_{R\text{-gmod-}R}(B, B') \to \mathrm{Hom}^\bullet_{\text{gmod-}R}(\Bbbk \otimes_R B, \Bbbk \otimes_R B')$$

is an isomorphism for any Soergel bimodules B, B'.

Here, $\operatorname{Hom}^\bullet$ denotes the graded Hom, which was defined for graded R-bimodules in (4.3). The definition for graded right R-modules is entirely analogous:

$$\operatorname{Hom}^\bullet_{\text{gmod-}R}(M, N) := \bigoplus_{d \in \mathbb{Z}} \operatorname{Hom}_{\text{gmod-}R}(M, N(d)).$$

Corollary 15.28 (Classification of Indecomposable Soergel Modules) *If W is finite and $\mathfrak{h}$ is reflection faithful, the Soergel modules $\overline{B}_w := \Bbbk \otimes_R B_w$ are indecomposable. Therefore, $\{\overline{B}_w : w \in W\}$ is a complete collection of indecomposable Soergel modules up to shift and isomorphism.*

Exercise 15.29 Deduce Corollary 15.28 from Proposition 15.27. (*Hint:* any quotient of a local ring is local.)

Remark 15.30 The assumptions that W is finite and $\mathfrak{h}$ is reflection faithful are crucial in the results above.

Recall that Soergel bimodules categorify the Hecke algebra $\mathrm{H} = \mathrm{H}(W)$: its split Grothendieck group $[\mathbb{S}\mathrm{Bim}(\mathfrak{h}, W)]_\oplus$ is naturally a $\mathbb{Z}[v, v^{-1}]$-algebra, and there is a $\mathbb{Z}[v, v^{-1}]$-algebra isomorphism

$$[\mathbb{S}\mathrm{Bim}(\mathfrak{h}, W)]_\oplus \simeq \mathrm{H} \tag{15.10}$$

determined by $[B_s] \mapsto b_s$ for $s \in S$. What do Soergel modules categorify?

The category $\overline{\mathbb{S}\mathrm{Bim}}(\mathfrak{h}, W)$ is additive and graded, like $\mathbb{S}\mathrm{Bim}(\mathfrak{h}, W)$, but it is no longer monoidal. However, we can still tensor over R on the right, so that $\overline{\mathbb{S}\mathrm{Bim}}(\mathfrak{h}, W)$ is naturally a *right module category* for $\mathbb{S}\mathrm{Bim}(\mathfrak{h}, W)$:

$$- \otimes_R - : \overline{\mathbb{S}\mathrm{Bim}}(\mathfrak{h}, W) \times \mathbb{S}\mathrm{Bim}(\mathfrak{h}, W) \to \overline{\mathbb{S}\mathrm{Bim}}(\mathfrak{h}, W). \tag{15.11}$$

This makes its split Grothendieck group $[\overline{\mathbb{S}\mathrm{Bim}}(\mathfrak{h}, W)]_\oplus$ into a right $[\mathbb{S}\mathrm{Bim}(\mathfrak{h}, W)]_\oplus$-module. Now, Corollary 15.28 says that, under the isomorphism (15.10), we may identify $[\overline{\mathbb{S}\mathrm{Bim}}(\mathfrak{h}, W)]_\oplus$ with the right regular representation of H, i.e. H acting on itself by right multiplication. Of course, there is nothing special about right modules here. One similarly defines left Bott–Samelson and Soergel modules, and the category of left Soergel modules categorifies the left regular representation of H.

For use in the next section, we also need to discuss modules over the *coinvariant algebra C* associated to the realization $(\mathfrak{h}, W)$. Recall from Definition 4.13 that this is the graded $\Bbbk$-algebra

$$C := R/I_W, \tag{15.12}$$

where $I_W \subset R$ is the homogeneous ideal generated by R^W_+, the graded subspace of R^W consisting of polynomials with constant term zero. Pull-back under the natural quotient map $R \twoheadrightarrow C$ is a full embedding

$$\text{gmod-}C \hookrightarrow \text{gmod-}R \tag{15.13}$$

identifying gmod-C as the full subcategory of graded right R-modules on which R_+^W acts by 0 on the right.

The W-action on R induces a W-action on C. The quotient map $R \twoheadrightarrow C$ is W-equivariant, hence restricts to a map of the s-invariants $R^s \to C^s$ for every $s \in S$. Thus for any $M \in$ gmod-C, we obtain a natural map

$$M \otimes_{R^s} R(1) \to M \otimes_{C^s} C(1). \tag{15.14}$$

Exercise 15.31 Assume that the characteristic of $\Bbbk$ is not equal to 2, and let $s \in S$. Show that $C^s = C^s \oplus \alpha_s C^s$. Conclude that the map (15.14) is an isomorphism, i.e. the full embedding (15.13) intertwines the endofunctors $- \otimes_{C^s} C(1)$ and $- \otimes_{R^s} R(1)$.

Now, on any (right) Bott–Samelson module, any element of R_+^W acts by 0 on the left and passes through each wall $\otimes_{R^s}$, hence also acts by 0 on the right. The same is therefore true of Soergel modules. Exercise 15.31 therefore allows one to equivalently define Bott–Samelson and Soergel modules as graded right C-modules: for any expression $(s_1, \ldots, s_m)$, we have a natural identification

$$\overline{\mathrm{BS}}(s_1, \ldots, s_m) \simeq \Bbbk \otimes_{C^{s_1}} C \otimes_{C^{s_2}} \cdots \otimes_{C^{s_m}} C(m). \tag{15.15}$$

15.4 Soergel's $\mathbb{V}$ Functor

Let us return to the Lie-theoretic setup of this chapter: $\mathfrak{g}$ is a semisimple complex Lie algebra with a triangular decomposition $\mathfrak{g} = \mathfrak{n}^- \oplus \mathfrak{h} \oplus \mathfrak{n}^+$, and $\mathcal{O}_0$ is the principal block of its BGG category $\mathcal{O}$. In this section, we define Soergel's $\mathbb{V}$ functor relating projective objects in $\mathcal{O}_0$ to Soergel modules.

We begin by clarifying the input data for Soergel modules. Consider the linear action of W on $\mathfrak{h}$ (not the dot action), and consider the associated Bott–Samelson and Soergel (bi)modules. Here, it is important that we take

$$R = \mathrm{Sym}(\mathfrak{h}) = \mathbb{C}[\mathfrak{h}^*]$$

with $\deg \mathfrak{h} = 2$, so that our (bi)modules are over $R = \mathrm{Sym}(\mathfrak{h})$ rather than the usual $\mathrm{Sym}(\mathfrak{h}^*)$. This reflects the fact that the Soergel-theoretic objects we consider are naturally associated to the *Langlands dual* of $\mathfrak{g}$.

Let us be more precise. Recall that the triangular decomposition determines a natural Coxeter structure (W, S) on the Weyl group, as well as simple roots $\alpha_s \in \mathfrak{h}^*$ and simple coroots $\alpha_s^\vee \in \mathfrak{h}$ indexed by simple reflections $s \in S$. Then the triple

$$(\mathfrak{h}, \{\alpha_s^\vee\} \subset \mathfrak{h}, \{\alpha_s\} \subset \mathfrak{h}^*)$$

defines a realization of (W, S) over $\mathbb{C}$, denoted simply by $\mathfrak{h}$. Let G be the connected simply-connected complex reductive with the same root system as $\mathfrak{g}$. Then $\mathfrak{h}$ is naturally identified with the base change to $\mathbb{C}$ of the root datum realization (see Example 5.41) of G.

Now, consider the dual realization $\mathfrak{h}^*$, i.e. the triple

$$(\mathfrak{h}^*, \{\alpha_s\} \subset \mathfrak{h}^*, \{\alpha_s^\vee\} \subset \mathfrak{h}). \tag{15.16}$$

That is, the simple roots (resp. simple coroots) of the realization $\mathfrak{h}^*$ are the simple coroots $\alpha_s^\vee$ (resp. simple roots α_s) of $\mathfrak{g}$. Then $\mathfrak{h}^*$ is naturally identified with the base change to $\mathbb{C}$ of the root datum realization of $G^\vee$, the Langlands dual complex reductive group.

Exercise 15.32 Show that, as W-representations, both $\mathfrak{h}$ and $\mathfrak{h}^*$ are isomorphic to the geometric representation over $\mathbb{C}$. (*Hint:* the Cartan matrix of a semisimple Lie algebra is symmetrizable.)

In particular, both $\mathfrak{h}$ and $\mathfrak{h}^*$ are reflection faithful, so the discussion of Bott–Samelson and Soergel modules from the previous section applies.

By Exercise 15.32, the realizations $\mathfrak{h}$ and $\mathfrak{h}^*$ lead to equivalent categories of Bott–Samelson and Soergel (bi)modules. However, these categories are not canonically equivalent, and we will see that the $\mathbb{V}$ functor naturally relates $\mathcal{O}_0$ for $\mathfrak{g}$ with Soergel modules associated to $\mathfrak{h}^*$. For this reason, we will consider Bott–Samelson and Soergel (bi)modules associated to the realization $(\mathfrak{h}^*, W)$, which are graded (bi)modules over $R = \mathrm{Sym}(\mathfrak{h})$.

We now define the $\mathbb{V}$ functor. Let w_0 be the longest element in W, and consider the antidominant[2] projective $P_{w_0} = P(w_0 \cdot 0)$. Soergel's $\mathbb{V}$ *functor* is defined by

$$\mathbb{V}(-) := \mathrm{Hom}_{\mathcal{O}_0}(P_{w_0}, -) : \mathcal{O}_0 \to \mathrm{mod}\text{-}\mathrm{End}_{\mathcal{O}_0}(P_{w_0}). \tag{15.17}$$

(For any $M \in \mathcal{O}_0$, the Hom space $\mathrm{Hom}_{\mathcal{O}_0}(P_{w_0}, M)$ is a right $\mathrm{End}_{\mathcal{O}_0}(P_{w_0})$-module under precomposition.)

Soergel's first result describes the endomorphism algebra of P_{w_0}. Let C be the coinvariant algebra associated to $(\mathfrak{h}^*, W)$, and let $q : R \to C$ be the quotient map. Consider two algebra maps. The first is the composition

$$Z(U(\mathfrak{g})) \xrightarrow{\gamma} R \xrightarrow{q} C, \tag{15.18}$$

where γ is the Harish-Chandra isomorphism (see Definition 14.31). The second is the action map

$$Z(U(\mathfrak{g})) \to \mathrm{End}_{\mathcal{O}_0}(P_{w_0}). \tag{15.19}$$

---

[2] A weight $\lambda \in \mathfrak{h}^*$ is called *antidominant* if $\langle \lambda + \rho, \alpha^\vee \rangle \notin \mathbb{Z}_{>0}$ for all $\alpha \in \Phi^+$.

15.4 Soergel's $\mathbb{V}$ Functor

Theorem 15.33 (Endomorphismensatz) *The algebra maps* (15.18) *and* (15.19) *are both surjective with the same kernel. Therefore, there is a canonical isomorphism*

$$C \simeq \mathrm{End}_{\mathcal{O}_0}(P_{w_0}). \tag{15.20}$$

From now on, we use (15.20) to view $\mathbb{V}$ as a functor

$$\mathbb{V} : \mathcal{O}_0 \to \mathrm{mod}\text{-}C \tag{15.21}$$

to *ungraded* right C-modules. Soergel proved the following properties of the $\mathbb{V}$ functor.

Theorem 15.34 *The $\mathbb{V}$ functor satisfies the following properties.*

1. (Struktursatz) *Let $M, Q \in \mathcal{O}_0$ with Q projective. Then the $\mathbb{V}$ functor induces an isomorphism*

$$\mathrm{Hom}_{\mathcal{O}_0}(M, Q) \xrightarrow{\sim} \mathrm{Hom}_{\mathrm{mod}\text{-}C}(\mathbb{V}(M), \mathbb{V}(Q)).$$

In particular, $\mathbb{V}$ is fully faithful on projectives.

2. *For every $s \in S$, there is a natural isomorphism*

$$\mathbb{V} \circ \Theta_s \simeq (- \otimes_{C^s} C) \circ \mathbb{V}. \tag{15.22}$$

Soergel's original proof of these results is rather complicated. There is a beautiful proof of the Struktursatz due to Beilinson–Bezrukavnikov–Mirkovic [21] using the notion of tilting objects.

The Struktursatz seems remarkable given that $\mathbb{V}$ is a very destructive functor and is far from being fully faithful on all of $\mathcal{O}_0$. Indeed, since P_w is the projective cover of L_w, we have

$$\dim_{\mathbb{C}} \mathbb{V}(M) = [M : L_{w_0}] \tag{15.23}$$

for any $M \in \mathcal{O}_0$. In particular, this can be used to show that

$$\mathbb{V}(L_w) \simeq \begin{cases} \mathbb{C} & \text{if } w = w_0, \\ 0 & \text{otherwise,} \end{cases} \qquad \mathbb{V}(\Delta_w) \simeq \mathbb{C} \quad \text{for all } w, \tag{15.24}$$

where $\mathbb{C}$ is the trivial module of C (i.e. image of R^+ acts by 0).

Example 15.35 Let us verify these results for $\mathfrak{g} = \mathfrak{sl}_2(\mathbb{C})$ by direct computation. To check the Endomorphismensatz, one first notices that $\dim \mathrm{End}(P_{w_0}) = 2$. As P_{w_0} is indecomposable, $\mathrm{End}(P_{w_0})$ is local of dimension 2, hence isomorphic to

$C \simeq \mathbb{C}[x]/(x^2)$. For instance, if we choose a basis of P_{w_0} with weight diagram

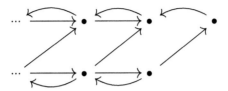

then x corresponds to the endomorphism that sends the vectors in the top row to zero and shifts the vectors in the bottom row to the corresponding ones on the top.

Exercise 15.36 Verify the rest of the results above for $\mathfrak{sl}_2(\mathbb{C})$.

The following is now straightforward. Soergel's results are usually phrased in terms of mod-C, but the discussion at the end of the previous section shows that one can just as well replace C with R.

Corollary 15.37 *The $\mathbb{V}$ functor restricts to an equivalence of $\mathbb{C}$-linear additive categories*

$$\mathbb{V} : \mathrm{Proj}(\mathcal{O}_0) \xrightarrow{\sim} \text{(ungraded Soergel modules)},$$

where $\mathrm{Proj}(\mathcal{O}_0)$ *denotes the full subcategory of* $\mathcal{O}_0$ *consisting of projectives, and the right hand side denotes the full subcategory of mod-R consisting of ungraded Soergel modules (i.e. Soergel modules with their gradings forgotten).*
For every $s \in S$, *we have* $\mathbb{V} \circ \Theta_s \simeq (- \otimes_R B_s) \circ \mathbb{V}$. *Moreover,*

$$\mathbb{V}(P_{\underline{x}}) \simeq \mathbb{C} \otimes_R \mathrm{BS}(\underline{x}), \qquad \mathbb{V}(P_x) \simeq \mathbb{C} \otimes_R B_x \qquad (15.25)$$

for any expression $\underline{x}$ and any $x \in W$.

Proof The $\mathbb{V}$ functor is fully faithful on projectives by the Struktursatz, so to show that it is an equivalence, it remains to show that it is essentially surjective. This will follow from the classification of indecomposable Soergel modules once we have shown that $\mathbb{V}(P_x) \simeq \mathbb{C} \otimes_R B_x$ for all $x \in W$.

The statement about wall-crossing functors is (15.22) (together with the discussion at the end of the previous section). Since $P_{\mathrm{id}} \simeq \Delta_{\mathrm{id}}$ we have $\mathbb{V}(P_{\mathrm{id}}) \simeq \mathbb{C}$ by (15.24). Then the claim about $P_{\underline{x}}$ follows from (15.22).

It therefore remains to show that $\mathbb{V}(P_x) \simeq \mathbb{C} \otimes_R B_x$. Choose a reduced expression $\underline{x}$ for x. Since $P_x \stackrel{\oplus}{\subset} P_{\underline{x}}$, we have $\mathbb{V}(P_x) \stackrel{\oplus}{\subset} \mathbb{C} \otimes_R \mathrm{BS}(\underline{x})$. We saw in the previous section that direct summands of $\mathbb{C} \otimes \mathrm{BS}(\underline{x})$ have the form $\mathbb{C} \otimes M$ where M is a direct summand of $\mathrm{BS}(\underline{x})$. As P_x is not a direct summand of $P_{\underline{w}}$ for any shorter expression $\underline{w}$ and $\mathbb{V}$ is fully faithful on projectives, the same is true for $\mathbb{V}(P_x)$. Therefore, $\mathbb{V}(P_x) = \mathbb{C} \otimes_R B_x$. □

Since $\mathcal{O}_0$ has finite homological dimension (see e.g. [86, Proposition 6.9]), there is a natural identification of $[\mathcal{O}_0]$ with the split Grothendieck group $[\mathrm{Proj}(\mathcal{O}_0)]_\oplus$ of the full subcategory of projectives. We saw in Corollary 15.22 that $\mathcal{O}_0$ equipped with the wall-crossing functors $\{\Theta_s\}_{s \in S}$ categorifies the abelian group $\mathbb{Z}[W]$ equipped with right multiplication by $\{1 + s\}_{s \in S}$.

Corollary 15.37 says that this can be upgraded: $\mathcal{O}_0$ equipped with the wall-crossing functors categorifies the right regular representation of $\mathbb{Z}[W]$ with the preferred generators $\{1 + s\}_{s \in S}$. What is more, there is a graded version of this story, where ungraded Soergel modules and bimodules are replaced with their ordinary graded analogues; thus in fact there is a graded version of $\mathcal{O}_0$ equipped with the wall-crossing functors that categorifies the right regular representation of the Hecke algebra $H(W)$ with the preferred generators $\{b_s\}_{s \in S}$. We will return to graded category $\mathcal{O}_0$ when we discuss Koszul duality in Chaps. 25 and 26, where this additional grading plays an essential role.

15.5 Soergel's Approach to the Kazhdan–Lusztig Conjecture

As before, let $\mathfrak{g}$ be a semisimple complex Lie algebra, and consider its BGG category $\mathcal{O}_0$. Recall the Kazhdan–Lusztig conjecture:

Conjecture 15.38 (Kazhdan–Lusztig) Let W be the Weyl group of $\mathfrak{g}$. For all $x, y \in W$, we have

$$[\Delta_y : L_x] = h_{y,x}(1).$$

Here, the right hand side denotes the value at $v = 1$ of the Kazhdan–Lusztig polynomial $h_{y,x} \in \mathbb{Z}[v]$.

We now explain how Soergel used the $\mathbb{V}$ functor to give a new proof to the Kazhdan–Lusztig conjecture, independent of the original proof by Beilinson–Bernstein and Brylinski–Kashiwara. By BGG reciprocity (Theorem 14.27), the Kazhdan–Lusztig conjecture is equivalent to the equality

$$(P_x : \Delta_y) = h_{y,x}(1)$$

for all $x, y \in W$. Using the identification $[\mathcal{O}_0] \simeq \mathbb{Z}[W]$ of the Grothendieck group, this can be phrased as

$$[P_x] = b_x|_{v=1} \tag{15.26}$$

for all x, where the right hand side denotes the specialization at $v = 1$ of the Kazhdan–Lusztig basis element $b_x \in H(W)$.

Recall that we say that Soergel's conjecture holds for the realization $(\mathfrak{h}, W)$ if we have

$$\operatorname{ch}(B_x) = b_x$$

for all $x \in W$.

Proposition 15.39 *Let $\mathfrak{g}$ be a semisimple complex Lie algebra. Then Soergel's conjecture for the geometric representation of its Weyl group implies the Kazhdan–Lusztig conjecture for $\mathfrak{g}$.*

Proof We will show (15.26) by induction on the Bruhat order. For $x = \mathrm{id}$, we have $P_{\mathrm{id}} = \Delta_{\mathrm{id}}$, so $[P_{\mathrm{id}}] = \mathrm{id} = b_{\mathrm{id}}|_{v=1}$.

Let $x > \mathrm{id}$. Choose $s \in S$ such that $xs < x$, and write $x = ws$ with $w < x$. Then

$$\Theta_s P_w \simeq P_x \oplus \bigoplus_{z<x} P_z^{\oplus m_z} \tag{15.27}$$

for some multiplicities m_z. Passing to the Grothendieck group, we get by induction

$$[P_x] = [\Theta_s P_w] - \sum_{z<x} m_z [P_z]$$

$$= [P_w](1+s) - \sum_{z<x} m_z [P_z] \tag{15.28}$$

$$= (b_w b_s - \sum m_z b_z)|_{v=1}.$$

On the other hand, applying $\mathbb{V}$ to (15.27), we get

$$\overline{B}_w B_s \simeq \overline{B}_x \oplus \bigoplus_{z<x} \overline{B}_z^{\oplus m_z}.$$

We saw that Soergel modules decompose exactly as Soergel bimodules do, so

$$B_w B_s \simeq B_x \oplus \bigoplus_{z<x} B_z^{\oplus m_z}.$$

Passing to characters and using Soergel's conjecture, we get

$$b_x = b_w b_s - \sum_{z<x} m_z b_z. \tag{15.29}$$

By (15.28) and (15.29), we get $[P_x] = b_x|_{v=1}$. □

For finite Weyl groups, Soergel proved his conjecture for the geometric representation by translating it into geometry and using the decomposition theorem, thereby

15.5 Soergel's Approach to the Kazhdan–Lusztig Conjecture

giving a new proof of the Kazhdan–Lusztig conjecture. Moreover, except for this last step, Soergel's proof is entirely algebraic. Here, it is crucial that we are working over a field of characteristic 0. We will say more about the decomposition theorem in Chap. 16.

Soergel conjectured that his conjecture continues to hold for an arbitrary Coxeter system. Recall from Sect. 5.6 that the geometric representation is not reflection faithful for a general Coxeter system. In this generality, Soergel defined a "Kac–Moody representation" that is in some sense the minimal reflection faithful representation.

Conjecture 15.40 (Soergel) Soergel's conjecture continues to hold for an arbitrary Coxeter system and its Kac–Moody representation.

In this generality, there is no geometry to rely on, and Soergel suggested that there should be an algebraic proof. In 2013, Elias and Williamson gave such a proof of Soergel's conjecture, thereby completing Soergel's program to give an algebraic proof of the Kazhdan–Lusztig conjecture.

Despite being written entirely in the language of Soergel (bi)modules, the Elias–Williamson proof is heavily inspired by geometry. In fact, they showed that Soergel modules satisfy a Hodge theory, behaving as though they arise as intersection cohomology of Schubert varieties of (possibly non-existent) flag varieties. The Elias–Williamson proof is the subject of Part IV of this book.

Chapter 16
Lightning Introduction to Perverse Sheaves

This chapter is based on expanded notes of a lecture given by the authors and taken by
Hang Huang and Kostiantyn Timchenko

Abstract We give an introduction to perverse sheaves. By focusing on the "table of stalks" of a perverse sheaf, one can study these objects concretely (under simplifying assumptions). We use the theory of perverse sheaves to explain why the two different descriptions of indecomposable Soergel bimodules should agree, at least for Weyl groups: that is, why the top summand of a Bott–Samelson bimodule should have a character matching the Kazhdan–Lusztig basis.

16.1 Motivation

Soergel's conjecture tells us that we have two descriptions of the indecomposable Soergel bimodule B_w:

1. For any reduced expression $\underline{w} = (s_1, \ldots, s_d)$ for w, B_w is the unique summand of $B_{\underline{w}} = B_{s_1} \cdots B_{s_d}$ that is not a summand of (any shift of) $B_{\underline{x}}$ for any shorter expression $\underline{x}$.
2. B_w is a Soergel bimodule such that $\mathrm{ch}(B_w) = b_w$. In other words, $\mathrm{ch}(B_w)$ is self-dual under the Kazhdan–Lusztig involution, and $\mathrm{ch}(B_w) = \delta_w + \sum_{y<w} h_{y,w} \delta_y$ for $h_{y,w} \in v\mathbb{Z}[v]$.

Why should these two characterizations be the same?

Let G be a complex reductive group with Borel subgroup B and Weyl group W. Let $B \times B$ act on G by left and right multiplication. There is a monoidal abelian category of $(B \times B)$-equivariant perverse sheaves on G with simple objects called IC_w, $w \in W$. These simples satisfy a condition analogous to condition 2,

---

H. Huang
University of Wisconsin Madison, Madison, WI, USA

K. Timchenko
Department of Mathematics, University of Notre Dame, Notre Dame, IN, USA

© The Editor(s) (if applicable) and The Author(s), under exclusive licence
to Springer Nature Switzerland AG 2020
B. Elias et al., *Introduction to Soergel Bimodules*, RSME Springer Series 5,
https://doi.org/10.1007/978-3-030-48826-0_16

essentially by definition, and a condition analogous to condition 1, thanks to a powerful result called the decomposition theorem. Soergel's observation was that the hypercohomology functor

$$H^\bullet_{B \times B} : \text{Perv}_{B \times B}(G) \to R\text{-gmod}_{qc}, \qquad (16.1)$$

is fully faithful, monoidal, and sends semisimple objects to Soergel bimodules. Thus $B_w = H^\bullet_{B \times B}(\text{IC}_w)$ has two descriptions as above.

Remark 16.1 The functor $H^\bullet_{B \times B}$ should be viewed as an analogue of Soergel's functor $\mathbb{V}$ from Chap. 15. However, instead of sending projective objects to Soergel bimodules, it sends semisimple objects to Soergel bimodules. This swap between projective objects and semisimple objects is an indication that Koszul duality is at work; see Chaps. 25 and 26 for more details.

In this chapter we will not give a definition of perverse sheaves. Instead, our focus will be to see how (at least under simplifying assumptions) one can compute with these objects without knowing a thing about what they are, as long as one is willing to accept several key properties and theorems.

16.2 Stratified Spaces and Examples

The setting for perverse sheaves is that of stratified spaces, and their resolutions of singularities.

Definition 16.2 Let X be a topological space. A *stratification* of X is a partially ordered set $(\Lambda, \leq)$ and a collection of locally closed subset $\{X_\lambda\}_{\lambda \in \Lambda}$ indexed by Λ, satisfying

1. $X = \bigsqcup_{\lambda \in \Lambda} X_\lambda$ and Λ is finite.
2. X_λ is a (smooth) connected complex manifold for each $\lambda \in \Lambda$.
3. $\overline{X_\lambda} = \bigsqcup_{\mu \leq \lambda} X_\mu$.

Remark 16.3 The results in this chapter do not apply to all stratifications, but require additional technical conditions. Roughly speaking, one would like X to be equally singular along each stratum (this is the origin of the notion of a *Whitney stratification*). For example, this condition holds if the diffeomorphisms of X act transitively on each stratum. In this chapter stratifications will always be given by the orbits of a group action on X, in which case these technical conditions are automatic. These subtleties will be ignored from now on.

Our main source of examples comes from a complex reductive group G acting on a complex algebraic variety, by considering the stratification by G-orbits. In what follows, by dimension we always mean complex dimension. We begin by recalling some standard constructions and results for complex reductive groups. The novice

16.2 Stratified Spaces and Examples

reader is encouraged to stick to type A (see Example 16.4 below). This will be the case used in examples throughout this chapter.

Let G be a complex reductive group.[1] Fix a Borel subgroup B containing a maximal torus T. Let $N(T)$ be the normalizer of T in G, and let $W = N(T)/T$, the Weyl group of (G, T). Then W has the natural structure of a Coxeter system (W, S). The quotient $X := G/B$ (the set of right B-cosets in G) can be given a topology, and even has the structure of a complex manifold or variety, called the *flag variety* of G.

Example 16.4 (Type A) Let $G = \mathrm{GL}_n(\mathbb{C})$, the group of invertible $n \times n$ complex matrices. Then we may choose B (resp. T) to be the subgroup of G consisting of upper-triangular (resp. diagonal) matrices:

$$B = \begin{pmatrix} * & & * \\ & * & \\ & & \ddots \\ 0 & & * \end{pmatrix}, \quad T = \begin{pmatrix} * & & 0 \\ & * & \\ & & \ddots \\ 0 & & * \end{pmatrix}. \tag{16.2}$$

In this case, the Weyl group $W = N(T)/T$ may be represented by permutation matrices, so that we have a natural identification $W = S_n$. Thus (W, S) is the type A_{n-1} Coxeter system.

Let $X = G/B$ be the flag variety. This name comes from the following concrete description of X in type A. Fix the vector space $\mathbb{C}^n$ with its standard basis $\{e_i\}_{i=1}^n$. A *(full) flag* in $\mathbb{C}^n$ is a collection of subspaces

$$V^\bullet = \left(0 = V^0 \subset V^1 \subset \cdots \subset V^{n-1} \subset V^n = \mathbb{C}^n\right),$$

where $\dim V^i = i$. The *standard flag* is the flag std, where std^k is the span of $\{e_i\}_{1 \leq i \leq k}$.

Exercise 16.5 The group $G = \mathrm{GL}_n(\mathbb{C})$ acts on $\mathbb{C}^n$. Show that it acts transitively on the space of flags (with $g \cdot V^i = g(V^i)$), and that the subgroup B is the stabilizer of the standard flag. Hence the quotient space G/B can be identified (at least as sets) with the space of flags, by sending $gB \in G/B$ to $g \cdot \mathrm{std}$.

Let us return to the setting of a general complex reductive group G. Since B is a subgroup of G, it acts on $X = G/B$ by left multiplication. The B-orbits are described by the following fundamental result.

Theorem 16.6 (Bruhat Decomposition) *We have*

$$G/B = \bigsqcup_{w \in W} BwB/B. \tag{16.3}$$

---

[1] By an abuse of notation, we also write G for its group of complex points, and similarly for B, T, etcetera.

Here, w denotes any lift of $w \in W = N(T)/T$ to $N(T)$. The set of cosets $BwB/B \subset G/B$ is independent of the choice of lift, and is called a **Schubert cell**. For each $w \in W$, BwB/B is isomorphic (as a complex variety) to the affine space $\mathbb{C}^{\ell(w)}$. Moreover, closure relation on these orbits is given by the Bruhat order $\leq$ on W:

$$\overline{BwB/B} = \bigsqcup_{x \leq w} BxB/B. \tag{16.4}$$

The closures $\overline{BwB/B}$ are called **Schubert varieties**.

Exercise 16.7 In type A, the Bruhat decomposition becomes a very concrete statement in linear algebra, which we explore in this exercise.

Using left and right multiplication by elements of B, we can realize certain of the row and column operations familiar from elementary linear algebra. More specifically, we can rescale a row, rescale a column, add a row to a higher row, or add a column to another column further to the right. (We cannot reorder the rows or columns.) Show that two matrices are in the same (B, B)-double coset if and only if they are related by these operations. Then show that any matrix can be reduced to a permutation matrix, and show that a generic matrix will reduce to the longest element of S_n.

Exercise 16.8 For any flag $V^\bullet$ in $\mathbb{C}^n$, we can define its *(standard) position* to be the matrix of integers f_{ij} for $1 \leq i, j \leq n$, where $f_{ij} = \dim(V^i \cap \mathrm{std}^j)$. Find a formula for the position of $w \cdot \mathrm{std}$, for a permutation $w \in S_n$. Show that the position of a flag is an invariant over B-orbits, and determines which Schubert cell a flag belongs to. Show that the numbers f_{ij} can only increase when passing from a flag in one Schubert cell to a flag in its closure.

Before beginning the next example, let us generalize the definition of a flag. Fix a sequence of positive integers $\underline{d} = (d_1, \ldots, d_n)$ with $\sum d_j = N$. A *partial flag (of type $\underline{d}$)* in $\mathbb{C}^N$ is a collection of subspaces

$$V^\bullet = \left(0 = V^0 \subset V^1 \subset \cdots \subset V^{n-1} \subset V^n = \mathbb{C}^N\right),$$

where $\dim V^k = \sum_{j=1}^k d_j$. In particular, $d_j = \dim V^j / V^{j-1}$. The corresponding standard partial flag $\mathrm{std}_{\underline{d}}$ is the partial flag where $\mathrm{std}_{\underline{d}}^k$ is spanned by e_i for $1 \leq i \leq \sum_{j=1}^k d_j$. The group $\mathrm{GL}_N(\mathbb{C})$ acts on $\mathbb{C}^N$ and acts transitively on the space of partial flags of fixed type, and the stabilizer of $\mathrm{std}_{\underline{d}}$ is the subgroup of block upper triangular matrices, where the blocks have size $d_1, \ldots, d_n$:

$$P = \begin{pmatrix} \mathrm{GL}_{d_1} & & & * \\ & \mathrm{GL}_{d_2} & & \\ & & \ddots & \\ 0 & & & \mathrm{GL}_{d_n} \end{pmatrix} \tag{16.5}$$

16.2 Stratified Spaces and Examples

A subgroup P of this form is called a *(standard) parabolic subgroup*.

Exercise 16.9 Verify that the Grassmannian $\mathrm{Gr}(k, N)$ of k-dimensional subspaces in $\mathbb{C}^N$ is isomorphic to the quotient of $\mathrm{GL}_N(\mathbb{C})$ by the parabolic subgroup corresponding to the sequence $d_1 = k$, $d_2 = N - k$.

Here is the next main example of a stratified space we will be looking at.

Example 16.10 Fix a sequence of positive integers $\underline{d} = (d_1, \ldots, d_n)$ with $\sum d_j = N$. Let $P = P_{\underline{d}}$ be the corresponding parabolic subgroup. Fix some $k < N$. Then as a subgroup of $\mathrm{GL}_N(\mathbb{C})$, P acts on $\mathrm{Gr}(k, N)$. We will consider $\mathrm{Gr}(k, N)$, stratified by its P-orbits.

Exercise 16.11 In analogy with Exercise 16.8, prove the following: for $W \in \mathrm{Gr}(k, N)$, its P-orbit is determined by the numbers $f_j = \dim(W \cap \mathrm{std}_{\underline{d}}^j)$, $1 \le j \le n$. Each f_j can only increase in the closure of an orbit.

Example 16.12 As a special case of Example 16.10, let $N = 4$, $d_1 = d_2 = 2$, and $k = 2$. Thus $P = P_{(2,2)}$ is the parabolic subgroup of $\mathrm{GL}_4(\mathbb{C})$ fixing V, the 2-dimensional span of e_1 and e_2 inside $\mathbb{C}^4$, and we consider the action of P on $X = \mathrm{Gr}(2, 4) = \{W \subset \mathbb{C}^4 \mid \dim W = 2\}$. By the previous exercise, this stratifies X into three complex manifolds:

$$X_i = \{W \in X \mid \dim W \cap V = i\} \quad \text{for } i = 0, 1, 2.$$

Note that the closure of each stratum is given by $\overline{X_i} = \{W \in X \mid \dim W \cap V \ge i\}$. So, for example, $\overline{X_0} = X$, while $\overline{X_2} = X_2$ is just a single point.

16.2.1 Stratified Resolutions and Schubert Varieties

If X is stratified, then $\overline{X_\lambda}$ will be stratified, for any stratum λ. We will be interested in (stratified) resolutions of singularities of $\overline{X_\lambda}$. Again, we make a slightly imprecise definition and ignore the thorny details.

Definition 16.13 A *(stratified) proper morphism* is a proper morphism $f : Y \to X$ of stratified spaces such that the image of every stratum of Y is a stratum of X, and such that for any stratum X_μ of X, the restriction $f|_{f^{-1}(X_\mu)} : f^{-1}(X_\mu) \to X_\mu$ is a topologically local trivial fiber bundle with compact fiber F_μ. A stratified proper morphism $f : Y \to \overline{X_\lambda}$ is a *(stratified) resolution of singularities* if in addition f is an isomorphism over each open stratum of X.

Note that the fiber $f^{-1}(x)$ over $x \in X$ only depends on the stratum containing x, i.e. $f^{-1}(x) = F_\mu$ when $x \in X_\mu$.

Our next goal is to produce stratified resolutions of singularities for Schubert varieties (we will do the same for Example 16.12 in Example 16.30). Let us first recall a useful construction.

Definition 16.14 Let H be a subgroup of G, and suppose H acts on a space X. The *induced space* $G \times^H X$ is the quotient of $G \times X$ by the free action of H given by the formula $h \cdot (g, x) = (gh^{-1}, hx)$.

Equivalently, one takes the quotient of $G \times X$ by the relation $(gh, x) \sim (g, hx)$ for $h \in H$. Thus the induced space is like a topological version of a tensor product.

Exercise 16.15 Show that $G \times^H X$ has a projection map to G/H with fiber the space X.

Let G be a complex reductive group, and let B, T, (W, S) be as before. For every simple reflection $s \in S$, one defines the *standard parabolic subgroup* $P_s := \langle B, n_s \rangle$, where $n_s \in N(T)$ is a preimage of $s \in W$. Then $P_s \supset B$ and $P_s/B \simeq \mathbb{P}^1_{\mathbb{C}}$.

Let us describe P_s concretely in type A. So let $G = \mathrm{GL}_n$, with B as before and $W = S_n$. If $s = (i, i+1) \in S_n$, then P_s is the standard parabolic subgroup corresponding to the sequence $(1, \ldots, 1, 2, 1, \ldots, 1)$, where 2 appears in the i-th place. That is, P_s consists of matrices in GL_n that are upper triangular, except possibly for an additional nonzero $(i+1, i)$ entry:

$$P_s = \begin{matrix} \\ \\ i \\ i+1 \\ \\ \end{matrix} \begin{pmatrix} * & & \overset{i\ \ i+1}{} & & * \\ & \ddots & & & \\ & & \boxed{\mathrm{GL}(2)} & & \\ & & & * & \\ & & & & \ddots \\ & 0 & & & * \end{pmatrix} \qquad (16.6)$$

Note that the quotient P_s/B is isomorphic to the quotient of GL_2 by its upper triangular matrices, which is isomorphic to $\mathrm{Gr}(1, 2) = \mathbb{P}^1_{\mathbb{C}}$.

We can now introduce our main example of resolutions of singularities.

Example 16.16 Let G be a complex reductive group, with B and (W, S) as before. Given an expression $\underline{w} = (s_1, \ldots, s_d)$ (i.e. a word in S), the corresponding *Bott–Samelson variety* $Y(\underline{w})$ is the space

$$Y(\underline{w}) = P_{s_1} \times^B P_{s_2} \times^B \cdots \times^B P_{s_d}/B. \qquad (16.7)$$

Since $P_{s_i}/B \simeq \mathbb{P}^1_{\mathbb{C}}$ for each i, it follows from Exercise 16.15 that $Y(\underline{w})$ is an iterated $\mathbb{P}^1_{\mathbb{C}}$-bundle over a point. Hence $Y(\underline{w})$ is smooth and projective of dimension d.

The multiplication map

$$\mathrm{mult} : P_{s_1} \times P_{s_2} \times \cdots \times P_{s_d} \to G$$
$$(p_1, p_2, \ldots, p_d) \mapsto p_1 p_2 \cdots p_d$$

16.2 Stratified Spaces and Examples

factors through $P_{s_1} \times^B P_{s_2} \times^B \cdots \times^B P_{s_d}$, hence induces a map

$$\text{mult} : Y(\underline{w}) = P_{s_1} \times^B P_{s_2} \times^B \cdots \times^B P_{s_d}/B \to G/B. \tag{16.8}$$

One can prove that, when $\underline{w}$ is a reduced expression for $w \in S_n$, the image of mult is precisely the Schubert variety $\overline{BwB/B}$, and the fiber over BwB/B is a point; in fact, the induced map

$$\text{mult} : Y(\underline{w}) \to \overline{BwB/B} \tag{16.9}$$

is a resolution of singularities, called a *Bott–Samelson resolution* of $\overline{BwB/B}$.

For the rest of this subsection, we focus on type A and write $s = (1, 2)$, $t = (2, 3)$, $u = (3, 4)$, ... for the simple reflections of $W = S_n$.

Example 16.17 In type A, the Bott–Samelson variety has a description using flags. A point in $Y(\underline{w})$ is a sequence of complete flags

$$(V_0^\bullet, V_1^\bullet, \ldots, V_d^\bullet),$$

where $V_0^\bullet = \text{std}$, and $V_i^\bullet$ and $V_{i+1}^\bullet$ agree except possibly in one location, determined by the simple reflection s_{i+1}. The map mult sends $(V_0^\bullet, V_1^\bullet, \ldots, V_d^\bullet) \mapsto V_d^\bullet$.

For example, consider $G = GL_3$. Then a point in $Y(s, t, s)$ is a quadruple of flags $(V_0^\bullet, V_1^\bullet, V_2^\bullet, V_3^\bullet)$, where:

- $V_0^\bullet = \text{std}$;
- $V_1^\bullet$ has the same plane as $V_0^\bullet$ but possibly a different line;
- $V_2^\bullet$ has the same line as $V_1^\bullet$ but possibly a different plane;
- $V_3^\bullet$ has the same plane as $V_2^\bullet$ but possibly a different line.

Exercise 16.18 Match up the flag description of the Bott–Samelson variety in type A in Example 16.17 with the description using induced spaces. (*Hint:* consider the map

$$P_{s_1} \times \cdots \times P_{s_d} \to (G/B)^{d+1}$$

$$(p_1, \ldots, p_d) \mapsto (V_0^\bullet, \ldots, V_d^\bullet),$$

where $V_0^\bullet = \text{std}$ and $V_i^\bullet = p_1 \cdots p_i \cdot \text{std}$ for $i \geq 1$.)

Exercise 16.19 Compute the fibers of the map mult : $Y(s, t, s) \to G/B$ over each stratum. The solution will be given in Example 16.28

Exercise 16.20 Compute the fibers of the map mult : $Y(s, s) \to G/B$ over each stratum.

16.2.2 Constructible Sheaves and Pushforwards

Let (X, Λ) be a stratified space.

> **Assumption 16.21** For simplicity, we assume henceforth that each stratum X_λ is simply-connected.

Definition 16.22 A *constructible sheaf* on a stratified space is _____.

We won't give the definition of constructible sheaf[2] here (see e.g. [46, §4.1] or [138] if you want to fill in the blank). Instead, we work with its combinatorial shadow, which will be all we need for computations.

To any sheaf $\mathcal{F}$ on any space X, we can associate a vector space $\mathcal{F}_x$ for each point $x \in X$. If $\mathcal{F}$ is a constructible sheaf on a stratified space (X, Λ), then $\mathcal{F}_x$ is a finite-dimensional vector space that depends only on the stratum X_λ containing x. We call this vector space $\mathcal{F}_\lambda$. Hence we can assign to $\mathcal{F}$ a collection of vector spaces $\{\mathcal{F}_\lambda \simeq \mathbb{C}^{n_\lambda}\}_{\lambda \in \Lambda}$ indexed by the strata. We visualize this collection as one column of a table (depicted here for $\Lambda = \{\lambda, \mu, \nu\}$):

$$\mathcal{F} \rightsquigarrow \begin{array}{c|c} \lambda & \mathbb{C}^{n_\lambda} \\ \hline \mu & \mathbb{C}^{n_\mu} \\ \hline \nu & \mathbb{C}^{n_\nu} \end{array} \qquad (16.10)$$

We note that this table does not determine the sheaf uniquely! The table is just the combinatorial shadow of the sheaf.

Definition 16.23 The Λ-*constructible derived category* $D_\Lambda(X)$ is _____.

Again, let us just mention a few properties, rather than giving a difficult definition (see e.g. [46, §4.1]). The objects of $D_\Lambda(X)$ are complexes of sheaves. One can take kernels and images of morphisms of sheaves, so there is a notion of cohomology for complexes. If $\mathcal{F} \in D_\Lambda(X)$, then each cohomology sheaf $h^i(\mathcal{F})$, for $i \in \mathbb{Z}$, is constructible. Thus to $\mathcal{F}$ we can now assign a collection of vector spaces $\{h^i(\mathcal{F})_\lambda\}_{(\lambda, i) \in \Lambda \times \mathbb{Z}}$ indexed by strata and cohomological degree. We visualize this collection as a table, called the *table of stalks*, where the rows correspond to strata, and columns correspond to cohomological degree:

---

[2]All our constructible sheaves will be sheaves of complex vector spaces.

16.2 Stratified Spaces and Examples

$$\mathcal{F} \rightsquigarrow \begin{array}{c|c|c|c} & -1 & 0 & 1 \\ \hline \lambda & & h^0(\mathcal{F})_\lambda & \\ \hline \mu & & & h^1(\mathcal{F})_\mu \\ \hline \nu & & & \end{array} \qquad (16.11)$$

Here and in the examples below, we have drawn the table when $\Lambda = \{\lambda, \mu, \nu\}$. Moreover, here we have only drawn the columns corresponding to cohomological degrees -1 through 1. There is a vector space in each entry of the table; we have only drawn two for illustration. Again, note that the table does not determine the complex uniquely.

Example 16.24 The constant sheaf $\underline{\mathbb{C}}_X$ on X (viewed as a complex supported in cohomological degree zero) has the following table:

$$\begin{array}{c|c|c} & 0 & 1 \\ \hline \lambda & \mathbb{C} & 0 \\ \hline \mu & \mathbb{C} & 0 \\ \hline \nu & \mathbb{C} & 0 \end{array} \qquad (16.12)$$

For any part of the table not pictured, the vector space in that spot is zero.

Example 16.25 The shifted constant sheaf $\underline{\mathbb{C}}_X[2]$ has the following table:

$$\begin{array}{c|c|c} & -2 & -1 \\ \hline \lambda & \mathbb{C} & 0 \\ \hline \mu & \mathbb{C} & 0 \\ \hline \nu & \mathbb{C} & 0 \end{array} \qquad (16.13)$$

We call the *support* of the table the set of entries where the vector space is nonzero.

Example 16.26 Let $\mathcal{F} = \underline{\mathbb{C}}_{\overline{X_\lambda}}$ be the constant sheaf on the closure of a stratum indexed by λ. We know that $\overline{X_\lambda} = \bigsqcup_{\mu \leq \lambda} X_\mu$. Therefore, $\mathcal{F}$ has the following table:

$$\begin{array}{c|c|c} & 0 & 1 \\ \hline \lambda & \mathbb{C} & 0 \\ \hline \mu < \lambda & \mathbb{C} & 0 \\ \hline \nu \not< \lambda & 0 & 0 \end{array} \qquad (16.14)$$

This table is supported on those strata μ with $\mu \leq \lambda$.

For a stratified morphism of stratified spaces $f : (Y, \Lambda_Y) \to (X, \Lambda_X)$, there is a (derived) pushforward functor

$$f_* : D_{\Lambda_Y}(Y) \to D_{\Lambda_X}(X). \qquad (16.15)$$

The following result follows from a fancy theorem called the "proper base change theorem" (see e.g. [99, Proposition 2.6.7]).

Proposition 16.27 *For any stratified proper map $f : Y \to X$, the pushforward $f_*\underline{\mathbb{C}}_Y$ of the constant sheaf has the following table:*

$$
\begin{array}{c|cccc}
 & 0 & 1 & 2 & \cdots \\ \hline
\lambda & H^*(F_\lambda) & & & \\
\mu & H^*(F_\mu) & & & \\
\nu & H^*(F_\nu) & & &
\end{array}
\tag{16.16}
$$

That is, the vector space in row λ and column i is $H^i(F_\lambda)$, the i-th cohomology (with coefficients in $\mathbb{C}$) of the fiber F_λ above the stratum X_λ.

Example 16.28 We continue Example 16.16 in type A_2, so that $G = \mathrm{GL}_3(\mathbb{C})$. The B-orbits on G/B are indexed by S_3. Set $s = (1, 2)$ and $t = (2, 3)$. Consider the Bott–Samelson resolution mult : $Y(s, t, s) \to G/B$, whose fibers were computed in Exercise 16.19. It follows from Proposition 16.27 that $\mathrm{mult}_* \underline{\mathbb{C}}_{Y(s,t,s)}[3]$ has the following table:

fiber		-3	-2	-1	0
$\{*\}$	sts	$\mathbb{C}$	0	0	0
$\{*\}$	ts	$\mathbb{C}$	0	0	0
$\{*\}$	st	$\mathbb{C}$	0	0	0
$\{*\}$	t	$\mathbb{C}$	0	0	0
$\mathbb{P}^1_{\mathbb{C}}$	s	$\mathbb{C}$	0	$\mathbb{C}$	0
$\mathbb{P}^1_{\mathbb{C}}$	id	$\mathbb{C}$	0	$\mathbb{C}$	0

(16.17)

The leftmost column of the table shows the fiber over each stratum. Each row contains the cohomology of the fiber, shifted down by 3. The meaning of the colored spots will be explained in the next subsection.

Exercise 16.29 Consider the Bott–Samelson resolution mult : $Y(s, s) \to G/B$, whose fibers were computed in Exercise 16.20. Using Proposition 16.27, compute the table of $\mathrm{mult}_* \underline{\mathbb{C}}_{Y(s,s)}[2]$.

Example 16.30 In this example we search for resolutions of singularities in Example 16.12. Recall that there were three P-orbits in $X = \mathrm{Gr}(2, 4)$, called X_0, X_1, and X_2. The closures of X_0 and X_2 are already smooth, so the identity map is a stratified resolution of singularities. We now seek a resolution of singularities for $\overline{X_1} = \{W \in X \mid \dim W \cap V \geq 1\}$. This space is singular precisely at the point where $W = V$ and $\dim W \cap V = 2$, which is the one-point orbit X_2.

16.2 Stratified Spaces and Examples

To find a resolution of singularities, we do not just assert that $W \cap V$ contains a line, but we choose this line. That is, consider the space

$$Y = \{L \subset W \subset \mathbb{C}^4 \mid L \subset W \cap V, \ \dim L = 1, \ \dim W = 2\}. \tag{16.18}$$

The forgetful map $f : Y \to X$ sending $(L, W) \mapsto W$ clearly has image equal to $\overline{X_1}$. The reader should compute that the fiber of f over a point in X_1 is a point, while the fiber over the point X_2 is $\mathbb{P}^1_{\mathbb{C}}$.

Meanwhile, there is also a forgetful map $g : Y \to \text{Gr}(1, V)$ sending $(L, W) \mapsto L$. Each fiber of g consists of all possible ways of choosing a two-dimensional subspace in $\mathbb{C}^4$ containing a fixed one-dimensional subspace, so this fiber is isomorphic to $\mathbb{P}^2_{\mathbb{C}}$. Hence Y is a $\mathbb{P}^2_{\mathbb{C}}$-bundle over $\text{Gr}(1, V) \simeq \mathbb{P}^1_{\mathbb{C}}$, and is smooth of dimension 3.

By Proposition 16.27, $f_* \underline{\mathbb{C}}_Y[3]$ has the following table:

fiber		−3	−2	−1	0
∅	X_0	0	0	0	0
{∗}	X_1	$\mathbb{C}$	0	0	0
$\mathbb{P}^1_{\mathbb{C}}$	X_2	$\mathbb{C}$	0	$\mathbb{C}$	0

(16.19)

Again, the color will be explained in the next subsection.

We will offer more exercises of a similar nature after we explain what you might do with this table.

16.2.3 Perverse Sheaves

Henceforth we write dim for the complex dimension of a complex manifold. We let $d_\lambda = \dim X_\lambda$ for a stratum λ.

Definition 16.31 A complex of sheaves $\mathcal{F} \in D_\Lambda(X)$ is called a *perverse sheaf*[3] if both of the following conditions are satisfied:

1. the table of $\mathcal{F}$ has the following form:

	$-d_\lambda$			$-d_\mu$			$-d_\nu$	
λ	∗	0	0	0	0	0	0	
μ	∗	∗	∗	∗	0	0	0	
ν	∗	∗	∗	∗	∗	∗	0	

(16.20)

---

[3] The name "perverse sheaf" is an abuse of terminology, as perverse sheaves are complexes of sheaves, not actual sheaves.

2. the same holds for $\mathbb{D}\mathcal{F}$, the Verdier dual of $\mathcal{F}$.

To be more precise about condition 1, on each table one can draw the *(twisted) diagonal*, which passes through the row λ in column $-d_\lambda$. We have shaded the spots lying on the diagonal in red. We require that the table of $\mathcal{F}$ is supported on and to the left of the diagonal. That is, $h^i(\mathcal{F})_\lambda = 0$ for $i > -d_\lambda$.

To be more precise about condition 2, we must define *(Poincaré–)Verdier duality*, which is a contravariant autoequivalence $\mathbb{D} : D_\Lambda(X) \to D_\Lambda(X)$. (To define this functor, we need some additional hypotheses on the stratified space (see Remark 16.3) which are satisfied for the all examples we consider in this chapter.) Again, we will not define this functor (see e.g. [99, Chapter III]). The key point is that if $\mathcal{F}$ is *(Verdier) self-dual*, meaning that $\mathbb{D}\mathcal{F} \simeq \mathcal{F}$, then one need only check condition 1 to show that $\mathcal{F}$ is perverse.

For our purposes, it will suffice to know the following properties of $\mathbb{D}$:

1. For any complex $\mathcal{F}$, we have $\mathbb{D}(\mathcal{F}[1]) \simeq \mathbb{D}(\mathcal{F})[-1]$.
2. If X is smooth, then $\mathbb{D}\mathbb{C}_X \simeq \mathbb{C}_X[2\dim_\mathbb{C} X]$. Combined with property 1, we see that $\mathbb{C}_X[\dim_\mathbb{C} X]$ is self-dual.
3. If f is proper, then $\mathbb{D}f_* = f_*\mathbb{D}$, so f_* preserves self-duality.

For example, in Examples 16.28 and 16.30, the resolution of singularities Y is smooth of dimension 3, so that $\mathbb{C}_Y[3]$ is self-dual. The resolution f is proper, so that $f_*\mathbb{C}_Y[3]$ is self-dual. By examining the tables in those examples, we see that both are perverse.

Exercise 16.32 In Exercise 16.29 you computed $\text{mult}_*\mathbb{C}_{Y(s,s)}[2]$. Verify that it is self-dual but not perverse.

Theorem 16.33 *The subcategory* $\text{Perv}(X) \subset D_\Lambda(X)$ *of perverse sheaves on X is abelian. Simple objects in* $\text{Perv}(X)$ *are called* intersection cohomology sheaves *and are in bijection with the poset Λ. We denote them by* IC_λ, *for* $\lambda \in \Lambda$.

Each IC_λ *is uniquely specified (up to a unique isomorphism) by the following two conditions:*

1. IC_λ *is self-dual, and*
2. IC_λ *has a table of the following form:*

		$-d_\lambda$			$-d_\mu$		$-d_\nu$	
$\not\leq \lambda$	0	0	0	0	0	0	0	0
λ	0	$\mathbb{C}$	0	0	0	0	0	0
μ	0	*	*	*	0	0	0	0
ν	0	*	*	*	*	*	0	0

(16.21)

More precisely, the table of IC_λ has support on the closure of X_λ (i.e. on strata $\leq \lambda$) and strictly to the left of the diagonal, except for the entry $\mathbb{C}$ on the diagonal in row λ. Though it takes more work to show, self-duality implies that IC_λ is also supported in cohomological degrees $\geq -d_\lambda$. In particular, the only term in row λ is the copy of $\mathbb{C}$ on the diagonal.

Theorem 16.33 does not specify what vector spaces can occur to the left of the diagonal, in rows $\mu < \lambda$. Part of the content of Theorem 16.33 is that there is exactly one self-dual perverse sheaf whose diagonal is just $\mathbb{C}$ in row λ, so that the vector spaces to the left of the diagonal are uniquely determined. Just what these vector spaces are is the big question. Heuristically, they measure the singularities for $\overline{X_\lambda}$ in a rather complicated and powerful way.

For example, the complex $f_*\underline{\mathbb{C}}_Y[3]$ in Example 16.30 is self-dual and has a table of the required form, so by uniqueness it must be the intersection cohomology sheaf of $\overline{X}_1$, $f_*\underline{\mathbb{C}}_Y[3] = IC_1$. Now, because we know the table of $f_*\underline{\mathbb{C}}_Y[3]$, we are able to determine the vector spaces to the left of the diagonal. Typically this is how we get a grasp on these unknown spaces: constructing IC_λ by some other means where we can compute its table more easily.

When $\overline{X_\lambda}$ is smooth, $\underline{\mathbb{C}}_{\overline{X_\lambda}}[d_\lambda]$ is self-dual and perverse, and satisfies the conditions to be IC_λ. This is the only case when the table of IC_λ is immediately clear.

Example 16.34 The complex $mult_*\underline{\mathbb{C}}_{Y(s,t,s)}[3]$ in Example 16.28 is not the intersection cohomology sheaf, because it has two nonzero entries on the diagonal. However, since the closure of the stratum $B(sts)B/B$ is all of G/B, it is smooth, so we know that IC_{sts} is $\underline{\mathbb{C}}_{G/B}[3]$.

Definition 16.35 As in any abelian category, a perverse sheaf of the form $\bigoplus IC_\lambda^{\oplus m_\lambda}$, $m_\lambda \in \mathbb{N}$, is called a *semisimple perverse sheaf*.

Finally, we can observe that a semisimple perverse sheaf is determined up to isomorphism by its table!

Lemma 16.36 *Given a semisimple perverse sheaf $\mathcal{F} = \bigoplus IC_\lambda^{\oplus m_\lambda}$, the multiplicities m_λ can be determined by reading down the dimensions along the diagonal. That is, $m_\lambda = \dim h^{-d_\lambda}(\mathcal{F})_\lambda$.*

For example, if we can prove that the complex in Example 16.28 is a semisimple perverse sheaf (which will follow from Theorem 16.43 below), then it must be $IC_{sts} \oplus IC_s$. If we already knew the table for IC_s, but didn't already know the table for IC_{sts}, this would give an effective way to compute it. Since the table of a direct sum is the direct sum of the tables, one could obtain the table of IC_{sts} by subtracting the known table of IC_s from the known table of $IC_{sts} \oplus IC_s$.

Definition 16.37 A complex is called a *semisimple complex* if it is isomorphic to a complex of the form $\bigoplus IC_\lambda[k]^{\oplus m_{\lambda,k}}$, $m_{\lambda,k} \in \mathbb{N}$.

Semisimple complexes are also determined up to isomorphism by their tables, although the computation of their multiplicities $m_{\lambda,k}$ is not quite as clear-cut, see the next exercise.

Exercise 16.38 This exercise gives an algorithm for decomposing a semisimple complex, if one already knows the tables of the IC sheaves. In the table of a complex of sheaves, the *k-th shifted diagonal* for $k \in \mathbb{Z}$ contains, for each row λ, the entry in column $-d_\lambda + k$.

Let $\mathcal{F}$ be a semisimple complex, and suppose that $\mathcal{F}$ is self-dual. Let k be the largest integer such that some nonzero vector space appears in the k-th shifted diagonal. If $k = 0$ then $\mathcal{F}$ is semisimple perverse and can be decomposed using Lemma 16.36. If $k > 0$ and a nonzero vector space appears in the k-th shifted diagonal in row λ, show that $\mathrm{IC}_\lambda[k] \oplus \mathrm{IC}_\lambda[-k]$ is a direct summand of $\mathcal{F}$.

Hence, if one already knows the table of IC_λ, then one can subtract $\mathrm{IC}_\lambda[k] \oplus \mathrm{IC}_\lambda[-k]$ from $\mathcal{F}$ to obtain the table of the complementary direct summand. This too is a semisimple complex, so we can continue to decompose it by the same algorithm. This algorithm works so long as one knows the tables of IC_λ for any summand which appears with a nonzero shift.

Exercise 16.39 Given that the complex in Exercise 16.32 is semisimple, verify that it is isomorphic to $\mathrm{IC}_s[1] \oplus \mathrm{IC}_s[-1]$.

Exercise 16.40 Compute the table of the pushforward of the constant sheaf on $Y(s, u, t, s, u)$ to the flag variety. Assuming that it is semisimple, decompose it into shifted IC sheaves. Use this to compute the table of IC_{sutsu}. (Along the way you may need to compute the tables of some other IC sheaves, which you can do by the same method.)

The reader can repeat this exercise for other expressions as desired. We recommend $Y(t, s, u, t)$ and $Y(s, t, s, u, t, s)$.

Exercise 16.41 Now repeat Example 16.30, but in a slightly harder setting ($N = 6$, $d_1 = d_2 = 3$, $k = 3$ in the notation of Example 16.10). Let V be a 3-dimensional subspace of $\mathbb{C}^6$, and let $Q \subset \mathrm{GL}_6(\mathbb{C})$ be its stabilizer in $\mathrm{Gr}(3, 6)$. Now we consider $X = \mathrm{Gr}(3, 6)$, stratified by its Q-orbits. It has four Q-orbits, X_i for $0 \le i \le 3$, where $i = \dim W \cap V$ for any $W \subset \mathbb{C}^6$ in the orbit. For each i there is a space

$$Y_i = \{Z \subset W \subset \mathbb{C}^6 \mid Z \subset W \cap V, \dim Z = i, \dim W = 3\}. \qquad (16.22)$$

To compute the dimension of Y_i and prove that it is smooth, consider the map $Y_i \to \mathrm{Gr}(i, V)$ which sends $(Z, W) \mapsto Z$. Now compute the fibers of the map $Y_i \to \mathrm{Gr}(3, 6)$ which sends $(Z, W) \mapsto W$. Can you deduce the dimension of X_i? Find the table of IC_i for all $0 \le i \le 3$.

Exercise 16.42 Let (X, Λ) be a stratified space. Fix $\lambda \in \Lambda$, and let $f : Y \to \overline{X_\lambda}$ be a (stratified) resolution of singularities. Can you formulate a condition on the dimension on the fibers F_μ, $\mu \in \Lambda$, of f that guarantees that $f_*\underline{\mathbb{C}}_Y[\dim Y]$ is a perverse sheaf? That it equals IC_λ? Now, look up the definitions of semismall and small maps.

16.2.4 The Decomposition Theorem

The decomposition theorem of Beilinson–Bernstein–Deligne–Gabber will be our tool to prove that complexes are semisimple. We do not state the most general form of this theorem.

Theorem 16.43 (Decomposition Theorem [19]) *If a morphism $f : Y \to X$ of algebraic varieties is proper and Y is smooth, then $f_*\underline{\mathbb{C}}_Y$ is a semisimple complex.*

Corollary 16.44 *Let $f : Y \to X$ be a (stratified) resolution of singularities of $\overline{X_\lambda}$. Then IC_λ is a direct summand of $f_*\underline{\mathbb{C}}_Y[\dim Y]$, and the remaining summands are (shifts of) IC_μ, where $\mu < \lambda$.*

We return to the question: what is the table of IC_λ? What are the mysterious spaces to the left of the diagonal? With the decomposition theorem in hand, there is an "algorithm" for answering this question.

Suppose one has a map $f_\lambda : Y(\lambda) \to X$ for each stratum λ, which is a resolution of singularities for its image $\overline{X_\lambda}$ (so in particular $\dim Y(\lambda) = \dim X_\lambda = d_\lambda$). Suppose one can compute the cohomology of the fibers of f_λ. Then one can compute the table of

$$\mathcal{F} = f_{\lambda*}\underline{\mathbb{C}}_{Y(\lambda)}[d_\lambda],$$

a self-dual semisimple complex. Using Corollary 16.44, this will be a direct sum of the desired sheaf IC_λ, together with shifts of IC_μ for various $\mu < \lambda$. By induction, we have already computed the table of IC_μ. So by Exercise 16.38, we can compute the decomposition of $\mathcal{F}$ into shifts of IC sheaves. By subtracting everything else, we can compute the table of IC_λ. This is effectively what was done in Exercise 16.40.

Remark 16.45 The conclusion of the decomposition theorem fails when, instead of working over $\mathbb{C}$, one considers coefficients in positive characteristic. The interested reader can learn more in [94, 95]. We also discuss the positive characteristic situation briefly in Sect. 27.4.

16.2.5 Connection to the Hecke Algebra

Let X be the flag variety, and W the corresponding Weyl group. If $\mathcal{F} \in D_\Lambda(X)$, then its table (or rather, the dimensions of the vector spaces in the table) can be encoded as an element of the Hecke algebra, which we call the *character* of $\mathcal{F}$. Namely, we set

$$\text{ch}(\mathcal{F}) = \sum_{w \in W, k \in \mathbb{Z}} \left(\dim H^{-\ell(w)-k}(\mathcal{F})_w \right) v^k \delta_w. \tag{16.23}$$

Said another way, the standard basis element δ_w of the Hecke algebra corresponds to a copy of $\mathbb{C}$ on the diagonal in row w, and multiplication by v moves the vector space to the left by one spot.

For semisimple complexes one has

$$\mathrm{ch}(\mathbb{D}(\mathcal{F})) = \overline{\mathrm{ch}(\mathcal{F})}, \qquad (16.24)$$

where $\overline{(-)}$ denotes the Kazhdan–Lusztig involution. Then the defining conditions of the simple perverse sheaf IC_w in Theorem 16.33 translate to the following conditions on its character:

1. $\overline{\mathrm{ch}(\mathrm{IC}_w)} = \mathrm{ch}(\mathrm{IC}_w)$,
2. $\mathrm{ch}(\mathrm{IC}_w) = \delta_w + \sum h_{y,w} \delta_y$ for $h_{y,w} \in v\mathbb{Z}[v]$.

But these are precisely the defining conditions of the Kazhdan–Lusztig basis element b_w (see Definition 3.9). By the uniqueness of the latter, we conclude that

$$\mathrm{ch}(\mathrm{IC}_w) = b_w. \qquad (16.25)$$

This matches description 2 of B_w from the motivational Sect. 16.1. As a bonus, each coefficient of $h_{y,w}$ is the dimension of some vector space, so we conclude that these Kazhdan–Lusztig polynomials live in $v\mathbb{N}[v]$! This is how the Kazhdan–Lusztig positivity conjecture was first proven, when W is a Weyl group.

Remark 16.46 It is not true for a general perverse sheaf that (16.24) holds. Thus, although Verdier duality and the Kazhdan–Lusztig involution are related, one needs to proceed with some care to connect them directly. That (16.24) holds for semisimple complexes is due to Brylinski, MacPherson, and Springer, see [167].

The interested reader can continue the story by learning about *convolution*. This equips semisimple B-equivariant complexes on X with a monoidal structure $\star$ (see the introduction to [59] and the references therein). After applying the character map ch, this intertwines with the multiplicative structure on the Hecke algebra. The equivariant hypercohomology functor identifies this geometric categorification with Soergel bimodules over $\mathbb{C}$.

From the definition of convolution one can conclude that

$$\mathrm{mult}_* \underline{\mathbb{C}}_{Y(s_1, s_2, \ldots, s_d)}[d] \simeq \mathrm{IC}_{s_1} \star \mathrm{IC}_{s_2} \star \cdots \star \mathrm{IC}_{s_d}. \qquad (16.26)$$

Moreover, Corollary 16.44 implies that IC_w is the unique new summand of this pushforward, when $(s_1, s_2, \ldots, s_d)$ is a reduced expression for w. This matches description 1 of B_w from the motivational Sect. 16.1.

In Exercise 16.48 below, we explain how one can prove that

$$\mathrm{ch}(\mathrm{mult}_*(\underline{\mathbb{C}}_{Y(s_1, s_2, \ldots, s_d)}[d])) = b_{s_1} b_{s_2} \cdots b_{s_d} \qquad (16.27)$$

without needing to use convolution.

16.2 Stratified Spaces and Examples

Exercise 16.47 Return to all the previous exercises involving the flag variety in this chapter, and see what statement about the Hecke algebra is being categorified.

Exercise 16.48 It is particularly easy to compute the cohomology of a stratified space X when every stratum is isomorphic to some affine space $\mathbb{C}^m$. In this case, $H^\bullet(X)$ is concentrated in even degrees, and $\dim H^{2m}(X)$ equals the number of strata of dimension m.

Assume we are in type A, and fix an expression $\underline{w} = (s_1, s_2, \ldots, s_d)$. In this exercise, we describe what we call the *Deodhar stratification* of $Y(\underline{w})$. Computing the cohomologies of the fibers of $\text{mult} : Y(\underline{w}) \to X$ by counting strata, one obtains a formula which precisely categorifies the Deodhar defect formula for $b_{s_1} b_{s_2} \cdots b_{s_d}$ from (3.36).

Let $\underline{e} \subset \underline{w}$ be a subexpression, and let $x_0 = \text{id}, x_1, \ldots, x_d = \underline{w}^{\underline{e}}$ denote the elements of W in the associated stroll (see Definition 3.32). Let us view $Y(\underline{w})$ as the set of sequences of flags

$$(V_0^\bullet, V_1^\bullet, \ldots, V_d^\bullet)$$

as in Example 16.17. Let $Y_{\underline{e}}$ denote the subset where the standard position of $V_i^\bullet$ is equal to x_i. Argue that $Y_{\underline{e}}$ is isomorphic to $\mathbb{C}^m$ for some m, by arguing that the choice of $V_i^\bullet$, given $V_{i-1}^\bullet$, is either a point or $\mathbb{C} = \mathbb{P}_\mathbb{C}^1 \setminus \text{pt}$. One should treat the four possibilities (where $e_i = U0, U1, D0, D1$) each individually. What is a formula for m? (It is not the defect exactly, but it is related to the defect.) Argue that (16.27) holds. What is the partial order on subexpressions of $\underline{w}$ induced by the closure relation in $Y(\underline{w})$?

Part IV
The Hodge Theory of Soergel Bimodules

Chapter 17
Hodge Theory and Lefschetz Linear Algebra

This chapter is based on expanded notes of a lecture given by the authors and taken by
Elijah Bodish and Jay Hathaway

Abstract The cohomology rings of smooth complex projective algebraic varieties satisfy a "package" of properties: Poincaré duality, weak Lefschetz, hard Lefschetz, and the Hodge–Riemann bilinear relations. These properties impose conditions on the structure of the cohomology ring, and give rise to interesting linear operators and bilinear forms on the ring. In several contexts this "package" yields important results about geometry, representation theory, or combinatorics. We discuss the above properties purely in terms of linear algebra, so that some of the techniques and intuition from geometry can be brought to bear on more general problems. Our focus is on the techniques which appear in the proof of Soergel's conjecture, but there have been successes in other fields as well (e.g. in the combinatorics of matroids and polytopes), and it seems there will be more to come.

17.1 Introduction

Let $\Bbbk$ be a field and let V be a finite-dimensional vector space over $\Bbbk$. A *bilinear form* on V is a function

$$(-,-) : V \times V \longrightarrow \Bbbk$$

which is $\Bbbk$-linear in each variable. We say that the form is *symmetric* if $(v, v') = (v', v)$ for all $v, v' \in V$, and *nondegenerate* if the mapping $v \mapsto (-, v)$ is an isomorphism between V and V^*. A fundamental question in mathematics is to understand the symmetric nondegenerate forms on a fixed vector space V.

E. Bodish · J. Hathaway
Department of Mathematics, University of Oregon, Fenton Hall, Eugene, OR, USA

© The Editor(s) (if applicable) and The Author(s), under exclusive licence to Springer Nature Switzerland AG 2020
B. Elias et al., *Introduction to Soergel Bimodules*, RSME Springer Series 5, https://doi.org/10.1007/978-3-030-48826-0_17

A bilinear form is defined independent of any fixed basis, but to see some of its properties we are inclined to choose an ordered basis of V, say $(v_1, \ldots, v_n)$. Then we may define the *Gram matrix* of the form with respect to this basis as the matrix $\bigl((v_i, v_j)\bigr)_{1 \leq i, j \leq n}$. This defines a bijection between bilinear forms on V and $(\dim V) \times (\dim V)$ matrices with entries in $\Bbbk$. The form is symmetric (resp. nondegenerate) if and only if the Gram matrix is a symmetric (resp. invertible) matrix.

There is an action of $\mathrm{GL}(V)$ on the set of all symmetric nondegenerate bilinear forms on V defined by

$$T \cdot (-, -) = (T(-), T(-)).$$

With respect to a fixed basis this action looks like $\mathrm{GL}_n(\Bbbk)$ acting on symmetric invertible $n \times n$ matrices by

$$M \mapsto A^t M A$$

The natural notion of equivalence for two forms then translates to their Gram matrices being in the same $\mathrm{GL}_n(\Bbbk)$ orbit. The nature of these orbits obviously depends on the dimension of V, but the most interesting behavior is observed when the field $\Bbbk$ varies.

Example 17.1 When $\Bbbk = \mathbb{C}$, all symmetric nondegenerate forms are equivalent.

Example 17.2 When $\Bbbk = \mathbb{F}_p$ with $p \neq 2$, symmetric nondegenerate forms are determined by the image of the determinant of the Gram matrix in $(\mathbb{F}_p^\times)/(\mathbb{F}_p^\times)^2$ (first think about this for the case of 1×1 matrices).

Example 17.3 When $\Bbbk = \mathbb{R}$ the Gram matrix is symmetric and invertible, hence it is conjugate, by an orthogonal change of basis, to a diagonal matrix with diagonal entries in $\mathbb{R}^\times$. Taking p to be the number of positive eigenvalues and q to be the number of negative eigenvalues, one says then that $p - q$ is the *signature* of the form. Sylvester's law of inertia is the statement that nondegenerate symmetric bilinear forms on a real vector space are classified up to equivalence by their signature.

The difference between the real and complex case should be thought of as a consequence of the difference between the topologies of $\mathbb{R}^\times$ and $\mathbb{C}^\times$. We will take advantage of the disconnectedness of $\mathbb{R}^\times$ in the following lemma as well.

Lemma 17.4 *Let V be a finite-dimensional $\mathbb{R}$-vector space. Suppose that B_λ is a continuous family of nondegenerate symmetric bilinear forms on V, parameterized by an interval $\lambda \in (a, b) \subset \mathbb{R}$. Then all B_λ have the same signature.*

Proof Since the forms are all symmetric, the eigenvalues of B_λ must remain real as $\lambda \in (a, b)$ varies. The signature of B_λ will only change if the sign of an eigenvalue changes. Since the forms vary continuously, the intermediate value theorem would then imply that there is some μ so that B_μ has 0 as an eigenvalue. Then B_μ is degenerate, a contradiction. □

Remark 17.5 In contrast to the proof above, when $\Bbbk = \mathbb{C}$ the eigenvalues are free to move around 0 in a continuous family. (Again thinking about the case of a form on a one-dimensional vector space is useful!) This gives some intuition as to why all symmetric nondegenerate forms over $\mathbb{C}$ are equivalent.

We will now restrict our attention to forms on $\mathbb{R}$-vector spaces. Any field of characteristic zero will suffice for the discussion of the hard Lefschetz property below, but once we reach the Hodge–Riemann bilinear relations the base field $\mathbb{R}$ will be essential, in order to have a notion of signature.

17.2 Hard Lefschetz

Fix a finite-dimensional graded $\mathbb{R}$-vector space

$$H = \bigoplus_{i \in \mathbb{Z}} H^i$$

and a symmetric nondegenerate *graded bilinear form*

$$\langle -, - \rangle : H \times H \longrightarrow \mathbb{R}.$$

By "graded" we mean that $\langle H^i, H^j \rangle = 0$ if $i \neq -j$.

It is immediate that $\langle -, - \rangle$ induces an isomorphism between H^{-i} and $(H^i)^*$. We say that the graded vector space H satisfies *Poincaré duality*. Thus, if $b_i = \dim(H^i)$ (the i-th *Betti number of* H), then $b_i = b_{-i}$ for all $i \in \mathbb{Z}$. Our convention is such that the mirror of Poincaré duality is in degree zero. We will use the notation $H^{\min}$ to denote the nonzero graded component of H of minimal degree, which is well defined as long as $H \neq 0$.

Definition 17.6 Let $L \in \mathrm{End}_\mathbb{R}(H)$ be a degree two operator

$$L : H^i \longrightarrow H^{i+2}.$$

1. We say L is a *Lefschetz operator* if $\langle La, b \rangle = \langle a, Lb \rangle$ for all $a, b \in H$.
2. If L is a Lefschetz operator, then it is said to satisfy *hard Lefschetz (hL)* if for all $i \geq 0$,

$$L^i : H^{-i} \longrightarrow H^i$$

is an isomorphism.

The following exercise requires some work, but is extremely worthwhile:

Exercise 17.7 Show that a degree two operator L on H satisfies hard Lefschetz if and only if there is an action of $\mathfrak{sl}_2(\mathbb{R})$ on H with e acting as L and $h \cdot v = kv$ for all $v \in H^k$. Moreover, this $\mathfrak{sl}_2(\mathbb{R})$-action is unique. In particular, deduce in this case that

$$b_0 \geq b_{-2} \geq \cdots \geq 0 \tag{17.1a}$$

and that

$$b_{-1} \geq b_{-3} \geq \cdots \geq 0. \tag{17.1b}$$

Example 17.8 The cohomology ring $H^*(\mathbb{CP}^n, \mathbb{R})$ of complex projective space is isomorphic as a graded ring to $\mathbb{R}[x]/(x^{n+1})$, where $\deg(x) = 2$. In order for the mirror of Poincaré duality to be in degree zero, we shift the grading on $H^*(\mathbb{CP}^n, \mathbb{R})$ down by n. (More generally, when we consider the cohomology ring of a projective variety over $\mathbb{C}$, we will shift the grading down by $\dim_{\mathbb{C}}(X)$.) Consider the graded bilinear pairing on this graded ring where (f, g) is the coefficient of x^n in the product fg. Observe that multiplication by x is a Lefschetz operator satisfying hard Lefschetz. (Geometrically, this bilinear pairing is the Poincaré pairing, and x is the first Chern class of a certain line bundle on $\mathbb{CP}^n$.)

When thinking about Lefschetz operators, it is helpful to have in mind the picture in Fig. 17.1.

The graded form on H restricts to the zero form on each graded component H^i when $i \neq 0$. Instead it pairs graded components in complementary degree. We are interested in obtaining a form on each graded component H^i, and to do this we will relate H^{-i} and H^i by applying L^i. This yields an identification $H^{-i} \xrightarrow{\sim} H^i$ (depending on L) exactly when L satisfies hard Lefschetz.

Fig. 17.1 This picture represents a Lefschetz operator satisfying hard Lefschetz. The $\star$'s represent basis vectors in each degree, and the arrows represent the action of L. If there is no arrow leaving a $\star$, then that basis vector is in the kernel of L

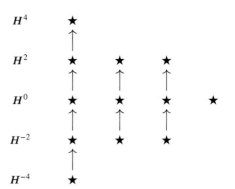

17.2 Hard Lefschetz

Definition 17.9 Let L be a Lefschetz operator. For each $i \geq 0$, define the *Lefschetz form* on H^{-i} with respect to L as

$$(a,b)_L^{-i} = \langle a, L^i b \rangle \tag{17.2}$$

for $a, b \in H^{-i}$.

The Lefschetz form is nondegenerate if and only if it induces an isomorphism between H^{-i} and $(H^{-i})^*$. Since this isomorphism factors as

$$H^{-i} \xrightarrow{L^i} H^i \xrightarrow[\sim]{x \mapsto \langle -, x \rangle} (H^{-i})^*, \tag{17.3}$$

the nondegeneracy of the Lefschetz form is equivalent to hard Lefschetz for L. Note that hard Lefschetz is satisfied trivially in degree zero for any Lefschetz operator L, and the Lefschetz form $(-,-)_L^0$ is independent of the choice of L.

Remark 17.10 If the Betti numbers of H make hard Lefschetz possible, i.e. (17.1) holds, then satisfying hard Lefschetz is a generic condition. Namely, the collection of degree two Lefschetz operators forms an affine space, and $L^i : H^{-i} \to H^i$ being an isomorphism is equivalent to the non-vanishing of a polynomial. Thus the operators satisfying hard Lefschetz form an open set in the Zariski topology of an affine space, and therefore are dense. In what follows, we will sometimes consider families of Lefschetz operators on a fixed graded vector space.

Suppose that L is a Lefschetz operator satisfying hard Lefschetz. The $\mathfrak{sl}_2(\mathbb{R})$ action specified in Exercise 17.7 is unique, and the weight spaces of H are the nonzero graded components. To decompose H into irreducible $\mathfrak{sl}_2(\mathbb{R})$ representations it suffices to determine the lowest weight spaces. In our context we introduce some new terminology to discuss the lowest weight spaces (without having to define a lowering operator). For all $i \geq 0$ set

$$P_L^{-i} = \ker(L^{i+1}) \cap H^{-i}. \tag{17.4}$$

We will refer to elements of this subspace as *primitive vectors* of degree $-i$, and call the subspace a *primitive subspace*.

Exercise 17.11 Check that P_L^{-i} is indeed the space of lowest weight vectors in H^{-i} (where we view H as an $\mathfrak{sl}_2(\mathbb{R})$-module via Exercise 17.7).

It is a consequence of the finite-dimensional representation theory of $\mathfrak{sl}_2(\mathbb{R})$ that

$$H = \bigoplus_{i>0} \bigoplus_{i>j>0} L^j \left(P_L^{-i} \right), \tag{17.5}$$

and that for $i \geq 0$,

$$H^{-i} = P_L^{-i} \oplus L\left(P_L^{-i-2}\right) \oplus L^2\left(P_L^{-i-4}\right) \oplus \cdots . \tag{17.6}$$

Observe that $L^j(P_L^{-i})$ is the intersection of the isotypic component of H of highest weight i and the $-i + 2j$ weight space of H. Can you see this decomposition in Fig. 17.1?

Exercise 17.12 Show that if L satisfies hard Lefschetz, then the primitive decomposition (17.6) is orthogonal with respect to the Lefschetz form.

Remark 17.13 A trivial but important formula which holds for any Lefschetz operator L is that

$$(La, Lb)_L^{-(i-2)} = (a, b)_L^{-i}. \tag{17.7}$$

Consequently, if L satisfies hard Lefschetz then the Lefschetz form is determined by its restriction to the primitive subspaces.

17.3 Hodge–Riemann Bilinear Relations

Recall that a symmetric $\mathbb{R}$-bilinear form is determined by its signature. In topology one is interested in what is sometimes called the *Poincaré form* on the cohomology ring of an even-dimensional manifold, defined by cupping complementary cocycles then integrating against the fundamental class. An interesting invariant arises when studying this form: the signature of its restriction to degree zero (again, we shift our grading down by half the dimension of the space), called the *signature* of the manifold.

When the manifold is a smooth complex projective variety, there are Lefschetz operators given by cupping with the first Chern class of ample line bundles (cf. [41]). These Lefschetz operators allow one to consider not just the signature of the Poincaré form in degree zero, but of the Lefschetz forms arising in all lower degrees as well. The signatures of these Lefschetz forms are also of interest, and the study of their properties is part of "Hodge theory."

Finite-dimensional representations of $\mathfrak{sl}_2(\mathbb{R})$ are partitioned into even and odd representations, so the same is true for Lefschetz operators satisfying hard Lefschetz. This means we can split H into its even and odd degree parts and study them separately. Recall that min is the most negative integer such that $H^{\min} \neq 0$.

Definition 17.14 Assume $H^{\mathrm{odd}} = 0$ or $H^{\mathrm{even}} = 0$ and that L is a Lefschetz operator satisfying hard Lefschetz. We say that $(H, \langle -, - \rangle, L)$ satisfies the *Hodge–Riemann bilinear relations (HR)* if the restriction of the Lefschetz form to the

17.3 Hodge–Riemann Bilinear Relations

primitive subspace

$$(-,-)_L^{\min+2i}|_{P_L^{\min+2i}} \quad (17.8)$$

is $(-1)^i$-definite.

Example 17.15 Consider the example pictured in Fig. 17.1, and suppose that it satisfies the Hodge–Riemann bilinear relations. The Lefschetz form must be positive definite on H^{-4}, since $-4 = \min$ and the whole space is primitive. By (17.7), the Lefschetz form is also positive definite on $L(H^{-4})$ and $L^2(H^{-4})$, i.e. on the two stars directly above the degree -4 star. Meanwhile, $\dim P_L^{-2} = 2$ and the Lefschetz form restricted to this space is negative definite. Thus the Gram matrix of the Lefschetz form on H^{-2} has one positive and two negative eigenvalues. Again by (17.7), these signs are copied to the image of L inside H^0, and the remaining part $P_L^0 \subset H^0$ is positive definite. Thus the Gram matrix of the Lefschetz form on H^0 has two positive and two negative eigenvalues.

If the Hodge–Riemann bilinear relations are satisfied, then each Lefschetz form is determined as follows. Let $i = \min +2j$. We first observe that the Lefschetz form restricted to the primitive subspace in H^{-i} is $(-1)^j$-definite. Then from the primitive decomposition

$$H^{-i} = \bigoplus_{j \geq k \geq 0} L^k \left(P_L^{-i-2k} \right) \quad (17.9)$$

we see that the Lefschetz form is $(-1)^{j+k}$-definite on $L^k \left(P_L^{-i-2k} \right)$, for all k.

Remark 17.16 The Hodge–Riemann bilinear relations tell us that the mixed signature of the Lefschetz forms is coming from some definiteness on each primitive subspace.

Exercise 17.17 Assume $H^{\text{odd}} = 0$. Find a formula for the signature of the Lefschetz form in degree zero in terms of the Betti numbers of H.

For context, let us state a classical result of Hodge theory (with a simplifying assumption).

Example 17.18 Let X be a smooth complex projective variety of complex dimension d, let $H^*(X)$ denote its cohomology ring with coefficients in $\mathbb{R}$, and let $\alpha \in H^2(X)$ be the first Chern class of an ample line bundle. Assume that in the Hodge decomposition of $H^*(X, \mathbb{C})$ only type (p, p) occurs, so in particular H is concentrated in even degree. (This is the case for any variety whose cohomology is generated by the fundamental classes of algebraic subvarieties. Examples include projective spaces, Grassmannians, flag varieties, and Schubert varieties.) Equip the cohomology ring of X with its Poincaré pairing $\langle -, - \rangle$. Then $(H^{*+d}(X), \langle -, - \rangle, \alpha \cup -)$ satisfies hard Lefschetz and the Hodge–Riemann bilinear relations.

When one has a nondegenerate form on a vector space, and a subspace of the vector space, it need not be the case that the restriction of the form to the subspace is nondegenerate. However, a (positive or negative) definite form will restrict to a definite form, and hence will remain nondegenerate! Here is an analogue of this crucial idea in Hodge theory.

Exercise 17.19 Suppose that $(H, \langle -, - \rangle, L)$ satisfies Hodge–Riemann and K is an L-invariant graded subspace of H satisfying Poincaré duality. Show that

$$(K, \langle -, - \rangle_K, L|_K)$$

satisfies Hodge–Riemann (up to a sign). In particular, the restriction of the Lefschetz form from H^{-i} to K^{-i} is nondegenerate, for each $i \geq 0$.

This elementary observation is a key ingredient in de Cataldo and Migliorini's proof of the decomposition theorem [41], and in Elias and Williamson's proof of the Soergel conjecture [57]. In both cases the main theorem will follow from the nondegeneracy of some form, but proving the nondegeneracy (say, by induction) is seemingly intractable. However, by proving a stronger "package" of results, through a complicated inductive argument, one can deduce nondegeneracy of a form by embedding the form in a larger one which has the Hodge–Riemann bilinear relations. Two important lemmas in this inductive proof will be presented in the next section.

Exercise 17.20 This exercise will explore $H^*(\text{Gr}(3, 6), \mathbb{R})$, the cohomology of the Grassmannian of 3-planes in $\mathbb{C}^6$, using a combinatorial model. Let $P(3, 6)$ denote the set of Young diagrams which fit inside a 3×3 rectangle. The *degree* of a partition will be -9 plus twice the number of boxes; for example the partition $(3,1,1)$ has degree $+1$. Two Young diagrams are *complementary* if one can be rotated 180 degrees in order to fill the full 3×3 rectangle with the other: for example $(3,1,0)$ and $(3,2,0)$ are complementary.

Let H be the graded vector space with basis $\{v_\lambda\}_{\lambda \in P(3,6)}$. Place a symmetric bilinear pairing on H, where $\langle v_\lambda, v_\mu \rangle = 1$ if λ and μ are complementary, and is equal to zero otherwise. Place an operator $L : H \to H(2)$ on this space, where $Lv_\lambda = \sum_\mu v_\mu$ is the sum over the Young diagrams $\mu \in P(3, 6)$ obtained from λ by adding one box.

1. Prove that L is a Lefschetz operator.
2. Prove that L satisfies hard Lefschetz. Compute a basis for each primitive subspace.
3. Prove that L satisfies the Hodge–Riemann bilinear relations.

This exercise may be made easier or harder by replacing $P(3, 6)$ with $P(k, n)$ for different values of k and n, and considering Young diagrams which fit inside an $k \times (n - k)$ box. The enterprising reader might try to find a formula for the number of different summands in the Lefschetz decomposition (i.e. the number of $i \geq 0$ such that $P_L^{-i} \neq 0$) as a function of k and n.

17.4 Lefschetz Lemmas

Classically, the "Kähler package" consists of Poincaré duality, hard Lefschetz, the Hodge–Riemann bilinear relations, as well as the Lefschetz hyperplane theorem (weak Lefschetz). We have not yet discussed the Lefschetz hyperplane theorem in this abstract setting. One reason for this omission is that there is not an obvious working candidate for a hyperplane section of a Soergel bimodule (see Sect. 19.1 for a further discussion). The analogue of the Lefschetz hyperplane theorem used in the proof of Soergel's Conjecture is the following lemma.

Lemma 17.21 (The Weak Lefschetz Argument) *Let V, W be graded vector spaces with graded nondegenerate symmetric bilinear forms $\langle -,-\rangle_V, \langle -,-\rangle_W$ and Lefschetz operators L_V, L_W. Let*

$$\sigma : V \longrightarrow W$$

be a degree one linear map such that

1. *σ is injective from negative degrees,*
2. *$\langle v, L_V v'\rangle_V = \langle \sigma v, \sigma v'\rangle_W$ for all $v, v' \in V$,*
3. *$\sigma L_V = L_W \sigma$.*

If L_W satisfies the Hodge–Riemann bilinear relations, then L_V satisfies hard Lefschetz.

Proof Let $i > 0$ (recall that hard Lefschetz always holds trivially in degree zero). We claim that

$$(\sigma(v), \sigma(v'))_{L_W}^{-(i-1)} = (v, v')_{L_V}^{-i} \tag{17.10}$$

for all $v, v' \in V^{-i}$. Indeed,

$$(\sigma(v), \sigma(v'))_{L_W}^{-(i-1)} = \langle \sigma(v), L_W^{i-1}\sigma(v')\rangle_W$$
$$= \langle \sigma(v), \sigma L_V^{i-1}(v')\rangle_W = \langle v, L_V^i(v')\rangle_V = (v, v')_{L_V}^{-i}. \tag{17.11}$$

Since the pairing $\langle -, -\rangle_V$ gives an isomorphism between V^{-i} and $(V^i)^*$, the Betti numbers of V satisfy $b_i = b_{-i}$. To prove that L_V satisfies hard Lefschetz, it therefore suffices to show that L_V^i is injective on V^{-i}. So let $v \in V^{-i}$ and suppose that $L_V^i(v) = 0$. Since $\sigma L_V = L_W \sigma$ it follows that $L_W^i(\sigma(v)) = \sigma(L_V^i(v)) = 0$, so

$$\sigma(v) \in \ker(L_W^i) \cap W^{-(i-1)} = P_{L_W}^{-(i-1)}. \tag{17.12}$$

By the Hodge–Riemann bilinear relations, the Lefschetz form $(-,-)_{L_W}^{-(i-1)}$ is positive or negative definite when restricted to the primitive subspace $P_{L_w}^{-(i-1)}$, so from

$$(\sigma(v), \sigma(v))_{L_W}^{-(i-1)} \stackrel{(17.10)}{=} (v, v)_{L_V}^{-i} = \langle v, L_V^i(v) \rangle_V = \langle v, 0 \rangle_V = 0 \qquad (17.13)$$

we may conclude that $\sigma(v) = 0$. Since σ is injective from negative degrees, $v = 0$. Thus L_V^i is injective on V^{-i}. □

Exercise 17.22 In the lemma above, what conditions would imply that σ sends primitive vectors to primitive vectors?

Lemma 17.23 (Limit Lemma) *Suppose that L_γ is a continuous family of Lefschetz operators on $(H, \langle -, - \rangle)$ parametrized by $\gamma \in [a, b] \subset \mathbb{R}$. Assume that L_γ satisfies hard Lefschetz for all γ (i.e. that all Lefschetz forms are nondegenerate). Then if one L_γ satisfies the Hodge–Riemann bilinear relations, they all do.*

Proof In each fixed degree the Hodge–Riemann bilinear relations are a statement about signatures of the Lefschetz forms. As the Lefschetz operators vary, the associated Lefschetz forms vary as well, but remain nondegenerate. Thus, if at least one form has eigenvalues distributed so that it satisfies the Hodge–Riemann bilinear relations, then by Lemma 17.4 every form will. □

Remark 17.24 Although the continuous family of Lefschetz forms all restrict on the primitive subspaces to definite forms, the primitive subspaces may change as L_γ varies. However, the primitive subspaces will remain on either the positive or the negative side of the isotropic cone as L_γ varies (the isotropic cone of a form $(-, -)$ on a vector space V is the union of the lines spanned by vectors $v \in V$ so that $(v, v) = 0$). The following example illustrates some of the subtlety present in this phenomenon.

Example 17.25 Consider $\overline{B_s B_s} = B_s B_s \otimes_R \mathbb{R}$ equipped with the global intersection form $\langle -, - \rangle$ (defined in Sect. 12.3), and consider the two-parameter family of Lefschetz operators

$$L_{a,b} := (a\rho \cdot -) \mathrm{id}_{B_s} + \mathrm{id}_{B_s}(b\rho \cdot -), \qquad (17.14)$$

where $(a, b) \in \mathbb{R}^2$ and $\rho = \frac{1}{2}\alpha_s$. This is left multiplication by $a\rho$ plus "middle multiplication" by $b\rho$.

The grading on the vector space $\overline{B_s B_s}$ can be seen explicitly in the decomposition

$$\overline{B_s B_s} = (\mathbb{R} \cdot c_{\mathrm{bot}}) \oplus (\mathbb{R} \cdot c_{01}) \oplus (\mathbb{R} \cdot c_{10}) \oplus (\mathbb{R} \cdot c_{\mathrm{top}}),$$

where we have written $c_{\underline{e}}$ (where $\underline{e} \subset (s, s)$, and $c_{\mathrm{bot}} = c_{00}$, $c_{\mathrm{top}} = c_{11}$) for the images of the 01-basis elements (see Sect. 12.1). We have $\deg c_{\mathrm{bot}} = -2$, $\deg c_{01} = \deg c_{10} = 0$, and $\deg c_{\mathrm{top}} = 2$.

17.4 Lefschetz Lemmas

With respect to this basis we have

$$L_{a,b} = \begin{bmatrix} 0 & 0 & 0 & 0 \\ (b-a) & 0 & 0 & 0 \\ a & 0 & 0 & 0 \\ 0 & a & (b+a) & 0 \end{bmatrix} \text{ and } L_{a,b}^2 = \begin{bmatrix} 0 & 0 & 0 & 0 \\ 0 & 0 & 0 & 0 \\ 0 & 0 & 0 & 0 \\ 2ab & 0 & 0 & 0 \end{bmatrix},$$

so $L_{a,b}$ satisfies hard Lefschetz if and only if a and b are nonzero. In this case the Lefschetz decomposition is

$$(\overline{B_s B_s})^{-2} = \mathbb{R} \cdot c_{\text{bot}} = P_{L_{a,b}}^{-2},$$

$$(\overline{B_s B_s})^0 = \mathbb{R} \cdot ((b-a) \cdot c_{01} + a \cdot c_{10}) \oplus \mathbb{R} \cdot ((b+a) \cdot c_{01} + (-a) \cdot c_{10})$$

$$= L_{a,b}\left(P_{L_{a,b}}^{-2}\right) \oplus P_{L_{a,b}}^0,$$

$$(\overline{B_s B_s})^2 = \mathbb{R} \cdot (2ab) \cdot c_{\text{top}} = L_{a,b}^2\left(P_{L_{a,b}}^{-2}\right).$$

The Gram matrix of the global intersection form with respect to the basis $(c_{\text{bot}}, c_{01}, c_{10}, c_{\text{top}})$ is

$$\begin{bmatrix} 0 & 0 & 0 & 1 \\ 0 & 0 & 1 & 0 \\ 0 & 1 & 2 & 0 \\ 1 & 0 & 0 & 0 \end{bmatrix}.$$

Using this we compute the Lefschetz form restricted to the primitive vectors:

$$(c_{\text{bot}}, c_{\text{bot}})_{L_{a,b}}^{-2} = 2ab,$$

$$((b+a) \cdot c_{01} + (-a) \cdot c_{10}, (b+a) \cdot c_{01} + (-a) \cdot c_{10})_{L_{a,b}}^0 = -2ab.$$

We see that $L_{a,b}$ satisfies the Hodge–Riemann bilinear relations if $2ab > 0$ and $-2ab < 0$, i.e. if a and b are both strictly positive or both strictly negative.

Exercise 17.26 Verify all the computations in this example using diagrammatics.

Remark 17.27 The decomposition $B_s B_s \simeq B_s(-1) \oplus B_s(1)$ is a decomposition of R-bimodules, so clearly each summand is preserved[1] by left (resp. right) multiplication by any element of R. So if L is left multiplication by a linear element

---

[1] Warning: direct summands are not preserved by arbitrary bimodule morphisms!

of R on $\overline{B_s B_s} = B_s B_s \otimes_R \mathbb{R}$ (a left R-module), then since

$$(\overline{B_s B_s})^{-2} \subset \overline{B_s(-1)} \quad \text{and} \quad (\overline{B_s B_s})^2 \subset \overline{B_s(1)},$$

the operator L^2 cannot restrict to an isomorphism between $(\overline{B_s B_s})^{-2}$ and $(\overline{B_s B_s})^2$. Thus no operator L induced by left multiplication can satisfy hard Lefschetz. This difficulty explains why we used "middle multiplication" when defining $L_{a,b}$ in Example 17.25.

Chapter 18
The Hodge Theory of Soergel Bimodules

This chapter is based on expanded notes of a lecture given by the authors and taken by
Libby Taylor and **Minh-Tam Trinh**

Abstract In this chapter, we survey Elias and Williamson's proof of Soergel's conjecture on the characters of Soergel bimodules. Their work actually establishes a "Hodge theory" for Soergel bimodules arising from arbitrary Coxeter systems, including purely algebraic versions of the hard Lefschetz theorem and the Hodge–Riemann bilinear relations, though not the weak Lefschetz theorem. The search for a substitute for the weak Lefschetz theorem will motivate the introduction of Rouquier complexes in the next chapter.

18.1 Introduction

This chapter is an introduction to Elias and Williamson's Hodge-theoretic proof of Soergel's conjecture for arbitrary Coxeter systems. We cover Sects. 1–5 of [57], leaving their Sect. 6 for Chaps. 19–20.

In Sect. 18.2, we explain why Soergel's conjecture implies the Kazhdan–Lusztig positivity conjecture and state the setup and results of [57]. Section 18.3 gives a broad outline of the Elias–Williamson proof. In Sect. 18.4, we discuss the absence of weak Lefschetz, setting the ground for Chaps. 19–20. In Sects. 18.5 and 18.6, we elaborate on two key parts of the proof: respectively, a theorem to deduce a "deformed" version of the Hodge–Riemann bilinear relations from the "undeformed" case in shorter Bruhat length [57, §5], and an embedding theorem relating a "local" intersection form to a "global" one [57, §4].

L. Taylor
Georgia School of Mathematics, Institute of Technology, Atlanta, GA, USA

M.-T. Trinh
Department of Mathematics, University of Chicago, Chicago, IL, USA

Section 18.7 surveys the development of a Hodge theory for matroids by Adiprasito, Huh, and Katz in [8], which was used to prove Rota's conjecture on the log-concavity of characteristic polynomials of matroids. We point out some similarities between their techniques and those of Elias–Williamson.

18.2 Overview and Preliminaries

18.2.1 The Conjectures of Soergel and Kazhdan–Lusztig

Fix a Coxeter system (W, S) and its Kac–Moody realization $\mathfrak{h}$, with base field $\mathbb{R}$. As usual, we write R for the ring $\mathrm{Sym}\,\mathfrak{h}^*$ and H for the Hecke algebra of (W, S) with parameter v. We write $\{\delta_x\}_{x \in W}$ for the standard basis of H and $\{b_x\}_{x \in W}$ for its Kazhdan–Lusztig basis.

Recall that $[\mathbb{S}\mathrm{Bim}]_\oplus$ denotes the split Grothendieck ring of $\mathbb{S}\mathrm{Bim}$. By the Soergel categorification theorem (Theorem 5.24), there is a ring isomorphism H $\to$ $[\mathbb{S}\mathrm{Bim}]_\oplus$ that sends $b_s \mapsto [B_s]$ for all $s \in S$ and sends v to the grading shift (1). In Chap. 5 we introduced a map $\mathrm{ch} : [\mathbb{S}\mathrm{Bim}]_\oplus \to$ H in the reverse direction, defined in terms of a standard filtration for each Soergel bimodule. Recall that a standard filtration on $B \in \mathbb{S}\mathrm{Bim}$ is an exhaustive R-bimodule filtration $0 = B^0 \subseteq B^1 \subseteq \cdots$ such that

$$B^i/B^{i-1} \simeq R_{x_i}^{\oplus h_{x_i}(B)} \qquad (18.1)$$

for some total ordering $x_1 < x_2 < \cdots$ of W which extends the Bruhat order. It turns out that the Laurent polynomials $h_{x_i}(B) \in \mathbb{Z}[v^{\pm 1}]$ depend on the element x_i but not on the choice of the total order, so that

$$\mathrm{ch}(B) = \sum_{x \in W} h_x(B)\delta_x \qquad (18.2)$$

is well-defined. We say that $\mathrm{ch}(B)$ is the *character of B*. Our objective is to prove the following conjecture.

Conjecture 18.1 (Soergel) $\mathrm{ch}(B_x) = b_x$ for all $x \in W$.

Since the Laurent polynomials $h_x(B)$ are graded ranks of modules, their coefficients are manifestly non-negative. Moreover, if Soergel's conjecture is true, then $h_y(B_x)$ equals the Kazhdan–Lusztig polynomial $h_{y,x}$. So in proving Soergel's conjecture, we get the Kazhdan–Lusztig positivity conjecture for free.

Conjecture 18.2 (Kazhdan–Lusztig) The coefficients of the Kazhdan–Lusztig polynomials for any Coxeter system are all non-negative.

18.2.2 Duality and Invariant Forms

In order to prove the Soergel conjecture, Elias–Williamson proved a stronger, Hodge-theoretic result which, while significantly more complicated, was amenable to an inductive proof. Our immediate goal is to develop the concepts and notation to state this stronger result properly.

Recall that, when B is a Soergel bimodule, the corresponding *Soergel module* is the right quotient $\overline{B} = B \otimes_R \mathbb{R}$. Unlike B, $\overline{B}$ is a finite-dimensional graded vector space. We would like to discuss hard Lefschetz and the Hodge–Riemann bilinear relations for certain operators acting on $\overline{B}$, but in order to do this, $\overline{B}$ must be equipped with a bilinear form $\langle -, - \rangle$. We now stage a digression to discuss forms on Soergel bimodules.

18.2.2.1 Morphisms Between Soergel Bimodules

For any $x \in W$, we write $S(x)$ to mean that Soergel's conjecture holds for x, i.e. that $\mathrm{ch}(B_x) = b_x$. We will frequently translate the hypotheses $S(-)$ into statements about Hom-spaces between Soergel bimodules using the Soergel Hom formula (Theorem 5.27), which we restate here.

Theorem 18.3 (Soergel Hom Formula) *For any Soergel bimodules B, B', we have*

$$\underline{\mathrm{rk}}\,\mathrm{Hom}^\bullet(B, B') = (\mathrm{ch}(B), \mathrm{ch}(B')), \tag{18.3}$$

where $(-, -)$ is the standard sesquilinear trace on H, and $\underline{\mathrm{rk}}$ denotes the graded rank of this graded Hom space as a free right R-module.

Since $(b_x, b_y) \in \delta_{x,y} + v\mathbb{Z}_{\geq 0}[v]$ (see Theorem 3.21), the Hom formula implies that if Soergel's conjecture holds, then the indecomposables B_x satisfy a kind of analogue of Schur's lemma.

Corollary 18.4 *Together, $S(x)$ and $S(y)$ imply*

$$\dim \mathrm{Hom}^i(B_x, B_y) = \begin{cases} \delta_{x,y} & \text{if } i = 0; \\ 0 & \text{if } i < 0. \end{cases}$$

18.2.2.2 Invariant Forms

Let $R\text{-gbim}_{\mathrm{free}}$ be the category of graded R-bimodules which are finitely generated and free as right R-modules.

Definition 18.5 Let B be any R-bimodule in R-gbim$_{\text{free}}$. An *invariant form* on B is a pairing $\langle \cdot, \cdot \rangle : B \times B \to R$ satisfying the following properties for all $b, b' \in B$ and $f \in R$:

1. $\deg(\langle b, b' \rangle) = \deg(b) + \deg(b')$ for homogeneous $b, b' \in B$,
2. $\langle fb, b' \rangle = \langle b, fb' \rangle$,
3. $\langle bf, b' \rangle = \langle b, b'f \rangle = \langle b, b' \rangle f$.

Remark 18.6 Note the difference between left and right multiplication: the form is bilinear for right multiplication by $f \in R$, but is only self-adjoint for left multiplication.

An invariant form on B induces an $\mathbb{R}$-valued bilinear form on $\overline{B} = B \otimes_R \mathbb{R}$.

Example 18.7 Recall from Sect. 12.3 (see Proposition 12.15) that a Bott–Samelson bimodule BS($\underline{x}$) for any expression $\underline{x}$ admits a nondegenerate invariant form $\langle \cdot, \cdot \rangle_{\underline{x}}$, an analogue of the intersection form on the cohomology of a projective variety. Explicitly,

$$\langle b_1, b_2 \rangle_{\underline{x}} = \mathrm{Tr}(b_1 b_2), \tag{18.4}$$

where the trace map Tr returns the coefficient of the 01-basis element c_{top}.

In ordinary linear algebra, a bilinear form on a finite-dimensional vector space V is the same data as a linear map $V \to V^*$, and the form is nondegenerate if and only if this map $V \to V^*$ is an isomorphism. It is not difficult to adapt these ideas to the category R-gbim$_{\text{free}}$.

Definition 18.8 The *(right) duality functor* is the contravariant autoequivalence $\mathbb{D}$ of R-gbim$_{\text{free}}$ given by

$$\mathbb{D}(-) = \mathrm{Hom}^{\bullet}_{-R}(-, R). \tag{18.5}$$

That is, $\mathbb{D}(B)$ is the space of right R-module maps from B to R. It is naturally an R-bimodule via $(r \cdot f \cdot r')(b) := rf(b)r'$ for $r, r' \in R$, $f \in \mathbb{D}(B)$ and $b \in B$.

We leave it to the reader to prove the following facts.

Proposition 18.9 *There is a canonical isomorphism of functors $\mathbb{D} \circ \mathbb{D} \simeq 1$. The data of an invariant form on $B \in R$-gbim$_{\text{free}}$ is the same as the data of an R-bimodule morphism $B \to \mathbb{D}(B)$. There is an isomorphism $\mathbb{D}(B(1)) \simeq \mathbb{D}(B)(-1)$ for any $B \in R$-gbim$_{\text{free}}$.*

Definition 18.10 We say that an invariant form on B is *nondegenerate* if the corresponding map $B \to \mathbb{D}(B)$ is an isomorphism.

Exercise 18.11 Match the definition of nondegeneracy to a more familiar notion (for every $b \in B$ there exists, etc.), but be careful. Prove that any nondegenerate invariant form descends to a nondegenerate form on the finite-dimensional vector space $\overline{B}$. If the induced form on $\overline{B}$ is nondegenerate, is the original invariant form on B nondegenerate?

Exercise 18.12 Prove that $\mathbb{D}(B_s) \simeq B_s$ as R-bimodules, in two ways.

1. Prove that the global intersection form on B_s is nondegenerate.
2. Find the isomorphism explicitly.

Exercise 18.13 For $B \in R\text{-gbim}_{\text{free}}$ show that one has a canonical isomorphism

$$\mathbb{D}(B \otimes_R B_s) \simeq \mathbb{D}(B) \otimes_R B_s. \qquad (18.6)$$

Deduce that any Bott–Samelson bimodule $\text{BS}(\underline{x})$ is self-dual. (*Hint:* see the proof of Proposition 5.9 in [165].)

18.2.2.3 Invariant Forms on Soergel Bimodules

It is a result of Soergel (see [165, Proposition 5.9] and its proof) that the category of Soergel bimodules (a full subcategory of $R\text{-gbim}_{\text{free}}$) is preserved by the duality functor $\mathbb{D}$.[1] One also has that $\mathbb{D}(B \otimes_R B') \simeq \mathbb{D}(B) \otimes \mathbb{D}(B')$ for all Soergel bimodules B, B'. Consequently $\mathbb{D}$ induces an algebra automorphism of the Grothendieck group $[\mathbb{S}\text{Bim}]_\oplus \simeq H$ sending v to v^{-1}. Since B_s is self-dual, this automorphism must agree with the Kazhdan–Lusztig bar involution, see Definition 3.8.

Soergel also proves that the indecomposable Soergel bimodules are all self-dual, that is, $\mathbb{D}(B_x) \simeq B_x$. This would follow immediately from the Soergel conjecture, since $\text{ch}(B_x) = b_x = \overline{b_x} = \text{ch}(\mathbb{D}(B_x))$, although Soergel proves it without knowing the Soergel conjecture. Thus any indecomposable Soergel bimodule has a nondegenerate invariant form.

Remark 18.14 One can prove these results for the diagrammatic Hecke category as well. The duality functor $\mathbb{D}$ is replaced by the functor from Chap. 11 which flips a diagram upside-down. The self-duality of Bott–Samelson bimodules follows by construction, and the self-duality of indecomposable objects is a general result about object-adapted cellular categories, see Exercise 11.43.

If one does assume $\text{S}(x)$, then $\dim \text{Hom}^0(B_x, \mathbb{D}(B_x)) = 1$ by Corollary 18.4. Thus the space of invariant forms on B_x is one-dimensional over $\mathbb{R}$, and any nonzero invariant form is nondegenerate. Let $\underline{x}$ be a reduced expression for x, and recall the nondegenerate (global) intersection form on $\text{BS}(\underline{x})$ from Example 18.7. For any embedding of B_x into $\text{BS}(\underline{x})$, the restriction of the intersection form from $\text{BS}(\underline{x})$ gives an invariant form on B_x. Here are two exercises which give different arguments to prove that this restriction is nonzero for any choice of embedding (and

---

[1] In fact, this will become clear in the exercises in a moment.

thus, if we know Soergel's conjecture, the restricted intersection form is the unique nondegenerate invariant form, up to scalar).

Exercise 18.15 As we saw in Proposition 18.9, a nondegenerate form on $BS(\underline{x})$ is the same data as an isomorphism

$$BS(\underline{x}) \xrightarrow{\sim} \mathbb{D}(BS(\underline{x})).$$

Use the Krull–Schmidt decomposition to deduce that for any choice of decomposition of $BS(\underline{x})$ into indecomposable Soergel bimodules, the induced map $B_x \to \mathbb{D}(B_x)$ is an isomorphism. Deduce that the intersection form on $BS(\underline{x})$ restricts to a nondegenerate form on B_x, and that any indecomposable Soergel bimodule is self-dual.

Exercise 18.16 Prove that, for any embedding of B_x inside $BS(\underline{x})$, the element $c_{\text{bot}} \in BS(\underline{x})$ actually lives inside the summand B_x. (*Hint:* Use standard filtrations and graded dimensions.) Deduce that $f c_{\text{bot}} \in B_x$, for any polynomial $f \in R$. Using Proposition 12.29, we know that $\rho^{\ell(x)} c_{\text{bot}}$ pairs nontrivially against c_{bot} when ρ is regular dominant. Deduce that the restriction of the global intersection form to B_x is nonzero.

Thus when $S(x)$ holds, it would be nice to say that there is a fixed nondegenerate invariant form on B_x, which we would denote by $\langle -, - \rangle_x$, and that this form agrees with the restriction of the global intersection form $\langle -, - \rangle_{\underline{x}}$ from $BS(\underline{x})$. We're not quite there yet: currently we know that $\langle -, - \rangle_x$ is well-defined up to an invertible scalar in $\mathbb{R}^\times$, but signs are important! If we rescale the embedding of B_x into $BS(\underline{x})$ by $\lambda \in \mathbb{R}^\times$, then this will rescale the restriction to B_x by the *positive* scalar λ^2, since $\langle \lambda b, \lambda b' \rangle_{\underline{x}} = \lambda^2 \langle b, b' \rangle_{\underline{x}}$. From this we might try to argue that $\langle -, - \rangle_x$ is well-defined up to a positive scalar, but there are issues here as well. Is the Hom space from B_x to $BS(\underline{x})$ only one-dimensional? Does the sign induced from $\underline{x}$ agree with the sign induced from a different reduced expression $\underline{x}'$? We will resolve these issues shortly in Corollary 18.18.

18.2.2.4 Lefschetz Forms and Positivity

The following properties were introduced in Chap. 17. Let B be a Soergel bimodule.

1. *Hard Lefschetz.* For any operator $L : B \to B(2)$, we write $hL(B, L)$ to mean that L^i induces an isomorphism $\overline{B}^{-i} \to \overline{B}^i$ for all $i \geq 0$.
2. *Hodge–Riemann.* For a nondegenerate invariant form $\langle \cdot, \cdot \rangle$ on B with Lefschetz operator L, we write $HR(B, \langle \cdot, \cdot \rangle, L)$ to mean that the *Lefschetz form*

$$(\alpha, \beta)_L^{-i} = \langle \alpha, L^i \beta \rangle \tag{18.7}$$

18.2 Overview and Preliminaries

is $(-1)^{(i-\min)/2}$-definite on the primitive subspace $P_L^{-i} \subset \overline{B}^{-i}$, where min is the degree of the lowest nonzero component of $\overline{B}$.

Note that $\mathrm{HR}(B, \langle \cdot, \cdot \rangle, L)$ implies $\mathrm{hL}(B, L)$.

Because the simple coroots $\alpha_s^\vee$ are linearly independent in $\mathfrak{h}$, we may choose an element $\rho \in \mathfrak{h}^*$ satisfying $\langle \alpha_s^\vee, \rho \rangle > 0$ for all $s \in S$.[2] Let L_0 denote the degree 2 operator given by left multiplication by ρ on any given Soergel bimodule. By property (2) of invariance (Definition 18.5), L_0 is self-adjoint for any invariant form, and thus is a Lefschetz operator.

Now let us discuss Proposition 12.29 in this language. Let $\underline{x}$ be a reduced expression, and consider the operator L_0 on the Soergel module $\overline{\mathrm{BS}(\underline{x})}$, which is a Lefschetz operator for the global intersection form. The number $N_{\underline{x}}(\rho)$ from Definition 12.28 can be reinterpreted as the value of the Lefschetz form

$$N_{\underline{x}}(\rho) = (c_{\mathrm{bot}}, c_{\mathrm{bot}})_{L_0}^{-\ell(\underline{x})} \tag{18.8}$$

on the one-dimensional degree $-\ell(\underline{x})$ part of $\overline{\mathrm{BS}(\underline{x})}$. Proposition 12.29 states that $N_{\underline{x}}(\rho) > 0$, implying that the Lefschetz form on $\overline{\mathrm{BS}(\underline{x})}$ is positive definite in the minimal degree, as would be required by the Hodge–Riemann bilinear relations. This is important enough to say again.

Corollary 18.17 *For a reduced expression $\underline{x}$, the Lefschetz form on $\overline{\mathrm{BS}(\underline{x})}$ induced by L_0 is positive definite in the minimal degree.*

We do not expect $\mathrm{hL}(\mathrm{BS}(\underline{x}), L_0)$ to hold for an arbitrary expression. We have already seen an example of a non-reduced expression where hard Lefschetz fails in Example 17.25, see also Remark 17.27. We do not even expect $\mathrm{hL}(\mathrm{BS}(\underline{x}), L_0)$ to hold for an arbitrary reduced expression. For a reduced expression where it fails, take $\underline{x} = (s, u, t, s, u)$ in the symmetric group S_4. Hard Lefschetz fails in these counterexamples for the same reason: if B is a Soergel bimodule which has an indecomposable summand of the form $B_y(k)$ for some $y \in W$ and some $k \neq 0$, then it is impossible for $\mathrm{hL}(B, L_0)$ to hold. Since L_0 is left multiplication by an element of R, it respects any R-bimodule decomposition. Any right quotient of an indecomposable bimodule $\overline{B}_y$ has a graded dimension which satisfies Poincaré duality, which implies that $\overline{B}_y(k)$ will not satisfy Poincaré duality for any $k \neq 0$.

However, when $\underline{x}$ is a reduced expression, the fact that $c_{\mathrm{bot}} \in \mathrm{BS}(\underline{x})$ actually lives in the direct summand B_x (see Exercise 18.16) combined with Corollary 18.17, implies that the Lefschetz form of L_0 is positive definite in minimal degree on B_x (that is, when B_x is equipped with the restriction of the intersection form from $\mathrm{BS}(\underline{x})$).

---

[2] There might not be such a ρ for the geometric representation, although it always exists for the dual geometric representation. For a general realization over $\mathbb{R}$, the existence of such a ρ can be called the *positivity assumption*.

Corollary 18.18 *For any reduced expression $\underline{x}$, and any embedding of B_x as a direct summand of $\mathrm{BS}(\underline{x})$, let $\langle -, - \rangle_{\underline{x}}$ denote the restriction of the intersection form from $\mathrm{BS}(\underline{x})$ to B_x. Then the Lefschetz form on $\overline{B_x}$ in minimal degree induced by L_0, namely*

$$\langle -, - \rangle_{L_0}^{-\ell(x)},$$

is positive definite. Consequently, if one assumes $S(x)$, then $\langle -, - \rangle_{\underline{x}}$ is nondegenerate and does not depend on the choice of reduced expression $\underline{x}$ or the choice of embedding, up to a positive scalar.

Definition 18.19 When $S(x)$ holds, we call $\langle -, - \rangle_x$ the *(global) intersection form* on $\overline{B_x}$. It is well-defined up to a positive scalar.

To justify the last sentence of the corollary, we remind the reader what we stated in Sect. 18.2.2.3, that $S(x)$ already implies that any nonzero form is nondegenerate, and that any two are equal up to a scalar (which in the setting of the corollary must be a positive scalar to preserve positive definiteness).

18.2.2.5 Key Statements in the Induction

We now introduce notation for the Hodge-theoretic statements we will prove. We write $\langle \cdot, \cdot \rangle_{\underline{x}}$ for the $\mathbb{R}$-valued intersection form on $\overline{\mathrm{BS}(\underline{x})}$. Now, for all $x \in W$, reduced expressions $\underline{x}$, and $s \in S$:

- We write $\mathrm{hL}(x)$ to mean $\mathrm{hL}(B_x, L_0)$.
- We write $\mathrm{HR}(\underline{x})$ to mean $\mathrm{HR}(B_x, \langle \cdot, \cdot \rangle_{\underline{x}}|_{B_x}, L_0)$ holds for every embedding of B_x into $\mathrm{BS}(\underline{x})$.
- We write $\mathrm{HR}(\underline{x}, s)$ to mean $\mathrm{HR}(B_x B_s, \langle \cdot, \cdot \rangle_{\underline{x}s}|_{B_x B_s}, L_0)$ holds for every embedding of B_x into $\mathrm{BS}(\underline{x})$.

As noted in Corollary 18.18, $S(x)$ implies that the restriction of the form from $\mathrm{BS}(\underline{x})$ to B_x is independent of the choice of $\underline{x}$ or the choice of embedding (up to positive scalar). Thus in the presence of $S(x)$, $\mathrm{HR}(\underline{x})$ is independent of $\underline{x}$. We let $\langle \cdot, \cdot \rangle_x$ denote the intersection form on $\overline{B_x}$, which only makes sense when $S(x)$ holds. Continuing our list of Hodge-theoretic statements:

- We write $\mathrm{HR}(x)$ to mean $S(x)$ and $\mathrm{HR}(B_x, \langle \cdot, \cdot \rangle_x, L_0)$.

Again, $S(x)$ and $\mathrm{HR}(\underline{x})$ imply $\mathrm{HR}(x)$.

We will write $S(< x)$, resp. $S(\leq x)$, to mean $S(y)$ holds for all $y < x$, resp. $y \leq x$, in the Bruhat order, and similarly with hL and HR.

18.2.2.6 Induced Forms

The statement HR($\underline{x}$, s), which is crucial for the inductive proof to come, is somewhat mysterious. It helps to have a better handle on $B_x B_s$ and its invariant form, which is the goal of this section.

Due to the existence of the 01-basis for BS($\underline{x}$) (see Sect. 12.1), one observes that the form $\langle \cdot, \cdot \rangle_{\underline{x}}$ can be built up in an inductive way solely from the forms $\langle \cdot, \cdot \rangle_{\underline{y}}$ for subexpressions $\underline{y}$. Something more general is true: If $s \in S$ and B is a Soergel bimodule equipped with a symmetric invariant form $\langle \cdot, \cdot \rangle_B$, then we can equip BB_s with an *induced form* $\langle \cdot, \cdot \rangle_{BB_s}$.

First, a right R-basis of B induces a right R-basis of BB_s as follows. Consider the maps $\xi_{\text{bot}}, \xi_s : B \to BB_s$ defined by

$$\xi_{\text{bot}}(b) = bc_{\text{bot}} \quad \text{and} \quad \xi_s(b) = bc_s. \tag{18.9}$$

If $\{e_i\}$ is a right basis of B, then $\{\xi_{\text{bot}}(e_i), \xi_s(e_i)\}$ is a right basis of BB_s. Pick an ordering on the basis $\{e_i\}$ of B, and let M_B be the Gram matrix of $\langle \cdot, \cdot \rangle_B$. Order the basis of BB_s so that all the elements $\xi_{\text{bot}}(e_i)$ come first. Then we define the induced pairing on BB_s by setting its Gram matrix M_{BB_s} to be

$$M_{BB_s} = \begin{pmatrix} \partial_s M_B & M_B \\ M_B & M_B \alpha_s \end{pmatrix}, \tag{18.10}$$

where $\partial_s M_B$ (resp. $M_B \alpha_s$) is the matrix obtained from M_B by applying ∂_s (resp. multiplication by α_s) to each entry. This choice ensures that the induced form remains invariant. Moreover, by the following exercise, when $B = \text{BS}(\underline{x})$ we get $\langle \cdot, \cdot \rangle_{BB_s} = \langle \cdot, \cdot \rangle_{\underline{x}s}$.

Exercise 18.20 Show that the form on BS($\underline{x}$) induced (in the sense above) inductively from the canonical form on R agrees with $\langle \cdot, \cdot \rangle_{\underline{x}}$.

Example 18.21 If $B = R$, then $BB_s = B_s$ and the ordered basis described above is just (c_{bot}, c_s) up to scaling. The recipe above gives the Gram matrix

$$M_{B_s} = \begin{pmatrix} 0 & 1 \\ 1 & \alpha_s \end{pmatrix}, \tag{18.11}$$

which is indeed the Gram matrix of $\langle \cdot, \cdot \rangle_s$ with respect to (c_{bot}, c_s).

Let's beat the point home. One can not determine the $\mathbb{R}$-valued global intersection form on $\overline{BB_s}$ just from the $\mathbb{R}$-valued global intersection forms on $\overline{B}$ and $\overline{B_s}$. The global intersection form on B instructs us, morally speaking, to force all polynomials to the far right of B and compute the coefficient of c_{top}. The $\mathbb{R}$-valued version on $\overline{B}$ instructs us further to forget any information about polynomials on the right of positive degree. Meanwhile, inside BB_s, the far right of B is not yet the far right of BB_s! Polynomials to the right of B must be pulled one step further, past the B_s, whence the $\partial_s M_B$ entry in (18.10). For this reason, one needs to retain information about the linear polynomials to the right of B in order to determine what happens in $\overline{BB_s}$.

18.2.3 The Main Theorem

In his original paper [161] on the Kazhdan–Lusztig conjecture, Soergel used the decomposition theorem (see Theorem 16.43) to show that for finite Weyl groups, $\overline{B_x}$ can be identified with the equivariant intersection cohomology of the Schubert variety corresponding to x (see Chap. 16 for a brief account), so that in this case, HR(x), hL(x), and S(x) follow from classical Hodge theory (for intersection cohomology). In [81], Härterich extended Soergel's result to the Weyl groups of symmetrizable Kac–Moody groups. The discovery of Elias–Williamson is that Hodge-theoretic behavior persists in arbitrary Coxeter systems, where no flag variety need exist.

Theorem 18.22 (Hodge Theory of Soergel Bimodules) S(x), hL(x), and HR(x) *hold for any x in any Coxeter system.*

This way of stating their main theorem is rather unenlightening, as it fails to convey the dynamic interplay of implications between Soergel's conjecture, the hard Lefschetz theorem, and the Hodge–Riemann bilinear relations that is the defining feature of the actual proof. Elias and Williamson's proof drew significant motivation from work of de Cataldo and Migliorini, who gave a Hodge theoretic proof of the Decomposition Theorem in complex algebraic geometry. In describing de Cataldo and Migliorini's proof, Williamson writes [39, 40]:

> Each ingredient is indispensable in the induction. One is left with the impression that the Decomposition Theorem is not a theorem by itself, but rather belongs to a family of statements, each of which sustains the others.

One could describe Elias and Williamson's proof in similar terms.

18.3 Outline of the Proof

The strategy is induction on the Bruhat order. The base cases HR(1) and HR(s) for $s \in S$ can be checked directly. Assume HR(x) for all x in some nonempty ideal $X \subseteq W$ (i.e. if $w \in X$ and $x \leq w$, then $x \in X$). A minimal element of the complement $W \setminus X$ can be written xs for some $x \in X$ and $s \in S$, in which case $xs > x$. Now the inductive step can be broken into two parts:

Step 1: HR($< xs$) $\implies$ HR($\underline{x}, s$).
Step 2: (HR($< xs$) and HR($\underline{x}, s$)) $\implies$ (S(xs) and HR($\underline{x}, s$)) $\implies$ HR(xs).

18.3.1 Step 1

18.3.1.1 Deforming the Lefschetz Operator

We know of no method to obtain HR($\underline{x}$, s) using only L_0. The approach of Elias–Williamson, following that of de Cataldo–Migliorini in [40], is to deform L_0 in a one parameter family of operators of constant signature and first establish Hodge–Riemann for a generic operator. This "signature" argument relies crucially on the choice of the base field $\mathbb{R}$ (as does the very statement of the Hodge–Riemann relations).

For any $\zeta \geq 0$, let $L_\zeta : B_x B_s \to B_x B_s$ be the endomorphism defined by

$$L_\zeta = (\rho \cdot -) \otimes \mathrm{id}_{B_s} + \mathrm{id}_{B_x} \otimes (\zeta \rho \cdot -). \tag{18.12}$$

Thus L_ζ is left multiplication by ρ plus "middle multiplication" by $\zeta \rho$ (cf. Example 17.25). Since $L_\zeta|_{\zeta \mapsto 0} = L_0$, the notation is justified.

One can think about L_ζ as

$$L_\zeta = L_0 + \zeta M$$

where M is middle multiplication by ρ. When computing the Lefschetz form, we will need to apply the operator $L_\zeta^k = (L_0 + \zeta M)^k$ in degree $-k$. By the binomial theorem, this is a sum of various operators $L_0^{k-b} M^b$ with known coefficients. When b is nonzero, the action of $L_0^a M^b$ on $\overline{B_x B_s}$ is not hard to compute. For example, consider a pure tensor of the form $m \otimes n$ for $m \in B_x$ and $n \in B_s$ both homogeneous. If $n = c_{\mathrm{top}}$ then $fn = nf$ for any $f \in R$, so the action of M on $m \otimes n$ agrees with right multiplication by ρ. (See Chap. 12 for the definition of c_{bot} and c_{top}.) In particular, M kills the image of $m \otimes c_{\mathrm{top}}$ inside $\overline{B_x B_s}$. If $n = c_{\mathrm{bot}}$ then (modulo right multiplication) M must be used to break the s strand, sending (the image of) $m \otimes c_{\mathrm{bot}}$ to a positive scalar multiple of (the image of) $m \otimes c_{\mathrm{top}}$.

Exercise 18.23 As a sanity check, prove that $L_0^a M^b$ acts by zero on $\overline{B_x B_s}$ when $b \geq 2$.

Exercise 18.24 Show that for any $a \geq 0$, the action of $L_0^a M$ on $\overline{B_x B_s}$ agrees with the action of $L_0' \otimes M'$ on $\overline{B_x} \otimes \overline{B_s}$, where L_0' (resp. M') denotes the operator of left multiplication by ρ on $\overline{B_x}$ (resp. $\overline{B_s}$). (Roughly speaking, the action of $L_0^a M$ on $\overline{B_x B_s}$ "decouples" into a tensor product of operators.) Consequently, prove that

$$\langle L_0^a M(m \otimes n), m' \otimes n' \rangle_{\overline{B_x B_s}} = \langle L_0^a m, m' \rangle_{\overline{B_x}} \cdot \langle Mn, n' \rangle_{\overline{B_s}}. \tag{18.13}$$

(*Hint:* Check that $L_0^a M$ sends $m \otimes c_{\mathrm{top}}$ to zero, and sends $m \otimes c_{\mathrm{bot}}$ to $m' \otimes c_{\mathrm{top}}$, where m' is any lift of $L_0^a m \in \overline{B_x}$.)

As noted in Sect. 18.2.2.6, the induced form on $B_x B_s$ is more subtle than the tensor product of the forms on B_x and B_s independently. There is a simple contribution where elements of B_x of complimentary degree pair off to be a scalar, and elements of B_s pair off to be a scalar, and these scalars are multiplied. But there is also a contribution where elements of B_x pair off to become a linear polynomial on the right, which then breaks the final B_s strand; call these the "nasty linear terms." Because of nasty linear terms, the Lefschetz form of L_0^k on $\overline{B_x B_s}^{-k}$ is quite complicated. However, as noted in the previous exercises, there are no nasty linear terms for $L_0^{k-b} M^b$ when $b > 0$. As ζ gets larger and larger, the contribution of the nasty linear terms from L_0^k gets swamped by the massive contribution of $\zeta L_0^{k-1} M$. As a result, in the limit as $\zeta \to \infty$, the Lefschetz pairing on $\overline{B_x B_s}$ is determined by the Lefschetz pairings on $\overline{B_x}$ and $\overline{B_s}$ independently; these tensor factors get "decoupled." This heuristic argument is proven more carefully in Sect. 18.5.

We extend earlier notation as follows:

1. We write $\mathrm{hL}(x, s)_\zeta$ to mean $\mathrm{hL}(B_x B_s, L_\zeta)$.
2. We write $\mathrm{HR}(\underline{x}, s)_\zeta$ to mean $\mathrm{HR}(B_x B_s, \langle \cdot, \cdot \rangle_{\underline{xs}}|_{B_x B_s}, L_\zeta)$ holds for every embedding of B_x into $\mathrm{BS}(\underline{x})$.

18.3.1.2 Flowchart for Step 1

We can depict the logic of Step 1 as follows:

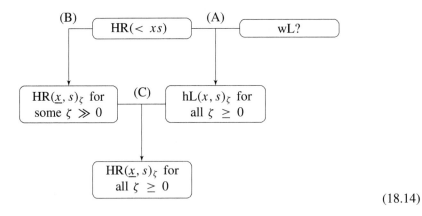

(18.14)

A: Here the argument involves deducing hard Lefschetz from Hodge–Riemann via a *weak-Lefschetz substitute* wL, as surveyed in Sect. 18.4 and further discussed in Chaps. 19–20.

B: This is the "decoupling argument," an explicit computation with Gram matrices to be found in Sect. 18.5.

18.3 Outline of the Proof

C: This is the following argument, which is specific to linear algebra over the real numbers. In the previous chapter, we saw (as Limit Lemma 17.4) the following fundamental fact from linear algebra:

$$\text{In a family of nondegenerate } \mathbb{R}\text{-valued symmetric forms,} \qquad (18.15)$$
$$\text{signature is constant.}$$

If $hL(x, s)_\zeta$ holds for all ζ, then the Lefschetz form corresponding to any value of ζ is nondegenerate, so (18.15) applies to the family $\{L_\zeta\}_{\zeta \geq 0}$. If $HR(\underline{x}, s)_\zeta$ holds for a particular ζ, then the Lefschetz form for that value of ζ has the right sign on each primitive subspace, so by (18.15), the same is true for every other ζ. One can compute the signature for very large ζ, so this argument goes through.

Thus we are able to deduce $HR(\underline{x}, s)_0 = HR(\underline{x}, s)$, as desired.

18.3.2 Step 2

18.3.2.1 The Local Intersection Form

To prove $HR(xs)$, we first prove $S(xs)$. We reduce $S(xs)$ to showing that for all $y < xs$, a certain symmetric form $(\cdot, \cdot)_y^{x,s}$ on $\mathrm{Hom}^0(B_y, B_x B_s)$ is nondegenerate.

Lemma 18.25 *Assume $HR(x)$ and $HR(y)$. Then there is an isomorphism*

$$\mathrm{Hom}^0(B_y, B_x B_s) \simeq \mathrm{Hom}^0(B_x B_s, B_y), \qquad (18.16)$$

under which $\phi : B_y \to B_x B_s$ corresponds to the adjoint $\phi^* : B_x B_s \to B_y$ defined by

$$\langle \phi(b_1), b_2 \rangle_{B_x B_s} = \langle b_1, \phi^*(b_2) \rangle_y \qquad (18.17)$$

for all $b_1 \in B_y$ and $b_2 \in B_x B_s$. Here $\langle \cdot, \cdot \rangle_{B_x B_s}$ is the form on $B_x B_s$ induced from $\langle \cdot, \cdot \rangle_x$.

Remark 18.26 The diagrammatic way to understand this isomorphism is the symmetry between the light leaves basis of $\mathrm{Hom}^0(B_x B_s, B_y)$ (in the quotient of the Hecke category by elements less than y) and the (upside-down) "light leaves basis" of $\mathrm{Hom}^0(B_y, B_x B_s)$, see Sect. 10.4.

Proof This follows from the nondegeneracy of both $\langle \cdot, \cdot \rangle_y$ and $\langle \cdot, \cdot \rangle_{B_x B_s}$. We get the former from $HR(y)$. We get the latter from $HR(x)$ together with the fact that a form induced from a nondegenerate form remains nondegenerate. See [57, Lemma 3.8] for details. □

Suppose $y < xs$. Recall that if $S(y)$ holds, then $\underline{\mathrm{rk}}\,\mathrm{End}^0(B_y) = 1$ by Corollary 18.4. In this case, the symmetric bilinear form

$$(\cdot,\cdot)_y^{x,s} = \mathrm{Hom}^0(B_y, B_x B_s) \times \mathrm{Hom}^0(B_y, B_x B_s) \to \mathrm{End}(B_y) \qquad (18.18)$$

that sends $(\phi, \psi) \mapsto \psi^* \circ \phi$ is called (by slight abuse of language) the *local intersection form on $B_x B_s$ at B_y* (or *at y*). In the flag variety setting, it arises from the local intersection form on the Schubert variety of xs along the stratum formed by the Schubert variety of y.

Lemma 18.27 *Assume $\mathrm{HR}(< xs)$. Then $S(xs)$ holds if and only if $(\cdot,\cdot)_y^{x,s}$ induces a nondegenerate $\mathbb{R}$-bilinear form for every $y < xs$.*

Proof Write

$$b_x b_s = b_{xs} + \sum_{y < xs} \mu(y, x; s) b_y, \qquad \mu(y, x; s) \in \mathbb{Z}, \qquad (18.19)$$

as in the inductive construction of the Kazhdan–Lusztig basis (see Sect. 3.3.2). For $y < xs$, the rank of $(\cdot,\cdot)_y^{x,s}$ equals the multiplicity of B_y as a summand of $B_x B_s$ (see Corollary 11.75), whereas by the Soergel Hom formula and $S(< xs)$, the rank of $\mathrm{Hom}^0(B_y, B_x B_s)$ equals $\langle b_y, b_x b_s \rangle = \mu(y, x; s)$. These ranks are the same for all y if and only if

$$B_x B_s \simeq B_{xs} \oplus \bigoplus_{y < xs} B_y^{\oplus \mu(y, x; s)}, \qquad (18.20)$$

which is equivalent (given $S(< xs)$) to $S(\leq xs)$. $\square$

18.3.2.2 $\mathrm{HR}(\underline{x}, s)$ Versus $\mathrm{HR}(\underline{x}s)$

Note that $\mathrm{HR}(\underline{x}, s)$ is the statement that $B_x B_s \stackrel{\oplus}{\subset} \mathrm{BS}(\underline{x}s)$ has the Hodge–Riemann bilinear relations for L_0, while $\mathrm{HR}(\underline{x}s)$ is the statement that the further summand $B_{xs} \stackrel{\oplus}{\subset} B_x B_s \stackrel{\oplus}{\subset} \mathrm{BS}(\underline{x}s)$ has the Hodge–Riemann bilinear relations for L_0.

Lemma 18.28 $\mathrm{HR}(\underline{x}, s)$ *and* $S(xs)$ *implies* $\mathrm{HR}(\underline{x}s)$.

Proof The restriction of the form on $B_x B_s$ to B_{xs} is nonzero (again, by the same argument as Exercise 18.15 or Exercise 18.16). Thus by $S(xs)$, the restricted form is nondegenerate. Since B_{xs} is an R-bimodule summand of $B_x B_s$, it is preserved by L_0. Now Exercise 17.19, applied to $\overline{B_{xs}} \subset \overline{B_x B_s}$, yields $\mathrm{HR}(\underline{x}s)$. $\square$

18.3.2.3 Flowchart for Step 2

We depict Step 2 as follows:

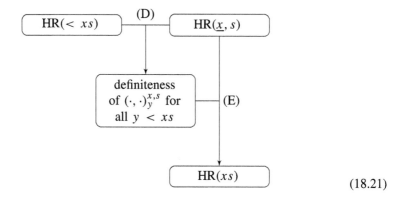

(18.21)

D: This step involves embedding $(\cdot, \cdot)_y^{x,s}$, up to scaling, within the (global) Lefschetz form on the primitive subspace of $(\overline{B_x B_s})^{-\ell(y)}$, which is definite by HR($\underline{x}$, s). This is the embedding theorem of Sect. 18.6.

E: This is the following argument. By Lemma 18.27, the nondegeneracy of $(\cdot, \cdot)_y^{x,s}$ for all $y < xs$ gives us S(xs). Lemma 18.28 gives HR($\underline{xs}$). Finally, S(xs) and HR($\underline{xs}$) together imply HR(xs), because End(B_{xs}) being of rank 1 implies the restriction of $\langle \cdot, \cdot \rangle_{\underline{xs}}$ to B_{xs} is just a (positive) rescaling of $\langle \cdot, \cdot \rangle_{xs}$. See also [57, Lemma 3.11] or Corollary 18.18.

18.4 The Weak Lefschetz Problem

The weak Lefschetz theorem in complex algebraic geometry states the following.

Theorem 18.29 (Weak Lefschetz) *Let X a smooth complex projective variety in $\mathbb{P}^n$ of dimension d and let $H \subseteq \mathbb{P}^n$ be a generic hyperplane. Then the pullback map*

$$H^i(X) \to H^i(X \cap H) \quad (18.22)$$

is injective for $i = d - 1$ and an isomorphism for $i < d - 1$.

The space $X \cap H$ is often called a *generic hyperplane section of X*. Because X and $X \cap H$ are both projective, one can also push forward cohomology classes along the inclusion $X \cap H \hookrightarrow X$ as follows: first apply Poincaré duality, then pushforward in homology, and then apply Poincaré duality again. Because the complex dimensions of $X \cap H$ and X differ by one, the "mirror" of Poincaré duality is different, and the pushforward will give a map $H^i(X \cap H) \to H^{i+2}(X)$. The weak Lefschetz theorem also implies that the pushforward map is an isomorphism for $i > d - 1$, and is surjective for $i = d - 1$. There is some section of some ample line bundle for which H is the zero set, and the composition $H^i(X) \to H^i(X \cap H) \to H^{i+2}(X)$ is equal to the cup product with the first Chern class of this line bundle; this is a Lefschetz operator L_H which is supposed to satisfy hard Lefschetz and the Hodge–Riemann bilinear relations.

The importance of the weak Lefschetz theorem is that it can be used to deduce the hard Lefschetz theorem for varieties of dimension d from the Hodge–Riemann bilinear relations for varieties of dimension $d - 1$. The key idea is to factor the Lefschetz operator L_H through the cohomology of the lower-dimensional space $X \cap H$. Then we can apply Lemma 17.21, an exercise in the abstract setting of Lefschetz linear algebra. (For more on this use of weak Lefschetz in geometry, see the appendix to Chap. 20.)

In our outline, Step 1(A) will emulate this idea from geometry. In the setting of Soergel bimodules, however, there is no reasonable analogue of a "generic hyperplane section." In order to circumvent this, Elias and Williamson use the first differential on a Rouquier complex as a substitute for (18.22).

Let us elaborate. Just as the classical Lefschetz operator L_H associated with H factors as

$$H^*(X) \to H^*(X \cap H) \to H^*(X)[2], \tag{18.23}$$

so can we factor the Lefschetz operator L_ζ on a Bott–Samelson module $\overline{\mathrm{BS}(\underline{w})}$ as

$$\overline{\mathrm{BS}(\underline{w})} \xrightarrow{\Omega} \bigoplus_i \overline{\mathrm{BS}(\underline{w_{\hat{i}}})}(1) \to \overline{\mathrm{BS}(\underline{w})}(2), \tag{18.24}$$

where $\underline{w_{\hat{i}}}$ is the expression formed by omitting the i-th element from $\underline{w}$. The map Ω occupies the same role as (18.22). For each word $\underline{w}$ there is an associated complex of Soergel bimodules called a Rouquier complex. The first differential in the Rouquier complex, up to some rescaling, agrees with the map Ω. The fact that the map in (18.22) is injective for $i = d - 1$ and an isomorphism for $i < d - 1$ is replaced by some analogous properties of Ω, which are deduced from the homological properties of Rouquier complexes. Thus the remarkable homological properties of Rouquier complexes serve as a satisfactory replacement for the weak Lefschetz theorem. This argument will be called the *weak Lefschetz substitute*.

Rouquier complexes are introduced in Chap. 19, and the weak Lefschetz substitute is the focus of Chap. 20. In the meantime, we just note that this weak Lefschetz substitute is the tool that gives us $\mathrm{hL}(x, s)_\zeta$ for all $\zeta \geq 0$.

18.5 From $\zeta = 0$ to $\zeta \gg 0$

In our outline, Step 1(B) follows from showing:

Theorem 18.30 *Let B be a summand of $\mathrm{BS}(\underline{x})$. If $\mathrm{HR}(B, \langle \cdot, \cdot \rangle_{\underline{x}}|_B, L_0)$ holds, then $\mathrm{HR}(BB_s, \langle \cdot, \cdot \rangle_{\underline{xs}}|_{BB_s}, L_\zeta)$ holds for all ζ large enough.*

Lemma 18.31 *Let V and W be finite-dimensional graded vector spaces equipped with respective Lefschetz data $(\langle \cdot, \cdot \rangle_V, L_V)$ and $(\langle \cdot, \cdot \rangle_W, L_W)$. Assume that:*

1. *W is only nonzero in even degrees, or W is only nonzero in odd degrees.*
2. *$\underline{\mathrm{rk}}\, V = (v + v^{-1})\,\underline{\mathrm{rk}}\, W.$*

If, for all $i \geq 0$, the signature of $(\cdot, \cdot)_{L_W}$ on $P_{L_W}^{-i+1}$ equals the signature of $(\cdot, \cdot)_{L_V}$ on all of V^{-i}, where we set $P_{L_W}^1 = 0$, then $\mathrm{HR}(V, \langle \cdot, \cdot \rangle_V, L_V)$ and $\mathrm{HR}(W, \langle \cdot, \cdot \rangle_W, L_W)$ are equivalent.

Exercise 18.32 Prove Lemma 18.31.

Example 18.33 Suppose that

$$W = W^{-2} \oplus W^0 \oplus W^2 \quad \text{and} \quad V = V^{-3} \oplus V^{-1} \oplus V^1 \oplus V^3. \tag{18.25}$$

Writing $p_{-i} = \dim P_{L_V}^{-i}$ and $q_{-i} = \dim P_{L_W}^{-i}$, we see that:

1. Hodge–Riemann holds for V if and only if $(\cdot, \cdot)_{L_V}$ has signature p_{-3} on V^{-3} and signature $p_{-3} - p_{-1}$ on V^{-1}.
2. Hodge–Riemann holds for W if and only if $(\cdot, \cdot)_{L_W}$ has signature q_{-2} on $P_{L_W}^{-2}$ and signature $-q_0$ on $P_{L_W}^0$ (and thus signature $q_{-2} - q_0$ on W^0).

The condition on graded ranks says that

$$p_{-3} = \dim V^{-3} = \dim W^{-2} = q_{-2}, \tag{18.26}$$

$$p_{-1} + p_{-3} = \dim V^{-1} = \dim W^0 + \dim W^{-2} = (q_0 + q_{-2}) + q_{-2}. \tag{18.27}$$

Thus $p_{-3} = q_{-2}$ and $p_{-3} - p_{-1} = -q_0$, which verifies the prediction of the lemma.

Proof *(Proof of Theorem 18.30)* We abbreviate the forms on B and BB_s to $\langle \cdot, \cdot \rangle_B$ and $\langle \cdot, \cdot \rangle_{BB_s}$, respectively.

By the lemma, we must show that the signature of $(\cdot, \cdot)_{L_0}^{-i+1}$ on $P_{L_0}^{-i+1} \subseteq \overline{B}^{-i+1}$ equals that of $(\cdot, \cdot)_{L_\zeta}^{-i}$ on all of $(\overline{BB_s})^{-i}$. As in Sect. 18.2.2.6, we will express everything in terms of explicit bases[3] Lift an (ordered) orthogonal basis of $\overline{B}^{-i-1}$

---

[3] All the bases discussed in this proof are technically ordered bases, but we found the notation for ordered sets to be distracting. The reader should be able to guess the ordering; we'll give a hint in the next footnote.

to a right R-basis $\{e_k\}$ of B^{-i-1}, and lift an (ordered) orthogonal basis of $P_{L_0}^{-i+1}$ to a basis $\{p_j\}$ of B^{-i+1}. Then:

1. $\{\rho e_k, p_j\}$ projects to an orthogonal basis of $\overline{B}^{-i+1}$.
2. $\{\xi_{\text{bot}}(\rho e_k), \xi_s(e_k), \xi_{\text{bot}}(p_j)\}$ projects to a basis[4] of $(\overline{BB_s})^{-i}$.

Consider the Gram matrix of the form $(\cdot, \cdot)_{L_\zeta}^{-i}$ on $(\overline{BB_s})^{-i}$ with respect to the basis in 2. Looking back at (18.10), one sees that the entries that involve the elements $\xi_s(e_i)_i$ are easier to handle, essentially because α_s vanishes when we base-change from R to $\mathbb{R}$. Ultimately, the Gram matrix of $(\cdot, \cdot)_{L_\zeta}^{-i}$ takes the form

$$M_\zeta = \begin{pmatrix} * & J & * \\ J & 0 & 0 \\ * & 0 & Q_\zeta \end{pmatrix} \qquad (18.28)$$

for some Q_ζ and nondegenerate diagonal matrix J. (See [57, p.1116] for details.) By inspection, M_ζ is nonsingular if and only if Q_ζ is nonsingular, and when this is the case, we can find a path in the space of nondegenerate symmetric matrices from M_ζ to

$$\begin{pmatrix} & J & \\ J & & \\ & & Q_\zeta \end{pmatrix}. \qquad (18.29)$$

Invoking (18.15) once more, we deduce that in this case, the signatures of M_ζ and Q_ζ coincide.

It remains to show that if $\zeta \gg 0$, then Q_ζ is nondegenerate of the same signature as the Gram matrix Q_0 of $(\cdot, \cdot)_{L_0}^{-i+1}$ with respect to $\{p_j\}$. We can assume $i > 0$. Writing $\text{Tr}_\mathbb{R}$ for the composition of Tr with the reduction $R \to R/R_+ = \mathbb{R}$, we compute

$$(\phi_{\text{bot}}(b_1), \phi_{\text{bot}}(b_2))_{L_\zeta}^{-i} = \text{Tr}_\mathbb{R}\left(\sum_{m=0}^i \binom{i}{m} \rho^{i-m}(b_1 b_2)(\zeta \rho)^m c_{\text{bot}}\right)$$
$$= \text{Tr}_\mathbb{R}(\rho^i (b_1 b_2) c_{\text{bot}}) + \zeta i \langle \rho, \alpha_s^\vee \rangle \text{Tr}_\mathbb{R}(\rho^{i-1}(b_1 b_2))$$
$$= \text{Tr}_\mathbb{R}(\rho^i (b_1 b_2) c_{\text{bot}}) + \zeta i \langle \rho, \alpha_s^\vee \rangle (b_1, b_2)_{L_0}^{-i+1}.$$
$$(18.30)$$

---

[4]The ordering on this basis has all the vectors $\xi_{\text{bot}}(\rho e_k)$ first, for various k, then the vectors $\xi_s(e_k)$ for various k, then the vectors $\xi_{\text{bot}}(p_j)$ for various j.

(The reduction to $\mathbb{R}$ is the reason why the terms with $m \geq 2$ vanish.) We conclude that $(1/\zeta)Q_\zeta \to i\langle \rho, \alpha_s^\vee\rangle Q_0$ as $\zeta \to \infty$. Note that $i\langle \rho, \alpha_s^\vee\rangle$ is a positive scalar. In particular, the signatures of Q_ζ and Q_0 match once ζ is sufficiently large. $\square$

18.6 From Local to Global Intersection Forms

Let $y < xs$ and fix a generator c_{bot} of $B_y^{-\ell(y)}$. Let

$$\iota = \iota_y^{x,s} : \text{Hom}(B_y, B_x B_s) \to (\overline{B_x B_s})^{-\ell(y)} \tag{18.31}$$

be the map that sends $\phi \mapsto \overline{\phi(c_{\text{bot}})}$. In our outline, Step 2(D) is completed by the following result.

Theorem 18.34 (Embedding) *The image of ι lands in $P_{L_0}^{-\ell(y)}$. If $\text{HR}(< xs)$ and $\text{HR}(x, s)$ hold, then ι is injective and satisfies*

$$(\iota(\phi_1), \iota(\phi_2))_{L_0}^{-\ell(y)} = N(\phi_1, \phi_2)_y^{x,s}, \tag{18.32}$$

where the scaling factor $N = \langle \rho^{\ell(y)} c_{\text{bot}}, c_{\text{bot}}\rangle_y$ is positive.

Proof The first claim holds because $\rho^{\ell(y)+1}$ annihilates $\overline{B_y}$ and ι commutes with ρ.

To show ι is injective: By precomposing with $\text{BS}(y) \to B_y$ and postcomposing with $\overline{B_x B_s} \to \overline{\text{BS}(xs)}$, we extend ϕ to a degree-0 morphism $\text{BS}(y) \to \overline{\text{BS}(xs)}$, which can be written as some linear combination of double light leaves. If $\overline{\phi(c_{\text{bot}})} = 0$, then every double light leaf that appears must factor through some $z < y$. We now argue that $\phi = 0$ using a "seesaw" argument:

1. From $S(y)$, we know $\text{ch}(B_y) \in \delta_y + \sum_{z<y} v\mathbb{Z}_{\geq 0}[v]\delta_z$. So the lower half of the double light leaf must have positive degree.
2. From $S(x)$, we know $\text{ch}(B_x B_s) \in b_x b_s + \sum_z \mathbb{Z}_{\geq 0}[v]\delta_z$. So the upper half of the double light leaf must have non-negative degree.

If the total degree of the double light leaf is 0, then these conditions cannot hold simultaneously. So $\phi = 0$ as desired.

For (18.32) we have:

$$(\iota(\phi_1), \iota(\phi_2))_{L_0}^{-\ell(y)} = \langle \rho^{\ell(y)} \overline{\phi_1(c_{\text{bot}})}, \overline{\phi_2(c_{\text{bot}})}\rangle_{\overline{B_x B_s}}$$
$$= \langle \rho^{\ell(y)} \phi_2^* \phi_1(c_{\text{bot}}), c_{\text{bot}}\rangle_y \tag{18.33}$$
$$= (\phi_1, \psi_2)_y^{x,s} \langle \rho^\ell(y) c_{\text{bot}}, c_{\text{bot}}\rangle_y$$

by definition.

Finally, the positivity result $\langle \rho^\ell(y) c_{\text{bot}}, c_{\text{bot}}\rangle_y > 0$ is the content of Proposition 12.29. $\square$

Exercise 18.35 In Exercise 11.80, several local intersection forms in the Soergel bimodule category were computed. Do the signatures agree with the Embedding Theorem 18.34 and the Hodge–Riemann bilinear relations?

Exercise 18.36 In Exercises 11.66 and 11.79, several local intersection forms in the Temperley–Lieb category were computed. How can you explain the signature of those forms using the Hodge theory of Soergel bimodules? You might also compute the following trivial intersection forms to help organize your thoughts: the local intersection forms on Hom(5, 5) and Hom(4, 4) (these are 1×1 matrices).

18.7 Hodge Theory of Matroids

Since the early 2000s, there have been many results that use analogues[5] of Hodge theory to solve combinatorial problems that were intractable without tools motivated by—but not making explicit use of—algebraic geometry [8, 15, 34, 57, 98, 142]. We highlight the most recent of these.

In a similar vein to Elias and Williamson's proof of Soergel's conjecture, Adiprasito, Huh, and Katz developed in [8] a Hodge theory for *matroids* that allowed them to prove Rota's log-concavity conjecture:

Conjecture 18.37 (Rota) If $\chi(t) = \sum_{i=0}^{r} w_{r-i} t^i$, where $w_0 = 1$, is the characteristic polynomial of an arbitrary matroid, then the sequence $\{w_i\}_i$ is log-concave:

$$w_i^2 \geq w_{i-1} w_{i+1} \qquad (18.34)$$

for $1 \leq w_i \leq r - 1$.

Like Elias–Williamson, they took inspiration from the work of de Cataldo and Migliorini on the decomposition theorem [39, 40]. The main theorem of Adiprasito–Huh–Katz is the following. For basic facts about matroids, including the definition of the Chow ring, we refer the reader to the excellent survey paper [13].

Theorem 18.38 (Hodge Theory of Matroids) *Let M be a matroid of rank $d + 1$. Let $A_{\mathbb{R}}^*$ be its Chow ring and let $\ell \in A_{\mathbb{R}}^1$ be ample. Then for $0 \leq i \leq d/2$, the following hold:*

1. Poincaré duality. *Multiplication is a perfect pairing*

$$A_{\mathbb{R}}^i \times A_{\mathbb{R}}^{d-i} \to A_{\mathbb{R}}^d \simeq \mathbb{R}. \qquad (18.35)$$

2. Hard Lefschetz. *Multiplication by ℓ^{d-2i} induces an isomorphism $A_{\mathbb{R}}^i \to A_{\mathbb{R}}^{d-i}$.*

---

[5] We should also mention the earlier results of [35, 168] which use classical Hodge theory to attack problems in combinatorics.

18.7 Hodge Theory of Matroids

3. Hodge–Riemann bilinear relations. *The symmetric bilinear form*

$$A_{\mathbb{R}}^i \times A_{\mathbb{R}}^i \to \mathbb{R} \qquad (18.36)$$

defined by $(a, b) \mapsto (-1)^i a \cdot \ell^i b$ *is positive definite on the kernel of* ℓ^{i+1}.

Let us point out certain parallels between the setting and techniques of Elias–Williamson and those of Adiprasito–Huh–Katz. The Bott–Samelson bimodule BS($\underline{x}$) for an expression $\underline{x}$ is analogous to the Bergman fan Σ_M of a matroid M. Taking subexpressions of $\underline{x}$ is analogous to contracting edges in the fan. In the course of the proof of Theorem 18.38, the authors leave the world of matroids by performing a sequence of flips on the Bergman fan. The intermediary objects are not actually Bergman fans of any matroid, although they are still balanced polyhedral fans. This can be viewed as being loosely analogous to Elias–Williamson's use of Rouquier complexes as a weak Lefschetz substitute, which leaves the world of Bott–Samelson bimodules. In both cases, the role of the intermediary objects is to supply a replacement for the target of the weak-Lefschetz map, i.e. for the geometric operation of taking a generic hyperplane section.

We emphasize that both of these results were motivated by classical algebraic geometry, yet neither of the proofs uses algebraic geometry itself. Without knowledge of the underlying geometric principles, though, it would be very difficult to motivate the techniques used in either the Soergel-bimodule setting or the matroid setting.

To end on a philosophical note, we emphasize the uncanny tendency of Poincaré duality, hard and weak Lefschetz, and Hodge–Riemann to come bundled together. This set of properties, having first been discovered in the context of compact Kähler manifolds, is often called the *Kähler package*.

Chapter 19
Rouquier Complexes and Homological Algebra

This chapter is based on expanded notes of a lecture given by the authors and taken by
Charles Cain and Ulrich Thiel

Abstract In the previous chapter, an outline of the proof of Hodge theory for Soergel bimodules was given. A key issue is the absence of the weak Lefschetz theorem. In the proof of Elias and Williamson, a substitute for this theorem is provided by the homological algebra of Rouquier complexes—remarkable complexes of Soergel bimodules which categorify the braid group. In this chapter, we briefly describe the underlying idea, review some necessary bits of homological algebra, and introduce Rouquier complexes formally. We also discuss the "diagonal miracle," which is a key step in establishing the hard Lefschetz theorem for Soergel bimodules.

19.1 Motivation

Suppose that $X \subset \mathbb{P}^N \mathbb{C}$ is a smooth projective variety. A *hyperplane section* of X is the intersection $X_H = X \cap H$ of X with a hyperplane $H \subset \mathbb{P}^N \mathbb{C}$. For generic hyperplanes, X_H is smooth and its complex dimension is one less than that of X. A recurring theme in algebraic geometry is to study X by means of its hyperplane sections. The following theorem, sometimes called the *weak Lefschetz theorem*, tells us that the topology of X is controlled to a large extent by the topology of its hyperplane sections.

C. Cain
University at Buffalo SUNY, Buffalo, NY, USA

U. Thiel
Department of Mathematics, University of Kaiserslautern, Kaiserslautern, Germany

© The Editor(s) (if applicable) and The Author(s), under exclusive licence to Springer Nature Switzerland AG 2020
B. Elias et al., *Introduction to Soergel Bimodules*, RSME Springer Series 5, https://doi.org/10.1007/978-3-030-48826-0_19

Theorem 19.1 (Lefschetz Hyperplane Theorem) *Let X be a smooth projective variety as above. For any hyperplane H the map induced by restriction*

$$H^k(X, \mathbb{Q}) \to H^k(X_H, \mathbb{Q})$$

is an isomorphism for $k < \dim_{\mathbb{C}} X - 1$ and is injective for $k = \dim_{\mathbb{C}} X - 1$.

There is a standard technique (see e.g. [183, §2.4]) for deducing the hard Lefschetz theorem for $H^*(X)$ from hard Lefschetz and the Hodge–Riemann bilinear relations for $H^*(X_H)$, where X_H denotes a smooth hyperplane section. This technique is essential to de Cataldo and Migliorini's Hodge-theoretic proof of the decomposition theorem [39, 40, 183].

As discussed in the previous chapter, the ideas of de Cataldo and Migliorini provided essential motivation for Elias and Williamson's proof of Soergel's conjecture. Whereas de Cataldo and Migliorini's idea to "deform the Lefschetz operator" has a transparent translation into the world of Soergel bimodules, the passage to a generic hyperplane section does not. Roughly speaking, the world of Soergel bimodules is rigid and combinatorial, and cannot accommodate generic objects like smooth hyperplane sections.

Let us try to be more precise. Suppose that W is a Weyl group and $\underline{w} = (s, t, \ldots, u)$ is an expression. We saw in Example 16.16 how to associate a smooth variety $Y(\underline{w})$ (the *Bott–Samelson variety*) to $\underline{w}$. This is a projective variety (in type A this follows from Exercises 16.17 and 16.18). A generic hyperplane section in $Y(\underline{w})$, for some choice of embedding, is not isomorphic to a Bott–Samelson variety—it is rather complicated. Here is a schematic picture (think of a cubic curve in $\mathbb{P}^2\mathbb{C}$):

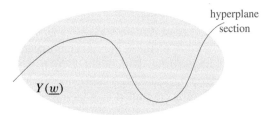

In particular, it is unreasonable to expect a Soergel bimodule description of a generic hyperplane section of a Bott–Samelson variety.

On the other hand, for any $1 \leq i \leq \ell(\underline{w})$ there is a smooth codimension one subvariety $Y(\underline{w})_i$ inside the Bott–Samelson variety $Y(\underline{w})$.[1] This subvariety is isomorphic to a Bott–Samelson variety $Y(\underline{x})$ for the expression obtained by omitting the ith term of $\underline{w}$, and thus its cohomology has a combinatorial description. Moreover, the restriction map on cohomology is easily understood—it is given by a "dot" map. Unfortunately, this subvariety is very special, and does not see much

---

[1] In the notation of Exercise 16.16, $Y(\underline{w})_i$ is given by replacing P_{s_i} by B in the product $P_{s_1} \times_B P_{s_2} \times_B \cdots \times_B P_{s_n}$.

19.1 Motivation

of the global topology of $Y(\underline{w})$. In particular no analogue of the weak Lefschetz theorem holds for restriction to the cohomology of this subvariety. We again give a schematic picture (think of a coordinate $\mathbb{P}^1\mathbb{C}$ inside $\mathbb{P}^2\mathbb{C}$):

However, if one adds up the cohomology groups of each $Y(\underline{w})_i$ over all $1 \le i \le \ell(\underline{w})$ we obtain a map

$$\overline{\mathrm{BS}(\underline{w})} \xrightarrow{\Omega} \bigoplus_i \overline{\mathrm{BS}(\underline{w}_{\hat{i}})}(1)$$

This is the map considered in (18.24). The analogue of the weak Lefschetz theorem is the fact that this map is injective in degrees below $\ell(\underline{w})$.[2] Again we give a schematic picture (think of the union of the three coordinate hyperplanes inside $\mathbb{P}^2\mathbb{C}$):

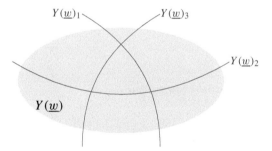

It is useful to think of the union of the subvarieties $Y(\underline{w})_i$ over all $1 \le i \le \ell(w)$ as something like a combinatorial version of a generic hyperplane section.

In order to adapt the standard technique alluded to above to prove the hard Lefschetz theorem for the indecomposable Soergel bimodule $\overline{B_w}$, one needs to understand the map Ω. The key observations are the following:

1. Ω is the first differential on a *Rouquier complex*;
2. one may use techniques of homological algebra (*minimal complexes*) to strip away many unnecessary terms from Ω.

---

[2]Notice that Ω is both worse and better than the restriction map occurring in the weak Lefschetz theorem: it is an isomorphism only in degree $-\ell(\underline{w})$ (worse), but is injective in more degrees (better).

These steps (in particular the second) demand technology from homological algebra. Roughly speaking, the missing generic hyperplane sections are made up for by a good dose of homological algebra. It is the goal of this chapter to introduce these techniques.

19.2 Some Homological Algebra

Rouquier complexes live in the homotopy category of Soergel bimodules. This section collects the homological algebra needed for their study: complexes and homotopies (Sect. 19.2.1), Gaussian elimination and minimal complexes (Sect. 19.2.2), and the Grothendieck group of the homotopy category (Sect. 19.2.3).

While the homological preliminary of this section will suffice for the discussion of Rouquier complexes in the rest of this chapter, some of the results (in particular in Sects. 19.5–19.6) are best understood in terms of more advanced homological algebra (triangulated categories and t-structures), which we briefly discuss in the appendix to this chapter. These topics will become more important in Part V of this book. In particular, it may help to know a little about triangulated categories in Chap. 23, and it will be essential for Chaps. 25–26.

Throughout this section, we work with an abstract additive category $\mathcal{A}$. The main example we have in mind is $\mathcal{A} = \mathbb{S}\text{Bim}$. For this reason, we will sometimes discuss what happens when $\mathcal{A}$ is monoidal or has a "grading shift" autoequivalence (1). When we discuss the Grothendieck group of $\mathcal{A}$, we assume that $\mathcal{A}$ is essentially small.

19.2.1 Complexes and Homotopies

Let $\mathcal{A}$ be an additive category. Let $C(\mathcal{A})$ be the category of (cochain) complexes in $\mathcal{A}$. An object in $C(\mathcal{A})$ is denoted by $(A^i, d_A^i)_{i \in \mathbb{Z}}$, where each A^i is an object of $\mathcal{A}$, and $d_A^i : A^i \to A^{i+1}$ are morphisms in $\mathcal{A}$ satisfying the familiar relations $d_A^{i+1} \circ d_A^i = 0$ for all $i \in \mathbb{Z}$. We often shorten this data to (A, d_A), or to just A. Given two complexes $A, B \in C(\mathcal{A})$, a morphism $f : A \to B$ in $C(\mathcal{A})$ consists of morphisms $(f^i : A^i \to B^i)_{i \in \mathbb{Z}}$ in $\mathcal{A}$ such that $d_B^i \circ f^i = f^{i+1} \circ d_A^i$ for all $i \in \mathbb{Z}$. We identify $\mathcal{A}$ with the full subcategory of $C(\mathcal{A})$ formed by the complexes concentrated in cohomological degree zero.

On $C(\mathcal{A})$ we have a *cohomological shift* autoequivalence [1] defined by

$$(A[1])^i := A^{i+1}, \quad d_{A[1]}^i := -d_A^{i+1} \tag{19.1}$$

for a complex A and by

$$(f[1])^i := f^{i+1} \tag{19.2}$$

19.2 Some Homological Algebra

for a morphism $f\colon A \to B$. If we draw a complex horizontally with differentials going from left to right, then [1] shifts the complex to the *left*.

A *homotopy* between a pair of morphisms $f, g\colon A \to B$ of complexes is a family $h = (h^i)_{i \in \mathbb{Z}}$ of morphisms $h^i\colon A^i \to B^{i-1}$ such that

$$f^i - g^i = h^{i+1} \circ d^i + d^{i-1} \circ h^i \tag{19.3}$$

for all $i \in \mathbb{Z}$. Here is a diagram which helps one remember this relation.

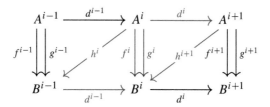

If a homotopy h between f and g exists, then f and g are called *homotopic* and we write $f \sim g$. This is clearly an equivalence relation on $\mathrm{Hom}_{C(\mathcal{A})}(A, B)$, whose equivalence classes are sometimes called *homotopy classes*. A *homotopy equivalence* is a morphism $f\colon A \to B$ of complexes for which there is a morphism $g\colon B \to A$ such that $f \circ g \sim \mathrm{id}_B$ and $g \circ f \sim \mathrm{id}_A$.

Exercise 19.2 Let $f, g\colon A \to B$ be homotopic morphisms. For any morphisms $i\colon A' \to A$ and $j\colon B \to B'$, prove that $f \circ i \sim g \circ i$ and $j \circ f \sim j \circ g$. Deduce that one can define composition of homotopy classes.

The *homotopy category* of $\mathcal{A}$, denoted by $K(\mathcal{A})$, is the category whose objects are complexes in $\mathcal{A}$, and whose morphisms are homotopy classes. Thus homotopy equivalences become isomorphisms in $K(\mathcal{A})$. Given two complexes A and B concentrated in degree zero, any family of morphisms $(h^i\colon A^i \to B^{i-1})_{i \in \mathbb{Z}}$ is automatically 0, so distinct morphisms $f, g\colon A \to B$ are never homotopic. Hence the fully faithful functor $\mathcal{A} \hookrightarrow C(\mathcal{A})$ induces a fully faithful functor $\mathcal{A} \hookrightarrow K(\mathcal{A})$.

Given complexes A, B, we write $A \simeq B$ for homotopy equivalence (isomorphism in $K(\mathcal{A})$) and $A \cong B$ for isomorphism of complexes (isomorphism in $C(\mathcal{A})$). A complex A is called *contractible* if $A \simeq 0$.

Exercise 19.3

1. Show that a complex A is contractible if and only if id_A is homotopic to the zero morphism on A.
2. Show that a complex of the form

$$\cdots \to 0 \to X \xrightarrow{\varphi} Y \to 0 \to \cdots$$

is contractible if and only if φ is an isomorphism.

Remark 19.4 In fact, the ideal of nullhomotopic morphisms (morphisms homotopic to the zero map) is equal to the ideal of morphisms which factor through contractible complexes. So it is valid to think that the homotopy category is obtained from the category of complexes by setting to zero all contractible complexes.

Given a morphism of complexes $f: A \to B$, its *cone* is the complex $\text{Cone}(f)$ with terms

$$\text{Cone}(f) := A[1] \oplus B, \tag{19.4}$$

i.e.,

$$\text{Cone}(f)^n = A^{n+1} \oplus B^n, \tag{19.5}$$

and with differentials

$$d^n_{\text{Cone}(f)} := \begin{pmatrix} d^n_{A[1]} & 0 \\ f[1]^n & d^n_B \end{pmatrix} = \begin{pmatrix} -d^{n+1}_A & 0 \\ f^{n+1} & d^n_B \end{pmatrix} : A^{n+1} \oplus B^n \to A^{n+2} \oplus B^{n+1}. \tag{19.6}$$

Exercise 19.5 Show that there is a short exact sequence in $C(\mathcal{A})$ of the form

$$0 \longrightarrow B \xrightarrow{i} \text{Cone}(f) \xrightarrow{p} A[1] \xrightarrow{h} 0. \tag{19.7}$$

When does this short exact sequence split?

Exercise 19.6 Show that $\text{Cone}(\text{id}_A)$ is contractible for any complex A via the homotopy

$$h^n: A^{n+1} \oplus A^n \to A^n \oplus A^{n-1}, \quad (a, a') \mapsto (a', 0).$$

Exercise 19.7 This exercise mixes the previous two exercises. For a map $f: A \to B$ of complexes, let i be the inclusion $B \to \text{Cone}(f)$. Find a homotopy equivalence between $\text{Cone}(i)$ and $A[1]$.

Exercise 19.8 Show that a morphism of complexes $f: A \to B$ is a homotopy equivalence if and only if $\text{Cone}(f)$ is contractible.

Exercise 19.9 Let $\mathcal{A}$ be $\Bbbk$-linear for some commutative ring $\Bbbk$, and let $\zeta \in \Bbbk^\times$. For any morphism $f: A \to B$ of complexes, prove that $\text{Cone}(f)$ and $\text{Cone}(\zeta f)$ are isomorphic.

Exercise 19.10 Show that any complex with only two nonzero terms can be realized as the cone of a map between complexes, each with only one nonzero term. Show that any complex with only three nonzero terms can be realized as the cone of a map from a one-term complex to a two-term complex. Generalize this result to construct any bounded complex as an iterated cone of one-term complexes.

19.2 Some Homological Algebra

Let $C^-(\mathcal{A})$, $C^+(\mathcal{A})$, and $C^b(\mathcal{A})$ be the full subcategory of $C(\mathcal{A})$ consisting of bounded above, bounded below, and bounded complexes, respectively. Similarly, let $K^-(\mathcal{A})$, $K^+(\mathcal{A})$, and $K^b(\mathcal{A})$ be the corresponding full subcategories of $K(\mathcal{A})$. If $\mathcal{A}$ is monoidal, so are $C^b(\mathcal{A})$ and $K^b(\mathcal{A})$ via the tensor product

$$(A \otimes B)^n = \bigoplus_{i+j=n} A^i \otimes B^j, \quad d(a \otimes b) = d(a) \otimes b + (-1)^{|a|} a \otimes d(b), \quad (19.8)$$

where $a \in A^i$, $b \in A^j$, and $|a| := i$. The tensor product of morphisms of complexes is defined similarly as for objects. The sign $(-1)^{|a|}$ in (19.8) ensures that $d \circ d = 0$.[3] The tensor product can also be defined on $K^+(\mathcal{A})$ and $K^-(\mathcal{A})$, since the direct sums in (19.8) are finite.

Remark 19.11 The subject of homological algebra is plagued by a proliferation of signs. This seems to be an inevitable fact of life, forced upon us by the following consideration: for the differential of a cone to satisfy $d \circ d = 0$, one needs to insert a sign somewhere. Where this sign goes is a mere convention, but one that demands consistency to avoid sign nightmares.

Here is one way to understand the particular choice of sign made in (19.6), or equivalently, the one in cohomological shift (19.1). Assume that $\mathcal{A}$ is a strict monoidal category, and define the cohomological shift by $A[1] := \mathbb{1}[1] \otimes A$, where $\mathbb{1}[1]$ is the complex consisting of the monoidal identity $\mathbb{1}$ in cohomological degree -1. Then the sign on $d_{A[1]}$ comes from the sign in (19.8). Setting $A[1] := A \otimes \mathbb{1}[1]$ leads to another, nonstandard sign convention, where $d_{A[1]}$ receives no sign and $d_{\text{Cone}(f)}$ receives a sign somewhere else.

Assume as in Sect. 4.1 that $\mathcal{A}$ consists of graded objects and has a grading shift autoequivalence (1). A complex in $\mathcal{A}$ is then bigraded, and we can imagine it as being arranged horizontally by cohomological degree and vertically by the grading in $\mathcal{A}$, which we call the *internal degree*. As with any autoequivalence of $\mathcal{A}$, the grading shift (1) induces an autoequivalence on each of the categories of complexes considered above, defined by applying (1) to each term and differential of a complex. We continue to denote these autoequivalences by (1). Thus each of these categories has two autoequivalences, [1] and (1), shifting each complex to the left and down, respectively.

19.2.2 Gaussian Elimination and Minimal Complexes

In the following exercise we will learn about a basic but extremely useful tool which allows us to simplify complexes in the homotopy category. We will use it many

---

[3] This sign comes from the so-called *Koszul sign rule*: when "commuting d past a," we pick up the sign $(-1)^{|d||a|}$, which equals $(-1)^{|a|}$ since d has degree 1.

times throughout the rest of the book. Recall from Exercise 19.3 that a complex of the form

$$\cdots \longrightarrow 0 \longrightarrow X \xrightarrow[\simeq]{\varphi} Y \longrightarrow 0 \longrightarrow \cdots \tag{19.9}$$

is contractible.

Exercise 19.12 (Gaussian Elimination (see e.g. [14])) Let $\mathcal{A}$ be an additive category. Suppose that $C \in C(\mathcal{A})$ has the form

$$\cdots \longrightarrow C^{i-1} \xrightarrow{\begin{pmatrix} a \\ b \end{pmatrix}} \widetilde{C}^i \oplus X \xrightarrow{\begin{pmatrix} c & d \\ e & \varphi \end{pmatrix}} \widetilde{C}^{i+1} \oplus Y \xrightarrow{\begin{pmatrix} f & g \end{pmatrix}} C^{i+2} \longrightarrow \cdots \tag{19.10}$$

where $\varphi : X \to Y$ is an isomorphism in $\mathcal{A}$.

1. Show that $C \simeq \widetilde{C} \oplus Z$ in $C(\mathcal{A})$, where

$$\widetilde{C}: \quad \cdots \longrightarrow C^{i-1} \xrightarrow{a} \widetilde{C}^i \xrightarrow{c-d\varphi^{-1}e} \widetilde{C}^{i+1} \xrightarrow{f} C^{i+2} \longrightarrow \cdots$$

$$Z: \quad \cdots \longrightarrow 0 \longrightarrow X \xrightarrow{\varphi} Y \longrightarrow 0 \longrightarrow \cdots \tag{19.11}$$

2. Deduce that the inclusion $\widetilde{C} \hookrightarrow C$ is an isomorphism in $K(\mathcal{A})$.

(*Hint:* First show that C is isomorphic as a complex to the complex below.)

$$\cdots \longrightarrow C^{i-1} \xrightarrow{\begin{pmatrix} a \\ b+\varphi^{-1}ea \end{pmatrix}} \widetilde{C}^i \oplus X \xrightarrow{\begin{pmatrix} c-d\varphi^{-1}e & 0 \\ 0 & \varphi \end{pmatrix}} \widetilde{C}^{i+1} \oplus Y \xrightarrow{\begin{pmatrix} f & g+fd\varphi^{-1} \end{pmatrix}} C^{i+2} \longrightarrow \cdots \tag{19.12}$$

The process of replacing C with the homotopy equivalent complex $\widetilde{C}$ is called[4] *Gaussian elimination of complexes*. Some explicit exercises in Gaussian elimination for complexes of Soergel bimodules can be found later in this chapter. One implication of Gaussian elimination is that all bounded contractible complexes are boring, and iterated Gaussian elimination is sufficient to reduce them to the zero complex.

---

[4]The name comes from the hint in the previous exercise.

Lemma 19.13 *Suppose that $\mathcal{A}$ is Karoubian (see Sect. 11.3.4). Then a bounded complex $C \in C(\mathcal{A})$ is contractible if and only if it is isomorphic to a finite direct sum of complexes of the form* (19.9).

Proof A direct sum of contractible complexes is clearly contractible. Conversely, suppose that C is contractible. Let d denote the differential on C. Being contractible means that the identity on C is homotopic to the zero map. Hence, there are maps $h^i : C^i \to C^{i-1}$ such that $\mathrm{id}_{C^i} = h^{i+1} d^i + d^{i-1} h^i$. Since C is bounded, there is $j \in \mathbb{Z}$ minimal with $C^j \neq 0$. We then have $\mathrm{id}_{C^j} = h^{j+1} d^j$. Let $e := d^j h^{j+1}$. This is an idempotent because $e^2 = d^j h^{j+1} d^j h^{j+1} = d^j \,\mathrm{id}\, h^{j+1} = e$. Since $\mathcal{A}$ is Karoubian, idempotents split and therefore d^j is an isomorphism from C^j onto a direct summand Y of C^{j+1}. Gaussian elimination now allows us to remove the summand $\cdots \to 0 \to C^j \to Y \to \cdots$ from C and we get a complex whose lowest nonzero term is in degree $> j$. We continue with this process and by induction we conclude that C is a finite direct sum of complexes of the form (19.9). $\square$

Exercise 19.14 This exercise serves as a warning to the careless Gaussian eliminator. Let A be the complex

$$\cdots \longrightarrow 0 \longrightarrow X \oplus X \xrightarrow{\begin{pmatrix} \mathrm{id}_X & \mathrm{id}_X \\ \mathrm{id}_X & \mathrm{id}_X \end{pmatrix}} X \oplus X \longrightarrow 0 \longrightarrow \cdots \tag{19.13}$$

concentrated in degrees 0 and 1. The careless eliminator might say "the map from the first X to the first X is an isomorphism, so I can eliminate these two terms. The map from the second X to the second X is also an isomorphism, so I can eliminate these terms as well. The result is the zero complex, so A is contractible." However, you are far more careful than that! Apply Gaussian elimination on the isomorphism from the first X to the first X. What complex remains? Is it contractible?

Assuming that $\mathcal{A}$ is Karoubian it follows from Lemma 19.13 that we can use Gaussian elimination to remove any contractible direct summand from a bounded complex. We thus say that a complex is *minimal* if it does not have any contractible direct summands. To turn a bounded complex into a minimal one by repeated application of Gaussian elimination, we have to ensure that this process terminates (i.e. that there are only finitely many contractible direct summands). This holds for example if $\mathcal{A}$ is Krull–Schmidt (see Sect. 11.3.4) since then each term has only finitely many direct summands.

Given a complex C, a *minimal subcomplex* is a direct summand $C_{\min}$ of C, such that $C_{\min}$ is minimal, and the inclusion $C_{\min} \hookrightarrow C$ is a homotopy equivalence. One might worry that Gaussian elimination might produce many different minimal subcomplexes, based on the order that one chooses to eliminate summands, but in fact minimal subcomplexes are unique.

Lemma 19.15 *Assume that $\mathcal{A}$ is Krull–Schmidt. Any homotopy equivalence between minimal bounded complexes is an isomorphism of complexes. In particular, any two minimal subcomplexes of a bounded complex are isomorphic as complexes.*

The proof is not difficult, but won't be given here. The key is the notion of the *radical* of an additive category (see [114, §2] and [57, §6.1]), analogous to the Jacobson radical of an algebra.

19.2.3 Grothendieck Groups

For an (essentially small) additive category $\mathcal{A}$ we have defined the split Grothendieck group $[\mathcal{A}]_\oplus$, see Definition 11.46. The categories $C^b(\mathcal{A})$ and $K^b(\mathcal{A})$ are also additive, so we could consider their split Grothendieck groups (although this will not be the right thing to do). For a bounded complex A we define its *Euler characteristic* as

$$\chi(A) := \sum_{i \in \mathbb{Z}} (-1)^i [A^i] \in [\mathcal{A}]_\oplus . \tag{19.14}$$

This gives a map $\chi \colon [C^b(\mathcal{A})]_\oplus \to [\mathcal{A}]_\oplus$. Clearly the Euler characteristic of any complex as in (19.9) is zero. Consequently, when $\mathcal{A}$ is Karoubian, Lemma 19.13 implies that all contractible complexes have zero Euler characteristic. In this case the Euler characteristic descends to a map

$$\chi \colon [K^b(\mathcal{A})]_\oplus \to [\mathcal{A}]_\oplus . \tag{19.15}$$

To elaborate: when $\mathcal{A}$ is Krull–Schmidt, any complex has the same Euler characteristic as its minimal complex. Since homotopy equivalent complexes have isomorphic minimal complexes (see Lemma 19.15) they also have the same Euler characteristic.

Note that, for any bounded complexes A, B and any chain map $f \colon A \to B$, we have

$$\chi(\mathrm{Cone}(f)) = \sum_{i \in \mathbb{Z}} (-1)^i [A^{i+1} \oplus B^i] = \chi(B) - \chi(A) . \tag{19.16}$$

As a special case when $B = 0$ (or immediately from (19.14)), we have

$$\chi(A[1]) = \sum_{i \in \mathbb{Z}} (-1)^i [A^{i+1}] = -\chi(A) . \tag{19.17}$$

19.2 Some Homological Algebra

The inclusion of categories $\mathcal{A} \hookrightarrow K^b(\mathcal{A})$ induces a map on their split Grothendieck groups which we denote by ϕ. Clearly

$$\chi(\phi([X])) = [X] \tag{19.18}$$

for any object X in $\mathcal{A}$.

Now let us put these ideas in context.

Definition 19.16 The *triangulated Grothendieck group* of $K^b(\mathcal{A})$, denoted $[K^b(\mathcal{A})]_\triangle$, is the quotient of $[K^b(\mathcal{A})]_\oplus$ by the relation

$$[\mathrm{Cone}(f)] = [B] - [A] \tag{19.19}$$

for any chain map $f : A \to B$.

In particular, when $B = 0$ we get that

$$[A[1]] = -[A] \tag{19.20}$$

inside $[K^b(\mathcal{A})]_\triangle$.

Remark 19.17 The homotopy category $K^b(\mathcal{A})$ is an example of a triangulated category, and the general definition of the triangulated Grothendieck group agrees with this definition for the homotopy category. See appendix for more details.

By (19.16), the Euler characteristic descends to a map

$$\chi : [K^b(\mathcal{A})]_\triangle \to [\mathcal{A}]_\oplus.$$

The map ϕ induces a map in the other direction.

Theorem 19.18 *Let $\mathcal{A}$ be an (essentially small) Karoubian category. Then the maps ϕ and χ define mutually inverse group isomorphisms*

$$\phi : [\mathcal{A}]_\oplus \simeq [K^b(\mathcal{A})]_\triangle : \chi. \tag{19.21}$$

The interesting part of the proof is found in the following exercise.

Exercise 19.19 If A is a one-term complex, then $\phi(\chi([A])) = [A]$ inside $[K^b(\mathcal{A})]_\triangle$, thanks to (19.17) and (19.20). By Exercise 19.10, any complex A with only two nonzero terms is a cone of two one-term complexes. Deduce that $\phi(\chi([A])) = [A]$. Continuing in this vein with Exercise 19.10, prove that $\phi(\chi([A])) = [A]$ for any bounded complex A.

Remark 19.20 The Karoubian assumption can be removed from Theorem 19.18. For any essentially small additive category, it is still true that $\chi(A) = 0$ for a bounded contractible complex A; see [158, Proposition 3.2]. The rest of the argument goes through unchanged.

Although $[\mathcal{A}]_\oplus$ and $[K^b(\mathcal{A})]_\triangle$ are isomorphic, there is a major philosophical difference. In $[\mathcal{A}]_\oplus$, only positive sums of indecomposable objects will lift to objects in the additive category $\mathcal{A}$. This means that one can prove positivity statements about elements of $[\mathcal{A}]_\oplus$ by lifting them to objects in $\mathcal{A}$. The other side of this same coin is that subtraction, while formally allowed in the Grothendieck group $[\mathcal{A}]_\oplus$, does not have a categorical meaning. Meanwhile, minus signs can appear naturally in the Euler characteristic of a complex in $K^b(\mathcal{A})$. If one wants to categorify $[B] - [A]$, the most "natural" way to accomplish this would be to find a chain map from A to B and take its cone. Thus both contexts have their advantages.

Remark 19.21 This points to a major difference in the categorical interpretation of addition and subtraction. When one sees a sum $[B] + [A]$, this might only indicate the direct sum $[B \oplus A]$. When one sees the difference $[B] - [A]$, this might indicate the cone of a morphism. This morphism is additional structure which is invisible on the Grothendieck group! There may be many different morphisms $A \to B$ with non-isomorphic cones, and picking the "correct categorification" of $[B] - [A]$ is a subtle question. Categorifying addition is easy, while categorifying subtraction requires that one produce extra structure.

Now, assume that $\mathcal{A}$ is equipped with a grading shift (1), inducing as before a shift (1) on $K^b(\mathcal{A})$. As noted in Sect. 11.2.1, this makes $[K^b(\mathcal{A})]_\triangle$ into a $\mathbb{Z}[v, v^{-1}]$-module with v acting by (1). It is clear that ϕ is then an isomorphism of $\mathbb{Z}[v, v^{-1}]$-modules.

Remark 19.22 Perhaps it is better to think that $C^b(\mathcal{A})$ is a category with two shifts, (1) and [1], which makes its split Grothendieck group into a $\mathbb{Z}[v, v^{-1}, t, t^{-1}]$-module, with t acting by [1]. In the passage from $[C^b(\mathcal{A})]_\oplus$ to $[K^b(\mathcal{A})]_\triangle$, one must impose the relation (19.20), which sets $t = -1$.

If $\mathcal{A}$ is monoidal, then by the definition of the induced monoidal structure on $K^b(\mathcal{A})$, see (19.8), it is clear that ϕ is a ring morphism, hence $[\mathcal{A}]_\oplus$ and $[K^b(\mathcal{A})]_\triangle$ are isomorphic as rings. In the presence of a shift functor as above, this is an isomorphism of $\mathbb{Z}[v, v^{-1}]$-algebras.

19.3 Rouquier Complexes and Categorification of the Braid Group

Let (W, S) be a Coxeter system, and let $\mathbb{S}\mathrm{Bim}$ be the associated category of Soergel bimodules (for the Kac–Moody realization). Recall that the basic motivation behind $\mathbb{S}\mathrm{Bim}$ is to categorify the Hecke algebra $H = H(W)$. Already in Chap. 4, we saw how the generating Bott–Samelson bimodule B_s satisfies the isomorphism

$$B_s B_s \simeq B_s(1) \oplus B_s(-1), \tag{19.22}$$

19.3 Rouquier Complexes and Categorification of the Braid Group

which categorifies the quadratic relation

$$b_s b_s = (v + v^{-1}) b_s. \tag{19.23}$$

More generally, Soergel's conjecture states that B_w categorifies the Kazhdan–Lusztig basis element b_w. This conjecture is the main goal of the present part of this book.

Let us take a step back and recall that H had another, easier basis: the standard basis $\{\delta_w : w \in W\}$, defined as products of the standard generators δ_s, $s \in S$. If the indecomposable Soergel bimodules categorify the Kazhdan–Lusztig basis, what categorifies the standard basis?

Recall that every Soergel bimodule B is isomorphic to a finite direct sum

$$B \simeq \bigoplus_{(w,d) \in W \times \mathbb{Z}} B_w(d)^{\oplus m_{w,d}}$$

for uniquely-defined multiplicities $m_{w,d} \geq 0$. Passing to the split Grothendieck group, we get

$$[B] = \sum_{(w,d) \in W \times \mathbb{Z}} m_{w,d} v^d [B_w].$$

Thus the class of every object in $\mathbb{S}\mathrm{Bim}$ is a $\mathbb{Z}_{\geq 0}[v, v^{-1}]$-linear combination of the classes of various B_w. Already for the type A_1 Coxeter system with simple reflection s, we know that

$$[R] = 1 \quad \text{and} \quad [B_s] = b_s = \delta_s + v, \tag{19.24}$$

so that

$$\delta_s = b_s - v = [B_s] - [R(1)], \tag{19.25}$$

which involves a minus sign, cannot be categorified by any Soergel bimodule.

We saw in Sect. 19.2.3 that minus signs can be categorified by passing to complexes, where -1 becomes the cohomological shift $[1]$. We therefore try to categorify δ_s by a complex of Soergel bimodules. If we moreover restrict to indecomposable complexes, we are led naturally[5] to the following candidate:

---

[5] The reader following Remark 19.21 might expect the cone of a morphism $R(1) \to B_s$. However, there are no nonzero morphisms $R(1) \to B_s$, and the cone of the zero morphism is decomposable and thus cannot be invertible (under tensor product). Since δ_s is invertible, we should expect its categorification to be invertible. Instead, the complex F_s defined in (19.26) is the shift $[-1]$ of the cone of a nonzero map $B_s \to R(1)$.

Later in this chapter we will define a complex F_s^{-1} which categorifies $\delta_s^{-1} = b_s - v^{-1}$. For F_s^{-1} the more naive approach works: taking a cone of a nonzero map $R(-1) \to B_s$.

$$F_s := \left(\cdots \longrightarrow 0 \longrightarrow \underline{B_s} \overset{\bullet}{\longrightarrow} R(1) \longrightarrow 0 \longrightarrow \cdots \right). \tag{19.26}$$

Here and in what follows, the zeroth term of a complex is marked by an underline. Thus F_s is a two-term complex of Soergel bimodules concentrated in cohomological degrees 0 and 1. We also use diagrammatics to represent morphisms between Soergel bimodules, see Theorem 10.20.

By definition, the generators δ_s, $s \in S$, satisfy the braid relations under multiplication in H. To deserve the title of a categorification of δ_s, the complexes F_s should satisfy the braid relations under categorified multiplication: the monoidal structure $\otimes_R$ on (bounded) complexes. It is a theorem of Rouquier that this is indeed the case *up to homotopy*.

Theorem 19.23 (Rouquier [159]) *The complexes F_s, $s \in S$, satisfy the braid relations up to homotopy. That is, for every $s, t \in S$ with $m_{st} < \infty$, the complexes of Soergel bimodules*

$$\underbrace{F_s F_t F_s \cdots}_{m_{st} \text{ terms}} \quad \text{and} \quad \underbrace{F_t F_s F_t \cdots}_{m_{st} \text{ terms}}$$

are homotopy equivalent.

Rouquier actually proved a much stronger statement, which we will see in Theorem 19.36 below.

For $w \in W$, choose a reduced expression $\underline{w} = (s_1, \ldots, s_m)$ for w, and set

$$F_w := F_{s_1} \cdots F_{s_m}. \tag{19.27}$$

By Matsumoto's theorem, Theorem 19.23 implies that F_w is well-defined up to isomorphism in $K^b\mathbb{S}\text{Bim}$, the bounded homotopy category of Soergel bimodules. We know from Theorem 19.18 and the final paragraphs of the previous section that the Euler characteristic gives a natural identification

$$[K^b\mathbb{S}\text{Bim}]_\triangle \simeq [\mathbb{S}\text{Bim}]_\oplus \simeq H \tag{19.28}$$

as $\mathbb{Z}[v, v^{-1}]$-algebras. We have thus obtained isomorphism classes F_w in $K^b\mathbb{S}\text{Bim}$ categorifying the standard basis elements δ_w.

Remark 19.24 Rouquier's proof of Theorem 19.23 in [159] first reduces the result to the finite dihedral group $W = \langle s, t \rangle$. In this case, it is shown that Rouquier complexes corresponding to $F_s F_t \cdots$ (m_{st} terms) have a particularly nice minimal complex that is manifestly symmetric in s and t. (You will see this in Exercise 19.31 below for $m_{st} = 3$.)

Rouquier's argument uses some techniques from Soergel's classical theory, which views Soergel bimodules as certain quasi-coherent sheaves (see Exercise 6.38

19.3 Rouquier Complexes and Categorification of the Braid Group

in Chap. 6 for an instance of this point of view). For a purely diagrammatic approach, see [5, Proposition 9.1.1].

Now let us use Gaussian elimination to play around with small examples of Rouquier complexes.

Exercise 19.25 As a warm-up in using Gaussian elimination to simplify complexes of Soergel bimodules, consider the tensor product

$$F_s F_s = \left(\underline{B_s} \longrightarrow R(1) \right) \otimes_R \left(\underline{B_s} \longrightarrow R(1) \right)$$

$$\cong \underline{B_s B_s} \begin{array}{c} \nearrow B_s(1) \searrow \\ \oplus \\ \searrow B_s(1) \nearrow \end{array} R(2). \quad (19.29)$$

Write down the differentials explicitly. There is an isomorphism

$$B_s B_s \simeq B_s(-1) \oplus B_s(1), \quad (19.30)$$

so that (19.29) is isomorphic to

$$\underline{B_s(-1) \oplus B_s(1)} \begin{array}{c} \nearrow B_s(1) \searrow \\ \oplus \\ \searrow B_s(1) \nearrow \end{array} R(2). \quad (19.31)$$

Choose your favorite projection and inclusion maps for the direct sum decomposition (19.30) (see e.g. Sect. 8.2.4 and (8.32)), and use this to find the differentials in (19.31). Deduce that the map from $B_s(1)$ in degree 0 to either copy of $B_s(1)$ in degree 1 is an isomorphism. For more practice, write down the chain maps which give the inverse isomorphisms between (19.29) and (19.31).

Now Gaussian eliminate $B_s(1)$ to deduce a homotopy equivalence of the form

$$F_s F_s \simeq \left(\underline{B_s(-1)} \longrightarrow B_s(1) \longrightarrow R(2) \right). \quad (19.32)$$

What are the differentials on the right hand side of (19.32)? For more practice, write down the chain maps which give the inverse homotopy equivalences between (19.32) and (19.29).

Exercise 19.26 Why does the homotopy equivalence (19.32) categorify the quadratic relation in H?

Exercise 19.27 Compute the minimal complex of $F_s B_s$.

For each $s \in S$, recall that δ_s is invertible in H. Consider the following two-term complex concentrated in cohomological degrees -1 and 0:

$$F_s^{-1} := \left(\cdots \longrightarrow 0 \longrightarrow R(-1) \xrightarrow{\downarrow} B_s \longrightarrow 0 \longrightarrow \cdots \right). \tag{19.33}$$

Since

$$\delta_s^{-1} = \delta_s + v - v^{-1} = b_s - v^{-1} = [B_s] - [R(-1)], \tag{19.34}$$

the complex F_s^{-1} categorifies δ_s^{-1} at least at the level of the Grothendieck group. The following exercise lifts this to the categorical level: we have homotopy equivalences

$$F_s F_s^{-1} \simeq R \simeq F_s^{-1} F_s, \tag{19.35}$$

where R denotes the complex consisting of R in cohomological degree 0. Since R is the monoidal identity of $K^b(\mathbb{S}\text{Bim})$, this justifies the notation F_s^{-1}.

Exercise 19.28 Use Gaussian elimination to verify the first isomorphism of (19.35). (The second isomorphism is similar.) That is, consider the tensor product

$$F_s F_s^{-1} = \left(B_s \longrightarrow R \right) \otimes_R \left(R(-1) \longrightarrow B_s \right)$$

$$\cong B_s(-1) \xrightarrow{\begin{array}{c} R \\ \oplus \\ B_s B_s \end{array}} B_s(1).$$

Use the isomorphism $B_s B_s \simeq B_s(1) \oplus B_s(-1)$, and Gaussian eliminate $B_s(1)$ and $B_s(-1)$. As in Exercise 19.25, you need to keep track of (at least some of) the differentials to justify each use of Gaussian elimination.

Exercise 19.29

1. Deduce from (19.35) that the complex $F_s F_s$ is indecomposable in the homotopy category.
2. Go back to the last step of Exercise 19.25: having deduced a homotopy equivalence of the form (19.32), show that the indecomposability of $F_s F_s$ determines the differentials uniquely up to a nonzero scalar, hence determines the minimal complex of $F_s F_s$.
3. Arguing similarly, find the minimal complex of F_s^n for all $n \geq 3$ as well.

Remark 19.30 In the previous exercise, the complexes F_s^n seem to stabilize as n goes to infinity. A similar statement holds for $(F_s^{-1})^n$. The limiting complex $F_s^{-\infty}$,

19.3 Rouquier Complexes and Categorification of the Braid Group

which is bounded above but not bounded below, is one of the projectors that will be constructed in Chap. 23.

Exercise 19.31 Show that for $m_{st} = 3$, the Rouquier complex

$$F_s F_t F_s \cong \begin{array}{c} B_s B_t B_s \end{array} \longrightarrow \begin{array}{c} B_s B_t(1) \\ \oplus \\ B_s B_s(1) \\ \oplus \\ B_t B_s(1) \end{array} \longrightarrow \begin{array}{c} B_s(2) \\ \oplus \\ B_t(2) \\ \oplus \\ B_s(2) \end{array} \longrightarrow R(3) \tag{19.36}$$

is homotopy equivalent to a complex of the following form:

$$\underline{B_{sts}} \longrightarrow \begin{array}{c} B_{st}(1) \\ \oplus \\ B_{ts}(1) \end{array} \longrightarrow \begin{array}{c} B_s(2) \\ \oplus \\ B_t(2) \end{array} \longrightarrow B_{\mathrm{id}}(3) \tag{19.37}$$

Find the differentials. Conclude that $F_s F_t F_s \simeq F_t F_s F_t$ in $K^b \mathbb{S}\text{Bim}$.

(Recall from Sect. 11.2.5 that one can draw morphisms in the Karoubi envelope by composing with the appropriate idempotents. In this case, the differentials out of B_{sts} can be drawn as morphisms out of $B_s B_t B_s$ precomposed with a box labeled $\text{JW}_{(s,t,s)}$.)

Rouquier originally proved Theorem 19.23 as an attempt to categorify the braid group.

Definition 19.32 Let (W, S) be a Coxeter system. The associated *(Artin) braid group* Br_W is the group defined by the presentation

$$\text{Br}_W = \langle \sigma_s, \, s \in S \mid \underbrace{\sigma_s \sigma_t \cdots}_{m_{st}} = \underbrace{\sigma_t \sigma_s \cdots}_{m_{st}} \text{ for all } s, t \in S \text{ with } m_{st} < \infty \rangle. \tag{19.38}$$

In other words, Br_W has the same presentation as W, minus the quadratic relations. Thus there is a natural quotient homomorphism

$$\text{Br}_W \twoheadrightarrow W : \sigma_s \mapsto s \text{ for all } s \in S. \tag{19.39}$$

The "braid" terminology comes from the case of symmetric groups, where $\text{Br}_n = \text{Br}_{S_n}$ can be viewed as the group of braids on n strands up to isotopy. We will examine this case further in Chap. 21 in the context of polynomial and homological invariants of knots and links.

By definition, every element of Br_W can be written as a product

$$\beta = \sigma_{s_1}^{e_1} \sigma_{s_2}^{e_2} \cdots \sigma_{s_m}^{e_m} \qquad (19.40)$$

for simple reflections s_i and $e_i \in \{\pm 1\}$. Such an expression is called a *braid word*, and we say that it *expresses* β. Braid words are analogous to expressions in the Coxeter group, except that they also have signs. We say that a braid word is *positive* if $e_i = +1$ for all i.

Definition 19.33 Given $\beta \in \mathrm{Br}_W$ as above, the associated *Rouquier complex* F_β is the complex

$$F_\beta := F_{s_1}^{e_1} F_{s_2}^{e_2} \cdots F_{s_m}^{e_m} \qquad (19.41)$$

in $K^b \mathbb{S}\mathrm{Bim}$.

For an expression $\underline{w} = (s_1, \ldots, s_m)$ in (W, S), we write $F_{\underline{w}}$ for the Rouquier complex associated to the positive lift of $\underline{w}$ to Br_W, and call it a *positive Rouquier complex*:

$$F_{\underline{w}} := F_{s_1} F_{s_2} \cdots F_{s_m}. \qquad (19.42)$$

If $\underline{w}$ is a reduced expression for w, then we write F_w for the isomorphism class of $F_{\underline{w}}$, which is independent of the reduced expression.

Theorem 19.23 and the equivalences (19.35) imply that F_β is well-defined up to isomorphism in $K^b \mathbb{S}\mathrm{Bim}$.

Definition 19.34 For a monoidal category $(\mathcal{C}, \otimes)$, the *Picard group* $\mathrm{Pic}(\mathcal{C}, \otimes)$ of $\mathcal{C}$ is the group of isomorphism classes of objects of $\mathcal{C}$ which are invertible under the tensor product.

Using this definition, Rouquier's result may therefore be phrased as follows:

Proposition 19.35 *There exists a homomorphism of groups*

$$f : \mathrm{Br}_W \to \mathrm{Pic}(K^b(\mathbb{S}\mathrm{Bim}), \otimes), \quad f(\sigma_s) = [F_s]. \qquad (19.43)$$

This is a *weak categorification* of the braid group. In fact, Rouquier showed that his complexes give a *strong categorification*, in the following sense.

Theorem 19.36 (Rouquier [159]) *The homomorphism f can be upgraded to a strict monoidal functor*

$$F : \Omega \mathrm{Br}_W \to K^b(\mathbb{S}\mathrm{Bim}), \quad F(\sigma_s) = F_s, \qquad (19.44)$$

where $\Omega \mathrm{Br}_W$ is the group Br_W considered as a monoidal category in the usual way. (The objects are elements of Br_W, and the only morphisms are the identity maps.)

Let us elaborate on this theorem. For two different braid words β and β' which express the same element of the braid group, the weak categorification result implies the existence of a homotopy equivalence $F_\beta \simeq F_{\beta'}$. The strong categorification result states that this homotopy equivalence can be chosen canonically, satisfying a number of compatibility statements. In other words, a Rouquier complex depends on the element in the braid group, not on the choice of braid word, up to unique isomorphism in the homotopy category. To give a concrete example of what this means, consider the reduced expression graph of an element $w \in W$, see Exercise 1.58 and Sect. 10.3. To any path in this graph there is a homotopy equivalence obtained by following the homotopy equivalences associated to each braid relation. The strong categorification result implies that any two paths give homotopic chain maps.

We conclude this discussion with an important open problem.

Conjecture 19.37 (Rouquier [159]) The map f is injective.

This has been proven for finite type A Coxeter groups (Khovanov–Seidel [113]), types ADE (Brav–Thomas [33]), and finite type (Jensen [90]).

19.4 Cohomology of Rouquier Complexes

The category $\mathbb{S}\text{Bim}$ is additive. One can define its homotopy category $K^b(\mathbb{S}\text{Bim})$, but one cannot define the usual cohomology functors $H^i = \text{Ker}\, d^i / \text{Im}\, d^{i-1}$, since there are no kernels or images. However, $\mathbb{S}\text{Bim}$ is a full subcategory of graded R-bimodules, an abelian category where we can take kernels and images. Thus we can use this ambient abelian category to define the cohomology of a complex of Soergel bimodules, although the answer will be a graded R-bimodule which is not necessarily a Soergel bimodule.

As it turns out, the cohomology of Rouquier complexes is quite simple.

In Sect. 5.3 we introduced the short exact sequence of R-bimodules which we called (Δ):

$$0 \to R_s(-1) \to B_s \to R(1) \to 0.$$

The second half of this short exact sequence is the complex F_s, from which we deduce that the cohomology of F_s is the shifted standard bimodule $R_s(-1)$.

$$H^i(F_s) = \begin{cases} 0 & \text{if } i \neq 0 \\ R_s(-1) & \text{if } i = 0. \end{cases} \tag{19.45}$$

Similarly, from the short exact sequence (∇) (also introduced in Sect. 5.3) we have

$$H^i(F_s^{-1}) = \begin{cases} 0 & \text{if } i \neq 0 \\ R_s(1) & \text{if } i = 0. \end{cases} \quad (19.46)$$

We claim that, as a consequence, the cohomology of a tensor product

$$F_{s_1}^{e_1} F_{s_2}^{e_2} \cdots F_{s_m}^{e_m}$$

(where $e_i = \pm 1$) is supported in degree zero, where it agrees with the tensor product of shifted standard bimodules

$$R_{s_1}(-e_1) \otimes R_{s_2}(-e_2) \otimes \cdots \otimes R_{s_m}(-e_m).$$

In other words, if β is a braid word which represents an element $w \in W$, then the cohomology of F_β is just $R_w(n)$ in degree zero, where $n = -\sum e_i$.

Remark 19.38 The reader should rightly worry that the cohomologies of a tensor product need not agree with the tensor product of the cohomologies! In this case, the problem disappears since all R-bimodules in sight are free as left and as right R-modules. The reader who is familiar with derived categories would replace F_s with the quasi-isomorphic complex $R_s(-1)$, and F_s^{-1} with the quasi-isomorphic complex $R_s(1)$, and would note that the derived tensor product agrees with the ordinary tensor product by the aforementioned freeness.

Remark 19.39 The reader familiar with derived categories should also note that the homotopy category $K^b(\mathbb{S}\text{Bim})$ is a much richer place to work in than the derived category of R-bimodules. In the derived category, F_s and F_s^{-1} are isomorphic (up to a grading shift), which is undesirable behavior! Rouquier's braid group categorification, when viewed as a homomorphism to the Picard group of the derived category of R-bimodules, would factor through the Coxeter group quotient of the braid group.

For future use, let us state the consequences of this computation of cohomology for positive Rouquier complexes. This is our analogue of the Lefschetz hyperplane theorem, see Sect. 19.1.

Lemma 19.40 *For any expression $\underline{w}$, the cohomology of the positive Rouquier complex $F_{\underline{w}}$ is $R_w(-\ell(\underline{w}))$, concentrated in homological degree 0. In particular, the first differential of the Rouquier complex $F_{\underline{w}}$ is injective in all graded degrees below $\ell(\underline{w})$.*

19.5 Perversity

Let X be a complex of Soergel bimodules. Each term X^i is isomorphic to a direct sum of shifts of $B_w(j)$ for various $w \in W$ and $j \in \mathbb{Z}$, possibly with multiplicity. By fixing such a decomposition and collecting terms of the form $B_w(j)$ for each $j \in \mathbb{Z}$, we obtain a decomposition

$$X^i = \bigoplus_{j \in \mathbb{Z}} X^i_j, \qquad X^i_j \simeq \bigoplus B_w(j), \qquad (19.47)$$

where w varies over W, possibly with multiplicity. It is helpful to visualize this decomposition on a planar grid, where X^i_j is placed at the coordinate $(i, -j)$:

$$\begin{array}{ccc} X^0_0 & X^1_0 & X^2_0 \\ X^0_1 & X^1_1 & X^2_1 \\ X^0_2 & X^1_2 & X^2_2 \end{array} \qquad (19.48)$$

In this picture, applying (1) (resp. [1]) to X shifts each entry down (resp. left) by one unit. We will refer to the dashed line going through the points $(i, -i)$ as the "diagonal"; we will see its importance shortly.

Let us draw some minimal Rouquier complexes on this planar grid. We have

$$F_s \quad = \quad \begin{array}{c} B_s \\ \searrow \\ R(1) \end{array} \qquad (19.49)$$

Here and in what follows, all empty components are 0, and the component at $(0, 0)$ is marked by an underline. We have

$$F_{st} \quad \simeq \quad \begin{array}{c} B_{st} \\ \searrow \\ B_s(1) \\ B_t(1) \\ \searrow \\ R(2) \end{array} \qquad (19.50)$$

and for $m_{st} = 3$, by Exercise 19.31,

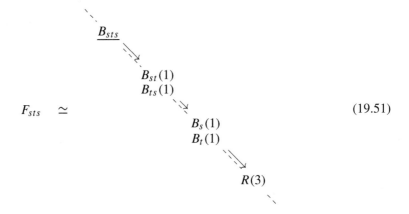
$$F_{sts} \simeq \begin{matrix} \underline{B_{sts}} \\ \searrow \\ B_{st}(1) \\ B_{ts}(1) \\ \searrow \\ B_s(1) \\ B_t(1) \\ \searrow \\ R(3) \end{matrix} \qquad (19.51)$$

where we have omitted the direct sum symbol. Meanwhile, by Exercise 19.25, we have

$$F_s F_s \simeq \begin{matrix} B_s(-1) \\ \searrow \\ \underline{0} \\ \searrow \\ B_s(1) \\ \searrow \\ R(2) \end{matrix} \qquad (19.52)$$

We see in these examples that there is some relationship between the cohomological degree and the shifts: the minimal Rouquier complexes F_s, F_{st}, F_{sts} associated to reduced expressions lie on the diagonal, while $F_s F_s$ at least lies on or above the diagonal.

We define two full subcategories

$$K^{\leq 0}, \; K^{\geq 0} \subset K^b \mathbb{S}\text{Bim} \qquad (19.53)$$

of $K^b \mathbb{S}\text{Bim}$, as follows. The category $K^{\leq 0}$ consists of complexes that are homotopy equivalent to complexes X which lie on or below the diagonal. In other words, X has a decomposition (19.47) satisfying $X^i_j = 0$ for all $j < i$. Equivalently, $K^{\leq 0}$ consists of complexes whose minimal complex has such a decomposition. Similarly, $K^{\geq 0}$ consists of complexes homotopy equivalent to complexes X with $X^i_j = 0$ for all $j > i$, i.e. X lies on or above the diagonal.

Similarly, one can define $K^{\leq 1}$ as those complexes (homotopy equivalent to complexes) which lie on or below the first off-diagonal. More generally, for $n \in \mathbb{Z}$,

define
$$K^{\leq n} := K^{\leq 0}[-n], \qquad K^{\geq n} := K^{\geq 0}[-n]. \tag{19.54}$$

For example, $K^{\geq 1}$ is the full subcategory consisting of those complexes (homotopy equivalent to complexes) which lie on or above the first off-diagonal. In other words, $X^i_j = 0$ for all $j > i - 1$.

We will not use it until Chap. 26, but organizing complexes by their diagonals in this fashion helps to prove some orthogonality results.

Proposition 19.41 *Assume the Soergel conjecture. Then for any $X \in K^{\leq 0}$ and $Y \in K^{\geq 1}$, we have $\mathrm{Hom}_{K^b \mathbb{S}\mathrm{Bim}}(X, Y) = 0$. This is often shortened to*

$$\mathrm{Hom}(K^{\leq 0}, K^{\geq 1}) = 0. \tag{19.55}$$

Proof This is an immediate consequence of

$$\mathrm{Hom}_{\mathbb{S}\mathrm{Bim}}(B_x, B_y(i)) = \begin{cases} \mathbb{C} \cdot \mathrm{id}_{B_x} & \text{if } i = 0 \text{ and } x = y, \\ 0 & \text{if } i = 0 \text{ and } x \neq y, \text{ or if } i < 0, \end{cases} \tag{19.56}$$

since this implies that any chain map between the minimal complexes of X and Y is the zero map. In turn, (19.56) follows from the Soergel Hom formula (Theorem 5.27) and the asymptotic orthonormality of the Kazhdan–Lusztig basis (Theorem 3.21), assuming the Soergel conjecture for B_x and B_y. □

Definition 19.42 We say that a complex X is *perverse* if X lies in $K^{\leq 0} \cap K^{\geq 0}$. In other words, the minimal complex of X lies on the diagonal.

As a consequence of Proposition 19.41, we deduce that if X and Y are perverse, then $\mathrm{Hom}(X, Y[n]) = 0$ for all $n < 0$.

19.6 The Diagonal Miracle

The examples above should have suggested several general results, which we address in this section.

Lemma 19.43 *Assume the Soergel conjecture. Then for any expression $\underline{w}$, we have $F_{\underline{w}} \in K^{\geq 0}$. That is, positive Rouquier complexes lie in $K^{\geq 0}$.*

Lemma 19.43 is an immediate consequence of the following proposition.

Proposition 19.44 *Let $s \in S$.*

1. The functors

$$(-) \otimes F_s^{-1}, \ F_s^{-1} \otimes (-) : K^b \mathbb{S}\mathrm{Bim} \to K^b \mathbb{S}\mathrm{Bim} \tag{19.57}$$

preserve $K^{\leq 0}$.

2. *The functors*

$$(-) \otimes F_s, \ F_s \otimes (-) : K^b \mathbb{S}\text{Bim} \to K^b \mathbb{S}\text{Bim} \tag{19.58}$$

preserve $K^{\geq 0}$.

We will prove this in the following series of exercises. First, as a special case, consider the complex $B_w(0)$ supported in cohomological degree zero, which is in both $K^{\geq 0}$ and $K^{\leq 0}$. What does the complex $F_s B_w$ look like? When $sw < w$, the minimal complex of $F_s B_w$ is isomorphic to $B_w(-1)$ supported in degree zero, by an argument almost identical to Exercise 19.27.

Exercise 19.45 Suppose now that $sw > w$. Using the Soergel conjecture, prove that $F_s B_w$ is perverse. (*Hint:* How does $b_s b_w$ decompose into the Kazhdan–Lusztig basis. How does $B_s B_w$ decompose into indecomposables, and what shifts appear? See Theorem 3.27.)

Exercise 19.46 Using the fact that $F_s B_w \in K^{\geq 0}$ for all $w \in W$, prove Proposition 19.44.

Now we have a much harder result.

Theorem 19.47 (Diagonal Miracle) *Assume the Soergel conjecture. Then for any reduced expression $\underline{w}$, the positive Rouquier complex $F_{\underline{w}}$ is perverse. Moreover, the minimal complex in degree zero is just $B_w(0)$. In other words, the minimal Rouquier complex $F_{\underline{w}}^{\min}$ is of the form*

$$B_w \to \bigoplus_{z < w} B_z^{\oplus n_z}(1) \to \bigoplus_{z' < w} B_{z'}^{\oplus n_{z'}}(2) \to \cdots, \qquad n_z, n_{z'} \in \mathbb{Z}_{\geq 0}.$$

To prove Theorem 19.47 from Lemma 19.43 is a somewhat technical argument, see [57, §6.5] and the next remark.

Remark 19.48 The gist of the argument is this. A paper of Libedinsky and Williamson [122] studied Rouquier complexes in the homotopy category of graded R-bimodules, which is larger than the homotopy category of Soergel bimodules; in particular, it contains standard bimodules. They prove a strong result about the associated graded of the standard filtration on a Rouquier complex for a reduced expression. If the Rouquier complex weren't perverse, one can take its associated graded and somehow deduce the existence of some impossible term (e.g. something appearing in negative homological degree, or below the diagonal).

The diagonal miracle really is a miraculous result. The Rouquier complex $F_{\underline{w}}$, even for a reduced expression, is a complicated mess of Bott–Samelson bimodules which have many nontrivial shifts in their direct summands. The diagonal miracle implies that all of these non-diagonal shifts can be Gaussian eliminated away, separating the wheat from the chaff and leaving only a nice diagonal complex. This

19.6 The Diagonal Miracle

is actually a very strong statement about morphisms, saying that sufficiently many pieces of the differential are isomorphisms.

Note that the diagonal miracle relies upon Soergel's conjecture, whereas in the next chapter we will use the diagonal miracle to prove Soergel's conjecture. In fact, one should actually prove both statements simultaneously with a double induction. When $\underline{w}$ is a reduced expression for w, proving that $F_{\underline{w}}$ is diagonal only uses $S(y)$ for $y \leq w$, and will be used to prove $S(ws)$ for some $s \in S$ with $ws > w$. For this book, we will not belabor this point, taking the diagonal miracle as a black box for our exposition of the Elias–Williamson proof of the Soergel conjecture.

Remark 19.49 The results of this and the previous section are best stated in terms of a homological gadget known as a "t-structure," which picks out an abelian category inside a triangulated one. Proposition 19.41 is one of the main things needed to prove that $(K^{\leq 0}, K^{\geq 0})$ is a t-structure, assuming Soergel's conjecture. Thus perverse complexes form an abelian category (although the notion of kernel and cokernel may not be the naive one). For more details see appendix below.

Appendix: More Homological Algebra

This appendix contains a brief discussion of triangulated categories, triangulated Grothendieck groups, and t-structures.

Triangulated Structure

A crucial aspect in homological algebra is that the homotopy category of an abelian category is *not* necessarily abelian. Instead, homotopy categories (even of an additive, not necessarily abelian category) admit the structure of a *triangulated category*, from which a lot of powerful homological algebra can be developed.

Triangulated categories were introduced by Verdier in his 1963 PhD thesis [173]. Useful (and non-French) references include [70, 77, 83, 139, 144], where the reader can find more details.

Let $\mathcal{T}$ be a category equipped with an autoequivalence [1]. A *triangle* in $\mathcal{T}$ is a diagram of the form

$$A \xrightarrow{f} B \xrightarrow{g} C \xrightarrow{h} A[1], \tag{19.59}$$

which is often shortened to $A \xrightarrow{f} B \xrightarrow{g} C \dashrightarrow^{h}$. Note that $f[1]$ gives a map from $A[1]$ to $B[1]$, and $h[-1]$ gives a map from $C[-1]$ to A, so that this diagram of maps can be extended in both directions indefinitely.

In brief, a *triangulated category* is an additive category $\mathcal{T}$ equipped with an autoequivalence [1] and a class of triangles that are called *distinguished*, subject to a number of axioms. In particular, any morphism $f: A \to B$ in $\mathcal{T}$ can be extended to a distinguished triangle

$$A \xrightarrow{f} B \xrightarrow{g} \mathrm{Cone}(f) \xrightarrow{h} A[1] \tag{19.60}$$

for an object $\mathrm{Cone}(f)$, called the *cone of f*, that is unique up to (non-unique) isomorphism.

Distinguished triangles play a role akin to that of short exact sequences in abelian categories, and cones can be viewed as a kind of combined kernel and cokernel. Said another way, the reader familiar with basic homological algebra should have seen the fact that a short exact sequence of complexes gives rise to a long exact sequence of cohomology groups. In a triangulated category there are no short exact sequences, but the distinguished triangles can be thought of as remembering the long exact sequences.

Let $\mathcal{A}$ be an additive category. The homotopy category $K(\mathcal{A})$ (as well as $K^*(\mathcal{A})$ for $* \in \{+, -, b\}$) can be given a triangulated structure as follows: the autoequivalence [1] is cohomological shift, and distinguished triangles are triangles that are isomorphic (in the obvious sense) to a triangle of the form

$$A \xrightarrow{f} B \xrightarrow{i} \mathrm{Cone}(f) \xrightarrow{p} A[1], \tag{19.61}$$

where i and p are the inclusion and projection.

For example, we have the following basic results, which follow quickly from the axioms of a triangulated category.

Lemma 19.50 *For any two objects A, B of a triangulated category, the triangle*

$$A \xrightarrow{i_A} A \oplus B \xrightarrow{p_B} B \xrightarrow{0} A[1]$$

is distinguished. Here, i_A is the inclusion and p_B is the projection.

One should think of this distinguished triangle as an analogue of a split short exact sequence.

Lemma 19.51 *For any object A of a triangulated category, the triangle*

$$A \longrightarrow 0 \longrightarrow A[1] \xrightarrow{\mathrm{id}_{A[1]}} A[1]$$

is distinguished.

Note that this is clear for the homotopy category $K(\mathcal{A})$ since $\mathrm{Cone}(A \to 0) = A[1]$.

19.6 The Diagonal Miracle

Lemma 19.52 *In any triangulated category, a morphism $f: A \to B$ is an isomorphism if and only if $\mathrm{Cone}(f) \simeq 0$.*

This was proved for $K(\mathcal{A})$ in Exercise 19.8.

Exercise 19.53 Assume that $\mathcal{A}$ is abelian. Show that it is *not* necessarily true that a short exact sequence $0 \to A \to B \to C \to 0$ in $\mathcal{A}$ induces a distinguished triangle $A \to B \to C \to A[1]$ in $K(\mathcal{A})$. Here is an example: let $\mathcal{A} = \mathrm{Ab}$ be the category of abelian groups and consider the short exact sequence $0 \to \mathbb{Z} \xrightarrow{\cdot 2} \mathbb{Z} \to \mathbb{Z}/2\mathbb{Z} \to 0$.

Exercise 19.54 Assume that $\mathcal{A}$ is abelian. Show that a short exact sequence $0 \to A \to B \to C \to 0$ of complexes does give rise to a distinguished triangle in $K(\mathcal{A})$ if the sequence "splits termwise"; i.e. for all i, the exact sequence $0 \to A^i \to B^i \to C^i \to 0$ splits.

Exercise 19.55 In this exercise, you will show that the natural functor

$$K^b \mathbb{B}\mathbb{S}\mathrm{Bim} \to K^b \mathbb{S}\mathrm{Bim}$$

is an equivalence of triangulated categories.

1. Observe that it is fully faithful.
2. Use induction along the Bruhat order to show that, for every $w \in W$, the complex consisting of B_w in cohomological degree 0 lies in the essential image.
3. Look up the definition of a *triangulated functor* (also called an *exact functor*) and of an equivalence of triangulated categories. Any functor between additive categories induces a triangulated functor of the homotopy categories. Deduce that the essential image of the functor $K^b \mathbb{B}\mathbb{S}\mathrm{Bim} \to K^b \mathbb{S}\mathrm{Bim}$ is closed under taking cones.
4. Deduce that the functor above is an equivalence of triangulated categories.

Triangulated Grothendieck Groups

For an (essentially small) additive category $\mathcal{A}$ we have defined the split Grothendieck group $[\mathcal{A}]_\oplus$, see Definition 11.46. If $\mathcal{A}$ is abelian, this group is exactly the quotient of the free abelian group on the isomorphism classes $[A]$ of objects A of $\mathcal{A}$ by the subgroup generated by the relations

$$[A] - [B] + [C] = 0 \tag{19.62}$$

whenever there is a *split* short exact sequence $0 \to A \to B \to C \to 0$ in $\mathcal{A}$, i.e. $B \simeq A \oplus C$. If we drop the "split" here, we obtain a quotient $[\mathcal{A}]$ of $[\mathcal{A}]_\oplus$ called the *Grothendieck group* of $\mathcal{A}$. The role of $[\mathcal{A}]$ is explained in the following exercise.

Exercise 19.56 An *additive function* on $\mathcal{A}$ is a function $\chi \colon \mathcal{A} \to G$ from the objects of $\mathcal{A}$ to an abelian group G such that

$$\chi(A) - \chi(B) + \chi(C) = 0 \tag{19.63}$$

whenever there is an exact sequence $0 \to A \to B \to C \to 0$. An additive map clearly descends to a group morphism $[\mathcal{A}] \to G$. Show that the obvious map $\chi \colon \mathcal{A} \to [\mathcal{A}]$ is the universal additive map: any other factors through it.

Exercise 19.57 Show that $[A] = [B]$ in $[\mathcal{A}]$ if and only if there are exact sequences

$$0 \to U \to A \oplus C \to V \to 0 \quad \text{and} \quad 0 \to U \to B \oplus C \to V \to 0.$$

See [125, Proposition 2.4.19] for help.

Pretending that distinguished triangles of a triangulated category are short exact sequences leads us to the definition of the *(triangulated) Grothendieck group* $[\mathcal{T}]_\triangle$ of a triangulated category $\mathcal{T}$: it is the quotient of the free abelian group on the isomorphism classes $[A]$ of objects A of $\mathcal{T}$ by the subgroup generated by the relations

$$[A] - [B] + [C] = 0 \tag{19.64}$$

whenever there is a distinguished triangle $A \to B \to C \to$ in $\mathcal{T}$. Note that this definition is consistent with the special case of the homotopy category in Definition 19.16.

It follows from Lemma 19.50 that the relation

$$[A \oplus B] = [A] + [B] \tag{19.65}$$

holds in $[\mathcal{T}]_\triangle$. Hence, we have a surjective group homomorphism

$$[\mathcal{T}]_\oplus \twoheadrightarrow [\mathcal{T}]_\triangle. \tag{19.66}$$

By Lemma 19.51, we have

$$[A[1]] = -[A] \tag{19.67}$$

in $[\mathcal{T}]_\triangle$. Note that (19.67) implies a special feature of the triangulated Grothendieck group: every element of $[\mathcal{T}]_\triangle$ is of the form $[A]$ for some $A \in \mathcal{T}$.

Exercise 19.58 Formulate and prove analogues of the statements in Exercises 19.56 and 19.57 for the triangulated Grothendieck group. (See [125, Corollary 2.6.16] for help with the latter.)

Perverse t-Structure

A t-structure on a triangulated category $\mathcal{T}$ is a homological gadget that allows one to cut out a full abelian subcategory of $\mathcal{T}$. This notion was introduced by Beilinson–Bernstein–Deligne(–Gabber) [19], where it was used to define the (abelian) category of perverse sheaves inside the (triangulated) constructible derived category of sheaves.

We only state the definition and first results, and refer the interested reader elsewhere for further details and proofs. Besides the original paper, some useful references include [46, Sect. 5.1], [99, Sect. 10.1].

Definition 19.59 A *t-structure* on a triangulated category $\mathcal{T}$ is a collection of strictly full subcategories

$$\mathcal{T}^{\leq n}, \mathcal{T}^{\geq n} \quad (n \in \mathbb{Z})$$

satisfying the following axioms:

1. $\mathcal{T}^{\leq n}[1] = \mathcal{T}^{\leq n-1}$ and $\mathcal{T}^{\geq n}[1] = \mathcal{T}^{\geq n-1}$ for all $n \in \mathbb{Z}$;
2. $\mathcal{T}^{\leq 0} \subseteq \mathcal{T}^{\leq 1}$ and $\mathcal{T}^{\geq 1} \subseteq \mathcal{T}^{\geq 0}$;
3. $\mathrm{Hom}(X, Y) = 0$ for all $X \in \mathcal{T}^{\leq 0}$ and $Y \in \mathcal{T}^{\geq 1}$;
4. Every object X in $\mathcal{T}$ fits into a distinguished triangle

$$X^{\leq 0} \to X \to X^{\geq 1} \to X^{\leq 0}[1]$$

with $X^{\leq 0} \in \mathcal{T}^{\leq 0}$ and $X^{\geq 1} \in \mathcal{T}^{\geq 1}$.

One often denotes a t-structure by $(\mathcal{T}^{\leq 0}, \mathcal{T}^{\geq 0})$, with the understanding that $\mathcal{T}^{\leq n}$ and $\mathcal{T}^{\geq n}$ for general $n \in \mathbb{Z}$ are defined by the first axiom.

Definition 19.60 The intersection $\mathcal{T}^{\leq 0} \cap \mathcal{T}^{\geq 0}$ is called the *heart* of the t-structure.

Let $\mathcal{A}$ be an abelian category, and consider its bounded derived category $\mathcal{T} = D^b(\mathcal{A})$, which is the quotient of $K^b(\mathcal{A})$ obtained by inverting quasi-isomorphisms. It turns out that the triangulated structure on $K^b(\mathcal{A})$ induces one on $D^b(\mathcal{A})$. (This is an example of *Verdier quotient* of a triangulated category.) The first example of a t-structure is the following.

Example 19.61 Let $\mathcal{T}^{\leq 0}$ (resp. $\mathcal{T}^{\geq 0}$) be the full subcategory of $D^b(\mathcal{A})$ consisting of complexes X such that $H^i(X) = 0$ only if $i > 0$ (resp. $i < 0$). Then $(\mathcal{T}^{\leq 0}, \mathcal{T}^{\geq 0})$ is a t-structure, called the *standard t-structure* on $D^b(\mathcal{A})$.

Exercise 19.62 Consider the functor $\mathcal{A} \to D^b(\mathcal{A})$ sending an object $X \in \mathcal{A}$ to the one-term complex X, supported in cohomological degree 0. Show that this functor induces an equivalence between $\mathcal{A}$ and the heart of the standard t-structure. Being an abelian category $\mathcal{A}$ has a notion of kernels and cokernels, while $D^b(\mathcal{A})$ only has a notion of cones. Given a map $f : X \to Y$, how does one recover the kernel or the cokernel of f from the cone of the corresponding morphism in $D^b(\mathcal{A})$? (*Hint:* use the last axiom for a t-structure.)

The previous exercise showed that the heart of the standard t-structure is a full abelian category inside the triangulated category $D^b(\mathcal{A})$. The first result about t-structures (and their *raison d'être*) states that this is true in general.

Theorem 19.63 *The heart of a t-structure is an abelian category.*

Now, fix a Coxeter system (W, S) and a realization $\mathfrak{h}$, and consider the associated category $\mathbb{S}\mathrm{Bim}$ of Soergel bimodules. In Sect. 19.5, we defined full subcategories $K^{\leq 0}$, $K^{\geq 0}$ of the homotopy category $K^b \mathbb{S}\mathrm{Bim}$.

Theorem 19.64 *If Soergel's conjecture holds for the realization* $(W, \mathfrak{h})$, *then* $(K^{\leq 0}, K^{\geq 0})$ *forms a t-structure on* $K^b \mathbb{S}\mathrm{Bim}$, *called the* perverse t-structure.

Proof (Sketch of Proof) We saw in Proposition 19.41 that $\mathrm{Hom}(K^{\leq 0}, K^{\geq 1}) = 0$.

Given $X \in K^b \mathbb{S}\mathrm{Bim}$, fix a decomposition (19.47) as before, then decompose each differential $d^i : X^i \to X^{i+1}$ accordingly, as

$$d^i = \bigoplus_{(j,k)\in \mathbb{Z}^2, k\geq 0} d^i_{j,k}, \qquad d^i_{j,k} : X^i_j \to X^{i+1}_{j+k}.$$

(Do you see why we only need components $k \geq 0$?) Now, let $X^{\leq 0}$ be the subcomplex

$$X^{\leq 0} = \ker(d^i_{i,0} : X^i_i \to X^{i+1}_i) \oplus \left(\bigoplus_{j>i} X^i_j\right).$$

In terms of the planar grid, $X^{\leq 0}$ consists of the direct sum of everything strictly below the diagonal, together with the kernels of the horizontal components of differentials out of the diagonal:

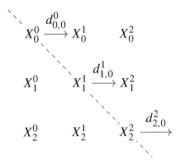

19.6 The Diagonal Miracle

One then shows that $X^{\leq 0}$ lies in $K^{\leq 0}$, and setting $X^{\geq 1} = \mathrm{Coker}(X^{\leq 0} \to X)$ for the natural inclusion, that $X^{\geq 1}$ lies in $K^{\geq 1}$ and $X^{\leq 0} \to X \to X^{\geq 1} \to X^{\leq 0}[1]$ is a distinguished triangle. □

Exercise 19.65 Use (19.56) to fill in the details in the proof sketch above. In particular, understand why kernels and cokernels in the argument above, which a priori are only graded R-bimodules, are in fact Soergel bimodules.

A complex of Soergel bimodules is said to be *perverse* if it lies in the heart of the perverse t-structure. Thus perverse complexes form a full abelian category inside $K^b \mathbb{S}\mathrm{Bim}$. Short exact sequences are those sequences $X \to Y \to Z$ with X, Y, Z perverse that can be completed (in a necessarily unique way) to a distinguished triangle in $K^b \mathbb{S}\mathrm{Bim}$.

Remark 19.66 We will return to the abelian category of perverse complexes (or more accurately, the analogue for right Soergel modules) in Chap. 26, where it will be understood as a Soergel-theoretic version of the graded category $\mathcal{O}_0$ of a semisimple Lie algebra. From this point of view, Rouquier complexes for reduced expressions (which are perverse by the diagonal miracle) will correspond to (graded lifts of) Verma modules.

Further discussion of the perverse t-structure involves a more geometric perspective on the Hecke category that is outside the scope of this book. We content ourselves here with a few (at this point perhaps cryptic) remarks to point the interested reader to the literature.

Remark 19.67 The perverse t-structure on $K^b \mathbb{S}\mathrm{Bim}$ is motivated by the similarly named perverse t-structure in geometry, or more precisely, in "mixed geometry." This is a geometric setting in which the category of sheaves has an additional grading, whose grading shift autoequivalence is called the *Tate twist*. We will return to the perverse t-structure in Chap. 26 in the context of Koszul duality, where this extra grading plays a crucial role. For more details on the connection to geometry, one may see [2].

Remark 19.68 Elias–Williamson deduced the inductive version of the diagonal miracle from results of Libedinsky–Williamson [122]. A more geometric perspective allows for a conceptual approach to results surrounding the perverse t-structure; see [2, 132]. For instance, Proposition 19.44 is [2, Proposition 4.6] (in the language of parity sheaves) and [132, Proposition 5.10]. These papers also handle cases where Soergel's conjecture fails (e.g. certain realizations over a field of positive characteristic). In such cases, one can still define a "perverse t-structure" on the bounded homotopy category of the Hecke category, but perverse complexes no longer coincide with those on the diagonal. See Sect. 26.5 for a related discussion.

Chapter 20
Proof of the Hard Lefschetz Theorem

This chapter is based on expanded notes of a lecture given by the authors and taken by
> **Gurbir Dhillon** and **Oscar Kivinen**

Abstract In this chapter, we outline the proof of Soergel's conjecture via versions of the hard Lefschetz theorem and Hodge–Riemann bilinear relations.

20.1 Introduction

We are finally in a position to complete the induction outlined in Chap. 18, thereby showing that the characters of the indecomposable Soergel bimodules are given by the Kazhdan–Lusztig polynomials. As discussed previously, the Elias–Williamson proof hinges on a positivity property (*the Hodge–Riemann bilinear relations*) and a nondegeneracy property (*hard Lefschetz*) of the operator L_0, which is simply left multiplication by the regular dominant element $\rho \in \mathfrak{h}^*$ (i.e. $\langle \alpha_s^\vee, \rho \rangle > 0$ for all $s \in S$). In the inductive proof of these properties, it is helpful to use a one-parameter deformation of this operator, as is developed below.

Depending on the reader's background and taste, it may be helpful when approaching these statements and their proofs to bear in mind their close historical and geometric counterparts. The appendix to this chapter has an orienting discussion along these lines, and could be read now as an extended introduction to this chapter. It does however require more background in algebraic geometry than we use in the remainder of the book. For this reason we leave this section as an optional appendix, intended for the interested and geometrically minded reader.

G. Dhillon
Department of Mathematics, Stanford University, Stanford, CA, USA

O. Kivinen
Department of Mathematics, UC Davis, Davis, CA, USA

© The Editor(s) (if applicable) and The Author(s), under exclusive licence to Springer Nature Switzerland AG 2020
B. Elias et al., *Introduction to Soergel Bimodules*, RSME Springer Series 5, https://doi.org/10.1007/978-3-030-48826-0_20

20.2 Preliminaries

As briefly discussed in the previous chapters, hard Lefschetz will be deduced from Hodge–Riemann and a weak Lefschetz substitute using Lemma 17.21. We repeat this lemma here for the reader's convenience.

Lemma 20.1 (The Weak Lefschetz Argument) *Let V, W be graded vector spaces with graded nondegenerate symmetric bilinear forms $\langle -, - \rangle_V, \langle -, - \rangle_W$ and Lefschetz operators L_V, L_W. Let*

$$\Omega : V \longrightarrow W$$

be a degree one linear map such that

1. *Ω is injective from negative degrees,*
2. *$\langle v, L_V v' \rangle_V = \langle \Omega v, \Omega v' \rangle_W$,*
3. *$\Omega L_V = L_W \Omega$.*

Then if L_W satisfies the Hodge–Riemann bilinear relations, it follows that L_V satisfies hard Lefschetz.

In Elias–Williamson's argument, the weak Lefschetz substitute Ω will be the first differential in a Rouquier complex. These complexes were introduced in Chap. 19. To apply Lemma 20.1, we will also need the so-called diagonal miracle.

Theorem 20.2 (Diagonal Miracle) *For any $w \in W$, the minimal Rouquier complex $F_w^{\min}$ is of the form*

$$B_w \to \bigoplus_{z<w} B_z^{\oplus n_z}(1) \to \bigoplus_{z'<w} B_{z'}^{\oplus n_{z'}}(2) \to \cdots, \qquad n_z, n_{z'} \in \mathbb{Z}_{\geq 0}. \tag{20.1}$$

That is, the zeroth term of $F_w^{\min}$ is B_w, and the i-th term is a direct sum of smaller indecomposable Soergel bimodules shifted down by i.

Recall that this was stated in Chap. 19 assuming Soergel's conjecture. This statement should in fact be incorporated into Elias–Williamson's inductive proof, but for simplicity we will take it as a black box.

In what follows, we simply write F_w for the minimal Rouquier complex $F_w^{\min}$. This should not be confused with the Rouquier complex $F_{\underline{w}} = \bigotimes_i (B_{s_i} \to R(1))$ for a reduced expression $\underline{w} = (s_1, \ldots, s_m)$ for w. By definition, F_w is the minimal complex of $F_{\underline{w}}$, and is defined up to a non-unique isomorphism.

Now, recall the induction outlined in Sect. 18.3. In this chapter, we explain Step 1(A).

> **Assumption 20.3** Throughout this chapter, let $x \in W$ and $s \in S$ be such that $xs > x$, and assume $\mathrm{HR}(w)$ for all $w < xs$. Our goal is to show $\mathrm{hL}(x, s)_\zeta$ for all $\zeta \geq 0$.

20.3 The Hodge–Riemann Relations for the Rouquier Complex

To apply Lemma 20.1, we need to know Hodge–Riemann on the target. For this, we first consider the following additional statement (Rouquier–HR) in our induction:

$$\mathrm{RoHR}(w) : \begin{array}{l}\text{the Hodge–Riemann bilinear relations hold} \\ \text{for } L_0 \text{ on the minimal Rouquier complex } F_w.\end{array}$$

Let us be more precise. Given a complex of Soergel bimodules C, denote its i-th term by C^i. The operator L_0 makes sense on any F_w^i, but we need to fix an inner product to talk about the Lefschetz form. We do this as follows. Given a reduced expression $\underline{w}$ for w, we can embed F_w as a summand inside $F_{\underline{w}}$. The i-th term of the latter is canonically a sum of Bott–Samelson bimodules corresponding to subexpressions of $\underline{w}$ of length $\ell(\underline{w}) - i$, all shifted down by i, so the summand F_w^i shifted up by i carries an induced invariant form. When we say that a Lefschetz operator on F_w satisfies some property, we mean that for any reduced expression $\underline{w}$ for w, there is an embedding of F_w as a summand in $F_{\underline{w}}$ such that the property holds for the induced inner product on each $\overline{F_{\underline{w}}^i}(-i)$.[1]

The goal of this section is to prove the following.

Proposition 20.4 *Assume* $\mathrm{HR}(w)$ *for all* $w < xs$. *Then* $\mathrm{RoHR}(x)$ *holds*.

Let us explain the content of Proposition 20.4 in down to earth terms. From the diagonal miracle we know that $F_x^i(-i)$ is a sum of Soergel bimodules $\bigoplus B_z^{\oplus n_z}$, and $\mathrm{HR}(z)$ is inductively known for any B_z occurring with nonzero multiplicity. Different isotypic components are automatically orthogonal for the Lefschetz form, by the Soergel Hom formula, and without loss of generality we may assume the direct sum decomposition $B_z^{\oplus n_z}$ is also orthogonal. A priori, if $n_z > 1$, in some of the summands the given form could be a positive multiple of the standard one, and in others a negative multiple. Proposition 20.4 is asserting that

---

[1] Actually, this is not quite enough. One can insist, with the same proof, that the Lefschetz properties hold after rescaling the inner products on each Bott–Samelson summand of $F_{\underline{w}}$ by a positive scalar, since we will use this extra flexibility later on in this chapter.

(i) this mixed signature situation never occurs.

Moreover, given our formulation of the Hodge–Riemann bilinear relations, it is further asserting that

(ii) the elements z for which $n_z > 0$ are either all even or all odd in length.

Finally, to have primitives from different isotypic components in the same graded degree be all positive or negative definite, it is asserting that

(iii) whether the induced form B_z is a positive or negative multiple of the standard form alternates according to $\ell(z) \mod 4$.

Before giving the proof of positivity, as a warm-up we invite the reader to think through the easier case of nondegeneracy:

Exercise 20.5 Show that hard Lefschetz holds for L_0 on F_x. That is, for any $i \geq 0$, deduce from our inductive hypotheses that hard Lefschetz holds for $\overline{F_x^i(-i)}$.

We now turn to the proof of Proposition 20.4, which will take up the rest of this section.

Proof Write $x = yt$, where $t \in S$ and $y < x$. Then F_x occurs as a summand in $F_y F_t$, and we may assume the embedding is an isometry. Recalling that F_t is the complex $B_t \to R(1)$ in degrees 0 and 1, we have

$$F_x^i \overset{\oplus}{\subset} (F_y F_t)^i = F_y^i B_t \oplus F_y^{i-1}(1), \tag{20.2}$$

where C^i denotes the i-th term of a complex C.

The action of L_0 preserves each summand, so writing $\mathcal{P}(-)$ for the $\mathbb{Z}_{\leq 0}$-graded subspace of primitives, we have:

$$\mathcal{P}\overline{(F_y F_t)^i} = \mathcal{P}(\overline{F_y^i B_t}) \oplus \mathcal{P}(\overline{F_y^{i-1}(1)}). \tag{20.3}$$

As $y < x$, by induction the inner product induced by L_0 on the second summand alternates between negative and positive definiteness.

A common approach to prove that a space has Hodge–Riemann is to embed it as an L-stable subspace inside another space with Hodge–Riemann, and it may look like that's what is happening here. We embed F_x^i using (20.2), and the second direct summand $F_y^{i-1}(1)$ has Hodge–Riemann. However, as we will see, the first direct summand $F_y^i B_t$ will not satisfy Hodge–Riemann! Thus we must proceed with care. The moral of the story will be that, while Hodge–Riemann does not hold on $F_y F_t$, it will hold on a big enough piece of $F_y F_t$ to guarantee that it holds on the direct summand F_x.

Let us turn to this first summand. When we tensor a Soergel bimodule B_z for $z \in W$ with B_t, either one of two things can happen, depending on whether $\ell(zt) =$

20.3 The Hodge–Riemann Relations for the Rouquier Complex

$\ell(z) + 1$ or $\ell(z) - 1$.[2] If $zt > z$, we have already seen that $B_z B_t$ is again a sum of unshifted Soergel bimodules. When $zt < z$, we get

$$B_z B_t \simeq B_z(1) \oplus B_z(-1). \tag{20.4}$$

Indeed, since we are in the regime where Soergel's conjecture holds, this follows from the corresponding equality at the level of characters (see Theorem 3.27).

Accordingly, we will write

$$F_y^i(-i) = \bigoplus_{z \leq y : zt > z} B_z^{\oplus n_z} \oplus \bigoplus_{z \leq y : zt < z} B_z^{\oplus n_z}. \tag{20.5}$$

Calling the first summand $B^\uparrow$, and the second summand $B^\downarrow$, we obtain

$$F_y^i(-i) B_t \simeq B^\uparrow B_t \oplus B^\downarrow(1) \oplus B^\downarrow(-1). \tag{20.6}$$

This is a decomposition as R-bimodules, and L_0 is left multiplication by an element of R, so L_0 preserves each direct summand.

For any $z \in W$, $\overline{B_z}$ is a finite-dimensional graded vector space satisfying Poincaré duality.[3] That is, the dimension in degree $-i$ is the same as the dimension in degree $+i$ for all i. Thus the direct sum $\overline{B^\downarrow}$ also satisfies Poincaré duality. In particular, $B^\downarrow(1)$ cannot possibly satisfy Poincaré duality, having a minimal degree min which is more negative than the maximal degree is positive. So $L_0^{\min}$ is zero on the minimal degree, because its target is zero, and the Lefschetz form on the lowest degree is identically zero. In this way, hard Lefschetz (and thus Hodge–Riemann) are guaranteed to fail on $B^\downarrow(1)$.

One should have the following picture in mind for the space $B^\downarrow(1) \oplus B^\downarrow(-1)$ (drawn when $B^\downarrow = B_t$):

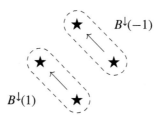

In this picture, clearly the Lefschetz operator squared is zero on the degree -2 piece.

---

[2] That is to say, when geometry is available, whether the convolution morphism is generically a fibration in points or curves.

[3] This is a general fact, which can be proven even when the Soergel conjecture fails. It is overkill, but one can also justify this statement for $z \leq y$ using the inductive assumption HR(z).

Before continuing, we first note that the vanishing of the Lefschetz form on $B^\downarrow(1)$ is not limited to its lowest degree component:

Lemma 20.6 *The Lefschetz form of L_0 on $\overline{B^\downarrow(1)}$ is identically zero.*

Proof We know inductively that the Lefschetz form is nondegenerate on $\overline{B_w}$ for all $w < xs$. It follows that the Lefschetz form on $\overline{B^\downarrow}$ is nondegenerate, and so the problems have to all stem from the shift down by one.

Indeed, the claim is easily seen from the $\mathfrak{sl}_2$ picture of hard Lefschetz (see Sect. 17.2). Namely, on the shifts of the primitives from B_w the form obviously vanishes (since one is raising a lowest weight vector from weight $-n$ to weight $n+2$), and self-adjointness of L_0 handles the rest, as in Remark 17.13. □

Remark 20.7 A word of warning. Our discussion of Lefschetz linear algebra always assumed we began with a space H equipped with a nondegenerate graded form $\langle -, - \rangle$. Now $B^\downarrow(1)$ is not such a space, and neither is $H(1)$ in general (as the shifted form will no longer be graded). It is $\overline{B^\downarrow(1)} \oplus \overline{B^\downarrow(-1)}$ which has a nondegenerate graded form, and we are only claiming that the restriction of the Lefschetz form to $\overline{B^\downarrow(1)}$ is zero. We do not claim that the graded form $\langle -, - \rangle$ restricts to the zero form on $\overline{B^\downarrow(1)}$, and it might not. □

We are now rather close. Namely, consider the embedding

$$\iota : F_x^i(-i) \to F_y^{i-1}(-i+1) \oplus B^\uparrow B_t \oplus B^\downarrow(1) \oplus B^\downarrow(-1), \tag{20.7}$$

where we still have $x = yt$. By the diagonal miracle and the fact that there are no maps between Soergel bimodules of negative degree, one component of ι is identically zero so we can rewrite this as

$$\iota : F_x^i(-i) \to F_y^{i-1}(-i+1) \oplus B^\uparrow B_t \oplus B^\downarrow(1). \tag{20.8}$$

Moreover, by Lemma 20.6, we can project away from $B^\downarrow(1)$, and the induced map

$$\iota' : \overline{F_x^i(-i)} \to \overline{F_y^{i-1}(-i+1)} \oplus \overline{B^\uparrow B_t} \tag{20.9}$$

will still be an isometry. We would be finished if (i) ι' were injective, and (ii) we have set things up so that primitive forms in the same degree across different $\overline{B_w}$, $w \in W$, appearing above have the same definiteness.

Both points are fairly mild. Point (ii) can be achieved with extra care and bookkeeping in the inductive procedure. Point (i) is handled by the following lemma (set $N = B^\downarrow(1)$), which concludes the proof of Proposition 20.4. □

Lemma 20.8 *Consider a split injection $(i, j) : S \to M \oplus N$, where S and M are sums of unshifted indecomposable Soergel bimodules. and N is a sum of Soergel bimodules with strictly positive shifts. Then $i : S \to M$ is a split injection.*

Proof By assumption there is a splitting map $M \oplus N \to S$, with components $p: M \to S$ and $q: N \to S$. Because there are no negative degree morphisms between indecomposable, self-dual Soergel bimodules,[4] the map q must be zero. Hence $\mathrm{id}_S = p \circ i + q \circ j = p \circ i$ and $i: S \to M$ is a split injection. □

20.4 Positivity of Breaking

In the rest of this chapter, we sketch the proof of $\mathrm{hL}(x, s)_\zeta$: the hard Lefschetz theorem for $B_x B_s$ and the deformed Lefschetz operators

$$L_\zeta : b_x \otimes b_s \mapsto \rho b_x \otimes b_s + b_x \otimes \zeta \rho b_s, \qquad \zeta \in \mathbb{R}_{\geq 0}. \tag{20.10}$$

Fix a reduced expression $\underline{x}$ for x. Let $\underline{xs}$ be the expression obtained by concatenation. For $1 \leq i \leq \ell(\underline{xs})$, let $\underline{xs}_{\hat{i}}$ denote the expression obtained from $\underline{xs}$ by removing the i-th simple reflection. Then the Rouquier complex $F_{\underline{x}} F_s$ starts as

$$\mathrm{BS}(\underline{xs}) \xrightarrow{\Omega} \bigoplus_{i=1}^{\ell(\underline{xs})} \mathrm{BS}(\underline{xs}_{\hat{i}})(1) \to \cdots . \tag{20.11}$$

Up to replacing $F_{\underline{x}} F_s$ by an isomorphic complex, we may sprinkle signs on its differentials and assume that the i-th component of Ω is

$$\left| \cdots \left| \begin{matrix} \bullet \\ | \end{matrix} \right| \cdots \right|,$$

where the dot map appears at the i-th place. We view Ω as a degree one morphism to a sum of unshifted Bott–Samelson bimodules. In particular, both the source and the target of Ω now carry canonical inner products.

Consider the induced map

$$\overline{\Omega} : \overline{\mathrm{BS}(\underline{xs})} \to \bigoplus_{i=1}^{\ell(\underline{xs})} \overline{\mathrm{BS}(\underline{xs}_{\hat{i}})} \tag{20.12}$$

on right quotients. The goal of this section is to prove Proposition 20.12 below, which says that with a little massaging, (20.12) can be put into the setup described in Lemma 20.1. By abuse of notation, for any $\zeta \geq 0$, we will also write L_ζ to denote

---

[4]This is not a priori clear, but follows from Soergel's conjecture. More carefully, to deduce this for a pair $B_{w_1}, B_{w_2}, w_1, w_2 \leq w_3$, one needs only Soergel's conjecture for all $w \leq w_3$. So, there is no circularity using this fact in the inductive argument.

the degree 2 endomorphism of BS($\underline{x}s$) which multiplies by ρ on the left, and by $\zeta\rho$ in the slot before s.

Remark 20.9 We will not be applying Lemma 20.1 to the first differential (20.12) of $\overline{F_{\underline{x}}F_s}$ itself, as one does not expect Bott–Samelson bimodules to satisfy Hodge–Riemann. We will complete the setup of Lemma 20.1 for (20.12), but to deduce hL$(x, s)_\zeta$ (in the next section), we will only apply Lemma 20.1 to the first differential of $\overline{F_{\underline{x}}F_s}$, which is a direct summand of $\overline{F_{\underline{x}}F_s}$.

We begin putting (20.12) into the setup of Lemma 20.1. The first condition holds even before passing to right quotients.

Lemma 20.10 *The first differential Ω of the Rouquier complex $F_{\underline{x}}F_s$ is injective from negative graded degrees.*

Proof This is an immediate consequence of the computation of the cohomology of Rouquier complexes in Lemma 19.40. In fact, Ω is injective in degrees less than $\ell(\underline{x}s)$. □

We also need a Lefschetz operator on $\bigoplus_{i=1}^{\ell(\underline{x}s)} \overline{BS(\underline{x}s_{\hat{i}})}$ which commutes with $\overline{\Omega}$ in the appropriate sense. In an ideal world, in order to prepare for an inductive proof, we want this Lefschetz operator to resemble "L_ζ for a smaller expression" on each direct summand (and we wish to minimize interaction between the different summands).

Lemma 20.11 *There are degree 2 endomorphisms*

$$L_i : \mathrm{BS}(\underline{x}s_{\hat{i}}) \to \mathrm{BS}(\underline{x}s_{\hat{i}})(2), \quad i = 1, \ldots, \ell(\underline{x}s),$$

such that:

1. *Their direct sum $L := \bigoplus_i L_i$ satisfies*

$$L \circ \Omega = \Omega \circ L_\zeta. \tag{20.13}$$

2. *On the right quotient, $\overline{L}$ is a Lefschetz operator.*

Proof The composition $\Omega \circ L_\zeta : \mathrm{BS}(\underline{x}s) \to \bigoplus_i \mathrm{BS}(\underline{x}s_{\hat{i}})(-2)$ has i-th component

$$\begin{cases} \boxed{\rho}\,\Big|\cdots\Big|\,\bullet\,\Big|\cdots\Big|\Big| + \Big|\cdots\Big|\,\bullet\,\Big|\cdots\Big|\,\boxed{\zeta\rho}\Big| & \text{if } 1 \le i < \ell(\underline{x}s), \\ \boxed{\rho}\,\Big|\ \cdots\ \Big|\,\bullet\,\Big| + \Big|\ \cdots\ \Big|\,\boxed{\zeta\rho}\,\bullet & \text{if } i = \ell(\underline{x}s), \end{cases} \tag{20.14}$$

where the dot map appears at the i-th place. (We have colored s red but left the simple reflections in $\underline{x}$ uncolored.) Let L_i denote this same map, but with the strand containing the dot removed. Then the first property is clear.

20.4 Positivity of Breaking

Moreover, for $1 \le i < \ell(\underline{x}s)$, L_i coincides with L_ζ for the shorter expression $\underline{x}s_{\hat{i}}$. For $i = \ell(\underline{x}s)$, the induced operator $\overline{L_i}$ coincides with $\overline{L_0}$ on $\overline{BS(\underline{x})}$ (since right multiplication by $\zeta\rho$ is zero in the right quotient). In particular, $\overline{L}$ is a Lefschetz operator. $\square$

One sees that the operators L_i are ripe for an inductive argument.

To complete the setup of Lemma 20.1, only the second condition remains:

$$\langle \overline{\Omega}v, \overline{\Omega}w \rangle \stackrel{?}{=} \langle \overline{L_\zeta}v, w \rangle. \tag{20.15}$$

Since only the right hand side depends on the parameter ζ, we should expect that some modification is necessary to get this to work.

It suffices to understand what happens to the 01-basis for Bott–Samelson bimodules introduced in Chap. 12. Given two elements v, w in the 01-basis of $BS(\underline{x}s)$, the pairing on the left hand side of (20.15) is the coefficient of c_{top} in

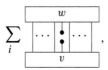

where in the i-th summand the i-th line is broken.

To understand the pairing on the right hand side, it will be helpful to rewrite L_ζ on $BS(\underline{x}s)$ using the polynomial forcing relation. For some scalars λ_i, $1 \le i \le \ell(\underline{x}s)$, and some linear polynomial f, we have

$$L_\zeta = \boxed{\rho}\bigg|\cdots\bigg|\bigg| + \bigg|\cdots\bigg|\boxed{\zeta\rho}$$

$$= \sum_i \lambda_i \bigg|\cdots\bigg|\substack{\bullet\\\bullet}\bigg|\cdots\bigg|\bigg| + \bigg|\cdots\bigg|\bigg|\boxed{f}.$$

The i-th strand is broken in the i-th term of the sum. It follows that the right hand side of (20.15) is the coefficient of c_{top} in

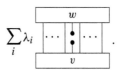

Therefore, if we rescale the standard inner product on $\overline{BS(\underline{x}s_{\hat{i}})}$ by λ_i, we obtain equality in (20.15).

We have now proven all parts of the following proposition except for the final statement.

Proposition 20.12 *With L_ζ, L_i, λ_i, and L defined as above, the first differential $\overline{\Omega}$, viewed as a degree $+1$ map*

$$\overline{\Omega} : \overline{\mathrm{BS}(\underline{x}s)} \to \bigoplus_{i=1}^{\ell(\underline{x}s)} \overline{\mathrm{BS}(\underline{x}s_{\hat{i}})},$$

satisfies the following properties. Equip $\overline{\mathrm{BS}(\underline{x}s_{\hat{i}})}$ with its usual intersection form rescaled by λ_i, and $\bigoplus_{i=1}^{\ell(\underline{x}s)} \overline{\mathrm{BS}(\underline{x}s_{\hat{i}})}$ with the direct sum of these rescaled forms. Then

1. $\overline{L} \circ \overline{\Omega} = \overline{\Omega} \circ \overline{L_\zeta}$,
2. $\overline{\Omega}$ *is injective from negative degrees, and*
3. $\langle \overline{\Omega} v, \overline{\Omega} v' \rangle = \langle \overline{L_\zeta} v, v' \rangle$ *for all $v, v' \in \overline{\mathrm{BS}(\underline{x}s)}$.*

Moreover, if $\underline{x}s$ is a reduced expression, we can choose all the scalars λ_i to be positive.

This final positivity statement is crucial. If a graded form (on a space with a Lefschetz operator) satisfies Hodge–Riemann, and one rescales it by a scalar λ, then the result satisfies Hodge–Riemann if and only if $\lambda > 0$. When $\lambda < 0$ one satisfies a version of Hodge–Riemann with opposite signs, which doesn't sound so bad at first. The issue arises when taking direct sums of multiple spaces. If the signs all match up appropriately, then the primitive subspace is a direct sum of spaces with positive definite forms, and hence it has a positive definite form. If the signs don't match, then the resulting form on the direct sum is nondegenerate but of mixed signature. Remember the special feature of positive definite forms: a nondegenerate form of mixed signature might become degenerate after restricting to a subspace, but a positive definite form restricts to a positive definite form. We used this argument to deduce the nondegeneracy of the local intersection form from Hodge–Riemann for the global intersection form, and it is also the foundation of the proof of Lemma 20.1.

The moral of the story is that *we need each λ_i to be strictly positive* in order to deduce any statements about hard Lefschetz for L_ζ from Hodge–Riemann for each L_i.

The positivity of λ_i has already been discussed and proven in appendix, see the Positive Breaking Lemma 12.34.

20.5 Sketch of the Proof of Hard Lefschetz

Now, by Lemma 20.1 applied to $\overline{\Omega}$, hard Lefschetz for L_0 on $B_x B_s$ would follow from Hodge–Riemann for L_0 on $(F_{\underline{x}} F_s)^1(-1)$. However, as already mentioned in Remark 20.9, one does not expect Hodge–Riemann to hold for the entire Bott–

20.5 Sketch of the Proof of Hard Lefschetz

Samelson bimodule.[5] A better hope is that the Hodge–Riemann relations hold for L_0 on the direct summand $(F_x F_s)^1(-1)$. This still can fail because of terms of the form $B_z B_s$, $zs < z$, as we saw in Lemma 20.6.

The key observation is that so long as $\zeta > 0$, so that the second ρ is "turned on," *we should expect Hodge–Riemann to hold for L_ζ on $B_z B_s$.* Thus for $\zeta > 0$, this will give hard Lefschetz for L_ζ on $B_x B_s$. Hard Lefschetz for L_0 will require slightly more care, but can be done. Finally, since we use it for $B_x B_s$, we will need to prove along the way that Hodge–Riemann holds for $B_z B_s$, when $zs < z$. Depending on one's taste, it may be helpful to think about the above statements in light of the discussion in appendix.

Exercise 20.13 Show by hand that hard Lefschetz fails for L_0 on $B_s B_s$. Similarly show that for L_ζ with $\zeta > 0$, both hard Lefschetz and Hodge–Riemann hold on $B_s B_s$.

Exercise 20.14 Continuing the previous exercise. If you know some geometry, work out what the line bundles are on the Bott–Samelson variety

$$Y(s,s) = P_s \overset{B}{\times} P_s / B \simeq \mathbb{P}^1 \times \mathbb{P}^1 \qquad (20.16)$$

which correspond to L_0 and L_1. (*Hint:* Calculate the degree of their restriction to the two Schubert curves. Compare these two curves to the divisors $\mathbb{P}^1 \times \text{pt}$, $\text{pt} \times \mathbb{P}^1$.)

Theorem 20.15 *If $xs > x$, then $\mathrm{hL}(x,s)_\zeta$ holds for $\zeta \geq 0$. If $xs < x$, then $\mathrm{hL}(x,s)_\zeta$ holds for $\zeta > 0$. (Note: both results rely upon certain inductive assumptions.)*

Proof There are three cases:

- $\zeta > 0$ and $xs > x$,
- $\zeta = 0$ and $xs > x$,
- $\zeta > 0$ and $xs < x$.

In the missing case $xs < x$ and $\zeta = 0$, $\mathrm{hL}(x,s)$ fails! The third case does not relate much to the other two. The first case is motivated by the deformation philosophy, whereas the second one is the actual goal. The reader is advised to take the first case as a warm-up to the second case.

Case 1 ($\zeta > 0$ and $xs > x$): By the diagonal miracle, F_x starts as $B_x \to F_x^1 \to \cdots$, so $F_x F_s$ starts as

$$B_x B_s \xrightarrow{\Phi = (\Phi_1, \Phi_2)} F_x^1 B_s \oplus B_x(1) \to \cdots . \qquad (20.17)$$

We wish to apply Lemma 20.1 to $\overline{\Phi}$. Supposing we have rescaled our intersection forms by λ_i, just as in the previous proposition we have $\langle \overline{\Phi}(v), \overline{\Phi}(v') \rangle =$

---
[5]There is a geometric reason for this: full Bott–Samelson resolutions are rarely semismall.

$\langle v, \overline{L_\zeta v'} \rangle$ for all $v, v' \in \overline{B_x B_s}$. The component Φ_1 commutes with L_ζ, whereas $\Phi_2(L_\zeta v) = \rho \cdot \Phi_2(v) + \Phi_2(v) \cdot \zeta \rho$ for all $v \in \overline{B_x B_s}$. Passing to Soergel modules kills the ζ term, so $\overline{\Phi}$ commutes with L_ζ. Moreover, $\overline{\Phi}$ is injective in negative degrees by the same homology argument as in Proposition 20.12.

So we almost have what we want. We still need Hodge–Riemann on the target of Φ, which on the second summand is already assumed. Meanwhile, F_x^1 is built from copies of B_z with $z < x$, and using the inductive assumption $\mathrm{HR}(z,s)_\zeta$ one can deduce that $\overline{F_x^1 B_s}$ has Hodge–Riemann.

Case 2 ($\zeta = 0$ and $xs > x$): Proceeding just as in Case 1, we rescale our intersection forms by λ_i. Since $\zeta = 0$ we cannot use exactly the same trick as before, because Hodge–Riemann will fail on $F_x^1 B_s$.

We may however decompose $F_x^1(-1) = B^\uparrow \oplus B^\downarrow$, so that the first two homological degrees of $F_x F_s$ are

$$B_x B_s \xrightarrow{\Phi} B^\uparrow B_s(1) \oplus B^\downarrow B_s(1) \oplus B_x(1), \tag{20.18}$$

and $B^\downarrow B_s(1) \simeq B^\downarrow(0) \oplus B^\downarrow(2)$. As before, our Lefschetz operator L_0 is left multiplication by an element in R, so any decomposition as R-bimodules is respected by L_0.

Now $F_x F_s \in K^{\geq 0}$ by Lemma 19.43,[6] so only negative shifts are allowed in its minimal complex. Hence it must be possible to Gaussian eliminate $B^\downarrow(2)$ against some direct summand in homological degree two. After doing this, the remaining complex, homotopy equivalent to $F_x F_s$, will have the form

$$B_x B_s \xrightarrow{d=(d_1,d_2,d_3)} B^\downarrow(0) \oplus B^\uparrow B_s(1) \oplus B_x(1), \tag{20.19}$$

and we now work with this complex instead. Our differential d is the sum of a degree zero map d_1 to an object $B^\downarrow$ satisfying hard Lefschetz, and a degree one map $d_2 + d_3$ to an object $B^\uparrow B_s \oplus B_x$ satisfying Hodge–Riemann.

We wish to show that $\overline{B_x B_s}$ satisfies hard Lefschetz, so choose a nonzero element v in degree $-k$. If $d_1(v) \neq 0$ then $L_0^k d_1(v) \neq 0$, using hard Lefschetz on $B^\downarrow$. Since d_1 is an R-bimodule map and commutes with L_0, we get $d_1(L_0^k v) \neq 0$, and thus $L_0^k v \neq 0$ as desired. If instead $d_1(v) = 0$, then $d = d_2 + d_3$ is a degree $+1$ map to a space satisfying Hodge–Riemann, and we can use the same argument as Lemma 20.1 to deduce that $L_0^k v \neq 0$.

Case 3 ($\zeta > 0$ and $xs < x$): According to the results on singular Soergel bimodules in [178], there is actually an (R, R^s)-bimodule B such that $B \otimes_{R^s} R \simeq B_x$. Thus the decomposition $B_x B_s \simeq B_x(-1) \oplus B_x(1)$ can be deduced in much the same way as the decomposition $B_s B_s \simeq B_s(-1) \oplus B_s(1)$ was deduced in Example 4.35. Morally speaking, there is not much difference between the case $x = s$ (which was Exercise 20.13) and the general case. Ultimately one

---

[6] As with the diagonal miracle, this statement should really be included in the inductive argument.

uses hL(x) to argue that $\overline{B_x}$ with its Lefschetz operator corresponds to some $\mathfrak{sl}_2$ representation X, and deduces that $\overline{B_x B_s}$ corresponds to the $\mathfrak{sl}_2$ representation $X \otimes \mathbb{R}^2$, the tensor product with the standard representation. We omit the details, see [57, Theorem 6.19].

This finishes the proof of hL$(x, s)_\zeta$. □

Appendix: Some Historical Context and Geometric Intuition for the Proof of Soergel's Conjecture

In this appendix we assume that the reader has some background in algebraic geometry (line bundles, ample cone, intersection theory ...) and the rudiments of Hodge theory. The material in this appendix is for motivation only, and is not used elsewhere in this book.

The Kazhdan–Lusztig Conjecture for Weyl Groups and the Decomposition Theorem

In this chapter we sketched why hard Lefschetz and the Hodge–Riemann bilinear relations hold for the global, whence local, intersection form. It is reasonable to wonder why one would anticipate these claims being true. In this appendix, we provide some historical and geometric motivation.

The first proofs of the Kazhdan–Lusztig conjecture went as follows. First, Kazhdan–Lusztig showed their polynomials gave the graded dimensions of the stalks of the intersection cohomology sheaf on a Schubert variety [109]. Second, Brylinski–Kashiwara and Beilinson–Bernstein made the connection between certain Lie algebra representations and regular holonomic D-modules on the flag variety [17, 38]. This was motivated by earlier work of Borel–Weil and Kempf, by which it was understood that many familiar objects of category $\mathcal{O}$ came from the local and global coherent cohomology of Schubert cells and varieties [111].

In the first step, much use is made of the fact that Schubert varieties are algebraic varieties, and not simply topological spaces. That is, one can use general theorems on the topology of algebraic varieties to control their intersection cohomology. Indeed, Kazhdan and Lusztig go so far as to write in [109]:

> Our paper contains only formal manipulations with deep unpublished results of Deligne, which he kindly explained to us.

In [109] Kazhdan and Lusztig deduce results about Schubert varieties $X_w(\mathbb{C})$ over the complex numbers by studying their counterparts over $\overline{\mathbb{F}}_q$. For finite q, extra control over the topology of $X_w(\overline{\mathbb{F}}_q)$ is given by the Galois group $\mathrm{Gal}(\overline{\mathbb{F}}_q/\mathbb{F}_q)$, namely use of the Frobenius and the associated theory of weights developed by Deligne.

In later developments, authors could avoid explicitly reducing to positive characteristic by using the extra control over the topology of $X_w(\mathbb{C})$ given by Saito's mixed Hodge modules, a theory of weights for perverse sheaves in complex algebraic geometry. In fact, it was long understood that the only consequence of the formalism one needs in proving the result of Kazhdan–Lusztig was the *decomposition theorem*, which roughly says pushing forward the constant (intersection cohomology) sheaf along a morphism between projective varieties is as semisimple as possible.[7]

Much more recently, de Cataldo–Migliorini gave a new proof of the decomposition theorem [39] that is very different from the original one by Beilinson–Bernstein–Deligne.[8] As they show, for semismall resolutions $\pi : X \to Y$ the decomposition theorem hinges on the nondegeneracy of local intersection forms on the locus with maximal (that is, middle) dimensional fibers. To show this directly, they pick an ample bundle $\mathcal{L}$ on the base. Since a generic hyperplane section pulled back from the base will miss these middle dimensional cycles, they are automatically primitive with respect to $\pi^*\mathcal{L}$.

If $\pi^*\mathcal{L}$ were ample, that single primitivity observation would be enough to finish the argument. Let us review why by revisiting some of the usual Hodge theory (recall X is smooth, and let us suppose X and Y are projective). From an ample class, one obtains a pairing on each cohomology group $H^i(X) := H^i(X, \mathbb{C})$ in the middle dimension and below. (This form is sometimes called the *Lefschetz form*.) For ease of imagination, take the class to be very ample, so that we have an associated embedding of X into a projective space. How does one pair the fundamental classes of subvarieties Z, Z' of the same dimension, by assumption at least half that of X? Well, we may imagine them to be in general position, intersect, and then record the *degree* of this intersection in the ambient projective space. Recall that the degree of a subvariety of a projective space is an integer, which e.g. for a hypersurface cut out by one homogeneous equation F is simply the degree of F as a polynomial. One knows this self-pairing, extended in a natural way to all of H^*, is nondegenerate (this is a consequence of *Poincaré duality* and the *hard Lefschetz theorem*). Notice that for middle dimensional cycles, the Lefschetz form and the usual intersection form agree. The Lefschetz form always is either positive or negative definite on primitive forms of type Hodge type (p, p) (this follows from the *Hodge–Riemann bilinear relations*), the desired nondegeneracy of the intersection pairing follows.

---

[7]This argument first appears in a Bourbaki talk of Springer [167], who in turn attributed it to Brylinski and MacPherson. One can also see this calculation using only the decomposition theorem explained in the notes of K. Rietsch [155]. In fact, a good exercise for the reader is to adapt the argument of [155] using only one step convolution (as opposed to the full Bott–Samelson resolution), as in the proof of Soergel's conjecture outlined here.

[8]The reader may wish to consult a survey of their work [183]. In fact, the relevant case for us is almost the case of a semismall resolution, as we will be simulating "one step convolution with a Schubert curve," which is a semismall morphism, albeit with a possibly singular source.

20.5 Sketch of the Proof of Hard Lefschetz

However, the pullback $\pi^*\mathcal{L}$ will rarely be ample. Surprisingly, de Cataldo–Migliorini show that the invoked properties in the preceding paragraph, namely hard Lefschetz and the Hodge–Riemann bilinear relations, still hold for $\pi^*\mathcal{L}$, which finishes the argument. (This part of the argument hinges on π being a semi-small map.) To show these facts about $\pi^*\mathcal{L}$, they use induction on the dimension along with multiplying $\pi^*\mathcal{L}$ by a *bona fide* ample class on X. Both these features show up in the proof of Soergel's conjecture, as we spell out below. Interestingly, de Cataldo–Migliorini write [39]:

> Furthermore, the direct proof of the nondegeneracy of these forms, with the new additional information on the signatures, neither relying on reduction to positive characteristic nor on Saito's theory of mixed Hodge modules, sheds light on the geometry underlying the decomposition theorem and gives some indications on its possible extensions beyond the algebraic category.

While they likely had complex analytic applications in mind, we saw in this chapter an example of more algebraic application of these ideas. Indeed, the overarching theme is that *Soergel bimodules for a general Coxeter group still behave as though they were the (equivariant intersection) cohomology of Schubert varieties, even when there is no flag variety available*. In particular, we want to apply the decomposition theorem, but in the absence of geometry this requires justification. Elias–Williamson realized that de Cataldo–Migliorini's approach to the decomposition theorem in terms of intersection forms, Hodge theory, and positivity can be nontrivially adapted to our setting.

We will spell out the previous paragraph in more detail in the remainder of the appendix. Namely, we will see that thinking about de Cataldo–Migliorini's argument leads us to guess both the statements of the theorems proved in this chapter as well as in many cases the structure of their proofs. Along the way, we will include gentle reminders about some of the geometric ideas which appear.

What We Need I: Hard Lefschetz in a Family

We were setting up an argument by induction on the length in W, which corresponds to the dimension of Schubert varieties when geometry is available. For $x \in W$ and $s \in S$ with $xs > x$, we saw that Soergel's conjecture for xs was the claim that as many B_z for $z < xs$ split off as possible from $B_x B_s$, which in turn was the assertion the local intersection form on $\mathrm{Hom}(B_z, B_x B_s)$ was nondegenerate. To show this, in the spirit of de Cataldo–Migliorini we will make use of the embedding $\mathrm{Hom}(B_z, B_x B_s) \to B_x B_s$, which intertwines the inner products (up to scalar). As in the geometric argument, it is easy to see the image lands in primitives, so again inspired by geometry we will hope that the Hodge Riemann bilinear relations hold in $B_x B_s$. This would provide an extremely transparent reason for the claimed nondegeneracy. Indeed, we thereby avoid the complication that for a general nondegenerate bilinear form, its restriction to a subspace can fail to be nondegenerate. (An example of such a failure is given by an isotropic line in a plane with an inner product of signature $+, -$.)

As in geometry, it is hard to see directly the definiteness on primitive forms for the Lefschetz form of L_0, which is left multiplication by ρ on $\overline{B_x B_s}$. Indeed, before considering the Hodge–Riemann relations it is even non-obvious that L_0 satisfies hard Lefschetz. To tackle the analogous geometric question, de Cataldo–Migliorini deformed $\pi^*\mathcal{L}$ to a ample class on X. Elias–Williamson accordingly deform L_0 to the 1-parameter family of operators:

$$L_\zeta : b_x \otimes b_s \mapsto \rho b_x \otimes b_s + b_y \otimes \zeta \rho b_s, \qquad \zeta \in \mathbb{R}_{\geq 0}. \tag{20.20}$$

Let us explain why this is a natural choice of deformation and how it is related to ampleness. We first explain a potentially confusing point. In the algebraic setting, we always consider the "right quotient" $\overline{B} \otimes_R \mathbb{R}$ of Soergel bimodules. In the language of equivariant cohomology, this corresponds to passage from $H^*_{B \times B}(G, \mathbb{R})$ to $H^*_{B \times \{1\}}(G, \mathbb{R}) = H^*(B \backslash G, \mathbb{R})$. Thus, in order to match to the above language of Soergel bimodules, it is necessary work with Schubert varieties in $B \backslash G$ rather than the more standard G/B.

Let us assume we are in the above setting ($xs > x$ with $s \in S$). Consider the convolution space $X_x \times_B P_s$, where $X_x \subset B \backslash G$ is the Schubert variety parameterized by $x \in W$. We have a commutative diagram of spaces

$$\begin{array}{ccccc} X_x \times^B P_s & \xrightarrow{i} & B \backslash G \times B \backslash P_s & = & B \backslash G \times \mathbb{P}^1 \\ \downarrow{m} & & \downarrow{p} & & \\ X_{xs} & \longrightarrow & B \backslash G & & \end{array} \tag{20.21}$$

The map i is the product of the multiplication and projection maps (a closed embedding); p is the projection; m is induced by the multiplication; and the unlabeled arrow is the inclusion. The map m defines a (singular variant of a) semi-small resolution of X_{xs}.

The Lefschetz operator L_0 is multiplication by the first Chern class of $m^*\mathcal{L}$, where $\mathcal{L}$ an ample line bundle on $B \backslash G$. This is not always ample on $X_x \times^B P_s$, because of the $\mathbb{P}^1$ which is contracted under the right hand map. When we instead take L_1, we add on the first Chern class of an ample line bundle on $\mathbb{P}^1$, which is ample on $B \backslash G \times \mathbb{P}^1$, and hence also on $X_x \times^B P_s$. In particular, by the usual (singular) Hodge theory, which we will review below, the intersection cohomology of $X_x \times^B P_s$ (alias $\overline{B_x B_s}$) will satisfy hard Lefschetz as well as the Hodge–Riemann bilinear relations with respect to the ample class L_1. Note that this ampleness of L_1 will still hold even when $xs < x$, which motivates the second claim of Theorem 20.15.

Exercise 20.16 Check the assertions made in the previous two paragraphs.

As explained in Chap. 18, for $\zeta \gg 0$, one can see directly the desired Hodge–Riemann bilinear relations, as is the case for de Cataldo–Migliorini. We want to deduce from this knowledge for $\zeta \gg 0$ that L_0 has the Hodge–Riemann bilinear relations. There are two a priori subtleties. First, for any graded piece $\overline{B_x B_s}^i \subset \overline{B_x B_s}$, $i \in \mathbb{Z}$, the primitive subspace with respect to L_ζ varies with the parameter ζ, and may even jump in dimension. Second, even if it did not vary, it is not at all true that the signature remains constant in a general family of symmetric bilinear forms. Indeed, the space of symmetric bilinear forms is a real vector space, in particular connected, so signatures can change.

However, as observed by de Cataldo–Migliorini, if we know the forms are all nondegenerate in our family, these issues disappear. Indeed, a dimension counting argument using nondegeneracy shows that the dimension of the primitive subspace cannot jump. Moreover, the connected components of the space of nondegenerate bilinear forms are given by the signatures without any zeroes. In particular, the signature will be constant in a family of nondegenerate symmetric bilinear forms. The upshot is *we want hard Lefschetz for all $\zeta \geq 0$*.

What We Need II: A Substitute for Hyperplane Sections

In trying to prove hard Lefschetz in our situation, it will be helpful to recall what is done classically, namely in Hodge theory for smooth projective varieties.[9]

There is an interesting approach to the hard Lefschetz theorem, by induction on the dimension.[10] In the inductive step, one takes a smooth projective d-fold $X \subset \mathbb{P}^N$. Intersecting it with a general hyperplane gives a smooth projective $(d-1)$-fold $X_H \subset \mathbb{P}^{N-1}$. It is a beautiful fact that the cohomology of X and X_H are extremely closely related. Roughly, *weak Lefschetz* says that cycles in X_H of degree less than or equal to the complex dimension of X_H are the same as hyperplane sections of cycles in X, except for possibly the middle degree, where X_H may have *more* cycles. Precisely, it says that the restriction map $i^* : H^*(X) \to H^*(X_H)$ gives isomorphisms

$$H^i(X) \simeq H^i(X_H), \qquad i < \dim_\mathbb{C} X_H, \tag{20.22}$$

---

[9] The careful reader will note that when geometry is available, the used versions of the Lefschetz theorems and the Hodge–Riemann relations will not be for the cohomology of smooth projective varieties, but rather for the intersection cohomology of irreducible projective varieties. We will ignore this issue in what follows.

[10] This approach actually fails in the setting of complex geometry, because the induction does not close (see [183, Remark 2.22] for further discussion). However it works in de Cataldo and Migliorini's setting (where they assume classical Hodge theory), and also in the setting of Soergel bimodules.

and an injection $H^{\dim_\mathbb{C} X_H}(X) \hookrightarrow H^{\dim_\mathbb{C} X_H}(X_H)$. Note that by Poincaré duality the remaining Betti numbers of X_H are determined by (20.22), so the only one not determined by $H^*(X)$ is the middle cohomology group dim $H^{\dim_\mathbb{C} X_H}(X_H)$.

Example 20.17 For the reader unfamiliar with this circle of ideas, it may be helpful to think about the simple case of a plane curve $C \subset \mathbb{P}^2$. This is cut out by a single equation, say of degree d. By taking the d-th Veronese embedding of $\mathbb{P}^2$, we can realize imposing this degree d equation as intersecting with a hyperplane. In this example, we have $X = \mathbb{P}^2$, $X_H = C$. Indeed $H^0(\mathbb{P}^2)$, which is generated by the fundamental class of $\mathbb{P}^2$, on the nose maps isomorphically onto $H^0(C)$, which is generated by the fundamental class of C. However, in the middle degree, we know $H^1(C)$ should be rank $2g$, if C is of genus g, but $H^1(\mathbb{P}^2) = 0$, so C has gained middle dimensional cohomology.

Let us return to the general discussion. Using weak Lefschetz, and hard Lefschetz for X_H, by pure linear algebra one obtains hard Lefschetz for X in all degrees below -1. In the middle degree 0, hard Lefschetz is just usual Poincaré duality, so it remains to understand degree -1. Here again a little linear algebra shows the problem is equivalent to the nondegeneracy of the Lefschetz pairing on primitive forms. To get further *we need to use a little more than hard Lefschetz for X_H*. Under i^*, it is easy to see primitive forms go to primitive forms, so to finish we can use the Hodge–Riemann bilinear relations. That is, Hodge–Riemann bilinear relations for a hyperplane section X_H imply hard Lefschetz for X. (This is the missing ingredient one needs in an inductive approach to the hard Lefschetz theorem.)

Therefore, *we want a substitute for a hyperplane section which satisfies Hodge–Riemann*. This is non-obvious in our context. The problem is that for the most part, we can only access analogues of the intersection cohomology of appropriately combinatorial subvarieties of $P_s \overset{B}{\times} X_x$. It is not a priori clear how to write down sections of $\mathcal{L}_\rho \boxtimes \mathcal{O}(1)$ whose vanishing loci would be of the desired form and comprise a family suitable for induction. A key observation of Elias–Williamson is that one can instead use the Rouquier complex, or more precisely its first differential, as a substitute for taking a hyperplane section.

Part V
Special Topics

Chapter 21
Connections to Link Invariants

This chapter is based on expanded notes of a lecture given by
You Qi
and taken by
Johannes Flake

Abstract We relate the quantum group of $\mathfrak{gl}_n$ and the Hecke algebra of the symmetric group S_m through a quantum version of Schur–Weyl duality and we show how both can be used to systematically generate link invariants. In the case of the quantum group, braided tensor category concepts lead to the construction of the Reshetikhin–Turaev quantum invariants, which include the Jones polynomial as a special case. Hecke algebras are employed together with Markov traces to obtain the celebrated HOMFLYPT polynomial in two variables, which can be specialized to give the quantum invariants as special cases. Finally, we describe how the HOMFLYPT polynomial is categorified using Rouquier complexes and Khovanov's triply-graded link homology.

21.1 Temperley–Lieb Algebra

We explore once more the Temperley–Lieb algebra. Previously we described the Temperley–Lieb category by generators and relations in Sect. 7.4, and generalized this with a formal variable δ in Sect. 9.2. The endomorphism ring of the object m is called the Temperley–Lieb algebra $\mathrm{TL}_{m,\delta}$, and we discussed its Jones–Wenzl projectors in Sect. 9.3, and its trace map in Exercise 9.28. We now give another exposition which does not rely on this previous material, describing $\mathrm{TL}_{m,\delta}$ itself by generators and relations.

Definition 21.1 (Temperley–Lieb Algebra) For each $m \in \mathbb{Z}_{\geq 1}$, the *Temperley–Lieb algebra* $\mathrm{TL}_{m,\delta}$ is the $\mathbb{Z}[\delta]$-algebra generated by $u_1, \ldots, u_{m-1}$ modulo the

---

Y. Qi
Division of Physics, Mathematics and Astronomy, Caltech, Pasadena, CA, USA

J. Flake
Department of Mathematics, Rutgers University, Piscataway, NJ, USA

© The Editor(s) (if applicable) and The Author(s), under exclusive licence to Springer Nature Switzerland AG 2020
B. Elias et al., *Introduction to Soergel Bimodules*, RSME Springer Series 5, https://doi.org/10.1007/978-3-030-48826-0_21

relations

$$u_i^2 = \delta u_i, \tag{21.1}$$

$$u_i u_j u_i = u_i \quad \text{if} \quad |i - j| = 1, \tag{21.2}$$

$$u_i u_j = u_j u_i \quad \text{if} \quad |i - j| \geq 2, \tag{21.3}$$

for all $1 \leq i, j \leq m - 1$.

Using graphical calculus, we can represent u_i by the diagram[1]

$$\left| \cdots \right| \underset{i\ \ i+1}{\overset{\smile}{\frown}} \left| \cdots \right|_m \tag{21.4}$$

and $1 \in \mathrm{TL}_{m,\delta}$ by the diagram $\left\| \cdots \right|$. Multiplication of diagrams is accomplished by stacking vertically and evaluating closed loops in the resulting diagrams to δ. The diagrams which can be obtained in this way are "crossingless matchings," and a general element in the Temperley–Lieb algebra is a $\mathbb{Z}[\delta]$-linear combination of crossingless matchings. The relation (21.1) evaluates particular closed loops to δ, while the other two relations imply that isotopic diagrams yield the same element of $\mathrm{TL}_{m,\delta}$.

The Temperley–Lieb algebra has a natural $\mathbb{Z}[\delta]$-linear *trace map*

$$\mathrm{Tr} \colon \mathrm{TL}_{m,\delta} \to \mathbb{Z}[\delta]. \tag{21.5}$$

For a single diagram, the trace is obtained by "closing up" the diagram and by evaluating each closed loop in the resulting closed diagram to δ. Graphically, for $f \in \mathrm{TL}_{m,\delta}$, this can be expressed as

$$\mathrm{Tr}(f) = \boxed{f} \tag{21.6}$$

---

[1] This element u_i is not to be confused with the diagram u from Sect. 7.4. We used the letter u previously because it looks like a cup. Here, we name the generators u_i to match the conventions in the literature.

21.1 Temperley–Lieb Algebra

Notice that, in particular,

$$\mathrm{Tr}(1) = \left(\cdots \bigcirc \cdots \right) = \delta^m. \tag{21.7}$$

A crucial feature of the trace map is that $\mathrm{Tr}(ab) = \mathrm{Tr}(ba)$ for any two elements $a, b \in \mathrm{TL}_{m,\delta}$, just as for the ordinary trace map on matrices.

$$\begin{array}{c}\boxed{a}\\ \boxed{b}\end{array} = \begin{array}{c}\boxed{b}\\ \boxed{v}\end{array} = \begin{array}{c}\boxed{b}\\ \boxed{a}\end{array} \tag{21.8}$$

Example 21.2 Consider the standard $\mathfrak{gl}_2$-module $V = \mathbb{C}^2$ with the standard basis e_1, e_2. Then on $V^{\otimes 2}$, switching the two tensor factors naturally intertwines the $\mathfrak{gl}_2$-action on the tensor product space (recall that $x \in \mathfrak{gl}_2$ naturally acts as $\Delta(x) = x \otimes 1 + 1 \otimes x$ on any tensor product of $\mathfrak{gl}_2$-modules). On the other hand, we can assign $\mathbb{C}$-linear maps to the "cup" and "cap" diagrams in the Temperley–Lieb category for $\delta = -2$:

$$\begin{array}{cccccc}
 & V \otimes V & e_1 \otimes e_2 - e_2 \otimes e_1 & & & \\
\cup : & \uparrow & \updownarrow & & & \\
 & \mathbb{C} & 1 & & & \\
 & \mathbb{C} & -1 & 1 & 0 & 0 \\
\cap : & \uparrow & \updownarrow & \updownarrow & \updownarrow & \updownarrow \\
 & V \otimes V & e_1 \otimes e_2 & e_2 \otimes e_1 & e_1 \otimes e_1 & e_2 \otimes e_2
\end{array} \tag{21.9}$$

It can be checked that these linear maps are, in fact, $\mathfrak{gl}_2$-module maps, where $\mathbb{C}$ is a $\mathfrak{gl}_2$-module via the trace of 2×2-matrices, and that the endomorphism of $V \otimes V$ sending $v \otimes w \mapsto w \otimes v$ can be realized as

$$\times := |\ | + \underset{\frown}{\smile}. \tag{21.10}$$

Consider the specialization $\mathrm{TL}_{2,-2} := \mathbb{C} \otimes_{\mathbb{Z}[\delta]} \mathrm{TL}_{2,\delta}$, where $\mathbb{Z}[\delta] \to \mathbb{C}$ sends $\delta \mapsto -2$. Then the assignment above yields an isomorphism of the group algebra $\mathbb{C}S_2$ with $\mathrm{TL}_{2,-2}$ that is compatible with the action of both algebras on $V \otimes V$.

We thus obtain a graphical way to compute the trace of any element from $\mathbb{C}S_2$ acting on $V^{\otimes 2}$, namely by closing up the corresponding Temperley–Lieb diagram and evaluating each closed loop to $\delta = -2$:

$$\mathrm{Tr}(\;|\;\;|\;) = 4 = (-2)^2 = \bigcirc\;,$$

$$\mathrm{Tr}(\;\times\;) = 2 = (-2)^2 - 2 = \bigcirc + \bigcirc\;.$$

The previous example can be generalized to arbitrary $m \in \mathbb{Z}_{\geq 1}$ by observing that

- S_m, and thus the group algebra $\mathbb{C}S_m$, acts on $V^{\otimes m}$ by permutation of tensor factors;
- $\mathrm{TL}_{m,-2}$ acts on $V^{\otimes m}$, where we now use $m - 1$ different cup and cap maps;
- there is a surjective algebra map $\mathbb{C}S_m \to \mathrm{TL}_{m,-2}$ sending the transposition $(i, i+1)$ to $1 + u_i$ for all $1 \leq i \leq m - 1$;
- the action of $\mathbb{C}S_m$ on $V^{\otimes m}$ factors through the action of $\mathrm{TL}_{m,-2}$, that is, we have the following commutative diagram:

$$\begin{array}{ccc} \mathbb{C}S_m & \cdots c \cdots & V^{\otimes m} \\ \downarrow & {\scriptstyle c} \nearrow & \\ \mathrm{TL}_{m,-2} & & \end{array} \qquad (21.11)$$

This generalization still applies to the case $\dim V = 2$ only. Finally, we can observe that the action of $\mathbb{C}S_m$ (or $\mathrm{TL}_{m,-2}$) on $V^{\otimes m}$ commutes with the action of $\mathfrak{gl}(V)$ on the same space.

Note however that the graphical trace Tr of a Temperley–Lieb diagram, viewed as an endomorphism of $V^{\otimes m}$, does not match the actual trace of the endomorphism, being off by a sign $(-1)^m$. After all, $\mathrm{Tr}(1) = (-2)^m$, while the dimension of $V^{\otimes m}$ is 2^m. This sign is discussed in the following remark.

Remark 21.3 The Temperley–Lieb algebras $\mathrm{TL}_{m,\delta}$ and $\mathrm{TL}_{m,-\delta}$ are isomorphic via the map sending $u_i \mapsto -u_i$, so it seems odd that we should choose to specialize $\delta = -2$ rather than $\delta = 2$. However, the Temperley–Lieb categories for δ and $-\delta$ are not isomorphic! For instance, scaling the cap by -1 and the cup by 1 would violate the isotopy relations. For the Temperley–Lieb category, it is only $\delta = -2$ (and not $\delta = 2$) which admits a functor to the category of $\mathfrak{gl}_2$-modules. People who study monoidal categories with duals are content to think that V actually has dimension -2, not dimension 2.

Exercise 21.4 Compute the trace of $s_1 s_2 \cdots s_{m-1}$ acting on $V^{\otimes m}$, where $V = \mathbb{C}^2$, in two ways: directly, and using the Temperley–Lieb algebra.

21.2 Schur–Weyl Duality

When $V = \mathbb{C}^n$ has dimension greater than two, we still want to study the action of S_m on $V^{\otimes m}$, together with the action of $\mathfrak{gl}(V)$. We will see that the actions are related by a double centralizer property duality, and in particular, the image of $\mathbb{C}S_m$ in $\mathrm{End}(V^{\otimes m})$ equals exactly the space of $\mathfrak{gl}(V)$-intertwiners.

We denote the universal enveloping algebra of $\mathfrak{gl}(V)$ by $U(\mathfrak{gl}(V))$. As an algebra, it is generated by the elements of $\mathfrak{gl}(V)$. The action of $\mathfrak{gl}(V)$ on tensor products is determined by the coproduct map $\Delta \colon U(\mathfrak{gl}(V)) \to U(\mathfrak{gl}(V)) \otimes U(\mathfrak{gl}(V))$, an algebra map which is defined by $\Delta(x) = x \otimes 1 + 1 \otimes x$ for all $x \in \mathfrak{gl}(V)$, see Sect. 21.4.

Theorem 21.5 (Schur–Weyl Duality) *Let V be an n-dimensional vector space over $\mathbb{C}$, and let $m \in \mathbb{Z}_{\geq 1}$. Then the images of $U(\mathfrak{gl}(V))$ and $\mathbb{C}S_m$ in $\mathrm{End}(V^{\otimes m})$ are centralizers of each other. Furthermore, we have a decomposition*

$$V^{\otimes m} \simeq \bigoplus_\lambda V_\lambda \otimes L_\lambda \tag{21.12}$$

as a $U(\mathfrak{gl}(V))$-module and $\mathbb{C}S_m$-module, where V_λ (resp. L_λ) are mutually non-isomorphic irreducible modules of $U(\mathfrak{gl}(V))$ (resp. $\mathbb{C}S_m$), which can be indexed by the Young diagrams λ with m boxes and at most $n = \dim V$ rows.

Example 21.6 For $n \geq 2$ and $m = 2$, there are two Young diagrams ▭ and ▯ which correspond to the trivial and the sign representation of S_2, respectively.

In fact, $V^{\otimes m} = V \otimes V$ decomposes into symmetric and anti-symmetric tensors on which the non-trivial element of S_2 acts (by swapping the tensor factors) as 1 or -1, respectively. The corresponding submodules are spanned by elements of the form $v \otimes w \pm w \otimes v$ for $v, w \in V$. Both submodules are irreducible under the action of $\mathfrak{gl}(V)$; they are isomorphic to $\mathrm{Sym}^2 V$ and $\wedge^2 V$, respectively.

Remark 21.7 (Sketch of Proof of Theorem 21.5) Consider $x \in \mathfrak{gl}(V)$, then x acts as

$$\Delta^{m-1}(x) = x \otimes \mathrm{id}^{\otimes(m-1)} + \mathrm{id} \otimes x \otimes \mathrm{id}^{\otimes(m-2)} + \cdots + \mathrm{id}^{\otimes(m-1)} \otimes x$$

in $\mathrm{End}(V^{\otimes m}) \simeq (\mathrm{End}(V))^{\otimes m}$ on $V^{\otimes m}$. This tensor product is invariant under any permutation of tensor factors, so it commutes with the action of any $\sigma \in S_m$ on $V^{\otimes m}$. Thus, the actions of $U(\mathfrak{gl}(V))$ and $\mathbb{C}S_m$ commute.

In particular, the image of $U(\mathfrak{gl}(V))$ in $\mathrm{End}(V^{\otimes m})$ is in the centralizer of the image of $\mathbb{C}S_m$. We even claim that it is the full centralizer, that is, the space of invariants

$$(\mathrm{End}(V^{\otimes m}))^{S_m} \simeq ((\mathrm{End}(V))^{\otimes m})^{S_m} \simeq \mathrm{Sym}^m(\mathrm{End}(V)).$$

It can be shown that the latter space is spanned by elements of the form $x^{\otimes m} = x \otimes x \otimes \cdots \otimes x$ for $x \in \mathrm{End}(V) = \mathfrak{gl}(V)$, and that every such $x^{\otimes m}$ is in the

Fig. 21.1 Link diagrams of the Hopf link, the trefoil and the (4,7)-torus knot

subalgebra generated[2] by $\Delta^{m-1}(x)$, $\Delta^{m-1}(x^2)$, ..., $\Delta^{m-1}(x^m)$, in particular in the image of $U(\mathfrak{gl}(V))$. Hence, the centralizer of the image of $\mathbb{C}S_m$ is the image of $U(\mathfrak{gl}(V))$.

Finally, by Maschke's theorem, $\mathbb{C}S_m$ is semisimple and $V^{\otimes m}$ is a semisimple $\mathbb{C}S_m$-module. Now by the double commutant theorem, the image of $\mathbb{C}S_m$ is its own double commutant, i.e. it is the centralizer of the image of $U(\mathfrak{gl}(V))$.

The decomposition statement is a corollary.

To reiterate, the images of $\mathbb{C}S_m$ and $U(\mathfrak{gl}(V))$ inside $\text{End}(V^{\otimes m})$ are commutants of each other, but neither the map $U(\mathfrak{gl}(V)) \to \text{End}(V^{\otimes m})$ nor the map $\mathbb{C}S_m \to \text{End}(V^{\otimes m})$ need be injective. When $n = 2$, the image of $\mathbb{C}S_m$ inside $\text{End}(V^{\otimes m})$ is precisely the Temperley–Lieb algebra, via the map $\mathbb{C}S_m \to \text{TL}_{m,-2}$ discussed above.

Exercise 21.8

1. Why does $(1 - s_i - s_{i+1} + s_i s_{i+1} + s_{i+1} s_i - s_i s_{i+1} s_i)$ in $\mathbb{C}S_n$ act as 0 on $V^{\otimes n}$ for $n \geq 3$ and $V = \mathbb{C}^2$?
2. What is the kernel of the action of $\mathbb{C}S_4$ on $V^{\otimes 4}$, when $V = \mathbb{C}^3$?

21.3 Trace and Link Invariants

We will discuss how the Temperley–Lieb algebra and its trace yield link invariants, including the celebrated Jones polynomial.

Definition 21.9 A *link* in $\mathbb{R}^3$ (or S^3) is a finite collection of smoothly embedded S^1 (circles). Links can be represented using *link diagrams*, that is, their projections onto suitable planes, where we allow only a finite number of crossing points, each of which corresponding to exactly two points of the link (Fig. 21.1). We will also discuss *oriented* links, where each embedded S^1 is given an orientation, indicated by an arrow in the link diagram.

---

[2] This follows from the statement that the complete homogeneous symmetric polynomials of degree 1 to m in m variables generate the algebra of symmetric polynomials in m variables.

21.3 Trace and Link Invariants

It is a well-known fact that two link diagrams describe isotopic links if and only if they are related by a finite sequence of *Reidemeister moves*:

$$(\text{RI}): \;\; \vcenter{\hbox{↷}} = \vcenter{\hbox{|}}, \qquad (\text{RII}): \;\; \vcenter{\hbox{⨯}} = \vcenter{\hbox{||}}, \qquad (\text{RIII}): \;\; \vcenter{\hbox{⨯⨯}} = \vcenter{\hbox{⨯⨯}}. \tag{21.13}$$

Note that these and similar pictures always refer to a small region of a possibly larger diagram, and we want to identify two diagrams if they are identical outside the small region, and related inside the small region as described.

Links can be obtained by taking closures of (type A) braid group elements. We recall this braid group from Chap. 19.

Definition 21.10 (Braid Group) The (type A) *braid group* on m strands is the group Br_m generated by $\sigma_1, \ldots, \sigma_{m-1}$ modulo the *braid relations*

$$\sigma_i \sigma_j = \sigma_j \sigma_i \quad \text{if} \quad |i - j| \geq 2, \tag{21.14}$$

$$\sigma_i \sigma_j \sigma_i = \sigma_j \sigma_i \sigma_j \quad \text{if} \quad |i - j| = 1, \tag{21.15}$$

for all $1 \leq i, j \leq m - 1$.

We represent σ_i and σ_i^{-1} diagrammatically by the following positive and negative simple crossing on m strands:

$$\sigma_i = \left| \cdots \right| \underset{i\; i+1}{\times} \left| \cdots \right|, \qquad \sigma_i^{-1} = \left| \cdots \right| \underset{i\; i+1}{\times} \left| \cdots \right|. \tag{21.16}$$

(The choice of upper strand at the crossing point is a mere convention.) Then an arbitrary element of Br_m is represented as an m-strand braid, in a way analogous to the strand diagram for the type A_{m-1} Coxeter group S_m (see Sect. 1.1.2). In fact, the assignment $\sigma_i \mapsto s_i = (i, i+1)$ defines a surjective group homomorphism

$$\mathrm{Br}_m \twoheadrightarrow S_m. \tag{21.17}$$

In this diagrammatics, the defining relations of Br_m become isotopy relations for braids. For example, (RII) says that $\sigma_i \sigma_i^{-1} = \mathrm{id}$. The relation (21.14) says that far-away crossings commute, while (21.15) becomes a version of (RIII) involving only positive crossings. Since two braids are isotopic if and only if they are related by a finite sequence of (RII) and (RIII) moves, it follows that we may view elements of Br_m as m-strand braids up to isotopy. The natural quotient map $\mathrm{Br}_m \twoheadrightarrow S_m$ sends a braid to the permutation induced on the strands.

The *closure* of a braid $b \in \mathrm{Br}_m$ is the link

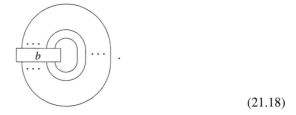

(21.18)

The following two results relate braids with links:

Theorem 21.11 (Alexander's Theorem) *Every (oriented) link is isotopic to the closure of an (oriented) braid.*

Theorem 21.12 (Markov) *The braid closures of $b' \in \mathrm{Br}_m$ and $b'' \in \mathrm{Br}_k$ are isotopic links if and only if b', b'' are related by a finite sequence of*

1. *conjugations in a fixed braid group: $b \mapsto aba^{-1}$;*
2. *Markov moves:*

$$\begin{array}{c}\boxed{b}\end{array} \mapsto \begin{array}{c}\boxed{b}\end{array}.$$

(21.19)

Hence, the problem of finding an invariant for (oriented) links is equivalent to finding an invariant for (oriented) braids up to conjugation and Markov moves. We note that conjugation invariance is the reason why we should be looking for trace maps on the braid group, i.e. maps Tr for which $\mathrm{Tr}(ab) = \mathrm{Tr}(ba)$.

Lemma 21.13 *The assignment*

$$\sigma_i \mapsto vu_i + v^{-1}, \qquad 1 \le i \le m-1, \tag{21.20}$$

defines a group homomorphism

$$\mathrm{Br}_m \to \mathrm{TL}^\times_{m,-v^2-v^{-2}} \tag{21.21}$$

from Br_m to the group of invertible elements in the Temperley–Lieb algebra at parameter $\delta = -v^2 - v^{-2}$.

Proof One easily checks using (21.1) that $vu_i + v^{-1}$ is invertible (with inverse $v^{-1}u_i + v$). We need to check that the elements $vu_i + v^{-1}$ satisfy the braid relations. The far-away braid relation (21.14) follows from the far-away commutation

21.3 Trace and Link Invariants

relation (21.3) in the Temperley–Lieb algebra. For the nearby braid relation (21.15), compute

$$(vu_i + v^{-1})(vu_{i+1} + v^{-1})(vu_i + v^{-1})$$
$$= v^3 u_i u_{i+1} u_i + v(u_i u_{i+1} + u_{i+1} u_i + u_i^2) + v^{-1}(2u_i + u_{i+1}) + v^{-3}$$
$$= v(u_i u_{i+1} + u_{i+1} u_i) + v^{-1}(u_i + u_{i+1}) + v^{-3} + \underbrace{(v^3 + v\delta + v^{-1})u_i}_{=0},$$

and note that the last expression is symmetric in u_i and u_{i+1}. Here, the last equality uses (21.1) and (21.2). □

Exercise 21.14 The assignment of Lemma 21.13 may be drawn diagrammatically as

$$\times \;\mapsto\; v\, \asymp +\, v^{-1}\, \big|\;\big|.$$

Verify the nearby braid relation for these images, now using diagrammatics. It may be helpful to use the computation of the inverse:

$$\times = (\times)^{-1} \;\mapsto\; v^{-1}\, \asymp +\, v\, \big|\;\big|.$$

Thus to any braid we have assigned an element of some Temperley–Lieb algebra. Composing with the trace map, we obtain a braid invariant that is moreover invariant under conjugation. However, the assignment is not invariant under the Markov move:

$$\times\!) \;\mapsto\; v\, \asymp\!) + v^{-1}\, \big|\; \big(\!\big) = -v^{-3}\, \big|,$$

$$\times\!) \;\mapsto\; v^{-1}\, \asymp\!) + v\, \big|\; \big(\!\big) = -v^3\, \big|.$$

To remedy this, we can pass to oriented braids and define:

$$\overset{\nearrow}{\times} \;\mapsto\; -v^3\left(v\, \asymp +\, v^{-1}\, \big|\;\big|\right), \qquad (21.22)$$

$$\overset{\nwarrow}{\times} \;\mapsto\; -v^{-3}\left(v^{-1}\, \asymp +\, v\, \big|\;\big|\right). \qquad (21.23)$$

Now both $\overset{\nearrow}{\times\!)}$ and $\overset{\nwarrow}{\times\!)}$ are mapped to the diagram $\big|$.

The constructed mapping assigns a diagram in $\text{TL}_{m,-v^2-v^{-2}}$ to every oriented braid. Applying the trace map we obtain a polynomial in δ, or a Laurent polynomial in v, which we denote by J.

By construction, J is invariant under conjugation, Markov moves and braid isotopy, thus, it is invariant under link isotopy of the braid closure. Furthermore, it follows from (21.22) and (21.23) that J satisfies the *skein relation*

$$q^2 J(\diagup\!\!\!\!\diagdown) - q^{-2} J(\diagdown\!\!\!\!\diagup) = (q - q^{-1}) J(\uparrow\ \uparrow) \tag{21.24}$$

with $q = -v^{-2}$.

The skein relation allows us to identify J as the celebrated *Jones polynomial*, which is the link invariant determined by this specific skein relation up to a normalization. The normalization of J as just constructed is given by

$$J(\bigcirc) = \delta = -v^2 - v^{-2} = q + q^{-1}. \tag{21.25}$$

There are variations P_n (for all $n \in \mathbb{Z}$) of the Jones polynomial with the more general skein relation

$$q^n P_n(\diagup\!\!\!\!\diagdown) - q^{-n} P_n(\diagdown\!\!\!\!\diagup) = (q - q^{-1}) P_n(\uparrow\ \uparrow) \tag{21.26}$$

and the normalization

$$P_n(\bigcirc) = \frac{q^n - q^{-n}}{q - q^{-1}}. \tag{21.27}$$

In this family of link invariants, P_0 corresponds to the *Alexander polynomial*, $P_1 \equiv 1$, and P_2 corresponds to the Jones polynomial. An even more general (two-parameter) version of this skein relation will be used to define the *HOMFLYPT polynomial* in Sect. 21.5.

References on link invariants and the Jones polynomial are Jones [93] and Kauffman [104].

21.4 Quantum Groups and Link Invariants

We recall that the module category of any Lie algebra $\mathfrak{g}$ is isomorphic to the module category of its universal enveloping algebra $U(\mathfrak{g})$. Since $U(\mathfrak{g})$ has the structure of a *cocommutative Hopf algebra*, the category of modules has the structure of a *symmetric tensor category*; in particular, the tensor product of two modules is

21.4 Quantum Groups and Link Invariants

naturally a module, where an element $x \in \mathfrak{g}$ acts as its image under the coproduct map

$$\Delta(x) = x \otimes 1 + 1 \otimes x \quad \in U(\mathfrak{g}) \otimes U(\mathfrak{g}), \tag{21.28}$$

and for any pair of modules (V, W), the *flip map*

$$\tau \colon V \otimes W \to W \otimes V, \quad v \otimes w \mapsto w \otimes v, \tag{21.29}$$

is a module map satisfying $\tau^2 = \mathrm{id}$.

Given a symmetric tensor category, we can assign a morphism in the category to any oriented braid if we label the strands of the braid with an object from the category, thus obtaining a tensor product object in the category, and translate every crossing to a morphism of the tensor product using the flip map for the corresponding pair of tensor factors. This does not distinguish crossing points with different upper strands, and as $\tau^2 = \mathrm{id}$, we have

$$\vcenter{\hbox{⤫}} = \vcenter{\hbox{↑↑}}. \tag{21.30}$$

In other words, if we pick an object V as the label for all strands, then for any $m \in \mathbb{Z}_{\geq 1}$, the symmetric group S_m will act on $V^{\otimes m}$, where the action of a transposition is given by τ applied to the corresponding pair of tensor factors, and the morphism associated to a braid is actually the morphism which corresponds to the image of the braid group element under the quotient morphism to the symmetric group.

To make things more interesting, we can consider the quantum group $U_q(\mathfrak{g})$, which is a one-parameter deformation of $U(\mathfrak{g})$ as a Hopf algebra.

Example 21.15 ($U_q(\mathfrak{sl}_2)$) For $\mathfrak{g} = \mathfrak{sl}_2$, the quantum group $U = U_q(\mathfrak{sl}_2)$ is the (unital) $\mathbb{Z}[q^{\pm}]$-algebra generated by E, F, K, K^{-1} subject to the relations

$$KK^{-1} = 1 = K^{-1}K, \quad KEK^{-1} = q^2 E, \quad KFK^{-1} = q^{-2}E, \quad [E, F] = \frac{K - K^{-1}}{q - q^{-1}},$$

together with a counit $\varepsilon \colon U \to \mathbb{Z}[q^{\pm}]$, a coproduct $\Delta \colon U \to U \otimes U$ and an antipode $S \colon U \to U$ defined as algebra maps by

$$\varepsilon(E) = 0, \quad \Delta(E) = K \otimes E + E \otimes 1, \quad S(E) = -K^{-1}E,$$
$$\varepsilon(F) = 0, \quad \Delta(F) = 1 \otimes F + F \otimes K^{-1}, \quad S(E) = -FK,$$
$$\varepsilon(K) = 1, \quad \Delta(K) = K \otimes K, \quad S(K) = K^{-1}.$$

It can be checked that this defines a Hopf algebra structure.

The coproduct of the Hopf algebra $U_q(\mathfrak{g})$ is not cocommutative (invariant under permutations of the tensor factors) any longer, and the flip τ will not be a module map anymore. However, $U_q(\mathfrak{g})$ is still a *quasitriangular Hopf algebra* making its module category a *braided tensor category*, that is, we have a *braiding map* $c\colon V \otimes W \to W \otimes V$ for any pair of modules (V, W), but $c \neq \tau$ and $c^2 \neq \mathrm{id}$, in general.

As before, we can assign a morphism to oriented braids, but now (generally)

$$\vcenter{\hbox{$\diagup\!\!\!\!\diagdown$}} \neq \uparrow\uparrow . \tag{21.31}$$

Since $c^2 \neq \mathrm{id}$, the symmetric group S_m will not act on $V^{\otimes m}$ for a module V; only the braid group Br_m, which lacks the quadratic relations, will act, where the braiding of adjacent tensor factors is given by c.

To generate actual link invariants, we also have to take care of the "cups" and "caps" in their diagrams, or equivalently, we have to find trace functions mapping the endomorphisms associated to braids to, say, polynomials in a way which is invariant under link isotopy. This is dealt with using *enhanced R-matrices* or *Ribbon categories*, and the link invariants obtained with this machinery (or its generalization to Lie superalgebras and their enveloping algebras) are sometimes called *Reshetikhin–Turaev* or *quantum invariants*.

Theorem 21.16 (Reshetikhin–Turaev, Kauffman–Saleur) *The invariants P_n from the end of Sect. 21.3 (including the Jones polynomial) are quantum invariants. For $n \geq 1$, P_n can be obtained as a quantum invariant from $U_q(\mathfrak{sl}_n)$.*

References for quantum groups and link invariants are Reshetikhin–Turaev [148], Turaev [170] and Kassel [102].

21.5 Ocneanu Trace and HOMFLYPT Polynomial

We fix $n, m \in \mathbb{Z}_{\geq 1}$. Then Schur–Weyl duality implies the existence of subalgebras $S(n, m)$, $Q(n, m)$ in $\mathrm{End}((\mathbb{C}^n)^{\otimes m})$ which are centralizers of each other such that the following diagram commutes:

$$\begin{array}{ccccc} U(\mathfrak{gl}_n) & \cdots\circlearrowright\cdots & (\mathbb{C}^n)^{\otimes m} & \cdots\circlearrowleft\cdots & \mathbb{C}S_m \\ \downarrow & \searrow & \circlearrowright & \swarrow & \downarrow \\ S(n,m) & \hookrightarrow & \mathrm{End}((\mathbb{C}^n)^{\otimes m}) & \hookleftarrow & Q(n,m) \end{array} \tag{21.32}$$

The algebra $S(n, m)$ is called the *Schur algebra*, the algebra $Q(n, m)$ is called the *n-row quotient* of $\mathbb{C}S_m$ for $n < m$. (Note that for $n \geq m$, the action map $\mathbb{C}S_m \to \mathrm{End}((\mathbb{C}^n)^{\otimes m})$ is injective.) As subalgebras of a finite-dimensional endomorphism

21.5 Ocneanu Trace and HOMFLYPT Polynomial

algebra, both algebras are finite-dimensional. Finally, note that, as we have seen above, $Q(2, m) \simeq \text{TL}_{m, -2}$, and that $\mathbb{C}S_m$ is the specialization of the Hecke algebra $\text{H}(S_m)$ (see Chap. 3) at $q = 1$.

Passing to the quantum case, $U(\mathfrak{gl}_n)$ is replaced by the quantum group $U_q(\mathfrak{gl}_n)$. We can consider $\mathbb{Q}(q)^n$ as a $U_q(\mathfrak{gl}_n)$-module. Then the braid group Br_m acts on $(\mathbb{Q}(q)^n)^{\otimes m}$, and the action factors through the Hecke algebra $\text{H}(S_m)$. Quantum Schur–Weyl duality now implies the existence of finite-dimensional subalgebras $S_q(n, m)$, $Q_q(n, m)$ in $\text{End}((\mathbb{Q}(q)^n)^{\otimes m})$ which are centralizers of each other such that the following diagram commutes:

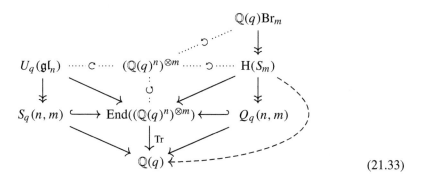

(21.33)

Here we have inserted the trace map on $\text{End}((\mathbb{Q}(q)^n)^{\otimes m})$, its restrictions to $S_q(n, m)$ and $Q_q(n, m)$, and an additional dashed arrow for later reference. The algebra $S_q(n, m)$ is sometimes called the *quantum Schur algebra*, the algebra $Q_q(n, m)$ is a Hecke algebra (q-)version of the n-row quotient (it will be discussed further in Chap. 22) and in particular, $Q_q(2, m) \simeq \text{TL}_{m, -[2]}$.

Remark 21.17 Technically, the algebras in the diagram should all be over $\mathbb{Q}(q)$, so they should be obtained by base change from the $\mathbb{Z}[q, q^{-1}]$ algebras $\text{H}(S_m)$ and $U_q(\mathfrak{gl}_n)$. However, every algebra in the diagram has an integral form, and the maps in the diagram descend to maps between these integral forms. It is the diagram with integral forms which is more amenable to categorification.

Both sides of the duality can be exploited to generate link invariants. As discussed in Sect. 21.4, we can use the braided tensor category of $U_q(\mathfrak{gl}_n)$-modules to construct the quantum link invariants. Alternatively we can use the Hecke algebra, generalizing the approach outlined in Sect. 21.3, where a link was first mapped to an element in the Temperley–Lieb algebra, whose trace then turned out to be a link invariant.

To generalize this idea, we fix a system of embeddings $S_1 \subset S_2 \subset \cdots$, and thus a system of embeddings $\text{H}(S_1) \subset \text{H}(S_2) \subset \cdots$, so we can make the following definition.

Definition 21.18 A *Markov trace* on $\bigcup_{m\geq 1} H(S_m)$ with parameter z is a linear map

$$\mathrm{Tr}: \bigcup_{m\geq 1} H(S_m) \to \mathbb{Z}[q^{\pm 1}, z] \qquad (21.34)$$

such that $\mathrm{Tr}(ab) = \mathrm{Tr}(ba)$ and $\mathrm{Tr}(M(b)) = z\,\mathrm{Tr}(b)$ for all $a, b \in \bigcup_{m\geq 1} H(S_m)$, where $M(b)$ is the modification of b according to the Markov move (21.19).[3]

Theorem 21.19 (Ocneanu) *There is a unique Markov trace on $\bigcup_{m\geq 1} H(S_m)$ with normalization* $\mathrm{Tr}(1) = 1$.

Of course, this unique Markov trace, when specialized at $z = 0$, agrees with the standard trace we introduced in Chap. 3, see Definition 3.12.

Remark 21.20 For a suitable specialization of q, z, this trace reproduces the trace on the n-row quotient algebra $Q_q(n, m)$ which comes from the endomorphism algebra $\mathrm{End}((\mathbb{Q}(q)^n)^{\otimes m})$. This is indicated by the dashed arrow in the commutative diagram (21.33).

Theorem 21.21 (HOMFLYPT[4]) *There is a unique $\mathbb{C}[t^{\pm 1}, x^{\pm 1}]$-valued invariant I of oriented links with the skein relation*

$$tI(\diagup\!\!\!\diagdown) - t^{-1}I(\diagdown\!\!\!\diagup) = xI(\uparrow \uparrow) \qquad (21.35)$$

and normalization $I(\bigcirc) = 1$.

Definition 21.22 This invariant is called the *HOMFLYPT polynomial*.

Comparing the skein relations, we note that the HOMFLYPT polynomial generalizes the quantum invariants P_n which we have seen in Sect. 21.3 and in Theorem 21.16. In this sense, it is the "one polynomial to rule them all."

It turns out that the HOMFLYPT polynomial can, indeed, be constructed using Hecke algebras and a generalization of the ideas from Sect. 21.3.

Theorem 21.23 (Jones) *Every Markov trace yields an invariant for oriented links. The Ocneanu trace yields the HOMFLYPT polynomial (up to a change of variables).*

Remark 21.24 The proof follows the key ideas of Sect. 21.3: For any oriented link, we pick an oriented braid whose closure is the given link. Next, we assign an element in the Hecke algebra $H(S_m)$ to the braid. Finally, we apply a modification of the Ocneanu trace map, which is still invariant under conjugation, and also invariant

---

[3] Alternatively, we can think of this as a family of trace maps which is compatible with the embeddings $H(S_m) \subset H(S_{m+1})$.

[4] HOMFLYPT is an acronym combining the initials of two independent groups of authors, Hoste–Ocneanu–Millett–Freyd–Lickorish–Yetter and Przytycki–Traczyk.

under the Markov move. This will yield a link invariant, and it remains to verify the skein relation.

It should be mentioned that the HOMFLYPT polynomial is not a complete invariant:

Theorem 21.25 (Kanenobu [97]) *There are infinitely many distinct links with the same HOMFLYPT polynomial.*

References for the HOMFLYPT polynomial and the Ocneanu trace are the original papers by HOMFLY [69] and Jones [92].

Exercise 21.26 Compute the Jones polynomial and the HOMFLYPT polynomial of the trefoil. The trefoil is the closure of σ_1^3 in the braid group on two strands.

21.6 Categorification of Braids and of the HOMFLYPT Invariant

21.6.1 Rouquier Complexes

We want to categorify link invariants and the HOMFLYPT polynomial. To this end, we start with the categorification of braid groups using Rouquier complexes.

Let us recall some material from Chap. 19. We fix $m \geq 1$ and consider the type A_{m-1} Coxeter system $W = S_m$ with generators $S = \{s_1, \ldots, s_{m-1}\}$. Let V be the geometric representation of W over $\Bbbk$. As usual, let $R := \mathrm{Sym}(V)$ be the symmetric algebra of V, and let $B_s := R \otimes_{R^s} R(1)$ for every $s \in S$. We consider the complexes of Soergel bimodules

$$F_s := \left(\cdots \longrightarrow 0 \longrightarrow \underline{B_s} \overset{\uparrow}{\longrightarrow} R(1) \longrightarrow 0 \longrightarrow \cdots \right)$$

$$F_s^{-1} := \left(\cdots \longrightarrow 0 \longrightarrow R(-1) \overset{\downarrow}{\longrightarrow} \underline{B_s} \longrightarrow 0 \longrightarrow \cdots \right) \tag{21.36}$$

for every $s \in S$, where the underline indicates that B_s is placed in cohomological degree 0 in both complexes.

As before, we denote the bounded homotopy category of Soergel bimodules by $K^b \mathbb{S}\mathrm{Bim}$, which is a monoidal category under tensor product of complexes. As discussed in Chap. 19, F_s and F_s^{-1} are inverse to each other in $K^b \mathbb{S}\mathrm{Bim}$. We have already seen the following theorem (see Theorem 19.36) in Chap. 19.

Theorem 21.27 (Rouquier) *The mapping* $F : \mathrm{Br}_m \to K^b \mathbb{S}\mathrm{Bim}$ *sending*

$$\sigma_{j_1}^{\epsilon_1} \cdots \sigma_{j_t}^{\epsilon_t} \mapsto F_{s_{j_1}}^{\epsilon_1} \otimes_R \cdots \otimes_R F_{s_{j_t}}^{\epsilon_t} \tag{21.37}$$

for all $t \geq 1$ and $j_i \in \{1, \ldots, m-1\}$, $\epsilon_i \in \{\pm 1\}$ *for all* $1 \leq i \leq t$ *is well-defined, i.e. it respects the braid relations, and it induces a group homomorphism from* Br_m *to the group of isomorphism classes of invertible objects in* $K^b\mathbb{S}\mathrm{Bim}$.

Hence, we can view Rouquier complexes, which are defined precisely to be the image of the above mapping, as a categorification of braid groups.

21.6.2 Hochschild Homology

In order to obtain link invariants as in Sect. 21.3, we need to find an analogue of the trace map for Rouquier complexes. We will see that the role of the trace map will be played by a combination of the Hochschild homology functor, the homology functor of chain complexes and the Euler characteristic.

For any $\Bbbk$-algebra A, we define $A^e := A \otimes_\Bbbk A^{\mathrm{opp}}$, the *enveloping algebra* of A. The category of A^e-modules is equivalent to the category of A-bimodules, but rephrasing bimodules in terms of A^e-modules allows one to use familiar ideas from homological algebra. For instance, it may not be obvious what a "projective resolution" of A-bimodules means, but it is obvious what a projective resolution of A^e-modules is. "Free A-bimodules" are A^e-modules that are isomorphic to a direct sum of copies of A^e itself.

Meanwhile, the regular A-bimodule, namely A itself, is not free as an A^e-module. Instead, it is the "universal" bimodule on which the left and right actions of A agree. That is, for any A-bimodule M, $\mathrm{Hom}_{A^e}(A, M)$ is isomorphic to the largest submodule on which the left and right actions agree, while $A \otimes_{A^e} M$ is the largest quotient on which the left and right actions agree, and agrees with the quotient of M by the space of all vectors $am - ma$ for $a \in A$ and $m \in M$. An A-bimodule on which the left and right actions agree is perhaps best thought of just as an A-module.

Definition 21.28 (Hochschild Homology) The functor $M \mapsto A \otimes_{A^e} M$, from A^e-modules to A-modules, is called the *A-coinvariants functor*, and is right exact. Its i-th left derived functor $M \mapsto \mathrm{Tor}_i^{A^e}(A, M)$ is called the *i-th Hochschild homology* of M and is denoted by $\mathrm{HH}_i(A, M)$. The *Hochschild homology* of M is defined as

$$\mathrm{HH}_*(A, M) := \bigoplus_{i \geq 0} \mathrm{HH}_i(A, M) . \tag{21.38}$$

Hochschild homology is a contravariant functor from the category of A-bimodules to the category of $\mathbb{Z}_{\geq 0}$-graded A-modules. Similarly, the *Hochschild cohomology* of M is

$$\mathrm{HH}^*(A, M) := \bigoplus_{i \geq 0} \mathrm{HH}^i(A, M) \tag{21.39}$$

21.6 Categorification of Braids and of the HOMFLYPT Invariant

where $\mathrm{HH}^i(A, M)$ is the i-th right derived functor of the left exact *A-invariants* functor $M \mapsto \mathrm{Hom}_{A^e}(A, M)$.

Let us note that these definitions also make sense for graded modules over graded rings.

If $(\cdots \to P_2 \to P_1 \to P_0 \to A \to 0)$ is a projective resolution of A as an A-bimodule, then the Hochschild homology of any A-bimodule M can be computed as the homology of the complex

$$\cdots \to P_2 \otimes_{A^e} M \to P_1 \otimes_{A^e} M \to P_0 \otimes_{A^e} M \to 0. \tag{21.40}$$

Example 21.29 (Koszul Resolution for Polynomial Rings, One Variable) Assume $A = \Bbbk[x]$ is a polynomial ring in one variable. Then $A^{\mathrm{opp}} = A$, $A^e = A \otimes A$ (both is true for any commutative algebra A), and A has the projective (Koszul) resolution

$$0 \longrightarrow A \otimes A \xrightarrow{(x\otimes 1 - 1 \otimes x)} A \otimes A \xrightarrow{\mu} A \longrightarrow 0, \tag{21.41}$$

where μ is the multiplication map.

Exercise 21.30 Suppose that M is a graded bimodule over $A = \Bbbk[x]$, where $\deg x = 2$. Deduce that $\mathrm{HH}_0(A, M)$ is the cokernel of the map $(x^l - x^r) \colon M(-2) \to M$, where x^l is the left action of x, and x^r is the right action of x. Similarly, $\mathrm{HH}_1(A, M)$ is the kernel of this map.

Example 21.31 (Koszul Resolution for Polynomial Rings, Many Variables) More generally, let A be the ring $R = \mathrm{Sym}(V)$ defined above; choosing a basis for V, we may identify R with a polynomial ring in $r = \dim V < \infty$ variables. Then the Koszul resolution of R as an R^e-module can be written in a basis-free way as

$$\Lambda^r V \otimes R^e \xrightarrow{d_r} \cdots \xrightarrow{d_3} \Lambda^2 V \otimes R^e \xrightarrow{d_2} V \otimes R^e \xrightarrow{d_1} R^e \xrightarrow{\mu} R \longrightarrow 0, \tag{21.42}$$

where for each $1 \leq k \leq r$, $\Lambda^k V$ denotes the k-th exterior power of V, and the differential $d_k \colon \Lambda^k V \otimes R^e \to \Lambda^{k-1} V \otimes R^e$ is determined by

$$d_k((r_1 \wedge \cdots \wedge r_k) \otimes (1 \otimes 1)) = \sum_{i=1}^{k} (-1)^{i+1} r_1 \wedge \cdots \widehat{r_i} \cdots \wedge r_k \otimes (r_i \otimes 1 - 1 \otimes r_i). \tag{21.43}$$

Exercise 21.32 As a warm-up, for $r \in \{1, 2, 3\}$, use a basis of V to write down (21.42) explicitly. Then show that (21.42) is a resolution for any $r > 1$.

Exercise 21.33 Let M be a graded bimodule over $R = \Bbbk[x_1, x_2]$, where $\deg x_i = 2$. Consider the complex

$$0 \longrightarrow M(-4) \xrightarrow{\begin{pmatrix} x_1^l - x_1^r \\ x_2^l - x_2^r \end{pmatrix}} \begin{matrix} M(-2) \\ \oplus \\ M(-2) \end{matrix} \xrightarrow{(x_2^l - x_2^r \ \ x_1^l - x_1^r)} \underline{M} \longrightarrow 0,$$

(21.44)

where the final M is in homological degree zero. Prove that $\mathrm{HH}_i(R, M)$ is the degree $-i$ cohomology of this complex.

Exercise 21.34 Let $R = \mathbb{R}[x_1, x_2]$, acted on by S_2, and define the Soergel bimodules R and B_s as usual. Apply the technique of the previous exercise to compute $\mathrm{HH}_*(R, R)$ and $\mathrm{HH}_*(R, B_s)$. For a future exercise, you should keep careful track of the grading shifts involved. (*Hint:* It is much easier to compute $\mathrm{HH}_*(R, B_s)$ if you think of R as $\mathbb{R}[x_1 + x_2, x_1 - x_2]$.)

For polynomial rings, there is a "Poincaré duality" relating Hochschild homology and cohomology. More precisely, if $A = R$ is a polynomial ring in d variables, then

$$\mathrm{HH}^i(R, M) \simeq \mathrm{HH}_{d-i}(R, M) \qquad (21.45)$$

for any R-bimodule M. Note that this isomorphism exchanges the gradings in the same way as Poincaré duality for a closed oriented manifold of dimension d.

Exercise 21.35 Let $A = R$ be a polynomial ring in d variables, and let M be any R-bimodule. Show that the complexes obtained by applying the functors $- \otimes_{R^e} M$ and $\mathrm{Hom}_{R^e}(-, M)$ to the Koszul resolution (21.42) can be identified up to a cohomological shift by d. Deduce the isomorphism (21.45).

(*Hint:* Fix an isomorphism $\Lambda^d V \xrightarrow{\sim} \Bbbk$. For any $0 \leq i \leq d$, multiplication defines a perfect pairing

$$(-) \wedge (-) : \Lambda^i V \times \Lambda^{d-i} V \to \Lambda^d V \xrightarrow{\sim} \Bbbk,$$

inducing isomorphisms $(\Lambda^i V)^* \simeq \Lambda^{d-i} V$. Use these isomorphisms to obtain the desired identification.)

Remark 21.36 This exercise shows that the isomorphism (21.45) arises from a "self-duality" in the Koszul resolution (21.42) of the polynomial ring itself. For a more general class of rings, there is a version of the isomorphism (21.45) due to van den Bergh [171, 172] that involves twisting the bimodule structure on one side by an automorphism.

Exercise 21.37 Let M be a graded bimodule over $R = \Bbbk[x_1, x_2]$, as in Exercise 21.33. Applying the appropriate grading shift and cohomological shift to the complex (21.44), one can obtain a complex whose cohomology in degree i matches

$\mathrm{HH}^i(R, M)$. Work this out explicitly, finding the correct shifts involved. Use this to compute $\mathrm{HH}^*(R, R)$ and $\mathrm{HH}^*(R, B_s)$ as in Exercise 21.34.

21.6.3 Categorifying the Standard Trace

We wish to argue that taking the Hochschild cohomology of a Soergel bimodule is the categorification of the standard trace map ϵ in the Hecke algebra. Recall that the standard trace could be expressed as $\epsilon(b) = (1, b)$, where $(-, -)$ is the standard form on the Hecke algebra (see Sect. 3.2.1). By the Soergel Hom formula (Theorem 5.27), $(-, -)$ categorifies to $\mathrm{Hom}(-, -)$. Thus the standard trace categorifies to $\mathrm{Hom}_{R^e}(R, -)$, whose derived functor is Hochschild cohomology.

The crucial feature of trace maps which we used in this chapter is that $\epsilon(ab) = \epsilon(ba)$. Since multiplication categorifies to $\otimes_A$, we want a relationship between $\mathrm{HH}^*(A, M \otimes_A N)$ and $\mathrm{HH}^*(A, N \otimes_A M)$ for any A-bimodules M and N. Let us instead consider Hochschild homology. At the start we have four A-actions: the left/right action on M/N, which we can think of as four A-actions on $M \otimes_\Bbbk N$. Taking the quotient $M \otimes_A N$ will "identify" the right action on M with the left action on N. Then applying the functor $\mathrm{HH}_0(A, -) = A \otimes_{A^e} (-)$ (or any derived functor HH_i) will identify the left action on M with the right action on N. Now two different A-actions remain. Meanwhile, computing $\mathrm{HH}_0(A, N \otimes_A M)$ will also identify the same actions, but in a different order. In other words, we can think of both $\mathrm{HH}_0(A, M \otimes_A N)$ and $\mathrm{HH}_0(A, N \otimes_A M)$ as the same quotient of $M \otimes_\Bbbk N$, where the inner actions are identified and the outer actions are identified, and these two spaces are isomorphic as $\Bbbk$-vector spaces! However, $\mathrm{HH}_0(A, M \otimes_A N)$ and $\mathrm{HH}_0(A, N \otimes_A M)$ are not isomorphic as A-modules: the A-module structure on $\mathrm{HH}_0(A, M \otimes_A N)$ comes from the outer action on $M \otimes_\Bbbk N$, while the A-module structure on $\mathrm{HH}_0(A, N \otimes_A M)$ comes from the inner action on $M \otimes_\Bbbk N$. A similar argument applies to Hochschild cohomology.

To reiterate, there is an isomorphism

$$\mathrm{HH}^*(A, M \otimes_A N) \simeq \mathrm{HH}^*(A, N \otimes_A M), \tag{21.46}$$

but only when viewing HH^* as a functor to graded $\Bbbk$-vector spaces, not as a functor to graded A-modules. This motivates viewing HH^* as a functor to vector spaces, rather than A-modules.

Exercise 21.38 Observe that for any Soergel bimodule M, the invariant subring R^W acts the same on the left and right of M. Deduce that for any Soergel bimodules M and N the isomorphism $\mathrm{HH}^*(R, M \otimes_R N) \simeq \mathrm{HH}^*(R, N \otimes_R M)$, which is not an isomorphism of R-modules, is at least an isomorphism of R^W-modules. Despite this, we continue to view $\mathrm{HH}^*(R, -)$ as a functor to vector spaces, to match the literature.

Now let us specialize $\Bbbk$ to $\mathbb{Q}$ and A to R. For any $\sigma \in \mathrm{Br}_m$, we have a (bounded) complex of Soergel bimodules

$$F(\sigma) = \Big(\cdots \to F^j(\sigma) \to F^{j+1}(\sigma) \to \cdots\Big).$$

After applying Hochschild cohomology, we obtain a complex of bigraded $\mathbb{Q}$-vector spaces

$$\mathrm{HH}^*(R, F(\sigma)) = \Big(\cdots \to \mathrm{HH}^*(R, F^j(\sigma)) \to \mathrm{HH}^*(R, F^{j+1}(\sigma)) \to \cdots\Big). \tag{21.47}$$

Note that there are three gradings: the *internal grading* of each graded R-bimodule, the *homological grading* in the complex, and the *Hochschild grading* which states which summand HH^i of HH^* is being considered. Taking homology of this complex yields a triply graded $\mathbb{Q}$-vector space which we call $\mathrm{HHH}(\sigma)$, the *triply graded link homology*. It categorifies the HOMFLYPT polynomial in the following sense.

Theorem 21.39 (Khovanov [112]) *Up to an overall grading shift, $\mathrm{HHH}(\sigma)$ is an invariant of oriented links, that is, it depends only on the closure of σ as an oriented link up to isotopy. Taking the Euler characteristic of $\mathrm{HHH}(\sigma)$ yields the HOMFLYPT polynomial, after some renormalization.*

The Euler characteristic turns the homological grading into a sign. Relating the two remaining gradings to the parameters t, x from (21.35) is actually quite difficult, as it passes through both a renormalization and a change of variable, but the precise relationship can be found in [52, Appendix].

Exercise 21.40 This exercise is surprisingly hard, which is why it is difficult to compute triply graded link homology by elementary techniques.

1. Consider the unknot, which is the closure of the identity braid on one strand. One should think of its triply graded link homology as the total Hochschild cohomology of $R_1 = \mathbb{Q}[x_1]$. Compute this triply graded vector space. (*Hint:* The graded dimension is $\frac{1+AQ^{-2}}{1-Q^2}$, where Q denotes the internal grading and A denotes the Hochschild grading.)
2. The unknot is also the closure of σ_1 in the braid group on two strands. Compute its triply graded link homology. Your base ring should be $R_2 = \mathbb{Q}[x_1, x_2]$, acted on by S_2 in the usual way. Your answer should agree with the previous one up to a shift.

Chapter 22
Cells and Representations of the Hecke Algebra in Type A

This chapter is based on expanded notes of a lecture given by the authors and taken by

Shotaro Makisumi

Abstract We describe the notion of cells in a Krull–Schmidt monoidal category. Applied to the category of Soergel bimodules in characteristic 0, Soergel's conjecture implies that this produces the Kazhdan–Lusztig cells in the Hecke algebra. The rest of the chapter focuses on type A. We recall the Robinson–Schensted correspondence and the description of the two-sided cells, which leads to a more explicit description of the k-row quotient of the Hecke algebra appearing in quantum Schur–Weyl duality. Finally, we discuss the relation between cells and representations of the Hecke algebra (in type A).

22.1 Cells

Recall the k-row quotient $Q_v(k, n)$ of the Hecke algebra $H(S_n)$ defined in the context of quantum Schur–Weyl duality in Chap. 21; for $k = 2$, $Q_v(2, n)$ is the Temperley–Lieb algebra on n strands. One of the main aims of this chapter is Theorem 22.33, which describes the kernel of the quotient map

$$H(S_n) \twoheadrightarrow Q_v(k, n). \qquad (22.1)$$

This description relies on the notion of Kazhdan–Lusztig cells, which is where we will begin.

S. Makisumi
Department of Mathematics, Columbia University, New York, NY, USA

© The Editor(s) (if applicable) and The Author(s), under exclusive licence to Springer Nature Switzerland AG 2020
B. Elias et al., *Introduction to Soergel Bimodules*, RSME Springer Series 5, https://doi.org/10.1007/978-3-030-48826-0_22

22.1.1 Cells for a Monoidal Category

Let $\mathcal{A}$ be a Krull–Schmidt (in particular, additive and Karoubian) monoidal category with monoidal ("tensor") product $\otimes$. (See Chap. 11 for a discussion of Krull–Schmidt categories.) Throughout this chapter, our main example will be the diagrammatic Hecke category $\mathcal{H}$ associated to a Coxeter system (W, S) and its geometric realization over $\mathbb{R}$.

> **Notation 22.1** Unlike elsewhere in this book, in this chapter we view $\mathcal{H}$ as a category whose Hom spaces are graded, not as a category with a grading shift endofunctor. In other words, $\mathcal{H} = (\text{Kar}(\mathcal{H}_{\text{BS}}^{\text{sh},\oplus}))^{\text{gr}}$ in the notation of Chap. 11.

Thus $\{B_w\}_{w \in W}$ is a complete list of indecomposable objects of $\mathcal{H}$ up to isomorphism (rather than up to isomorphism and grading shift). By analogy, for general $\mathcal{A}$ we continue to denote by

$$\{B_w\}_{w \in W} \tag{22.2}$$

a complete list of indecomposable objects up to isomorphism.

Definition 22.2 A *two-sided 2-ideal*

$$\mathcal{I} \subset \mathcal{A} \tag{22.3}$$

is a collection of morphisms in $\mathcal{A}$ that is closed under the following operations: (pre- and post-)composition with any morphism, and (right and left) tensor product with the identity morphism of any object.

This is a natural definition for a monoidal category: a two-sided 2-ideal $\mathcal{I}$ is exactly what is needed for a monoidal structure to be induced on the quotient category

$$\mathcal{A}/\mathcal{I}. \tag{22.4}$$

Since

$$f \otimes g = (f \otimes \text{id}_{Y'}) \circ (\text{id}_X \otimes g) \tag{22.5}$$

for any morphisms $f \colon X \to Y$ and $g \colon X' \to Y'$ in a monoidal category, a two-sided 2-ideal is in fact closed under tensor product with any morphism. Viewing a monoidal category as living in a plane as in Chap. 7, we can say that $\mathcal{I}$ is closed under four-sided composition: top and bottom for actual composition, and right and left for the tensor product.

22.1 Cells

Example 22.3 Let $\mathcal{I} \subset \mathcal{H}$ be a two-sided 2-ideal. Suppose that $\mathcal{I}$ contains id_{B_s} for some $s \in S$, i.e. a vertical strand colored s. Using vertical composition with the dot morphisms, we see that $\mathcal{I}$ contains the s-colored barbell

$$\stackrel{\bullet}{\underset{\bullet}{|}} = \alpha_s. \tag{22.6}$$

Let $t \in S$. Using horizontal composition with id_{B_t} on either side, followed by vertical composition with t-colored trivalent vertices, we see that $\mathcal{I}$ contains

$$\diamondsuit = \partial_t(\alpha_s) \Big|. \tag{22.7}$$

(See Chap. 7 for this calculation.) If $m_{st} \neq 2$, then $\partial_t(\alpha_s) \in \mathbb{R}$ is invertible, so $\mathcal{I}$ also contains id_{B_t}.

Example 22.4 Morphisms in $\mathcal{H}$ in the image of R_+^W (symmetric polynomials of strictly positive degree) acting on the right form a two-sided 2-ideal. Indeed, this collection of morphisms is clearly closed under top, bottom, and left composition, and it is closed under right composition since W-invariant polynomials slide through all strands. (We restrict to polynomials of positive degree in order to obtain a proper 2-ideal.)

Definition 22.5 A *left 2-ideal* (resp. *right 2-ideal*) is a collection of morphisms that is closed under top, bottom, and left (resp. right) composition.

Example 22.6 Morphisms in $\mathcal{H}$ in the image of R_+ on the right form a left 2-ideal.

For $w \in W$, let

$$\mathcal{I}_{w,LR} \subset \mathcal{A} \tag{22.8}$$

be the two-sided 2-ideal generated by (i.e. the smallest one containing) id_{B_w}, or equivalently the collection of linear combinations of morphisms that factor through $M \otimes B_w \otimes N$ for some objects M, N. Similarly, let

$$\mathcal{I}_{w,L} \quad (\text{resp. } \mathcal{I}_{w,R}) \tag{22.9}$$

be the left (resp. right) 2-ideal generated by id_{B_w}, or equivalently the collection of linear combinations of morphisms that factor through $M \otimes B_w$ (resp. $B_w \otimes M$) for some object M.

We now come to the main definitions of this subsection.

Definition 22.7 Define the *two-sided preorder* $\le_{LR}$ on W by $z \le_{LR} w$ if

$$\mathrm{id}_{B_z} \in \mathcal{I}_{w,LR}. \tag{22.10}$$

This is a preorder, and its equivalence classes are called *two-sided cells*. Similarly define the *left preorder* $\le_L$ and *left cells* using the left 2-ideal $\mathcal{I}_{w,L}$, and the *right preorder* $\le_R$ and *right cells* using the right 2-ideal $\mathcal{I}_{w,R}$.

By (left, right, two-sided) cells in a Coxeter group, we mean cells for the associated Soergel category $\mathcal{H}$ (for the geometric realization over $\mathbb{R}$).

Exercise 22.8 Check that the (left, right, two-sided) preorder really is a preorder (i.e. it is reflexive and transitive).

We will use the following description of these preorders.

Exercise 22.9 Show that $z \le_{LR} w$ if and only if

$$B_z \overset{\oplus}{\subset} M \otimes B_w \otimes N \tag{22.11}$$

for some objects M, N (and the analogous statements for the left and right preorders).

Example 22.10 For $s, t \in S$ in any Coxeter system (W, S), the calculation in Example 22.3 shows that $t \le_{LR} s$ whenever $m_{st} \ne 2$. One also sees this from the inclusion $B_t \overset{\oplus}{\subset} B_t B_s B_t$.

Example 22.11 Let $W = \langle s, t \rangle$ with $m_{st} = 3$. Then $\mathcal{H}$ has six indecomposable objects up to isomorphism. In the diagram below, a red (resp. blue) arrow $B_x \to B_y$ indicates an inclusion $B_y \overset{\oplus}{\subset} B_x B_s$ (resp. $B_y \overset{\oplus}{\subset} B_x B_t$):

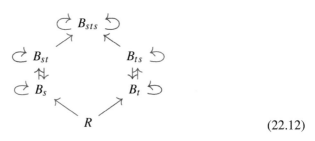

$$\tag{22.12}$$

Since B_s, B_t generate $\mathcal{H}$ in the appropriate sense, this diagram completely encodes the right preorder: $x \le_R y$ if and only if one can reach B_x from B_y by following

22.1 Cells

arrows (of any color). The partition of W into the right cells is therefore as follows:

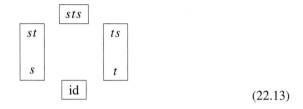

(22.13)

Repeating this for tensoring on the left, one obtains the left cells $\{id\}$, $\{s, ts\}$, $\{t, st\}$, and $\{sts\}$. The two-sided preorder combines the arrows of both preorders, so the two-sided cells are $\{id\}$, $\{s, t, st, ts\}$, $\{sts\}$. For instance, the two-sided 2-ideal generated by id_s contains the identity morphism of all indecomposable objects except for R.

Remark 22.12 Warning: do not confuse the cell order with the Bruhat order! From small examples one may get the impression that $w <_{LR} x$ implies that $w < x$ in the Bruhat order, but this is false in general.

Exercise 22.13

1. Verify the following description of cells for the Coxeter group $W = \langle s, t \rangle$ with $m_{st} = 4$.

   - Left cells: $\{id\}$, $\{s, ts, sts\}$, $\{t, st, tst\}$, $\{w_0\}$.
   - Right cells: $\{id\}$, $\{s, st, sts\}$, $\{t, ts, tst\}$, $\{w_0\}$.
   - Two-sided cells: $\{id\}$, $\{s, t, st, ts, sts, tst\}$, $\{w_0\}$.

   Better yet, draw a colored graph analogous to (22.12). Generalize this to arbitrary dihedral type.
2. For any Coxeter group W, show that the maximal (left, right, or two-sided) cell is $\{id\}$. For finite W, show that the minimal cell is $\{w_0\}$.
3. Compute the cells in S_4. (You have already computed the direct sum decompositions of $B_x B_s$ and $B_s B_x$ in Exercise 3.29.)

Exercise 22.14

1. Prove that elements of a Coxeter group in the same left cell have the same right descent set.
2. Find an example of two different left cells in a Coxeter group with the same right descent set.

Exercise 22.15 Let (W, S) be an irreducible Coxeter system. Let C be the set of non-identity elements of W with a unique reduced expression, and for each $s \in S$, let $C_s = \{w \in C \mid ws < w\}$.

1. Show that C_s is a contained in a single left cell, for all $s \in S$. (*Hint:* If $w \in C$ with $\ell(w) \geq 2$, then $w = tux$ for some $t, u \in S$ and $x \in C \cup \{\mathrm{id}\}$. What do you know about m_{tu}? Show that $w \leq_L ux$ (easy). Show that $ux \leq_L w$ (use a dihedral calculation).)
2. Show that C is contained in a single two-sided cell.

Remark 22.16 In fact, C_s is a left cell, and C is a two-sided cell. This was proven by Lusztig in [128, Proposition 3.8].

22.1.2 Cell Module Categories

Given a two-sided 2-ideal $\mathcal{I} \subset \mathcal{A}$, let $\mathcal{A}_\mathcal{I}$ be the full subcategory of $\mathcal{A}$ spanned by objects X such that $\mathrm{id}_X \in \mathcal{I}$. Equivalently, $\mathcal{A}_\mathcal{I}$ is the biggest full subcategory of $\mathcal{A}$ with all morphisms contained in $\mathcal{I}$. Whereas $\mathcal{A}/\mathcal{I}$ was a monoidal quotient category, $\mathcal{A}_\mathcal{I}$ is a monoidal subcategory. In fact, we have

$$\mathcal{A} \otimes \mathcal{A}_\mathcal{I} \otimes \mathcal{A} \subset \mathcal{A}_\mathcal{I}, \tag{22.14}$$

so that $\mathcal{A}_\mathcal{I}$ is a left and right module category over $\mathcal{A}$.

For $w \in W$ and a two-sided cell $\lambda \subset W$, we write $w \leq_{LR} \lambda$ (resp. $w <_{LR} \lambda$) to mean that the two-sided cell μ containing w satisfies $\mu \leq_{LR} \lambda$ (resp. $\mu <_{LR} \lambda$).

Definition 22.17 Given a two-sided cell λ, let

$$\mathcal{I}_{\leq \lambda} \quad (\text{resp. } \mathcal{I}_{<\lambda}) \tag{22.15}$$

be the two-sided 2-ideal consisting of all linear combinations of morphisms factoring through some B_w, $w \leq_{LR} \lambda$ (resp. $w <_{LR} \lambda$). Write $\mathcal{A}_{\leq \lambda}$ instead of $\mathcal{A}_{\mathcal{I}_{\leq \lambda}}$. The *cell module category* is the quotient

$$\mathcal{A}_\lambda := \mathcal{A}_{\leq \lambda}/\mathcal{I}_{<\lambda}. \tag{22.16}$$

This is a module category for $\mathcal{A}$, an example of a categorical representation.

Example 22.18 Consider $\mathcal{H}$ for W finite. The minimal cell $\lambda_{\min}$ consists of just w_0, so $\mathcal{I}_{<\lambda_{\min}} = 0$, and $\mathcal{H}_{\lambda_{\min}}$ is the full subcategory generated by B_{w_0} (i.e. finite direct sums of shifts of B_{w_0}).

Meanwhile, the maximal cell $\lambda_{\max}$ consists of just R, so

$$\mathcal{H}_{\lambda_{\max}} := \mathcal{H}_{\leq \lambda_{\max}}/\mathcal{I}_{<\lambda_{\max}} = \mathcal{H}/\mathcal{I}_{<\lambda_{\max}}. \tag{22.17}$$

This is the quotient of $\mathcal{H}$ by morphisms factoring through any object besides R (for instance α_s for any $s \in S$). Hence $\mathcal{H}_{\lambda_{\max}}$ has one nonzero indecomposable object up to isomorphism, the image of the object R, and its endomorphism ring is $R/R_+ = \mathbb{R}$ (when working with the geometric realization).

22.1.3 Cells for a Based Algebra

Cells are usually defined at the level of the Grothendieck group. We briefly recall the analogous notions in this setting.

Let $(A, \{b_w\}_{w \in W})$ be a *based algebra*, i.e. A is an algebra that is free as a module over some base ring $\Bbbk$, together with a fixed basis $\{b_w\}_{w \in W}$. For $z, w \in W$, write

$$z \xleftarrow{LR} w \quad (\text{resp. } z \xleftarrow{L} w, \quad \text{resp. } z \xleftarrow{R} w) \tag{22.18}$$

if there exist $m, n \in A$ such that b_z appears with nonzero coefficient when $mb_w n$ (resp. mb_w, resp. $b_w n$) is written in the given basis. Unlike the analogous notion for a monoidal category, this relation is not necessarily transitive. Let

$$\leq_{LR} \quad (\text{resp. } \leq_L, \quad \text{resp. } \leq_R) \tag{22.19}$$

be the transitive closure, the *two-sided* (resp. *left*, resp. *right*) *preorder*. Its equivalences classes are the *two-sided* (resp. *left*, resp. *right*) cells for $(A, \{b_w\}_{w \in W})$.

Given a left cell λ, consider the left ideals

$$A_{\leq_L \lambda} := \bigoplus_{w \leq_L \lambda} \Bbbk b_w, \qquad A_{<_L \lambda} := \bigoplus_{w <_L \lambda} \Bbbk b_w \tag{22.20}$$

of A, and define the left *cell module*

$$A_\lambda := A_{\leq_L \lambda} / A_{<_L \lambda}. \tag{22.21}$$

One similarly defines the right and two-sided cell modules, which are right A-modules and A-bimodules.

Soergel's conjecture implies that the tensor decomposition of Soergel bimodules is controlled by the Kazhdan–Lusztig basis. This implies that cells for $\mathcal{H}$ are exactly the *Kazhdan–Lusztig cells*, i.e. the cells for the Hecke algebra with respect to the Kazhdan–Lusztig basis.

Remark 22.19 The Grothendieck group of a Krull–Schmidt monoidal category comes with a distinguished basis, the class of the indecomposable objects. To define cells for an algebra, we must specify this basis as additional data.

For $\mathcal{H}$ defined using a realization over a field of characteristic $p > 0$, Soergel's conjecture no longer holds. The classes of the indecomposable objects give a new basis of the Hecke algebra, the *p-canonical basis* (or *p-Kazhdan–Lusztig basis*) (see Chap. 27). Cells with respect to this basis are called *p-Kazhdan–Lusztig cells*.

Exercise 22.20 For each based algebra, find the left, right, and two-sided cells. What do the cell modules look like?

1. A matrix algebra, with its basis of matrix entries.
2. A product of matrix algebras, with its basis of matrix entries for each term in the product.
3. A polynomial ring $\mathbb{C}[x]$, with the basis $\{x^k\}$.
4. A polynomial ring $\mathbb{C}[x]$, with the basis $\{1, x, x^2 - 1, x^3 - 2x, \ldots\}$. To interpret this basis, consider the ring $\mathbb{C}[q, q^{-1}]$, and the subring of invariants under $q \mapsto q^{-1}$. This subring is a polynomial ring generated by $x = [2]$, and the basis described is $\{[n]\}$.

22.2 Cells in Type A

For the symmetric group, a theorem of Kazhdan and Lusztig (see Theorem 22.26 below) gives a description of the cells. To state it, we need to recall some combinatorics of tableaux.

22.2.1 Young Diagrams and Tableaux

Recall that a partition λ of n (written $\lambda \vdash n$) can be represented by a *Young diagram*. For example, the partition $(4, 3, 1) \vdash 8$ is represented by the following diagram having 4, 3, and 1 box(es) in the first, second, and third row:

We identify a partition λ with its Young diagram and speak of "shape λ" and "boxes of λ." Given a partition $\lambda \vdash n$, a *Young tableau (of shape λ)* is a filling of the boxes of λ by positive integers. A *standard (Young) tableau* is one in which these numbers are 1 through n, each occurring once, and strictly increase going down any column and to the right in any row. Here is an example of a standard tableau of shape $(4, 3, 1)$:

$$T = \begin{array}{|c|c|c|c|} \hline 1 & 2 & 5 & 8 \\ \hline 3 & 6 & 7 \\ \cline{1-3} 4 \\ \cline{1-1} \end{array}$$

(22.22)

22.2 Cells in Type A

The set of standard tableaux of shape λ is denoted by $\mathrm{SYT}(\lambda)$.

We will need the following statistics on Young diagrams and tableaux. Given a partition λ, number its rows from top to bottom and its columns from left to right, both starting at 0. Given a box in λ, its *row number* (resp. *column number*) is the number assigned to the row (resp. column) containing it, and its *content* is its column number minus its row number. (Thus the content indicates which diagonal the box lies on.) The *total row number*, *total column number*, and *total content* of λ, denoted by

$$r(\lambda), \quad c(\lambda), \quad x(\lambda), \tag{22.23}$$

is the sum of the respective statistic over all boxes of λ. For example, here are the row number, column number, and content of each box in $\lambda = (4, 3, 1)$:

$$
\begin{array}{|c|c|c|c|}\hline 0 & 0 & 0 & 0 \\\hline 1 & 1 & 1 \\\cline{1-3} 2 \\\cline{1-1}\end{array}, \quad
\begin{array}{|c|c|c|c|}\hline 0 & 1 & 2 & 3 \\\hline 0 & 1 & 2 \\\cline{1-3} 0 \\\cline{1-1}\end{array}, \quad
\begin{array}{|c|c|c|c|}\hline 0 & 1 & 2 & 3 \\\hline -1 & 0 & 1 \\\cline{1-3} -2 \\\cline{1-1}\end{array}.
\tag{22.24}$$

Thus $r(\lambda) = 5$, $c(\lambda) = 9$, and $x(\lambda) = 4 = c(\lambda) - r(\lambda)$.

Given a standard tableau T of shape $\lambda \vdash n$, let

$$r(\boxed{k}), \quad c(\boxed{k}), \quad x(\boxed{k}) \quad \text{for } 1 \leq k \leq n \tag{22.25}$$

denote the row number, column number, and content of the k-box in T, the (unique) box in T containing k. For example, for the standard tableau T in (22.22), we have $r(\boxed{7}) = 1$, $c(\boxed{7}) = 2$, $x(\boxed{7}) = 1$.

22.2.2 The Robinson–Schensted Correspondence

The description of cells in type A uses the following correspondence between permutations in S_n and pairs of standard Young tableaux of the same shape.

Theorem 22.21 (Robinson–Schensted Correspondence) *For each $n \geq 1$, there exists an explicit bijection*

$$S_n \xleftrightarrow{1:1} \{(P, Q, \lambda) \mid \lambda \vdash n \text{ and } P, Q \in \mathrm{SYT}(\lambda)\}. \tag{22.26}$$

In the triple (P, Q, λ), P is called the *insertion tableau* or *P-symbol*, and Q is called the *recording tableau* or *Q-symbol*. The forward map is given by the *Schensted's bumping algorithm*, which we illustrate in an example. The map in reverse is harder to describe, but given a triple (P, Q, λ), one can unravel this algorithm to deduce the corresponding permutation.

Example 22.22 Consider the permutation

$$\sigma = 4\,6\,2\,7\,5\,3\,1\,8 \in S_8$$

written in one-line notation. That is, $\sigma(1) = 4$, $\sigma(2) = 6$, and so on. To find the associated triple (P, Q, λ), start with P, Q being empty. We read σ from left to right, modifying P, Q at each step.

The first number is 4. Add it to P, then record this by adding a 1 to Q:

$$P = \boxed{4}, \qquad Q = \boxed{1}.$$

The next number, 6, can be added to the right of 4 while keeping P a standard tableau. Add a corresponding box 2 to Q:

$$P = \boxed{4\ 6}, \qquad Q = \boxed{1\ 2}.$$

The next number, 2, cannot be added to the right of 6. Instead, we place 2 where 4 is, *bumping* 4 to the second row. (There is a unique number one can bump to keep the first row strictly increasing.) Place a 3 in Q to record the new box:

$$P = \begin{array}{|c|c|}\hline 2 & 6 \\\hline 4 \\\cline{1-1}\end{array}, \qquad Q = \begin{array}{|c|c|}\hline 1 & 2 \\\hline 3 \\\cline{1-1}\end{array}.$$

7 can be placed on the top row:

$$P = \begin{array}{|c|c|c|}\hline 2 & 6 & 7 \\\hline 4 \\\cline{1-1}\end{array}, \qquad Q = \begin{array}{|c|c|c|}\hline 1 & 2 & 4 \\\hline 3 \\\cline{1-1}\end{array}.$$

5 bumps 6, which can be added to the right in the second row:

$$P = \begin{array}{|c|c|c|}\hline 2 & 5 & 7 \\\hline 4 & 6 \\\cline{1-2}\end{array}, \qquad Q = \begin{array}{|c|c|c|}\hline 1 & 2 & 4 \\\hline 3 & 5 \\\cline{1-2}\end{array}.$$

22.2 Cells in Type A

3 bumps 5, which bumps 6, which starts the third row:

$$P = \begin{array}{|c|c|c|} \hline 2 & 3 & 7 \\ \hline 4 & 5 \\ \cline{1-2} 6 \\ \cline{1-1} \end{array} \quad , \quad Q = \begin{array}{|c|c|c|} \hline 1 & 2 & 4 \\ \hline 3 & 5 \\ \cline{1-2} 6 \\ \cline{1-1} \end{array} \quad .$$

1 bumps 2, which bumps 4, which bumps 6:

$$P = \begin{array}{|c|c|c|} \hline 1 & 3 & 7 \\ \hline 2 & 5 \\ \cline{1-2} 4 \\ \cline{1-1} 6 \\ \cline{1-1} \end{array} \quad , \quad Q = \begin{array}{|c|c|c|} \hline 1 & 2 & 4 \\ \hline 3 & 5 \\ \cline{1-2} 6 \\ \cline{1-1} 7 \\ \cline{1-1} \end{array} \quad .$$

Finally, 8 can be added to the right on the first row:

$$P = \begin{array}{|c|c|c|c|} \hline 1 & 3 & 7 & 8 \\ \hline 2 & 5 \\ \cline{1-2} 4 \\ \cline{1-1} 6 \\ \cline{1-1} \end{array} \quad , \quad Q = \begin{array}{|c|c|c|c|} \hline 1 & 2 & 4 & 8 \\ \hline 3 & 5 \\ \cline{1-2} 6 \\ \cline{1-1} 7 \\ \cline{1-1} \end{array} \quad .$$

This defines the insertion tableau P and the recording tableau Q, which are clearly of the same shape.

Exercise 22.23 Pick a few random elements of S_{10} and apply Schensted's bumping algorithm to find the corresponding triple (P, Q, λ).

The Robinson–Schensted correspondence satisfies the following properties.

1. We have

$$(P, Q, \lambda)^{-1} = (Q, P, \lambda). \tag{22.27}$$

In particular, $w \in S_n$ is involution if and only if $P = Q$.

2. Consider the Dynkin diagram automorphism

$$S_n \to S_n : s_i \mapsto s_{n-i} \quad \text{for } 1 \leq i \leq n. \tag{22.28}$$

Diagrammatically, this is a horizontal flip. It can also be described as conjugation by the longest element: $w \mapsto w_0 w w_0$. In terms of triples, this automorphism acts as

$$(P, Q, \lambda) \mapsto (P^\vee, Q^\vee, \lambda). \tag{22.29}$$

Here, $P^\vee$ is the *Schützenberger dual* of P. We will not recall the *Schützenberger involution* $P \mapsto P^\vee$ on $\mathrm{SYT}(\lambda)$ (defined for instance using *jeu de taquin*, see e.g. [71]) other than to note that it is an involution. For instance, we have

$$P = \begin{array}{|c|c|c|} \hline 1 & 3 & 4 \\ \hline 2 \\ \cline{1-1} \end{array}, \qquad P^\vee = \begin{array}{|c|c|c|} \hline 1 & 2 & 3 \\ \hline 4 \\ \cline{1-1} \end{array}. \qquad (22.30)$$

Exercise 22.24

1. Prove that the number of rows in the partition associated to a permutation $w \in S_n$ under the Robinson–Schensted correspondence is the length of the longest decreasing sequence in the one-row notation of w: $i_1 < i_2 < \cdots < i_k$ such that $w(i_1) > w(i_2) > \cdots > w(i_k)$.
2. (Harder) The number of rows is also the size of the first column. Find a formula for the sum of the sizes of the first two columns. (Be careful, it is easy to get this slightly wrong!)

Exercise 22.25 Redraw the graph (22.12), but replacing each element of S_3 with its corresponding triple (P, Q, λ) under the Robinson–Schensted correspondence. Do you notice any relationship between the right cells and the triples (P, Q, λ)? What about left cells? Two-sided cells?

22.2.3 Cells in Type A

We can now state the description of cells in type A, which already appears in the original paper that introduced the Kazhdan–Lusztig basis. In the following theorem, we use the Robinson–Schensted correspondence to associate an insertion tableau, a recording tableau, and a shape to each permutation.

Theorem 22.26 (Kazhdan–Lusztig [108]) *The two-sided cells of S_n are in bijection with partitions $\lambda \vdash n$. The two-sided cell indexed by λ consists of permutations with shape λ:*

$$\{(-,-,\lambda)\}. \qquad (22.31)$$

Both the left cells and the right cells contained in this two-sided cell are in bijection with standard tableaux of shape λ. Each left (resp. right) cell consists of permutations with a fixed recording (resp. insertion) tableau:

$$\{(-, Q, \lambda)\}, \qquad (\textit{resp. } \{(P, -, \lambda)\}). \qquad (22.32)$$

22.2 Cells in Type A

Remark 22.27 Since it is common to read this result dyslexically the first time, let us reiterate: the left cell is determined by the Q-symbol, and the right cell by the P-symbol, not vice versa.

Theorem 22.26 and properties of the Robinson–Schensted correspondence imply several important facts about cells. First, w and w^{-1} always lie in the same two-sided cell, but they lie in the same left (or right) cell if and only if w is an involution. Second, each two-sided cell in type A is a "square" since its elements are indexed by a pair (P, Q). If one arranges the elements of a two-sided cell in a matrix with rows P and columns Q, then the left cells are the columns, right cells are the rows, and the intersection of a left cell with a right cell (both contained in this two-sided cell) is exactly one element. Moreover, the two-sided cell indexed by λ has size $(\#\mathrm{SYT}(\lambda))^2$.

Remark 22.28 In Exercise 22.13 the cells in type B_2 were computed, and the middle two-sided cell had size 6, which is not a square number. Moreover, the intersection of a left cell and a right cell need not have size 1. Thus the features discussed in the previous paragraph are very special to type A.

Remark 22.29 This square behavior of cells occurs in any *cellular algebra*. We will not define this notion, but we can see this behavior within the more refined notion of an object-adapted cellular category (see Chap. 11). If X is an object in such a category, then $\mathrm{End}(X)$ has a basis $\{c^\lambda_{S,T}\}_{\lambda \in \Lambda}$ where $T \in E(X, \lambda)$ and $S \in \overline{E}(X, \lambda)$. The terms in cell λ form a square, since $E(X, \lambda)$ and $\overline{E}(X, \lambda)$ are in bijection, just like the rows and columns of a square matrix are in bijection.

Again, the Kazhdan–Lusztig basis is a cellular basis for the Hecke algebra only in type A.[1] However, for finite W, the Hecke algebra is cellular for some modified basis; see [75].

Remark 22.30 Recall that the group algebra $\mathbb{C}[S_n]$, being semisimple, is isomorphic to a product of matrix algebras indexed by $\lambda \vdash n$. Philosophically, one should think of the two-sided cell filtration of the Hecke algebra $\mathrm{H}(S_n)$ (which is no longer semisimple in general) as an analogue of this direct product decomposition. In particular, the subquotients of this filtration, while not being matrix algebras, have the same size as the corresponding matrix algebras in $\mathbb{C}[S_n]$.

To obtain actual matrix algebras as subquotients, one needs to modify the multiplication. This is Lusztig's J-ring [126].

The following result was proved by Graham in his thesis.

Theorem 22.31 (Graham [79]) *In* $\mathrm{H}(S_n)$, *we have*

$$\delta_{w_0} b_{(P,Q,\lambda)} = (-1)^{c(\lambda)} v^{x(\lambda)} b_{(P^\vee, Q, \lambda)} + (lower\ terms). \tag{22.33}$$

---

[1]This is in contrast to the fact that, on the next categorical level, the Hecke category $\mathcal{H}$ is an object-adapted cellular category for all Coxeter groups W.

Here, $b_{(P,Q,\lambda)}$ denotes the Kazhdan–Lusztig basis element for the permutation corresponding to (P, Q, λ) under the Robinson–Schensted correspondence, and lower terms lie in the span of $b_{(P',Q',\lambda')}$ for $\lambda' \leq_{LR} \lambda$.

Thus δ_{w_0} acts by a permutation matrix on Kazhdan–Lusztig cell modules, up to a sign and a power of v. (This result has been generalized to arbitrary finite type by Mathas [135] and to unequal parameters by Lusztig [130].)

Exercise 22.32 Check Graham's theorem for S_3 by direct computation.

22.2.4 The k-row Quotient of the Hecke Algebra

We can now state the description of the k-row quotient of $H(S_n)$ promised at the start of this chapter.

Theorem 22.33 *For $k \geq 1$, let $\lambda_k = (n-k, 1, 1, \ldots, 1) \vdash n$, the $(k+1)$-row hook partition of n. Then*

$$H(S_n)/\mathcal{I}_{\leq \lambda_k} \xrightarrow{\sim} Q_v(k, n). \qquad (22.34)$$

In other words, the kernel of the quotient map $H(S_n) \twoheadrightarrow Q_v(k, n)$ *is spanned by the Kazhdan–Lusztig basis elements b_w for permutations $w \in S_n$ whose shape under the Robinson–Schensted correspondence has $> k$ rows.*

Remark 22.34 The authors are not aware of a good reference for this theorem. It should follow (with some effort) from the results of [74].

Example 22.35 Let $n = 3$, $k = 2$. The only partition of 3 with more than 2 rows is $\lambda_3 = (1, 1, 1)$, which corresponds to the minimal cell $\{w_0\}$ of S_3. Thus the two-sided ideal $\mathcal{I}_{\leq \lambda_3}$ is one-dimensional, spanned by

$$b_{w_0} = \delta_{w_0} + v\delta_{st} + v\delta_{ts} + v^2\delta_s + v^2\delta_t + v^3\delta_{\text{id}}. \qquad (22.35)$$

Theorem 22.33 implies that $\dim Q_v(2, 3) = 6 - 1 = 5$, which indeed equals $\dim \text{TL}_3$.

Exercise 22.36 Repeat this dimension check for S_4. That is, use Theorem 22.33 to compute the dimension of $Q_v(2, 4)$ (by computing the sizes of appropriate two-sided cells), and check that this agrees with the number of crossingless matchings on 4 strands. Can you generalize this to S_n?

Remark 22.37 In Example 22.35, observe that the element of $\mathbb{C}[S_3]$ that acts as 0 on $(\mathbb{C}^2)^{\otimes 3}$ is $w_0 - st - ts + s + t - 1$. This is the specialization of b_{w_0} at $v = -1$ rather than $v = 1$. For better agreement with our conventions, one should include a twist by the sign representation in quantum Schur–Weyl duality.

22.3 Representations of the Hecke Algebra in Type A

Assumption 22.38 In this section, our Hecke algebra will be over the base field $\mathbb{Q}(v)$, where v is an indeterminate or a complex number that is not a root of unity.

Recall from Chap. 21 the braid group Br_n on n strands, generated by the simple positive crossings

$$\sigma_i = \Big| \cdots \Big| \underset{i\ i+1}{\times} \Big| \cdots \Big|, \qquad 1 \leq i \leq n-1. \tag{22.36}$$

The i-th *Jucys–Murphy braid* $j_i \in \mathrm{Br}_n$ is defined by

$$j_1 = \mathrm{id}, \qquad j_i = \sigma_{i-1}\sigma_{i-2}\cdots\sigma_2\sigma_1\sigma_1\sigma_2\cdots\sigma_{i-1} \quad \text{for } 2 \leq i \leq n. \tag{22.37}$$

Diagrammatically,

$$j_i = \quad \tag{22.38}$$

One easily sees that j_i commutes with $\mathrm{Br}_{i-1} \subset \mathrm{Br}_n$, and in particular with j_k for $k < i$. Hence Jucys–Murphy braids pairwise commute.

Since the standard generators $\delta_{(i,i+1)}$ of $\mathrm{H}(S_n)$ satisfy the braid relation, the assignment $\sigma_i \mapsto \delta_{(i,i+1)}$ extends to a group homomorphism

$$\mathrm{Br}_n \to \mathrm{H}(S_n)^\times. \tag{22.39}$$

The i-th *Jucys–Murphy element* is the image of j_i under this homomorphism. We also denote it by j_i by an abuse of notation. An easy induction yields an explicit formula for these images in the standard basis:

$$j_1 = 1, \qquad j_i = 1 + \sum_{1 \leq k < i} (v^{-1} - v)\delta_{(k,i)}, \quad i \geq 2. \tag{22.40}$$

The Jucys–Murphy elements generate a maximal commutative subalgebra of $H(S_n)$ called the *Gelfand–Zetlin* algebra. The following result says that this subalgebra is large enough to detect the irreducible representations of $H(S_n)$.

Theorem 22.39 (See [87, 145]) *Isomorphism classes of irreducible representations of $H(S_n)$ are in bijection with partitions of n. Let V_λ be the irreducible representation corresponding to $\lambda \vdash n$. Then V_λ has a Young basis $\{e_T\}_{T \in SYT(\lambda)}$, a simultaneous eigenbasis for the Jucys–Murphy elements $\{j_1, j_2, \ldots, j_n\}$, satisfying*

$$j_i e_T = v^{2x(\boxed{i})} e_T. \tag{22.41}$$

Here, $x(\boxed{i})$ denotes the content of the i-box in T.

This theorem describes the possible spectrum (simultaneous eigenvalues) of the Jucys–Murphy elements on finite-dimensional representations of $H(S_n)$. For instance, $x(\boxed{1}) = 0$ in any standard tableau, consistent with $j_1 = \delta_{\mathrm{id}}$. Writing the quadratic relation in $H(S_n)$ as $(\delta_{s_i} - v^{-1})(\delta_{s_i} + v) = 0$, we see that any eigenvalue of δ_{s_i} must be either v^{-1} or $-v$. Then any eigenvalue of $j_2 = \delta_{s_1}^2$ is $v^{\pm 2}$, consistent with the fact that $x(\boxed{2}) = \pm 1$ in any standard tableau.

Example 22.40 Since there is only one standard tableau T of shape $(1, \ldots, 1) \vdash n$, the irreducible representation $V_{(1,1,\ldots,1)}$ is one-dimensional, and $j_i e_T = v^{2(i-1)} e_T$ for every $1 \leq i \leq n$. Inducting on i, one sees that all δ_{s_i} acts as $-v$. This is the *sign representation* of $H(S_n)$; setting $v = 1$, we obtain the sign representation of S_n. Similarly, the one-dimensional representation $V_{(n)}$ is the *trivial representation*, in which every δ_{s_i} acts as v^{-1}.

Remark 22.41 One interpretation of Theorem 22.39 is that by simultaneously diagonalizing the Jucys–Murphy elements, one can not only project from an arbitrary representation to its isotypic components, but also find a special basis of each irreducible representation. Because the simultaneous eigenbasis for $\{j_1, \ldots, j_n\}$ is also a simultaneous eigenbasis for $\{j_1, \ldots, j_{n-1}\}$, this basis behaves nicely under the restriction functor from $H(S_n)$-representations to $H(S_{n-1})$-representations. This idea is explored further in [145].

The following exercises give an idea of the proof of Theorem 22.39.

Exercise 22.42 By the definition of j_i, we have $\delta_{s_i} j_i \delta_{s_i} = j_{i+1}$ in $H(S_n)$ for $1 \leq i \leq n-1$. Deduce that

$$\delta_{s_i} j_i = j_{i+1} \delta_{s_i} + X \tag{22.42}$$

for some element $X \in H(S_n)$. Find X.

22.3 Representations of the Hecke Algebra in Type A

Exercise 22.43 Let A denote the subalgebra of $\mathrm{H}(S_n)$ generated by j_i, j_{i+1}, and $H = \delta_{s_i}$. Let x be an eigenvector for both j_i and j_{i+1} in some $\mathrm{H}(S_n)$-module, with eigenvalues λ_i and λ_{i+1}, respectively. Equation (22.42) implies that the subspace $A \cdot x$ is at most 2-dimensional, spanned by x and Hx.

1. Suppose that x and Hx are linearly dependent. What relation does this imply between the eigenvalues λ_i, λ_{i+1}? (There are two cases.)
2. Suppose that x and Hx are linearly independent. Find another eigenvector y for j_i and j_{i+1} (linearly independent from x), and compute its eigenvalues.
3. (Optional) Find the matrix for H in the basis $\{x, y\}$, assuming that $\lambda_i = v^a$ and $\lambda_{i+1} = v^b$. Up to scalars, this is what is called the *Young seminormal form*.

The spectral approach to the representation theory of S_n using Jucys–Murphy elements is the topic of [145], but in a slightly different context. The following remark is intended to help the reader contrast Theorem 22.39 with what is found in [145].

Remark 22.44 The elements $\{j_i\}$ in the Hecke algebra are sometimes called *multiplicative* Jucys–Murphy elements. This is to contrast them with the *additive* Jucys–Murphy elements $\{L_i\}$, obtained by taking the "derivative" of the multiplicative Jucys–Murphy elements.

$$L_i := \frac{1}{v^{-2}-1}(j_i - 1) \overset{(22.40)}{=} \sum_{1 \le j < i} v\delta_{(j,i)}. \tag{22.43}$$

Thus $\{e_T\}$ is a simultaneous eigenbasis for both the multiplicative and additive Jucys–Murphy elements, and the eigenvalue of L_i on e_T is a renormalization of the quantum number of $x(\boxed{i})$. Both the additive and multiplicative versions are often simply called Jucys–Murphy elements in the literature.

The multiplicative Jucys–Murphy elements are more obviously connected to the braid group, but they are all pure braids (i.e. they induce the trivial permutation on the braids), so after the specialization $v \mapsto 1$ they are all sent to the identity element of $\mathbb{Q}[S_n]$. Meanwhile, the additive Jucys–Murphy elements specialize to the so-called Jucys–Murphy or Young–Jucys–Murphy elements in $\mathbb{Q}[S_n]$, which are sums of certain transpositions. In [145] they work in $\mathbb{Q}[S_n]$. It might be interesting for the reader to adapt the previous exercises to work directly with $\{L_i\}$ instead of $\{j_i\}$. After setting $v \mapsto 1$, this would match the results proven in [145].

The *full twist* on n strands is the element

$$\mathrm{ft}_n := j_1 j_2 \cdots j_n \in \mathrm{Br}_n. \tag{22.44}$$

Diagrammatically, for example,

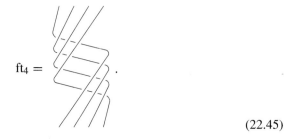

$$\text{ft}_4 = \qquad . \qquad (22.45)$$

As with the Jucys–Murphy elements, we continue to write ft_i for the image of ft_i in $H(S_n)$ under the map (22.39). It is easily seen that the subalgebra of $H(S_n)$ generated by $1 = \text{ft}_1, \text{ft}_2, \ldots, \text{ft}_n$ agrees with the subalgebra generated by $1 = j_1, j_2, \ldots, j_n$. Note that ft_n is central in Br_n (anything done to the bottom of ft_n twists right to the top), hence ft_n is central in $H(S_n)$. This is an advantage of working with full twists instead of Jucys–Murphy elements.

Note that $\text{ft}_n = \delta_{w_0}^2$, which can be seen by cutting the diagram (22.45) along a slant, and observing that each half lifts (a different reduced expression for) the longest element of S_n. Thus Theorem 22.31 implies that

$$\text{ft}_n b_{(P,Q,\lambda)} = b_{(P,Q,\lambda)} (-1)^{2c(\lambda)} v^{2x(\lambda)} + \text{(lower terms)}. \qquad (22.46)$$

Of course $(-1)^{2c(\lambda)} = 1$, but we keep this sign around since it becomes relevant in Chap. 23.

Together with Theorem 22.39, we deduce the following result.

Corollary 22.45 *The left cell representation $V_{(-,Q,\lambda)}$ is isomorphic to V_λ. The central full twist ft_n acts on V_λ by multiplication by $(-1)^{2c(\lambda)} v^{2x(\lambda)}$.*

In particular, cell modules give all irreducible representations of $H(S_n)$.

The central full twist ft_n almost distinguishes between distinct irreducible representations. For instance, $H(S_3)$ has three irreducible representations, corresponding to the partitions

$$\square\square\square, \quad \square\!\square, \quad \square, \qquad (22.47)$$

and ft_3 acts on them by the scalars v^6, v^0, and v^{-6}, respectively. Since an arbitrary representation of $H(S_3)$ is a direct sum of some copies of the irreducible representations, projection to an eigenspace of ft_3 is the same as projecting to an isotopic component. Unfortunately, for $n \geq 6$, the full twist alone cannot tell apart

22.3 Representations of the Hecke Algebra in Type A

certain irreducible representations; the partitions

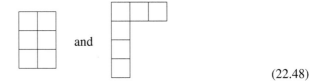

(22.48)

have the same total content -3. To properly distinguish between these irreducible representations, one could simultaneously diagonalize the full twists $\mathrm{ft}_1, \mathrm{ft}_2, \ldots, \mathrm{ft}_n$.

The considerations of this chapter suggest that a spectral approach to constructing categorical representations of the Hecke category may be fruitful. The next chapter discusses such a categorified spectral approach.

Chapter 23
Categorical Diagonalization

This chapter is based on expanded notes of a lecture given by the authors and taken by
Alex Chandler, **Nachiket Karnick**, and **Dmitry Vagner**

Abstract In classical linear algebra, given a diagonalizable operator on a vector space, Lagrange interpolation produces an idempotent decomposition of the identity corresponding to the projections to eigenspaces. In this chapter, we explain a categorical analogue of this procedure, due to Elias and Hogancamp: given a "diagonalizable" functor acting on a monoidal homotopy category, we produce idempotent functors which project to "eigencategories." The main application is to the full twist Rouquier complex, acting on the homotopy category of Soergel bimodules.

This chapter surveys the theory of categorical diagonalization developed by Elias and Hogancamp [50, 51].

23.1 Classical Linear Algebra

Recall the following situation from linear algebra. Let $\Bbbk$ be a field and V be a $\Bbbk$-vector space of dimension $n \in \mathbb{N}$, and consider a linear endomorphism

$$f : V \to V.$$

A. Chandler
Department of Mathematics, North Carolina State University, Raleigh, NC, USA

N. Karnick
Department of Mathematics, Indiana University-Bloomington, Bloomington, IN, USA

D. Vagner
Department of Mathematics, Duke University, Durham, NC, USA

© The Editor(s) (if applicable) and The Author(s), under exclusive licence to Springer Nature Switzerland AG 2020
B. Elias et al., *Introduction to Soergel Bimodules*, RSME Springer Series 5, https://doi.org/10.1007/978-3-030-48826-0_23

If $\kappa \in \mathbb{k}$ is a scalar and $m \in V$ is a nonzero vector such that

$$fm = \kappa m,$$

then m is called an *eigenvector* of f with *eigenvalue* κ, or simply a κ-*eigenvector*. The subspace V_κ of κ-eigenvectors is called the κ-*eigenspace* of f.

Let $\{\kappa_i\}_{i=0}^r$ denote the set of (distinct) eigenvalues of f. If V has a basis of eigenvectors (an *eigenbasis*) of f, then f is commonly called diagonalizable. The matrix for f with respect to an eigenbasis is a diagonal matrix with entries κ_i with multiplicity $\dim V_{\kappa_i}$.

You knew all this already. However, we would like to discuss diagonalizability without the mention of eigenvectors. Eigenvectors are often not easy to find! This problem is accentuated further in the categorification, and the idea that f is diagonalizable if it has a basis of eigenvectors is not easy to categorify.

Remark 23.1 In (22.33) it was argued that the action of δ_{w_0} on $H(S_n)$ is block upper triangular with respect to the Kazhdan–Lusztig basis, where the blocks are the two-sided cells, and upper-triangularity is with respect to the cell order. Moreover, comparable blocks have distinct eigenvalues. By familiar arguments from linear algebra, this implies that δ_{w_0} is diagonalizable. However, there are no known formulas for the eigenvectors in the literature. The same argument applies to the full twist ft_n by (22.46).

Exercise 23.2 Continuing your work from Exercise 22.32, compute the eigenvectors of δ_{w_0} acting on $H(S_3)$.

Consider the direct sum decomposition

$$V = \bigoplus_{i=0}^r V_{\kappa_i}.$$

The composition of the projection map $V \to V_{\kappa_i}$ with the inclusion map $V_{\kappa_i} \to V$ is an idempotent endomorphism

$$p_i : V \to V_{\kappa_i} \to V,$$

which we call the *projection* to V_{κ_i}. One can verify that the set $\{p_i\}_{i=0}^r$ is a *complete collection of orthogonal idempotents* or an *idempotent decomposition (of identity)*. That is, it satisfies the following relations:

$$p_i^2 = p_i, \qquad p_i p_j = 0 \text{ for } i \neq j, \qquad \sum_i p_i = \mathrm{id}_V. \tag{23.1}$$

Since the idempotent p_i projects to the κ_i-eigenspace, we have the formula

$$f \circ p_i = \kappa_i p_i = p_i \circ f. \tag{23.2}$$

These idempotents give the basis-free explanation for why f can be written as a block matrix, where each block is a scalar multiple of the identity.

23.1 Classical Linear Algebra

Definition 23.3 We say that a linear map $f: V \to V$ is *diagonalizable* if there exists a finite set I, an idempotent decomposition $(p_i)_{i \in I}$ of the identity, and scalars $(\kappa_i)_{i \in I}$ in $\Bbbk$, such that (23.2) holds for each $i \in I$. This data is called a *diagonalization* of f.

In Definition 23.3 we have not assumed that the scalars κ_i are distinct. In particular, the idempotent decomposition may give a finer decomposition of V than merely the splitting into eigenspaces. In this sense, Definition 23.3 is slightly more general than the discussion above. The underlying set (without multiplicity) $\{\kappa_i \mid p_i \neq 0\}$ is called the *spectrum* of f.

For a scalar $\kappa \in \Bbbk$, it is often useful to rephrase properties of eigenvectors in terms of the operator $f - \kappa = f - \kappa \mathrm{id}_V$. For instance, m is a κ-eigenvector if and only if

$$(f - \kappa)m = 0,$$

and if p_κ is the projection to the κ-eigenspace, then

$$(f - \kappa)p_\kappa = 0 = p_\kappa(f - \kappa). \tag{23.3}$$

If f is diagonalizable with spectrum $\{\kappa_j\}_{j=0}^r$, one has the relation

$$(f - \kappa_0)(f - \kappa_1) \cdots (f - \kappa_r) = 0. \tag{23.4}$$

In fact, an operator f satisfies (23.4) if and only if it is diagonalizable, as we recall in the next proposition. For sake of clarity (and because the categorical situation is different), we refer to an operator which satisfies (23.4) as being *prediagonalizable*, with respect to the set of distinct scalars $\{\kappa_j\}_{j=0}^r$. Prediagonalizability comes with a minimality condition: $\Pi(f - \kappa_i) \neq 0$ when the product is taken over any proper subset of $\{0, \ldots, r\}$.

Proposition 23.4 *If $f \in \mathrm{End}(V)$ is diagonalizable with spectrum $\{\kappa_j\}_{j=0}^r$, then it is prediagonalizable with respect to $\{\kappa_j\}_{j=0}^r$. Conversely, if $f \in \mathrm{End}(V)$ is prediagonalizable with respect to $\{\kappa_j\}_{j=0}^r$, then f is diagonalizable with spectrum $\{\kappa_j\}_{j=0}^r$.*

Proof We only prove the second statement, leaving the first as an exercise. Suppose that f is prediagonalizable with respect to distinct scalars $\{\kappa_j\}_{j=0}^r$. We construct the idempotents $p_i, i \in I = \{0, \ldots, r\}$, via *Lagrange interpolation*:

$$p_i = p_i(f) = \prod_{j \neq i} \frac{f - \kappa_j}{\kappa_i - \kappa_j}. \tag{23.5}$$

It follows immediately from (23.4) that $(f - \kappa_i)p_i = 0 = p_i(f - \kappa_i)$. It remains to show that $\{p_i\}_{i \in I}$ constitutes an idempotent decomposition.

Letting $\Bbbk[f]$ denote the subring of $\mathrm{End}(V)$ generated by f, there is a surjection $Q = \Bbbk[x]/\langle(x - \kappa_0) \cdots (x - \kappa_r)\rangle \to \Bbbk[f]$ which sends x to f. The formula (23.5) is actually the formula for a degree r polynomial $p_i(x) \in \Bbbk[x]$, after replacing each instance of f with x. We also denote the image of p_i in Q by the symbol p_i. We will show that p_i are a complete list of orthogonal idempotents already in the ring Q, which will imply the corresponding statement in $\Bbbk[f]$.

Every element of Q is the image of an element of $\Bbbk[x]$ of degree $\leq r$. The element $p_i \in \Bbbk[x]$ satisfies $p_i(\kappa_i) = 1$ and $p_i(\kappa_j) = 0$ when $i \neq j$. For any scalars $a_i \in \Bbbk$, the polynomial $L = \sum_i a_i p_i$ satisfies $L(\kappa_i) = a_i$. Since any polynomial of degree $\leq r$ is determined by its values at $r + 1$ distinct points, we see that L is the unique polynomial of degree $\leq r$ with these values. In other words, any $L \in Q$ satisfies $L = \sum_i L(\kappa_i) p_i$. In particular, letting $L = 1$ yields completeness $1 = \sum_i p_i$, letting $L = p_i^2$ yields idempotence $p_i^2 = p_i$, and letting $L = p_i p_j$ yields orthogonality $p_i p_j = 0$ when $i \neq j$. □

Exercise 23.5 Prove the first statement in the proposition.

The following section is dedicated to categorifying this construction.

Remark 23.6 The definition of (pre)diagonalizability referred not to V itself but only to its endomorphism algebra $\mathrm{End}(V)$. These notions still make sense for an element f of any $\Bbbk$-algebra. Proposition 23.4 applies even when the base ring $\Bbbk$ is not a field, so long as $\kappa_i - \kappa_j$ is invertible for all $i \neq j$.

23.2 Categorified Linear Algebra

23.2.1 Eigenobjects

We want to categorify everything in sight. We replace the operator f acting on the vector space V with an endofunctor F of an additive category $\mathcal{V}$. Analogous to Remark 23.6, categorified linear algebra can be phrased entirely within the monoidal category $\mathrm{End}(\mathcal{V})$ of endofunctors of $\mathcal{V}$, and we therefore assume this more general setup from the start. Thus let $\mathcal{A} = (\mathcal{A}, \otimes, \mathbb{1})$ be a monoidal additive category acting (by additive functors) on an additive category $\mathcal{V}$:

$$(-) \otimes (-) : \mathcal{A} \times \mathcal{V} \to \mathcal{V}.$$

Fix some object $F \in \mathcal{A}$, which we identify with the functor $F \otimes (-) : \mathcal{V} \to \mathcal{V}$. We may also consider situations where $\mathcal{A}$ and $\mathcal{V}$ are abelian or triangulated, in which case we assume that the action map $\mathcal{A} \times \mathcal{V} \to \mathcal{V}$ is biexact.

The first interesting question is what should categorify scalar multiplication. We fix a monoidal subcategory $\mathcal{K} \subset \mathcal{A}$ of *scalar objects*, which induce *scalar endofunctors* $\lambda \otimes - : \mathcal{V} \to \mathcal{V}$ for each scalar object $\lambda \in \mathcal{K}$. As a typical example, if $\mathcal{V}$ has a natural notion of "grading shift" autoequivalences and $\mathcal{A} = \mathrm{End}(\mathcal{V})$, then the scalar objects might consist of (direct sums of) grading shifts. We call $\lambda \in \mathcal{K}$ *invertible* if there exists $\lambda^{-1} \in \mathcal{K}$ such that $\lambda \otimes \lambda^{-1} \simeq \mathbb{1} \simeq \lambda^{-1} \otimes \lambda$.

What would it mean to categorify the equation $fm = \kappa m$? A naive guess would be to consider a nonzero object $M \in \mathcal{V}$ and a scalar object $\lambda \in \mathcal{K}$ such that there is an isomorphism $F \otimes M \simeq \lambda \otimes M$. We call M a *weak λ-eigenobject* for F in this case. For such a structure to be sufficiently amenable to categorical machinery, however, we must fix this isomorphism in a natural way.

Definition 23.7 Suppose $\lambda \in \mathcal{K}$, $\alpha : \lambda \to F$ in $\mathcal{A}$, and $M \in \mathcal{V}$ is nonzero. We call M an *eigenobject of F with eigenmap α*, or an *α-eigenobject*, if the map

$$\alpha \otimes \mathrm{id}_M : \lambda \otimes M \to F \otimes M$$

is an isomorphism. In this situation, we call α a *forward eigenmap*. We define the *α-eigencategory* $\mathcal{V}_\alpha$ to be the smallest full subcategory of $\mathcal{V}$ containing the α-eigenobjects (and zero).

Remark 23.8 We may also consider a *backward eigenmap* $\beta : F \to \lambda$, defined similarly. A given functor may have many forward eigenmaps and few backward eigenmaps, or vice versa. See Exercise 23.33 for an example.

The reader would be well served by following an interesting example as they learn about these new concepts, and we encourage the simultaneous reading of Sect. 23.3.

The notion of an eigenobject has some considerable subtleties which are missing in linear algebra. For example, in linear algebra, eigenspaces for distinct eigenvalues intersect trivially. Similarly, eigencategories for nonisomorphic scalar functors intersect in the zero object (under the assumption that only 0 is fixed by a nontrivial scalar functor). However, two distinct eigenmaps with the same scalar functor might lead to distinct eigencategories that do intersect nontrivially. Taking linear combinations of these eigenmaps may give a new eigenmap.

Example 23.9 Let $\mathcal{A}$ be the monoidal category of $\mathbb{Z}$-modules, whose monoidal identity is $\mathbb{1} = \mathbb{Z}$. Let $F = \lambda = \mathbb{Z}$, and consider the map $\alpha_k : \mathbb{Z} \to \mathbb{Z}$ given by multiplication by $k \in \mathbb{Z}$. Then an eigenobject for α_k is a $\mathbb{Z}$-module for which k acts invertibly. For instance, $\mathbb{Q}$ is an eigenobject for α_k for all $k \neq 0$.

Remark 23.10 The study of eigencategories sharing a given scalar functor has a great overlap with algebraic geometry. We refer the reader to [50, Chapter 2] for more discussion.

23.2.2 Prediagonalizability

We now consider categorified analogues of prediagonalizability and diagonalizability. To do so, we must categorify the difference $f - \kappa$. As we saw in Sect. 19.2.3, this can be done via the mapping cone construction, so long as we work in the triangulated setting. Briefly, for any map $\phi\colon X \to X'$ in a triangulated category, there is an object $\mathrm{Cone}(\phi)$ and a distinguished triangle

$$X \xrightarrow{\phi} X' \to \mathrm{Cone}(\phi) \to X[1].$$

The (triangulated) Grothendieck group of a triangulated category has the relation $[C_1] + [C_3] = [C_2]$ for every distinguished triangle $C_1 \to C_2 \to C_3 \to C_1[1]$, thus yielding the relation

$$[\mathrm{Cone}(\phi)] = [X'] - [X].$$

Thus, cones can be used to "categorify subtraction."

The cone of a morphism is well-defined up to isomorphism, but not up to unique isomorphism, and is not functorial in general. To avoid this problem, we restrict our attention to homotopy categories of additive categories, where cones can be defined canonically.

> **Assumption 23.11** Henceforth, let $\mathcal{A}$ be a *monoidal homotopy category* $(\mathcal{A}, \otimes, \mathbb{1})$, i.e. we assume that there exists an additive monoidal category $(\mathcal{B}, \otimes, \mathbb{1})$ such that $\mathcal{A}$ is a full triangulated monoidal subcategory of the homotopy category $K(\mathcal{B})$.

Remark 23.12 The condition that $\mathcal{A}$ is monoidal typically forces one to restrict to complexes which are either bounded above or bounded below. When tensoring complexes (see (19.8)) which are infinite in both directions, an infinite direct sum appears in each homological degree, which may cause problems.

For clarity, we often write $X \simeq Y$ to denote when two complexes are homotopy equivalent, and hence isomorphic in the homotopy category.

With these ideas in hand, we can now categorify prediagonalizability.

Lemma 23.13 *Let $F \in \mathcal{A}$, and let $\alpha \colon \lambda \to F$ be a morphism from a scalar object. Then $M \in \mathcal{V}$ is an α-eigenobject if and only if $\mathrm{Cone}(\alpha) \otimes M \simeq 0$.*

Proof From the distinguished triangle

$$\lambda \otimes M \xrightarrow{\alpha \otimes \mathrm{id}_M} F \otimes M \longrightarrow \mathrm{Cone}(\alpha \otimes \mathrm{id}_M) \longrightarrow \lambda \otimes M[1]$$

and Lemma 19.52, we see that $\alpha \otimes \mathrm{id}_M$ is an isomorphism if and only if $0 \simeq \mathrm{Cone}(\alpha \otimes \mathrm{id}_M) \simeq \mathrm{Cone}(\alpha) \otimes M$. □

23.2 Categorified Linear Algebra

Definition 23.14 We say $F \in \mathcal{A}$ is *categorically prediagonalizable* if there is a finite set of maps $\{\alpha_i : \lambda_i \to F\}_{i \in I}$, with λ_i scalar objects, such that

$$\bigotimes_{i \in I} \operatorname{Cone}(\alpha_i) \simeq 0.$$

We also assume that the tensor product over a proper subset of I is nonzero. We call the set $\{\alpha_i\}_{i \in I}$ a *prespectrum* for F.

Remark 23.15 Unlike in $\Bbbk[f]$ where the elements $f - \kappa_i$ and $f - \kappa_j$ commute, the functors $\operatorname{Cone}(\alpha_i)$ and $\operatorname{Cone}(\alpha_j)$ need not tensor-commute. We will ignore this issue in this survey, and assume these cones "strongly commute" (see [50, Definition 6.12] for the definition).

The prespectrum of a functor F is not unique, and even the number of eigenmaps in a prespectrum can change! We refer the reader to [50, Chapter 2] for further discussion.

Example 23.16 We provide (without background) one example of non-uniqueness, which arises in algebraic geometry. We work in the monoidal category of coherent sheaves on $\mathbb{P}^r$. Let $\mathcal{O}(1)$ denote its defining ample line bundle. Then any basis of $\operatorname{Hom}(\mathcal{O}, \mathcal{O}(1))$ will be a prespectrum for the endofunctor $\mathcal{O}(1) \otimes (-)$.

23.2.3 Twisted Complexes

Just as in the decategorified case, we would like to use prediagonalizability to construct an idempotent decomposition via a sort of Lagrange interpolation. We first need to define what a categorical idempotent decomposition is. In order to do this, we need to think a bit harder about categorifying addition. In an additive category there is only one way to add two objects (take the direct sum), but in a triangulated (or abelian) setting there can be many. We now introduce the language of twisted complexes, which formalize iterated mapping cones.

Definition 23.17 A *twisted complex* $(C_i, d_{ij})_{i,j \in I}$ is the data of a poset $(I, \leq)$, chain complexes (C_i, d_{C_i}), and maps $d_{ij} : C_j \to C_i$ (not necessarily chain maps) of homological degree $+1$ for $j \leq i$ such that

1. $d_{ii} = d_{C_i}$, and
2. $\sum_{k \in [j,i]} d_{ik} d_{kj} = 0$.

We assume that $(I, \leq)$ is *interval finite*, i.e. that $[j, i] = \{k \mid j \leq k \leq i\}$ is finite for all $j \leq i$, in order that the sum above will make sense.

Bicomplexes are a special case of twisted complexes, where $I = \mathbb{Z}$ and $d_{ij} = 0$ unless $i = j + 1$ or $i = j$. Given a twisted complex, we can obtain a chain complex called the convolution, which generalizes the notion of a total complex of a bicomplex.

Definition 23.18 Given a twisted complex $(C_i, d_{ij})_{i,j \in I}$, its *convolution* or *total complex* is the chain complex given by

$$\text{Tot}(C_i, d_{ij})_{i,j \in I} = \left(\bigoplus_{i \in I} C_i, \sum_{(i,j) \in I^2} d_{ij} \right). \quad (23.6)$$

Remark 23.19 Questions like "is this convolution well-defined" and "what if we used the direct product rather than the direct sum" will be ignored in this survey. Ultimately, we will assume that certain scalar objects are small (see Definition 23.22), which will take care of such technical issues.

Note that the convolution is distinct from simply taking the direct sum of complexes (C_i, d_{C_i}) in that it provides a more interesting "twisted" differential. Thus a convolution is potentially a more interesting categorification of the sum of complexes. We use this to categorify the notion of an idempotent decomposition.

23.2.4 Diagonalizability

Definition 23.20 Let $(\mathcal{A}, \otimes, \mathbb{1})$ be a monoidal homotopy category and $\mathbf{P} = (\mathbf{P}_i, d_{ij})_{i,j \in I}$ a twisted complex in $\mathcal{A}$, with finite index set I. We say $\mathbf{P}$ is an *idempotent decomposition* of $\mathbb{1}$ if

1. $\mathbf{P}_i \neq 0$ for all i,
2. $\mathbf{P}_i \otimes \mathbf{P}_i \simeq \mathbf{P}_i$ for all i,
3. $\mathbf{P}_i \otimes \mathbf{P}_j \simeq 0$ whenever $i \neq j$, and
4. $\text{Tot}(\mathbf{P}) \simeq \mathbb{1}$.

This gives us the language to define what it means for a functor $F \in \mathcal{A}$ to be diagonalizable.

Definition 23.21 Let F be an object of the homotopy monoidal category $(\mathcal{A}, \otimes, \mathbb{1})$, $(I, \leq)$ a finite poset, $\{\lambda_i\}_{i \in I} \subset \mathcal{K}$ a set of scalar objects, $\{\alpha_i : \lambda_i \to F\}_{i \in I}$ a set of morphisms in $\mathcal{A}$, and $\mathbf{P} = (\mathbf{P}_i, d_{ij})_{i,j \in I}$ a twisted complex in $\mathcal{A}$. We say $(\mathbf{P}_i, \alpha_i)_{i \in I}$ is a *diagonalization* of F if $\mathbf{P}$ is an idempotent decomposition of $\mathbb{1} \in \mathcal{A}$, and for all $i \in I$,

$$\text{Cone}(\alpha_i) \otimes \mathbf{P}_i \simeq 0 \simeq \mathbf{P}_i \otimes \text{Cone}(\alpha_i). \quad (23.7)$$

Note that (23.7) is a direct categorification of (23.3). However, there is a mysterious extra piece of data in the categorical definition: a partial order on the set I of eigenvalues.

23.2.5 Smallness

The following condition is needed to ensure that certain infinite convolutions we will consider are well-defined.

Definition 23.22 We call $\lambda \in \mathcal{K}$ *small* if, for all objects $B \in \mathcal{A}$, the infinite direct sum $\bigoplus_{n \geq 0} \lambda^{\otimes n} \otimes B$ exists in $\mathcal{A}$ and is isomorphic to the direct product.

If λ categorifies the scalar κ, then smallness of λ ensures that the infinite sum

$$1 + \kappa + \kappa^2 + \cdots$$

can be categorified. Typically, if λ is small, then λ^{-1} is not small.

Example 23.23 When $\mathcal{A}$ is the bounded above homotopy category, then $[1]$ is small, and $[-1]$ is not.

23.2.6 Lagrange Interpolation

We would now like to categorify Lagrange interpolation (23.5): given a prediagonalizable functor $F \in \mathcal{A}$, we wish to construct a diagonalization. Letting $c_{ji} = \frac{f - \kappa_j}{\kappa_i - \kappa_j}$, we have that $p_i = \prod_{j \neq i} c_{ji}$. We will then hope to define

$$\mathbf{P}_i = \bigotimes_{j \neq i} \mathbf{C}_{ji}, \qquad (23.8)$$

where $\mathbf{C}_{ji}$ categorifies c_{ji}. To categorify c_{ji}, we make use of the infinite geometric series expansion

$$c_{ji} = \frac{f - \kappa_j}{\kappa_i - \kappa_j} = \kappa_i^{-1} \left(\frac{f - \kappa_j}{1 - \frac{\kappa_j}{\kappa_i}} \right) \qquad (23.9)$$

$$= \kappa_i^{-1}(f - \kappa_j)(1 + (\tfrac{\kappa_j}{\kappa_i}) + (\tfrac{\kappa_j}{\kappa_i})^2 + \cdots)$$

$$= \tfrac{1}{\kappa_i}(f - \kappa_j) + \tfrac{\kappa_j}{\kappa_i^2}(f - \kappa_j) + \cdots.$$

We could naively categorify this sum via $\bigoplus_{n \geq 0} \lambda_i^{-1} \operatorname{Cone}(\alpha_j) \otimes (\lambda_j \lambda_i^{-1})^{\otimes n}$, so long as $\lambda_j \lambda_i^{-1}$ is small. This sum only uses the eigenmap $\alpha_j : \lambda_j \to F$. We will, however, want to also keep track of the eigenmap $\alpha_i : \lambda_i \to F$. This is where the convolution will again play a role, allowing for a categorification of sums with more interesting differentials. We will set $\mathbf{C}_{ji} = \mathbf{C}_{\alpha_j, \alpha_i}$ using Definition 23.24 below.

Definition 23.24 Let $\alpha : \lambda \to F$ and $\beta : \mu \to F$ be maps from invertible scalar objects λ and μ, such that $\lambda\mu^{-1}$ is small. We then define the complex

$$\mathbf{C}_{\alpha,\beta} = \mathrm{Tot}\left(\begin{array}{ccccccc} & \frac{1}{\mu}\alpha & \frac{\lambda}{\mu} & -\frac{\lambda}{\mu^2}\beta & \frac{\lambda}{\mu^2}\alpha & \frac{\lambda^2}{\mu^2} & -\frac{\lambda^2}{\mu^3}\beta \\ & \searrow & & \searrow & & \searrow & \\ \frac{1}{\mu}F & & & \frac{\lambda}{\mu^2}F & & & \cdots \end{array} \right). \qquad (23.10)$$

Note that if we remove the arrows pointing southeast, we obtain the naive direct sum mentioned above.

Since $\mu\lambda^{-1}$ is unlikely to also be small, the complex $\mathbf{C}_{\beta,\alpha}$ will be defined by a different formula. Note the homological shift [1].

$$\mathbf{C}_{\beta,\alpha}[1] = \mathrm{Tot}\left(\begin{array}{ccccccccc} \mathbb{1} & & -\frac{1}{\mu}\beta & \frac{1}{\mu}\alpha & \frac{\lambda}{\mu} & -\frac{\lambda}{\mu^2}\beta & \frac{\lambda}{\mu^2}\alpha & \frac{\lambda^2}{\mu^2} & -\frac{\lambda^2}{\mu^3}\beta \\ & \searrow & & \searrow & & \searrow & & \searrow & \\ & & \frac{1}{\mu}F & & & \frac{\lambda}{\mu^2}F & & & \cdots \end{array} \right). $$

$$(23.11)$$

Remark 23.25 The signs on β in (23.10) and (23.11) are not truly important, and one would obtain an isomorphic complex had one defined $\mathbf{C}_{\alpha,\beta}$ without the signs. The reason for the signs is to simplify another aspect of the theory which we do not discuss at all in this survey, see [50, Proposition 7.23 and Remark 7.24] for details.

Exercise 23.26 Equation (23.9) involved power series in $\frac{\kappa_j}{\kappa_i}$. Give the corresponding expansion of c_{ji} using power series in $\frac{\kappa_i}{\kappa_j}$, and match this to (23.11).

Exercise 23.27 Note that $c_{ji} = 1 - c_{ij}$. Categorify this statement with a distinguished triangle, using (23.10) and (23.11).

We are finally ready to categorify Lagrange interpolation, and hence present our main theorem.

Theorem 23.28 (Diagonalization Theorem) *Consider a monoidal homotopy category* $(\mathcal{A}, \otimes, \mathbb{1})$*, and a categorically prediagonalizable object F with prespectrum* $\{\alpha_i : \lambda_i \to F\}_{i \in I}$*. Suppose that* $(I, \leq)$ *is a finite totally ordered set, such that each* λ_i *is invertible and* $\lambda_i \lambda_j^{-1}$ *is small whenever* $j < i$*. Then the following categorified projectors constitute an idempotent decomposition of* $\mathbb{1}$:

$$\mathbf{P}_i = \bigotimes_{j \neq i} \mathbf{C}_{ji}. \qquad (23.12)$$

Here, $\mathbf{C}_{ji} = \mathbf{C}_{\alpha_j, \alpha_i}$, where we use (23.10) when $j > i$ and (23.11) when $i > j$.

23.3 A Toy Example

Note that this theorem introduces a new wrinkle into the mix: a total order on the scalar functors, corresponding to which ratios of scalar functors are small.

Remark 23.29 Suppose that the scalar objects λ_i have distinct homological shifts. Then it is typically the case that ordering the λ_i by their homological shifts will provide a suitable total order on I in the bounded above homotopy category, and the opposite order will be suitable in the bounded below homotopy category.

For a proof of this theorem, consult [50, §8].

23.3 A Toy Example

The example in this section (when $n = 2$) serves as a toy model for the category of Soergel bimodules in type A_1, which we will see below in Sect. 23.4.1.

Let $A = \mathbb{Z}[C_n]$ be the group algebra of $C_n = \{1, x, \ldots, x^{n-1}\}$, the cyclic group of order n. Consider the homotopy category $K^b(A\text{-mod})$ as a monoidal category under $\otimes_\mathbb{Z}$, and let $\mathcal{A} = K^b(A\text{-mod})$ act on the left on $\mathcal{V} = \mathcal{A}$. Our subcategory of scalar objects will consist of direct sums of homological shifts of $\mathbb{Z}$, the monoidal identity.

Consider the complex

$$F = \left(0 \longrightarrow \underline{A} \xrightarrow{1-x} A \xrightarrow{\epsilon_{x \to 1}} \mathbb{Z} \longrightarrow 0 \right) \qquad (23.13)$$

concentrated in cohomological degrees 0 through 2. Here and in what follows, we adopt the following conventions: an element $a \in A$ indicates multiplication by a; the map $\epsilon_{x \to 1}$ sends x to 1; and the underlined term $\underline{A}$ in the complex lies in cohomological degree zero.

We will diagonalize the functor

$$K^b(A\text{-mod}) \to K^b(A\text{-mod}) : M \mapsto F \otimes_\mathbb{Z} M, \qquad (23.14)$$

again denoted by F. Note that, by Morita theory, natural transformations between functors given by tensoring with a complex are specified by chain maps between the corresponding complexes.

Suppose we had an eigenmap $\alpha_0 : \mathbb{Z} \to F$, given by a single nonzero component α_0^0:

$$\begin{array}{c} 0 \longrightarrow \underline{A} \longrightarrow A \longrightarrow \mathbb{Z} \longrightarrow 0. \\ \uparrow\alpha_0^0 \\ \mathbb{Z} \end{array} \qquad (23.15)$$

Let us be greedy and expect an eigenobject concentrated in homological degree 0, i.e. of the form $M = (0 \to \underline{M} \to 0)$. Then the condition $\text{Cone}(\alpha_0) \otimes_{\mathbb{Z}} M \simeq 0$ implies that

$$0 \longrightarrow M \xrightarrow{\alpha_0^0(1)} \underline{M[C_n]} \xrightarrow{1-x} M[C_n] \xrightarrow{\epsilon_{x \to 1}} M \longrightarrow 0 \qquad (23.16)$$

has zero homology and thus $\text{Ker}(1-x) = \text{Im}\,\alpha_0^0(1)$. This forces α_0^0 to be the map

$$\alpha_0^0 : 1 \mapsto 1 + x + x^2 + \cdots + x^{n-1} = \frac{x^n - 1}{x - 1}. \qquad (23.17)$$

The reader can check in this case that for $M = A$, the complex $\text{Cone}(\alpha_0) \otimes A$ is indeed homotopic to 0, and thus A is an α_0-eigenobject for F.

An eigenmap of the form $\alpha : \mathbb{Z}[-1] \to F$ is not possible, since all chain maps

$$\begin{array}{c} 0 \longrightarrow \underline{A} \longrightarrow A \longrightarrow \mathbb{Z} \longrightarrow 0, \\ \uparrow \\ \mathbb{Z} \end{array} \qquad (23.18)$$

are nullhomotopic.

Suppose that some morphism $\alpha_1 : \mathbb{Z}[-2] \to F$, given by a chain map

$$\begin{array}{c} 0 \longrightarrow \underline{A} \longrightarrow A \longrightarrow \mathbb{Z} \longrightarrow 0, \\ \alpha_1^2 \uparrow \\ \mathbb{Z} \end{array} \qquad (23.19)$$

were an eigenmap with an eigenobject of the form $M = (0 \to \underline{M} \to 0)$. Then

$$0 \longrightarrow \underline{M[C_n]} \xrightarrow{\begin{bmatrix} 1-x \\ 0 \end{bmatrix}} M[C_n] \oplus M \xrightarrow{\begin{bmatrix} \epsilon_{x \to 1} & \alpha_1^2(1) \end{bmatrix}} M \longrightarrow 0 \qquad (23.20)$$

must have zero homology. However, there is no way to choose α_1^2 so that there is no homology in degree 1. Thus we are forced to consider a more complicated eigenobject (one that is not concentrated in a single homological degree). Luckily, since there are only two possible eigenmaps, if we are to diagonalize F then we must have

$$\text{Cone}(\alpha_1) \otimes_{\mathbb{Z}} \text{Cone}(\alpha_0) \simeq 0.$$

23.3 A Toy Example

Thus $\mathrm{Cone}(\alpha_0)$ must be an eigenobject corresponding to α_1. With the choice $\alpha_1^2 = 1$ we will argue that this is indeed the case. Now $\mathrm{Cone}(\alpha_0)$ looks like

$$0 \longrightarrow A \xrightarrow{\begin{bmatrix}1-x\\0\end{bmatrix}} A \oplus \mathbb{Z} \xrightarrow{\begin{bmatrix}\epsilon_{x\to 1} & 1\end{bmatrix}} \mathbb{Z} \longrightarrow 0. \quad (23.21)$$

Applying Gaussian elimination to eliminate the two copies of $\mathbb{Z}$, we obtain the homotopy-equivalent complex

$$0 \longrightarrow \underline{A} \xrightarrow{1-x} A \longrightarrow 0 \longrightarrow 0. \quad (23.22)$$

Since we already have $\mathrm{Cone}(\alpha_1) \otimes A \simeq 0$, a short exact sequence argument will show that $\mathrm{Cone}(\alpha_1) \otimes (-)$ annihilates any finite complex built from copies of A. In particular

$$\mathrm{Cone}(\alpha_1) \otimes \mathrm{Cone}(\alpha_0) \simeq 0.$$

Thus we have constructed two eigenmaps, and shown that F is prediagonalizable. We now use α_0 and α_1 to construct the categorified projectors. The projector $\mathbf{P}_0 = \mathbf{C}_{\alpha_1,\alpha_0}$ is the total complex of the twisted complex in Fig. 23.1 below. Applying Gaussian elimination to eliminate all the copies of $\mathbb{Z}$ in the total complex, we find that

$$\mathbf{P}_0 \simeq \quad \cdots \xrightarrow{\frac{x^n-1}{x-1}} A \xrightarrow{1-x} A \xrightarrow{\frac{x^n-1}{x-1}} A \xrightarrow{1-x} \underline{A}. \quad (23.23)$$

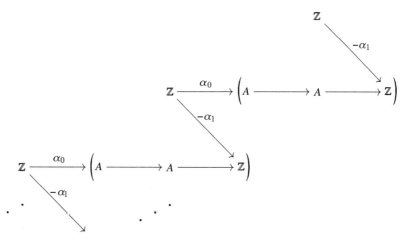

Fig. 23.1 To get the projector $\mathbf{P}_0$ we take the total complex of this twisted complex

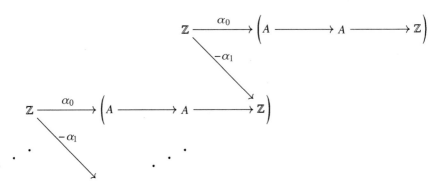

Fig. 23.2 To get the projector $\mathbf{P}_1$ we take a total complex of this twisted complex

Similarly, $\mathbf{P}_1 = \mathbf{C}_{\alpha_0, \alpha_1}$ is the total complex of the twisted complex shown in Fig. 23.2 below. After applying Gaussian elimination (to eliminate all but one copy of $\mathbb{Z}$), we get

$$\mathbf{P}_1 \simeq \cdots \xrightarrow{\frac{x^n-1}{x-1}} A \xrightarrow{1-x} A \xrightarrow{\frac{x^n-1}{x-1}} A \xrightarrow{1-x} A \xrightarrow{\epsilon_{x \to 1}} \mathbb{Z}. \quad (23.24)$$

We claim that $\mathbf{P}_1 \otimes (-)$ annihilates A, which we leave to the reader. This can be bootstrapped to see that $\mathbf{P}_1 \otimes \mathbf{P}_0 \simeq 0$, since $\mathbf{P}_0$ is built from copies of A. Similarly, $\mathbf{P}_1$ agrees with $\mathbb{Z}$ up to copies of A, so that $\mathbf{P}_1 \otimes \mathbf{P}_1 \simeq \mathbf{P}_1 \otimes \mathbb{Z} \simeq \mathbf{P}_1$.

Remark 23.30 This argument may look sketchy, because $\mathbf{P}_0$ is built from *infinitely many* copies of A. One must make such arguments carefully, see [50, §4.5].

There is a quotient map of complexes $\mathbf{P}_1 \to \mathbf{P}_0[1]$, with kernel $\mathbb{Z}$. This leads to the distinguished triangle

$$\mathbf{P}_0 \to \mathbb{Z} \to \mathbf{P}_1 \to \mathbf{P}_0[1]$$

which categorifies $1 = [\mathbb{Z}] = [\mathbf{P}_0] + [\mathbf{P}_1]$.

Exercise 23.31 Find the twisted complex $(\mathbf{P}_0 \oplus \mathbf{P}_1, d)$ whose convolution is homotopy equivalent to $\mathbb{Z}$.

Altogether, we have shown that F is categorically diagonalizable with spectrum $\{\alpha_0, \alpha_1\}$ in the category of bounded above complexes.

Exercise 23.32 Now show that F is categorically diagonalizable with the same spectrum, in the category of bounded below complexes. This will reverse the total order on $I = \{0, 1\}$, so it will change the constructions of $\mathbf{P}_0$ and $\mathbf{P}_1$.

The following exercises emphasize a difference between forward and backward eigenmaps.

Exercise 23.33 We have just found a categorical diagonalization of F using forward eigenmaps. Does F have any backward eigenmaps? Can it be categorically diagonalized using backward maps? (*Hint:* the answer to the second question is no!)

Exercise 23.34 Now consider the complex

$$F' = \left(0 \longrightarrow \mathbb{Z} \xrightarrow{\alpha_0^0} A \xrightarrow{1-x} \underline{A} \longrightarrow 0 \right). \tag{23.25}$$

Find the forward and backward eigenmaps of F'. Show that it can be categorically diagonalized using backward eigenmaps.

23.4 Diagonalizing the Full Twist

Let (W, S) be a Coxeter system, and let $\mathbb{S}\text{Bim}$ be the category of Soergel bimodules associated to its geometric representation over $\mathbb{R}$. Consider $\mathcal{A} = K^b(\mathbb{S}\text{Bim})$ acting on the left on $\mathcal{V} = \mathcal{A}$. We saw in Sect. 19.3 that the complexes

$$F_s = \underline{B_s} \xrightarrow{\uparrow} R(1)$$

for $s \in S$ categorify the elements $\delta_s \in H(W)$, and that their tensor products, called Rouquier complexes, categorify the products of these standard generators.

In this section, we apply categorical diagonalization to the full twist Rouquier complex of finite Coxeter systems.

23.4.1 Type A_1

Let us start with the case of type A_1, with $S = \{s\}$. Here, the full twist complex $\text{FT}_2 = F_s F_s$ categorifies the full twist element $\text{ft}_2 = \delta_s \delta_s$. Since the quadratic relation of the Hecke algebra can be written as

$$(\delta_s - v^{-1})(\delta_s + v) = 0,$$

we see that δ_s has eigenvalues v^{-1} and $-v$, hence $\text{ft}_2 = \delta_s^2$ has eigenvalues $(-1)^2 v^2$ and v^{-2}. This suggests that FT_2 should have two categorical eigenvalues, which will end up being the shifts $\mathbb{1}(2)[-2]$ and $\mathbb{1}(-2)[0]$. Note that the monoidal identity $\mathbb{1}$ is just the complex consisting of R in cohomological degree 0.

We computed in Exercise 19.25 that

$$\mathrm{FT}_2 \simeq \underline{B_s(-1)} \xrightarrow{\;\begin{smallmatrix}|&-&|\\\end{smallmatrix}\;} B_s(1) \xrightarrow{\;|\;} R(2). \qquad (23.26)$$

Tensoring on the right by B_s (in cohomological degree 0) and using $B_s B_s = B_s(1) \oplus B_s(-1)$, we get

$$\mathrm{FT}_2 \otimes B_s \simeq \underline{B_s B_s(-1)} \xrightarrow{\;|\;|-|\;|\;} B_s B_s(1) \xrightarrow{\;|\;} B_s(2)$$

$$\simeq \underline{B_s(-2) \oplus B_s(0)} \longrightarrow B_s(0) \oplus B_s(2) \longrightarrow B_s(2). \qquad (23.27)$$

If one keeps track of the differentials, one sees that one can Gaussian eliminate $B_s(0)$ and $B_s(2)$ from the last complex, so that

$$\mathrm{FT}_2 \otimes B_s \simeq B_s(-2) = \mathbb{1}(-2)[0] \otimes B_s. \qquad (23.28)$$

Thus B_s is a weak categorical eigenobject with eigenvalue $\mathbb{1}(-2)[0]$.

Exercise 23.35 Keep track of the differentials in the computation above and verify that Gaussian elimination applies as claimed.

In fact, B_s is an eigenobject for the eigenmap $\alpha_\square : \mathbb{1}(-2)[0] \to F$ given by the following chain map:

$$\begin{array}{ccccccc}
0 & \longrightarrow & \underline{B_s(-1)} & \longrightarrow & B_s(1) & \longrightarrow & R(-2) & \longrightarrow & 0. \\
& & \big\uparrow & & & & & & \\
& & \underline{R(-2)} & & & & & &
\end{array} \qquad (23.29)$$

There is a second eigenmap $\alpha_\square : \mathbb{1}(-2)[-2] \to F$, given by the following chain map, which is entirely analogous to the eigenmap α_1 from the toy example in Sect. 23.3:

$$\begin{array}{ccccccc}
0 & \longrightarrow & \underline{B_s(-1)} & \longrightarrow & B_s(1) & \longrightarrow & R(-2) & \longrightarrow & 0. \\
& & & & & & \big\uparrow{\scriptstyle 1} & & \\
& & & & & & R(-2) & &
\end{array} \qquad (23.30)$$

23.4 Diagonalizing the Full Twist

Because $\text{Cone}(\alpha_{\square\!\square}) \otimes B_s \simeq 0$, and because $\text{Cone}(\alpha_{\square\!\square})$ is "built" from shifts of B_s it will follow that

$$\text{Cone}(\alpha_{\square\!\square}) \otimes \text{Cone}(\alpha_{\square\!\square}) \simeq 0. \tag{23.31}$$

Thus FT_2 is categorically prediagonalizable with eigenvalues $\mathbb{1}(-2)[0]$ and $\mathbb{1}(2)[-2]$, categorifying

$$(\text{ft}_2 - v^2)(\text{ft}_2 - v^{-2}) = 0. \tag{23.32}$$

Exercise 23.36 Verify the statements above about the eigenmaps $\alpha_{\square\!\square}$ and $\alpha_{\square\!\square}$

Now let us compute the projectors. At the decategorified level, the eigenvalue v^{-2} gives rise (via Lagrange interpolation) to the projector

$$p_{\square\!\square} = \frac{\text{ft}_2 - v^2}{v^{-2} - v^2} = \frac{b_s}{v + v^{-1}}, \tag{23.33}$$

and the eigenvalue v^2 gives rise to

$$p_{\square\!\square} = \frac{\text{ft}_2 - v^{-2}}{v^2 - v^{-2}} = 1 - \frac{b_s}{v + v^{-1}}. \tag{23.34}$$

We now use the eigenmaps $\alpha_{\square\!\square}$ and $\alpha_{\square\!\square}$ to construct the categorified projectors $P_{\square\!\square}$ and $P_{\square\!\square}$ which diagonalize FT_2. To get $P_{\square\!\square}$ we take the total complex of the following twisted complex:

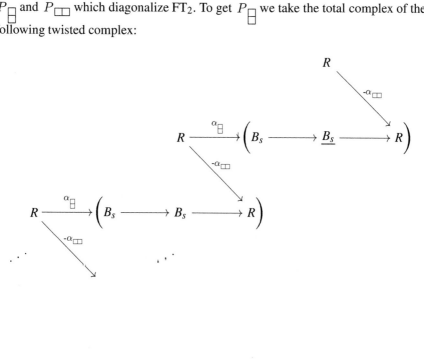

Applying Gaussian elimination to this complex, we find that

$$P_{\square} \simeq \cdots \longrightarrow B_s \longrightarrow B_s \longrightarrow B_s \longrightarrow B_s \longrightarrow \underline{B_s}.$$

Similarly, $P_{\square\square}$ is the total complex of the following twisted complex:

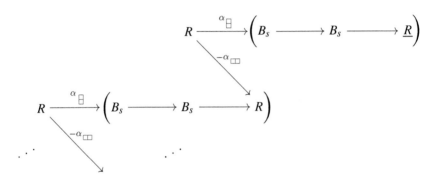

Applying Gaussian elimination yields

$$P_{\square\square} \simeq \cdots \longrightarrow B_s \longrightarrow B_s \longrightarrow B_s \longrightarrow B_s \longrightarrow \underline{R}. \tag{23.35}$$

The homotopy equivalence

$$\mathrm{Cone}\left(P_{\square\square}[-1] \to P_{\square}\right) \simeq \mathbb{1} \tag{23.36}$$

decategorifies to the idempotent decomposition

$$p_{\square\square} + p_{\square} = 1. \tag{23.37}$$

23.4.2 Type A

Recall from Chap. 22 that the Hecke algebra $H(S_n)$ in type A_{n-1} has a filtration by 2-sided cell ideals $I_{\leq \lambda}$, indexed by partitions λ of n. The subquotient $I_{\leq \lambda}/I_{<\lambda}$ is isotypic, being a direct sum of copies of the cell module V_λ. These cell modules are in bijection with the irreducible representations of $H(S_n)$. Moreover, the full twist $\mathrm{ft}_n = \delta_{w_0}^2$ acts on V_λ by the scalar $(-1)^{2c(\lambda)}v^{2x(\lambda)}$ (see Corollary 22.45). Thus

23.4 Diagonalizing the Full Twist

the full twist $\text{ft}_n \in H(S_n)$ is diagonalizable, with eigenvalues associated to each partition of n.

We do not say "with eigenvalues in bijection with partitions of n," because as noted in the end of Chap. 22, there are distinct partitions which give rise to the same eigenvalue. Thus projection from $H(S_n)$ to its isotypic components is a finer decomposition of identity than the eigenspace decomposition of ft_n. Still, this decomposition of identity is a diagonalization of ft_n, in the sense of Definition 23.3.

In [51], it is proven that FT_n, the Rouquier complex lifting ft_n, is categorically diagonalizable. One begins by proving that it is prediagonalizable.

Theorem 23.37 ([51, Theorem 1.6]) *For each partition λ of n, there is an eigenmap*

$$\alpha_\lambda : \mathbb{1}(2x(\lambda))[2c(\lambda)] \to \text{FT}_n.$$

One has

$$\bigotimes \text{Cone}(\alpha_\lambda) \simeq 0. \tag{23.38}$$

For the case of FT_3 we get the picture shown in Fig. 23.3. An ambitious reader might try to work out what the projectors look like in this case, but be ready for a truly massive calculation.

From here we may wish to apply Theorem 23.28 to construct categorical projections to eigencategories. This works for $n < 6$. However, the existence of distinct partitions with the same eigenvalue (or even the same homological shift) for

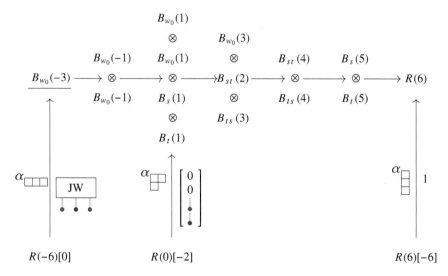

Fig. 23.3 The complex FT_3 and its eigenmaps

$n \geq 6$ causes the assumptions of Theorem 23.28 to fail. Nonetheless, the following theorem is proven in [51].

Theorem 23.38 ([51, Theorem 1.8]) *The functor* FT_n *is categorically diagonalizable with one projection for each partition* λ *of n, relative to the eigenmaps* α_λ *of the previous theorem.*

Remark 23.39 To prove this result, [51] constructs categorical projections inductively by simultaneously diagonalizing the commuting Rouquier complexes FT_1, $FT_2, \ldots, FT_n$ associated to the tower of Coxeter groups $S_1 \subset S_2 \subset \cdots \subset S_n$. This is morally the same as simultaneously diagonalizing the Jucys–Murphy elements, as discussed in Theorem 22.39. Note, however, that while the Jucys–Murphy braids do have associated Rouquier complexes, these Jucys–Murphy complexes are not categorically diagonalizable (they do not admit enough eigenmaps)!

Remark 23.40 The full twist FT_n does not admit enough backward eigenmaps to be categorically diagonalizable. Meanwhile, the inverse full twist FT_n^{-1}, which is the Rouquier complex for ft_n^{-1}, is categorically diagonalizable with backward eigenmaps, but not with forward eigenmaps. This is the underlying reason why the Jucys–Murphy complexes $FT_i FT_{i-1}^{-1}$ are not categorically diagonalizable using either forward or backward eigenmaps.

23.4.3 The General Case

For any finite Coxeter group W, one can define the *full twist* element $ft_W = \delta_{w_0}^2$. Lifting this element to the braid group of W, one can define the corresponding Rouquier complex FT_W. In addition, the Hecke algebra $H(W)$ has a cell filtration by 2-sided ideals, although the subquotients are not isotypic. There is a generalization of the eigenvalue formula (22.31) due to Mathas [135], which states that ft_W is diagonalizable, with eigenvalues associated to the 2-sided cells of W.

Conjecture 23.41 For each finite Coxeter group W, the full twist FT_W is categorically diagonalizable, with one eigenmap for each 2-sided cell of W.

See [50, Conjecture 3.5] for a more precise version of this conjecture. Theorem 23.38 shows that this conjecture holds in type A. It has also been proven by Elias–Hogancamp for dihedral groups, in forthcoming work.

Chapter 24
Singular Soergel Bimodules and Their Diagrammatics

This chapter is based on expanded notes of a lecture given by the authors and taken by
Ben Elias

Abstract Instead of considering the ring R and its bimodules, one can consider the family of rings $\{R^I\}$, where R^I denotes the invariants in R under the action of a finite parabolic subgroup $W_I \subset W$. Using the functors of induction and restriction between these various rings, one obtains the 2-category of singular Soergel bimodules. These bimodules were first studied in Williamson's thesis, where it was proven that they categorify the Hecke algebroid. One advantage of working with singular Soergel bimodules is that one can "zoom in" on a bimodule: for example, instead of studying the Bott–Samelson bimodule $B_s = R \otimes_{R^s} R(1)$, which is a composition of restriction and induction bimodules, one can study the induction or restriction bimodule itself. After introducing singular Soergel bimodules, we give a diagrammatic calculus for describing bimodule morphisms.

24.1 The Classical Theory of Singular Soergel Bimodules

This material follows Elias [47], Elias–Snyder–Williamson [62] and forthcoming work of Elias–Williamson. For an introduction to this material in the literature, see [49, Chapter 5] and [56, Chapter 2].

B. Elias
Department of Mathematics, University of Oregon, Fenton Hall, Eugene, OR, USA

24.1.1 Bimodules and Functors

In Chap. 7, Example 7.6, a 2-category $\mathcal{B}$im was defined. This 2-category is the setting for many other algebraic categories and 2-categories, in that they are sub-2-categories of $\mathcal{B}$im. For convenience, we repeat the definition here.

Definition 24.1 Let $\mathcal{B}$im denote the following 2-category. The objects are rings. For two rings R and S, the category $\operatorname{Hom}_{\mathcal{B}\mathrm{im}}(R, S)$ is the category of (S, R)-bimodules.

Let us expand on this definition. A 1-morphism from R to S is an (S, R)-bimodule M, which we may write as $_S M_R$ to emphasize which rings act where. The 2-morphisms from $_S M_R$ to $_S N_R$ are the bimodule morphisms from M to N. Moreover, given $_S M_R$ and $_T N_S$, the 1-morphism composition is the tensor product $_T (N \otimes_S M)_R$.

The 2-category $\mathcal{B}$im is a sub-2-category of $\mathcal{C}$at (see Example 7.5), as we now explain. Instead of the objects being rings R, we can think of them as representing the category of (left) R-modules. One can think of the bimodule $_S M_R$ as representing the functor $_S M \otimes_R (-)$, which is a functor from R-modules to S-modules. Each natural transformation between such functors is realized by a bimodule morphism, and composition of functors is realized by taking the tensor product.

In the converse direction, it is often fruitful to think of functors between module categories as being represented by bimodules. For example, let $A \subset B$ be a ring extension. Then the functor of induction, from A-modules to B-modules, agrees with the functor $_B B \otimes_A (-)$, and thus can be represented by the bimodule $_B B_A$. Similarly, the functor of restriction, from B-modules to A-modules, agrees with the functor $_A B \otimes_B (-)$, and is represented by the bimodule $_A B_B$.

24.1.2 Singular Soergel Bimodules

We begin by defining a combinatorial category of bimodules (analogous to Bott–Samelson bimodules). Afterwards, we will define singular Soergel bimodules by taking direct summands.

Definition 24.2 Recall that a subset $I \subset S$ is called *finitary* if the corresponding parabolic subgroup W_I is finite. In this case, we let w_I denote the longest element of W_I, and let $\ell(I)$ denote the length of w_I.

Definition 24.3 For any subset $I \subset S$, let $R^I = R^{W_I}$ denote the invariants in R under the action of the parabolic subgroup W_I. We sometimes use shorthand for R^I when I is small. For example, when $I = \{s, t\}$, we write $R^{s,t}$ instead of $R^{\{s,t\}}$.

24.1 The Classical Theory of Singular Soergel Bimodules

Note that when $I \subset J$, we have $R^J \subset R^I$. As we will discuss in Sect. 24.2.1 below, this is actually a graded Frobenius extension of degree $\ell(J) - \ell(I)$ (the special case when $I = \emptyset$ and $J = \{s\}$ is familiar). As a consequence, induction and restriction functors between these rings will be biadjoint functors (up to shift).

Definition 24.4 Let $\mathcal{SBS}\text{Bim}$, the 2-category of *singular Bott–Samelson bimodules*, be the full sub-2-category of $\mathcal{B}\text{im}$ defined as follows. The objects are finitary subsets $I \subset S$, identified with the rings R^I. The 1-morphisms are tensor-generated by induction and (shifted) restriction between R^I and R^J when these parabolic subgroups differ by a single simple reflection. More precisely, they are generated by the bimodules

$$\text{Ind}_J^I = {}_{R^I}R^I_{R^J}, \qquad \text{Res}_J^I = {}_{R^J}R^I_{R^I}(\ell(J) - \ell(I)), \tag{24.1}$$

where $I \subsetneq J$, $J = I \cup \{s\}$ for some $s \in S$.

In analogy with Bott–Samelson bimodules, the category $\mathcal{SBS}\text{Bim}$ is not closed under direct sums, direct summands, or grading shifts, but its morphism spaces are taken to be the graded vector space $\text{Hom}^\bullet(M, N) := \bigoplus_{k \in \mathbb{Z}} \text{Hom}(M, N(k))$.

Remark 24.5 Note that restriction has a built-in grading shift. In similar fashion, the Bott–Samelson bimodule B_s was defined as $R \otimes_{R^s} R(1)$, rather than $R \otimes_{R^s} R$. In both cases, the grading shift is included to match the graded Frobenius structure, see the end of Sect. 24.2.1.

More concretely, the 1-morphisms of $\mathcal{SBS}\text{Bim}$ can be associated with sequences $(I_1, I_2, \ldots, I_d)$ of finitary subsets, where I_k and I_{k+1} differ by a single simple reflection for each k. To this sequence we associate the corresponding composition of induction and restriction bimodules. For example, in the special case when $I_1 \supset I_2 \subset I_3 \supset I_4 \subset \cdots \supset I_{d-1} \subset I_d$, the corresponding bimodule is

$$_{R^{I_1}}R^{I_2} \otimes_{R^{I_3}} R^{I_4} \otimes \cdots \otimes R^{I_{d-1}}_{R^{I_d}}(N),$$

where N is determined by adding up the grading shifts for each restriction bimodule. Note that this bimodule is a 1-morphism in $\text{Hom}(I_d, I_1)$.

One might ask why we restrict ourselves to inductions and restrictions that add just one simple reflection at a time. After all, for any I and J with $I \subset J$, $R^J \subset R^I$ is a Frobenius extension, so restriction and induction should be nice objects. The answer is that any such induction can be obtained by composing the induction functors which add one simple reflection at a time, since

$$\text{Ind}_K^I \simeq \text{Ind}_J^I \otimes_{R^J} \text{Ind}_K^J, \tag{24.2}$$

for any $I \subset J \subset K$.

The singular Bott–Samelson bimodules in $\text{Hom}(\emptyset, \emptyset)$, that is, the (R, R)-bimodules, include some familiar faces as well as some new friends.

Example 24.6 For the sequence $(\emptyset, s_1, \emptyset, s_2, \ldots \emptyset, s_d, \emptyset)$, the corresponding singular Bott–Samelson bimodule is

$$R \otimes_{R^{s_1}} R \otimes_{R^{s_2}} \cdots \otimes_{R^{s_d}} R(d),$$

which is none other than the ordinary Bott–Samelson bimodule $BS(s_1, s_2, \ldots, s_d)$.

Example 24.7 For the sequence $(\emptyset, s, st, s, \emptyset)$, the corresponding singular Bott–Samelson bimodule is isomorphic to

$$R \otimes_{R^{s,t}} R(m_{st}). \tag{24.3}$$

Note that $m_{st} < \infty$, or this is not a sequence of finitary subsets. In fact, the bimodule in (24.3) is isomorphic to the indecomposable Soergel bimodule $B_{w_{s,t}}$, where $w_{s,t}$ is the longest element of the parabolic subgroup generated by $\{s, t\}$. However, this bimodule is not an ordinary Bott–Samelson bimodule.

Example 24.8 For the sequence $(\emptyset, s, st, stu, st, s, su, s, \emptyset)$, the corresponding singular Bott–Samelson bimodule is (up to grading shift) isomorphic to

$$R \otimes_{R^{s,t,u}} R^s \otimes_{R^{s,u}} R,$$

which is a new sort of beast.

Of course, we should also study (R^I, R^J)-bimodules for various I and J.

Example 24.9 Fix $s, t \in S$ with $m_{st} < \infty$. Then $R^s \otimes_{R^{s,t}} R^t(m_{st} - 1)$ and ${}_{R^s}R_{R^t}(1)$ are singular Bott–Samelson bimodules in $\text{Hom}(t, s)$.

Definition 24.10 Let $\mathcal{SS}\text{Bim}$, the 2-category of *singular Soergel bimodules*, be the graded Karoubi envelope of $\mathcal{SBS}\text{Bim}$ (see Sect. 11.2.3 of Chap. 11). In other words, it is obtained from $\mathcal{SBS}\text{Bim}$ by adjoining all direct sums, grading shifts, and most importantly, all direct summands.

Exercise 24.11 Show that in type A_1 with $S = \{s\}$, the indecomposable singular Soergel bimodules (up to grading shift) are:

- The (R, R)-bimodules R and B_s.
- The (R^s, R)-bimodule $R(1)$.
- The (R, R^s)-bimodule R.
- The (R^s, R^s)-bimodule R^s.

A natural question is whether the category of Soergel bimodules $\mathcal{S}\text{Bim}$ agrees with the category $\text{Hom}(\emptyset, \emptyset)$ inside the 2-category $\mathcal{SS}\text{Bim}$. It is obvious that $\mathcal{S}\text{Bim} \subset \text{Hom}(\emptyset, \emptyset)$. After all, as seen in Example 24.6, Bott–Samelson bimodules are objects in $\text{Hom}(\emptyset, \emptyset)$, and the category is closed under taking direct summands. The other inclusion is far less obvious. For example, is every direct summand of the bimodule in Example 24.8 actually a summand of an ordinary Bott–Samelson bimodule? The answer is yes.

24.1 The Classical Theory of Singular Soergel Bimodules

Theorem 24.12 *The category* $\mathrm{Hom}_{\mathcal{S}\mathbb{S}Bim}(\emptyset, \emptyset)$ *is equal to the category* $\mathbb{S}\mathrm{Bim}$, *as subcategories of* (R, R)-*bimodules*.

This is a consequence of the categorification theorems below, which classify the indecomposable bimodules in each category $\mathrm{Hom}_{\mathcal{S}\mathbb{S}Bim}(I, J)$.

24.1.3 Categorification Theorems

The Grothendieck "group" of a 2-category is a 1-category, just as the Grothendieck group of a monoidal category is a ring. As it turns out, the (split) Grothendieck group of singular Soergel bimodules is a category called the *Hecke algebroid*, which is a "smearing out" of the Hecke algebra over various parabolic subgroups. We wish to avoid introducing the Hecke algebroid and its technicalities, so we will necessarily be vague.

Remark 24.13 Let A be a ring, and $\{e_i\}$ a collection of idempotents in A. Let $\mathcal{C}$ denote the full subcategory of right A-modules whose objects are the right ideals $e_i A$. Left multiplication by an element of $e_j A e_i$ gives a morphism of right A-modules from $e_i A$ to $e_j A$, and in fact $e_j A e_i \simeq \mathrm{Hom}_\mathcal{C}(e_i A, e_j A)$. Composition of morphisms is given by multiplying elements of $e_k A e_j$ and $e_j A e_i$ inside A. Thus, to give an explicit construction of $\mathcal{C}$, one can consider the subsets $e_i A e_j$ of A for various idempotents e_i and e_j. This matches precisely with the formal construction of the (partial) Karoubi envelope, see Sect. 11.2.

Now consider the case when the elements e_i are not idempotents but *quasi-idempotents*, so that $e_i^2 = c_i e_i$ for some (not necessarily invertible) element c_i. Then one can still model the full subcategory $\mathcal{C}$ using subsets of A, but the construction is slightly more technical. For example, the Hom spaces will be given by the intersections $e_j A \cap A e_i$, and composition will be a renormalized multiplication.

This is effectively how the Hecke algebroid is constructed. The quasi-idempotents in question are the Kazhdan–Lusztig basis elements b_{w_I} associated with longest elements of finite parabolic subgroups.

We now state the Soergel–Williamson categorification theorem, proven by Williamson in his thesis [178, Theorems 1 and 2]. For comparison, see the Soergel categorification theorem, Theorems 5.24 and 11.1.

Theorem 24.14 (Soergel–Williamson Categorification Theorem) *There is a bijection between indecomposable singular Soergel bimodules (up to isomorphism and grading shift) in* $\mathrm{Hom}(I, J)$, *and the set of double cosets* $W_J \backslash W / W_I$. *The split Grothendieck group of* $\mathcal{S}\mathbb{S}\mathrm{Bim}$ *is isomorphic to the Hecke algebroid via a natural map* c, *which sends the generators of the Hecke algebroid to the (classes of the) generating 1-morphisms of* $\mathcal{SBSS}\mathrm{Bim}$.

Remark 24.15 There is also a *Soergel–Williamson Hom formula*. If B and B' are two singular Soergel bimodules in $\mathrm{Hom}(I, J)$, then $\mathrm{Hom}(B, B')$ is free as both a left R^J-module and a right R^I-module, of different ranks. The Hom formula states that these ranks are controlled by a particular pairing between $[B]$ and $[B']$ in the Hecke algebroid.

Remark 24.16 There is also a theory of standard modules and standard filtrations for singular Soergel bimodules. This constructs a character map, which is the inverse to the isomorphism c given in Theorem 24.14.

Remark 24.17 For ordinary Soergel bimodules, not only is there a bijection between W and the indecomposable Soergel bimodules, but one can "construct" B_w by choosing a reduced expression $\underline{w}$, and considering the unique indecomposable summand in $BS(\underline{w})$ which has not already appeared in smaller Bott–Samelson bimodules. A similar statement can be made for singular Soergel bimodules, once one has developed the theory of expressions and reduced expressions for double cosets. This theory is introduced in (the original, unpublished version of) Williamson's thesis [177, §1.3], and is currently a topic of exploration.

24.2 One-Color Singular Diagrammatics

Now we introduce a diagrammatic calculus which efficiently encodes the morphisms between singular Bott–Samelson bimodules as planar diagrams. To warm up, we begin with type A_1, which is really the story of the Frobenius extension $R^s \subset R$.

24.2.1 Diagrammatics for a Frobenius Extension

Let $A \subset B$ be a Frobenius extension of commutative rings. We denote the inclusion map by $\iota \colon A \to B$. Recall from Sect. 8.1.4 that the Frobenius extension is the structure of an A-linear map $\partial \colon B \to A$, for which there exist dual bases $\{b_i\}$ and $\{b_i^*\}$ of B (as a finite rank free A-module) such that $\partial(b_i b_j^*) = \delta_{ij}$. Associated to this Frobenius extension are four bimodule maps: the A-bimodule maps

$$\iota \colon A \to B, \tag{24.4a}$$

$$\partial \colon B \to A, \tag{24.4b}$$

and the B-bimodule maps

$$m \colon B \otimes_A B \to B, \tag{24.4c}$$

$$\Delta \colon B \to B \otimes_A B. \tag{24.4d}$$

24.2 One-Color Singular Diagrammatics

Here, m is the multiplication map, and the coproduct map Δ sends 1 to $\sum b_i \otimes b_i^*$.

We now give diagrammatic notation for these four maps in the 2-category $\mathcal{B}\text{im}$. We denote the induction and restriction bimodules as follows.

$$\text{Ind} = \underline{B \uparrow A}, \qquad \text{Res} = \underline{A \downarrow B}. \tag{24.5}$$

For example, the picture $\underline{A \downarrow B \uparrow A \downarrow B}$ represents the bimodule ${}_A B \otimes_A B_B$.

We represent the identity morphisms of these bimodules as oriented vertical lines, downward for restriction and upward for induction:

$$\text{id}_{\text{Ind}} = \; B \uparrow A \;, \qquad \text{id}_{\text{Res}} = \; A \downarrow B \;. \tag{24.6}$$

Note that an empty diagram, whose sole region is labeled A, represents the identity morphism of the A-bimodule A (the monoidal identity in (A, A)-bimodules).

Remark 24.18 Recall that, when drawing 2-morphisms in a 2-category like $\mathcal{B}\text{im}$, the regions are labeled by objects; in this case, by rings. The data of the orientations is redundant, as the labeling of the regions will determine whether the interface between two regions represents induction or restriction. Redundant or no, the orientations are extremely helpful for organizing the diagrammatics, so we include them. Note that, conversely, the orientations determine the region labels. When "walking along the oriented strand," the larger ring B is always to one's left.

Now we can draw the four Frobenius maps from (24.4) as oriented cups and caps.

$$m = \begin{array}{c} B \\ \uparrow \\ A \end{array}, \quad \iota = \begin{array}{c} B \\ \uparrow \\ A \\ A \end{array} \quad {}_A B_A \atop A \tag{24.7}$$

$$\Delta = \begin{array}{c} B \otimes_A B \\ \uparrow \\ B \end{array} {}_{B \otimes_A B} \atop B, \quad \partial = \begin{array}{c} A \\ A \\ \uparrow \\ B \end{array} \quad {}_A B_A$$

This notation is justified by the isotopy relations

$$\text{(diagrams)} \qquad (24.8)$$

which indicate that the functors of induction and restriction are biadjoint. The reader should review the discussion of isotopy, biadjunction, and cyclicity from Sect. 7.5.

Exercise 24.19 Verify the isotopy relations, directly from the definitions of the cups and caps.

There are additional bimodule morphisms which we should hope to encode diagrammatically. Namely, for any $f \in A$, multiplication by f is an A-bimodule map $A \to A$. Similarly, for any $g \in B$, multiplication by g is a B-bimodule map $B \to B$. These morphisms are drawn as follows, and as before, we call these maps *boxes*:

$$(\text{mult. by } f \in A) = \boxed{f}^{\,A} \quad , \quad (\text{mult. by } g \in B) = \boxed{g}^{\,B} \, . \qquad (24.9)$$

With these conventions in place, we have a host of additional relations which are satisfied for any Frobenius extension. For example, one has

$$\boxed{f}\Big|_{A\ B} \;=\; \Big|_{A\ B}\boxed{f} \qquad (24.10)$$

for $f \in A$ (perhaps one should write $\iota(f)$ on the right hand side), and

$$\boxed{f}\,\boxed{g}_{\,A} \;=\; \boxed{fg}_{\,A} \qquad (24.11)$$

for $f, g \in A$ (with a similar relation for $f, g \in B$).

24.2 One-Color Singular Diagrammatics

Here are some more interesting relations. Consider the clockwise circle. This is an endomorphism of B, so it is multiplication by some $\mu \in B$.

$$\left(A\right)_B = \boxed{\mu}_B \qquad (24.12)$$

Here, $\mu = \sum_i b_i b_i^*$ is called the *product-coproduct element*, because $\mu = m(\Delta(1))$. For example, for the Frobenius extension $R^s \subset R$, the dual bases can be chosen as $\{\alpha_s/2, 1\}$ and $\{1, \alpha_s/2\}$, so that $\mu = \alpha_s$.

Lemma 24.20 *One has* $\partial(\mu) = n$, *where n is the rank of B as a free A-module.*

Proof This is immediate from the fact that $\partial(b_i b_i^*) = 1$. □

Consider a counterclockwise circle, with a polynomial $g \in B$ inside. One has the following relation.

$$\left(\boxed{g}\right)_B \,_A = \boxed{\partial(g)}\,_A \qquad (24.13)$$

The final relation expresses the fact that B is free as an A-bimodule. It writes the identity morphism of B as a sum of orthogonal idempotents which factor though A. See Sect. 8.2.4 for more on what it means to express a direct sum decomposition diagrammatically.

$$A \,|\, B \,|\, A = \sum_i \boxed{b_i} \atop \boxed{b_i^*} \qquad (24.14)$$

Rotating by 90°, one has the following equivalent relation.

$$\begin{array}{c} A \\ B \\ A \end{array} = \sum_i \boxed{b_i}\;\boxed{b_i^*} \atop B\;A\;B \qquad (24.15)$$

Exercise 24.21 Verify the relations above.

In fact, it is proven in [62, Claim 2.1] that the generators and relations above are sufficient to express all morphisms between bimodules obtained from induction and restriction.

Theorem 24.22 *Let* Frob$(A \subset B)$ *denote the diagrammatic category generated by oriented cups and caps and by boxes, modulo the relations listed above. Then* Frob$(A \subset B)$ *is equivalent to the full subcategory of* $\mathcal{B}$im *generated by induction and restriction bimodules.*

We remind the reader of the definition of a graded Frobenius extension from Sect. 8.1.4. When A and B are graded rings, and ∂ is homogeneous of degree -2ℓ, we call this a *graded Frobenius extension of degree* ℓ. In this case, it is more convenient to renormalize our generating bimodules. We let

$$\operatorname{Ind}_A^B = {}_B B_A, \qquad \operatorname{Res}_A^B = {}_A B_B(\ell). \tag{24.16}$$

We continue to draw this (shifted) restriction bimodule using a downward strand. With this convention, one can check that the clockwise cups and caps represent homogeneous maps of degree $+\ell$, while the counterclockwise maps represent homogeneous maps of degree $-\ell$. Of course, if $f \in A$ is homogeneous of degree k, then so is multiplication by f.

24.2.2 Relationship to the One-Color Soergel Calculus

Fix a simple reflection s, and consider the graded Frobenius extension $R^s \to R$ of degree 1, with trace map ∂_s. We refer to the category Frob$(R^s \subset R)$ as the *singular Soergel calculus* in one color. Previously in Sect. 8.2, we gave a diagrammatic calculus for Bott–Samelson bimodules in one color, which encoded morphisms between tensor products of B_s. Now, this monoidal generator B_s can be expressed as a composition of two smaller pieces, restriction followed by induction. The following theorem indicates how the original one-color calculus can be expressed in terms of our new singular diagrammatic calculus.

Theorem 24.23 *Let* (W, S) *have type* A_1. *There is a fully faithful monoidal functor* $\mathcal{F}$ *from* $\mathcal{H}_{\mathrm{BS}}$ *to* Frob$(R^s \subset R)$, *defined as follows. On objects, it is defined by*

$$B_s = \quad\underline{\quad\bullet\quad}\quad \mapsto \quad\underline{\quad\boxed{}\quad}. \tag{24.17}$$

Here and henceforth, a colorless region represents the ring R, *while a red region represents the ring* R^s. *On morphisms,* $\mathcal{F}$ *is defined by*

24.2 One-Color Singular Diagrammatics

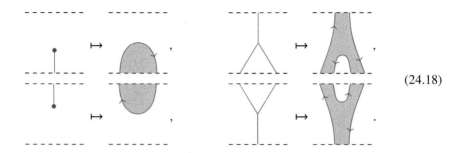

(24.18)

The relationship between the original Soergel calculus and the singular Soergel calculus can be visualized as a "deformation retract." By retracting all the regions labeled R^s into univalent and trivalent graphs, one obtains an inverse to $\mathcal{F}$ (on the 1-morphisms in its image).

Exercise 24.24 Confirm that the functor $\mathcal{F}$ is well-defined, i.e. the one-color relations hold in $\mathrm{Frob}(R^s \subset R)$ after applying the functor.

One important aspect of singular Soergel calculus is that the relations of ordinary Soergel calculus are implied by smaller, more "local" relations. For example, associativity

is implied by locality

while the needle relation

$$\bigcirc = 0$$

is implied by the counterclockwise circle relation

$$\bigcirc = \boxed{\partial(1)} = 0.$$

This idea, that one can "zoom in" on a Soergel diagram and explain its features more locally will become even more important when we introduce more colors. For example, the all-important and mysterious $2m$-valent vertex is no more than a composition of much simpler singular diagrams, as we shall see.

24.2.3 Relationship with the Temperley–Lieb 2-Category

Now fix (W, S) and choose two (distinct) simple reflections $s, t \in S$. Inside $S\mathbb{BS}$Bim, we can consider those bimodules obtained by induction and restriction between the rings R, R^s, and R^t, without ever going to $R^{s,t}$. The diagrammatics established above for Frobenius extensions can already be used to draw many such bimodules and the morphisms between them. For example, 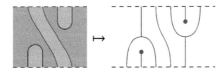 denotes ${}_{R^t}R \otimes_{R^s} R(2)$, where again we use red for s and blue for t. Universal Soergel diagrams (built from univalent and trivalent vertices, blue or red) can be sent, by a "functor" we also call $\mathcal{F}$, to singular Soergel diagrams built from caps and cups (blue or red).[1] This "functor" $\mathcal{F}$ has an "inverse" given by deformation retraction, just as in Theorem 24.23.

Let us assume for simplicity that our realization has a symmetric Cartan matrix, so that $\partial_s(\alpha_t) = a_{st} = a_{ts} = \partial_t(\alpha_s)$.

In Sect. 9.2 we introduced the Temperley–Lieb 2-category $2\mathcal{TL}_\delta$. Write $2\mathcal{TL}_{a_{st}}$ for the specialization where δ is equal to a_{st}. We gave a *non-monoidal* functor $\Sigma \colon 2\mathcal{TL}_{a_{st}} \to \mathbb{BS}$Bim. Namely, given a colored crossingless matching, we applied a deformation retract to each colored region to obtain a universal diagram.

This is a functor, in that composition (vertical stacking) is preserved. However, this functor was not monoidal, meaning that horizontal concatenation is not preserved. This is because of the issues inherent in deformation retraction near the right and left edge of the picture. For example, we have

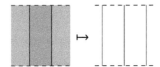

---

[1] Note that $\mathcal{F}$ is not actually a functor, as it is not defined on the $2m_{st}$-valent vertex. It is a functor, and is fully faithful, in the case when $m_{st} = \infty$.

24.2 One-Color Singular Diagrammatics

and

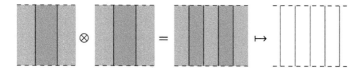

whereas

A simpler example is that

Deformation retraction sends the identity 1-morphism of red to something which is not a monoidal identity, leading to the "duplication" problem we see above.

It is obviously unnatural to consider non-monoidal functors between monoidal categories. The problem was merely that $\mathbb{BSB}$im was not the correct setting for the target of this functor. Instead, we upgrade Σ to a genuine 2-functor

$$2\Sigma : 2\mathcal{TL}_{a_{st}} \to \mathcal{S}\mathbb{BS}\text{Bim}. \tag{24.19}$$

Recall that the objects of $2\mathcal{TL}$ are the two colors, red and blue. We send red to the parabolic subset $\{s\}$, and blue to $\{t\}$. The generating 1-morphisms of $2\mathcal{TL}$ are a map from red to blue, and a map from blue to red. These are sent by 2Σ to the bimodules ${}_{R^t}R_{R^s}(1)$ and ${}_{R^s}R_{R^t}(1)$, respectively. That is,

$$\begin{array}{c}\includegraphics[height=1em]{} \mapsto \includegraphics[height=1em]{}\end{array}, \tag{24.20}$$

and similarly with red and blue switched. The (non-oriented) cups and caps of $2\mathcal{TL}$ are "expanded" to become pairs of oriented cups and caps in $\mathcal{S}\mathbb{BS}$Bim.

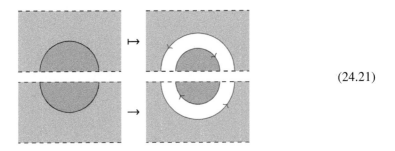
(24.21)

For example, we apply 2Σ to the Jones–Wenzl projector on two strands.

(24.22)

Exercise 24.25 Verify that 2Σ is a monoidal 2-functor. Draw the image of the Jones–Wenzl projector on three strands.

The utility of the functor Σ was that the idempotent decompositions in $2\mathcal{TL}$ could be used to provide idempotent decompositions of alternating Bott–Samelson bimodules like $B_s B_t B_s B_t B_s$ (in this example, we would need $m_{st} \geq 5$). In particular, Jones–Wenzl projectors could be used to project to indecomposable Soergel bimodules. The 2-functor 2Σ can be used in the same way to construct projections to indecomposable singular Soergel bimodules.

Let us briefly discuss the connection between Σ and 2Σ. The following is a consequence of the Soergel–Williamson categorification theorem.

Proposition 24.26 *Let I, J be finitary parabolic subsets. Consider the functor $\iota_{J,I}$ from (R^J, R^I)-bimodules to (R, R)-bimodules, where*

$$\iota_{J,I}(M) = R \otimes_{R^J} M \otimes_{R^I} R(\ell(I)). \tag{24.23}$$

That is, $\iota_{J,I}$ is just composition with $\operatorname{Res}_I^\emptyset$ on the right, and $\operatorname{Ind}_J^\emptyset$ on the left. Then $\iota_{J,I}$ induces a functor from $\operatorname{Hom}(I, J)$ to $\operatorname{Hom}(\emptyset, \emptyset)$ in $\mathcal{SSBim}$, and this functor is faithful.

When $I = J$, $\operatorname{Hom}(I, I)$ is a monoidal category, as is $\operatorname{Hom}(\emptyset, \emptyset)$, but $\iota_{I,I}$ is not a monoidal functor (it does not send the monoidal identity to the monoidal identity).

The functor Σ is just the composition of the genuine 2-functor 2Σ with the non-monoidal functors $\iota_{s,s}, \iota_{s,t}, \iota_{t,s}, \iota_{t,t}$. For example, the Jones–Wenzl projector on 2 strands is sent by $\iota_{s,s}$ to

24.3 Singular Soergel Diagrammatics in General

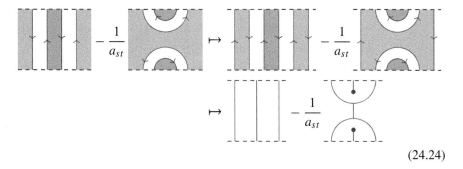

(24.24)

The second map above is the deformation retraction inverse to the "functor" $\mathcal{F}$.

24.3 Singular Soergel Diagrammatics in General

Just as the Frobenius extension structure on $R^s \subset R$ controls the diagrammatics for induction and restriction bimodules, the Frobenius extension structures for the extensions $R^J \subset R^I$ (when $I \subset J$) will control the diagrammatics for all of $S\mathbb{B}\mathbb{S}$Bim. We discuss these Frobenius extensions now.

24.3.1 The Upgraded Chevalley Theorem, Part I

In this section, R will denote the polynomial ring of the geometric representation of a Coxeter system (W, S), over the base ring $\mathbb{R}$. The Chevalley theorem states that R is free as an R^I-module, and R^I is a polynomial ring generated in various degrees, so long as I is finitary. The upgraded Chevalley theorem[2] states that $R^I \subset R$ is actually a graded Frobenius extension of degree $\ell(I)$, when I is finitary. We state the result precisely in Theorem 24.36 below, and generalize it in Theorem 24.40. A lot of this material can be generalized to other realizations, but great care is required, and we postpone that discussion until Sect. 24.3.6.

Remark 24.27 The upgraded Chevalley theorem, and the other statements made in this section, are standard and relatively well-known results, too much so to be called "folklore." They follow quickly from results of Demazure [44] and Bernstein–Gelfand–Gelfand [24]. However, it is difficult to find the results stated in the literature as we state them here. Often different results have to be pieced together. Also, some proofs use geometry, and thus work only for Weyl groups, while other proofs work more generally but are stated only for Weyl groups. Almost all the

---

[2]There is no name for this theorem in the literature; "upgraded Chevalley theorem" is not common terminology.

results we use can all be found explicitly in various parts of Williamson's PhD thesis [177]. To improve the state of the literature, and to give a one-stop shop for learning this material, we will give proofs or sketches here.

Recall from Sect. 4.3 that for each $s \in S$ there is a Demazure operator $\partial_s \colon R \to R$ of degree -2, defined by

$$\partial_s(f) = \frac{f - sf}{\alpha_s}. \tag{24.25}$$

Some of its key properties are:

- $\operatorname{Ker} \partial_s = R^s$, which is evident from the definition.
- ∂_s is R^s-linear, in that $\partial_s(fg) = f \partial_s(g)$ for $f \in R^s$.
- The image of ∂_s is R^s. This follows from R^s-linearity and the existence of a linear polynomial f with $\partial_s(f) = 1$ (by Demazure surjectivity).
- For simple reflections $s, t \in S$ with $m_{st} < \infty$, the operators ∂_s and ∂_t satisfy the braid relation.

In particular, $\partial_s^2 = 0$. For $w \in W$ we define ∂_w to be the composition $\partial_{\underline{w}} = \partial_{s_1} \partial_{s_2} \cdots \partial_{s_d}$ for any reduced expression $\underline{w} = (s_1, s_2, \ldots, s_d)$. Consequently, for $x, y \in W$, we have

$$\partial_x \partial_y = \partial_{xy} \qquad \text{if } \ell(xy) = \ell(x) + \ell(y), \tag{24.26}$$

$$\partial_x \partial_y = 0 \qquad \text{if } \ell(xy) < \ell(x) + \ell(y). \tag{24.27}$$

For a finitary subset $I \subset S$, let us write ∂_I for ∂_{w_I}. We claim that the image of ∂_I is contained in R^I. If $s \in I$ then $\partial_s \partial_I = 0$, because $\ell(sw_I) < \ell(w_I)$. Hence, the image of ∂_I is contained in the kernel of ∂_s, which is R^s. Since this is true for all $s \in I$, the image of ∂_I is contained in R^I. It is also easy to see that ∂_I is R^I-linear. A more difficult statement to prove is that the image of ∂_I is precisely R^I, which will follow by linearity from the existence of a polynomial $f \in R$ for which $\partial_I(f)$ is an invertible scalar (we will construct one below). Note that ∂_I kills any polynomial of degree $< \ell(I)$, for degree reasons. Ultimately, ∂_I will be the Frobenius trace map for the extension $R^I \subset R$, as we will prove in this section.

A *root* of W_I is an element of $\mathfrak{h}^*$ of the form $w(\alpha_s)$ for $s \in I$ and $w \in W_I$. A *positive root* has the additional constraint that $\ell(ws) = \ell(w) + 1$. Positive roots are in bijection with reflections, so there are $\ell(I)$ positive roots of W_I, see [85, §5.4] for more details. We let Φ_I^+ denote the set of positive roots of W_I, and we define the polynomial $\mu_I \in R$ by the formula

$$\mu_I = \prod_{\alpha \in \Phi_I^+} \alpha. \tag{24.28}$$

24.3 Singular Soergel Diagrammatics in General

Any simple reflection $s \in I$ will permute the set $\Phi_I^+ \setminus \alpha_s$, and will negate α_s, so that

$$s(\mu_I) = -\mu_I. \tag{24.29}$$

Exercise 24.28 When $m_{st} = 2$, compute that $\partial_s \partial_t (\alpha_s \alpha_t) = 4$. When $m_{st} = 3$, compute that $\partial_s \partial_t \partial_s (\alpha_s \alpha_t (\alpha_s + \alpha_t)) = 6$.

Below we will prove that $\partial_I(\mu_I) = |W_I|$, implying that $\partial_I \colon R \to R^I$ is surjective. The proof is slightly simplified using an alternate description of ∂_I due to Demazure, which is independently useful.

Lemma 24.29 *Let $R^{-I} \subset R$ denote the set of I-antiinvariant polynomials,*

$$R^{-I} = \{f \in R \mid s(f) = -f \text{ for all } s \in I\}. \tag{24.30}$$

Then R^{-I} is a free R^I-module of rank 1, generated by μ_I.

For a proof, see [85, Proposition 3.13]. The basic idea is this. For any reflection t there is a corresponding positive root $\alpha_t \in \mathfrak{h}^*$, and a hyperplane $H_t \subset \mathfrak{h}$ which is fixed by t and killed by α_t. If f is a polynomial satisfying $tf = -f$ then f vanishes on H_t, meaning that α_t must divide it (see [85, Lemma 3.3] if this is not clear to you). In particular, any antiinvariant polynomial must be divisible by every positive root, and hence by μ_I. The ratio f/μ_I is W_I-invariant.

Now, let $A_I \colon R \to R$ denote the *antisymmetrization operator*

$$A_I(f) = \sum_{w \in W} (-1)^{\ell(w)} w(f). \tag{24.31}$$

Note that A_I is an R^I-linear map of degree zero, whose image is contained inside R^{-I}. In particular, for any polynomial f of degree $< \ell(I)$, $A_I(f) = 0$, since R^{-I} is generated in degree $\ell(I)$. Also note that $A_I(\mu_I) = |W_I| \cdot \mu_I$, since μ_I is antiinvariant.

Lemma 24.30 (Demazure's Formula) *When I is finitary, we have*

$$\partial_I(f) = \frac{A_I(f)}{\mu_I}. \tag{24.32}$$

This also implies that

$$\partial_I(\mu_I) = |W_I|. \tag{24.33}$$

Exercise 24.31 Prove (24.32) directly when $I = S$ in types A_1 (trivial), $A_1 \times A_1$, A_2, and B_2.

Proof (Following Demazure [44, Proposition 1 and Lemme 4].) By the Chevalley theorem, we know that R is a free R^I-module with a particular graded rank. In particular, it has an R^I-basis of homogeneous polynomials, all but one of which has degree $< \ell(I)$, and one of which has degree exactly $\ell(I)$. Thus A_I kills all basis elements except for the last. Since A_I is not the zero operator, it must send the last basis element to a nonzero scalar multiple of μ_I. Since $A_I(\mu_I) \neq 0$, we see that μ_I has a nonzero coefficient in this last basis element. Consequently, we can replace the last basis element with μ_I and still get a basis of R over R^I.

Now an arbitrary polynomial $f \in R$ can be written as

$$f = \sum a_i p_i + a\mu_I$$

where $a_i, a \in R^I$ and $\deg p_i < \ell(I)$. Both A_I and ∂_I will kill p_i, so by R^I-linearity, they will also kill $a_i p_i$. Thus $\partial_I(f) = a\partial_I(\mu_I)$ and $A_I(f) = aA_I(\mu_I)$. Since $\partial_I(\mu_I)$ is a scalar and $A_I(\mu_I)$ is a scalar multiple of μ_I, we can deduce that $\partial_I(f)$ and $\frac{A_I(f)}{\mu_I}$ agree up to scalar, and it remains to determine what that scalar is.

Let us only sketch the remainder of the proof. When looking at the composition $\partial_s \partial_t$ for two simple distinct simple reflections, we get

$$\partial_s \partial_t(f) = \partial_s\left(\frac{f - tf}{\alpha_t}\right) = \frac{\frac{f-tf}{\alpha_t} - \frac{sf-stf}{s(\alpha_t)}}{\alpha_s} = \frac{s(\alpha_t)(f - tf) - \alpha_t(sf - stf)}{\alpha_s \alpha_t s(\alpha_t)}.$$

More generally, using induction one can write ∂_w as a sum of expressions of the form $\frac{\pm g \cdot x(f)}{h}$, where $x \in W$, and g and h are products of positive roots. Any such expression is unique, using the linear independence of the operators $x \in W$. One can prove inductively that $x \leq w$ in any such expression, and that the leading term where $x = w$ has a particular form. Using this one can prove that ∂_I has leading term $\frac{(-1)^{\ell(w_I)} w_I(f)}{\mu_I}$, agreeing with the corresponding term in $\frac{A_I}{\mu_I}$. Thus the scalar of proportionality must be 1. $\square$

Our final goal is to prove the existence of dual bases for R over R^I, with respect to the pairing $(f, g) \mapsto \partial_I(fg)$. However, finding a dual basis is more difficult than proving one exists! Let us set the stage with some general results.

Lemma 24.32 *Let R be a commutative ring and let $M \simeq R^{\oplus n} \simeq N$ be free R-modules of the same rank n. Let $(-, -) \colon M \times N \to R$ be an R-bilinear pairing, and let $A = \{e_1, \ldots, e_n\} \subset M$ and $B = \{f_1, \ldots, f_n\} \subset N$ be dual sets in that*

$$(e_i, f_j) = \delta_{ij}. \tag{24.34}$$

Then A is a basis for M, B is a basis for N, and the pairing is nondegenerate, inducing an isomorphism of R-modules $M \to \mathrm{Hom}_R(N, R)$.

Proof Choose arbitrary bases for M and N, and let G be the Gram matrix of the pairing with respect to those bases. Let X be the matrix whose i-th column expresses e_i in terms of the basis for M, and Y be the matrix whose i-th column expresses f_i in terms of the basis for N. These are $n \times n$ matrices with entries in R. Then the "Gram

24.3 Singular Soergel Diagrammatics in General

matrix" for the pairing of A and B is $X^t G Y$. By (24.34) we deduce that $X^t G Y$ is the identity matrix. Therefore X and Y are invertible, so A and B are bases. The rest is standard. □

So one need not prove spanning and linear independence of the suspected dual bases, one need only show they are dual sets. In the next exercise, we argue that one need only show that they are "almost" dual sets. We will apply the exercise in the case $R = R^I$.

Exercise 24.33 Let R be a graded commutative ring, supported in non-negative degrees, where the degree zero part $R_0 = \Bbbk$ is a field. Let R_+ denote the positively-graded part of R, its unique graded maximal ideal. Let M and N be finitely-generated free graded R-modules of the same graded rank. Let $(-,-): M \times N \to R$ be a graded R-bilinear pairing of some degree, and let $A = \{e_1, \ldots, e_n\} \subset M$ and $B = \{f_1, \ldots, f_n\} \subset N$ be *almost dual sets* in that the elements e_i, f_i are all homogeneous, and

$$(e_i, f_j) \in \delta_{ij} + R_+. \tag{24.35}$$

Then A is a basis for M, B is a basis for N, and the pairing is nondegenerate, inducing an isomorphism of graded R-modules $M \to \operatorname{Hom}_R(N, R)$.

Hint In the proof of the previous lemma, it was not important that $X^t G Y$ was the identity matrix, only that $X^t G Y$ was invertible. Order the basis elements by degree, and use degree arguments to deduce that the Gram matrix of A paired against B is upper triangular, with ones on the diagonal, and elements of R_+ above the diagonal.

Remark 24.34 The above exercise can be used even without the assumption that M and N are free, but only that their rank (in the sense of commutative algebra) is n. One proves that the span of A is a free submodule of M which can not be "enlarged," i.e. it is not properly contained inside a free submodule of the same rank.

The following lemma, when W is a Weyl group and $I = S$, is the foundation of the field of Schubert calculus, which studies a particular basis of the coinvariant ring $C = R/(R^W)_+$. The basis in question is called the *Schubert basis*.

Lemma 24.35 *When I is finitary, the elements $\{\partial_w(\mu_I) \mid w \in W_I\}$ form an R^I-basis for R. For any $x, y \in W_I$ we have*

$$\partial_I(\partial_x(\mu_I) \cdot \partial_y(\mu_I)) = |W_I|^2 \delta_{xw_I, y} + (R^I)_+. \tag{24.36}$$

The pairing $(f, g) := \partial_I(fg)$ is nondegenerate.

Proof We need only prove (24.36), which will imply that $A = \{\partial_x(\mu_I)\}_{x \in W_I}$ and $B = \{\partial_{xw_I}(\mu_I)\}_{x \in W_I}$ are almost dual sets for R over R^I. Since the Chevalley theorem implies that R is free over R^I with the appropriate graded rank, we can apply Exercise 24.33 to deduce that A is a basis and that the pairing is nondegenerate.

Let $x, y \in W_I$ and consider $\partial_I(\partial_x(\mu_I)\partial_y(\mu_I))$. This is an element of R^I of degree $\ell(I) - \ell(x) - \ell(y)$, so if $\ell(x) + \ell(y) > \ell(I)$ then the result is zero, and if $\ell(x) + \ell(y) < \ell(I)$ then the result is in R_+^I. So we need only consider the case when $\ell(x) + \ell(y) = \ell(I)$.

The operator ∂_s is R^s-linear and has image in R^s, from which one deduces that

$$\partial_s(\partial_s(f)g) = \partial_s(f)\partial_s(g) = \partial_s(f\partial_s(g)).$$

Since ∂_I has an expression ending in ∂_s for any $s \in I$, we can use this to transfer any ∂_s from the beginning of ∂_x to the beginning of ∂_y. Thus

$$\partial_I(\partial_x(\mu_I)\partial_y(\mu_I)) = \partial_I(\mu_I \cdot \partial_{x^{-1}}\partial_y(\mu_I)).$$

If $\ell(x^{-1}) + \ell(y) \neq \ell(x^{-1}y)$ then $\partial_{x^{-1}}\partial_y = 0$. Otherwise, $\ell(x^{-1}y) = \ell(I)$ so that $x^{-1}y = w_I$ and $y = xw_I$. The result is then

$$\partial_I(\partial_x(\mu_I)\partial_y(\mu_I)) = \partial_I(\mu_I\partial_I(\mu_I)) = \partial_I(\mu_I)^2 = |W_I|^2,$$

as required. □

The upgraded Chevalley theorem (part I) is a repackaging of Lemma 24.35.

Theorem 24.36 *The inclusion map $R^I \to R$ and the trace map $\partial_I \colon R \to R^I$ equip R with the structure of a graded Frobenius algebra over R^I.*

Remark 24.37 Lemma 24.35 produces a pair of almost dual bases for R over R^I, from which one deduces the existence of a pair of dual bases. One could find the actual dual bases explicitly using linear algebra (say, by finding the inverse to the Gram matrix). The question remains: is there a nice combinatorial description (using roots and Demazure operators) of dual bases for R over R^I? Said another way, what is the dual basis to the Schubert basis? We like to think of this as a basic question in *equivariant Schubert calculus*, which (as far as the authors are aware) is unsolved.

24.3.2 The Upgraded Chevalley Theorem, Part II

More generally, consider two nested finitary subsets $I \subset J \subset S$, so that $R^J \subset R^I$. Consider the element $w_J^I = w_J w_I^{-1} \in W$. Of course, $w_I^{-1} = w_I$, but we write the inverse here in order to emphasize how the lengths add up. The corresponding Demazure operator will be denoted

$$\partial_J^I := \partial_{w_J^I}.$$

24.3 Singular Soergel Diagrammatics in General

Also define the following element of R, a product of positive roots:

$$\mu_J^I = \prod_{\alpha \in \Phi_J^+ \setminus \Phi_I^+} \alpha. \tag{24.37}$$

Our notation matches the previous notation, in that $\partial_J^\emptyset = \partial_J$ and $\mu_J^\emptyset = \mu_J$.

By construction, it is clear that whenever $I \subset J \subset K$ are finitary subsets of S, we have

$$\partial_K^I = \partial_K^J \circ \partial_J^I. \tag{24.38}$$

Exercise 24.38 When $m_{st} = 3$, we have $\partial_{s,t}^s = \partial_s \partial_t$. Compute that $\partial_{s,t}^s(\mu_{s,t}^s) = \partial_{s,t}^s(\alpha_t(\alpha_s + \alpha_t)) = 3$.

Lemma 24.39 *The operator ∂_J^I sends R^I to R^J surjectively. One has that $\mu_J^I \in R^I$, and*

$$\partial_J^I(\mu_J^I) = |W_J/W_I|. \tag{24.39}$$

We also have

$$\partial_I(\mu_J) = |W_I|\mu_J^I. \tag{24.40}$$

Proof First we show that ∂_J^I sends R^I to R^J. But R^I is the image of ∂_I, and $\partial_J^I \circ \partial_I = \partial_J$ since $w_J^I w_I = w_J$ and the lengths add up. Hence ∂_J^I sends R^I to the image of ∂_J, which is R^J. Moreover, it is surjective, since ∂_J is surjective.

The only positive roots which W_I can send to negative roots live in Φ_I^+. Thus W_I preserves the set $\Phi_J^+ \setminus \Phi_I^+$, and fixes the polynomial μ_J^I. Now

$$\partial_I(\mu_J) = \partial_I(\mu_J^I \mu_I) = \mu_J^I \partial_I(\mu_I) = |W_I|\mu_J^I,$$

proving (24.40). In this computation we used the fact that ∂_I is R^I-linear and $\mu_J^I \in R^I$, and we used (24.33). Continuing, we have

$$|W_J| = \partial_J(\mu_J) = \partial_J^I \circ \partial_I(\mu_J) = |W_I|\partial_J^I(\mu_J^I), \tag{24.41}$$

from which we deduce (24.39). □

Here is the rest of the upgraded Chevalley theorem.

Theorem 24.40 *Let $I \subset J$ be finitary subsets of S. Then $R^J \subset R^I$ is a graded Frobenius extension of degree $\ell(J) - \ell(I)$, with trace ∂_J^I, and product-coproduct element μ_J^I. In particular, $R^J \subset R$ is a graded Frobenius extension of degree $\ell(J)$, with trace ∂_J, and product-coproduct element μ_J.*

Like the previous lemma, the idea is merely to bootstrap facts about the Frobenius extensions $R^J \subset R$ and $R^I \subset R$ to deduce facts about the extension $R^J \subset R^I$.

Proof Let Y be the set of minimal length coset representatives for the left cosets $W_I \setminus W_J$, and X be the set of minimal length coset representatives for the right cosets W_J/W_I. The map $x \mapsto x^{-1}$ gives a length-preserving bijection between X and Y. There is another bijection between X and Y: for each $y \in Y$ there exists a unique $x \in X$ such that $xw_I y = w_J$ and $\ell(x) + \ell(w_I) + \ell(y) = \ell(w_J)$. To see this, note that $w_I y \in W_J$ has a complement, i.e. an element $z \in W_J$ such that $zw_I y = w_J$ and $\ell(z) + \ell(w_I y) = \ell(w_J)$. Clearly z is a minimal length right coset representative, or else $\ell(zw_I) < \ell(z) + \ell(w_I)$. Composing these two bijections we get a bijection $Y \to Y$, denoted $y \mapsto y^\circ$, where y° is the unique element such that $y^{-1} w_I y^\circ = w_J$. In other words,

$$y^\circ = w_I y w_J.$$

Let $A = \{\partial_{w_I y}(\mu_J)\}_{y \in Y}$. Let $B = \{\partial_{w_I y^\circ}(\mu_J)\}_{y \in Y}$. All these elements live in R^I, since they are in the image of ∂_I. The goal is to prove that A and B are almost dual sets for R^I over R^J, up to a factor of $|W_J|^2$. Now, observe that

$$\partial_J^I(\partial_{w_I y}(\mu_J)\partial_{w_I z}(\mu_J)) = \partial_J^I(\partial_I(\partial_y(\mu_J))\partial_{w_I z}(\mu_J)))$$
$$= \partial_J^I(\partial_I(\partial_y(\mu_J)\partial_{w_I z}(\mu_J))), \qquad (24.42)$$

where the last equality holds since $\partial_{w_I z}(\mu_J) \in R^I$ and ∂_I is R^I-linear. But $\partial_J^I \circ \partial_I = \partial_J$, so we get

$$\partial_J^I(\partial_{w_I y}(\mu_J)\partial_{w_I z}(\mu_J)) = \partial_J(\partial_y(\mu_J)\partial_{w_I z}(\mu_J)) = |W_J|^2 \delta_{yw_J, w_I z} + R_+^J \qquad (24.43)$$

by (24.36). But we have $\delta_{yw_J, w_I z} = \delta_{z, y^\circ}$. This proves our goal.

It remains to show that the product-coproduct elements are as stated. We follow the proof from [47, Theorem 3.32]. For the moment let ϕ_J^I denote the product-coproduct element for the Frobenius extension $R^J \subset R^I$.

In [62] some general facts are proven about chains of Frobenius extensions. One implication is that, whenever $I \subset J \subset K$ we have

$$\phi_K^I = \phi_J^I \phi_K^J. \qquad (24.44)$$

In particular, if we can prove that $\phi_I = \mu_I$, then this will imply that $\phi_J^I = \frac{\phi_I}{\phi_J} = \mu_J^I$. A general fact about Frobenius extensions is that when you apply the Frobenius trace to the product-coproduct element, you get an integer which is the dimension of the extension. Since $\partial_I(\mu_I) = |W_I| = \partial_I(\phi_I)$, it suffices to prove that ϕ_I and μ_I are colinear. By Lemma 24.29, it suffices to prove that ϕ_I is I-antiinvariant.

Using (24.44), for any $s \in I$ one has $\phi_I = \phi_s \phi_I^s$. Since $\alpha_s = \phi_s$ is s-antiinvariant and $\phi_I^s \in R^s$, this implies that ϕ_I is s-antiinvariant. Thus ϕ_I is I-antiinvariant, so for the reasons just discussed $\phi_I = \mu_I$. □

Remark 24.41 Again, as far as the authors are aware, finding combinatorial dual bases for the Frobenius extension $R^J \subset R^I$ is an open problem.

24.3.3 Diagrammatics for a Cube of Frobenius Extensions

All this structure can be encapsulated in what we call the *Soergel (hyper)cube of Frobenius extensions*. When $|S| = n$, the poset of subsets of S forms an n-dimensional hypercube. The subposet of finitary subsets may not be a hypercube but we abusively refer to it as a cube anyway. For each $I \subset S$ finitary, we associate the ring R^I. For each edge of the cube (corresponding to $I \subsetneq J$, with $J = I \cup \{s\}$), we associate the Frobenius extension $R^J \subset R^I$, with its fixed Frobenius trace map ∂_J^I. These Frobenius extensions are *compatible* in the sense that (24.38) holds; the consequence is that for any two vertices of the cube $I \subset K$ which are related in the partial order, the Frobenius extension $R^K \subset R^I$ is uniquely defined. Altogether, we have a *compatible Frobenius hypercube*, as introduced in the paper [62].

In [62], a diagrammatic calculus is described which gives generators and relations for an arbitrary cube of Frobenius extensions. More precisely, it provides certain diagrammatic generators representing bimodule morphisms which exist for any Frobenius cube, and relations which hold for any Frobenius cube, though no claim is made that these generators describe all the morphisms, or that these are all the relations. Let us demonstrate these diagrammatics for the Soergel cube.

Let us write Is as shorthand for $I \cup \{s\}$, when $s \notin I$. The identity 2-morphisms of the induction and restriction bimodules between R^I and R^{Is} are drawn as follows.

$$\begin{array}{cc} \underset{s}{\underline{Is \;\downarrow\; I}} & \underset{s}{\underline{I \;\uparrow\; Is}} \end{array} \qquad (24.45)$$

While the regions are labeled by I and Is, the strand itself is colored by the simple reflection s. In particular, the diagrammatics for Frobenius extensions from the previous section applies, giving four cups and caps with degrees as indicated below:

$$(24.46)$$

$\ell(Is) - \ell(I) \qquad \ell(Is) - \ell(I) \qquad \ell(I) - \ell(Is) \qquad \ell(I) - \ell(Is)$

Note that the clockwise cup and cap have positive degrees, and the counterclockwise cup and cap have negative degrees.

Now suppose that $s, t \notin I$, and $Ist = I \cup \{s, t\}$ is finitary. Then induction from R^{Ist} to R^I (which is just the bimodule $_{R^I}R^I_{R^{Ist}}$) can be described in two different and isomorphic ways using the generators of $S\mathbb{BS}\text{Bim}$: as the composition $\text{Ind}^I_{Is} \text{Ind}^{Is}_{Ist}$, or the composition $\text{Ind}^I_{It} \text{Ind}^{It}_{Ist}$. The natural isomorphism between these two compositions (which is just the identity map on the underlying bimodule R^I) is drawn as follows, and has degree zero. It is called the *crossing*:

$$\begin{array}{c} It \\ I \quad \times \quad Ist \\ Is \end{array} \qquad (24.47)$$

Note that both the induction from Is to I, and the induction from Ist to It, are drawn using an s-colored strand, though they represent different functors entirely (as evidenced by the different region labels). The convention where these different functors are drawn the same way is quite useful; as a consequence, this picture looks like a transverse crossing of two 1-manifolds. The following exercise implies that planar diagrammatics involving these crossings are isotopy-invariant.

Exercise 24.42 The crossing is cyclic. Its rotation by 180° is equal to the natural isomorphism (the identity map on the underlying bimodule) between the two ways of describing restriction from R^I to R^{Ist}. (*Hint:* the easiest approach is to define the 180° rotated crossing to be the identity map, and check compatibility of the two crossings under half-twists, see [62, Equations (1.2) and (1.3)].)

Unlike in the diagrammatic category $\mathcal{H}_{\text{BS}}$, rotating a singular diagram will change the degree of the diagram, because the cups and caps do not have degree zero! Rotating the crossing by 90° gives the *sideways crossings*

$$\begin{array}{cc} \begin{array}{c} I \\ Is \quad \times \quad It \\ Ist \end{array} & \begin{array}{c} Ist \\ It \quad \times \quad Is \\ I \end{array} \end{array} \qquad (24.48)$$

with degree $\ell(Ist) - \ell(Is) - \ell(It) + \ell(I)$.

Exercise 24.43 Verify that the degree of the sideways crossings is as stated, and is always ≥ 0. Write down a formula for how each crossing acts on the corresponding bimodules. (One formula will be easy, one formula will be hard.)

Putting it all together, one can define a diagrammatic 2-category Frob with objects given by finitary subsets $I \subset S$. The generating 1-morphisms are given by following an edge of the Soergel cube, and are drawn with a colored oriented strand. The generating 2-morphisms are given by oriented cups and caps, crossings, and multiplication by a polynomial $f \in R^I$ inside a region labeled I. As earlier, 2-

24.3 Singular Soergel Diagrammatics in General

morphisms are invariant up to isotopy. Consequently, a diagram in this category can be depicted as a collection of colored 1-manifolds with boundary, properly embedded in the planar strip, where the different colors can cross each other transversely. Not every such collection of 1-manifolds will give rise to a well-defined morphism: there must be an appropriate labeling of the regions as well.

In addition to the relations from Sect. 24.2.1, which hold for each Frobenius extension, one also has a number of relations which look like "oriented Reidemeister moves."

$$\text{(24.49)}$$

$$\text{(24.50)}$$

$$\text{(24.51)}$$

$$\text{(24.52)}$$

$$\text{(24.53)}$$

The empty boxes in these relations represent certain polynomials or linear combinations of polynomials, determined by the Frobenius structures. For example, the polynomial in (24.50) is $\frac{\mu_{Ist}}{\mu_{Is}\mu_{It}}$. For more details, see [62].

As noted above, the 2-category Frob is not yet equivalent to $\mathcal{SBS}$Bim. It only has the relations which hold for any cube of Frobenius extensions, while we need special relations for our particular case. We will let $\mathcal{SH}_{BS}$ denote the quotient of Frob by certain additional diagrammatic relations, some of which we spell out below.

24.3.4 The Jones–Wenzl Relation

One of the most important families of additional relations is the Jones–Wenzl relation.

Fix $s, t \in S$ with $m = m_{st} < \infty$. For $k \geq 1$, let v_k denote the following singular diagram, whose boundary repeats $\emptyset, s, \emptyset, t, \ldots$ exactly k times (when read around the circle).

$$\begin{array}{ccc} k=1 & k=2 & k=3 \\ & k=4 & \end{array} \tag{24.54}$$

Here we use the color purple to indicate the region labeled $\{s, t\}$.

Exercise 24.44 Check that the degree of v_k is exactly $2(m - k - 1)$, as pictured above. Now rotate the picture so that $\emptyset$ is the region on the right and left. Check that the degree of this rotation is $2(m - k)$. (Again, the degree of a singular picture is not rotation invariant.)

In particular, v_{m-1} has degree 0. Another way to obtain a morphism of degree 0 is to take a 2-morphism in the image of the 2-functor 2Σ. Taking the negligible Jones–Wenzl projector JW_{m-1} and applying 2Σ yields the linear combination of diagrams we denote JW_{m-1}. The *singular Elias–Jones–Wenzl relation* states that $v_{m-1} = JW_{m-1}$. Here are the examples when $m = 2, 3$.

$$m = 2 \qquad \qquad = \qquad \tag{24.55}$$

$$m = 3 \qquad \qquad = \qquad + \qquad$$

24.3 Singular Soergel Diagrammatics in General

Note that when $m = 2$, this agrees with (24.50).

Exercise 24.45 Write down the singular Elias–Jones–Wenzl relation for $m = 4$.

An implication of the singular Elias–Jones–Wenzl relation is that any instance of v_k for $k < m$ can be rewritten as a diagram without any purple regions. This is true for v_{m-1} because morphisms in the image of 2Σ have no purple regions. For $k < m - 1$, one can obtain v_k by applying cups and caps to v_{m-1}. However, the diagram v_m cannot be simplified any further. When rotating v_m so that its rightmost and leftmost regions are $\emptyset$, this diagram has degree zero.

Theorem 24.46 *Let (W, S) have dihedral type. The functor $\mathcal{F}$ from Theorem 24.23 extends to a monoidal functor from $\mathcal{H}_{BS}$ to $\mathcal{SH}_{BS}$, sending the $2m$-valent vertex to (the appropriate rotation of) v_m.*

For dihedral groups it turns out that the singular Elias–Jones–Wenzl relation is the only additional relation one needs to get a category equivalent to $\mathcal{SBS}$Bim. For a proof, see [47, Chapter 6].

The reader should be comforted by the following idea. The $2m$-valent vertex, which was rather mysterious, becomes easier to understand when seen as the singular Soergel diagram v_m: it is composed of numerous smaller pieces, each of which is straightforward (cups, caps, crossings). The complicated relations of Soergel calculus are mostly consequences of more local relations in singular Soergel calculus, and most of these local relations seem topological, like the Reidemeister moves.

24.3.5 Additional Relations, Applications, and Future Work

Unfortunately, for larger Coxeter groups there are more relations still. Unlike the ordinary Soergel calculus, which only had relations coming from ranks 1, 2, and 3, singular Soergel calculus will have new relations in any rank. A number of additional relations are known in type A, due to the author and Williamson; see [56, Chapter 2.5] for some examples.

In forthcoming work, the author and Williamson will give a list of relations which are known to hold in $\mathcal{SBS}$Bim in type A, and which are conjectured to be sufficient to produce a diagrammatic category $\mathcal{SH}_{BS}$ which is equivalent to $\mathcal{SBS}$Bim in type A and $\widetilde{A}$. This conjecture is proven in special cases of small rank (the dihedral group [47], type $\widetilde{A}_2$ [49, Appendix], type A_n for small n).

In forthcoming work, the author and Libedinsky will produce a family of "singular light leaves" and a "singular double leaves basis" (in type A and in dihedral type). We will prove that double leaves form a basis for $\mathcal{SBS}$Bim, using the Soergel–Williamson Hom formula. Once a diagrammatic proof is found that double leaves form a basis for $\mathcal{SH}_{BS}$, this will prove the conjecture that the relations are correct.

The double leaves basis makes $\mathcal{SBS}$Bim and $\mathcal{SH}_{BS}$ into a *fibered cellular category*, which is a slight variation on an object-adapted cellular category (see Chap. 12). See [54] for details.

For existing applications of singular Soergel diagrammatics, we point the reader to [49] and [56].

24.3.6 Other Realizations

It is not true that $R^I \subset R$ is a Frobenius extension for any realization. For the sake of the literature, we discuss the status of the upgraded Chevalley theorem for other realizations.

The proof of Theorem 24.36 frequently used the fact that $|W_I|$ was invertible in $\mathbb{R}$. The most serious problem for other realizations is that $|W_I|$ might be zero in finite characteristic, and might not be invertible if we work over $\mathbb{Z}$. This will clearly ruin the entire argument: the set $A = \{\partial_w(\mu_I)\}$ can not possibly be a basis because $\partial_{w_I}(\mu_I) = |W_I|$ is the unique degree zero element of the basis. We discussed this problem already in type A_1 (in characteristic 2, or over $\mathbb{Z}$). There we argued that $R^s \subset R$ was still a Frobenius extension for some larger realization (not spanned by the roots), so long as it contains some linear polynomial ϖ which satisfies $\partial_s(\varpi) = 1$.

In fact, Demazure was aware of this problem when he wrote [44]. For most of the paper he works over $\mathbb{Z}$, and he proves Demazure's formula in this context.[3] Then he proves the strongest results [44, Théorème 2] only after assuming the existence[4] of an element $a_I \in R$ for which $A_I(a_I) = \mu_I$. In particular, by Demazure's formula this is equivalent to the condition $\partial_I(a_I) = 1$. Under this assumption, he proves that the set

$$X = \{\partial_w(a_I)\}_{w \in W_I} \tag{24.56}$$

is a basis (and half of a dual basis) for R over R^I. It is easy to adapt the arguments used in the proof of Theorem 24.36 to prove that (other issues aside) $R^I \subset R$ is a graded Frobenius extension if and only if ∂_I is surjective.

Exercise 24.47 Consider the realization defined over $\Bbbk = \mathbb{Z}$ which comes from the action of S_n on the ring $\mathbb{Z}[x_1, \ldots, x_n]$. Let $a = x_1^{n-1} x_2^{n-2} \cdots x_{n-1}^1 x_n^0$. Prove that $\partial_{w_0}(a) = 1$, and that $\sum_{w \in W}(-1)^{\ell(w)} w(a) = \prod(x_i - x_j)$.

---

[3] Demazure restricts his attention to Weyl groups, whence the existence of a realization over $\mathbb{Z}$. His arguments apply equally well to Coxeter groups with other commutative base rings of characteristic zero, and their (potentially finite characteristic) specializations.

[4] In fact, Demazure in [44, §5] determines explicitly for certain standard realizations over $\mathbb{Z}$ which primes need to be inverted to guarantee the existence of such an element.

24.3 Singular Soergel Diagrammatics in General

So, to summarize, the condition that $|W_I|$ is invertible is a red herring: the correct condition to assume is that ∂_I is surjective. Let us call *finitary Demazure surjectivity* the condition that, for all $I \subset S$ finitary, the map $\partial_I : R \to R^I$ is surjective. This is a necessary assumption for the results of this chapter to work. We have already assumed all along (this was Demazure surjectivity, Assumption 5.48) that each ∂_s is surjective.

Finitary Demazure surjectivity can be achieved when $|W_I|$ is invertible for all finitary I, of course, but can also be achieved by using realizations which are not spanned by the roots.

Exercise 24.48 Consider the geometric representation in type A_2 (spanned by the roots), but defined over some arbitrary commutative ring $\Bbbk$. Prove that ∂_{w_0} is surjective if and only if 3 is invertible.

Exercise 24.49 Let $I = \{s, t\}$ for distinct simple reflections $s, t \in S$, with $m_{st} = m < \infty$. Let ϖ_t be a linear polynomial satisfying $\partial_s(\varpi_t) = 0$ and $\partial_t(\varpi_t) = 1$. Prove that $\partial_I^s(\varpi_t^{m-1}) = [m-1]!$, where the quantum numbers are defined as in Exercise 5.51. Thus (assuming also that ∂_s is surjective) we see that $[m-1]!$ being invertible implies that ∂_I is surjective. Verify that the set $\{\varpi_t^k\}_{0 \le k < m}$ is almost dual to itself.

Remark 24.50 Continuing the setup of Exercise 24.49, we see that $[m-1]!$ being invertible implies that ∂_I is a Frobenius trace. Note from Exercise 5.51 that if $[k] = 0$ for some $0 < k < m$ then $(st)^k$ acts trivially on the span of α_s and α_t. So long as the realization is not particularly degenerate, $[m-1]!$ being nonzero is equivalent to the action of this particular dihedral group being faithful.

Now, here is a list of the other features of the geometric realization that we used in the proof of Theorem 24.36. First we have the fundamentals:

- The Demazure operators satisfy the braid relations.
- The existence of a finite set of positive roots for any finite parabolic subgroup I, having size $|W_I|$, and satisfying certain properties.

Then we have the deductions:

- The W_I-antiinvariants are generated over R^I by μ_I.
- Demazure's formula: $\partial_I = \frac{A_I}{\mu_I}$. (In the proof we used the linear independence of the operators $w \in W_I$ acting on the fraction field of R.)
- The ordinary Chevalley theorem: R is free over R^I of a particular graded rank.

Let us examine these one by one (starting with the deductions, then moving to the fundamentals), and ask which ones hold for other realizations under the assumption of finitary Demazure surjectivity. An important point to note is that (with minor exceptions) these properties are preserved by base change of realizations. For example, if Demazure's formula holds for a realization, it will continue to hold after base change. Hence we can assume that $\Bbbk$ is not more complicated than it needs to be for its Cartan matrix to provide a realization of (W, S). For example, we can assume that $\Bbbk$ is a UFD.

In characteristic 2, the notion of antiinvariants makes no sense (or it agrees with the notion of invariants, in which case they are clearly not generated by μ_I). Here is an alternate proof that μ_I divides any antiinvariant polynomial f, so long as $2 \neq 0$. Let $s \in I$, so that $sf = -f$. Now $\partial_s(f)\alpha_s = f - sf = 2f$. Since a polynomial ring over a UFD is a UFD, we see that α_s divides f. This is true for all $s \in I$. Now $w(f)$ is a unit multiple of f for any $w \in W_I$, so $w(\alpha_s)$ divides f as well. Hence all the roots of W_I divide f. However, if the fundamentals are wrong and there are fewer than $\ell(w_I)$ positive roots, then this result will face some issues, which we examine later.

Demazure [44, Lemme 4] proved Demazure's formula integrally, so it is true in any specialization, even in characteristic 2. His proof assumes the theory of positive roots and the faithfulness of the W_I-action. The proof that we used above can be easily adapted (by replacing μ_I with a) to the situation where ∂_I is surjective, so long as the positive roots behave in the familiar fashion.

Similarly, Demazure [44, Théorème 2 and 3] proved the ordinary Chevalley theorem as a corollary of the other results, under the assumption that ∂_I is surjective. To sketch another proof, the set X from (24.56) has a dual set defined in the analogous way, meaning that it spans a free R^I-module. Note that X contains $1 = \partial_I(a)$. It is annoying but possible to prove directly that the span of X is all of R. The result having been proven by Demazure, we will not belabor the point.

Let us mention that Theorem 24.40 creates no additional subtleties, and the proof can be adapted without difficulty.

Now we ask what happens when the fundamentals fail.

Throughout this book we have often been using the simplifying assumption that the realization is *odd-balanced*, see [47, Definition 3.6]. When the realization is odd-balanced, the Demazure operators satisfy the braid relations and there is a consistent notion of positive roots. However, when a realization is not odd-balanced, then Demazure operators only satisfy the braid relations up to an invertible scalar! Such realizations are actually useful, see [49]. In particular, one can only define an operator ∂_w for $w \in W$ up to scalar, but not on the nose. Similarly, the "positive roots" are only well-defined up to scalars.

Example 24.51 Suppose that $\partial_s(\alpha_t) = \partial_t(\alpha_s) = 1$. Then $m_{st} = 3$, and $\partial_s\partial_t\partial_s = -\partial_t\partial_s\partial_t$. Note that $s(\alpha_t) = -t(\alpha_s)$, so that it is not clear whether $s(\alpha_t)$ or $t(\alpha_s)$ should be the positive root. If we set $\mu = \alpha_s\alpha_t s(\alpha_t) = -\alpha_s\alpha_t t(\alpha_s)$ then $s(\mu) = -\mu$ and $t(\mu) = -\mu$.

Even in the odd-unbalanced case, there is a finite list of roots such that every root is obtained from this list by multiplication by an invertible scalar; the image of the roots in projective space is a finite set. If the size of this set is $\ell(I)$, then the element μ_I is well-defined up to scalar, and it is still W_I-antiinvariant! After all, μ_I is a product of a scalar, α_s, and a collection of roots which split into s-orbits. Ultimately, the odd-unbalanced case will not cause any real problems in the proof of Theorem 24.36; one must simply choose some scalar to pin down the trace map and the product-coproduct element compatibly. The almost dual bases will remain almost dual up to some invertible scalars, which is no problem.

24.3 Singular Soergel Diagrammatics in General

To construct a compatible Frobenius hypercube in the non-balanced case, one must keep track of additional scalars, which is a notational nightmare. Still, it poses no intractable problems, and Theorem 24.40 can be adapted to the non-balanced case as well. For details on how to treat the general case, see [49, Chapter 5, and specifically 5.5]. For additional details see [47, Appendix A.3 and following].

Now let us ask the thornier problem: what if the theory of roots does not work at all? What if the number of roots (counted in projective space) is fewer than $\ell(I)$? This is what you should expect when the realization is not reflection faithful, as two distinct reflections will share the same reflecting hyperplane.

Example 24.52 Any realization of type A_2 (with Coxeter generators s, t) can also be viewed as a realization of type G_2 by inflation. (In characteristic 2, every realization of G_2 factors through its quotient A_2.) The subrings R^W and R^{-W} inside R do not depend on whether W is thought of as type A_2 or type G_2, they only care about the action of the generators s and t. As a consequence of the type A_2 theory, there are only three distinct roots (up to sign), and their product is anti-invariant. Chevalley's theorem holds for A_2 and thus fails for G_2 (the graded rank is incorrect). Note that $\partial_s \partial_t \partial_s = \partial_t \partial_s \partial_t$, and hence $\partial_s \partial_t \partial_s \partial_t \partial_s \partial_t = 0$. Thus this operator $\partial_{w_0} = 0$ can clearly not be a Frobenius trace for type G_2! Of course, finitary Demazure surjectivity also fails.

One should not expect the Chevalley theorem to hold when the realization is not a faithful representation of W_I. Let us call a realization *finitary reflection faithful* if it is reflection faithful when restricted to each finite parabolic subgroup. We do not know of any examples of realizations which are not finitary reflection faithful, but for which finitary Demazure surjectivity still holds, and we do not know whether the conditions are independent. The following conjecture is not found in the literature, and while we expect it to follow from the techniques used above without much difficulty, we do not feel this book is the appropriate place to prove it.

Conjecture 24.53 The upgraded Chevalley theorem 24.40 holds for any realization which satisfies finitary Demazure surjectivity and is finitary reflection faithful.

Finally, let us discuss the diagrammatic category for other realizations. To use the diagrammatic machinery of [62] we must have a Frobenius hypercube, so one needs the upgraded Chevalley theorem. The results of [62] are proven under a technical assumption on the realization, called (∗) in that paper, which is an analogue of finitary reflection faithfulness. Let $I \subset S$ be finitary, and let $s, t \in S \setminus I$ be such that $J = Is$ and $K = It$ are both finitary. Then (∗) is the condition that R^I has dual bases over R^J where one of the bases is entirely contained in R^K. In particular, this implies that R^J and R^K generate R^I as a ring.

For example, when $I = \emptyset$, this implies that R is generated by R^s and R^t whenever $m_{st} < \infty$. This could fail when $R^s = R^t$, i.e. when $s = t$ as operators on $\mathfrak{h}^*$, which can happen for $m_{st} < \infty$ in some finite characteristics. If (∗) holds then there is an element ϖ_t for which $\partial_s(\varpi_t) = 0$ and $\partial_t(\varpi_t) = 1$. As noted in Exercise 24.49, this is enough to prove that $\partial_{s,t}$ is a Frobenius trace map. We only need to assume (∗) for each dihedral parabolic subgroup in order to use all the diagrammatic technology of this chapter for dihedral groups, which was tacitly (and possibly unnecessarily) used in the construction of the functor $\mathcal{F}$ from the non-singular diagrammatic category $\mathcal{H}_{BS}$ to the bimodule category $\mathbb{BSBim}$ (see Sect. 10.2.6).

Chapter 25
Koszul Duality I

This chapter is based on expanded notes of a lecture given by the authors and taken by
Noah Arbesfeld and Visu Makam

Abstract In 1979, Bernstein, Gelfand, and Gelfand described a duality between certain parts of the derived categories of the symmetric and exterior algebras. This type of duality, called Koszul duality, has since been observed in various contexts. After introducing the necessary homological algebra, we formulate this phenomenon as a differential graded version of Morita theory, and present Koszul duality for a polynomial ring in one variable. We then present some evidence (on the level of characters) of such a self-duality in the principal block of category $\mathcal{O}$.

25.1 Introduction

Roughly speaking, a Fourier transform is an operation which collects a signal and decomposes the signal into its component frequencies; in particular, a Fourier transform exchanges certain frequencies with delta functions, and conversely. In many important cases, a Fourier transform can reveal hidden structure. The fact that decomposition into component frequencies involves no loss of information is implied by the fact that applying the Fourier transform twice to a function returns the original function (up to a constant and a sign).

Fourier transform-like operations appear in geometry and algebra. Roughly speaking, in geometry, such operations often exchange vector bundles with skyscraper sheaves, and in algebra, they exchange projective objects with simple objects. In algebraic settings, such transforms are often instances of *Koszul duality*, originally described by Bernstein, Gelfand, and Gelfand [25] as a duality relating the symmetric and exterior algebras. Some other examples may be found in [6], [20], [32], [133], and [153].

N. Arbesfeld
Department of Mathematics, Columbia University, New York, NY, USA

V. Makam
Department of Mathematics, University of Michigan, Ann Arbor, MI, USA

© The Editor(s) (if applicable) and The Author(s), under exclusive licence to Springer Nature Switzerland AG 2020
B. Elias et al., *Introduction to Soergel Bimodules*, RSME Springer Series 5, https://doi.org/10.1007/978-3-030-48826-0_25

25.1.1 Morita Theory

We present Koszul duality as a differential graded version of Morita theory, and hence we begin with a quick overview of Morita theory; one source for this material is [16, Ch. II].

For a ring R, we let mod-R denote the category of finitely generated right R-modules. Morita theory tells us two things: first, when a given abelian category $\mathcal{A}$ is equivalent to a category of the form mod-R, and second, when two categories mod-R and mod-R' are equivalent.

For a projective object P in an abelian category $\mathcal{A}$, consider $\langle P \rangle$, the full subcategory generated by P, i.e. an object belongs to $\langle P \rangle$ if it has a presentation $P^{\oplus m} \to P^{\oplus n}$. Loosely speaking, Morita theory asserts that the subcategory $\langle P \rangle$ and End(P) determine each other. This is not surprising: the data of a presentation can be written in a matrix of endomorphisms of P.

To be precise, let $\mathcal{A}$ be an abelian category. We make the important additional assumption that each object of $\mathcal{A}$ has finite length.[1] Recall that a *projective object* is an object $P \in \mathcal{A}$ for which the functor $\text{Hom}_{\mathcal{A}}(P, -)$ is exact.

Lemma 25.1 *The functor* $\text{Hom}_{\mathcal{A}}(P, -)$ *defines an equivalence*

$$\langle P \rangle \xrightarrow{\sim} \text{mod-}\text{End}_{\mathcal{A}}(P),$$

where mod-$\text{End}_{\mathcal{A}}(P)$ *denotes the category of finitely-generated right* $\text{End}_{\mathcal{A}}(P)$-*modules.*

Exercise 25.2 Prove the lemma. (*Hint:* First show that $h = \text{Hom}(P, -)$ is fully-faithful, using only that P is projective. Then use our assumptions on $\mathcal{A}$ to deduce that it is essentially surjective.)

A projective object P is called a *projective generator* if $\langle P \rangle = \mathcal{A}$.

Corollary 25.3 *If P is a projective generator, then* $\text{Hom}_{\mathcal{A}}(P, -)$ *gives an equivalence*

$$\mathcal{A} \xrightarrow{\sim} \text{mod-}\text{End}_{\mathcal{A}}(P).$$

The other part of Morita theory characterizes rings with equivalent module categories in terms of projective generators.

Theorem 25.4 *Let R and R' be two rings. Then, the categories* mod-R *and* mod-R' *are equivalent if and only if R' is isomorphic to* $\text{End}_R(P)$ *for some finitely-generated projective generator* $P \in$ mod-R.

---

[1] This assumption can be dropped at the expense of additional technicalities, see [16, Ch. II].

Two rings R and R' with equivalent module categories are said to be *Morita equivalent*. The rings R and $\operatorname{End}_R(P)$ have isomorphic centers; in particular, two Morita-equivalent commutative rings must also be isomorphic.

Example 25.5 (Finite-Dimensional Algebras) Consider a finite-dimensional algebra A over an algebraically closed field $\Bbbk$. The Krull–Schmidt theorem (Proposition 11.50) implies that the indecomposable projective A-modules are direct summands of A itself. Let $P_1, \ldots, P_n$ be a set of indecomposable projectives up to isomorphism, and let $P = \bigoplus_{i=1}^n P_i$. Then, P is a projective generator, and hence we have an equivalence between the categories mod-A and mod-$\operatorname{End}_A(P)$. It is easy to see that $\operatorname{End}_A(P)$ is a finite-dimensional basic algebra, i.e. all its simple modules are one dimensional.

Any finite-dimensional basic algebra is isomorphic to the path algebra of some quiver with relations. Hence, the above application of Morita theory allows us to understand the category mod-A by studying a corresponding quiver with relations. This latter object can sometimes be easier to compute with.

However, we remark that there is no known explicit quiver algebra description for category $\mathcal{O}$, and no nice description is expected! Instead, our approach in this book is to form a large projective generator P comprised of Soergel (or more precisely Bott–Samelson) modules, and work with the algebra $\operatorname{End}_{\mathcal{O}_0}(P)$ explicitly.

25.2 dg-Algebras

Our next aim is to formulate a version of Morita theory for the derived category of modules over a dg-algebra. In this section we recall the requisite homological algebra.

Definition 25.6 A *differential graded algebra*, or *dg-algebra*, is a $\mathbb{Z}$-graded algebra $A = \bigoplus_{i \in \mathbb{Z}} A^i$ equipped with a degree 1 differential $d : A^i \to A^{i+1}$ (so $d \circ d = 0$) satisfying the Leibniz rule $d(ab) = d(a)b + (-1)^{|a|} a d(b)$ for homogeneous elements a.

The following construction produces an important family of examples of dg-algebras. Let $\mathcal{A}$ be an additive category and X be a chain complex in $\mathcal{A}$. Then, form a dg-algebra $\mathcal{E}\mathrm{nd}(X) = \bigoplus_{i \in \mathbb{Z}} \mathcal{E}\mathrm{nd}^i(X)$ as follows. Set

$$\mathcal{E}\mathrm{nd}^i(X) = \prod_{k \in \mathbb{Z}} \operatorname{Hom}(X^k, X^{k+i}), \qquad (25.1)$$

and let multiplication be given by composition of morphisms. We equip $\mathcal{E}\mathrm{nd}^i(X)$ with the following differential: for $f = (f^k)_{k \in \mathbb{Z}} \in \mathcal{E}\mathrm{nd}^i(X)$, set $df = ((df)^k)_{k \in \mathbb{Z}}$, where

$$(df)^k = d \circ f^k - (-1)^i f^{k+1} \circ d. \qquad (25.2)$$

Exercise 25.7 Check that with these definitions $\mathcal{E}\mathrm{nd}(X)$ is a dg-algebra.

Even for seemingly innocuous complexes X, the dg-algebra $\mathcal{E}\mathrm{nd}(X)$ can be unwieldy, as we see in the following exercise.

Exercise 25.8 Let $\Lambda = \Bbbk[\epsilon]/(\epsilon^2)$, the algebra of *dual numbers*, and consider the chain complex

$$X = (\cdots \to \Lambda \xrightarrow{\epsilon} \Lambda \xrightarrow{\epsilon} \Lambda \to 0 \to 0 \to \cdots)$$

in Λ-mod.

1. Show that X gives a resolution of the trivial Λ-module $\Bbbk$, and hence that

$$\mathrm{Ext}^i_{\Lambda\text{-mod}}(\Bbbk, \Bbbk) = \Bbbk \quad \text{for all } i \geq 0.$$

   (A little harder: show that $\mathrm{Ext}^\bullet(\Bbbk, \Bbbk) = \Bbbk[x]$ as graded algebras, where $\deg x = 1$.)
2. Show that $\mathcal{E}\mathrm{nd}(X)$ consists of a countable direct product of copies of Λ in every degree.
3. If you are feeling brave: show explicitly that the cohomology of $(\mathcal{E}\mathrm{nd}(X), d)$ is isomorphic to $\Bbbk[x]$ above. Can you explain why this has to be the case without calculating?

We will be interested in modules over dg-algebras. The appropriate notion of a module in this setting is as follows.

Definition 25.9 A (right) *dg-module* over a dg-algebra A is a graded right A-module $M = \bigoplus_{i \in \mathbb{Z}} M^i$ with differential $d : M^i \to M^{i+1}$ such that $d(m \cdot a) = d(m) \cdot a + (-1)^{|m|} m \cdot d(a)$ for homogeneous elements m.

For example, if A is concentrated in degree zero, then its differential d must be zero. Consequently, a dg-module over such an algebra A is simply a complex of right A-modules whose differential is a homomorphism of right A-modules.

A more complicated class of examples can be constructed as follows. Given two complexes $X, Y \in C(\mathcal{A})$, we can define $\mathcal{H}\mathrm{om}(X, Y)$ as above; namely, set

$$\mathcal{H}\mathrm{om}^i(X, Y) = \prod_{k \in \mathbb{Z}} \mathrm{Hom}(X^k, Y^{k+i}), \tag{25.3}$$

and define the differential just as for $\mathcal{E}\mathrm{nd}(X)$ above. Then, given any $Y \in C(\mathcal{A})$, the complex $\mathcal{H}\mathrm{om}(X, Y)$ is a right dg-module for $\mathcal{E}\mathrm{nd}(X)$; this is a manifestation of the general fact that, in appropriate settings, $\mathrm{Hom}(X, Y)$ is a right $\mathrm{End}(X)$-module.

Exercise 25.10 Check that with these definitions $\mathcal{H}\mathrm{om}(X, Y)$ is a right dg-module. (*Hint:* This should be easy if you have done Exercise 25.7.)

To construct the derived category of dg-modules, we need the notion of a quasi-isomorphism.

Definition 25.11 A *quasi-isomorphism* of dg-algebras $\phi : A \to A'$ is a homomorphism of dg-algebras such that the induced map on cohomology $H^\bullet(A) \to H^\bullet(A')$ is an isomorphism of graded rings. A quasi-isomorphism of dg-modules is defined analogously.

Exercise 25.12 Continuing Exercise 25.8, find a quasi-isomorphism between $\Bbbk[x]$ (viewed as a dg-algebra with $\deg x = 1$ and $d = 0$) and $(\mathcal{E}\mathrm{nd}(X), d)$.

Let dg-$C(A)$ denote the category of right dg-modules over A; by regarding a dg-module as a complex, we may then define the homotopy and derived categories in the usual manner. Namely, define the homotopy category dg-$K(A)$ to be the category obtained by identifying homotopic maps of dg-modules, and define the derived category dg-$D(A)$ to be the category obtained from dg-$K(A)$ by inverting quasi-isomorphisms. Both dg-$K(A)$ and dg-$D(A)$ are triangulated categories (see Chap. 19 for a discussion of triangulated categories).

Given quasi-isomorphic dg-algebras, one has the following relationship between their derived dg-module categories.

Proposition 25.13 *A quasi-isomorphism* $\phi : A \to A'$ *induces a triangulated equivalence of derived categories*

$$\text{dg-}D(A') \xrightarrow{\sim} \text{dg-}D(A). \tag{25.4}$$

There are some subtleties in this proposition that we are sweeping under the rug. The reader is referred to [110, 154] and [169, Tag 09S5] for a detailed treatment.

25.3 dg-Morita Theory

Recall that Morita theory loosely states that for a projective object P in an abelian category $\mathcal{A}$, the subcategory $\langle P \rangle$ and the ring $\mathrm{End}(P)$ determine each other. Following [153, Section 1.3], we present a derived version of such a statement.

Given an abelian category $\mathcal{A}$ and an object X, a naive hope would be that the full triangulated subcategory $\langle X \rangle_\Delta$ generated by X inside the derived category $D(\mathcal{A})$ may be recovered from the algebra

$$\mathrm{End}^\bullet(X) = \bigoplus_{m \in \mathbb{Z}} \mathrm{Hom}_{D(\mathcal{A})}(X, X[m]); \tag{25.5}$$

here, generation is in a triangulated category, so shifts of X are allowed. Roughly speaking, in analogy with the abelian case discussed above, one thinks of objects $A \in \langle X \rangle_\Delta$ as fitting into distinguished triangles

$$\bigoplus_{i=1}^{j} X[m_i] \to \bigoplus_{j=1}^{k} X[n_j] \to A \to \left(\bigoplus_{i=1}^{j} X[m_i] \right)[1]. \tag{25.6}$$

More precisely, one builds objects in $\langle X \rangle_\Delta$ by taking successive cones of maps between sums of X and its shifts, and hence one might first hope that knowing $\text{End}^\bullet(X)$ is enough.

In general, however, more information is needed; the basic issue is that the cone appearing in (25.6) is not functorial. Such information can sometimes be contained in the dg-algebra $\mathcal{E}\text{nd}(X)$. Namely, for certain complexes X, one may formulate a differential graded analogue of Morita equivalence using the homological algebra introduced above.

A first basic difficulty is that morphisms in $D(\mathcal{A})$ are rather difficult to pin down. Here we make the following definition, which allows us to change ambient category from the derived category $D(\mathcal{A})$ to the homotopy category $K(\mathcal{A})$:

Definition 25.14 A complex X is called *end-acyclic* if, for all $m \in \mathbb{Z}$, the natural homomorphism

$$\text{Hom}_{K(\mathcal{A})}(X, X[m]) \to \text{Hom}_{D(\mathcal{A})}(X, X[m]) \qquad (25.7)$$

is an isomorphism.

Remark 25.15 One can think of end-acyclic as the correct notion in the derived setting which replaces "finitely-generated projective" in classical Morita theory.

Exercise 25.16 Show that bounded above complexes of projectives and bounded below complexes of injectives are end-acyclic. Find an example of a complex of projectives which is not end-acyclic. (*Hint:* Let Λ be the algebra of dual numbers appearing in Exercise 25.8, and now consider infinite complex

$$\cdots \to \Lambda \xrightarrow{\epsilon} \Lambda \xrightarrow{\epsilon} \Lambda \to \cdots \text{.)}$$

If X is end-acyclic then the quotient functor from the homotopy category to the derived category induces an equivalence

$$K(\mathcal{A}) \supset \langle X \rangle_\Delta \xrightarrow{\sim} \langle X \rangle_\Delta \subset D(\mathcal{A}).$$

This can be proven by induction, assuming the result for two objects in a distinguished triangle and deducing it for the third (do it!).

Let X be an end-acyclic complex, and let us set A to be the dg-algebra $\mathcal{E}\text{nd}(X)$. Then, we have a functor

$$\mathcal{H}\text{om}(X, -) : C(\mathcal{A}) \to \text{dg-}C(A) \qquad (25.8)$$

sending $X \mapsto A$. This functor descends to a functor between homotopy categories $\mathcal{H}\text{om}(X, -) : K(\mathcal{A}) \to \text{dg-}K(A)$, yielding the diagram of functors

$$D(\mathcal{A}) \leftarrow K(\mathcal{A}) \xrightarrow{\mathcal{H}\text{om}(X,-)} \text{dg-}K(A) \to \text{dg-}D(A). \qquad (25.9)$$

25.3 dg-Morita Theory

Then, dg-Morita theory describes how the category $\langle X \rangle_\Delta \subset K(\mathcal{A})$ behaves under these functors.

Theorem 25.17 *Let $\mathcal{A}$ be an abelian category, and let $X \in C(\mathcal{A})$ be an end-acyclic complex. Then, there is an equivalence of triangulated categories*

$$D(\mathcal{A}) \supset \langle X \rangle_\Delta \xrightarrow{\sim} \langle A \rangle_\Delta \subset \text{dg-}D(A). \tag{25.10}$$

Remark 25.18 A more general version of this statement is presented in [153, Section 1.3], where X is replaced by a finite family of complexes that pairwise satisfy the condition for end-acyclicity above.

Loosely speaking, dg-Morita theory can be summarized as the statement that, in algebraic settings, triangulated subcategories are described by dg-algebras. In practice, it is difficult to work with all of $\mathcal{E}\text{nd}(X)$. The following condition will enable us to pass from $\mathcal{E}\text{nd}(X)$ to its cohomology, which will be a simpler object.

Definition 25.19 A dg-algebra A is said to be *formal* if A is quasi-isomorphic to its cohomology $\text{H}^\bullet(A)$.

For a general dg-algebra A, the cohomology $\text{H}^\bullet(A)$ is a graded ring, and can be viewed as a dg-algebra with trivial differential. The following exercise relates the cohomology ring of the dg-algebra $\mathcal{E}\text{nd}(X)$ with maps in the homotopy category between X and its shifts.

Exercise 25.20 For any complex $X \in C(\mathcal{A})$, one has

$$\text{H}^\bullet(\mathcal{E}\text{nd}(X)) = \bigoplus_{i \in \mathbb{Z}} \text{Hom}_{K(\mathcal{A})}(X, X[i]). \tag{25.11}$$

In the particular case where X is end-acyclic, we obtain

$$\text{H}^\bullet(\mathcal{E}\text{nd}(X)) = \bigoplus_{i \in \mathbb{Z}} \text{Hom}_{D(\mathcal{A})}(X, X[i]) = \text{End}^\bullet(X). \tag{25.12}$$

As a quasi-isomorphism of dg-algebras induces an equivalence of derived categories, we deduce the following corollary.

Corollary 25.21 *If X is end-acyclic, and $\mathcal{E}\text{nd}(X)$ is formal, we obtain a triangulated equivalence of derived categories*

$$\text{dg-}D(\mathcal{E}\text{nd}(X)) \xrightarrow{\sim} \text{dg-}D(\text{End}^\bullet(X)), \tag{25.13}$$

where $\text{End}^\bullet(X)$ is viewed as a dg-algebra with trivial differential.

25.4 Koszul Duality for Polynomial Rings

Koszul duality is an equivalence of derived categories of graded modules over certain types of graded rings. We illustrate this phenomenon with the following extended example.

Let V be a 1-dimensional vector space over a field $\Bbbk$, and let $\Lambda = \Lambda V = \Bbbk \oplus V$ be its exterior algebra, considered as a graded algebra with V in degree -1, and $\Bbbk$ in degree 0. Let $S = \operatorname{Sym}(V^*) = \Bbbk[x]$ be the symmetric algebra of its linear dual, considered as a graded algebra with V^* in degree 1.

Consider the trivial module $\Bbbk$ in the category Λ-gmod, i.e. the category of graded and finitely-generated Λ-modules. As usual, we denote the shift of grading functor by (1) (see Sect. 4.1 for our conventions).

The following equivalence will be our first example of Koszul duality.

Theorem 25.22 *There is a triangulated equivalence*

$$D^b(\Lambda\text{-gmod}) \supset \langle \Bbbk \rangle_\Delta \xrightarrow[\kappa]{\sim} \langle S \rangle_\Delta \subset D^b(S\text{-gmod}). \tag{25.14}$$

Here, $\langle \Bbbk \rangle_\Delta$ means the graded *triangulated category generated by $\Bbbk$ (and similarly for $\langle S \rangle_\Delta$), i.e. $\langle \Bbbk \rangle_\Delta$ contains $\Bbbk[m](n)$ for all $m,n \in \mathbb{Z}$. Under this equivalence we have:*

$$\kappa : \Bbbk \mapsto S,$$

$$\kappa : \Lambda \mapsto \Bbbk(1)[-1].$$

Exercise 25.23 Show that the natural inclusions

$$\langle \Bbbk \rangle_\Delta \subset D^b(\Lambda\text{-gmod}) \quad \text{and} \quad \langle S \rangle_\Delta \subset D^b(S\text{-gmod})$$

are in fact equivalences, i.e. $\Bbbk$ and S generate the respective graded triangulated categories. (Recall that Λ-gmod and S-gmod denote finitely-generated graded modules.) Thus in the setting of the theorem Koszul duality is an equivalence of derived categories

$$D^b(\Lambda\text{-gmod}) \xrightarrow[\kappa]{\sim} D^b(S\text{-gmod}). \tag{25.15}$$

Remark 25.24 The second part of the theorem ensures that κ sends interchanges simple (point like) and projective (fundamental frequency like) modules. It is for this reason that κ can be thought of as akin to a Fourier transform.

Exercise 25.25 Consider for a moment the categories Λ-mod and S-mod of *ungraded* (finitely-generated) modules. Show that $D^b(\Lambda\text{-mod})$ and $D^b(S\text{-mod})$ are not equivalent as triangulated categories. (*Hint:* Show that $\operatorname{Hom}(X, X[i])$ is non-zero for infinitely many i for some $X \in D^b(\Lambda\text{-mod})$, whereas this is not the

25.4 Koszul Duality for Polynomial Rings

case for any $Y \in D^b(S\text{-mod})$.) The moral is that the presence of the grading in Theorem 25.22 is essential.

The functor κ is a composition of two equivalences, which we now describe.

The first equivalence is dg-Morita theory. Note that $D^b(\Lambda\text{-gmod})$ has two grading shift functors: cohomological shift [1], and internal (module) grading shift (1). We need some preparation to handle this extra grading. For $X \in C(\Lambda\text{-gmod})$, define

$$\mathcal{E}\mathrm{nd}(X) := \bigoplus_{i,j} \mathcal{E}\mathrm{nd}^{i,j}(X) = \bigoplus_{i,j} \prod_{k \in \mathbb{Z}} \mathrm{Hom}_{\Lambda\text{-gmod}}(X^k, X^{k+i}(j)). \qquad (25.16)$$

Then $\mathcal{E}\mathrm{nd}(X)$ is a dgg-algebra, a dg algebra with an extra internal grading, where the differential has bidegree $(1, 0)$ (i.e. cohomological degree 1). Similarly, we can define a bigraded algebra

$$\mathrm{End}^{\bullet,\bullet}(X) := \bigoplus_{i,j} \mathrm{Hom}_{D(\Lambda\text{-gmod})}(X, X[i](j)). \qquad (25.17)$$

Note that the module $\Bbbk$ admits a graded projective resolution

$$\cdots \to \Lambda(2) \to \Lambda(1) \to \Lambda \to \Bbbk. \qquad (25.18)$$

Let $\widetilde{S} = \mathrm{Sym}(V^*) = \Bbbk[\widetilde{x}]$ be the symmetric algebra, bigraded with V^* in degree $(1, 1)$.

Exercise 25.26 Using the resolution (25.18), prove the following:

1. There is an isomorphism of bigraded algebras $\mathrm{End}^{\bullet,\bullet}(\Bbbk) \xrightarrow{\sim} \widetilde{S}$.
2. The dg-algebra $\mathcal{E}\mathrm{nd}(P)$ is formal.

(*Hint*: This exercise is a graded version of Exercises 25.8 and 25.12.)

Remark 25.27 The above exercise is an illustration of a general principle that bigraded dg-algebras with cohomology concentrated on a diagonal (in this case $i = j$) are formal. For more on this technique see [153, §1.5].

Applying a straightforward extension of dg-Morita theory to include the extra grading, we obtain a triangulated equivalence

$$D^b(\Lambda\text{-gmod}) \supset \langle \Bbbk \rangle_\Delta \xrightarrow{\sim} \langle \widetilde{S} \rangle_\Delta \subset \mathrm{dgg}\text{-}D^b(\widetilde{S}), \qquad (25.19)$$

which intertwines the internal shift (1) on $D^b(\Lambda\text{-gmod})$ with the shift on $\mathrm{dgg}\text{-}D^b(\widetilde{S})$ that shifts the second grading down. Here, $\widetilde{S}$ is viewed as a dgg-algebra with zero differential.

A second equivalence is needed to pass from the dgg-derived category of $\widetilde{S}$ to the ordinary derived category of graded S-modules; this equivalence is a shearing operation that intertwines the internal grading with a combination of cohomological and internal gradings. We briefly indicate this process. First, note that a dgg-module $M = \bigoplus M^{i,j}$ over the dgg-algebra $\widetilde{S}$ consists of a complex of vector spaces, equipped with chain maps $d : M^{i,j} \to M^{i+1,j}$ and $x : M^{i,j} \to M^{i+1,j+1}$, as depicted below:

$$\begin{array}{ccccc}
M^{0,2} & \xrightarrow{d} & M^{1,2} & \xrightarrow{d} & M^{2,2} \\
& \nearrow^x & & \nearrow^x & \\
M^{0,1} & \xrightarrow{d} & M^{1,1} & \xrightarrow{d} & M^{2,1} \\
& \nearrow^x & & \nearrow^x & \\
M^{0,0} & \xrightarrow{d} & M^{1,0} & \xrightarrow{d} & M^{2,0}.
\end{array} \qquad (25.20)$$

Redrawing the picture, we can view the dg-module M as a complex of graded modules over S:

$$\begin{array}{ccccc}
M^{0,2} & \xrightarrow{d} & M^{1,2} & \xrightarrow{d} & M^{2,2} \\
\uparrow^x & & \uparrow^x & & \\
M^{0,1} & \xrightarrow{d} & M^{1,1} & \xrightarrow{d} & M^{2,1} \\
\uparrow^x & & \uparrow^x & & \\
M^{0,0} & \xrightarrow{d} & M^{1,0} & \xrightarrow{d} & M^{2,0}.
\end{array} \qquad (25.21)$$

Here the columns can be viewed as graded S-modules. Hence, after a shift in grading, M may be regarded as a complex of graded S-modules, whose differentials are the horizontal maps in the picture. This grading shift is called shearing, and furnishes a triangulated equivalence

$$\text{dgg-}D^b(\widetilde{S}) \supset \langle \widetilde{S} \rangle_\Delta \xrightarrow{\sim} \langle S \rangle_\Delta \subset D^b(S\text{-gmod}), \qquad (25.22)$$

which now intertwines (1) with $(1)[-1]$.

Combining dg-Morita theory and shearing, we obtain the equivalence κ. In particular, for $M \in D^b(\Lambda\text{-gmod})$, we have

$$\kappa(M(i)) = \kappa(M)(i)[-i]. \qquad (25.23)$$

A similar equivalence, proved by Bernstein, Gelfand and Gelfand in [25], holds when V is replaced by any finite-dimensional vector space; in this case, the so-called *Koszul resolution* plays the role of the resolution (25.18). The general class of graded rings for which Koszul duality can be formulated is called Koszul rings; the definition may be found, for example, in [20, Def. 1.2.1].

Exercise 25.28 Think about running the above argument the other way: start with the trivial S module, and use dg-Morita theory and shearing to get an equivalence

$$D^b(S\text{-gmod}) \supset \langle \Bbbk \rangle_\Delta \xrightarrow[\kappa']{\sim} \langle \Lambda \rangle_\Delta \subset D^b(\Lambda\text{-gmod}).$$

Prove that $\langle \Bbbk \rangle_\Delta \subset D^b(S\text{-gmod})$ consists of complexes with finite-dimensional total cohomology. (In particular, this version involves "smaller" subcategories than the version discussed above.)

25.5 Review of the Kazhdan–Lusztig Conjecture

Our goal in the next two sections is to explain some considerations which led Beilinson and Ginzburg to their conjectures on Koszul duality for the principal block of category $\mathcal{O}$.[2] However, before we get to this we need to recall the Kazhdan–Lusztig conjecture, and deduce an equivalent formulation. In this and the following section, we assume the reader is on good terms with Chaps. 14 and 15.

Let $\mathfrak{g}$ be a complex semisimple Lie algebra. We consider $\mathcal{O}_0$, the principal block of category $\mathcal{O}$. As in Sect. 15.2.5 we use the dot action to relabel our simple, standard, costandard, indecomposable projective, and indecomposable injective objects in $\mathcal{O}_0$ by the Weyl group W. That is, for all $x \in W$ we set

$$L_x := L(x \cdot 0), \quad \Delta_x := \Delta(x \cdot 0), \quad \nabla_x := \nabla(x \cdot 0), \quad P_x := P(x \cdot 0), \quad I_x := I(x \cdot 0).$$

The Grothendieck group $[\mathcal{O}_0]$ carries a bilinear pairing $\langle -, - \rangle$, called the *Euler form*, defined by

$$\langle [M], [N] \rangle = \sum_i (-1)^i \dim \operatorname{Ext}^i(M, N). \quad (25.24)$$

As $\mathcal{O}$ has finite homological dimension, this pairing is well-defined. Recall from Theorem 14.23 (see also e.g. [86, Theorems 3.3(c) and 6.12]) that we have ("fundamental vanishing")

$$\dim \operatorname{Ext}^i(\Delta_x, \nabla_y) = \begin{cases} 1 & \text{if } i = 0 \text{ and } x = y, \\ 0 & \text{otherwise.} \end{cases} \quad (25.25)$$

---

[2] All we know for a fact is that the Kazhdan–Lusztig inversion formula led Beilinson and Ginzburg to their conjecture. How exactly this occurred is not known to the authors. Thus there is some guessing involved below.

This implies that $\{[\Delta_x]\}$ and $\{[\nabla_y]\}$ are dual bases of $[\mathcal{O}_0]$. The properties defining projective covers ensure that $\{[P_x]\}$ and $\{[L_y]\}$ are dual bases, as are $\{[L_x]\}$ and $\{[I_y]\}$.

We begin with the Kazhdan–Lusztig conjecture (see Conjecture 15.38 and (15.26))

$$[P_x] = \sum_{z \in W} h_{z,x}(1)[\Delta_z]. \tag{25.26}$$

Let us fix x and pair both sides of (25.26) with $[\nabla_z]$ using the Euler form. Using (25.25) and the fact that P_x is the projective cover of L_x, we deduce

$$[\nabla_z : L_x] = h_{z,x}(1),$$

where the left hand sides denotes the multiplicity of L_z in ∇_x. (Alternatively, this can be seen as an instance of BGG reciprocity, see Theorem 14.27.) Since Δ_z and ∇_z have the same character, we deduce that $[\Delta_z : L_x] = h_{z,x}(1)$. In other words,

$$[\Delta_z] = \sum_{x} h_{z,x}(1)[L_x] \quad \text{in } [\mathcal{O}_0]. \tag{25.27}$$

We would like to invert this formula to find an expression for simple modules in terms of Verma modules. To do so we need to invert the matrix of Kazhdan–Lusztig polynomials appearing on the right hand side of (25.27). This is achieved by the following remarkable observation, whose proof is purely combinatorial (but will not be given here):

Theorem 25.29 (Kazhdan–Lusztig Inversion Formula [108]) *For all $x, y \in W$, we have*

$$\sum_{z \in W} (-1)^{\ell(z)+\ell(y)} h_{z,x} h_{zw_0, yw_0} = \begin{cases} 1 & \text{if } x = y, \\ 0 & \text{otherwise.} \end{cases} \tag{25.28}$$

We now proceed to invert (25.27). For fixed $y \in W$ we have

$$\sum_{z \in W} (-1)^{\ell(z)+\ell(y)} h_{zw_0, yw_0}(1)[\Delta_z] = \sum_{z,x \in W} (-1)^{\ell(z)+\ell(y)} h_{zw_0, yw_0}(1) h_{z,x}(1)[L_x] = [L_y].$$

Using the substitutions $y \mapsto yw_0$ and $x \mapsto xw_0$ this equality becomes

$$[L_{yw_0}] = \sum_{z \in W} (-1)^{\ell(z)+\ell(y)} h_{z,y}(1)[\Delta_{zw_0}]. \tag{25.29}$$

Remark 25.30 The above considerations show that in fact (25.29) is equivalent to the Kazhdan–Lusztig conjecture. It was in the form (25.29) that the Kazhdan–Lusztig conjecture was first made (see [108, Conjecture 1.5]).

Exercise 25.31 Prove (25.29) for $y = \text{id}$ and $y = w_0$. (*Hint:* One case should be very easy; for the other the Weyl character formula or BGG resolution (see Theorem 13.2) is helpful.)

25.6 Evidence of Koszul Duality in Category $\mathcal{O}$

A basic observation is that (up to signs and a w_0) the original form of the Kazhdan–Lusztig conjecture (25.26) and the equivalent form (25.29) look remarkably similar, although they involve the classes of very different modules. This simple observation turns out to have a deep explanation via Koszul duality. In the following we sketch why one might hope to explain this similarity via Koszul duality; in the next chapter we will roll up our sleeves and actually explain the form that Koszul duality takes in our context.

The example of Koszul duality for polynomial rings (see Sect. 25.4) was meant to highlight the following features:

1. One often needs to introduce a grading for Koszul duality to become visible (see e.g. Exercise 25.25).
2. As an equivalence of triangulated categories, Koszul duality commutes with the triangulated shift functor [1]. However, it *does not* commute with the shift of grading functor (1). Rather, one has the basic relation (see Sect. 25.4)

$$(i) \mapsto (i)[-i]. \tag{25.30}$$

3. In analogy with the Fourier transform, Koszul duality exchanges "delta functions" (e.g. simple modules) and "fundamental frequencies" (e.g. projective modules), up to a shift.

Let us first discuss the issue of introducing the grading. It is natural to expect that the principal block $\mathcal{O}_0$ admits a graded version $\widetilde{\mathcal{O}}_0$, together with a "forgetting the grading functor"

$$\textsf{For} : \widetilde{\mathcal{O}}_0 \to \mathcal{O}_0.$$

(We defer until Sect. 26.2 in the next chapter exactly what we mean by "admits a graded version." A good basic model to have in mind is the following: Suppose our category is equivalent to modules over a finite-dimensional algebra A. Then a graded version of our category would be given by considering *graded* modules over our algebra, for some grading on A; in which case **For** is the functor of forgetting the grading.)

We do not expect all modules in $\widetilde{\mathcal{O}}_0$ lift to $\widetilde{\mathcal{O}}_0$ (just as not all ungraded modules over a graded ring are gradable). However we do expect lifts of all the objects considered in the previous section. We denote these objects by a tilde: i.e. L_x lifts to $\widetilde{L}_x$, Δ_x lifts to $\widetilde{\Delta}_x$, etc.

Aside from Koszul duality, another reason to expect the existence of such a grading is in order to "explain" the v in Kazhdan–Lusztig polynomials. That is, if we write the coefficients of Kazhdan–Lusztig polynomials as

$$h_{z,x} = \sum h_{z,x}^i v^i$$

we might hope that the Kazhdan–Lusztig conjecture (25.26) is refined to an equality

$$[\widetilde{P}_x] = \sum_{z \in W, i \in \mathbb{Z}} h_{z,x}^i [\widetilde{\Delta}_z(i)]. \qquad (25.31)$$

In other words, the v appearing in Kazhdan–Lusztig polynomials measures *graded* multiplicities. If this is the case then the same argument as above (this time taking account of the grading throughout) would give us the following graded analogue of (25.32):

$$[\widetilde{L}_{yw_0}] = \sum (-1)^{\ell(z)+\ell(y)} h_{z,y}^i [\widetilde{\Delta}_{zw_0}(i)]. \qquad (25.32)$$

Now let us dream for a moment that there is a Koszul duality equivalence

$$\kappa : D^{\mathrm{b}}(\widetilde{\mathcal{O}}_0) \xrightarrow{\sim} D^{\mathrm{b}}(\widetilde{\mathcal{O}}_0)$$

sending

$$\widetilde{\Delta}_z \mapsto \widetilde{\Delta}_{zw_0}. \qquad (25.33)$$

Via the crucial property (25.30), we deduce

$$\kappa : \widetilde{\Delta}_z(i) \mapsto \widetilde{\Delta}_{zw_0}(i)[-i]$$

which gives

$$\kappa : [\widetilde{\Delta}_z(i)] \mapsto (-1)^i [\widetilde{\Delta}_{zw_0}(i)] \qquad (25.34)$$

on Grothendieck groups. Using (25.31) we deduce that on the classes of indecomposable projectives κ acts via

$$[\widetilde{P}_x] \mapsto \sum (-1)^i h_{z,x}^i [\widetilde{\Delta}_{zw_0}(i)] = \sum (-1)^{\ell(z)+\ell(x)} h_{z,x}^i [\widetilde{\Delta}_{zw_0}(i)] = [\widetilde{L}_{xw_0}].$$

25.6 Evidence of Koszul Duality in Category $\mathcal{O}$

(For the first equality we have used that $h^i_{z,x} \neq 0$ only if i is of the same parity as $\ell(z)+\ell(x)$. For the second we have used (25.32).) Similarly, we deduce that on the classes of simples we get

$$[\widetilde{L}_{yw_0}] \mapsto \sum (-1)^{\ell(z)+\ell(y)}(-1)^i h^i_{z,y}[\widetilde{\Delta}_z(i)] = \sum h^i_{z,y}[\widetilde{\Delta}_z(i)] = [\widetilde{P}_y].$$

Thus the simple prescription (25.33) exchanges the classes of simple and indecomposable projective modules! In the following chapter we will develop these ideas in more detail, where (a variant of) κ will be seen to exist.

Remark 25.32 One important point to keep in mind is that Koszul duality will turn out less symmetric than the above discussion appears to suggest. For example, it will send graded lifts of Verma modules to *dual* Verma modules. The reason that this is compatible with the above is that dual modules (see Sect. 14.3) have the same class in the Grothendieck group. Thus distinctions between e.g. Verma modules and their duals, or between indecomposable projective and injective modules, are only visible once we work on the level of categories.

Chapter 26
Koszul Duality II

This chapter is based on expanded notes of a lecture given by the authors and taken by
Shotaro Makisumi

Abstract We formulate a Koszul duality for the principal block of the BGG category $\mathcal{O}$ in the language of Soergel bimodules, and give some indication of the proof.

26.1 Introduction

As discussed in Chap. 25, Beilinson and Ginzburg realized that there should be a self-duality of the principal block $\mathcal{O}_0$ of category $\mathcal{O}$ sending simple objects to projective objects. To explain the Kazhdan–Lusztig inversion formula, this duality should involve two gradings: first, being a derived equivalence, it involves the cohomological grading; but it should also involve a new "mixed" grading, reflecting the fact that the Kazhdan–Lusztig polynomial is a polynomial rather than a number.

In this chapter, we introduce a Soergel-theoretic graded category $\mathcal{O}_0$, which gives a way to work with graded $\mathcal{O}_0$ using complexes of Soergel modules. We will refer to this category as *Soergel* $\mathcal{O}_0$ (read aloud as "Soergel category O"). We formulate a precise statement of Koszul duality for Soergel $\mathcal{O}_0$ in Theorem 26.26 and discuss the main ingredients of the proof, as in [133].

The relation to geometry is outside the scope of this book. In particular, we will discuss neither the work of Bezrukavnikov and Yun [26] nor of Achar and Riche [2], which served as important motivation in both the formulation and the proof of Theorem 26.26.

S. Makisumi
Department of Mathematics, Columbia University, New York, NY, USA

© The Editor(s) (if applicable) and The Author(s), under exclusive licence to Springer Nature Switzerland AG 2020
B. Elias et al., *Introduction to Soergel Bimodules*, RSME Springer Series 5, https://doi.org/10.1007/978-3-030-48826-0_26

26.2 Graded Category $\mathcal{O}_0$

26.2.1 Desiderata

We begin by explaining what properties we want from graded category $\mathcal{O}_0$.

Let $\mathfrak{g}$ be a semisimple complex Lie algebra with Borel and Cartan subalgebras $\mathfrak{b} \supset \mathfrak{h}$, and consider the principal block $\mathcal{O}_0$ of its category $\mathcal{O}$. We retain the notation of Chap. 14, with the following exception:

> **Assumption 26.1** Throughout this chapter, we index structural objects (L_w, P_w, I_w, Δ_w, ∇_w) in $\mathcal{O}_0$ using the antidominant weight, $w_0 \cdot 0 = -2\rho$. For example, $L_w = L(w \cdot (-2\rho)) = L(ww_0 \cdot 0)$.

This differs from the convention in Chap. 14 by w_0.

Since $\mathcal{O}_0$ is a finite length abelian category with enough projectives, Morita theory tells us that there is an equivalence

$$\mathrm{Hom}(P, -) : \mathcal{O}_0 \xrightarrow{\sim} A\text{-mod},$$

where P is a projective generator of $\mathcal{O}_0$, $A := \mathrm{End}_{\mathcal{O}_0}(P)^{\mathrm{opp}}$, and A-mod denotes the category of (ungraded) finitely-generated left modules over A. For example, one might let $P := \bigoplus_{w \in W} P_w$. Now suppose there was a way to equip A with a grading $A = \bigoplus_{n \in \mathbb{Z}} A_n$. Then we can consider the category A-gmod of finitely-generated graded modules and graded (degree zero) maps, which has a grading shift endofunctor $\langle 1 \rangle$ and a forget-the-grading functor to A-mod.

The category A-gmod models what we want from graded category $\mathcal{O}_0$.

Desiderata 26.2 *Graded category* $\mathcal{O}_0$, denoted by $\mathcal{O}_0^{\mathrm{mix}}$, should have the following properties:

- it is a $\mathbb{C}$-linear finite length abelian category equipped with an exact endofunctor $\langle 1 \rangle$ called *Tate twist*;
- there exists an exact "forgetful" functor $\mathsf{For} : \mathcal{O}_0^{\mathrm{mix}} \to \mathcal{O}_0$ and a natural isomorphism $\nu : \mathsf{For} \circ \langle 1 \rangle \simeq \mathsf{For}$;
- For is a *degrading functor*: for any $M, N \in \mathcal{O}_0^{\mathrm{mix}}$ and $i \in \mathbb{Z}_{\geq 0}$, the natural map

$$\bigoplus_{n \in \mathbb{Z}} \mathrm{Ext}^i_{\mathcal{O}_0^{\mathrm{mix}}}(M, N\langle n \rangle) \to \mathrm{Ext}^i_{\mathcal{O}_0}(\mathsf{For}(M), \mathsf{For}(N)) \tag{26.1}$$

induced by For and ν is an isomorphism.

26.2 Graded Category $\mathcal{O}_0$

In addition, structural objects in $\mathcal{O}_0$ admit "graded lifts" to $\mathcal{O}_0^{\text{mix}}$:

- for every $X \in \{L, P, I, \Delta, \nabla\}$ and $w \in W$, there exists an object $X_w^{\text{mix}} \in \mathcal{O}_0^{\text{mix}}$ satisfying $\mathsf{For}(X_w^{\text{mix}}) \simeq X_w$.
- for every $w \in W$, there exist morphisms

$$P_w^{\text{mix}} \to \Delta_w^{\text{mix}} \to L_w^{\text{mix}} \to \nabla_w^{\text{mix}} \to I_w^{\text{mix}} \qquad (26.2)$$

in $\mathcal{O}_0^{\text{mix}}$ making P_w^{mix} (resp. I_w^{mix}) a projective cover (resp. injective hull) of L_w^{mix}.

Let us clarify the analogy with graded modules. For $M, N \in \mathcal{O}_0^{\text{mix}}$, a morphism $M \to N\langle n \rangle$ in $\mathcal{O}_0^{\text{mix}}$ should be thought of as a "graded map $M \to N$ of degree n," so that the "graded Hom space" $\bigoplus_{n \in \mathbb{Z}} \mathrm{Hom}_{\mathcal{O}_0^{\text{mix}}}(M, N\langle n \rangle)$ consists of "graded maps $M \to N$ of all degrees." The isomorphism (26.1) for $i = 0$ says that this gives a grading on the "ungraded Hom space" $\mathrm{Hom}_{\mathcal{O}_0}(\mathsf{For}(M), \mathsf{For}(N))$, which consists of all "(ungraded) maps $M \to N$."

Graded lifts are not unique. For example, if some object $L_w^{\text{mix}} \in \mathcal{O}_0^{\text{mix}}$ satisfies $\mathsf{For}(L_w^{\text{mix}}) \simeq L_w$, then $\mathsf{For}(L_w^{\text{mix}}\langle n \rangle) \simeq L_w$ for any $n \in \mathbb{Z}$. However, once we have chosen an object L_w^{mix} (i.e. once we have "fixed a choice of grading shift"), this pins down the choice of grading shift on $P_w^{\text{mix}}, \Delta_w^{\text{mix}}$, etc. This is because each morphism in (26.2) lives in a one-dimensional graded Hom space by (26.1), so there is a unique grading shift where the nonzero morphism has "degree zero."

Of course, the trivial grading of A, placing everything in degree zero, yields an uninteresting graded category A-gmod containing no more information than A-mod. The grading on $\mathcal{O}_0^{\text{mix}}$ will be a special one that reveals the duality conjectured by Beilinson and Ginzburg.

Theorem 26.3 (Koszul Duality for $\mathcal{O}_0$, Beilinson–Ginzburg–Soergel [20])
There exists a triangulated equivalence

$$\kappa^{\text{BGS}} : D^b \mathcal{O}_0^{\text{mix}} \xrightarrow{\sim} D^b \mathcal{O}_0^{\text{mix}}$$

satisfying $\kappa^{\text{BGS}} \circ \langle 1 \rangle \simeq [1]\langle -1 \rangle \circ \kappa^{\text{BGS}}$ *and*

$$\kappa^{\text{BGS}}(\Delta_w^{\text{mix}}) \simeq \nabla_{w^{-1}w_0}^{\text{mix}}, \qquad \kappa^{\text{BGS}}(P_w^{\text{mix}}) \simeq L_{w^{-1}w_0}^{\text{mix}}$$

for all $w \in W$.

Being a triangulated functor, κ^{BGS} intertwines the cohomological shifts $[1]$ (up to natural isomorphism). However, since κ^{BGS} does not intertwine Tate twists, one cannot complete the following diagram to a commutative square:

$$\begin{array}{ccc} D^b \mathcal{O}_0^{\text{mix}} & \xrightarrow{\kappa^{\text{BGS}}} & D^b \mathcal{O}_0^{\text{mix}} \\ \downarrow {\mathsf{For}} & & \downarrow {\mathsf{For}} \\ D^b \mathcal{O}_0 & & D^b \mathcal{O}_0. \end{array}$$

Thus Koszul duality is a hidden self-duality of $\mathcal{O}_0$, one that was not visible without the graded lift $\mathcal{O}_0^{\mathrm{mix}}$.

The goal of this chapter is to formulate an analogous result in the setting of Soergel $\mathcal{O}_0$.

Exercise 26.4 A fact closely related to the Kazhdan–Lusztig conjecture is that $\mathcal{O}_0^{\mathrm{mix}}$ is *Koszul* (see also Exercise 26.11): for all $x, y \in W$, $i \in \mathbb{Z}_{\geq 0}$, and $n \in \mathbb{Z}$,

$$\mathrm{Ext}^i_{\mathcal{O}_0^{\mathrm{mix}}}(L_x^{\mathrm{mix}}, L_y^{\mathrm{mix}}\langle n \rangle) = 0 \quad \text{unless } i = -n. \tag{26.3}$$

Show that (26.3) together with Theorem 26.3 implies an isomorphism of algebras

$$\mathrm{End}_{\mathcal{O}_0}(P) \simeq \mathrm{Ext}^{\bullet}_{\mathcal{O}_0}(L, L),$$

where $P := \bigoplus_{w \in W} P_w$ and $L := \bigoplus_{w \in W} L_w$. (This is an isomorphism of ungraded algebras that comes from an isomorphism of graded algebras in $\mathcal{O}_0^{\mathrm{mix}}$.)

26.2.2 Motivation from Soergel's $\mathbb{V}$ Functor

Rather than directly define $\mathcal{O}_0^{\mathrm{mix}}$, we will first define a triangulated category that should be viewed as its bounded derived category, then define $\mathcal{O}_0^{\mathrm{mix}}$ as the heart of a certain t-structure.

To motivate the definitions, recall Soergel's $\mathbb{V}$ functor (see Sect. 15.4):

$$\mathbb{V} = \mathrm{Hom}_{\mathcal{O}_0}(P_{\mathrm{id}}, -) : \mathcal{O}_0 \to \mathrm{End}(P_{\mathrm{id}})^{\mathrm{opp}}\text{-mod}.$$

Let W be the Weyl group of $\mathfrak{g}$ (with respect to our fixed Cartan $\mathfrak{h}$), acting naturally on $\mathfrak{h}$. Soergel showed that there is a canonical isomorphism of $\mathrm{End}(P_{\mathrm{id}})$ with the coinvariant algebra $C = \mathrm{Sym}(\mathfrak{h})/(\mathrm{Sym}(\mathfrak{h})_+^W)$, and that via this isomorphism,

$$\mathbb{V}|_{\mathrm{Proj}\,\mathcal{O}_0} : \mathrm{Proj}\,\mathcal{O}_0 \xrightarrow{\sim} (\text{ungr. Soergel mod.}). \tag{26.4}$$

Here, the right hand side denotes the additive category of Soergel modules associated to the realization $\mathfrak{h}^*$ of W, viewed as ungraded C-modules.

Since $\mathcal{O}_0$ has finite homological dimension (see e.g. [86, Proposition 6.9]), the natural functor $K^b \mathrm{Proj}\,\mathcal{O}_0 \to D^b \mathcal{O}_0$ is an equivalence. Composing with $K^b \mathbb{V}|_{\mathrm{Proj}\,\mathcal{O}_0}$, we obtain an equivalence

$$D^b \mathcal{O}_0 \simeq K^b(\text{ungr. Soergel mod.}).$$

26.2 Graded Category $\mathcal{O}_0$

This suggests that the bounded homotopy category of (the usual graded) Soergel modules should play the role of $D^b \mathcal{O}_0^{\text{mix}}$. Moreover, this definition makes sense for an arbitrary Coxeter system.

26.2.3 Definition of Soergel Category $\mathcal{O}_0$

For various reasons (some purely expository, some to ensure that certain of the statements in this chapter are actually true!), we assume the following:[1]

> **Assumption 26.5** In this chapter, let (W, S) be a finite Coxeter system, and let $\mathfrak{h}$ be the geometric realization over $\mathbb{C}$.

As usual, let $R = \text{Sym}(\mathfrak{h}^*) = \mathbb{C}[\alpha_s : s \in S]$, graded with $\deg \mathfrak{h}^* = 2$, and let $\mathbb{S}\text{Bim}$ be the category of Soergel bimodules associated to the realization $(W, \mathfrak{h})$ (see Sect. 5.7). Rather than consider actual Soergel modules, it will be convenient to introduce a category $\overline{\mathbb{S}\text{Bim}}$ obtained from $\mathbb{S}\text{Bim}$ as follows.

Definition 26.6 Let $\overline{\mathbb{S}\text{Bim}}$ be the category having the same objects as $\mathbb{S}\text{Bim}$ and graded Hom spaces

$$\text{Hom}^\bullet_{\overline{\mathbb{S}\text{Bim}}}(B, B') := \mathbb{C} \otimes_R \text{Hom}^\bullet_{\mathbb{S}\text{Bim}}(B, B').$$

Thus we have a natural quotient functor

$$q : \mathbb{S}\text{Bim} \to \overline{\mathbb{S}\text{Bim}}, \tag{26.5}$$

whose induced maps on graded Homs are surjective with kernel spanned by morphisms of the form $\lambda \cdot f$ for $\lambda \in R_+$. Here, R_+ is the ideal of R consisting of polynomials with no constant term.

The grading shift (1) on $\mathbb{S}\text{Bim}$ induces a grading shift on $\overline{\mathbb{S}\text{Bim}}$, still denoted by (1). Note that $\overline{\mathbb{S}\text{Bim}}$ is graded $\mathbb{C}$-linear and right R-linear, but no longer left R-linear.

Example 26.7 For $s \in S$, we have

$$\text{Hom}^\bullet_{\mathbb{S}\text{Bim}}(B_s, B_s) = R \,\Big|\, \oplus R \,\Big|\, \simeq R \oplus R(-2)$$

---

[1] See, however, Sect. 26.5 for a brief discussion of Soergel $\mathcal{O}_0$ for more general realizations.

as a graded left R-module, so

$$\operatorname{Hom}^{\bullet}_{\overline{\mathbb{S}\mathrm{Bim}}}(B_s, B_s) = \mathbb{C}\,\big|\, \oplus \mathbb{C}\, {\textstyle\frac{\bullet}{\bullet}} \simeq \mathbb{C} \oplus \mathbb{C}(-2),$$

as a graded left $\mathbb{C}$-vector space. Here, we have written $\big|$ for $1 \otimes \big|$ and ${\textstyle\frac{\bullet}{\bullet}}$ for $1 \otimes {\textstyle\frac{\bullet}{\bullet}}$.

Remark 26.8 The functor $\mathbb{C} \otimes_R (-)$ from $\mathbb{S}\mathrm{Bim}$ to the category of (actual) Soergel modules factors through $\overline{\mathbb{S}\mathrm{Bim}}$ (by the definition of $\overline{\mathbb{S}\mathrm{Bim}}$). The induced functor from $\overline{\mathbb{S}\mathrm{Bim}}$ to Soergel modules is an equivalence by a result of Soergel (see Proposition 15.27). For this reason, we will use the term "Soergel modules" to refer to objects in $\overline{\mathbb{S}\mathrm{Bim}}$.

Recall from Theorem 19.64 that Soergel's conjecture implies the existence of a "perverse" t-structure on $K^b \mathbb{S}\mathrm{Bim}$, where the perverse complexes (complexes lying in the heart) are those homotopic to "diagonal" complexes. This is also true for $K^b \overline{\mathbb{S}\mathrm{Bim}}$.

Theorem 26.9 *If Soergel's conjecture holds for $\mathbb{S}\mathrm{Bim}$ associated to the realization $(W, \mathfrak{h})$, then the full subcategories*

$$K^{\leq 0} := \langle B_x[-i](m) \mid m \geq i \rangle \quad \text{and} \quad K^{\geq 0} := \langle B_x[-i](m) \mid m \leq i \rangle$$

define a t-structure on $K^b \overline{\mathbb{S}\mathrm{Bim}}$, called the perverse t-structure.

Definition 26.10 Soergel-theoretic graded category $\mathcal{O}_0$ (or simply *Soergel category* $\mathcal{O}_0$ or *Soergel* $\mathcal{O}_0$) is the heart of the perverse t-structure on $K^b \overline{\mathbb{S}\mathrm{Bim}}$:

$$\mathcal{O}_0^{\mathrm{mix}} = \mathcal{O}_0^{\mathrm{mix}}(W, \mathfrak{h}) := K^{\leq 0} \cap K^{\geq 0}.$$

Recall that $K^b \overline{\mathbb{S}\mathrm{Bim}}$ is a triangulated category whose distinguished triangles are triangles isomorphic to those of the form $A \xrightarrow{f} B \to \mathrm{Cone}(f) \to A[1]$ (see Chap. 19 for a brief review). Short exact sequences in the $\mathbb{C}$-linear abelian category $\mathcal{O}_0^{\mathrm{mix}}$ are those sequences $A \to B \to C$ with A, B, C lying in $\mathcal{O}_0^{\mathrm{mix}}$ that can be completed (in a necessarily unique way) to a distinguished triangle in $K^b \overline{\mathbb{S}\mathrm{Bim}}$.

Define a new shift on $K^b \overline{\mathbb{S}\mathrm{Bim}}$, called the *Tate twist*, by

$$\langle 1 \rangle := [1](-1) : K^b \overline{\mathbb{S}\mathrm{Bim}} \to K^b \overline{\mathbb{S}\mathrm{Bim}}.$$

Since $\langle 1 \rangle$ preserves the class of diagonal complexes, it sends perverse complexes to perverse complexes, hence restricts to an exact endofunctor of $\mathcal{O}_0^{\mathrm{mix}}$.

For $w \in W$, the indecomposable Soergel bimodule B_w can be viewed as a complex supported in cohomological degree 0. This gives an object of $K^b \overline{\mathbb{S}\mathrm{Bim}}$ living in the heart of the perverse t-structure, which we continue to denote by B_w. One can show that these objects B_w are exactly the simple objects in $\mathcal{O}_0^{\mathrm{mix}}$ up to Tate twist and isomorphism. (In fact, by a result of [65], this fact is equivalent to

26.2 Graded Category $\mathcal{O}_0$

Soergel's conjecture; see [132, §5.6].) We also write L_w^{mix} for B_w:

$$\{ \text{simples in } \mathcal{O}_0^{\text{mix}} \}/\langle 1 \rangle, \simeq \xleftrightarrow{1:1} W$$

$$L_w^{\text{mix}} = B_w \longleftrightarrow w.$$

Thus this definition of Soergel $\mathcal{O}_0$ is "already Koszul dual" to Soergel's description of the Lie-theoretic $\mathcal{O}_0$ using $\mathbb{V}$, in which (ungraded) Soergel modules corresponded to projectives.

In the case that the realization $(W, \mathfrak{h})$ arises from a semisimple complex Lie algebra $\mathfrak{g}$, we wish to construct a forgetful functor

$$\text{For} : \mathcal{O}_0^{\text{mix}}(W, \mathfrak{h}) \to \mathcal{O}_0(\mathfrak{g})$$

as in Desiderata 26.2. From Definition 26.10, however, it is not at all obvious how to define such a functor. Indeed, one uses Koszul duality (to be formulated more precisely in Sect. 26.4 below) to define For and verify that $\mathcal{O}_0^{\text{mix}}$ satisfies the desiderata!

Nevertheless, let us assume the existence of For and ν as in Desiderata 26.2, and see for $\mathfrak{g} = \mathfrak{sl}_2(\mathbb{C})$ how Soergel $\mathcal{O}_0$ provides a nontrivial grading on $\mathcal{O}_0$. A more systematic overview of basic homological properties of $\mathcal{O}_0^{\text{mix}}$ follows in Sect. 26.3.

Exercise 26.11 Show that Soergel $\mathcal{O}_0$ is Koszul (see Exercise 26.4).

26.2.4 Example: Soergel $\mathcal{O}_0$ in Type A_1

Let $W = S_2 = \langle s \rangle$, $\mathfrak{h} = \mathbb{C}\alpha_s^\vee$, and define $\mathcal{O}_0^{\text{mix}}$ as in Definition 26.10. Then $\mathcal{O}_0^{\text{mix}}$ has two simple objects $L_{\text{id}}^{\text{mix}} = R$ and $L_s^{\text{mix}} = B_s$ up to Tate twist and isomorphism. Let $\mathcal{O}_0 = \mathcal{O}_0(\mathfrak{sl}_2(\mathbb{C}))$, and assume that $\mathcal{O}_0^{\text{mix}}$ satisfies Desiderata 26.2. In particular, we have $\text{For}(L_{\text{id}}^{\text{mix}}) \simeq L_{\text{id}}$ and $\text{For}(L_s^{\text{mix}}) \simeq L_s$.

In $\mathcal{O}_0$, there is a nonsplit short exact sequence

$$0 \to L_{\text{id}} \to \Delta_s \to L_s \to 0.$$

In fact, we have $\dim \text{Ext}^1_{\mathcal{O}_0}(L_s, L_{\text{id}}) = 1$. By (26.1), we deduce that there exists some $n_0 \in \mathbb{Z}$ with

$$\dim \text{Ext}^1_{\mathcal{O}_0^{\text{mix}}}(L_s^{\text{mix}}, L_{\text{id}}^{\text{mix}}\langle n \rangle) = \begin{cases} 1 & \text{if } n = n_0; \\ 0 & \text{if } n \neq n_0. \end{cases}$$

We seek a lift $\Delta_s^{\text{mix}} \in \mathcal{O}_0^{\text{mix}}$ of Δ_s fitting into a nonsplit short exact sequence

$$0 \to L_{\text{id}}^{\text{mix}}\langle n_0 \rangle \to \Delta_s^{\text{mix}} \to L_s^{\text{mix}} \to 0 \tag{26.6}$$

in $\mathcal{O}_0^{\mathrm{mix}}$. From the description of short exact sequences in $\mathcal{O}_0^{\mathrm{mix}}$ (see the paragraph following Definition 26.10), one sees that there is only one such sequence up to isomorphism:

$$\begin{array}{ccccccccc}
0 & \longrightarrow & R(1) & \xrightarrow{\mathrm{id}} & R(1) & \longrightarrow & 0 & \longrightarrow & 0 \\
\uparrow & & \uparrow & & \uparrow\!\!\bullet & & \uparrow & & \uparrow \\
\bullet\, 0 & \longrightarrow & 0 & \longrightarrow & B_s & \xrightarrow{\mathrm{id}} & B_s & \longrightarrow & 0.
\end{array}$$

Here, each column denotes a complex in $K^b\overline{\mathbb{S}\mathrm{Bim}}$ supported in cohomological degrees $[0, 1]$; the bottom left bullet marks the bottom row as degree 0. The horizontal arrows are (components of) maps of complexes representing morphisms in $K^b\overline{\mathbb{S}\mathrm{Bim}}$. The corresponding distinguished triangle in $K^b\overline{\mathbb{S}\mathrm{Bim}}$ is a shift and a rotation of

$$B_s \xrightarrow{\bullet} R(1) \to \mathrm{Cone}(\underset{\bullet}{\bullet}) \to .$$

Since $R(1)[-1] = R\langle -1\rangle$, we have $n_0 = -1$ in (26.6), and

$$\Delta_s^{\mathrm{mix}} \simeq \left(\begin{array}{c} R(1) \\ \bullet\uparrow \\ \bullet\, B_s \end{array} \right).$$

This value of n_0 is consistent with the Koszulity of $\mathcal{O}_0^{\mathrm{mix}}$ (see Exercises 26.4 and 26.11).

Similarly, the short exact sequence $0 \to L_s^{\mathrm{mix}} \to \nabla_s^{\mathrm{mix}} \to L_{\mathrm{id}}^{\mathrm{mix}} \to 0$ in $\mathcal{O}_0$ lifts to $0 \to L_s^{\mathrm{mix}} \to \nabla_s^{\mathrm{mix}} \to L_{\mathrm{id}}^{\mathrm{mix}}\langle 1\rangle \to 0$ given by

$$\begin{array}{ccccccccc}
\bullet\, 0 & \longrightarrow & B_s & \xrightarrow{\mathrm{id}} & B_s & \longrightarrow & 0 & \longrightarrow & 0 \\
\uparrow & & \uparrow & & \downarrow\!\!\bullet & & \uparrow & & \uparrow \\
0 & \longrightarrow & 0 & \longrightarrow & R(-1) & \xrightarrow{\mathrm{id}} & R(-1) & \longrightarrow & 0,
\end{array}$$

where now the top row sits in cohomological degree 0, and

$$\nabla_s^{\mathrm{mix}} \simeq \left(\begin{array}{c} \bullet\, B_s \\ \downarrow\!\!\bullet \\ R(-1) \end{array} \right).$$

26.2 Graded Category $\mathcal{O}_0$

Remark 26.12 Observe that Δ_s^{mix} (resp. ∇_s^{mix}) is the image of the Rouquier complex F_s (resp. F_s^{-1}) (see Sect. 19.3) under the natural functor $K^b\mathbb{S}\text{Bim} \to K^b\overline{\mathbb{S}\text{Bim}}$. We will return to this point in Sect. 26.3.

This accounts for four of the five indecomposable objects in $\mathcal{O}_0$: $L_{\text{id}} (\simeq \Delta_{\text{id}} \simeq \nabla_{\text{id}} \simeq I_{\text{id}})$, L_s, $\Delta_s (\simeq P_s)$, and $\nabla_s (\simeq I_s)$. The last indecomposable $P_{\text{id}} \simeq I_{\text{id}}$ is characterized by either of the non-split short exact sequences

$$0 \to \Delta_s \to P_{\text{id}} \to L_{\text{id}} \to 0, \qquad 0 \to L_{\text{id}} \to I_{\text{id}} \to \nabla_s \to 0,$$

lifting to

$$0 \to \Delta_s^{\text{mix}}\langle ?\rangle \to P_{\text{id}}^{\text{mix}} \to L_{\text{id}}^{\text{mix}} \to 0, \qquad 0 \to L_{\text{id}}^{\text{mix}} \to I_{\text{id}}^{\text{mix}} \to \nabla_s^{\text{mix}}\langle ?\rangle \to 0.$$

Having already determined $\Delta_s^{\text{mix}}, \nabla_s^{\text{mix}} \in \mathcal{O}_0^{\text{mix}}$, similar considerations as before determine these sequences up to isomorphism:

$$\begin{array}{ccccccccc}
0 & \to & R(2) & \xrightarrow{\text{id}} & R(2) & \to & 0 & \to & 0 \\
\uparrow & & \uparrow & & \uparrow & & \uparrow & & \uparrow \\
0 & \to & B_s(1) & \xrightarrow{\text{id}} & B_s(1) & \to & 0 & \to & 0 \\
\uparrow & & \uparrow & & \downarrow\uparrow & & \uparrow & & \uparrow \\
\bullet\, 0 & \to & 0 & \to & R & \xrightarrow{\text{id}} & R & \to & 0
\end{array}$$

and

$$\begin{array}{ccccccccc}
\bullet\, 0 & \to & R & \xrightarrow{\text{id}} & R & \to & 0 & \to & 0 \\
\uparrow & & \uparrow & & \uparrow & & \uparrow & & \uparrow \\
0 & \to & 0 & \to & B_s(-1) & \xrightarrow{\text{id}} & B_s(-1) & \to & 0 \\
\uparrow & & \uparrow & & \downarrow\uparrow & & \uparrow & & \uparrow \\
0 & \to & 0 & \to & R(-2) & \xrightarrow{\text{id}} & R(-2) & \to & 0.
\end{array}$$

Thus

$$P_{\text{id}}^{\text{mix}} \simeq \begin{pmatrix} R(2) \\ \uparrow \\ B_s(1) \\ \downarrow \\ \bullet\, R \end{pmatrix}, \qquad I_{\text{id}}^{\text{mix}} \simeq \begin{pmatrix} \bullet\, R \\ \uparrow \\ B_s(-1) \\ \downarrow \\ R(-2) \end{pmatrix},$$

and $P_{\mathrm{id}}^{\mathrm{mix}}\langle 2\rangle \simeq I_{\mathrm{id}}^{\mathrm{mix}}$. Note that $\downarrow \circ \uparrow = \alpha_s \cdot \mathrm{id}_R$, which is 0 in $\overline{\mathbb{S}\mathrm{Bim}}$ but not in $\mathbb{S}\mathrm{Bim}$. The complexes $P_{\mathrm{id}}^{\mathrm{mix}}$ and $I_{\mathrm{id}}^{\mathrm{mix}}$ do not lie in the essential image of the natural functor $K^b\mathbb{S}\mathrm{Bim} \to K^b\overline{\mathbb{S}\mathrm{Bim}}$.

The following perverse complex plays an important role in Koszul duality:

$$T_s^{\mathrm{mix}} = \begin{pmatrix} R(1) \\ \uparrow \\ \bullet\ B_s \\ \uparrow \\ R(-1) \end{pmatrix}. \qquad (26.7)$$

Thus $T_s^{\mathrm{mix}} \simeq P_{\mathrm{id}}^{\mathrm{mix}}\langle 1\rangle \simeq I_{\mathrm{id}}^{\mathrm{mix}}\langle -1\rangle$. One may characterize T_s^{mix} up to isomorphism by the two short exact sequences

$$0 \to \Delta_s^{\mathrm{mix}} \to T_s^{\mathrm{mix}} \to \Delta_{\mathrm{id}}^{\mathrm{mix}}\langle 1\rangle \to 0, \qquad 0 \to \nabla_{\mathrm{id}}^{\mathrm{mix}}\langle -1\rangle \to T_s^{\mathrm{mix}} \to \nabla_s^{\mathrm{mix}} \to 0.$$

An object having such "standard" and "costandard" filtrations is said to be *tilting*; we will give a precise definition in Sect. 26.3.2.

Exercise 26.13 Compute the algebras $\mathrm{End}_{\mathcal{O}_0}(P_{\mathrm{id}} \oplus P_s)$ and $\mathrm{Ext}^\bullet_{\mathcal{O}_0}(L_{\mathrm{id}} \oplus L_s, L_{\mathrm{id}} \oplus L_s)$ by working in $\mathcal{O}_0^{\mathrm{mix}}$ and using (26.1).

Exercise 26.14 Based on the computations in this subsection, can you guess how to define a contravariant duality functor $\mathbb{D} : K^b\overline{\mathbb{S}\mathrm{Bim}} \to K^b\overline{\mathbb{S}\mathrm{Bim}}$ that restricts to an exact functor on $\mathcal{O}_0^{\mathrm{mix}}$, lifting the duality in $\mathcal{O}_0$ (see Chap. 14)? How does it interact with Tate twist?

26.3 Homological Properties of Soergel $\mathcal{O}_0$

To state Koszul duality, we first need a few basic homological properties of Soergel $\mathcal{O}_0$, analogous to those of the Lie-theoretic $\mathcal{O}_0$. Since the proofs of some of these results use technology that is outside the scope of this book (see Sect. 26.5 for a brief discussion), we merely give an overview of the main results.

26.3.1 Highest Weight Structure

We observed in Remark 26.12 that Δ_s^{mix}, ∇_s^{mix} in the type A_1 Soergel $\mathcal{O}_0$ are images of the Rouquier complexes F_s, $F_s^{-1} \in K^b\overline{\mathbb{S}\mathrm{Bim}}$ (see Sect. 19.3). More generally,

26.3 Homological Properties of Soergel $\mathcal{O}_0$

Rouquier complexes associated to reduced expressions play the role of Vermas and dual Vermas in $\mathcal{O}_0^{\text{mix}}$. Namely, for each $w \in W$, choose a reduced expression $(s_1, \ldots, s_k)$ and define

$$\Delta_w^{\text{mix}} := F_{s_1} \cdots F_{s_k}, \qquad \nabla_w^{\text{mix}} := F_{s_1}^{-1} \cdots F_{s_k}^{-1}.$$

Since Rouquier complexes satisfy braid relations up to homotopy (Theorem 19.23), these objects are defined up to isomorphism in $K^b\mathbb{S}\text{Bim}$. Denote also by $\Delta_w^{\text{mix}}, \nabla_w^{\text{mix}}$ their images under the natural functor $K^b\mathbb{S}\text{Bim} \to K^b\overline{\mathbb{S}\text{Bim}}$. The diagonal miracle (Theorem 19.47) asserts that these complexes lie in $\mathcal{O}_0^{\text{mix}}$.

As the notation suggests, Δ_w^{mix} and ∇_w^{mix} are graded lifts of Vermas and dual Vermas and satisfy analogous properties. For example, observe that there are canonical maps

$$\Delta_w^{\text{mix}} \to L_w^{\text{mix}} \to \nabla_w^{\text{mix}}$$

in $\mathcal{O}_0^{\text{mix}}$.

Theorem 26.15 *The canonical map $\Delta_w^{\text{mix}} \to L_w^{\text{mix}}$ is a projective cover in $(\mathcal{O}_0^{\text{mix}})_{\leq w}$. The canonical map $L_w^{\text{mix}} \to \nabla_w^{\text{mix}}$ is an injective envelope in $(\mathcal{O}_0^{\text{mix}})_{\leq w}$.*

Here, $(\mathcal{O}_0^{\text{mix}})_{\leq w}$ denotes the Serre subcategory of $\mathcal{O}_0^{\text{mix}}$ generated by $L_x^{\text{mix}}\langle n \rangle$ for $x \leq w, n \in \mathbb{Z}$.

Theorem 26.16 ("Fundamental Vanishing" [122]) *For $x, y \in W$, we have*

$$\text{Ext}^i_{\mathcal{O}_0^{\text{mix}}}(\Delta_x^{\text{mix}}, \nabla_y^{\text{mix}}) = \begin{cases} \mathbb{C} & \text{if } x = y \text{ and } i = 0; \\ 0 & \text{otherwise.} \end{cases}$$

These properties almost amount to the following.

Theorem 26.17 *The standard objects $\{\Delta_w^{\text{mix}} \mid w \in W\}$ and costandard objects $\{\nabla_w^{\text{mix}} \mid w \in W\}$ give a (graded) highest weight structure on $(\mathcal{O}_0^{\text{mix}}, \langle 1 \rangle)$ with respect to the partially ordered set $(W, \leq)$.*

A *(graded) highest weight category* is an axiomatization of highest weight categories such as category $\mathcal{O}_0$ appearing commonly in representation theory. For more details in the graded setting, see [2, Appendix A], which instead uses the terminology "graded quasihereditary."

The axioms of a (graded) highest weight category imply various homological properties of $\mathcal{O}_0^{\text{mix}}$ analogous to those of $\mathcal{O}_0$, such as the following.

Corollary 26.18 *$\mathcal{O}_0^{\text{mix}}$ has finite length and has enough projectives and injectives. Hence every L_w^{mix} admits a projective cover and an injective envelope $P_w^{\text{mix}} \twoheadrightarrow L_w^{\text{mix}} \hookrightarrow I_w^{\text{mix}}$.*

26.3.2 Tilting Objects and the Realization Functor

The results in this subsection are essentially lifted from [2, Appendix A] (where they are stated for a general graded highest weight category) and specialized to the case of $\mathcal{O}_0^{\mathrm{mix}}$.

Definition 26.19 An object X in $\mathcal{O}_0^{\mathrm{mix}}$ admits a *standard* (resp. *costandard*) *filtration* if there exists a filtration

$$0 = F^0 X \subset F^1 X \subset \cdots \subset F^k X = X$$

such that every $F^i X / F^{i-1} X$, $0 < i \leq k$, is isomorphic to some $\Delta_w^{\mathrm{mix}}\langle n \rangle$ (resp. $\nabla_w^{\mathrm{mix}}\langle n \rangle$). We say that X is *tilting* if it admits both a standard and a costandard filtration.

If X admits a standard (resp. costandard) filtration, for any $w \in W$ and $n \in \mathbb{Z}$ we write

$$(X : \Delta_w \langle n \rangle) \qquad (\text{resp. } (X : \nabla_w \langle n \rangle)) \tag{26.8}$$

for the number of subquotients in the filtration that are isomorphic to $\Delta_w \langle n \rangle$ (resp. $\nabla_w \langle n \rangle$). It is easy to see using Theorem 26.16 that these multiplicities are independent of the choice of filtration.

Theorem 26.20 ([2, Appendix A] (Ringel [156] in the Ungraded Setting)) *For every $w \in W$, there exists a unique indecomposable tilting object T_w^{mix} contained in $(\mathcal{O}_0^{\mathrm{mix}})_{\leq w}$ and satisfying $(T_w^{\mathrm{mix}} : \Delta_w^{\mathrm{mix}}) = (T_w^{\mathrm{mix}} : \nabla_w^{\mathrm{mix}}) = 1$. Every indecomposable tilting object in $\mathcal{O}_0^{\mathrm{mix}}$ is isomorphic to $T_w^{\mathrm{mix}}\langle n \rangle$ for some $w \in W$ and $n \in \mathbb{Z}$.*

Let $\mathrm{Tilt} \subset \mathcal{O}_0^{\mathrm{mix}}$ denote the full additive subcategory of tilting objects.

Lemma 26.21 *There are triangulated equivalences*

$$K^{\mathrm{b}} \mathrm{Tilt} \xrightarrow{\sim} D^{\mathrm{b}} \mathcal{O}_0^{\mathrm{mix}} \xrightarrow{\sim} K^{\mathrm{b}} \overline{\mathbb{S}\mathrm{Bim}}.$$

The first functor is the natural one; the fact that it is an equivalence follows from the vanishing of higher extensions between tilting objects. The second equivalence (via a "realization" functor) confirms our original motivation from the end of Sect. 26.2.2, that $K^{\mathrm{b}} \overline{\mathbb{S}\mathrm{Bim}}$ should play the role of the bounded derived category of $\mathcal{O}_0^{\mathrm{mix}}$.

Remark 26.22 The second equivalence of Lemma 26.21 more subtle than it appears at first sight. For an abstract triangulated category $\mathcal{T}$ and a t-structure with heart $\mathcal{A}$, it is in general not true that $D^{\mathrm{b}} \mathcal{A} \simeq \mathcal{T}$. Even worse, one cannot even construct a realization functor $D^{\mathrm{b}} \mathcal{A} \to \mathcal{T}$ without additional structure on $\mathcal{T}$, and even when one can, the functor may not be an equivalence.

26.3.3 Ringel Duality

Let w_0 be the longest element of W. Since w_0 is an involution, we have $\Delta_{w_0}^{\mathrm{mix}} \otimes_R \nabla_{w_0}^{\mathrm{mix}} \simeq R \simeq \nabla_{w_0}^{\mathrm{mix}} \otimes_R \Delta_{w_0}^{\mathrm{mix}}$ in $K^b \overline{\mathbb{S}\mathrm{Bim}}$. It follows that $(-) \otimes_R \Delta_{w_0}^{\mathrm{mix}}$ is an autoequivalence of $K^b \overline{\mathbb{S}\mathrm{Bim}}$ with quasi-inverse $(-) \otimes_R \nabla_{w_0}^{\mathrm{mix}}$. (Since $\overline{\mathbb{S}\mathrm{Bim}}$ is a left quotient, one can still tensor on the right over R.)

Theorem 26.23 (Ringel Self-duality for Soergel $\mathcal{O}_0$ [132] (Following [2])) *The triangulated equivalence*

$$\mathrm{Rin} := (-) \otimes_R \Delta_{w_0}^{\mathrm{mix}} : K^b \overline{\mathbb{S}\mathrm{Bim}} \xrightarrow{\sim} K^b \overline{\mathbb{S}\mathrm{Bim}}$$

satisfies $\mathrm{Rin}(\nabla_x^{\mathrm{mix}}) \simeq \Delta_{xw_0}^{\mathrm{mix}}$ *and* $\mathrm{Rin}(T_x^{\mathrm{mix}}) \simeq P_{xw_0}^{\mathrm{mix}}$ *for all* $x \in W$.

Unlike Koszul duality, this duality can be seen without the mixed grading.

Exercise 26.24 Show that $\mathrm{Rin}(\nabla_x^{\mathrm{mix}}) \simeq \Delta_{xw_0}^{\mathrm{mix}}$ for all $x \in W$.

Exercise 26.25 For type A_1 (see Sect. 26.2.4), verify that $\mathrm{Rin}(T_{\mathrm{id}}^{\mathrm{mix}}) \simeq P_s^{\mathrm{mix}}$ and $\mathrm{Rin}(T_s^{\mathrm{mix}}) \simeq P_{\mathrm{id}}^{\mathrm{mix}}$.

26.4 Koszul Duality

26.4.1 Statement

In the form originally conjectured by Beilinson–Ginzburg, Koszul duality for $\mathcal{O}_0$ sends $\Delta_x^{\mathrm{mix}} \mapsto \nabla_{x^{-1}w_0}^{\mathrm{mix}}$ and $P_x^{\mathrm{mix}} \mapsto L_{x^{-1}w_0}^{\mathrm{mix}}$. We instead consider its composition with Ringel duality, as suggested later also by Beilinson–Ginzburg [18]:

$$\nabla_x^{\mathrm{mix}} \xmapsto{\mathrm{Ringel}} \Delta_{xw_0}^{\mathrm{mix}} \xmapsto{\mathrm{Koszul}} \nabla_{x^{-1}}^{\mathrm{mix}}$$

$$T_x^{\mathrm{mix}} \longmapsto P_{xw_0}^{\mathrm{mix}} \longmapsto L_{x^{-1}}^{\mathrm{mix}}$$

We call this composition Koszul duality, although "Koszul–Ringel" is more accurate. Among the various reasons to consider this composition are that it eliminates w_0, and that somewhat miraculously, it also behaves nicely on Vermas and simples:

$$\Delta_x^{\mathrm{mix}} \xmapsto{\text{"Koszul–Ringel"}} \Delta_{x^{-1}}^{\mathrm{mix}}$$

$$L_x^{\mathrm{mix}} \longmapsto T_{x^{-1}}^{\mathrm{mix}}$$

Given a realization $\mathfrak{h} = (\mathfrak{h}, \{\alpha_s^\vee \in \mathfrak{h}\}, \{\alpha_s \in \mathfrak{h}^*\})$ of (W, S), the *dual realization* $\mathfrak{h}^* = (\mathfrak{h}^*, \{\alpha_s \in \mathfrak{h}^*\}, \{\alpha_s^\vee \in \mathfrak{h}\})$ is obtained by exchanging the roles of roots and coroots. For reasons we will see, Koszul duality naturally relates graded categories $\mathcal{O}_0$ associated to dual realizations.

Let (W, S) be a finite Coxeter system, and let $\mathfrak{h}$ be its geometric realization over $\mathbb{C}$. Let $R = \operatorname{Sym} \mathfrak{h}^*$ and $R^\vee = \operatorname{Sym} \mathfrak{h}$, graded with $\deg \mathfrak{h}^* = 2$ and $\deg \mathfrak{h} = 2$. Let $\mathbb{S}\mathrm{Bim}$ be the category of Soergel bimodules associated to $(W, \mathfrak{h})$, and let $\overline{\mathbb{S}\mathrm{Bim}}$ be its left quotient as in Definition 26.6. Let $\mathbb{S}\mathrm{Bim}^\vee$ be the category of Soergel bimodules associated to the dual realization $(W, \mathfrak{h}^*)$, and let $\underline{\mathbb{S}\mathrm{Bim}^\vee}$ be its *right* quotient, i.e. take quotients of (graded) morphisms spaces of $\mathbb{S}\mathrm{Bim}^\vee$ by morphisms of the form $f \cdot h$, where h lies in the augmentation ideal of $R^\vee$.

The heart of the perverse t-structure on $K^b\overline{\mathbb{S}\mathrm{Bim}}$ (resp. $K^b\underline{\mathbb{S}\mathrm{Bim}^\vee}$) is graded highest weight; denote the structural objects by $L_w^{\mathrm{mix}} = B_w, \Delta_w^{\mathrm{mix}}, \nabla_w^{\mathrm{mix}}, P_w^{\mathrm{mix}}, I_w^{\mathrm{mix}}$, T_w^{mix} (resp. $L_w^{\mathrm{mix},\vee} = B_w^\vee, \Delta_w^{\mathrm{mix},\vee}, \nabla_w^{\mathrm{mix},\vee}, P_w^{\mathrm{mix},\vee}, I_w^{\mathrm{mix},\vee}, T_w^{\mathrm{mix},\vee}$). Left quotient and right quotient are related by horizontal reflection of diagrams, which exchanges $X_w^{\mathrm{mix}} \leftrightarrow X_{w^{-1}}^{\mathrm{mix}}$ for all structural objects $X \in \{L = B, \Delta, \nabla, P, I, T\}$. This explains the absence of inverses in the indices in the following formulation of Koszul duality.

Theorem 26.26 (Koszul Duality for Soergel $\mathcal{O}_0$ [133]) *There exists a triangulated equivalence*

$$\kappa : K^b\overline{\mathbb{S}\mathrm{Bim}} \xrightarrow{\sim} K^b\underline{\mathbb{S}\mathrm{Bim}^\vee}$$

satisfying $\kappa \circ \langle 1 \rangle \simeq (1) \circ \kappa$ *and*

$$\kappa(\Delta_w^{\mathrm{mix}}) \simeq \Delta_w^{\mathrm{mix},\vee}, \quad \kappa(\nabla_w^{\mathrm{mix}}) \simeq \nabla_w^{\mathrm{mix},\vee}, \quad \kappa(T_w^{\mathrm{mix}}) \simeq B_w^\vee, \quad \kappa(B_w) \simeq T_w^{\mathrm{mix},\vee}.$$

The remainder of this section explains some of the ingredients that go into the proof of Theorem 26.26.

26.4.2 Monodromy Action

Let $h \in R^\vee$ be homogeneous of degree d. Since $\underline{\mathbb{S}\mathrm{Bim}^\vee}$ is obtained from $\mathbb{S}\mathrm{Bim}^\vee$ by modding out morphism spaces on the right, left multiplication by h defines a morphism $m(h)_B : B \to B(d)$ for any $B \in \underline{\mathbb{S}\mathrm{Bim}^\vee}$, and more generally a morphism

$$m(h)_{\mathcal{F}} : \mathcal{F} \to \mathcal{F}(d)$$

for any complex $\mathcal{F} \in K^b\underline{\mathbb{S}\mathrm{Bim}^\vee}$. These morphisms define a natural transformation $m(h) : \mathrm{id} \to (d)$ of endofunctors of $K^b\underline{\mathbb{S}\mathrm{Bim}^\vee}$, and varying $h \in R^\vee$, we obtain an

26.4 Koszul Duality

algebra map

$$m : R^\vee \to \bigoplus_{d\in\mathbb{Z}} \text{Hom}(\text{id}, (d)).$$

Thus $R^\vee$ acts functorially on every object in $K^b\underline{\mathbb{S}\text{Bim}}^\vee$ in a graded way compatible with (1). For the purpose of these paragraphs, let us say that $R^\vee$ acts on the category with shift $(K^b\underline{\mathbb{S}\text{Bim}}^\vee, (1))$.

Similarly, R acts on $(K^b\overline{\mathbb{S}\text{Bim}}, (1))$ via right multiplication. But Theorem 26.26 implies an additional action: $R^\vee$ should act on $(K^b\overline{\mathbb{S}\text{Bim}}, \langle 1\rangle)$ via some algebra map

$$\mu : R^\vee \to \bigoplus_{d\in\mathbb{Z}} \text{Hom}(\text{id}, \langle d\rangle). \tag{26.9}$$

Concretely, for $h \in R^\vee$ homogeneous of degree d, we expect morphisms

$$\mu(h)_\mathcal{F} : \mathcal{F} \to \mathcal{F}\langle d\rangle,$$

functorial in $\mathcal{F} \in K^b\overline{\mathbb{S}\text{Bim}}$.

A key ingredient in the proof of Theorem 26.26 is to find this hidden action (26.9), called the *monodromy action*. We illustrate this action by way of examples.

Example 26.27 Recall the tilting complex T_s^{mix} in type A_1 (26.7). Since the natural functor $q : \mathbb{S}\text{Bim} \to \overline{\mathbb{S}\text{Bim}}$ is surjective on morphisms, we can (arbitrarily) lift each component of the differential of T_s^{mix} to a morphism in $\mathbb{S}\text{Bim}$, obtaining for example the sequence

$$(\widetilde{T_s^{\text{mix}}}, \widetilde{d}) \simeq \begin{pmatrix} R(1) \\ \uparrow \\ \bullet\ B_s \\ \downarrow\uparrow \\ R(-1) \end{pmatrix}.$$

Since $\uparrow\circ\downarrow = \alpha_s \cdot \text{id}$, this is not a complex in $\mathbb{S}\text{Bim}$. However, since it becomes a genuine complex after applying q, each component of $\widetilde{d} \circ \widetilde{d}$ is a sum of morphisms of the form $\lambda \cdot f$, $\lambda \in \mathfrak{h}^*$:

$$R(-1) \rightsquigarrow R(1) : \alpha_s \cdot \text{id}_{R(-1)}.$$

For any $X \in \mathfrak{h}$, we define $\mu(X)_{T_s^{\mathrm{mix}}}$ by "contracting X with $\tilde{d} \circ \tilde{d}$," as follows:

$$\mu(X)_{T_s^{\mathrm{mix}}} : T_s^{\mathrm{mix}} = \begin{array}{c} R(1) \quad\quad R(-1) \\ \uparrow \quad\quad \nearrow\uparrow \\ \bullet\; B_s \quad \alpha_s(X)\cdot\mathrm{id} \; B_s(-2) \\ \uparrow \quad\quad \uparrow \\ R(-1) \quad\quad R(-3) \end{array} = T_s^{\mathrm{mix}}\langle 2\rangle.$$

Note that $\tilde{d} \circ \tilde{d}$ has cohomological degree 2 and contraction decreases the Soergel bimodule degree by 2 (so that $\alpha_s(X) \in \mathbb{C}$), giving the combined shift $[2](-2) = \langle 2 \rangle$.

Given $X, Y \in \mathfrak{h}$, one defines

$$\mu(XY)_{T_s^{\mathrm{mix}}} := \mu(X)_{T_s^{\mathrm{mix}}}\langle 2\rangle \circ \mu(Y)_{T_s^{\mathrm{mix}}} = \mu(Y)_{T_s^{\mathrm{mix}}}\langle 2\rangle \circ \mu(X)_{T_s^{\mathrm{mix}}} : T_s^{\mathrm{mix}} \to T_s^{\mathrm{mix}}\langle 4 \rangle;$$

see Exercise 26.30 below. For T_s^{mix}, we have $\mu(h)_{T_s^{\mathrm{mix}}} = 0$ for $h \in (\mathfrak{h})^2$, for degree reasons.

Example 26.28 Let $W = S_3 = \langle s, t \rangle$, and consider the complex

$$\mathcal{F} = \left(R \xrightarrow{\bullet} B_s(1) \xrightarrow{\bullet} B_t(3) \xrightarrow{\bullet} R(4) \right)$$

in $K^b\overline{\mathbb{S}\mathrm{Bim}}$. Lift $(\mathcal{F}, d)$ to a sequence $(\tilde{\mathcal{F}}, \tilde{d})$ in $\mathbb{S}\mathrm{Bim}$ given by the same formula. Then $\tilde{d} \circ \tilde{d}$ has components

$$R \rightsquigarrow B_t(3) : \;\bullet\!\circ\!\bullet = \alpha_s \cdot \bullet, \qquad B_s(1) \rightsquigarrow R(4) : \;\bullet\!\circ\!\bullet = \alpha_t \cdot \bullet.$$

Given $X \in \mathfrak{h}$, the corresponding components of $\mu(X)_{\mathcal{F}}$ are computed by contraction with X:

$$\mu(X)_{\mathcal{F}} : \mathcal{F} = \begin{array}{c} R(4) \quad\quad R(2) \\ \uparrow \quad\quad \nearrow\uparrow \\ B_t(3) \;\; \alpha_t(X)\uparrow \; B_t(1) \;\bullet \\ \uparrow \quad\quad \uparrow \\ B_s(1) \;\; \alpha_s(X)\downarrow \; B_s(-1) \\ \uparrow \quad\quad \uparrow \\ \bullet\; R \quad\quad R(-2) \end{array} = \mathcal{F}\langle 2 \rangle.$$

26.4 Koszul Duality

In general, given $\mathcal{F} \in K^b\mathbb{S}\text{Bim}$ and $X \in \mathfrak{h}$, one produces a *monodromy map* $\mu(X)_{\mathcal{F}} : \mathcal{F} \to \mathcal{F}\langle 2\rangle$ by choosing a lift $\tilde{\mathcal{F}}$ to a sequence in $\mathbb{S}\text{Bim}$ and contracting the square of this "differential" as in the examples above. For degree reasons, the monodromy action on any object in $K^b\overline{\mathbb{S}\text{Bim}}$ is nilpotent, as it must be if it is to correspond under Koszul duality to the multiplication action on $K^b\mathbb{S}\text{Bim}^\vee$.

Exercise 26.29 Show that $\mu(X)_{\mathcal{F}} : \mathcal{F} \to \mathcal{F}\langle 2\rangle$ commutes with the differentials and does not depend up to homotopy on the choice of lift $\tilde{\mathcal{F}}$, i.e. it is a well-defined morphism in $K^b\overline{\mathbb{S}\text{Bim}}$. (*Hint:* use the fact that graded Homs between Soergel bimodules are graded free as left R-modules.)

Exercise 26.30 For any morphism $f : \mathcal{F} \to \mathcal{G}$ in $K^b\overline{\mathbb{S}\text{Bim}}$ and $X \in \mathfrak{h}$, show that $\mu(X)_{\mathcal{G}} \circ f = f \circ \mu(X)_{\mathcal{F}}$. Thus the morphisms $\mu(X)_{\mathcal{F}}$ define a natural transformation $\mu(X) : \text{id} \to \langle 2\rangle$ between endofunctors of $K^b\overline{\mathbb{S}\text{Bim}}$. Deduce that the assignment $X \mapsto \mu(X)$ extends to a graded algebra map (26.9).

26.4.3 Wall-Crossing Functors

Theorem 26.26 also implies an equivalence of additive categories

$$\kappa' : \text{Tilt} \xrightarrow{\sim} \underline{\mathbb{S}\text{Bim}}^\vee$$

satisfying $\kappa' \circ \langle 1\rangle \simeq (1) \circ \kappa'$ and $\kappa'(T_w^{\text{mix}}) \simeq B_w^\vee$. We know that the endofunctors

$$\theta_s^\vee := B_s^\vee \otimes_{R^\vee} (-) \simeq R^\vee \otimes_{(R^\vee)^s} (-)(1) : \underline{\mathbb{S}\text{Bim}}^\vee \to \underline{\mathbb{S}\text{Bim}}^\vee, \quad s \in S,$$

"generate" all Soergel modules: B_w appears as the unique "biggest summand"

$$B_w \overset{\oplus}{\subset} \theta_{s_1}^\vee \cdots \theta_{s_k}^\vee(B_{\text{id}}), \text{ for any reduced expression } \underline{w} = (s_1, \ldots, s_k). \quad (26.10)$$

Another key step in the proof of Theorem 26.26 is to define the Koszul dual of these functors: endofunctors $\xi_s : \mathcal{O}_0^{\text{mix}} \to \mathcal{O}_0^{\text{mix}}$, called *wall-crossing functors*, such that T_w^{mix} appears as the unique biggest summand

$$T_w^{\text{mix}} \overset{\oplus}{\subset} \xi_{s_1} \cdots \xi_{s_k}(T_{\text{id}}^{\text{mix}}), \text{ for any reduced expression } \underline{w} = (s_1, \ldots, s_k).$$

In fact, [133] constructs triangulated functors $\xi_s : K^b\overline{\mathbb{S}\text{Bim}} \to K^b\overline{\mathbb{S}\text{Bim}}$ that restrict to such exact endofunctors of $\mathcal{O}_0^{\text{mix}}$.

Let us comment briefly on this construction. By analogy with $\theta_s^\vee$, one may hope to define ξ_s as $T_s^{\text{mix}} \otimes_R (-)$. However, this only defines a functor $K^b\mathbb{S}\text{Bim} \to K^b\mathbb{S}\text{Bim}$; since a complex $\mathcal{F} \in K^b\overline{\mathbb{S}\text{Bim}}$ does not in general lift to $K^b\mathbb{S}\text{Bim}$, tensoring over R on the left does not make sense. This problem is solved in [133] as follows: first, one chooses a lift $\tilde{\mathcal{F}}$ as in the definition of monodromy, so that

$T_s^{\mathrm{mix}} \otimes_R \widetilde{\mathcal{F}}$ makes sense, although it is not a genuine complex in $K^{\mathrm{b}}\overline{\mathbb{S}\mathrm{Bim}}$. One then adds certain components to the differential of $T_s^{\mathrm{mix}} \otimes_R \widetilde{\mathcal{F}}$, depending on the monodromy of $\mathcal{F}$, to turn it into a genuine complex $\xi_s(\mathcal{F})$ in $K^{\mathrm{b}}\overline{\mathbb{S}\mathrm{Bim}}$.

Remark 26.31 We can also consider the functors

$$\theta_s := (-) \otimes_R B_s : \overline{\mathbb{S}\mathrm{Bim}} \to \overline{\mathbb{S}\mathrm{Bim}}, \quad s \in S,$$

and the induced triangulated functors on $K^{\mathrm{b}}\overline{\mathbb{S}\mathrm{Bim}}$, which we also denote by θ_s. Thus we now have two sets of triangulated endofunctors of $K^{\mathrm{b}}\overline{\mathbb{S}\mathrm{Bim}}$ indexed by S:

$$\xi_s \circlearrowright K^{\mathrm{b}}\overline{\mathbb{S}\mathrm{Bim}}. \circlearrowleft \theta_s$$

One can show that $\xi_s \circ \theta_t \simeq \theta_t \circ \xi_s$ for all $s, t \in S$ (not necessarily distinct). We know that the functors θ_s make $[K^{\mathrm{b}}\overline{\mathbb{S}\mathrm{Bim}}]_\Delta \simeq [\overline{\mathbb{S}\mathrm{Bim}}]_\oplus \simeq [\mathbb{S}\mathrm{Bim}]_\oplus$ into the right regular representation of the Hecke algebra; if we are to believe in Koszul duality, then ξ_s must give the left regular representation. Koszul duality may thus be viewed as the symmetry between these two commuting Hecke actions on $D^{\mathrm{b}}\mathcal{O}_0^{\mathrm{mix}} \xrightarrow{\sim} K^{\mathrm{b}}\overline{\mathbb{S}\mathrm{Bim}}$.

By definition, the right action θ_s comes from the action of the monoidal category $\mathbb{S}\mathrm{Bim}$ on $K^{\mathrm{b}}\overline{\mathbb{S}\mathrm{Bim}}$. In fact, the left Hecke action can also be categorified. In [5], a monoidal category playing the Koszul dual role of $\mathbb{S}\mathrm{Bim}$ is constructed. This category of "free-monodromic tilting sheaves" acts on the left of $K^{\mathrm{b}}\overline{\mathbb{S}\mathrm{Bim}}$, and ξ_s becomes the action of a particular object (a certain enhancement of the tilting complex T_s^{mix}). In fact, this category, defined in terms of $\mathbb{S}\mathrm{Bim}$, is equivalent as a monoidal category to $\mathbb{S}\mathrm{Bim}^\vee$. This monoidal Koszul duality is the subject of [6]. For a gentle survey, see [134].

Remark 26.32 In Lie-theoretic $\mathcal{O}_0$, we saw in Chap. 15 that the wall-crossing functors Θ_s "generate" (in the same sense as above) the projectives in $\mathcal{O}_0$. One can show that Θ_s commutes with Ringel duality, so that Θ_s similarly generate the tiltings in $\mathcal{O}_0$.

Recall from the discussion following Definition 26.10 that our definition of Soergel $\mathcal{O}_0^{\mathrm{mix}}$ was "already Koszul dual" to Soergel's description of $\mathcal{O}_0$. In particular, the functors $\theta_s^\vee$ considered above are graded lifts of the Koszul dual of Θ_s, which are known as derived Zuckerman functors. It is instead the functors ξ_s that lift Θ_s. This explains why we call ξ_s rather than $\theta_s^\vee$ the wall-crossing functors.

26.4.4 Outline of the Proof of Theorem 26.26

With the tools from Sect. 26.4.2 and Sect. 26.4.3, we can outline the proof of Theorem 26.26. The two sides of Koszul duality will be related using a graded analogue of Soergel's $\mathbb{V}$ functor. Let

$$\mathbb{V}' := \bigoplus_{d \in \mathbb{Z}} \mathrm{Hom}(P_{\mathrm{id}}^{\mathrm{mix}}, -\langle d \rangle) : K^b\overline{\mathbb{S}\mathrm{Bim}} \to A^{\mathrm{opp}}\text{-gmod},$$

$$\text{where } A := \bigoplus_{d \in \mathbb{Z}} \mathrm{Hom}(P_{\mathrm{id}}^{\mathrm{mix}}, P_{\mathrm{id}}^{\mathrm{mix}}\langle d \rangle).$$

Then we define

$$\mathbb{V} = \mu_{P_{\mathrm{id}}^{\mathrm{mix}}}^* \circ \mathbb{V}' : K^b\overline{\mathbb{S}\mathrm{Bim}} \to R^\vee\text{-gmod},$$

where $\mu_{P_{\mathrm{id}}^{\mathrm{mix}}}^*$ denotes pullback under the algebra map $\mu_{P_{\mathrm{id}}^{\mathrm{mix}}} : R^\vee = (R^\vee)^{\mathrm{opp}} \to A^{\mathrm{opp}}$ induced by the monodromy action.

As mentioned in Remark 26.8, $\underline{\mathbb{S}\mathrm{Bim}}^\vee$ is equivalent to actual Soergel modules in $R^\vee$-gmod. One shows that $\mathbb{V} \circ \xi_s \simeq \theta_s^\vee \circ \mathbb{V}$ for all $s \in S$, so that $\mathbb{V}$ restricts to a functor $\kappa' : \mathrm{Tilt} \to \underline{\mathbb{S}\mathrm{Bim}}^\vee$. That this is an equivalence relies on a study of the socle of the standard objects, using an argument from the beautiful paper [21]. Finally, one defines κ as the composition

$$K^b\overline{\mathbb{S}\mathrm{Bim}} \xrightarrow[\sim]{\text{Lemma 26.21}} K^b\mathrm{Tilt} \xrightarrow[\sim]{K^b\kappa'} K^b\underline{\mathbb{S}\mathrm{Bim}}^\vee.$$

26.5 Some Odds and Ends

We have left out many details as well as other aspects of this story. As a partial remedy and to orient the interested readers, we describe in this section how the results above generalize to other realizations.

The proof that Soergel $\mathcal{O}_0$ is graded highest weight relies on more geometric/local incarnations of the Hecke category: Borel-equivariant parity sheaves on flag varieties, or Braden–MacPherson (BMP) sheaves on Bruhat moment graphs, both of which have Soergel bimodules as precisely their (Borel-equivariant) global sections. Using these objects, one may define a "perverse" t-structure on $K^b\overline{\mathbb{S}\mathrm{Bim}}$ even for some realizations for which Soergel's conjecture fails (for instance, realizations over positive characteristic) and Definition 26.9 does not define a t-structure.

For realizations arising from geometry, one can use parity sheaves to obtain a graded highest weight Soergel $\mathcal{O}_0$ ("mixed modular Bruhat-constructible perverse sheaves on flag varieties"); see [2]. Analogous results were obtained (with much the same proof) in [132] for BMP sheaves, which make sense for a general Coxeter system, but under the quite restrictive assumption that the realization is reflection faithful (see Sect. 5.6). Very roughly, what allows these results is that Schubert cells are more visible from parity sheaves or BMP sheaves than from Soergel bimodules.

The perverse t-structure is then obtained by glueing the standard t-structure on each cell, as in [19, §1.4] ("recollement").

For finite W and $\mathfrak{h}$ reflection faithful, Koszul duality as stated in Theorem 26.26 still holds, with the same proof [133]. Note that Theorem 26.26 is phrased as switching tiltings and Soergel modules, not tiltings and simples. Only the former description generalizes; in positive characteristic, B_w is in general no longer perverse (see [123]) and may not be simple even when it is perverse. Moreover, unlike with the Lie-theoretic $\mathcal{O}_0$, which is determined up to equivalence by its Weyl group, the distinction between the duals $\mathbb{S}\mathrm{Bim}$ and $\mathbb{S}\mathrm{Bim}^\vee$ matters in general.

More recently, the perverse t-structure and a graded highest weight Soergel $\mathcal{O}_0$ have been defined in [7] in the setting of the diagrammatic Hecke category for an arbitrary Coxeter system and realization. The expected Koszul duality in this generality is a work in progress.

Chapter 27
The *p*-Canonical Basis

This chapter is based on expanded notes of a lecture given by the authors and taken by
    Gordon C. Brown, **Christopher Chung**, and **Christopher Leonard**

Abstract Given a generalized Cartan matrix and an associated realization over $\mathbb{Z}$, one obtains by base change a realization of the Weyl group over a field of any characteristic p. The resulting diagrammatic Hecke category yields a basis of the Hecke algebra called the *p-canonical basis*, a positive characteristic analogue of the Kazhdan–Lusztig basis.

In this chapter, we define the *p*-canonical basis and describe how it is computed in practice using the local intersection form. We present examples that demonstrate how it differs from the Kazhdan–Lusztig basis, and the advantages of the diagrammatic Hecke category over Soergel bimodules in positive characteristic. Finally, we indicate the relation of the *p*-canonical basis to parity complexes on the flag variety and its relevance to modular representation theory.

27.1 Introduction

Whereas the Hecke algebra only depends on a Coxeter system, we have seen that its categorification depends on the additional data of a realization, which can be defined over a field of any characteristic or even integrally. In this book, we have focused primarily on the geometric and Kac–Moody realizations over $\mathbb{R}$ or $\mathbb{C}$, in which case the indecomposable objects of the Hecke category categorify the Kazhdan–Lusztig basis (Soergel's conjecture). However, the Soergel categorification theorem holds for more general realizations, hence still yields a self-dual basis of the Hecke algebra with positive structure constants that, in general, differs from the Kazhdan–Lusztig basis.

G. C. Brown
Department of Mathematics, University of Oklahoma, Norman, OK, USA

C. Chung · C. Leonard
Department of Mathematics, University of Virginia, Charlottesville, VA, USA

Among these bases, the most important ones are obtained as follows. Start with the root datum realization of a complex connected reductive group, or more generally the Kac–Moody root datum of a complex Kac–Moody group. This realization is defined over $\mathbb{Z}$, hence by base change over a field of any characteristic p. The basis obtained from the associated Hecke category only depends on the root system and the characteristic p, and is called the *p-canonical* (or *p-Kazhdan–Lusztig*) basis. The entries of the base change matrix from the standard to the p-canonical basis are the *p-Kazhdan–Lusztig* polynomials, which are now in general only Laurent polynomials.

This chapter is an introduction to the p-canonical basis. In Sect. 27.2, we define the p-canonical basis and state its first properties. In Sect. 27.3, we explain how the p-canonical basis can be computed using the local intersection form. The standard reference for the material in these two sections is [91], to which the reader is referred for more details.

Just as the Kazhdan–Lusztig basis is related to characteristic 0 representation theory, there is an emerging philosophy (due to Soergel, Fiebig, Williamson, Riche–Williamson, ...) that the p-canonical basis controls modular (i.e. characteristic p) representation theory. In the final Sects. 27.4 and 27.5, we explain how the p-canonical basis is related to both geometry (parity complexes) and to representation theory (algebraic representations of reductive groups, Soergel's modular category $\mathcal{O}$).

27.2 Definition of the *p*-Canonical Basis

The starting data for defining the p-canonical basis is the following.

Definition 27.1 A *generalized Cartan matrix* (GCM) is a square matrix $A = (a_{ij})$, with rows and columns indexed by a finite set I, having integer entries and such that

1. $a_{ii} = 2$ for all i,
2. $a_{ij} \leq 0$ if $i \neq j$, and
3. $a_{ij} = 0$ if and only if $a_{ji} = 0$.

A generalized Cartan matrix A is *decomposable* if there exists a decomposition $I = I_1 \sqcup I_2$ such that $a_{ij} = 0$ if $i \in I_1$ and $j \in I_2$; it is *indecomposable* otherwise.

Remark 27.2 The notion of a GCM should not be confused with that of a Cartan matrix associated to a realization as defined in Sect. 5.7. Unlike the latter, GCMs must have entries in $\mathbb{Z}$. This assumption is needed to ensure that, given a GCM, there exists a realization over $\mathbb{Z}$ having this GCM as its Cartan matrix (in the sense of Sect. 5.7).

27.2 Definition of the p-Canonical Basis

Definition 27.3 Given a GCM A, its *Weyl group* is the Coxeter system (W, S) with S in natural bijection with I, determined by

$$m_{st} = 2, 3, 4, 6, \text{ or } \infty \qquad \text{according as} \qquad a_{ij}a_{ji} = 0, 1, 2, 3, \text{ or } \geq 4$$

for distinct $s, t \in S$.

A GCM is of *finite type* if all of its principal minors are positive. Finite type GCMs form an important class of GCMs, as they are exactly the classical Cartan matrices (of root systems of semisimple complex Lie algebras). Indecomposable finite type GCMs therefore admit the familiar Killing–Cartan classification into types (A_n, B_n, C_n, D_n, E_6, E_7, E_8, F_4, G_2). Their associated Weyl groups are the Coxeter systems of the same type (where $C_n = B_n$ for this purpose), which are exactly the finite crystallographic Coxeter systems (see Definition 1.36).

Example 27.4 An example of a GCM that is not finite type is $\begin{pmatrix} 2 & -7 \\ -409 & 2 \end{pmatrix}$.

From now on, fix a GCM A, and let (W, S) be its Weyl group. By construction, (W, S) is a crystallographic Coxeter system. Moreover, fix a realization $\mathfrak{h}$ of (W, S) over $\mathbb{Z}$ (see Definition 5.37) with Cartan matrix A. Recall that this consists of a free finite-rank $\mathbb{Z}$-module $\mathfrak{h}$ together with subsets

$$\{\alpha_s^\vee\}_{s \in S} \subset \mathfrak{h}, \qquad \{\alpha_s\}_{s \in S} \subset \mathfrak{h}^* = \mathrm{Hom}_\mathbb{Z}(\mathfrak{h}, \mathbb{Z})$$

of simple coroots $\alpha_s^\vee$ and simple roots α_s, satisfying:

1. $A = (\langle \alpha_s^\vee, \alpha_t \rangle)_{s, t \in S}$,
2. the assignment

$$s(v) = v - \langle v, \alpha_s \rangle \alpha_s^\vee$$

for $s \in S$ and $v \in \mathfrak{h}$ extends to an action of W on $\mathfrak{h}$,

and a technical condition we will not recall.

For finite type GCMs, such integral realizations can be obtained as root datum realizations of split reductive groups of the same type.

Example 27.5 Let A denote a Cartan matrix of type A_{n-1}, so that W is the symmetric group S_n and S is its simple transpositions. We discuss two important and different realizations of (W, S) with the same Cartan matrix.

For the first, set

$$\mathfrak{h}_{\mathrm{GL}_n} = \bigoplus_{i=1}^n \mathbb{Z} e_i \tag{27.1}$$

and let $\{e_i^*\}$ denote the dual basis of $\mathfrak{h}_{\mathrm{GL}_n}^*$. For $i = 1, \ldots, n-1$ let

$$\alpha_i^\vee := e_i - e_{i+1} \quad \text{and} \quad \alpha_i := e_i^* - e_{i+1}^*. \tag{27.2}$$

Then $(\mathfrak{h}_{\mathrm{GL}_n}, \{\alpha_i\}, \{\alpha_i^\vee\})$ provides a realization of $W = S_n$, which we call the GL_n *realization*.

On the other hand, we can instead take

$$\mathfrak{h}_{\mathrm{SL}_n} = \left\{ \sum \lambda_i e_i \in \mathfrak{h}_{\mathrm{GL}_n} \mid \sum \lambda_i = 0 \right\}, \tag{27.3}$$

in which case we can identify $\mathfrak{h}_{\mathrm{SL}_n}^* = \mathfrak{h}_{\mathrm{GL}_n}^* / (\sum e_i^*)$. The images of $\{\alpha_i\}$ (resp. $\alpha_i^\vee$) in $\mathfrak{h}_{\mathrm{SL}_n}^*$ (resp. $\mathfrak{h}_{\mathrm{SL}_n}$) provide another realization, called the SL_n *realization*. (The terminology comes from algebraic groups, where these realizations arise as the root data of GL_n and SL_n respectively.)

From $\mathfrak{h}$, we obtain by base change a realization over any field $\Bbbk$: $\mathfrak{h}_\Bbbk := \mathfrak{h} \otimes_\mathbb{Z} \Bbbk$. In what follows, we write p for the characteristic of $\Bbbk$ (possibly $p = 0$).

Exercise 27.6 Let

$$\mathfrak{h}_{B_n} = \bigoplus_{i=1}^n \mathbb{Z} e_i, \quad \mathfrak{h}_{B_n}^* = \bigoplus_{i=1}^n \mathbb{Z} e_i^* \tag{27.4}$$

and set

$$\begin{aligned}\Delta &= \{e_i^* - e_{i+1}^* \mid 1 \le i \le n-1\} \cup \{e_n^*\}, \\ \Delta^\vee &= \{e_i - e_{i+1} \mid 1 \le i \le n-1\} \cup \{2e_i \mid 1 \le i \le n\}.\end{aligned} \tag{27.5}$$

Show that $(\mathfrak{h}_{B_n}, \Delta, \Delta^\vee)$ provides a realization of (W, S) of type B_n. Similarly, interchanging the roles of $\mathfrak{h}_{B_n}$ and $\mathfrak{h}_{B_n}^*$ (and Δ and $\Delta^\vee$) provides another realization of W. For $\Bbbk$ any field, show that these two realizations are isomorphic over $\Bbbk$ if and only if $p \ne 2$. (These two realizations arise as the root data of algebraic groups of types B_n and C_n.)

We assume throughout that $\mathfrak{h}_\Bbbk$ satisfies Demazure surjectivity: for every $s \in S$, the maps $\alpha_s : \mathfrak{h}_\Bbbk \to \Bbbk$ and $\alpha_s^\vee : \mathfrak{h}_\Bbbk^* \to \Bbbk$ are surjective.

Remark 27.7 Demazure surjectivity is automatically satisfied if $p \ne 2$, or if the Coxeter system (W, S) is of simply-laced type of rank at least 2. By enlarging $\mathfrak{h}$, one can always extend a realization to one satisfying Demazure surjectivity. For instance, for the type A_1 Cartan matrix, the SL_2 realization satisfies Demazure surjectivity if and only if $p \ne 2$, whereas the GL_2 realization satisfies Demazure surjectivity for all p.

Let $\mathcal{H}$ be the diagrammatic Hecke category associated to $\mathfrak{h}_\Bbbk$ (see Chaps. 10 and 11). The (analogue of the) Soergel categorification theorem still holds for $\mathcal{H}$ (see

27.2 Definition of the p-Canonical Basis

Theorems 11.1 and 11.26, or see [59, Lemma 6.24, Theorem 6.25, and Corollary 6.26]).

Theorem 27.8 *We have the following:*

1. $\mathcal{H}$ *is a Krull–Schmidt category.*
2. *For every $w \in W$ there exists a unique, indecomposable object $B_w \in \mathcal{H}$, which is a direct summand of $\underline{w}$ for any reduced expression $\underline{w}$ of w, and which is not isomorphic to any grading shift of any direct summand of any reduced expression $\underline{x}$ for $x < w$. Up to isomorphism, the object B_w does not depend on the reduced expression $\underline{w}$ of w.*
3. *The set $\{B_w \mid w \in W\}$ is a complete set of representatives of the isomorphism classes of indecomposable objects in $\mathcal{H}$ up to grading shift.*
4. *There is a unique isomorphism* $\mathrm{H} \to [\mathcal{H}]_\oplus$ *of $\mathbb{Z}[v, v^{-1}]$-algebras such that $b_s \mapsto [B_s]$ for all $s \in S$. The inverse map* $\mathrm{ch} \colon [\mathcal{H}]_\oplus \to \mathrm{H}$ *is called the* character map.

Remark 27.9 In fact the above theorem holds when $\Bbbk$ is any complete local ring such as the p-adic integers $\mathbb{Z}_p$. That $\Bbbk$ be complete local is required for $\mathcal{H}$ to be Krull–Schmidt (see Theorem 11.26), hence we cannot work over $\mathbb{Z}$ here.

Remark 27.10 Warning! For $p > 0$, the above theorem does *not* necessarily extend to the case when $\mathcal{H}$ is replaced with the category $\mathbb{S}\mathrm{Bim}$ of Soergel bimodules (see Example 27.17 below). Indeed, it is known that a realization of an infinite Coxeter system whose Cartan matrix is a GCM is never reflection faithful, so Soergel's classical theory as reviewed in Chap. 4 does not apply.

Definition 27.11 The p-*canonical basis* of H is $\{{}^p b_w \mid w \in W\}$ where

$$ {}^p b_w := \mathrm{ch}([B_w]). $$

(It is a basis by Theorem 27.8.) The p-*Kazhdan–Lusztig polynomials* ${}^p h_{y,x} \in \mathbb{Z}[v, v^{-1}]$ are defined by

$$ {}^p b_x = \delta_x + \sum_{y < x} {}^p h_{y,x} \delta_y. $$

That ${}^p b_w$ depends only[1] on p, and not on $\Bbbk$ (hence the name "p-canonical basis"), is a consequence of the following result (see [91, Lemma 4.1]).

Lemma 27.12 *For every $w \in W$, B_w is absolutely indecomposable; that is, it remains indecomposable after any change of base.*

We conclude this section by stating some first properties of the p-canonical basis (see [91, Proposition 4.2]).

---

[1] But it *does* depend on the chosen realization, as we will see below.

Proposition 27.13 *For all $x, y \in W$, we have the following:*

1. $^0b_x = b_x$, *i.e. the 0-canonical basis is the Kazhdan–Lusztig basis.*
2. ^pb_x *is self-dual:* $\overline{^pb_x} = {^pb_x}$.
3. $^pb_x = \delta_x + \sum_{y<x} {^ph_{y,x}} \delta_y$ *with* $^ph_{y,x} \in \mathbb{Z}_{\geq 0}[v, v^{-1}]$.
4. $^pb_x = b_x + \sum_{y<x} {^pm_{y,x}} b_y$ *with self-dual* $^pm_{y,x} \in \mathbb{Z}_{\geq 0}[v, v^{-1}]$.
5. $^pb_x {^pb_y} = \sum_{z \in W} {^p\mu^z_{x,y}} {^pb_z}$ *with self-dual* $^p\mu^z_{x,y} \in \mathbb{Z}_{\geq 0}[v, v^{-1}]$.
6. $^pb_x = b_x$ *for $p \gg 0$, i.e. for a fixed x, there exist only finitely many primes for which $^pb_x \neq b_x$.*

Point 1 is Soergel's conjecture, which for Weyl groups relied on the decomposition theorem, whereas the remaining points follow from relatively simple considerations in the Hecke category. In each of points 3, 4, and 5, the positivity of the coefficients, traditionally a "mysterious" phenomenon, is immediate since they all measure certain ranks or multiplicities in $\mathcal{H}_{BS}$.

27.3 Computing the *p*-Canonical Basis

As in the previous section, we continue to let $\mathcal{H}$ be the diagrammatic Hecke category associated to a realization $\mathfrak{h}_\Bbbk := \Bbbk \otimes_\mathbb{Z} \mathfrak{h}$, where $\Bbbk$ is a field of characteristic p. In practice we compute the *p*-canonical basis using the local intersection form, which was the subject of Sect. 11.3.4. Let us recall this notion, specialized to the situation at hand.

For $X, Y \in \mathcal{H}$ and $w \in W$, let $\mathrm{Hom}^\bullet_{<w}(X, Y)$ denote the set of morphisms in $\mathrm{Hom}^\bullet(X, Y)$ that factor through some shift of an indecomposable Soergel bimodule B_x for some $x < w$ (or a Bott–Samelson bimodule for a reduced expression of some $x < w$). Define the graded $\Bbbk$-module

$$\mathrm{Hom}^\bullet_{\not< w, \Bbbk}(X, Y) := \Bbbk \otimes_R \left(\mathrm{Hom}^\bullet(X, Y) / \mathrm{Hom}^\bullet_{<w}(X, Y) \right)$$

where $\Bbbk \otimes_R -$ kills the subspace spanned by the action of positive degree polynomials acting on the left. We also write $\mathrm{End}^\bullet_{\not< w, \Bbbk}(X)$ for $\mathrm{Hom}^\bullet_{\not< w, \Bbbk}(X, X)$. In particular, we have a canonical isomorphism

$$\mathrm{End}^\bullet_{\not< w, \Bbbk}(B_w) \simeq \Bbbk \qquad (27.6)$$

sending the class of id_{B_w} to 1. That this assignment is an isomorphism follows from the double leaves basis (Sect. 10.4).

For an expression $\underline{x}$ and an element $w \in W$, the *local intersection pairing of $B_{\underline{x}}$ at w* is the $\Bbbk$-valued pairing

$$I_{\underline{x}, w} : \mathrm{Hom}^\bullet_{\not< w, \Bbbk}(B_w, B_{\underline{x}}) \times \mathrm{Hom}^\bullet_{\not< w, \Bbbk}(B_{\underline{x}}, B_w) \to \mathrm{End}^\bullet_{\not< w, \Bbbk}(B_w) \stackrel{(27.6)}{\simeq} \Bbbk \qquad (27.7)$$

27.3 Computing the p-Canonical Basis

induced by composition. For any $d \in \mathbb{Z}$, the *d-th graded piece* of the local intersection form is the restriction

$$I_{\underline{x},w}^d : \mathrm{Hom}^d_{\not< w, \Bbbk}(B_w, B_{\underline{x}}) \times \mathrm{Hom}^{-d}_{\not< w, \Bbbk}(B_{\underline{x}}, B_w) \to \Bbbk. \tag{27.8}$$

By Corollary 11.75, the graded rank of $I_{\underline{x},w}$ equals the graded multiplicity of B_w in $B_{\underline{x}}$.

These ranks can be computed using the light leaves maps (Sect. 10.4). Choose a reduced expression $\underline{w}$ for w. Then for each of the quotient spaces $\mathrm{Hom}_{\not< w, \Bbbk}$, we may freely replace every instance of B_w with $B_{\underline{w}}$; this only results in a canonically isomorphic space. In particular, this identifies the isomorphism (27.6) with the isomorphism

$$\varphi_{\underline{w}} : \mathrm{End}^\bullet_{\not< w, \Bbbk}(B_{\underline{w}}) \simeq \Bbbk \tag{27.9}$$

sending the class of an endomorphism of $B_{\underline{w}}$ to the constant term of the coefficient of $\mathrm{id}_{B_{\underline{w}}}$ when it is written in the double leaves basis. Moreover, let $E(\underline{x}, w)$ be the set of subexpressions of $\underline{x}$ that express w, and for each $\underline{e} \in E(\underline{x}, w)$, fix a corresponding light leaf $LL_{\underline{x},\underline{e}} : B_{\underline{x}} \to B_{\underline{w}}$, with upside down light leaf $\overline{LL}_{\underline{x},\underline{e}} : B_{\underline{w}} \to B_{\underline{x}}$. Then these light leaves maps give a $\Bbbk$-basis for $\mathrm{Hom}^\bullet_{\not< w, \Bbbk}(B_{\underline{w}}, B_{\underline{x}})$ and $\mathrm{Hom}^\bullet_{\not< w, \Bbbk}(B_{\underline{x}}, B_{\underline{w}})$, respectively, so the local intersection form can be viewed as the matrix

$$\left(\varphi_{\underline{w}}(LL_{\underline{x},\underline{e}} \circ \overline{LL}_{\underline{x},\underline{f}}) \right)_{\underline{e},\underline{f} \in E(\underline{x},w)}, \tag{27.10}$$

with rows and columns indexed by $E(\underline{x}, w)$. The d-th graded piece is the submatrix corresponding to rows and columns indexed by $E^d(\underline{x}, w)$, the subset of $E(\underline{x}, w)$ consisting of subexpressions of defect d.

In our present situation of the Hecke category associated to the realization $\mathfrak{h}_\Bbbk$, we can say more. The computation above can done in the diagrammatic Bott–Samelson category $\mathcal{H}_{BS}$ (i.e. before taking Karoubi envelope), which is already defined for the realization $\mathfrak{h}$ over $\mathbb{Z}$. This produces matrices with entries in $\mathbb{Z}$, and the local intersection form associated to $\mathfrak{h}_\Bbbk$ is then obtained by reducing these integer matrices modulo p.

From this discussion, it follows that the graded ranks of local intersection forms only depend on the characteristic $p \geq 0$ of $\Bbbk$. Denote these graded ranks by

$$^p n_{\underline{x},w} := \sum_{d \in \mathbb{Z}} \mathrm{rk}(I_{\underline{x},w}^d) v^d \in \mathbb{Z}_{\geq 0}[v, v^{-1}]. \tag{27.11}$$

Then Corollary 11.75 implies the following (cf. [91, Lemma 3.4]).

Lemma 27.14 *In* $\mathcal{H}$, *we have a direct sum decomposition*

$$B_{\underline{x}} \simeq \bigoplus_{w \in W} B_w^{\oplus \, {}^p n_{\underline{x},w}} \qquad (27.12)$$

so that in H,

$$b_{\underline{x}} = \sum_{w \in W} {}^p n_{\underline{x},w} {}^p b_w. \qquad (27.13)$$

Since reducing an integer matrix modulo $p > 0$ can only decrease its rank, it also follows from this discussion that we can only have nonzero multiplicity in positive characteristic if we have nonzero multiplicity in characteristic 0. So to determine the decomposition of $B_{\underline{x}}$ in positive characteristic, we only need to check those w such that b_w occurs with nonzero coefficient in the expansion of $b_{\underline{x}}$ in terms of the 0-canonical (i.e. Kazhdan–Lusztig) basis.

Example 27.15 Let us compute the p-canonical basis in type B_2. Take the Cartan matrix $\begin{pmatrix} 2 & -1 \\ -2 & 2 \end{pmatrix}$ and let $S = \{s, t\}$, so

$$\langle \alpha_s^\vee, \alpha_t \rangle = -2, \qquad \langle \alpha_t^\vee, \alpha_s \rangle = -1.$$

For a reduced expression $\underline{x}$ with $\ell(\underline{x}) \leq 2$, we know that $b_{\underline{x}} = b_x$, so ${}^p b_x = b_x$ for all p. Now take $\underline{x} = (s, t, s)$. In H,

$$b_{(s,t,s)} = b_{sts} + b_s.$$

We know that ${}^p b_{sts}$ appears in $b_{(s,t,s)}$ with multiplicity 1, so to determine the decomposition of $B_{(s,t,s)}$, it remains to compute the multiplicity of B_s. The relevant subexpressions are 100 and 001 with defects 0 and 2 respectively. For degree reasons, the only possible non-trivial pairing is 100 with itself. The corresponding light leaf is

using the coloring s and t. Flipping this and composing with itself we get

$$\left(\text{diagram} \right) = \partial_s(\alpha_t) = -2$$

27.3 Computing the p-Canonical Basis

so $I_{\underline{x},w} = \begin{pmatrix} -2 & 0 \\ 0 & 0 \end{pmatrix}$ and

$$\text{gr rk}(I_{\underline{x},w}) = \begin{cases} 1 & \text{if } p \neq 2 \\ 0 & \text{if } p = 2. \end{cases}$$

Therefore

$$b_{(s,t,s)} = \begin{cases} {}^p b_{sts} + {}^p b_s & \text{if } p \neq 2 \\ {}^2 b_{sts} & \text{if } p = 2 \end{cases}$$

$$\Rightarrow {}^2 b_{sts} = \begin{cases} b_{sts} & \text{if } p \neq 2 \\ b_{sts} + b_s & \text{if } p = 2. \end{cases}$$

Similarly, $I_{(t,s,t),t} = \begin{pmatrix} -1 & 0 \\ 0 & 0 \end{pmatrix}$ and so ${}^p b_{tst} = b_{tst}$ for all p.

The p-canonical basis for C_2 can be deduced by swapping s and t. Note that if $p = 2$, then this gives a different basis for H, despite the fact that B_2 and C_2 have the same Weyl group. So the p-canonical basis really does depend on the Cartan matrix, not just the Coxeter system. The difference is even more pronounced for B_3 vs. C_3 (see [91, §5.4]).

Exercise 27.16 In type B_2 (or C_2), show that ${}^p b_{stst} = b_{stst}$ for all p. (*Hint:* Recall that $b_{(s,t,s,t)} = b_{stst} + 2b_{st}$. Thus one must show that two copies of B_{st} will split off of $B_{(s,t,s,t)}$ in any characteristic, or equivalently, that the local intersection form at st has rank 2. See also Exercise 11.67.)

Example 27.17 Type G_2: take the Cartan matrix $\begin{pmatrix} 2 & -1 \\ -3 & 2 \end{pmatrix}$ and write $S = \{s, t\}$ with

$$\langle \alpha_s^\vee, \alpha_t \rangle = -3, \qquad \langle \alpha_t^\vee, \alpha_s \rangle = -1.$$

Take $p = 2$. Then

$$ {}^2 b_{stst} = b_{stst} + b_{st}, \qquad {}^2 b_{ststs} = b_{ststs} + b_s,$$

These equations still hold after interchanging s and t as the Cartan matrix is symmetric when $p = 2$. For all other $x \in W$, we have ${}^2 b_x = b_x$.

There is an obvious map $W(G_2) \twoheadrightarrow W(A_2)$ sending generators to generators. Since the Cartan matrices for G_2 and A_2 are identical when $p = 2$, any realization of A_2 induces a realization of G_2. Soergel bimodules can't tell the difference between these and so (worryingly) the Soergel categorification theorem says that indecomposables in $\mathbb{S}\text{Bim}_{G_2}$ are in bijection with $W(A_2)$. The diagrammatic category $\mathcal{H}$ however is still well-behaved, illustrating the mantra that $\mathcal{H}$ is the correct replacement for $\mathbb{S}\text{Bim}$ when the latter behaves poorly.

Exercise 27.18 Check the computations in Example 27.17.

The computation in Example 27.15 shows more generally that entries of the Cartan matrix also appear in local intersection forms. Thus it is not surprising that p-torsion (that is, a difference between the 0-canonical and p-canonical bases) should appear in types B and G, where numbers divisible by p occur in the Cartan matrix itself. It takes a little longer to manifest, but p-torsion also occurs in type A. To measure the p-torsion that appears, let

$$T_n = \{p \mid \exists\, x \in S_n \text{ such that } {}^p b_x \neq b_x\}$$

for any $n \in \mathbb{N}$.

Example 27.19 Type A_n: if $n < 7$ then ${}^p b_x = b_x$ for all p and x [185, §5.1], but ${}^p b_x \neq b_x$ for $p = 2$ and 38 of the 40320 elements of $S_8 = W(A_7)$. Thus $T_n = \emptyset$ for $n < 7$, and $T_8 = \{2\}$.

Exercise 27.20 Writing i in place of $s_i = (i, i+1)$, the expression $\underline{x} = (4, 3, 5, 2, 4, 6, 1, 3, 5, 7, 2, 4, 6, 3, 5, 4)$ in S_8 has a unique subexpression $\underline{e}$ with defect 0 and $w = \underline{x}^{\underline{e}} = 13435437$, the longest element in the parabolic subgroup corresponding to the subset $\{1, 3, 4, 5, 7\}$ of simple reflections. Calculate local intersection form $I^0_{\underline{x}, w}$ and hence show that ${}^2 b_x \neq b_x$. (This is a hard exercise. See Exercise 10.42 for a related calculation and some hints.)

Theorem 27.21 ([141, 147]) *For any prime p, $\exists\, x \in S_{4p}$ such that ${}^p b_x \neq b_x$.*

The results above may make p-torsion seem rare in type A, but the reality is quite the opposite.

Theorem 27.22 (Torsion Explosion, Williamson [182]) *The size of T_n, and thus also its largest element, grows at least exponentially with n.*

Remark 27.23 Currently, the only known methods to compute p-canonical bases involve computing local intersection forms in the diagrammatic category to determine how objects decompose. This is a computationally intensive and challenging process. Perhaps this highlights further how wonderful the Soergel (and Kazhdan–Lusztig) conjectures are! These conjectures state that, in characteristic zero, one need not do difficult computations in the category itself, but can find the answer only by examining the Grothendieck group. Both procedures are algorithmic, but an algorithm in the Hecke algebra is a lot easier than an algorithm in the Hecke category.

Remark 27.24 In [82] He and Williamson reduce the calculation of certain entries in the local intersection form to a simple formula in the nil Hecke ring. See that paper and [91, §3] for more details. This allows one to compute many examples of p-torsion even in high rank.

27.4 Geometric Incarnation of the Hecke Category

We saw in Chap. 16 that semisimple complexes on flag varieties give a geometric categorification of the Hecke algebra. Indeed, this geometric categorification is older than Soergel bimodules, going back to the birth of Kazhdan–Lusztig theory.

In this section we describe the Juteau–Mautner–Williamson theory of *parity sheaves* [95], which generalizes this geometric categorification to positive characteristic coefficients. For crystallographic Coxeter groups and their Cartan realizations, this gives a geometric interpretation to the associated p-canonical basis as graded stalks of indecomposable parity complexes (see Corollary 27.36 below).

27.4.1 Parity Complexes on Flag Varieties

Let us assume that we are in the setting of Sect. 16.2; that is, X is a variety equipped with a suitably nice stratification $X = \bigsqcup_{\lambda \in \Lambda} X_\lambda$. We consider the derived category $D^b_\Lambda(X, \Bbbk)$ of Λ-constructible sheaves with coefficients in a field $\Bbbk$. Parity complexes can be defined as soon as the strata X_λ satisfy certain conditions on their cohomology and fundamental groups (see [95]). The simplest case is when all strata are isomorphic to affine spaces. We will assume this from now on.

Remark 27.25 The key example is when $X = G/B$ is the flag variety of a complex reductive group and Λ is the stratification by B-orbits (see Sect. 16.2). In this case we denote $D^b_\Lambda(G/B, \Bbbk)$ by $D^b_{(B)}(G/B, \Bbbk)$.

Recall that given any Λ-constructible complex of sheaves $\mathcal{F}$ we considered in Sect. 16.2.2 its table of stalks.

Definition 27.26 A constructible complex of sheaves $\mathcal{F} \in D^b_\Lambda(X, \Bbbk)$ is a *parity complex* if the columns of the table of stalks of $\mathcal{F}$ and its Verdier dual $\mathbb{D}\mathcal{F}$ vanish in all even or odd degrees. An indecomposable parity complex is called a *parity sheaf*. We denote the full additive category of parity sheaves by $\mathrm{Parity}_\Lambda(X, \Bbbk)$.

Remark 27.27 The attentive reader will note that (at least with $\mathbb{Q}$-coefficients) the degree i column of the table of stalks of IC_w is zero if i is not of the same parity as $\ell(w)$.[2] Because IC_w is Verdier self-dual, it follows that the same is true for the table of stalks of $\mathbb{D} \mathrm{IC}_w$. Thus IC_w is a parity complex, and (being indecomposable) is even a parity sheaf. However, when we take intersection cohomology complexes of Schubert varieties with coefficients in positive characteristic, they will no longer be parity sheaves in general.

The main observation of [95] is the following:

---

[2] In terms of Kazhdan–Lusztig polynomials, this translates into the fact (easily proved by induction) that $v^{-(\ell(w)-\ell(x))} h_{x,w} \in \mathbb{Z}[v^{-2}]$ for all $x, w \in W$, see Exercise 3.28.

Theorem 27.28 *For a fixed stratum $X_\lambda \subset X$, there exists (up to shift and isomorphism) at most one parity sheaf whose support is $\overline{X_\lambda}$.*

If the parity sheaf of the theorem exists, then it is necessarily self-dual up to a shift. We denote the unique indecomposable parity sheaf by $\mathcal{E}_\lambda$ and normalize it so that it is self-dual.

Remark 27.29 It follows from Theorem 27.28 and Remark 27.27 that in the case of flag varieties and $\Bbbk = \mathbb{Q}$, every stratum $X_w = BwB/B$ supports a parity sheaf, and we have $\mathcal{E}_w = \mathrm{IC}_w$. In this setting Theorem 27.28 gives yet another characterization of intersection cohomology complexes.

We now specialize to the case of the flag variety G/B of a complex reductive group. Recall that W denotes the Weyl group and $S \subset W$ its simple reflections. For each simple reflection $s \in S$, let P_s denote the corresponding standard parabolic subgroup and its partial flag variety G/P_s. The natural quotient map

$$\pi_s : G/B \to G/P_s \tag{27.14}$$

is projective (in particular proper) and is a stratified map for the Schubert stratifications, hence induces (derived) pushforward and pullback functors between the Schubert-constructible derived categories of sheaves of $\Bbbk$-vector spaces:

$$\pi_{s*} : D^b_{(B)}(G/B, \Bbbk) \rightleftarrows D^b_{(B)}(G/P_s, \Bbbk) : \pi_s^*. \tag{27.15}$$

Both π_{s*} and π_s^* preserve parity complexes (see [95, Proposition 4.10]). Thus, for any expression $\underline{w} = (s_1, \ldots, s_m)$, the *Bott–Samelson complex* $\mathcal{E}_{\underline{w}}$ defined by

$$\mathcal{E}_{\underline{w}} := \pi_{s_m}^* \pi_{s_m *} \cdots \pi_{s_1}^* \pi_{s_1 *} \underline{\Bbbk}_{X_{\mathrm{id}}}[m]. \tag{27.16}$$

is a parity complex. One may check (see [164, Lemma 3.2.1], [95, Proposition 4.11]) that $\mathcal{E}_{\underline{w}}$ is isomorphic to the direct image of the (shifted) constant sheaf $\underline{\Bbbk}_{Y(\underline{w})}[m]$ under a Bott–Samelson morphism (see Example 16.16)

$$\mathrm{mult} : Y(\underline{w}) = P_{s_1} \times^B P_{s_2} \times^B \cdots \times^B P_{s_m}/B \to G/B.$$

Because mult is a resolution of singularities if $\underline{w}$ is a reduced expression, it follows that parity sheaves exist on every stratum.

Thus one may characterize $\mathrm{Parity}_{(B)}(G/B, \Bbbk)$ as follows: it is the smallest strictly full subcategory of $D^b_{(B)}(G/B, \Bbbk)$ that contains the skyscraper sheaf $\underline{\Bbbk}_{X_{\mathrm{id}}}$ at the point stratum and is closed under finite direct sum $\oplus$, cohomological shift [1], push-pull $\pi_s^* \pi_{s*}$ for each $s \in S$, and direct summands $\overset{\oplus}{\subset}$.

Remark 27.30 The varieties here are defined over $\mathbb{C}$, and it is rather the coefficient field $\Bbbk$ that has characteristic $p > 0$. Roughly speaking, we are in the realm of algebraic topology, not algebraic geometry!

Let us compare and contrast parity sheaves with intersection cohomology sheaves. Recall that intersection cohomology complexes of Schubert varieties (with $\mathbb{Q}$-coefficients) have two definitions: the first is via self-duality and a degree bound on their stalks; and the second is as the unique summand with maximal support of a complex obtained via direct image from a Bott–Samelson resolution. The equivalence between these two characterizations is provided by the decomposition theorem. The first is intrinsic; the second inductive. The first has the advantage of uniquely determining the stalks, which amounts to the uniqueness of the Kazhdan–Lusztig basis. In the setting of Soergel bimodules, we use the second characterization to define Soergel bimodules, and Soergel's conjecture amounts to showing that the two characterizations agree.

In the setting of parity sheaves we again have two definitions: the first via conditions on the stalks and costalks; and the second as the unique summand with maximal support of a complex. However here the decomposition theorem is missing, and the degree bound condition is replaced by a much weaker parity vanishing condition. It follows that the conditions on the stalks and costalks are not enough to determine the stalks of parity sheaves. Thus (in the absence of a significant new idea) we have to perform calculations in the diagrammatic category or in the geometry of flag varieties to determine this basis.

27.4.2 Parity Complexes and the Hecke Category

The reader probably already suspects that there is a connection between the indecomposable parity sheaves $\mathcal{E}_w$ and the indecomposable objects in the diagrammatic category. However to make this precise we need to work in an equivariant setting, where we have machinery of convolution at our disposal. (We have already touched on this at the end of Chap. 16.) For this we need to pass to the equivariant derived category $D_B^b(G/B, \Bbbk)$ and consider equivariant versions of parity sheaves. We will not dwell on these technical issues and instead refer the reader to [95, 184]. The theory of the previous section carries over to the equivariant setting, and we denote by $\mathcal{E}_w \in D_B^b(G/B, \Bbbk)$ the corresponding parity sheaf.

Recall our complex reductive group G, Borel subgroup B, maximal torus $B \supset T$, Weyl group W, and simple reflections S from earlier. The structure theory of split reductive groups classifies such $G \supset B \supset T$ by a combinatorial data called a based root datum, which may be viewed as a realization $\mathfrak{h}$ of (W, S) over $\mathbb{Z}$ with Cartan matrix C (the Cartan matrix of our root system).

As before, for a field $\Bbbk$, denote by $\mathfrak{h}_\Bbbk$ the base change realization $\Bbbk \otimes_\mathbb{Z} \mathfrak{h}$ of (W, S).

Proposition 27.31 *There is a canonical isomorphism*

$$\mathbb{H}_B^\bullet(\mathrm{pt}; \Bbbk) \simeq R.$$

It follows that the total $B \times B$-equivariant cohomology of a space or equivariant sheaf may be viewed as a graded R-bimodule. This is also true if we take the B-equivariant cohomology of equivariant sheaves on G/B, as we may view this as the $B \times B$-equivariant cohomology of equivariant sheaves on G. The following result is essentially due to Soergel [164].

Theorem 27.32 *Let $\Bbbk$ be a field whose characteristic is $p \geq 5$. Then the total cohomology functor restricts to a $\Bbbk$-linear monoidal equivalence*

$$\mathbb{H} : \mathrm{Parity}_B(G/B, \Bbbk) \to \mathbb{S}\mathrm{Bim}(\mathfrak{h}_\Bbbk, W)$$

sending $\mathbb{H}(\mathcal{E}_w) \simeq B_w$ and satisfying $\mathbb{H} \circ [1] \simeq (1) \circ \mathbb{H}$.

Remark 27.33 The assumption that $p \geq 5$ is to ensure that $\mathfrak{h}_\Bbbk$ is reflection faithful, so that the classical theory of Soergel bimodules applies (see [121, Appendix]).

The discussion above generalizes to Kac–Moody groups, which are certain infinite-dimensional generalizations of connected reductive groups. Let A be a GCM with Weyl group (W, S). Then a realization $\mathfrak{h}$ over $\mathbb{Z}$ of (W, S) with Cartan matrix A, as in Sect. 27.2, may also be viewed as a *Kac–Moody root datum* associated to A. To the latter, one associates a complex Kac–Moody group $\mathcal{G}$ with a canonical Borel subgroup and maximal torus $\mathcal{B} \supset \mathcal{T}$.

The Kac–Moody flag variety $\mathcal{G}/\mathcal{B}$ (in general a complex ind-variety) has a combinatorial structure much like flag varieties associated to reductive groups (e.g. Schubert decomposition, maps to partial flag varieties). In particular, one can still define an additive category $\mathrm{Parity}_\mathcal{B}(\mathcal{G}/\mathcal{B}, \Bbbk)$ of $\mathcal{B}$-equivariant parity complexes with any coefficient field $\Bbbk$, monoidal under $\mathcal{B}$-convolution.

Although Soergel bimodules can still be defined for this realization, it is expected that Soergel bimodules in general no longer categorify the Hecke algebra. However, parity complexes always categorify the Hecke algebra. In fact, parity complexes always agree with the diagrammatic Hecke category.

Theorem 27.34 (Riche–Williamson [152, Part 3]) *Assume that $\mathfrak{h}_\Bbbk$ satisfies Demazure surjectivity. Then there exists a $\Bbbk$-linear monoidal equivalence*

$$\Delta : \mathcal{H}(\mathfrak{h}_\Bbbk, W) \xrightarrow{\sim} \mathrm{Parity}_\mathcal{B}(\mathcal{G}/\mathcal{B}, \Bbbk) \tag{27.17}$$

sending $\Delta(w) \simeq \mathcal{E}_w$ for all $w \in W$ and satisfying $[1] \circ \Delta \simeq \Delta \circ (1)$.

Remark 27.35 Just as with the functor to bimodules from Chap. 10 (see Theorem 10.20), one first defines a monoidal functor Δ^{BS} from $\mathcal{H}_{\mathrm{BS}}(\mathfrak{h}_\Bbbk, W)$ to "Bott–Samelson parity complexes" by generators and relations, then passes to the Karoubi envelope to get Δ. For example, $\Delta^{\mathrm{BS}}(s) := \mathcal{E}_s$ for $s \in S$, and the image of

27.4 Geometric Incarnation of the Hecke Category

the s-colored dots are the adjunction maps

$$\Delta^{\mathrm{BS}}\left(\,\begin{array}{c}\bullet\\|\end{array}\,\right) := \left(\mathcal{E}_s \to i_*i^*\mathcal{E}_s \simeq \mathcal{E}_{\mathrm{id}}[1]\right),$$
$$\Delta^{\mathrm{BS}}\left(\,\begin{array}{c}|\\\bullet\end{array}\,\right) := \left(\mathcal{E}_{\mathrm{id}}[-1] \simeq i^!i_!\mathcal{E}_s \to \mathcal{E}_s\right),$$
(27.18)

where $i : X_{\mathrm{id}} \hookrightarrow \mathcal{G}/\mathcal{B}$ is the inclusion of the point stratum.

To show that these assignments extend to a monoidal functor Δ^{BS}, one must check that the images of the generating morphisms satisfy the defining relations of $\mathcal{H}_{\mathrm{BS}}(\mathfrak{h}_\Bbbk, W)$. But how is this done? Unlike with bimodules, calculating directly with parity complexes involves potentially very complicated computations in the constructible derived category. For instance, how would one possibly verify the Zamolodchikov relations?

The trick is to observe that each relation comes from a finitary parabolic subgroup. Using this, Riche–Williamson reduced the verification of each relation to the same verification for an analogous functor involving this finitary parabolic subgroup and the corresponding complex reductive group. In this case, by Theorem 27.32 one can further compose with the $\mathbb{H}$ functor, hence verify the relations in bimodules, which was already done by Elias–Williamson.

With a little more work, Theorem 27.34 implies the following result.

Corollary 27.36 *In the setting above, p-Kazhdan–Lusztig polynomials are the Poincaré polynomials of stalks of indecomposable parity complexes. More precisely, for any $x, w \in W$, we have*

$$^p h_{x,w} = \sum_{i \in \mathbb{Z}} \dim H^{-\ell(x)-i}(\mathcal{E}_w|_x) \cdot v^i.$$
(27.19)

Sketch of Proof Theorem 27.28 implies that the map

$$\mathrm{ch}(\mathcal{E}) := \sum_{x \in W} \left(\sum_{i \in \mathbb{Z}} \dim H^{-\ell(x)-i}(\mathcal{E}|_x) \cdot v^i\right) \delta_x$$

defines an isomorphism between the split Grothendieck group of parity sheaves and the Hecke algebra. Now [185, Lemma 3.3] implies that $\mathrm{ch}(\mathcal{E} \star \mathcal{E}_s) = \mathrm{ch}(\mathcal{E}) b_s$ for any parity sheaf $\mathcal{E}$ and $s \in S$, where $\star$ denotes the convolution product (see Sect. 16.2.5). It follows that ch agrees, under the equivalence of Theorem 27.34, with the character map on the diagrammatic category, hence the corollary. □

This is analogous to (16.25), which describes Kazhdan–Lusztig polynomials in terms of the stalks of intersection cohomology complexes.

Remark 27.37 Historically, parity sheaves were defined earlier (first by Soergel in the setting of finite flag varieties [164], and by Juteau et al. [95] in the generality discussed here) than the diagrammatic Hecke category. Depending on one's taste, one might prefer to take Corollary 27.36 as the definition of the *p*-canonical basis.

27.5 Modular Representation Theory of Reductive Groups

We close the book by surveying a topic that has played a central role in the development of the entire theory of Soergel bimodules, from Soergel's classical approach to parity complexes and finally to the diagrammatic Hecke category. This is the modular representation theory of reductive groups.

We make no attempt here to give precise definitions or complete explanations. Rather, our goal is to tell a story and point out the connections to the mathematics covered in this book, with the hope of orienting the interested reader for further exploration. The basic results of the theory mentioned below can be found in Jantzen's book [88]. Some of the more recent advances below are exposited in [3, 66, 149, 180, 184].

Let G be a connected reductive group over an algebraically closed field $\Bbbk$. For an algebraic group G, a natural class of representations to study is what are called *algebraic representations*:[3] representation

$$\rho : G \to \mathrm{GL}(V)$$

such that ρ is a morphism of algebraic groups.[4] Let $\mathrm{Rep}(G)$ denote the abelian category of finite-dimensional algebraic representations of G.

Fix a Borel subgroup and a maximal torus $T \subset B$. Let X be the character lattice of T, and let X^+ be the set of dominant characters. Then classical results give a bijection

$$X^+ \xleftrightarrow{1:1} \{ \text{ simple modules } \}/ \simeq \qquad (27.20)$$
$$\lambda \longleftrightarrow L(\lambda).$$

---

[3] In the literature, one more often encounters the term "rational representations."

[4] In particular, this forces V to be a $\Bbbk$-vector space. Another fascinating area of research (which we do not consider here) is the representation theory of finite groups of Lie type, like $\mathrm{GL}_n(\mathbb{F}_q)$ or $\mathrm{Sp}_{2n}(\mathbb{F}_q)$, where $q = p^k$ for a prime p. These finite groups can act on a vector space V over any field, possibly of characteristic p (the equal characteristic case), possibly of characteristic 0, possibly of characteristic $\ell \neq p$ (the unequal characteristic case). These three possibilities have dramatically different behavior. Meanwhile, algebraic representations are always in "equal characteristic."

27.5 Modular Representation Theory of Reductive Groups

The next question is to determine the characters of the simple modules $L(\lambda)$. When $\Bbbk$ has characteristic 0, the answer is classical and given by the Weyl character formula. When $\Bbbk$ has positive characteristic, however, this fundamental question remains unsolved in general.

Let us assume henceforth that $\Bbbk$ is an algebraically closed field of characteristic $p > 0$. Much like with the BGG category $\mathcal{O}$ of a semisimple complex Lie algebra, $\mathrm{Rep}\,G$ can be studied using translation functors defined using tensor products, and this reduces the question of simple characters to those simple modules in the *principal block* $\mathrm{Rep}_0(G)$. Unlike the BGG regular block $\mathcal{O}_0$, the combinatorics of $\mathrm{Rep}_0(G)$ is controlled by the (dual) *affine*[5] Weyl group W^{aff}, or more precisely the set ${}^f W^{\mathrm{aff}}$ of elements in W^{aff} which are minimal in their right coset for the finite Weyl group $W_f \subset W^{\mathrm{aff}}$.

Again just like $\mathcal{O}_0$, the principal block $\mathrm{Rep}_0(G)$ is a highest weight category and comes with the following distinguished isomorphism classes of objects: simple objects $L(w \cdot_p 0)$, standard objects $\Delta(w \cdot_p 0)$, costandard objects $\nabla(w \cdot_p 0)$, and tilting objects $T(w \cdot_p 0)$, where $w \in {}^f W^{\mathrm{aff}}$ and $\cdot_p$ is a certain W^{aff}-action on X that depends on the characteristic p.[6]

Recall the Kazhdan–Lusztig conjecture for category $\mathcal{O}$, in the form stated in (13.13):

$$[L(y \cdot 0)] = \sum_{y \leq x} (-1)^{\ell(y)+\ell(x)} h_{xw_0, yw_0}(1) [\Delta(x \cdot 0)]. \qquad (27.21)$$

The analogue of this conjecture for algebraic groups is Lusztig's conjecture:

Conjecture 27.38 (Lusztig [127]) *In the Grothendieck group of* $\mathrm{Rep}_0(G)$ *we have*

$$[L(y \cdot_p 0)] = \sum (-1)^{\ell(y)+\ell(x)} h_{w_f x, w_f y}(1) [\Delta(x \cdot_p 0)]$$

where the sum runs over $x \in {}^f W^{\mathrm{aff}}$, *and* w_f *denotes the longest element in* W_f.

There are some important constraints on the prime and highest weight in this conjecture:

1. The prime p should not be "too small." Lusztig's original conjecture was equivalent to requiring that $p \geq 2h - 3$, where h is the Coxeter number of G (e.g. $h = n$ for GL_n). Following work of Kato [103] it was widely believed for some time that $p \geq h$ might be the correct bound.
2. The weight $L(y \cdot_p 0)$ should not be "too distant from 0." (Roughly speaking, one wants at most one iteration of Frobenius to play a role in the character of

---

[5]This appearance of the affine Weyl group can be explained by the Frobenius map of G, which only exists in positive characteristic.

[6]Recall that $\mathrm{Rep}_0(G)$ consists only of finite-dimensional algebraic representations. In this case (in contrast to category $\mathcal{O}_0$) $\mathrm{Rep}_0(G)$ has neither injective nor projective objects. However injective (resp. projective) objects do exist in the ind- (resp. pro-) completion of $\mathrm{Rep}_0(G)$.

$L(y \cdot_p 0)$.) One formulation of the conjecture requires all digits of the highest weight $y \cdot_p 0$ to be $\leq p - 1$, when expanded in fundamental weights.

For a detailed discussion of these issues, see [180, §1.12–1.15].

A priori, Lusztig's conjecture only provides a character formula for a finite list of highest weights. However, the Steinberg tensor product theorem allows one to deduce the characters of all simple modules, once these characters are known.

A first proof of Lusztig's conjecture for large p was obtained in the 1990s, by combining work of Kazhdan and Lusztig [105–107], Lusztig [129], Kashiwara and Tanisaki [100, 101], and Andersen et al. [9]. The idea is to first establish an analogue of Lusztig's conjecture for quantum groups at a root of unity, and then compare the representation of the quantum group (an object in characteristic 0) with the representation theory of G (an object in characteristic p). The character formula for the quantum group was established by combining the above works of Kazhdan–Lusztig, Kashiwara–Tanisaki, and Lusztig, and the comparison between the quantum group and the algebraic group was performed by Andersen–Jantzen–Soergel. A disadvantage of this approach is that there is a loss of control when passing to characteristic p. The result was that Lusztig's conjecture was known to hold for large p (depending on the root system), without an explicit bound on p. More recently, Fiebig established a direct bridge between parity sheaves and modular representations via moment graphs and was able to establish an explicit (though enormous) bound on p [67].

27.5.1 Soergel's Modular Category $\mathcal{O}$

In his 2000 paper [164], Soergel proposed an approach to understand a small part of Lusztig's conjecture. He defined a *modular category* $\mathcal{O}$, a certain subquotient of $\text{Rep}_0(G)$, whose combinatorics is controlled by the finite Weyl group rather than the affine Weyl group. Like the BGG category $\mathcal{O}_0$, modular category $\mathcal{O}$ admits an analogous $\mathbb{V}$ functor to Soergel modules, now over $\Bbbk$.

Combined with the $\mathbb{H}$ functor (Theorem 27.32), Soergel thus related modular category $\mathcal{O}$ with parity sheaves on Schubert varieties in the complex (finite) flag variety Langlands dual to G. The following diagram gives a schematic summary of Soergel's approach:

$$\left\{ \begin{array}{c} \Bbbk\text{-parity complexes} \\ \text{on flag variety} \end{array} \right\} \xrightarrow{\mathbb{H}} \{\text{Soergel mod.}/\Bbbk\} \xleftarrow{\mathbb{V}} \left\{ \begin{array}{c} \text{projectives} \\ \text{in modular } \mathcal{O} \end{array} \right\}. \qquad (27.22)$$

This is analogous to his approach to the Kazhdan–Lusztig conjecture (the reader should compare this diagram with the approach of Soergel outlined in Sect. 13.4). Soergel wrote in the introduction to [164]:

> The goal of this article is to forward [the problem of simple multiplicities of Weyl modules] to the topologists or geometers.

Soergel thus related a small part of the Lusztig's conjecture to Soergel's conjecture for the realization $\mathfrak{h}_\Bbbk$ of the finite Weyl group. In particular, if Lusztig's conjecture holds for p, then the p-canonical basis for $\mathfrak{h}_\Bbbk$ agrees with the Kazhdan–Lusztig basis.

It is through this connection that Williamson produced his celebrated counterexamples to Lusztig's conjecture. As a consequence of his Theorem 27.22, Williamson showed that, already in type A, any lower bound on p for which Lusztig's conjecture is true must be at least exponential in the rank![7]

Remark 27.39 Under the relation to parity complexes, Williamson's discovery corresponds to the existence of p-torsion in the $\mathbb{Z}$-intersection cohomology of type A Schubert varieties. This explains the name "torsion explosion." While Williamson's first construction of p-torsion was diagrammatic, he also gave an entirely geometric proof [181]. While the entire proof can therefore be phrased without the diagrammatics, it is important to note that extensive calculations with the diagrammatics were how Williamson discovered these counterexamples in the first place.

27.5.2 The Riche–Williamson Conjecture

There is something unsatisfactory about the happenings of the previous section. One begins with a category $\mathrm{Rep}_0(G)$ of interest, whose combinatorics are related to the affine Weyl group. In order to study it, one looks at a very small subquotient, a toy model, namely Soergel's modular category $\mathcal{O}$. Soergel proves that this toy model is governed by Soergel bimodules (and hence the Hecke category) for the finite Weyl group, which is a very small part of the affine Weyl group. What would be more satisfactory is a connection between the entire category $\mathrm{Rep}_0(G)$ and the Hecke category for the affine Weyl group. This would enable one to compute multiplicities in $\mathrm{Rep}_0(G)$ (rather than in the toy model) using p-Kazhdan–Lusztig polynomials for the affine Weyl group.

After Williamson's discovery of counterexamples to Lusztig's conjecture, a new conjecture was proposed by Riche and Williamson. Like with Soergel's approach to the Kazhdan–Lusztig conjecture, the idea is to understand $\mathrm{Rep}_0(G)$ via the action of similarly defined wall-crossing functors, which are now indexed by the simple reflections of W^{aff}.

---

[7]More precisely, Williamson was able to reduce this statement to a difficult question concerning the prime divisors of entries of products of fixed length of certain 2×2 integral matrices. Kontorovich, McNamara and Williamson were able to understand the growth of these prime divisors, using recent results in number theory (see the appendix to [182]).

Conjecture 27.40 (Categorical Riche–Williamson Conjecture [152] (Rough Statement)) Wall-crossing functors give an action of the diagrammatic Hecke category for the affine Weyl group on $\text{Rep}_0(G)$.

A bit more precisely, the conjecture posits the existence of a monoidal functor from the Hecke category (for a particular realization of the affine Weyl group in characteristic p) to the monoidal category of endofunctors of $\text{Rep}_0(G)$, sending each generating object (s) to the wall-crossing functor labeled by s.

Riche–Williamson proved that their categorical conjecture implies a description of $\text{Rep}_0(G)$ as the "anti-spherical quotient" (see [124]) of the Hecke category, as well as:

Conjecture 27.41 (Numerical Riche–Williamson Conjecture [152] (Rough Statement)) There are character formulas for indecomposable modules involving certain p-Kazhdan–Lusztig polynomials, for $p > h$ (the Coxeter number). By an observation of Andersen, this also implies (for $p > 2h - 1$) character formulas for simple modules in terms of p-Kazhdan–Lusztig polynomials.

Let us comment on the status of these conjectures. In type A, Conjecture 27.40 (and hence Conjecture 27.41) has been proved by Riche–Williamson (in the same paper) and by Elias and Losev [56]. While the two approaches are different, they both rely on a phenomenon special to type A: the connection between the Hecke category and the Khovanov–Lauda–Rouquier algebra. Notably, the Elias–Losev proof uses diagrammatics for singular Soergel bimodules, which is discussed in Chap. 26.

While the categorical Riche–Williamson conjecture remains open in other types, the numerical conjecture has been proved using a different approach. This proof was completed by Achar et al. [6], combining a positive characteristic Koszul duality for affine flag varieties with a recent string of advances in modular geometric representation theory (Achar and Rider [4], Mautner and Riche [137], Achar and Riche [1]). (These results are a positive characteristic version of work of Bezrukavnikov and collaborators [10, 11, 26].) For more on this development, the reader is referred to the introduction to [6], Achar–Riche's survey [3], Williamson's surveys [180, 184], and Riche's habilitation thesis [149].

Remark 27.42 As we noted above, in principle these results give character formulas for simple modules, under reasonable bounds on the characteristic. However the formulas that one obtains in practice are very unwieldy. Recently, Riche and Williamson have explained that the above solution of the numerical Riche–Williamson conjecture implies a remarkably simple character formula for simple modules in terms of p-Kazhdan–Lusztig polynomials [151]. It turns out that another formulation of Lusztig's conjecture (Lusztig's so-called periodic conjecture) holds if one replaces Kazhdan–Lusztig polynomials by p-Kazhdan–Lusztig polynomials.

27.5.3 This Is Not the End

Thanks to these recent developments, we are starting to learn that the p-canonical basis gives an answer to some long-standing questions in Lie-theoretic modular representation theory, analogous to the role that the Kazhdan–Lusztig basis plays in characteristic zero Lie theory. Whereas the Kazhdan–Lusztig basis can be computed entirely in the Hecke algebra, the p-canonical basis requires a computation in the Hecke category: one needs to compute the ranks of local intersection forms. While this diagrammatic computation can be made algorithmic, it is far more computationally intensive than computing the Kazhdan–Lusztig basis.

What still remains is the important task of better understanding the p-canonical bases, and especially those of affine Weyl groups. At present, very little is known beyond (affine) SL_2, where the p-canonical basis can be computed explicitly for every p. An exciting recent development is a conjecture of Lusztig and Williamson [131] that the p-canonical basis for (affine) SL_3 is controlled by a certain dynamical system that they describe as "billiards bouncing in alcoves."[8] Their conjecture also shows how beautiful yet complicated the situation seems to be. According to Williamson [184] the calculation of the p-canonical basis in affine Weyl groups for large p "seems to me to be one of the most interesting [open] problems in representation theory."

---

[8] One can watch this billiard in action on Williamson's YouTube channel: https://www.youtube.com/watch?v=Ru0Zys1Vvq4.

References

1. P.N. Achar, S. Riche, Modular perverse sheaves on flag varieties I: tilting and parity sheaves. Ann. Sci. Éc. Norm. Supér. (4) **49**(2), 325–370 (2016). With a joint appendix with Geordie Williamson. https://doi.org/10.24033/asens.2284
2. P.N. Achar, S. Riche, Modular perverse sheaves on flag varieties, II: Koszul duality and formality. Duke Math. J. **165**(1), 161–215 (2016). https://doi.org/10.1215/00127094-3165541
3. P.N. Achar, S. Riche, *Dualité de Koszul formelle et théorie des représentations des groupes algébriques réductifs en caractéristique positive* (2018). arXiv:1807.08690.
4. P.N. Achar, L. Rider, The affine Grassmannian and the Springer resolution in positive characteristic. Compos. Math. **152**(12), 2627–2677 (2016). https://doi.org/10.1112/S0010437X16007661
5. P.N. Achar, S. Makisumi, S. Riche, G. Williamson. *Free-monodromic mixed tilting sheaves on flag varieties* (2017). arXiv:1703.05843.
6. P.N. Achar, S. Makisumi, S. Riche, G. Williamson, Koszul duality for Kac-Moody groups and characters of tilting modules. J. Am. Math. Soc. **32**(1), 261–310 (2019). https://doi.org/10.1090/jams/905
7. P.N. Achar, S. Riche, C. Vay, Mixed perverse sheaves on flag varieties for Coxeter groups. Can. J. Math. **72**(1), 1–55 (2020). https://doi.org/10.4153/cjm-2018-034-0
8. K. Adiprasito, J. Huh, E. Katz, Hodge theory for combinatorial geometries. Ann. Math. (2) **188**(2), 381–452 (2018). https://doi.org/10.4007/annals.2018.188.2.1
9. H.H. Andersen, J.C. Jantzen, W. Soergel, Representations of quantum groups at a pth root of unity and of semisimple groups in characteristic p: independence of p. Astérisque **220**, 321 (1994)
10. S. Arkhipov, R. Bezrukavnikov, Perverse sheaves on affine flags and Langlands dual group. Israel J. Math. **170**, 135–183 (2009). With an appendix by Bezrukavnikov and Ivan Mirković. https://doi.org/10.1007/s11856-009-0024-y
11. S. Arkhipov, R. Bezrukavnikov, V. Ginzburg, Quantum groups, the loop Grassmannian, and the Springer resolution. J. Am. Math. Soc. **17**(3), 595–678 (2004). https://doi.org/10.1090/S0894-0347-04-00454-0
12. G. Azumaya, On maximally central algebras. Nagoya Math. J. **2**, 119–150 (1951). http://projecteuclid.org/euclid.nmj/1118764746
13. M. Baker, Hodge theory in combinatorics. Bull. Am. Math. Soc. (N.S.) **55**(1), 57–80 (2018). https://doi.org/10.1090/bull/1599
14. D. Bar-Natan, Fast Khovanov homology computations. J. Knot Theory Ramifications **16**(3), 243–255 (2007). https://doi.org/10.1142/S0218216507005294

15. G. Barthel, J.-P. Brasselet, K.-H. Fieseler, L. Kaup, Combinatorial duality and intersection product: a direct approach. Tohoku Math. J. (2) **57**(2), 273–292 (2005). http://projecteuclid.org/euclid.tmj/1119888340
16. H. Bass, *Algebraic K-theory* (W. A. Benjamin, Inc., New York/Amsterdam, 1968), pp. xx+762
17. A. Beilinson, J. Bernstein, Localisation de g-modules. C. R. Acad. Sci. Paris Sér. I Math. **292**(1), 15–18 (1981)
18. A. Beilinson, V. Ginzburg, Wall-crossing functors and $\mathcal{D}$-modules. Represent. Theory **3**, 1–31 (1999). https://doi.org/10.1090/S1088-4165-99-00063-1
19. A.A. Beilinson, J. Bernstein, P. Deligne, Faisceaux pervers, in *Analysis and Topology on Singular Spaces, I (Luminy, 1981)*, vol. 100. Astérisque (Société mathématique de France, Paris, 1982), pp. 5–171
20. A. Beilinson, V. Ginzburg, W. Soergel, Koszul duality patterns in representation theory. J. Am. Math. Soc **9**(2), 473–527 (1996). https://doi.org/10.1090/S0894-0347-96-00192-0
21. A. Beilinson, R. Bezrukavnikov, I. Mirković, Tilting exercises. Mosc. Math. J. **4**(3), 547–557, 782 (2004). https://doi.org/10.17323/1609-4514-2004-4-3-547-557
22. I.N. Bernstein, I.M. Gelfand, S.I. Gelfand, Structure of representations that are generated by vectors of highest weight. Funckcional. Anal. i Priložen. **5**(1), 1–9 (1971)
23. I.N. Bernstein, I.M. Gelfand, S.I. Gelfand, Differential operators on the base affine space and a study of g-modules, in *Lie Groups and Their Representations. Summer School of the Bolyai János Mathematical Society*, ed. by I.M. Gelfand (Halsted Press, New York, 1975), pp. 21–64
24. I. Bernstein, I.M. Gelfand, S. Gelfand, Schubert cells and cohomology of the spaces *G/P*. Russ. Math. Surv. **23**, 1–26 (1973)
25. I.N. Bernstein, I.M. Gelfand, S.I. Gelfand, Algebraic vector bundles on $\mathbf{P}^n$ and problems of linear algebra. Funktsional. Anal. i Prilozhen **12**(3), 66–67 (1978). English trans.: Funct. Anal. Appl. **12**, 212–214 (1979)
26. R. Bezrukavnikov, Z. Yun, On Koszul duality for Kac-Moody groups. Represent. Theory **17**, 1–98 (2013). https://doi.org/10.1090/S1088-4165-2013-00421-1
27. A. Björner, Some combinatorial and algebraic properties of Coxeter complexes and Tits buildings. Adv. Math. **52**(3), 173–212 (1984). https://doi.org/10.1016/0001-8708(84)90021-5
28. R.E. Block, The irreducible representations of the Lie algebra $\mathfrak{sl}(2)$ and of the Weyl algebra. Adv. Math. **39**(1), 69–110 (1981). https://doi.org/10.1016/0001-8708(81)90058-X
29. N. Bourbaki, *Lie Groups and Lie Algebras. Chapters 4–6*. Elements of Mathematics (Berlin). Translated from the 1968 French original by Andrew Pressley (Springer, Berlin, 2002), pp. xii+300. https://doi.org/10.1007/978-3-540-89394-3
30. T. Braden, Remarks on the combinatorial intersection cohomology of fans. Pure Appl. Math. Q. **2**(4), 1149–1186 (2006). Special Issue: In honor of Robert D. MacPherson. Part 2. https://doi.org/10.4310/PAMQ.2006.v2.n4.a10
31. T. Braden, R. MacPherson, From moment graphs to intersection cohomology. Math. Ann. **321**(3), 533–551 (2001). https://doi.org/10.1007/s002080100232
32. T. Braden, A. Licata, N. Proudfoot, B. Webster, Gale duality and Koszul duality. Adv. Math. **225**(4), 2002–2049 (2010). https://doi.org/10.1016/j.aim.2010.04.011
33. C. Brav, H. Thomas, Braid groups and Kleinian singularities. Math. Ann. **351**(4), 1005–1017 (2011). https://doi.org/10.1007/s00208-010-0627-y
34. P. Bressler, V.A. Lunts, Hard Lefschetz theorem and Hodge-Riemann relations for intersection cohomology of nonrational polytopes. Indiana Univ. Math. J. **54**(1), 263–307 (2005). https://doi.org/10.1512/iumj.2005.54.2528
35. M. Brion, The structure of the polytope algebra. Tohoku Math. J. (2) **49**(1), 1–32 (1997). https://doi.org/10.2748/tmj/1178225183
36. M. Broué, G. Malle, J. Michel, *Towards Spetses. I*, vol. 4, pp. 2–3. Dedicated to the memory of Claude Chevalley (1999), pp. 157–218
37. K.S. Brown, *Buildings* (Springer, New York, 1989), pp. viii+215. https://doi.org/10.1007/978-1-4612-1019-1
38. J.-L. Brylinski, M. Kashiwara, Kazhdan-Lusztig conjecture and holonomic systems. Invent. Math. **64**(3), 387–410 (1981). https://doi.org/10.1007/BF01389272

39. M.A.A. de Cataldo, L. Migliorini, The hard Lefschetz theorem and the topology of semismall maps. Ann. Sci. École Norm. Sup. (4) **35**(5), 759–772 (2002). https://doi.org/10.1016/S0012-9593(02)01108-4
40. M.A.A. de Cataldo, L. Migliorini, The Hodge theory of algebraic maps. Ann. Sci. École Norm. Sup. (4) **38**(5), 693–750 (2005). https://doi.org/10.1016/j.ansens.2005.07.001
41. M.A.A. de Cataldo, L. Migliorini, The decomposition theorem, perverse sheaves and the topology of algebraic maps. Bull. Am. Math. Soc. (N.S.) **46**(4), 535–633 (2009). https://doi.org/10.1090/S0273-0979-09-01260-9
42. C. Chevalley, Invariants of finite groups generated by reflections. Am. J. Math. **77**, 778–782 (1955). https://doi.org/10.2307/2372597
43. C.W. Curtis, I. Reiner, *Representation Theory of Finite Groups and Associative Algebras*. Pure and Applied Mathematics, vol. XI (Interscience Publishers, a division of John Wiley & Sons, New York/London, 1962), pp. xiv+685
44. M. Demazure, Invariants symétriques entiers des groupes de Weyl et torsion. Invent. Math. **21**, 287–301 (1973). https://doi.org/10.1007/BF01418790
45. V.V. Deodhar, Local Poincaré duality and nonsingularity of Schubert varieties. Commun. Algebra **13**(6), 1379–1388 (1985). https://doi.org/10.1080/00927878508823227
46. A. Dimca, *Sheaves in Topology*. Universitext (Springer, Berlin, 2004), pp. xvi+236. https://doi.org/10.1007/978-3-642-18868-8
47. B. Elias, The two-color Soergel calculus. Compos. Math. **152**(2), 327–398 (2016). https://doi.org/10.1112/S0010437X15007587
48. B. Elias, Thicker Soergel calculus in type A. Proc. Lond. Math. Soc. (3) **112**(5), 924–978 (2016). https://doi.org/10.1112/plms/pdw012
49. B. Elias, Quantum Satake in type A. Part I. J. Comb. Algebra **1**(1), 63–125 (2017). https://doi.org/10.4171/JCA/1-1-4
50. B. Elias, M. Hogancamp, *Categorical Diagonalization* (2017). arXiv:1707.04349
51. B. Elias, M. Hogancamp, *Categorical Diagonalization of Full Twists* (2017). arXiv:1801.00191
52. B. Elias, M. Hogancamp, On the computation of torus link homology. Compos. Math. **155**(1), 164–205 (2019)
53. B. Elias, M. Khovanov, Diagrammatics for Soergel categories. Int. J. Math. Math. Sci. 58 (2010). Art. ID 978635. https://doi.org/10.1155/2010/978635
54. B. Elias, A.D. Lauda, Trace decategorification of the Hecke category. J. Algebra **449**, 615–634 (2016). https://doi.org/10.1016/j.jalgebra.2015.11.028
55. B. Elias, N. Libedinsky, Indecomposable Soergel bimodules for universal Coxeter groups. Trans. Am. Math. Soc. **369**(6), 3883–3910 (2017). With an appendix by Ben Webster. https://doi.org/10.1090/tran/6754
56. B. Elias, I. Losev, *Modular Representation Theory in Type A Via Soergel Bimodules* (2017). arXiv:1701.00560
57. B. Elias, G. Williamson, The Hodge theory of Soergel bimodules. Ann. Math. (2) **180**(3), 1089–1136 (2014). https://doi.org/10.4007/annals.2014.180.3.6
58. B. Elias, G. Williamson, *Relative Hard Lefschetz for Soergel Bimodules* (2016). arXiv:1607.03271
59. B. Elias, G. Williamson, Soergel calculus. Represent. Theory **20**, 295–374 (2016). https://doi.org/10.1090/ert/481
60. B. Elias, G. Williamson, Diagrammatics for Coxeter groups and their braid groups. Quantum Topol. **8**(3), 413–457 (2017). https://doi.org/10.4171/QT/94
61. B. Elias, G. Williamson, Localized calculus for the Hecke category (in preparation)
62. B. Elias, N. Snyder, G. Williamson, On cubes of Frobenius extensions, in *Representation Theory—Current Trends and Perspectives*. EMS Series of Congress Reports European Mathematical Society, Zürich (2017), pp. 171–186
63. P.I. Etingof, A.A. Moura, Elliptic central characters and blocks of finite dimensional representations of quantum affine algebras. Represent. Theory **7**, 346–373 (2003). https://doi.org/10.1090/S1088-4165-03-00201-2

64. P. Etingof, S. Gelaki, D. Nikshych, V. Ostrik, *Tensor Categories*. Mathematical Surveys and Monographs, vol. 205 (American Mathematical Society, Providence, RI, 2015), pp. xvi+343
65. P. Fiebig, The combinatorics of Coxeter categories. Trans. Am. Math. Soc. **360**(8), 4211–4233 (2008). https://doi.org/10.1090/S0002-9947-08-04376-6
66. P. Fiebig, Lusztig's conjecture as a moment graph problem. Bull. Lond. Math. Soc. **42**(6), 957–972 (2010). https://doi.org/10.1112/blms/bdq058
67. P. Fiebig, An upper bound on the exceptional characteristics for Lusztig's character formula. J. Reine Angew. Math. **673**, 1–31 (2012). http://dx.doi.org/10.1515/CRELLE.2011.170
68. P. Fiebig, G. Williamson, Parity sheaves, moment graphs and the p-smooth locus of Schubert varieties. Ann. Inst. Fourier (Grenoble) **64**(2), 489–536 (2014). http://aif.cedram.org/item?id=AIF_2014__64_2_489_0
69. P. Freyd, D. Yetter, J. Hoste, W.B.R. Lickorish, K. Millett, A. Ocneanu, A new polynomial invariant of knots and links. Bull. Am. Math. Soc. (N.S.) **12**(2), 239–246 (1985). https://doi.org/10.1090/S0273-0979-1985-15361-3
70. T. Fritz, *Notes on Triangulated Categories* (2014). arXiv:1407.3765.
71. W. Fulton, *Young Tableaux: With Applications to Representation Theory and Geometry*. London Mathematical Society Student Texts, vol. 35 (Cambridge University Press, Cambridge, 1997), pp. x+260
72. W. Fulton, J. Harris, *Representation Theory: A First Course, Readings in Mathematics*. Graduate Texts in Mathematics, vol. 129 (Springer, New York, 1991), pp. xvi+551. https://doi.org/10.1007/978-1-4612-0979-9
73. D. Gaitsgory, *Geometric Representation Theory*. Lecture Notes (2005). Available at http://www.math.harvard.edu/~gaitsgde/267y/catO.pdf
74. A.M. Garsia, T.J. McLarnan, Relations between Young's natural and the Kazhdan-Lusztig representations of S_n. Adv. Math. **69**(1), 32–92 (1988). https://doi.org/10.1016/0001-8708(88)90060-6
75. M. Geck, Hecke algebras of finite type are cellular. Invent. Math. **169**(3), 501–517 (2007). https://doi.org/10.1007/s00222-007-0053-2
76. M. Geck, G. Pfeiffer, *Characters of Finite Coxeter Groups and Iwahori-Hecke Algebras*. London Mathematical Society Monographs. New Series, vol. 21 (The Clarendon Press, Oxford University Press, New York, 2000), pp. xvi+446
77. S.I. Gelfand, Y.I. Manin, *Methods of Homological Algebra*, 2nd edn. Springer Monographs in Mathematics (Springer, Berlin, 2003), pp. xx+372. https://doi.org/10.1007/978-3-662-12492-5
78. E. Gorsky, A. Neguţ, J. Rasmussen, *Flag Hilbert Schemes, Colored Projectors and Khovanov-Rozansky Homology* (2016). arXiv:1608.07308
79. J.J. Graham, *Modular Representations of Hecke Algebras and Related Algebras*. PhD thesis. University of Sydney, 1995.
80. J.J. Graham, G.I. Lehrer, Cellular algebras. Invent. Math. **123**(1), 1–34 (1996). https://doi.org/10.1007/BF01232365
81. M. Härterich, *Kazhdan-Lusztig-Basen, unzerlegbare Bimoduln und die Topologie der Fahnenmannigfaltigkeit einer Kac-Moody-Gruppe*. PhD thesis. Albert-Ludwigs-Universität Freiburg, 1999. https://freidok.uni-freiburg.de/data/18
82. X. He, G. Williamson, Soergel calculus and Schubert calculus. Bull. Inst. Math. Acad. Sin. (N.S.) **13**(3), 317–350 (2018)
83. T. Holm, P. Jørgensen, Triangulated categories: definitions, properties, and examples, in *Triangulated Categories*. London Mathematical Society Lecture Note Series, vol. 375 (Cambridge University Press, Cambridge, 2010), pp. 1–51. https://doi.org/10.1017/CBO9781139107075.002
84. J.E. Humphreys. *Introduction to Lie Algebras and Representation Theory*. Graduate Texts in Mathematics, vol. 9. Second printing, revised (Springer, New York/Berlin, 1978), pp. xii+171
85. J.E. Humphreys, *Reflection Groups and Coxeter Groups*. Cambridge Studies in Advanced Mathematics, vol. 29 (Cambridge University Press, Cambridge, 1990), pp. xii+204. https://doi.org/10.1017/CBO9780511623646

86. J.E. Humphreys, *Representations of Semisimple Lie Algebras in the BGG Category O*. Graduate Studies in Mathematics, vol. 94 (American Mathematical Society, Providence, RI, 2008), pp. xvi+289. https://doi.org/10.1090/gsm/094
87. A.P. Isaev, O.V. Ogievetsky, On representations of Hecke algebras. Czechoslovak J. Phys. **55**(11), 1433–1441 (2005). https://doi.org/10.1007/s10582-006-0022-9
88. J.C. Jantzen, *Representations of Algebraic Groups*. Mathematical Surveys and Monographs, 2nd edn., vol. 107 (American Mathematical Society, Providence, RI, 2003), pp. xiv+576
89. J.C. Jantzen, Moment graphs and representations, in *Geometric Methods in Representation Theory. I*, vol. 24. Sémin. Congr. Soc. Math. France, Paris (2012), pp. 249–341
90. L.T. Jensen, The 2-braid group and Garside normal form. Math. Z. **286**(1–2), 491–520 (2017). https://doi.org/10.1007/s00209-016-1769-8
91. L.T. Jensen, G. Williamson, The p-canonical basis for Hecke algebras, in *Categorification and Higher Representation Theory*. Contemporary Mathematics, vol. 683 (American Mathematical Society, Providence, RI, 2017), pp. 333–361
92. V.F.R. Jones, Hecke algebra representations of braid groups and link polynomials. Ann. Math. (2) **126**(2), 335–388 (1987). https://doi.org/10.2307/1971403
93. V.F.R. Jones, A polynomial invariant for knots via von Neumann algebras. Bull. Am. Math. Soc. (N.S.) **12**(1), 103–111 (1985). https://doi.org/10.1090/S0273-0979-1985-15304-2
94. D. Juteau, C. Mautner, G. Williamson, Perverse sheaves and modular representation theory, in *Geometric Methods in Representation theory. II*, vol. 24. Sémin. Congr. Soc. Math. France, Paris (2012), pp. 315–352
95. D. Juteau, C. Mautner, G. Williamson, Parity sheaves. J. Am. Math. Soc. **27**(4), 1169–1212 (2014). https://doi.org/10.1090/S0894-0347-2014-00804-3
96. L. Kadison, New Examples of Frobenius Extensions. University Lecture Series, vol. 14 (American Mathematical Society, Providence, RI, 1999), pp. x+84. https://doi.org/10.1090/ulect/014
97. T. Kanenobu, Infinitely many knots with the same polynomial invariant. Proc. Am. Math. Soc. **97**(1), 158–162 (1986). https://doi.org/10.2307/2046099
98. K. Karu, Hard Lefschetz theorem for nonrational polytopes. Invent. Math. **157**(2), 419–447 (2004). https://doi.org/10.1007/s00222-004-0358-3
99. M. Kashiwara, P. Schapira, *Sheaves on Manifolds*, vol. 292. Grundlehren der Mathematischen Wissenschaften [Fundamental Principles of Mathematical Sciences]. With a chapter in French by Christian Houzel, Corrected reprint of the 1990 original (Springer, Berlin, 1994), pp. x+512
100. M. Kashiwara, T. Tanisaki, Kazhdan-Lusztig conjecture for affine Lie algebras with negative level. Duke Math. J. **77**(1), 21–62 (1995). http://dx.doi.org/10.1215/S0012-7094-95-07702-3
101. M. Kashiwara, T. Tanisaki, Kazhdan-Lusztig conjecture for affine Lie algebras with negative level. II. Nonintegral case. Duke Math. J. **84**(3), 771–813 (1996). http://dx.doi.org/10.1215/S0012-7094-96-08424-0
102. C. Kassel, *Quantum Groups*. Graduate Texts in Mathematics, vol. 155 (Springer, New York, 1995), pp. xii+531. https://doi.org/10.1007/978-1-4612-0783-2
103. S. Kato, On the Kazhdan-Lusztig polynomials for affine Weyl groups. Adv. Math. **55**(2), 103–130 (1985). https://doi.org/10.1016/0001-8708(85)90017-9
104. L.H. Kauffman, State models and the Jones polynomial. Topology **26**(3), 395–407 (1987). https://doi.org/10.1016/0040-9383(87)90009-7
105. D. Kazhdan, G. Lusztig, Tensor structures arising from affine Lie algebras. I. J. Am. Math. Soc. **6**(4), 905–947 (1993). https://doi.org/10.2307/2152745
106. D. Kazhdan, G. Lusztig, Tensor structures arising from affine Lie algebras. II. J. Am. Math. Soc. **6**(4), 949–1011 (1993). https://doi.org/10.1090/S0894-0347-1993-1186962-0
107. D. Kazhdan, G. Lusztig, Tensor structures arising from affine Lie algebras. III. J. Am. Math. Soc. **7**(2), 335–381 (1994). https://doi.org/10.2307/2152762
108. D. Kazhdan, G. Lusztig, Representations of Coxeter groups and Hecke algebras. Invent. Math. **53**(2), 165–184 (1979). https://doi.org/10.1007/BF01390031

109. D. Kazhdan, G. Lusztig, Schubert varieties and Poincaré duality, in *Geometry of the Laplace Operator (Proceedings of Symposia in Pure Mathematics, University of Hawaii, Honolulu, HI, 1979)*. Proceedings of Symposia in Pure Mathematics, vol. XXXVI (American Mathematical Society, Providence, RI, 1980), pp. 185–203
110. B. Keller, Deriving DG categories. Ann. Sci. École Norm. Sup. (4) **27**(1), 63–102 (1994). http://www.numdam.org/item?id=ASENS_1994_4_27_1_63_0
111. G. Kempf, The Grothendieck-Cousin complex of an induced representation. Adv. Math. **29**(3), 310–396 (1978). https://doi.org/10.1016/0001-8708(78)90021-X
112. M. Khovanov, Triply-graded link homology and Hochschild homology of Soergel bimodules. Int. J. Math. **18**(8), 869–885 (2007). https://doi.org/10.1142/S0129167X07004400
113. M. Khovanov, P. Seidel, Quivers, Floer cohomology, and braid group actions. J. Am. Math. Soc. **15**(1), 203–271 (2002). https://doi.org/10.1090/S0894-0347-01-00374-5
114. H. Krause, Krull-Schmidt categories and projective covers. Expo. Math. **33**(4), 535–549 (2015). https://doi.org/10.1016/j.exmath.2015.10.001
115. J. Kübel, From Jantzen to Andersen filtration via tilting equivalence. Math. Scand. **110**(2), 161–180 (2012). https://doi.org/10.7146/math.scand.a-15202
116. J. Kübel, Tilting modules in category $\mathcal{O}$ and sheaves on moment graphs. J. Algebra **371**, 559–576 (2012). https://doi.org/10.1016/j.jalgebra.2012.09.008
117. T.Y. Lam, *A First Course in Noncommutative Rings*. Graduate Texts in Mathematics, 2nd edn., vol. 131 (Springer, New York, 2001), pp. xx+385. https://doi.org/10.1007/978-1-4419-8616-0
118. R.G. Larson, M.E. Sweedler, An associative orthogonal bilinear form for Hopf algebras. Am. J. Math. **91**, 75–94 (1969). https://doi.org/10.2307/2373270
119. A.D. Lauda, An introduction to diagrammatic algebra and categorified quantum $\mathfrak{sl}_2$. Bull. Inst. Math. Acad. Sin. (N.S.) **7**(2), 165–270 (2012)
120. N. Libedinsky, Sur la catégorie des bimodules de Soergel. J. Algebra **320**(7), 2675–2694 (2008). https://doi.org/10.1016/j.jalgebra.2008.05.027
121. N. Libedinsky, Light leaves and Lusztig's conjecture. Adv. Math. **280**, 772–807 (2015). https://doi.org/10.1016/j.aim.2015.04.022
122. N. Libedinsky, G. Williamson, Standard objects in 2-braid groups. Proc. Lond. Math. Soc. (3) **109**(5), 1264–1280 (2014). https://doi.org/10.1112/plms/pdu022
123. N. Libedinsky, G. Williamson, A non-perverse Soergel bimodule in type A. C. R. Math. Acad. Sci. Paris **355**(8), 853–858 (2017). https://doi.org/10.1016/j.crma.2017.07.011
124. N. Libedinsky, G. Williamson, *The Anti-spherical Category* (2017). arXiv:1702.00459
125. W. Lu, A.K. McBride, *Algebraic Structures on Grothendieck Groups* (2013). Available at https://alistairsavage.ca/pubs/Lu-McBride-Alg-Str-on-Groth-Grps.pdf
126. G. Lusztig, *Hecke Algebras with Unequal Parameters*. CRM Monograph Series, vol. 18 (American Mathematical Society, Providence, RI, 2003), pp. vi+136
127. G. Lusztig, Some problems in the representation theory of finite Chevalley groups. In: *The Santa Cruz Conference on Finite Groups (University of California, Santa Cruz, CA, 1979)*. Proceedings of Symposia in Pure Mathematics, vol. 37 (American Mathematical Society, Providence, RI, 1980), pp. 313–317.
128. G. Lusztig, Some examples of square integrable representations of semisimple p-adic groups. Trans. Am. Math. Soc. **277**(2), 623–653 (1983). https://doi.org/10.2307/1999228
129. G. Lusztig, Monodromic systems on affine flag manifolds. Proc. Roy. Soc. Lond. Ser. A **445**(1923), 231–246 (1994). http://dx.doi.org/10.1098/rspa.1994.0058
130. G. Lusztig, Action of longest element on a Hecke algebra cell module. Pac. J. Math. **279**(1–2), 383–396 (2015). https://doi.org/10.2140/pjm.2015.279.383

131. G. Lusztig, G. Williamson, Billiards and tilting characters for SL_3, in *SIGMA Symmetry Integrability and Geometry Methods and Applications*, vol. 14 (2018). Paper No. 015, 22. https://doi.org/10.3842/SIGMA.2018.015
132. S. Makisumi, *Mixed Modular Perverse Sheaves on Moment Graphs* (2017). arXiv:1703.01571
133. S. Makisumi, *Modular Koszul Duality for Soergel Bimodules* (2017). arXiv:1703.01576
134. S. Makisumi, On monoidal Koszul duality for the Hecke category. Rev. Colombiana Mat. **53**(suppl.), 195–222 (2019). https://doi.org/10.15446/recolma.v53nsupl.84084
135. A. Mathas, On the left cell representations of Iwahori-Hecke algebras of finite Coxeter groups. J. Lond. Math. Soc. (2) **54**(3), 475–488 (1996). https://doi.org/10.1112/jlms/54.3.475
136. H. Matsumoto, Générateurs et relations des groupes de Weyl généralisés. C. R. Acad. Sci. Paris **258**, 3419–3422 (1964)
137. C. Mautner, S. Riche, Exotic tilting sheaves, parity sheaves on affine Grassmannians, and the Mirković-Vilonen conjecture. J. Eur. Math. Soc. **20**(9), 2259–2332 (2018). https://doi.org/10.4171/JEMS/812
138. L. Maxim, *Intersection Homology & Perverse Sheaves: With Applications to Singularities*. Graduate Texts in Mathematics, vol. 281 (Springer, New York, 2019)
139. J. May, *The Axioms for Triangulated Categories* (2005). Available%20at%20%5Curl%7Bhttp://math.uchicago.edu/~may/MISC/Triangulate.pdf%7D.
140. P. McMullen, On simple polytopes. Invent. Math. **113**(2), 419–444 (1993). https://doi.org/10.1007/BF01244313
141. P. McNamara, G. Williamson, Tame torsion (in preparation)
142. G. Mikhalkin, I. Zharkov, Tropical eigenwave and intermediate Jacobians. *Homological Mirror Symmetry and Tropical Geometry*. Lecture Notes of the Unione Matematica Italiana, vol. 15 (Springer, Cham, 2014), pp. 309–349. https://doi.org/10.1007/978-3-319-06514-4_7
143. C. Năstăsescu, F. Van Oystaeyen, *Methods of Graded Rings*. Lecture Notes in Mathematics, vol. 1836 (Springer, Berlin, 2004), pp. xiv+304
144. A. Neeman, *Triangulated Categories*. Annals of Mathematics Studies, vol. 148 (Princeton University Press, Princeton, NJ, 2001), pp. viii+449. https://doi.org/10.1515/9781400837212
145. A. Okounkov, A. Vershik, A new approach to representation theory of symmetric groups. Selecta Math. (N.S.) **2**(4), 581–605 (1996). https://doi.org/10.1007/PL00001384
146. P. Polo, Construction of arbitrary Kazhdan-Lusztig polynomials in symmetric groups. *Represent. Theory* **3**, 90–104 (1999). https://doi.org/10.1090/S1088-4165-99-00074-6
147. P. Polo, Examples of torsion in intersection cohomology. Unpublished
148. N.Y. Reshetikhin, V.G. Turaev, Ribbon graphs and their invariants derived from quantum groups. Commun. Math. Phys. **127**(1), 1–26 (1990). http://projecteuclid.org/euclid.cmp/1104180037
149. S. Riche, *Geometric Representation Theory in Positive Characteristic*. Mémoire d'habilitation, Université Blaise Pascal (Clermont Ferrand 2) (2016). https://tel.archives-ouvertes.fr/tel-01431526
150. S. Riche, La théorie de Hodge des bimodules de Soergel (d'apres Soergel et Elias-Williamson). Séminaire Bourbaki, no. 1339, Astérisque (2017, to appear). arXiv:1711.02464
151. S. Riche, G. Williamson, A simple character formula (2018). arXiv:1801.00896
152. S. Riche, G. Williamson, Tilting modules and the p-canonical basis. Astérisque **397**, ix+184 (2018)
153. S. Riche, W. Soergel, G. Williamson, Modular Koszul duality. Compos. Math. **150**(2), 273–332 (2014). https://doi.org/10.1112/S0010437X13007483
154. J. Rickard, Morita theory for derived categories. J. Lond. Math. Soc. (2) **39**(3), 436–456 (1989). https://doi.org/10.1112/jlms/s2-39.3.436
155. K. Rietsch, An introduction to perverse sheaves, in *Representations of Finite Dimensional Algebras and Related Topics in Lie Theory and Geometry*. Fields Institute Communications, vol. 40 (American Mathematical Society, Providence, RI, 2004), pp. 391–429.

156. C.M. Ringel, The category of modules with good filtrations over a quasi-hereditary algebra has almost split sequences. Math. Z. **208**(2), 209–223 (1991). https://doi.org/10.1007/BF02571521
157. M. Ronan, *Lectures on Buildings*. Updated and revised. University of Chicago Press, Chicago, IL, 2009, pp. xiv+228
158. D.E.V. Rose, A note on the Grothendieck group of an additive category. Vestn. Chelyab. Gos. Univ. Mat. Mekh. Inform. **3**(17), 135–139 (2015)
159. R. Rouquier, Categorification of $\mathfrak{sl}_2$ and braid groups, in *Trends in Representation Theory of Algebras and Related Topics*. Contemporary Mathematics, vol. 406 (American Mathematical Society, Providence, RI, 2006), pp. 137–167. https://doi.org/10.1090/conm/406/07657
160. G.C. Shephard, J.A. Todd, Finite unitary reflection groups. Can. J. Math. **6**, 274–304 (1954). https://doi.org/10.4153/cjm-1954-028-3
161. W. Soergel, Kategorie $\mathcal{O}$, perverse Garben und Moduln über den Koinvarianten zur Weylgruppe. J. Am. Math. Soc. **3**(2), 421–445 (1990). https://doi.org/10.2307/1990960
162. W. Soergel, The combinatorics of Harish-Chandra bimodules. J. Reine Angew. Math. **429**, 49–74 (1992). https://doi.org/10.1515/crll.1992.429.49
163. W. Soergel, Kazhdan-Lusztig polynomials and a combinatoric[s] for tilting modules. Represent. Theory **1**, 83–114 (1997). https://doi.org/10.1090/S1088-4165-97-00021-6
164. W. Soergel, On the relation between intersection cohomology and representation theory in positive characteristic. J. Pure Appl. Algebra **152**(1–3) (2000). Commutative Algebra, Homological Algebra and Representation Theory (Catania/Genoa/Rome, 1998), pp. 311–335. https://doi.org/10.1016/S0022-4049(99)00138-3
165. W. Soergel, Kazhdan-Lusztig-Polynome und unzerlegbare Bimoduln über Polynomringen. J. Inst. Math. Jussieu **6**(3), 501–525 (2007). https://doi.org/10.1017/S1474748007000023
166. W. Soergel, Andersen filtration and hard Lefschetz. Geom. Funct. Anal. **17**(6), 2066–2089 (2008). https://doi.org/10.1007/s00039-007-0640-9
167. T.A. Springer, Quelques applications de la cohomologie d'intersection, in *Bourbaki Seminar, Vol. 1981/1982*. Astérisque. Soc. Math. France, Paris, vol. 92 (1982), pp. 249–273
168. R.P. Stanley, Log-concave and unimodal sequences in algebra, combinatorics, and geometry, in *Graph Theory and Its Applications: East and West (Jinan, 1986)*. Annals of the New York Academy of Sciences, vol. 576 (New York Academy of Sciences, New York, 1989), pp. 500–535. https://doi.org/10.1111/j.1749-6632.1989.tb16434.x
169. The Stacks Project Authors, *Stacks Project*. https://stacks.math.columbia.edu
170. V.G. Turaev, *Quantum Invariants of Knots and 3-Manifolds*. De Gruyter Studies in Mathematics, 3rd edn., vol. 18 (De Gruyter, Berlin, 2016), pp. xii+596. MR1292673. https://doi.org/10.1515/9783110435221
171. M. van den Bergh, A relation between Hochschild homology and cohomology for Gorenstein rings. Proc. Am. Math. Soc. **126**(5), 1345–1348 (1998). https://doi.org/10.1090/S0002-9939-98-04210-5
172. M. van den Bergh, Erratum to: "A relation between Hochschild homology and cohomology for Gorenstein rings" [Proc. Am. Math. Soc. **126**(5), 1345–1348 (1998)]. Proc. Am. Math. Soc. **130**(9), 2809–2810 (2002). https://doi.org/10.1090/S0002-9939-02-06684-4
173. J.-L. Verdier, Des catégories dérivées des catégories abéliennes. Astérisque **239** (1996). With a preface by Luc Illusie, Edited and with a note by Georges Maltsiniotis, xii+253 pp. (1997)
174. D.-N. Verma, Structure of certain induced representations of complex semisimple Lie algebras. Bull. Am. Math. Soc. **74**, 60–166 (1968). https://doi.org/10.1090/S0002-9904-1968-11921-4
175. H. Wenzl, On sequences of projections. C. R. Math. Rep. Acad. Sci. Canada **9**(1), 5–9 (1987)
176. B.W. Westbury, Invariant tensors and cellular categories. J. Algebra **321**(11), 3563–3567 (2009). https://doi.org/10.1016/j.jalgebra.2008.07.004
177. G. Williamson, *Singular Soergel bimodules*. PhD thesis. Albert-Ludwigs-Universität Freiburg, 2008
178. G. Williamson, Singular Soergel bimodules. Int. Math. Res. Not. IMRN **20**, 4555–4632 (2011). https://doi.org/10.1093/imrn/rnq263

179. G. Williamson, Local Hodge theory of Soergel bimodules. Acta Math. **217**(2), 341–404 (2016). https://doi.org/10.1007/s11511-017-0146-8
180. G. Williamson, Algebraic representations and constructible sheaves. Jpn. J. Math. **12**(2), 211–259 (2017). https://doi.org/10.1007/s11537-017-1646-1
181. G. Williamson, On torsion in the intersection cohomology of Schubert varieties. J. Algebra **475**, 207–228 (2017). https://doi.org/10.1016/j.jalgebra.2016.06.006
182. G. Williamson, Schubert calculus and torsion explosion. J. Am. Math. Soc. **30**(4), 1023–1046 (2017). With a joint appendix with Alex Kontorovich and Peter J. McNamara. https://doi.org/10.1090/jams/868
183. G. Williamson, The Hodge theory of the decomposition theorem. Astérisque **390** (2017). Séminaire Bourbaki, vol. 2015/2016. Exposés 1104–1119, Exp. No. 1115, 335–367
184. G. Williamson, Parity sheaves and the Hecke category. in *Proceedings of the International Congress of Mathematicians 2018) (2018, to appear)*. arXiv:1801.00896
185. G. Williamson, T. Braden, Modular intersection cohomology complexes on flag varieties. Math. Z. **272**(3–4), 697–727 (2012). https://doi.org/10.1007/s00209-011-0955-y

Index

Symbols
$\star$, 330
$(-,-)$, 46
$\langle-,-\rangle$, 337
$(-,-)_L^{-i}$, 339
$[-]_\oplus$, 75
$[-]$, 395
$[-]_\Delta$, 379, 396
$\overline{(-)}$, 44
$\mathcal{A}$, 28
$\overline{\mathcal{A}}$, 28
A^e, 436
B_s, 67
B_w, 87, 202
$\overline{B}_w$, 306
$\text{BS}(\underline{w})$, 68
Br_W, 385
$\mathbb{BSBim}$, 70
$\mathbb{BSBim}_Q$, 198
$\mathbb{BSBim}(V)$, 89
$\mathcal{B}\text{im}$, 121, 482
$\mathcal{C}$, 64, 267
$\mathcal{C}_{\min}$, 377
$\mathcal{C}\text{at}$, 121
$\text{Cone}(f)$, 374
$\mathbb{D}$, 280, 326, 350, 538
Δ-filtration, 82
$\Delta(\lambda)$, 260, 274
Δ_w, 301, 530
F_β, 386
H, 39
H_φ, 40
$\text{H}(W)$, 39

$\mathcal{H}$, 202
$\mathcal{H}_{\text{BS}}(\mathfrak{h}, W)$, 187
$\mathcal{H}_{\text{BS}}$, 152, 164, 175, 182
$\mathcal{H}_{\text{BS}}(s)$, 148
$\mathbb{H}$, 269, 562
$\text{Hom}^\bullet$, 60
I_w, 530
JW_n, 158
$\text{JW}_{\underline{w}}$, 159
Kar, 202, 209
$L(\lambda)$, 260, 276
L_w, 301, 530
LL, 214
$\overline{LL}_{\underline{w},\underline{e}}$, 191
$\text{LL}^x_{f,\underline{e}}$, 192
$\overline{LL}$, 214
$\mathcal{L}(w)$, 17
$\mathcal{O}$, 277
$\mathcal{O}_0$, 301
$\mathcal{O}_0^{\text{gr}}$, 534
$\mathcal{O}_\lambda$, 286, 294
Φ, 272
$P(\lambda)$, 282
P_w, 530
Q, 84, 198
Q_w, 85, 198
R, 91
R^I, 62, 482
R_+^W, 64
R_+, 255
R^s, 62
R_x, 78
R-gbim, 67

$\mathcal{R}(w)$, 17
$\mathbb{S}$Bim, 69
$\mathbb{S}$Bim(V), 89
StdBim, 78
StdBim(V), 89
$\mathcal{S}\mathbb{B}\mathbb{S}$Bim, 483
$\mathcal{S}\mathcal{H}_{BS}$, 505
$\mathcal{S}\mathbb{S}$Bim, 484
Σ, 155
$\mathcal{TL}$, 128
$TL_{n,\delta}$, 158
T_λ^μ, 298
Θ_s, 301, 546
$U_q(\mathfrak{sl}_2)$, 431
$\mathbb{V}$, 267, 532, 566
W_I, 37
$Z(\mathfrak{g})$, 283
α_s, 91
$\alpha_s^\vee$, 91
a_{st}, 93
b_x, 44
c_{bot}, 240
$c_{\underline{e}}$, 240
c_{id}, 69
c_s, 69
c_{top}, 240
can_x, 248
ch, 83, 116, 329, 563
ch_Δ, 82
ch_∇, 83
χ_λ, 283
$\text{defect}(\underline{e})$, 56
δ_x, 40
∂_I, 496
∂_s, 64
∂_w, 66
ϵ, 46
ft_n, 457
$\mathfrak{h}$, 90, 260
$\mathfrak{h}^*$, 90, 93, 308, 542
$h_{y,x}$, 44
$\ell(I)$, 482
$\ell(w)$, 14
m_{st}, 3
$[n]$, 95
∇-filtration, 82
$\nabla(\lambda)$, 280
∇_w, 530
ω, 44
ρ, 252, 285
w_I, 482
$\underline{w}$, 4
1-tensor, 68
2Σ, 493

$2m_{st}$-valent vertices, 153, 182
2-ideal, 442

A
Additive closure, 208
Affine
    reflection, 26
        group, 26
    transformation, 25
Alcove, 28
    fundamental, 29
Alexander polynomial, 430
Algebra
    based, 447
Antiinvariant, 64
Associated category with shift, 206
Associated graded category, 205
Asymptotically orthonormal, 47

B
Barbell, 146
    relation, 176
Bar involution, 44
Basis
    Kazhdan–Lusztig, 44
    standard, 41
Bernstein–Gelfand–Gelfand (BGG)
    reciprocity, 268, 282
    resolution, 261
Betti number, 337
Biadjoint, 130
Biadjointness, 46
Bilinear form, 335
    graded, 337
    nondegenerate, 335
    symmetric, 335
Block decomposition, 283, 286
Bott–Samelson
    bimodule, 68
        singular, 483
    category, 70
        diagrammatic, 175
    complex, 560
    module, 305
    resolution, 321
    variety, 320, 370
Box, 146, 175
Braden–MacPherson sheaf, 108
Braid
    closure, 428
    -ed tensor category, 432
    group, 385, 427

Index

relation, 4, 40
Bruhat
  decomposition, 265, 317
  graph, 20, 100
  order, 20

C

Canonical subexpression, 58, 248
Cap, 126, 136, 183
Cartan matrix, 93
  generalized, 550
Casimir element, 283, 287
Category
  essentially small, 225
  indecomposable, 283
  Karoubian, 227
  Krull–Schmidt, 227
  ribbon, 432
  triangulated, 393
  with shift, 204
Category $\mathcal{O}$, 262, 277
  Soergel-theoretic, 534
Cell, 444
  Kazhdan–Lusztig, 447
  module, 447
  module category, 446
  preorder, 444
Cellular
  category, 203
  ideal, 215
  pairing, 217
  structure, 213
Character, 83, 329
  of Braden–MacPherson sheaves, 116
  central, 284
  of constructible sheaves, 329, 563
  of Soergel bimodules, 82
Chevalley
  involution, 279
  theorem, 62
    upgraded, 495
Coinvariant algebra, 64, 267, 306
Column number, 449
Comonoid object, 134
Complex
  contractible, 373
  minimal, 377
  Rouquier, 386
  total, 468
  twisted, 467
Composition
  form, 230
  pairing, 230

  local, 235
Cone, 374, 394
Constructible
  derived category, 322
  sheaf, 322
Content, 449
Convolution, 330, 468
Coroot, 274
Coxeter
  complex, 35, 48
  graph, 4
  group, 4
  system, 3
    crystallographic, 14
    rank, 3
    universal, 153
Crossingless matchings, 127, 422
Crystallographic, 14
Cup, 126, 136, 183
Cyclicity, 131
  of dot maps, 137, 184
  of $2m_{st}$-valent vertex, 162, 184

D

Death by pitchfork, 165
Decomposition theorem, 329, 413
Defect, 56
Deletion condition, 18, 31
Demazure
  operator, 64, 66
  –'s formula, 497
  surjectivity, 93, 187, 552
    finitary, 509
Deodhar
  defect, 56
  –'s lemma, 56
Descent set
  left, 17
  right, 17
Dg-
  algebra, 515
    formal, 519
  module, 516
Diagonalization, 468
Diagonal miracle, 392, 402, 539
Diagram
  linear, 119
  planar, 121
  string, 121
  universal, 151
Diagrammatic
  Bott–Samelson category, 175
  Hecke category, 175

Diagrammatic Hecke category, 175, 182
　one-color, 148
　two-color, 164
　universal two-color, 152
Dihedral group, 10
　infinite, 11
Domain
　fundamental, 33
Dot action, 261, 285
Dots, 135, 182
Double leaves, 192
Dual
　module, 280
　realization, 93, 542
　sets, 230
　Verma module, 280
Duality
　for bimodules, 350
　Langlands, 93
　Poincaré, 337
　Verdier, 326

E
Eigenmap, 465
Eigenobject, 465
Element
　Casimir, 283
　longest, 19
Elias–Jones–Wenzl relation, 163, 178
　singular, 506
Enveloping algebra, 436
Euler
　characteristic, 378
　form, 523
Exchange condition, 17, 31
Expression, 4
　reduced, 14
　sub-, 21, 43

F
Face, 28
Facet, 28
　dimension, 28
　support, 28
False orthonormality, 47
Filtration
　∇-, 82
　Δ-, 82
　standard, 281
Finitary, 62, 482
Flag variety, 265, 317
Formal, 519

Frobenius
　algebra, 48
　algebra object, 134, 135
　associativity, 134, 136, 176
　extension, 140, 142, 486
　reciprocity, 142
　trace, 140
　unit, 135, 176
Full twist, 457, 475, 480
Functor
　triangulated, 395

G
Gaussian elimination, 376
Gelfand–Zetlin algebra, 456
Geometric
　realization, 92
　representation, 12, 91
Global
　intersection form, 243
Glueing, 35
Graded
　category, 204
　direct summand, 60
　free, 61
　Hom, 60
　module, 60
　pieces, 59
　ring, 60
　shift, 60
　submodule, 60
　vector space, 59
Grading
　Hochschild, 440
　homological, 440
　internal, 440
　shift, 60
Gram matrix, 336
Graph
　Bruhat, 20, 100
　moment, 104
　rex, 20
Grothendieck group
　abelian, 262, 395
　split, 75, 225
　triangulated, 379, 396

H
Half-space, 28
Hard Lefschetz, 337, 352
Harish-Chandra homomorphism, 285
　twisted, 286

Heart, 397
Hecke
    algebra, 39
    algebroid, 485
    category
        diagrammatic, 175
Hensel's lemma, 228
Henselian ring, 228
$\mathbb{H}$ functor, 269, 562
Highest weight
    module, 274
    vector, 262
Hochschild
    cohomology, 436
    homology, 436
Hodge–Riemann bilinear relations, 340, 352, 402
HOMFLYPT polynomial, 430, 434
Homotopic, 373
Homotopy, 373
    category, 373
    equivalence, 373
Hyperplane
    reflecting, 25
    section, 369

I
Ideal, 215
Idempotent
    decomposition, 468
    split, 227
Indecomposable, 70
    category, 283
    object, 224
Internal degree, 375
Intersection cohomology, 266, 326
Intersection form
    global, 243, 365
    local, 230, 236, 360, 365
Intersection pairing
    local, 236, 554
Invariant subrings, 62
Inversion, 15
Involution
    bar, 44
    Kazhdan–Lusztig, 44
Isotopy, 174, 183
    linear, 120
    presentation, 140, 174
    rectilinear, 123
    true, 138

J
Jones polynomial, 430
Jones–Wenzl
    Elias–, 163, 178
    projector, 158
    relation, 163, 178
Jucys–Murphy, 455

K
Kac–Moody
    group, 562
    representation, 90
Kähler package, 367
Karoubian, 227
    closure, 209
Kazhdan–Lusztig
    anti-involution, 44
    basis, 44
    cell, 447
    conjecture, 264, 311
    inversion formula, 264, 524
    involution, 44
    polynomial, 44, 264
Killing form, 273
Koszul
    sign rule, 375
Koszul duality, 269
    for category $\mathcal{O}$, 525, 542
Krull–Schmidt
    category, 74, 227, 442
    theorem, 226

L
Lagrange interpolation, 463
Langlands dual, 93, 307
Laurent polynomial ring, 39
Lefschetz
    form, 339, 414
    operator, 337
Leibniz rule
    twisted, 65
Length, 4, 14
Light leaves, 190
Limit lemma, 344, 359
Linear
    diagrams, 119
    isotopy, 120
Link, 426
    oriented, 426
Linked, 285
Local
    composition pairing, 235

decomposition, 226
  intersection form, 230, 236
  intersection pairing, 236
Localization, 198
  functor, 85, 115
Locally finite, 277
Lusztig's conjecture, 565

M
Markov
  move, 428
  trace, 434
Mates, 131
Matsumoto's theorem, 19, 33
Maximal vector, 274
Minimal complex, 377, 402
Mirror picture, 9
Moment graph, 104
  GKM, 104
Monoid object, 133
Morita equivalence, 515
Multiplicity, 230

N
Needle relation, 147, 176
NilCoxeter algebra, 66

O
Object
  comonoid, 134
  Frobenius algebra, 134
  indecomposable, 224
  monoid, 133
  projective, 514
  scalar, 464
One-color relations, 176
Order
  path dominance, 58

P
Pairing
  cellular, 217
Parabolic subgroup, 37, 62, 319
Parity
  complex, 171, 559
  sheaf, 559
Path dominance order, 58
p-canonical basis, 553
Perfect pairing, 66
Perverse
  complex of Soergel bimodules, 391
  sheaf, 266, 325
    semisimple, 327
  t-structure, 398, 534
Picard group, 386
Pitchfork, 165
p-Kazhdan–Lusztig polynomial, 553
Poincaré duality, 337
Polynomial forcing relation, 176
Positive breaking lemma, 253
Prediagonalizable, 463
Presentation, 174
  isotopy, 140, 174
  of $\mathcal{H}_{\mathrm{BS}}$, 182
Primitive
  subspace, 339
  vector, 339
Principal block, 264, 287
Projective
  cover, 282
  generator, 514
  object, 514
Projectivization, 225
Pseudoreflection, 63

Q
Quadratic
  relation, 39
Quantum
  invariant, 432
  number, 95, 160, 254
  Schur algebra, 433
Quasi-
  idempotent, 485
  isomorphism, 517

R
Radical, 378
Rank, 3
Realization, 90
  balanced, 97, 162, 510
  dual, 93
  geometric, 92
  restriction of, 92
Reduced expression graph, 20
Reflecting hyperplane, 25
Reflection, 4, 11, 25
  affine, 26
  faithful, 89
  pseudo, 63
  simple, 4
Reflection group

Index 587

affine, 26
crystallographic, 35
Regular dominant, 252
Reidemeister moves, 427
Relation
  barbell, 146, 176
  biadjunction, 183
  braid, 4, 40
  cyclicity, 162
  death by pitchfork, 165
  Elias–Jones–Wenzl, 163
  Frobenius associativity, 176
  Frobenius unit, 176
  fusion, 146
  Jones–Wenzl, 163, 178
  keyhole, 146
  multiplication, 146
  needle, 176
  polynomial forcing, 69, 147, 176
  polynomial slide, 147
  quadratic, 4, 39
  rotation of $2m_{st}$-valent vertices, 184
  rotation of trivalent vertices, 184
  rotation of univalent vertices, 184
  singular Elias–Jones–Wenzl, 506
  skein, 430
  two-color associativity, 163, 177
  two-color dot contraction, 163, 178
  Zamolodchikov, 178
Representation
  geometric, 12
  Kac–Moody, 90
  sign, 15
Reshetikhin–Turaev invariant, 432
Riche–Williamson conjecture, 568
Ring
  Henselian, 228
  semiperfect, 227
Robinson–Schensted correspondence, 449
Root, 272, 496
  hyperplane, 294
  lattice, 274
  positive, 95, 99, 273, 496
  simple, 91
  space, 272
  system, 273
  vector, 272
Rouquier complex, 386, 436, 538
Row number, 449

S

Satake equivalence, 160
Schützenberger involution, 452

Schubert
  basis, 499
  cell, 318
  variety, 265, 318
Schur algebra, 432
  quantum, 433
Schur–Weyl duality, 425
Semiperfect ring, 227
Semisimple
  complex, 269, 327
  perverse sheaf, 327
Sheaf
  Braden–MacPherson, 108
  perverse, 325
Shift
  cohomological, 372
  functor, 204
Signature, 336, 340, 357
Sign representation, 15
Simple
  coroots, 91
  roots, 91, 273
Singular
  Bott–Samelson bimodule, 483
  Bruhat graph, 102
  Elias–Jones–Wenzl relation, 506
  Soergel bimodule, 484
Skein relation, 430
Soergel
  bimodule, 69
  categorification theorem, 75, 86, 202
  Hom formula, 88, 349
  module, 267, 305, 534
  –'s conjecture, 88, 269, 312
  $\mathbb{V}$ functor, 267, 308, 532, 566
  –Williamson categorification theorem, 485
Specialization, 40
Standard
  basis, 41
  bimodule, 78, 198
  filtration, 281
  form, 46
  multiplicity, 281
  parabolic subgroup, 37, 62, 319
  trace, 46
  Young tableau, 448
Strand diagram, 5
  dotted, 8
  generic, 7
Stratification, 316
  Deodhar, 331
  Whitney, 316
Strictification, 122
Stroll, 31, 55

length, 31
reduced, 31
Struktursatz, 309
Subexpression, 21, 43
    canonical, 58, 248
Subgroup
    parabolic, 37
Support, 323
Symmetric
    algebra, 61
    category, 124
    tensor category, 430
Symmetric group
    even-signed, 10
    signed, 8

T
Tate twist, 399, 530, 534
Temperley–Lieb
    algebra, 158, 421
    category, 128
    category (generic), 154
    two-colored, 2-category, 155
Three-color relations, 178
Trace
    on Bott–Samelson bimodule, 243
    on Hecke algebra, 46
    on Temperley–Lieb algebra, 161, 422
Translation, 25
    functor, 298
Triangle, 393
    distinguished, 394
Triangulated
    category, 393
    functor, 395
Triply graded link homology, 440
Trivalent vertices, 135, 182
t-structure, 397
    perverse, 398, 534
    standard, 397
Two-color
    associativity, 163, 177
    dot contraction, 163, 178
    relations, 162, 177

U
Unique decomposition, 224
Univalent vertices, 135, 182
Universal enveloping algebra, 274

V
Vector
    highest weight, 262
Verdier
    duality, 326
    quotient, 397
Verma
    flag, 281
    module, 260, 274
$\mathbb{V}$ functor, 267, 308, 532, 566

W
Wall, 28
Wall-crossing functor, 301, 545
Weak Lefschetz, 361
    substitute, 371, 402, 410
Weight, 23, 272
    antidominant, 308
    compatible, 297
    dominant integral, 274
    integral, 274
    lattice, 274
    module, 262, 272
    space, 272
    vector, 272
Weyl
    chamber, 294
    character formula, 261
    group, 273
        associated to a GCM, 551

Y
Young diagram and tableau, 448

Z
Zamolodchikov equations, 178